WATER PURIFICATION

WATER PURIFICATION

Nanotechnology in the Agri-Food Industry, Volume 9

Edited by

ALEXANDRU MIHAI GRUMEZESCU

Department of Science and Engineering of Oxide Materials and Nanomaterials, Faculty of Applied Chemistry and Materials Science, University Politehnica of Bucharest, Bucharest, Romania

ELSEVIER

ACADEMIC PRESS

An imprint of Elsevier

elsevier.com

Academic Press is an imprint of Elsevier
125 London Wall, London EC2Y 5AS, United Kingdom
525 B Street, Suite 1800, San Diego, CA 92101-4495, United States
50 Hampshire Street, 5th Floor, Cambridge, MA 02139, United States
The Boulevard, Langford Lane, Kidlington, Oxford OX5 1GB, United Kingdom

Library of Congress Cataloging-in-Publication Data
A catalog record for this book is available from the Library of Congress

British Library Cataloguing-in-Publication Data
A catalogue record for this book is available from the British Library

ISBN: 978-0-12-804300-4

Publisher: Nikki Levy
Acquisition Editor: Patricia Osborn
Editorial Project Manager: Jaclyn Truesdell
Production Project Manager: Caroline Johnson
Designer: Mark Rogers

Typeset by Thomson Digital

CONTENTS

LIST OF CONTRIBUTORS

Ana S. Abreu
University of Minho, Department of Polymer Engineering (DEP), Institute of Polymers and Composites (IPC), Institute of Nanostructures, Nanomodelling and Nanofabrication (i3N), Campus of Azurém, Guimarães, Portugal

Ashraf F. Ali
National Research Centre, Inorganic Chemistry Department, Dokki, Cairo, Egypt

Ali Reza Amani-Ghadim
Azarbaijan Shahid Madani University, Applied Chemistry Research Laboratory, Department of Chemistry, Faculty of Science, Tabriz, Iran

Justinah S. Amoko
Department of Chemistry, Adeyemi College of Education, Ondo, Obafemi Awolowo University, Ile-Ife, Nigeria

Priscila Anadão
University of São Paulo, Cidade Universitária, Department of Metallurgical and Materials Engineering, Engineering School, São Paulo, SP, Brazil

Ecaterina Andronescu
University POLITEHNICA of Bucharest, Faculty of Applied Chemistry and Materials Science; Department of Science and Engineering of Oxide Materials and Nanomaterials; University POLITEHNICA of Bucharest, Faculty of Applied Chemistry and Materials Science; National Research Center for Micro and Nanomaterials, Bucharest, Romania

Sahar M. Atwa
University of Louisiana at Monroe, School of Sciences, Monroe, LA, United States; Cairo University, Department of Chemistry, Faculty of Science, Giza, Egypt

Paul Cătălin Balaure
University "Politehnica" of Bucharest, Faculty of Applied Chemistry and Materials Science, "Costin Nenişescu" Department of Organic Chemistry, Bucharest, Romania

Lerato Baloyi
Tshwane University of Technology, Department of Environmental, Water and Earth Sciences, Faculty of Science, Pretoria, South Africa

Arijit Basu
The Hebrew University of Jerusalem, Department of Medicinal Chemistry and Natural Products, Institute for Drug Research, School of Pharmacy—Faculty of Medicine, Jerusalem, Israel

Rajani Bharati
Green Catalysis and Process Technology Research Lab, Department of Chemical Engineering, MANIT, Bhopal, Madhya Pradesh, India

Chiranjib Bhattacharjee
Jadavpur University, Chemical Engineering Department, Kolkata, West Bengal, India

Mariana Carmen Chifiriuc
University of Bucharest, Research Institute of the University of Bucharest, Bucharest, Romania

Doina Maria Cirstea
National Institute of Public Health, Institute of Biology, Bucharest, Romania

Ani Ioana Cotar
University of Bucharest, Research Institute of the University of Bucharest, Bucharest, Romania

Carmen Curutiu
University of Bucharest, Research Institute of the University of Bucharest, Bucharest, Romania

Arsénio de Sá
University of Minho, Department of Polymer Engineering (DEP), Institute of Polymers and Composites (IPC), Institute of Nanostructures, Nanomodelling and Nanofabrication (i3N), Campus of Azurém, Guimarães, Portugal

Abraham J. Domb
The Hebrew University of Jerusalem, Department of Medicinal Chemistry and Natural Products, Institute for Drug Research, School of Pharmacy—Faculty of Medicine, Jerusalem, Israel

Mir Saeed Seyed Dorraji
University of Zanjan, Applied Chemistry Research Laboratory, Department of Chemistry, Faculty of Science, Zanjan, Iran

Emad M. El-Giar
University of Louisiana at Monroe, School of Sciences, Monroe, LA, United States

Shady Farah
The Hebrew University of Jerusalem, Department of Medicinal Chemistry and Natural Products, Institute for Drug Research, School of Pharmacy—Faculty of Medicine, Jerusalem, Israel

Ezekiel O. Fehintola
Department of Chemistry, Adeyemi College of Education, Ondo, Obafemi Awolowo University, Ile-Ife, Nigeria

Anton Ficai
University POLITEHNICA of Bucharest, Faculty of Applied Chemistry and Materials Science; Department of Science and Engineering of Oxide Materials and Nanomaterials; University POLITEHNICA of Bucharest, Faculty of Applied Chemistry and Materials Science; National Research Center for Micro and Nanomaterials, Bucharest, Romania

Denisa Ficai
University POLITEHNICA of Bucharest, Faculty of Applied Chemistry and Materials Science; Department of Inorganic Chemistry; University POLITEHNICA of Bucharest, Faculty of Applied Chemistry and Materials Science; National Research Center for Micro and Nanomaterials, Bucharest, Romania

Alberto Figoli
Institute on Membrane Technology (ITM CNR), Ronde, Italy

Denisa Alexandra Florea
Politehnica University of Bucharest, Department of Biomaterials and Medical Devices, Faculty of Medical Engineering, Bucharest, Romania

Elvis Fosso-Kankeu
North-West University, School of Chemical and Minerals Engineering, Faculty of Engineering, Potchefstroom, South Africa

Mariana Oana Mihaela Fufă
Department of Science and Engineering of Oxide Materials and Nanomaterials, Faculty of Applied Chemistry and Materials Science, University Politehnica of Bucharest; National Institute for Laser, Plasma and Radiation Physics, Lasers Department, Magurele, Bucharest, Romania

Bin Gao
University of Florida, Department of Agricultural and Biological Engineering, Gainesville, FL, United States

Irina Gheorghe
University of Bucharest, Research Institute of the University of Bucharest, Bucharest, Romania

Alexandru Mihai Grumezescu
Department of Science and Engineering of Oxide Materials and Nanomaterials, Faculty of Applied Chemistry and Materials Science, University Politehnica of Bucharest, Bucharest, Romania

Dragoş Gudovan
University "Politehnica" of Bucharest, Faculty of Applied Chemistry and Materials Science, "Costin Neniţescu" Department of Organic Chemistry, Bucharest, Romania

Iulia Alexandra Gudovan
University "Politehnica" of Bucharest, Faculty of Applied Chemistry and Materials Science, "Costin Neniţescu" Department of Organic Chemistry, Bucharest, Romania

Alina Maria Holban
Department of Science and Engineering of Oxide Materials and Nanomaterials, Faculty of Applied Chemistry and Materials Science, University Politehnica of Bucharest; University of Bucharest, Department of Botanic-Microbiology, Faculty of Biology, Bucharest, Romania

Abubakar Ismail
Department of Water Resources and Environmental Engineering, Ahmadu Bello University, Zaria, Nigeria

Ilunga Kamika
University of South Africa, Department of Environmental Science, College of Agriculture and Environmental Sciences, Florida, South Africa

Ozcan Konur
Yildirim Beyazit University, Faculty of Engineering and Natural Sciences, Department of Materials Engineering, Kecioren, Ankara, Turkey

Konda Reddy Kunduru
The Hebrew University of Jerusalem, Department of Medicinal Chemistry and Natural Products, Institute for Drug Research, School of Pharmacy—Faculty of Medicine, Jerusalem, Israel

Veronica Lazăr
University of Bucharest, Research Institute of the University of Bucharest, Bucharest, Romania

Salihu Lukman
Department of Civil Engineering, University of Hafr Al-Batin, Hafr Al-Batin, Saudi Arabia; Department of Water Resources and Environmental Engineering, Ahmadu Bello University, Zaria, Nigeria

Ana Vera Machado
University of Minho, Department of Polymer Engineering (DEP), Institute of Polymers and Composites (IPC), Institute of Nanostructures, Nanomodelling and Nanofabrication (i3N), Campus of Azurém, Guimarães, Portugal

Florica Marinescu
National Institute for Research and Development in Environmental Protection, Institute of Biology, Bucharest, Romania

Ajay Kumar Mishra
College of Engineering, Science and Technology, University of South Africa, Nanotechnology and Water Sustainability Research Unit, Johannesburg, South Africa

Magda Mihaela Mitache
Titu Maiorescu University, Faculty of Medicine; Public Health Department Bucharest, Bucharest, Romania

Maggy NB Momba
Tshwane University of Technology, Department of Environmental, Water and Earth Sciences, Faculty of Science, Pretoria, South Africa

Isabel Moura
University of Minho, Department of Polymer Engineering (DEP), Institute of Polymers and Composites (IPC), Institute of Nanostructures, Nanomodelling and Nanofabrication (i3N), Campus of Azurém, Guimarães, Portugal

Lizzy Mpenyana-Monyatsi
Tshwane University of Technology, Department of Environmental, Water and Earth Sciences, Faculty of Science, Pretoria, South Africa

Michael Nazarkovsky
The Hebrew University of Jerusalem, Department of Medicinal Chemistry and Natural Products, Institute for Drug Research, School of Pharmacy—Faculty of Medicine, Jerusalem, Israel

Liliana Neagu
University of Bucharest, Research Institute of the University of Bucharest; Politehnica University of Bucharest, Department of Biomaterials and Medical Devices, Faculty of Medical Engineering, Bucharest, Romania

Isaiah A. Oke
Department of Civil Engineering, Obafemi Awolowo University, Ile-Ife, Nigeria

Sukanchan Palit
University of Petroleum and Energy Studies, Department of Chemical Engineering, Energy Acres, Dehradun, Uttarakhand, India

Kostantinos Pantapasis
University Politehnica of Bucharest, Department of Biomaterials and Medical Devices, Faculty of Medical Engineering, Bucharest, Romania

Rajendra P. Pawar
Deogiri College, Department of Chemistry, Aurangabad, Maharashtra, India

Roxana Cristina Popescu
Department of Science and Engineering of Oxide Materials and Nanomaterials, Faculty of Applied Chemistry and Materials Science, University Politehnica of Bucharest; "Horia Hulubei", National Institute of Physics and Nuclear Engineering, Department of Life and Environmental Physics, Magurele, Bucharest, Romania

Ankur Sarkar
Jadavpur University, Chemical Engineering Department, Kolkata, West Bengal, India

Santanu Sarkar
Environment Research Group, R&D, Tata Steel Ltd, Jamshedpur, Jharkhand, India

Suresh Sundaramurthy
Green Catalysis and Process Technology Research Lab, Department of Chemical Engineering, MANIT, Bhopal, Madhya Pradesh, India

Ratna G. Suthar
University of Florida, Department of Agricultural and Biological Engineering, Gainesville, FL, United States

Chandrakant Thakur
Department of Chemical Engineering, NIT, Raipur, Chhattisgarh, India

SERIES FOREWORD

The emergence of nanotechnology has reached impressive heights in recent years and the development of special nanodevices and nanomaterials has found intriguing applications in agriculture and food sector. Most of the investigated nanotechnological approaches initially aimed to solve evolving problems in the agri-food industry in order to impact on the economic potential. Soon after the implementation of new technologies and approaches that were using nanostructured materials, the worldwide concern was rapidly extended to numerous applications that could be developed by using the science of nanosized materials. Smart materials, biosensors, packaging materials, nutraceuticals, and nanodevices have been designed to address numerous agri-food related issues with direct impact in health, economy, ecology, and industry. As the engineering of nanostructures has constantly progressed and extended its applications, there is virtually unlimited potential in this sector. However, the widely differing opinions on the applicability and usefulness of nanotechnology between both specialists and the general public has hampered progress. The main concern manifested by people is related to the potential risk for health and the environmental impact of the recently developed nanoengineered materials and devices. Therefore, current approaches are strictly considering these concerns when designing nanotechnological solutions for agriculture and food sectors.

This multivolume series was developed by the constant need to discover current inquiries and approaches on the field of agri-food science and also to learn about the most recent progress, approaches, and applications that have emerged through nanotechnology.

As agriculture is the backbone of most developing countries, nanotechnology has the potential to revolutionize the agriculture and food sector by promoting productivity through genetic improvement of plant and animal foods. It can also ensure the delivery of drugs, genes, and pesticides to specific sites at cellular levels in targeted plants and animals, by limiting side effects. Nanotechnology can be used to evaluate gene expression under different stress condition for both plant and animal foods through the development of nanoarray-based gene-technologies. Additionally, this technology can detect fertilizers, pesticides with high precision by smart nanosenors for an adequate management of the natural resources. Moreover, numerous industrial-related applications with direct impact on economy have emerged. For example, nano- and micro-structured arrays can detect the early presence of pathogens, contaminants, and food spoilage factors. Other applications for this technology are smart integration systems for food processing and packaging, as well as nanoemulsion-based decontaminants for food equipment and storage compartments, and nanoparticles that facilitate the bioavailability and delivery of nutrients directly to cells.

The potential benefits of nanotechnology for agriculture, food, fisheries, and aquaculture were identified and supported by many countries, which invested a significant amount of money in the development of applications. Also, numerous campaigns are currently trying to increase awareness on the developing process and recent technologies in order to influence the acceptance of customers. Although nanoagri-food industrialized concept could help to find a sustainable solution for the current global food crisis, the offered advantages should balance the concerns regarding soil, water, environment, and health related issues that such approach could bring.

The series entitled *Nanotechnology in the Agri-Food Industry* brings comprehensive and recent knowledge regarding the impact of the science of nanometer-sized materials on the field of agriculture and food industry, but also discuss the current inquiries regarding risks of these applications in all relevant fields such as environment and health, aiming to increase awareness to a wider amount of readers.

Dr Alina Maria Holban
Department of Microbiology and Immunology, Faculty of Biology, University of Bucharest, Romania

Dr Monica Cartelle Gestal
Harvill Lab, University Of Georgia, Athens, GA, United States

SERIES PREFACE

About the Series (Volumes I–X)

In a permanently changing society, health and well being remain the key drivers for the food industry. Despite the technological progress made in the agri-food industry, a true food crisis emerges in several areas of the globe. This can be explained by insufficient food but mostly by inadequate food for a very distinct range of consumers. In this context, innovative technologies represent the core throughout the whole food chain from raw materials/ingredient sourcing, food processing, quality control of finished products, and packaging. Nanotechnology, coupled with novel interdisciplinary approaches and processing methods, has enabled some important advances recently flourishing in many of these areas. The science of nanosized materials can improve and even resolve the huge challenges faced by the food and bioprocessing industries for developing and implementing systems that can produce qualitative and quantitative foods that are safe, sustainable, environment friendly, and efficient. This emerging tool finds its applications in various fields and represents an endless approach for the development of innovative strategies in food development, processing, and packaging.

This multivolume set aims to bring together the most recent and innovative applications of nanotechnology in the agri-food industry, but also to present the future perspectives in the design of new or alternative foods.

The series contains 200 chapters organized in 10 volumes, prepared by outstanding research groups that made significant impacts on the field of nanotechnology and food-related research sectors. This comprehensive set represents an updated and highly structured material for undergraduate and postgraduate students in food science, biotechnological, engineering fields, but also a valuable resource of recent scientific progress, along with most known applications of nanomaterials on the food industry to be used by researchers, engineers, and academia. Moreover, novel opportunities and ideas for developing or improving technologies in the agri-food industry by innovative companies, biotechnological industries, and other economical structures are highlighted and their potential is widely dissected. This series may be also valuable for the wide audience interested in recent nanotechnological progress in the agri-food field worldwide.

These 10 volumes cover almost all aspects related to the applications of *Nanotechnology in the Agri-Food Industry* and are named as:

Volume I Novel Approaches
Volume II Encapsulations
Volume III Emulsions
Volume IV Nutraceuticals
Volume V Nutrient Delivery
Volume VI Food Preservation
Volume VII Food Packaging
Volume VIII NanoBioSensors
Volume IX Water Purification
Volume X New Pesticides and Soil Sensors

Each volume contains 20 chapters, which were carefully composed and illustrated to highlight the most innovative and intensively investigated applications of nanotechnology on particular wide interest domains of the agri-food industry field.

Alexandru Mihai Grumezescu
University Politehnica of Bucharest,
Department of Science and Engineering
of Oxide Materials and Nanomaterials,
Faculty of Applied Chemistry and
Materials Science, Bucharest, Romania
http://grumezescu.com/

VOLUME PREFACE

About Volume IX: Water Purification

Improving water quality provides social and economic benefits both to developed and developing countries. Adequate water purification approaches could bring lower costs, a safer environment, and industry and also a huge benefit for human health.

In recent years, the use of nanostructured materials in developing novel water purification approaches bring numerous advantages over conventional methods previously employed for water purification. From the design of nanoporous membranes to functional nanoparticles and nanosensors, the science of nanosized materials offers now a sustainable alternative for the removal of main polutants, such as arsenic, bacteria, organic materials, nitrates, salinity, and viruses from most water sources.

The present book highlights the recent progress in the field of water purification using nanomaterials and also discusses the main advantages of novel approaches based on nanosized materials in the design of ecological and personalized water purification processes.

The fact that most nanobased approaches involve chlorine-free water purification methods, represents a particular interest since carcinogenic disinfection by-products may be formed when components of natural water interact with chloramines or chlorine. These aspects support a safer and ecofriendly water purification approach.

Volume IX contains 20 chapters, prepared by outstanding international researchers from USA, Brazil, Portugal, India, Italy, Romania, Nigeria, South Africa, Egypt, Israel, and Iran.

In Chapter 1, *Recent Advances in Using Magnetic Materials for Environmental Applications*, Denisa Ficai et al. present the most used classes of nanomaterials for environmental applications, the mechanisms of water purification, as well as the advances in this field induced by the synergic use of more components, especially magnetic materials, metals, metal oxides, carbon-based materials, and silicates.

Konda Reddy Kunduru, in Chapter 2, *Nanotechnology for Water Purification: Applications of Nanotechnology Methods in Waste Water Treatment*, presents an up-to-date overview about nanotechnology-enabled water treatment processes, showing how they transform our water supply and wastewater treatment. The following topics are discussed in this chapter: different nanomaterials, properties, mechanisms, advantages compared to existing methods, limitations, research needs for commercialization, and toxicity of nanomaterials.

Chapter 3, prepared by Ratna G. Suthar et al., *Nanotechnology for Drinking Water Purification* discusses the applications of nanotechnology in the areas of adsorption and membrane separation, particularly with respect to how nanotechnology and engineered nanomaterials are applied to improve the efficiency, as well as to reduce the cost and environmental footprint of the conventional water purification methods.

Chapter 4, *Application of Nanotechnology in Drinking Water Purification*, prepared by Alberto Figoli et al. focuses on nanotechnology-enabled water treatment processes utilizing silver nanoparticles, zerovalent iron nanoparticles, nanostructured photocatalysts, membranes-based nanotechnology, and nanoadsorbents. In the case of each technology, it has been tried to survey the necessity and efficacy with considering the important challenges.

Maggy NB Momba et al., in Chapter 5, *Nanotechnology-Based Filters for Cost-Effective Drinking Water Purification in Developing Countries*, discuss the capability of disinfection to remove microbial contaminants and about the cost effectiveness of the technology applicable for domestic use in rural communities.

Chapter 6, *Development of Magnetic Nanoparticles for Fluoride and Organic Matter Removal From Drinking Water*, prepared by Ashraf F. Ali et al. highlights the synthesis and use of magnetic nanoparticle compositions in the treatment of surface, ground, and wastewaters; with emphasis on the elimination of fluoride ions and organic matter. The chapter also covers the properties, advantages, and the challenges associated with the use of nanomaterials in the drinking water systems.

Mariana Oana Mihaela Fufă et al. in Chapter 7, *Microorganisms: New Trends in Environmental-Friendly and Energy-Saving Water Purification*, highlight the impressive potential of both nonpathogenic and pathogenic microorganisms in water self-purification and water purification approaches.

Arsénio de Sá et al. in Chapter 8, *Polymeric Materials for Metal Sorption From Hydric Resources*, provide an overview about polymeric materials that are being reported as adsorbents of metal ion, with high emphasis to those derived from natural wastes in waters.

E. Fosso-Kankeu et al. in Chapter 9, *Photocatalytic Degradation and Adsorption Techniques Involving Nanomaterials for Biotoxins Removal From Drinking Water*, cover the limitations of conventional water treatment techniques in the removal of biotoxins and review the potential of nanomaterials-based treatment in the removal of toxins and toxic materials.

Santanu Sarkar et al. in Chapter 10, *Nanotechnology-Based Membrane Separation Process for Drinking Water Purification*, reveal and discuss several aspects of next generation nanomaterials-based membrane treatment process for water purification. Three major categories of membranes are approached: nanoparticle based ceramic membranes, inorganic–organic membranes, and biologically inspired membranes. Each development in the field of nanomembrane has shown the promising pathway of improved energy efficient water purification process.

Florica Marinescu et al. in Chapter 11, *Fate of Antibiotics During Water Treatment: Impact on Antimicrobial Resistance in Environmental and Clinical Strains*, give an up-to-date overview about the origin, transfer and accumulation of antibiotic resistance genes in the wastewater treatment plants in the aquatic environment. This chapter offers important information regarding a worldwide key problem, antibiotic resistance and its management.

Roxana Cristina Popescu et al., in Chapter 12, *Nanostructurated Membranes for the Microbiological Purification of Drinking Water*, focused on the removal of microorganisms from contaminated water by using different filtering membranes and discuss some examples of nanofiltration membranes which are under current use or research.

Chapter 13, *Gold Nanoparticles: Advances in Water Purification Approaches*, prepared by Kostantinos Pantapasis et al., reviews the advantages of gold nanoparticles in water purification and discusses current progress regarding synthesis and characterization of such nanoparticles suitable for obtaining pure water sources.

In Chapter 14, *Nanomaterials and Food-Processing Wastewater*, prepared by Rajani Bharti et al., the focus is on green nanomaterials used for the removal of pollutants in the food processing wastewater. Moreover, various parameters utilized to prepare nanoparticles from waste stream, which can be used for numerous applications, are also revealed and discussed.

Chapter 15 prepared by Priscila Anadão et al., *Nanocomposite Filtration Membranes for Drinking Water Purification*, presents the theory related to membrane process, such as micro-, ultra-, nanofiltration, and reverse osmosis. Also, the most recent scientific and technological advances in the development of nanocomposite membranes are presented in this work.

Paul Cătălin Balaure et al., in Chapter 16, *Nanotechnology Depollution of Heavy Metals Present in Potable Water*, discuss about integrated solutions in the nanodepollution of waters, taking them from the laboratory scale to industrial applications. Some solutions involve "carbonaceous" materials, such as carbon nanotubes, graphene oxide, nanostructured carbon, but also metal oxides, magnetic metal oxides, natural polymers, waste materials, and others.

Sukanchan Palit in Chapter 17, *Application of Nanotechnology, Nanofiltration, and Drinking and Wastewater Treatment—a Vision for the Future*, provides a cogent insight about the varied application of nanotechnology and nanofiltration in drinking water treatment. The author also delineates the recent scientific endeavor in the field of nanofiltration and other environmental engineering and chemical engineering separation processes with specific importance to water treatment.

Chapter 18, *Microbial Biofilms From the Aquatic Ecosystems and Water Quality*, prepared by Liliana Neagu et al., gives an overview on the characteristics of microbial biofilms found in the aquatic ecosystems with a special focus on measuring their negative and beneficial roles for the water treatment and quality.

Isaiah A. Oke et al., in Chapter 19, *Removal of Lead Ions From Water and Wastewaters Electrochemically*, present interesting details about the treatment technique in removing lead ion concentration from typical and synthetic industrial wastewaters as promising tool in metal recovery techniques and purification.

Chapter 20, prepared by Ozcan Konur, *Scientometric Overview Regarding Water Nanopurification*, presents a scientometric overview of the research in the water purification nanomaterials and recently designed processes, while revealing the strong public policy implications providing important incentives for the key stakeholders involved in this research field.

Alexandru Mihai Grumezescu
University Politehnica of Bucharest,
Department of Science and Engineering
of Oxide Materials and Nanomaterials,
Faculty of Applied Chemistry and
Materials Science, Bucharest, Romania
http://grumezescu.com/

1

RECENT ADVANCES IN USING MAGNETIC MATERIALS FOR ENVIRONMENTAL APPLICATIONS

Denisa Ficai*,†, Anton Ficai,†, Ecaterina Andronescu**,†**

**University POLITEHNICA of Bucharest, Faculty of Applied Chemistry and Materials Science; Department of Inorganic Chemistry, Bucharest, Romania; **University POLITEHNICA of Bucharest, Faculty of Applied Chemistry and Materials Science; Department of Science and Engineering of Oxide Materials and Nanomaterials, Bucharest, Romania; †University POLITEHNICA of Bucharest, Faculty of Applied Chemistry and Materials Science; National Research Center for Micro and Nanomaterials, Bucharest, Romania*

1 Introduction

The environmental protection as well as food safety and energy independence are current needs of the humankind. Worldwide, to overpass these needs, the researchers are trying to improve existent or to develop new strategies (Ghasemzadeh et al., 2014). In order to take practical steps to reduce and combat the negative phenomena induced by pollution, even the extinction of species of plants and animals, the pollutant factors must be correctly identified and combated. The development of solutions and technical measures to avoid (refers to control the sources of pollution of the various components of the environment: water, air, soil, fauna, and flora) or to combat pollution or annihilate the effect of these pollutants or even to remove these pollutants will assure a durable development (Pham et al., 2009; Morin-Crini and Crini, 2013).

Water is a basic component of the environment in direct contact with soil and atmosphere. Consequently, water pollution is of great concern being involved also in soil and even air contamination. It is well known that, especially near some industrial areas and cities, the water quality is affected. Nowadays due to the awareness of the importance of the role of water quality many methods of industrial wastewater purification were improved or

Water Purification. http://dx.doi.org/10.1016/B978-0-12-804300-4.00001-0

developed (Lipnizki and Field, 2002). Three main water treatment routes are applied: chemical, physical, and biological but, many times, two or even all the three routes are combined in order to obtain the required quality of the water (Pham et al., 2009; Hanna et al., 2010; Horikoshi and Serpone, 2009) .Contamination of different classes of water (surface water, groundwater, etc.) is a serious problem worldwide caused by the rapid population growth and industrial expansion as well as some climate changes, such as long droughts (Cundy et al., 2008; Chong et al., 2010; Zeng et al., 2011). Because of the involved risks, strict, and carefully control of the polluting sources are required because once discharged into the environment (heavy metals, bacteria, viruses, organic contaminants from industry, inorganic compounds) through wastewater they are harmful to humans and the environment.

Starting from theses premises, over time, a number of physical, chemical, and biological method for removal of these contaminants from wastewater (Jiang et al., 2006; Huang et al., 2010; Pang et al., 2011a), such as photocatalytic oxidation, adsorption/separation processing, and bioremediation (Pang et al., 2011a,b; Huang et al., 2006; Zelmanov and Semiat, 2008; Long et al., 2011) were studied and some of them are currently exploited at industrial level. The choose of a particular method of removing contaminants takes into account several factors, such as processing efficiency, operational method, energy requirements, and economic benefit. Another and perhaps most important factor in choosing the method of purifying wastewater is the type of material used. Recently, nanomaterials (NMS) have been suggested as efficient, cost-effective alternative to existing and environmental friendly materials for wastewater treatment (Dimitrov, 2006; Friedrich et al., 1998; Dastjerdi and Montazer, 2010).

In recent years, the synthesis, characterization, and use of magnetic nanoparticles (MNP) with spinel structure has attracted much attention as decontamination agent of wastewater. Also, MNPS functionalization with different materials (Polimet network silica, amino acids, etc.) leads to advanced materials with core–shell structure successfully used in wastewater decontamination. The advantages that recommend these structures are especially related to the improved dispersion and stability of core shell nanostructures during application and, the shell can also act as an active layer which can bind, selectively, desired species. Although organosilanes are compounds with very low toxicity, high biocompatibility, high capacity for adsorbing poisonous polymers, and are used widely in the chemical industry, only few works concerning Fe_3O_4/organosilanes/polymer core–shell nanostructures for removal of environmental pollutants are published in the

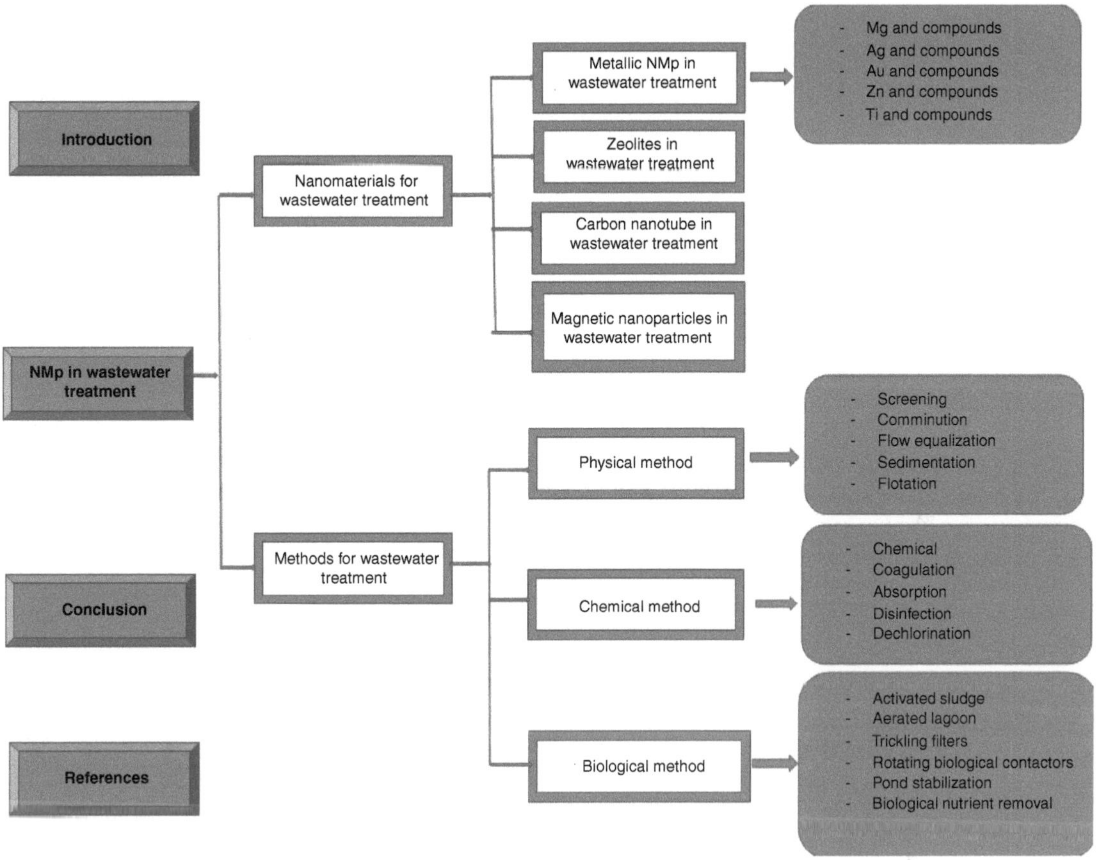

Figure 1.1. Schematic illustration of the main nanomaterials (NMS) used wastewater treatment.

scientific literature meaning that the perspectives in this field is very promising (Stoimenov et al., 2002).

In this chapter the importance of nanoparticles (especially those magnetic) in wastewater decontamination of sewage and wastewater decontamination are presented. The structure of this chapter is illustrated in Fig. 1.1.

2 Nanomaterials for Wastewater Treatment

Given the nanometric size of these materials, the physical, chemical, mechanical, and biological properties are different comparing with the bulk materials. This very low dimensionality of the materials, conferred by high specific surface area (SSA), high reactivity, strong sorption capability, sensing, and detection make these materials very promising candidate for environmental applications as sorbent of various organic and inorganic pollutants as well as catalyst for degradation of organic components (Rickerby and Morrison, 2007; Vaseashta et al., 2007; Qu et al., 2013).

Today some classes of NMS are extensively tested in water purification process: metal and metal-oxide nanoparticles (Ag, Fe_3O_4, ZnO, TiO_2...), carbon nanotubes and nanofibres, graphene, zeolites, and dendrimers (Tiwari et al., 2008).

2.1 Metal Nanoparticles in Wastewater Treatment

Metal nanoparticles show two unique properties that make them promising candidates as sorbents. First of all, metal nanoparticles have much larger SSAs than bulk particles, and secondly they can be functionalized with various functional groups in order to increase their affinity towards different target compounds. Experimental tests also found that due to the unique properties of these metal nanoparticles they exhibit high absorption capacity (Tiwari et al., 2008). By using modern characterization techniques the interaction of metal nanoparticles with various toxic substances (metal cations, bacteria ...) from sewage and industrial wastewater was shown. In the case of microorganisms, using microscopic techniques (atomic force microscopy, transmission electron microscopy and confocal laser microscopy) it was found that the integrity of cell membranes suffer considerable changes that lead to their degradation and consequently microorganisms death (Tiwari et al., 2008). Photocatalytic nanoparticles are used successfully to destroy pesticides, antibiotics or industrial solvents, for instance (Tiwari et al., 2008).

A comprehensive mini-database containing various metallic nanoparticles and their action in wastewater treatment is presented in Table 1.1.

2.1.1 Mg Derivate Compounds (Tiwari et al., 2008)

Magnesium and magnesium oxide nanoparticles are very effective biocides against Gram-positive (*Escherichia coli*) and Gram-negative bacteria (*Bacillus megaterium*) as well as bacterial spores (*Bacillus subtillus*) (Stoimenov et al., 2002).

Magnesium oxide nanoparticles (MgO) are oxides with spinel structure with high SSA particles of 25–50 m^2/g, and 5–100 nm in diameter. Magnesium nanoparticles, are spherical black particles with 20–60 nm diameter and SSA of 30–70 m^2/g.

Table 1.1 Various MNP used in Wastewater Treatment

Type of MNP	Type of Pollutant Remove	Observation	References
Fe_3O_4/dimercapto-succinic acid (DMSA)	Toxic cations: Pb^{2+}, Hg^{2+}, Cd^{2+}, Tl^{+}, Ag^{+}	• These heavy metals form coordinative binding with DMSA • The Fe_3O_4 can be easily removed from solution by using a 12T magnet • The chemical affinity, capacity, kinetics, and stability of the MNPs to Fe_3O_4/DMSA was compared with conventional resin based sorbents (GT-73), AC, and nanoporous silica (SAMMS) with similar surface area in river water, groundwater, seawater, and human blood and plasma. As a conclusion, the Fe_3O_4/DMSA core–shell system exhibited a capacity of 227 mg of Hg^{2+}/g, 30 times higher than GT-73	Yantasee et al. (2007)
Fe_3O_4/succinic acid Fe_3O_4/ ethylenediamine Fe_3O_4/2,3 dimercapto-succinic acid	Cations: Pb^{2+}, Cu^{2+}, Cr^{3+}, Ni^{2+}, Co^{2+}, As^{3+}, Cd^{2+} Bacterial pathogens (*E. coli*) from water	Remove the metal cations from water by forming chelate complexes, ion exchange process or electrostatic interaction The antibacterial efficiency is strongly dependent on the concentration of NMP and their contact time, the surface functionality of Fe_3O_4 and pH Also the surface of MNPs are highly effective, efficient, and economically viable in comparison to the existing industrial purification processes	Singh et al. (2011)
Fe_3O_4/Gum kondagogu (GK)	Cd^{2+}, Cu^{2+}, Pb^{2+}, Ni^{2+}, Zn^{2+}, Hg^{2+} from aqueous solution	Spherical iron oxide nanoparticles with diameters in the range of 8–15 nm, magnetic saturation value of ~60 emu/g were obtained the removal efficiencies for the metal cations by the GK–MNP at a pH of 5.0 and at a temperature of 30.0°C were 106.8 mg/g for Cd^{2+} and 35.07 mg/g for the Hg^{2+}; the removal efficiency decreasing as follow: $Cd^{2+} > Cu^{2+} > Pb^{2+} > Ni^{2+} > Zn^{2+} > Hg^{2+}$	Saravanana et al. (2011)
Fe_3O_4/APS/AA/CA	Heavy metals: Cd^{2+}, Zn^{2+}, Pb^{2+}, Cu^{2+}, from aqueous solution	Fe_3O_4 MNPs modified with 3-aminopropyltriethoxysilane (APS) and copolymers of acrylic acid (AA) and crotonic acid (CA) The adsorption capacity was tested at different pH, contact time and cation concentration	Ge et al. (2012)

(*Continued*)

Table 1.1 Various MNP used in Wastewater Treatment (*cont.*)

Type of MNP	Type of Pollutant Remove	Observation	References
Amino-modified Fe_3O_4	Cu^{2+}, Cr^{3+}	The adsorption capacity was 12.43 mg/g for Cu^{2+} ions and 11.24 mg/g for Cr^{3+} ions	Huang and Chen (2009)
Fe_3O_4/carboxymethyl-β-cyclodextrin (CM-β-CD)	Cu^{2+}, from aqueous solution	The absorption capacity of Cu^{2+} was enchanted by grafting the carboxymethyl-β-cyclodextrin on the surface on the Fe_3O_4 nanoparticles. Cu^{2+} removal is realized by forming coordinative bond with the hydroxyl and carboxyl groups of CM-β-CD structure. It proves that Cu^{2+} retention onto Fe_3O_4/(CM-β-CD) is dependent on pH and temperature. FTIR and XPS reveal that Cu(II) is retained onto CMCD-MNPs surface, this process being reversible and can be desorbed from CMCD-MNPs by citric acid solution. The desorption efficiency was 96.2% and the CMCD-MNPs exhibit good recyclability	Badruddoza et al. (2011)
Fe_3O_4/HA	Cations, such as Hg^{2+}, Pb^{2+}, Cd^{2+}, and Cu^{2+} from water	The absorption capacities was over 99% for Hg^{2+} and Pb^{2+}, and over 95% for Cu^{2+} and Cd^{2+} in natural and tap water at optimized pH	Liu et al. (2008)
Fe_3O_4	As^{5+}	The removal of As^{5+} and its specific species resulted in different sorption processes were investigated on magnetite nanoparticles It was found that As^{5+} could form complexes at the surfaces on Fe_3O_4 and could be progressively incorporated in their structure These results shown that the As^{5+}–magnetite interactions are essential for some developing effective water treatment technologies for arsenic	Wang et al. (2011)
Fe_3O_4/multiwall boron nitride nanotubes (BNNTs)	As^{5+} from water solutions	Adsorption capacity of As^{5+} was studied at neutral pH (6.9) and room temperature The Langmuir, Freundlich, and Dubinin–Radushkevich adsorption isotherms were tested for As^{5+} at various concentrations (from 1 to 40 mg/L) under the same conditions Experimental results showed that the absorption was achieved both through chemical complexation and through physical electrostatic interactions with a slight preference for chemisorption	Chen et al. (2011)
Magnetic iron–nickel oxide	Cr^{6+}	The capacity adsorption of the magnetic iron–nickel oxide was of 30 mg/g for Cr(VI)	Wei et al. (2009)

Table 1.1 Various MNP used in Wastewater Treatment (*cont.*)

Type of MNP	Type of Pollutant Remove	Observation	References
Fe_3O_4/MCP (activated maize cob powder)	MB	Tan et al. (2012) shown that the Fe_3O_4–MCP can be used as potential adsorbents for the removal of MB from dye rich wastewater, and the adsorption of MB dye was pH dependent and the higher efficiency of dye concentration removal was at pH 6.0 Also they showed that the MNPs present high surface area and large pore size for dye molecules to be adsorbed on the surface of Fe_3O_4–MCP	Tan et al. (2012)
Fe_3O_4/SiO_2/thiol	Mercury	The adsorption capacity of the Fe_3O_4/SiO_2/thiol is 148.8 mg/g at pH 6.5, due to the strong interaction between mercury and the thiol groups. Fe_3O_4/SiO_2/thiol can be reused because mercury loaded on the sorbents could be easily desorbed with 1 mol/L HCl containing 3 wt.% of thiourea, and the sorbents showed good reusability. The adsorption capacities were still kept over 110 mg/g in tested natural water samples, showing practical significance in remediation of mercury-contaminated water	Zhang et al. (2013)
Fe_3O_4 wheat straw	MB	The adsorption capacity is dependent by pH and temperature and the removal efficiency increases with increases of temperature	Pirbazari et al. (2014)
Fe_3O_4 nanoparticles stabilized with carboxy (succinic acid), amino (ethylenediamine) and thiol (2,3-dimercaptosuccinic acid) containing agents	Cr^{3+}, Co^{2+}, Ni^{2+}, Cu^{2+}, Cd^{2+}, Pb^{2+}, and As^{3+} and bacterial pathogens (*E. coli*) from water	It has been observed that magnetite nanoparticles of average size of about 6–10 nm bearing carboxyl or thiol exhibit good removal capacity of a wide range of cations. However, each amine functionalized Fe_3O_4 is of 40 nm because of the high tendency of agglomeration due to the possibility of forming H bonds between the amino groups, at the usually wastewater treatment conditions Depending upon the surface functionality (COOH, NH_2, or SH), magnetic nanoadsorbents capture metal ions either by forming chelate complexes or ion exchange process or electrostatic interaction. It has been observed that the antibacterial efficiency is strongly dependent on the concentration of nanoadsorbents and their inoculation time. Furthermore, these nanoadsorbents can be used as highly efficient separable and reusable materials for removal of toxic metal ions	Singh et al. (2011)
Fe_3O_4/mesostructured SiO_2–NH_2	Cu(II)	The adsorption capacity of Fe_3O_4/mesostructured SiO_2–NH_2 was 0.5 mmol/g (31.75 mg/g) for Cu(II)	Kim et al. (2003)
PEI coated Fe_3O_4/MNPs–NH_2	Cr(VI)	The maximum adsorption capacity for Cr(VI) was 83.3 mg/g	Pang et al. (2011b)

2.1.2 Silver and Ag(I) Compounds (Vaseashta et al., 2007; Ni et al., 2007)

Silver and Ag(I) compounds are successfully used as antimicrobial compounds for coliform bacteria found in wastewater (Jain and Pradeep, 2005). The high antibacterial activity induced by the size and shape, low toxicity, and microbial inactivation in water make these nanoparticles very attractive biocides with increasing applicability in both industrial, environmental as well as medical applications (Furno et al., 2004; Morones et al., 2005; Zhao and Stevens, 1998; Inoue et al., 2002; Kumar et al., 2004; Nedelcu et al., 2014a; Patrascu et al., 2015). Silver nanofibers can be used as first aid bandages, plastics, soap, and textiles, in treatment of certain viruses, in self-cleaning fabrics, as conductive filler. Jain and Pradeep (2005) and Sons et al. (2004) shown that the Ag nanoparticles can be used as active biocides against Gram-positive and Gram-negative bacteria including *E. coli, Staphylococcus aureus, Klebsiella pneumoniae,* and *Pseudomonas aeruginosa.* Silver nanoparticles, nanodots, or nanopowder are spherical or flake-like with a dimension of about 1–40 nm and SSA larger than 1 m^2/g. Even the most silver precursors (silver nitrate, silver chloride, silver acetate) present biocides activities, the mechanism of action being different and involve Ag^+ as active agent (Sondi and Salopek-Sondi, 2004; Baker et al., 2005; Panacek et al., 2006; Kim et al., 2007; Shrivastava et al., 2007). The antibacterial activity is strongly dependent by the particle size and shape (Morones et al., 2005; Makhluf et al., 2005; Nedelcu et al., 2014b). It was found that smaller Ag nanoparticles (8 nm) present heigher bactericide activity comparing with larger particles (11–23 nm) (Makhluf et al., 2005), the truncated triangular silver nanoplates exhibiting the higher antibacterial effects especially comparing with the spherical and rod-shaped particles (Pal et al., 2007).

A number of specialty papers showed that the antibacterial mechanism involves bacterial membranes damage with the formation of free radicals (Xiu et al., 2011, 2012), interactions with DNA and membrane properties by changing the cell surface adhesion (Kim et al., 2007; Liau et al., 1997; Danilczuk et al., 2006).

Although Ag nanoparticles have been used efficiently for inactivating bacteria and viruses as well as reducing membrane biofouling, their long-term efficacy against membrane biofouling has not been reported mainly due to loss of silver ions over the time. So, further work to reduce this loss of silver ions is required for long-term control of membrane biofouling. Alternatively, doping of Ag nanoparticles with other metallic nanoparticles or its composites with metal-oxide nanoparticles can solve the issue and this could also lead to the simultaneous removal of inorganic/organic compounds from water/wastewater.

Silver nanocrystals incorporated in different polymers was successfully used as antibacterial (against both Gram-positive and Gram-negative bacteria) antifungal, antiviral, antiinflammatory agent (Savage and Diallo, 2005). For instance, Ag nanoparticles incorporated in cellulose acetate fibers (Son et al., 2004) were found to be effective against both Gram-positive and Gram-negative bacteria (Savage and Diallo, 2005). Ag nanoparticles incorporated into poly (ε-caprolactone)-based polyurethane nanofiber mats (Balogh et al., 2001; Chen et al., 2003) exhibited very good antimicrobial properties (Lala et al., 2007; Chen and Chiang, 2008; Vimala et al., 2009) and good antibacterial properties against *E. coli* (Jain and Pradeep, 2005). Ag nanoparticles, incorporated in polysulfone membranes was used as water filtration membranes (Zodrow et al., 2009), and shown good antibacterial properties against *E. coli, Pseudomonas,* and so forth (Chou et al., 2005; Lee et al., 2007).

2.1.3 Gold and Gold-Based Compounds

Gold nanoparticles, nanodots, and nanopowder are brown metal particles, usually with spherical morphology, characterized by high surface area 1–3 m^2/g and particle size of about 20–100 nm. From the environmental protection point of view it was demonstrated that the gold nanoparticles coated with palladium was successfully used as catalysts for removing tri-chloroethane (TCE) from groundwater. The remove capacity was ~2,200 times better than palladium.

2.1.4 Zn and Zinc-Based Compounds

Zinc oxide nanoparticles was successfully used to remove arsenic from water, even though bulk zinc oxide cannot absorb arsenic (Nichols et al., 2006).

2.1.5 Ti and Ti-Based Compounds

TiO_2 with Fe_3O_4 nanoparticles are most promising photocatalysts for wastewater purification (Adesina, 2004; Li et al., 2008a) due to good photoactivity and nontoxicity (Hashimoto et al., 2005). The basic mechanism of TiO_2 photocatalysts involves the production of highly reactive oxidants, such as OH radicals, for water disinfection of microorganisms, bacteria, fungi, algae, and viruses (Hashimoto et al., 2005; Einaga et al., 1999; Fujishima et al., 2000;

Shephard et al., 2002; Ibanez et al., 2003; Liu and Yang, 2003; Cho et al., 2004, 2005; Hajkova et al., 2007; Zan et al., 2007). TiO_2 nanoparticles reduce the viability of some pathogens, such as protozoa, fungi, *E. coli*, and *P. aeruginosa* after only 8 h of exposure to simulated solar radiation, the photocatalytic activity being activated by light (Lonnen et al., 2005).

Based on the literature data, it was demonstrated that photocatalytic activity for inactivation of *E. coli* is size and shape dependent and thus the antibacterial activity of the spherical, rod-like and films of TiO_2 against *E. coli* is very different (Yu et al., 2002; Joo et al., 2005; Shieh et al., 2006). TiO_2 nanocomposites containing multiwalled CNTs showed the complete inactivation of *Bacillus cereus*

In a recent study (Amin et al., 2014), the disinfection efficiency of TiO_2 by inactivation of *Fecalis coliforms* under sunlight (Gelover et al., 2006) was evaluated. Ni et al. (2007) and Fujishima et al. (2008) have improved the photocatalytic capability of TiO_2, by extending its optical absorbance in the visible light region. In order to improve the photocatalytic capability they has doping TiO_2 with transition metals (Subramanian et al., 2001), nitrogen (Asahi et al., 2001; Burda et al., 2003; Irie et al., 2003; Yang and Gao, 2004; Di Valentin et al., 2004; Livraghi et al., 2005, 2006; Liu et al., 2006), carbon (Khan et al., 2002; Sakthivel and Kisch, 2003; Noworyta and Augustynski, 2004; Lin et al., 2005; Rincon et al., 2005; Wang and Lewis, 2005; Mitoraj et al., 2007), sulfur (Umebayashi et al., 2002; Ohno et al., 2003; Yamamoto et al., 2004), and fluorine (Yu et al., 2002). TiO_2 nanoparticles doped with sulfur, and iron have present strong antibacterial effects against *E. coli* (Yu et al., 2005; Egerton et al., 2006). TiO_2 nanoparticles doped with nitrogen was successfully used as catalysts for degradation of microbial contaminants in water (Chaturvedi et al., 2012).

Nanostructured TiO_2 photocatalysts doped with palladium was successfully used as catalysts for disinfection efficiency against *E. coli* (Gram-negative bacterium), *P. aeruginosa*, and *Bacillus subtilis* spores due to photocatalytic oxidation under visible-light illumination (Wu et al., 2008, 2009). By doping of TiO_2 with Ag nanoparticles has improved bacterial inactivation of *E. coli* (Sokmen et al., 2001; Sung-Suh et al., 2004; Vamathevan et al., 2004; Zhang et al., 2005; Page et al., 2007; Liga et al., 2011).

TiO_2 nanoparticles can be incorporated in thin films or membrane filters (Belhacova et al., 1999), in order to use them in the process of decontamination of water against microorganisms and organic pollutants under UV and visible light irradiation. Also, TiO_2 nanoparticle films successfully inactivated *E. coli K12* in distilled water (Alrousan et al., 2009). TiO_2 nanoparticles incorporated in polypropylene polymeric matrix present higher biocidal activity against *Enterococcus faecalis* and *P. aeruginosa* (Kubacka et al., 2009).

2.2 Zeolites in Wastewater Treatment

Zeolites are effective sorbents and ion exchange media for metal ions from acid mine wastewaters.

An example of this type of zeolites is NaP1 zeolites with structural formula $Na_6Al_6 \cdot Si_{10}O_{32}$, $12H_2O$ which have a high density of Na^+ ion exchange sites. Alvarez-Ayuso et al. (2003) use the synthetic NaP1 zeolites in order to remove Cr(III), Ni(II), Zn(II), Cu(II), and Cd(II) from metal electroplating wastewater. This kind of zeolite was easy synthesized by hydrothermal activation of fly ash with low Si/Al ratio at 150°C in 1.0–2.0 M NaOH solutions.

2.3 Carbon Nanotubes and Nanofibres in Wastewater Treatment

Another type of NMS which can be successfully used antimicrobial characteristics agents is CNTs. Numerous studies have shown that the CNTs possess good antimicrobial and antifungal activity against some bacteria, such as *E. coli* and *Salmonella* (Zhu et al., 2006; Deng et al., 2008; Nepal et al., 2008; Upadhyayula et al., 2008; Akasaka and Watari, 2009) and viruses (Brady-Estevez et al., 2008; Mostafavi et al., 2009) due to their special structural characteristics [their fibrous shape (Kang et al., 2007, 2008), the size and length of the tubes, and number of layers] (Li et al., 2008a; Wick et al., 2007) and the physical, chemical, and cytotoxic properties (Upadhyayula et al., 2009). Due to the outstanding physicochemical properties, large SSA, external diameter of CNTs, large composition of mesoporous volume, the adsorption of cyanobacterial toxins on CNTs is also higher when is compared activated carbon (AC) (Yan et al., 2004; Ye et al., 2007; de Albuquerque et al., 2008; Long and Yang, 2001).

The CNTs posses a high capacity to adsorb heavy metals (Rao et al., 2007), such as Cd^{2+} (Li et al., 2003; Di et al., 2006), Cr^{3+} (Di et al., 2006), Pb^{2+} (Li et al., 2005), and Zn^{2+} (Lu et al., 2006) and metalloids (arsenic compounds) (Peng et al., 2005). Also, Fe and cerium oxide nanoparticles supported on CNTs present a high capacity to remove heavy metal ions (Lu et al., 2005;

Lu and Chiu, 2006; Sharma et al., 2009), and cerium oxide nanoparticles supported on CNTs are used mostly to adsorb arsenic (Di et al., 2006). Fast adsorption kinetics of CNTs is mainly due to the highly accessible adsorption sites and the short intraparticle diffusion distance (Rao et al., 2007).

3 Magnetic Nanoparticles in Wastewater Treatment

Recent researches have shown that many of the problems caused by poor water quality could be solved or ameliorated using MNPs or other products of nanotechnology development (Xu et al., 2012; Schulte and Dutta, 2005).

The use of iron-based MNPs in water treatment processes has gained special attention due to their special properties, being highly profitable as adsorbents and for filtration purpose. Some of the most important characteristics which recommend iron-based MNPs in decontamination processes are: high SSA can be obtained by controlling the synthesis conditions, there are superparamagnetic particles, nontoxic and can easily removed by using adequate magnetic field (Ficai et al., 2013; McHenry and Laughlin, 2000; Afkhami et al., 2010; Pan et al., 2010). Also due to the high biocompatibility and chemical inertness iron oxide MNPs can be used in various biotechnological applications (Huang et al., 2003; Roco, 2003; Gupta and Gupta, 2005). The magnetic nanoparticle can be easily synthesized via various synthesis methods and by using adequate stabilizing agents. In function of the synthesis conditions the size distribution, morphology, magnetic properties, and surface chemistry of NMS (Boyer et al., 2010; Dias et al., 2011) can be easily controlled. The literature data presents the synthesis methods and the properties and potential applications of various MNPs via coprecipitation, sol–gel methods, hydrothermal decomposition, and so on (Jeong et al., 2007; Machala et al., 2007; Ficai et al., 2012).

Due to the low chemical stability, and to avoid aggregation magnetite nanoparticles are often treated with surface modification agents (organic or inorganic polymers, such as polymeric materials or SiO_2, and organic multifunctional molecules) in order to increase the functionality, mobility, and chemical and thermal stability (Ficai et al., 2013; Schrick et al., 2004; Kanel et al., 2007; Tiraferri et al., 2008; Huang and Chen, 2009). Bare or functionalized MNPs can be successfully used in wastewater treatment. Depending on the nature of the functional groups attached to the polymer component, these systems can be optimized for retention of heavy metals from various wastewater (functionalization of polymeric shell is made with groups that can complex cations heavy metals) or to decontaminate samples of organic compounds, nitrites, fluoride, and arsenic (functionalization is achieved by immobilization of enzymes, is preferred by covalent immobilization to the detriment of the asset through electrostatic interactions as in this case extends the lifetime of functionalized magnetic system by immobilization of enzymes) (Table 1.1). Thus, the MNPs has being also used as auto catalysts to retain Pb^{2+} from the exhausted gases.

4 Wastewater Treatment Method

Contamination of water with different toxic cations Pb^{2+}, Cu^{2+}, Cr^{3+}, Ni^{2+}, Co^{2+}, Cr^{6+}, Ag^{+}, As^{3+}, As^{5+}, and the presence of various microorganisms (*E. coli*, *Sarcina lutea*, and *S. aureus*, etc.) represent a very delicate and sever environment and public health issue (Zhou et al., 2009a,b). In order to remove these cations and microorganisms from water, various physical, chemical, and biological methods were developed and applied, the most representative methods being: reverse osmosis, absorption, membrane filtration, flotation, ion exchange, precipitation. etc.) (Wei et al., 2009).

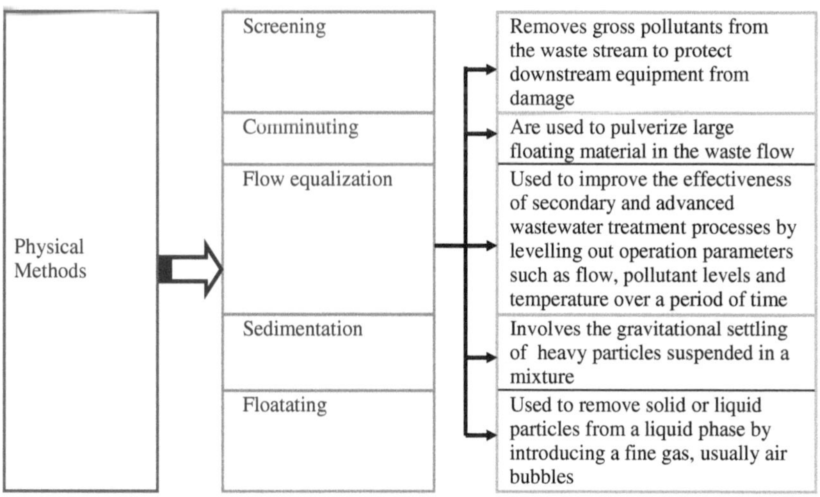

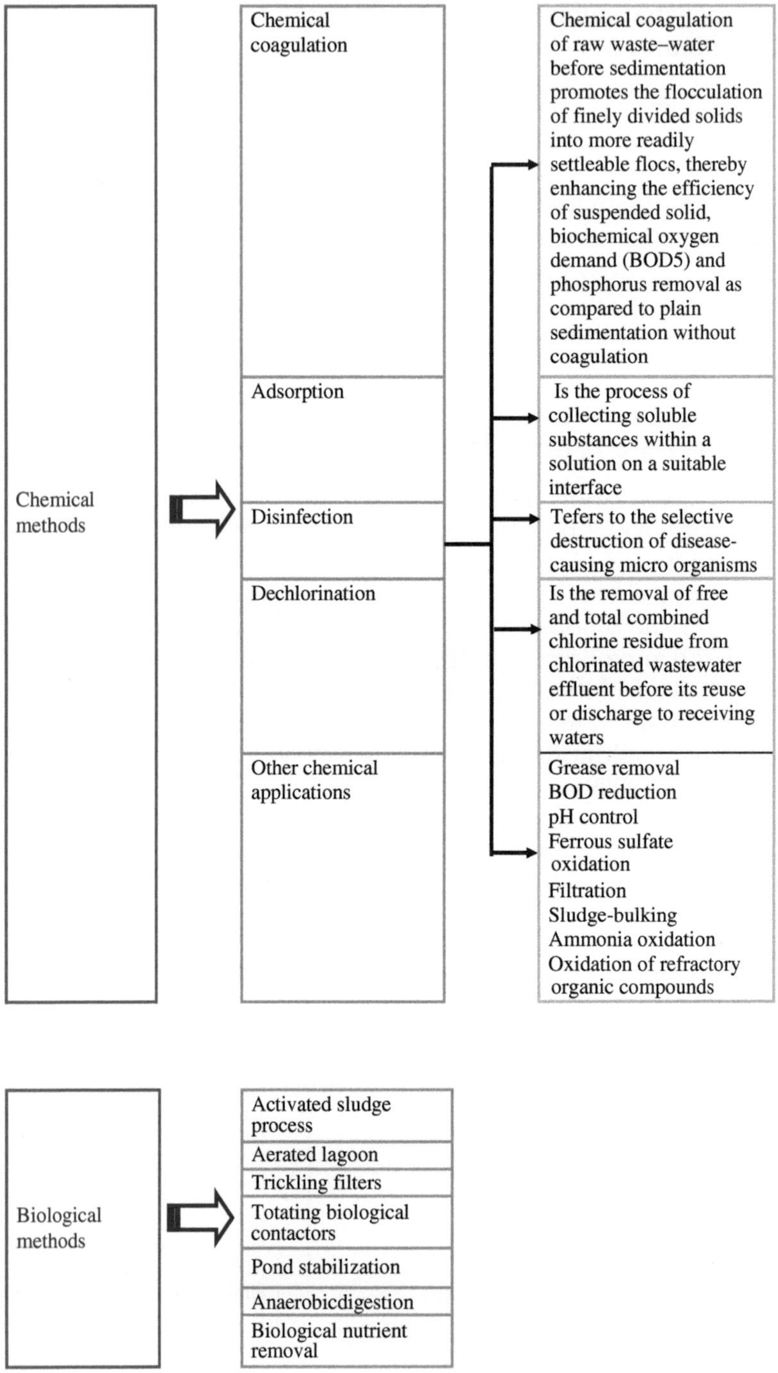
Chemical methods
Chemical coagulation
Chemical coagulation of raw waste–water before sedimentation promotes the flocculation of finely divided solids into more readily settleable flocs, thereby enhancing the efficiency of suspended solid, biochemical oxygen demand (BOD5) and phosphorus removal as compared to plain sedimentation without coagulation
Adsorption
Is the process of collecting soluble substances within a solution on a suitable interface
Disinfection
Tefers to the selective destruction of disease-causing micro organisms
Dechlorination
Is the removal of free and total combined chlorine residue from chlorinated wastewater effluent before its reuse or discharge to receiving waters
Other chemical applications
Grease removal
BOD reduction
pH control
Ferrous sulfate oxidation
Filtration
Sludge-bulking
Ammonia oxidation
Oxidation of refractory organic compounds
Biological methods
Activated sludge process
Aerated lagoon
Trickling filters
Totating biological contactors
Pond stabilization
Anaerobicdigestion
Biological nutrient removal

The most efficient technique for removing the toxic cation is the absorption due to its easy handling, low cost, and high efficiency.

4.1 Absorption Methods

To establish the optimal method and material for wastewater treatment is a sever environment and public health issue, which depends on a number of factors, such as the standards to be respected, the absorption capacity, and the cost (Huang et al., 2008; Oller et al., 2011).

Over the time many researchers from various research areas (environmental, chemical, biotechnological, and material science) have collaborated together to obtain new absorbents that can solve wastewater treatment problems (Reddy and Yuna, 2016; Reddy and Lee, 2014, 2013a).

In many published papers the preparation, characterization, and application of a range of adsorbents including: carbon nanotubes, magnetic multifunctional composites (MCs) (Reddy and Lee, 2013b), ordered mesoporous materials (Wu et al., 2011), organic/inorganic nanoadsorbents (Ray and Shipley, 2015), graphene (Yu et al., 2015), ACs (Hadi et al., 2015), zeolites, (Figueiredo and Quintelas, 2014), and dendrimers has been described. Although AC was the most used material in wastewater treatment, being still present in commercial applications but presents some disadvantages which limit its applications.

- After extensive usage, pollutant loaded AC must be separated from the aqueous solution and regenerated by filtration or centrifugation process (Clifford et al., 1983).
- Due to its nonpolar surface, it is inappropriate for inorganic contaminant removal.
- Also exhibits a low adsorption affinity for low molecular weight polar organic pollutants (Qu et al., 2013).

To be used as absorbent the materials must meet a series of conditions: high performance, high absorption capacity, rapid absorption kinetic, environmentally nontoxic, high SSA, reusability, and ease of separation.

Nowadays, MNPs with spinel structure (Reddy and Yuna, 2016) are considered promising candidates for industrial wastewater treatment by absorption method, due to strong adsorption capacity, easy separation by applying a magnetic field (Fan et al., 2012; Hu et al., 2005a; Carabante et al., 2009), enhanced stability, high specific surface area assured by the use of the adequate stabilizing agents, superparamagnetic properties, high thermal, mechanical, chemical, and corrosive stability (Liu et al., 2000a, 2000b).

Table 1.2 Size—Absorption Capacity of MNPs

Type of Absorbent	Size of Nanoparticles (nm)	Pollutant Removed	Observation	References
$MnFe_2O_4$	4–6 (TEM)	Pb^{2+}	pH = 6.8 ± 0.4	Bateer et al. (2014)
	20–40 (TEM)	Cr^{6+}	pH = 2	Ahalya et al. (2014)
	~11 (XRD)	Pb^{2+}	pH = 5; desorption and regeneration was made in presence of 0.2 M HCl	Kumar et al. (2014)
		As^{3+}	pH = 6.5; desorption and regeneration was made in presence of 1 M NaOH	
		As^{5+}	pH = 4; desorption and regeneration was made in presence of 1 M NaOH	
$CuFe_2O_4$	<100 (SEM)	Pb^{2+}	—	Bakhshayesh and Dehghani (2014)
	20–90 (SEM)	Cd^{2+}	pH = 6	Tu et al. (2012)
	60 (SEM)	As^{3+}	pH = 4.2; desorption and regeneration was made in presence of 0.1 M H_3PO_4	Tu et al. (2013)

The absorption capacity of MNPs is also strongly influenced by particle size and shape and the nature of the shell. There have been several attempts to prepare MNPs that have a particular size and/or shape in order to be used in a specific application (Reddy and Yuna, 2016) (Table 1.2).

Hu et al. (2005b) tested the $MnFe_2O_4$ nanoparticles (with a diameter about of 10 nm and a surface area of 208 m^2/g) for remove Cr(VI) from wastewater. They shown that these nanoparticles exhibit sorption capacity of 31.6 mg/g (Hu et al., 2005b). Duan et al. (2015) prepared $Co_{0.6}Fe_{2.4}O_4$ microparticles using a thermal decomposition method with an adsorption capacity of 80.3 mg Pb(II) g^{-1}. Wei et al. (2014) used cobalt ferrite hollow spheres with a surface area of 230 m^2/g in order to remove U(VI) from aqueous solutions. They show that the binding capacity was 170 mg/g and the equilibrium was obtained after 180 min. Jia et al. (2015) used hierarchical one-dimensional $ZnFe_2O_4$ nanorods (100–200 nm) for Cr(VI) removal with a maximum adsorption capacity of 34.2 mg/g, the equilibrium being quickly reached after 10 min.

The sorption performance of MNPs and their derivatives can be evaluated using several isotherm and kinetic equations:

Table 1.3 Isoterm and Kinetic Equation used for Evaluated Sorbtion Performance

	Isotherm Equation			Kinetics Equation	
	Linear	**Nonlinear**		**Linear**	**Nonlinear**
Langmuir (Badruddoza et al., 2011)	$C_e/q_e = (1/K_L Q_m) + (C_e/Q_m)$	$q_e = Q_m K_{Lce}/(1 + K_{Lce})$	Pseudo first order (Wang et al., 2011)	$\log(q_e - q_t) = \log(q_e) - k_1/2.303t$	$q_t = q_e(1 - e^{-k_1 t})$
	C_e = equilibrium concentration of solute (mg/L), q_e = solute amount adsorbed/unit weight at equilibrium (mg/g), Q_m = maximum monolayer Langmuir sorption capacity, K_L = Langmuir constant			k_1 = the first order rate equation, q_t = the amount adsorbed at time "t", q_e = the amount adsorbed at equilibrium	
Freundlich (Liu et al., 2008)	$\log q_e = \log K_F + (1/n)\log C_e$	$q_e = K_F C_e^{1/n}$	Pseudo second order (Chen et al., 2011; Wei et al., 2009)	$t/q_t = 1/k_2 q_e^2 + t/q_e$	$q_t = q_e^2 k_2 t/(1 + q_e k_2 t)$
	C_e = the equilibrium concentration of adsorbate (mg/L), q_e = the adsorbate amount adsorbed per unit weight at equilibrium (mg/g), K_F = Freundlich constant, $1/n$ = the adsorption intensity			k_2 = the second order rate equation, q_t = the amount of adsorbate adsorbed at time t, q_e = the amount of adsorbate adsorbed at and equilibrium	

Langmuir (1918) and Freundlich (1906), pseudo-first order (PFO) (Lagergren, 1898) and pseudo-second order (PSO) (Ho and McKay, 1999; Ho, 2006) (Table 1.3).

4.2 Photocatalysis Methods

Photocatalysis is another wastewater purification technique consisting in degradation of organic pollutants under the influence of light (Akhavan and Azimirad, 2009). Although it is one of the advanced technologies of water purification technology that shows some disadvantages in the use of MNPs in photocatalysis process of toxic compounds. One of the most important advantage of this method is related to the easy and cost-effective separation of the photocatalyst after the treatment (limited labor costs, fast precipitation without costly chemicals), usually by centrifugation or decanting but the low yield of the treatment process limits the kinetics and efficiency (Bandara et al., 2007) can be considered a strong disadvantage.

In order to increase the catalytic activity the researchers have tried to synthesize metal nanoparticles as small as possible in order to increase the SSA, also the photocatalytic nanoparticles are combined with other new metal nanoparticles (Zhang and Fang, 2010).

Over the years a number of scientific experiments showed that the iron oxide with core–shell structure shows a good photocatalytic activity in visible light, due to generation of electron–hole pairs through the narrow band-gap (Bandara et al., 2007). Because it can be removed easily after completion of the photocatalytic reaction, the core–shell NMS are used successfully in photochemical processes of heterogeneous catalysis degradation of organic contaminants from wastewater (Senapati et al., 2012; Tian et al., 2011).

Singh et al. (2015) synthesized Fe_3O_4–Cr_2O_3(MNCs) core/shell structure with an average size of 50 ± 5 nm using ultrasonication assisted coprecipitation method in order to use this material as photocatalytic agent in degradation of 4-chlorophenol in aqueous solution under UV irradiation. The degradation of the of 4-chlorophenol was approximately 20% after the first 0.5 h, 50% after 1.5 h, and finally nearly 100% after 2.5 h irradiation with UV light due to high band-gap energy (2.94 eV) and the formation of hetero-junction (p–n junction) between the Fe_3O_4 and Cr_2O_3 interface (Chen et al., 2008).

Li et al. (2015) prepared by a sol–gel method, a photocatalyst of magnetic-modified sepiolite supported with nanosized titanium dioxide (TiO_2/MMS). The TiO_2 nanoparticles used for synthesis of TiO_2/MMS was mainly anatase phase. By the experimental data they show that, in comparison with TiO_2 nanoparticles, the TiO_2/MMS system exhibited enhanced higher photocatalytic activity and adsorption ability, after 120 min of UV irradiation 90.9% of the total 2,4-dichlorophenol (DCP) was degraded. In cases of the simultaneous photocatalytic treatment of Cr(VI) and DCP the removal efficiency of Cr(VI) and DCP was almost 100% and 96.9%, respectively. The obtained results revealed that the magnetic sepiolite supported TiO_2 can be successfully used as photocatalyst for simultaneous decontamination of Cr(VI) and DCP.

Zhanga et al. (2016) has synthesized PANI (polianiline)/Fe_3O_4/ZnO photocatalysts microspheres with core–shell structure using different PANI:ZnO ratios by Pickering emulsion route, in one step. The PANI/Fe_3O_4/ZnO microspheres were obtained by incorporation of Fe_3O_4-OA (oleic acid modified Fe_3O_4) nanoparticles and tunable ZnO in PANI thin shell. They tested the catalytic activity of the PANI/Fe_3O_4/ZnO microspheres by the degradation of Methylen Blue (MB) under visible light irradiation and they proved that the PANI/Fe_3O_4/ZnO photocatalyst present enhanced photocatalytic activity owing to fast separation of photo-generated electron-hole pairs. The PANI/Fe_3O_4/ZnO microspheres photocatalysts can be reused for seven cycles without modified the stability and degradation efficiency. The photocatalytic activity of PANI/Fe_3O_4/ZnO was tested also on Rhodamine B (RhB) under UV light in aqueous environment and the obtained results shows excellent efficiency toward the photo-degradation of RhB as well as high stability and recycling.

Zhang et al. (2016) prepared Fe_3O_4/TiO_2/Ag composite microspheres in two steps via a facile chemical route. In the first step the Fe_3O_4/TiO_2 microspheres were prepared by the solvothermal method, while in the second step Ag nanoparticles were anchored onto the out-layer of TiO_2 by the tyrosine-reduced method. The photocatalytic activity of the Fe_3O_4/TiO_2 microspheres was tested for the photocatalytic degradation of MB. The results indicate that the photocatalytic activity of Fe_3O_4/TiO_2/Ag composite microspheres is superior to that of Fe_3O_4/TiO_2 due to the dual effects of the enhanced light absorption and reduction of photoelectron-hole pair recombination in TiO_2 with the introduction of Ag/NPs. Moreover, these magnetic Fe_3O_4/TiO_2/Ag composite microspheres can be completely removed from the dispersion with the help of magnetic separation and reused with little or no loss of catalytic activity.

Hu et al. (2011) prepared TiO_2/SiO_2@Fe_3O_4 photocatalysts with core–shell structure (TiO_2/Fe_3O_4 core and a porous TiO_2 shell) by sol–gel. The photocatalytic activity of TiO_2/SiO_2@Fe_3O_4 composite was tested for the degradation of methyl orange and MB dyes. The experimental results proved that the two different dyes were completely decolorized in 60 min under UV irradiation and also shown that after recycling for 6 times by using external magnetic field the amount of catalyst and the photocatalytic activity were almost not decrease. Also, TiO_2/SiO_2/Fe_3O_4 (TSF) electrode, obtained by dispersing of the TSF nanoparticles in ethanol followed by their immobilization on a graphite electrode surface, in order to be used for photoelectrocatalytic (PEC) exhibit good photodegradation of diclofenac. They showed that 45 min PEC treatment, 95.3% of diclofenac was degraded on the magnetically attached TSF electrode.

4.3 Desalination

Desalination is another alternative wastewater treatment for obtaining fresh water. Though expensive, membrane based desalination processes cover most of the desalination capability out of which only reverse osmosis accounts for 41% (Humplik et al., 2011).

NMS are very useful in developing membrane filters for water/wastewater treatment and desalination (Srivastava et al., 2004) due to they offer opportunities to control the cost of desalination process and increase its energy efficiency and among these are CNTs (Holt et al., 2006; Fornasiero et al., 2008), zeolites (Li et al., 2004, 2008b), and more recently graphene (Sint et al., 2008; Jiang et al., 2009; Bai et al., 2010).

Thin film nanocomposite membranes containing Ag and TiO_2 nanoparticles was used successfully because they exhibit good salt rejection (Lee et al., 2007, 2008) depending on the number of coatings in TiO_2/Al_2O_3 (aluminiumoxide) composite ceramic membranes coated by iron. By functionalization the surface on the CNTs with carboxyl groups, membranes have better selective rejection of some components but this has resulted in reduced permeability rendering CNTs incapable for desalination (Nednoor et al., 2005; Mauter and Elimelech, 2008). In conclusion, a uniform CNT diameter of less than 0.8 nm confer high salt rejection (Hinds, 2012).

5 Conclusions

Nanotechnology is rapidly penetrating the most of the markets, including the environmental one. Metal and metal-oxides, carbon based materials (including active carbon, carbon nanotubes, but also graphene and graphene oxide), zeolites and so on are increasingly studied for wastewater treatment especially because nowadays they are economically and technically accessible, exhibit low environmental toxicity and the national and international protocols impose strict standards.

Magnetic materials and especially magnetite prove high interest for high-tech applications for industrial, environmental, and medical applications. Moreover, the remarkable properties of the magnetite can be further improved by developing proper core–shell structures, the end groups of the shell molecules being able to interact with various species, including hazardous ions, organic molecules (dyes, antibiotics, pesticides), microorganisms, and so on. Amino, carboxyl, and thiol end groups are extremely exploited in order to remove hazardous agents form wastewaters. The most important mechanism for removal of the hazardous species seems to be the absorption because of the high affinity of these groups to many ions as well as organic compounds. The high interest for magnetic materials is revealed due to the easy removal of these NMS after the removal of the hazardous species and, sometimes, these can be regenerate and reuse for more times.

Acknowledgement

The present work was possible due to the EU-funding grant POSCCE-A2-O2.2.1-2013-1/Axa prioritara 2, Project No. 638/12.03.2014, cod SMIS-CSNR 48652.

References

Adesina, A.A., 2004. Industrial exploitation of photocatalysis: progress, perspectives and prospects. Catal. Surv. Asia 8, 265–273.

Afkhami, A., Saber-Tehrani, M., Bagheri, H., 2010. Modified maghemite nanoparticles as an efficient adsorbent for removing some cationic dyes from aqueous solution. Desalination 263, 240–248.

Ahalya, K., Suriyanarayanan, N., Sangeetha, S., 2014. Effect of pH and annealing temperatures on structural, magnetic, electrical, dielectric and adsorption properties of manganese ferrite nano particles. Mat. Sci. Semicon. Process. 27, 672–681.

Akasaka, T., Watari, F., 2009. Capture of bacteria by flexible carbon nanotubes. Acta Biomater 5, 607–612.

Akhavan, O., Azimirad, R., 2009. Photocatalytic property of Fe_2O_3 nanograin chains coated by TiO_2 nanolayer in visible light irradiation. Appl. Catal. A-Gen. 369, 77–82.

Alrousan, D.M.A., Dunlop, P.S.M., McMurray, T.A., Byrne, J.A., 2009. Photocatalytic inactivation of *E. coli* in surface water using immobilised nanoparticle TiO_2 films. Water Res. 43, 47–54.

Alvarez-Ayuso, E., Garcia-Sanchez, A., Querol, X., 2003. Purification of metal electroplating wastewaters using zeolites. Water Res. 37, 4855–4862.

Amin, M.T., Alazba, A.A., Manzoor, U., 2014. A review of removal of pollutants from water/wastewater using different types of nanomaterials. Adv. Mater. Sci. Eng. 2014.

Asahi, R., Morikawa, T., Ohwaki, T., Aoki, K., Taga, Y., 2001. Visible-light photocatalysis in nitrogen-doped titanium oxides. Science 293, 269–271.

Badruddoza, A.Z.M., Tay, A.S.H., Tan, P.Y., Hidajat, K., Uddin, M.S., 2011. Carboxymethyl-beta-cyclodextrin conjugated magnetic nanoparticles as nano-adsorbents for removal of copper ions: synthesis and adsorption studies. J. Hazard. Mater. 185, 1177–1186.

Bai, J.W., Zhong, X., Jiang, S., Huang, Y., Duan, X.F., 2010. Graphene nanomesh. Nat. Nanotechnol. 5, 190–194.

Baker, C., Pradhan, A., Pakstis, L., Pochan, D.J., Shah, S.I., 2005. Synthesis and antibacterial properties of silver nanoparticles. J. Nanosci. Nanotechno. 5, 244–249.

Bakhshayesh, S., Dehghani, H., 2014. Nickel and cobalt ferrites nanoparticles: synthesis, study of magnetic properties and their use as magnetic adsorbent for removing lead(II) ion. J. Iran. Chem. Soc. 11, 769–780.

Balogh, L., Swanson, D.R., Tomalia, D.A., Hagnauer, G.L., McManus, A.T., 2001. Dendrimer-silver complexes and nanocomposites as antimicrobial agents. Nano Lett. 1, 18–21.

Bandara, J., Klehm, U., Kiwi, J., 2007. Raschig rings-Fe_2O_3 composite photocatalyst activate in the degradation of 4-chlorophenol and Orange II under daylight irradiation. Appl. Catal. B Environ. 76, 73–81.

Bateer, B., Tian, C.G., Qu, Y., Du, S.C., Yang, Y., Ren, Z.Y., et al., 2014. Synthesis, size and magnetic properties of controllable $MnFe_2O_4$ nanoparticles with versatile surface functionalities. Dalton T 43, 9885–9891.

Belhacova, L., Krysa, J., Geryk, J., Jirkovsky, J., 1999. Inactivation of microorganisms in a flow-through photoreactor with an immobilized TiO_2 layer. J. Chem. Technol. Biotechnol. 74, 149–154.

Boyer, C., Whittaker, M.R., Bulmus, V., Liu, J.Q., Davis, T.P., 2010. The design and utility of polymer-stabilized iron-oxide nanoparticles for nanomedicine applications. NPG Asia Mater. 2, 23–30.

Brady-Estevez, A.S., Kang, S., Elimelech, M., 2008. A single-walled-carbon-nanotube filter for removal of viral and bacterial pathogens. Small 4, 481–484.

Burda, C., Lou, Y.B., Chen, X.B., Samia, A.C.S., Stout, J., Gole, J.L., 2003. Enhanced nitrogen doping in TiO_2 nanoparticles. Nano Lett. 3, 1049–1051.

Carabante, I., Grahn, M., Holmgren, A., Kumpiene, J., Hedlund, J., 2009. Adsorption of as (V) on iron oxide nanoparticle films studied by in situ ATR-FTIR spectroscopy. Colloid Surf. A 346, 106–113.

Chaturvedi, S., Dave, P.N., Shah, N.K., 2012. Applications of nano-catalyst in new era. J. Saudi Chem. Soc. 16, 307–325.

Chen, C.Y., Chiang, C.L., 2008. Preparation of cotton fibers with antibacterial silver nanoparticles. Mater Lett. 62, 3607–3609.

Chen, Y., Wang, L., Jiang, S., Yu, H.J., 2003. Study on novel antibacterial polymer materials (I) preparation of zeolite antibacterial agents and antibacterial polymer composite and their antibacterial properties. J. Polym. Mater. 20, 279–284.

Chen, S.F., Zhang, S.J., Liu, W., Zhao, W., 2008. Preparation and activity evaluation of p–n junction photocatalyst NiO/TiO_2. J. Hazard. Mater. 155, 320–326.

Chen, R.Z., Zhi, C.Y., Yang, H., Bando, Y., Zhang, Z.Y., Sugiur, N., et al., 2011. Arsenic (V) adsorption on Fe_3O_4 nanoparticle-coated boron nitride nanotubes. J. Colloid Interface Sci. 359, 261–268.

Cho, M., Chung, H., Choi, W., Yoon, J., 2004. Linear correlation between inactivation of *E-coli* and OH radical concentration in TiO_2 photocatalytic disinfection. Water Res. 38, 1069–1077.

Cho, M., Chung, H.M., Choi, W.Y., Yoon, J.Y., 2005. Different inactivation Behaviors of MS-2 phage and *Escherichia coli* in TiO_2 photocatalytic disinfection. Appl. Environ. Microb. 71, 270–275.

Chong, M.N., Jin, B., Chow, C.W.K., Saint, C., 2010. Recent developments in photocatalytic water treatment technology: a review. Water Res. 44, 2997–3027.

Chou, W.L., Yu, D.G., Yang, M.C., 2005. The preparation and characterization of silver-loading cellulose acetate hollow fiber membrane for water treatment. Polym. Adv. Technol. 16, 600–607.

Clifford, D., Chu, P., Lau, A., 1983. Thermal regeneration of powdered activated carbon (pac) and pac-biological sludge mixtures. Water Res. 17, 1125–1138.

Cundy, A.B., Hopkinson, L., Whitby, R.L.D., 2008. Use of iron-based technologies in contaminated land and groundwater remediation: a review. Sci. Total Environ. 400, 42–51.

Danilczuk, M., Lund, A., Sadlo, J., Yamada, H., Michalik, J., 2006. Conduction electron spin resonance of small silver particles. Spectrochim. Acta A 63, 189–191.

Dastjerdi, R., Montazer, M., 2010. A review on the application of inorganic nano-structured materials in the modification of textiles: focus on anti-microbial properties. Colloid Surf. B 79, 5–18.

de Albuquerque, E.C., Mendez, M.O.A., Coutinho, A.D., Franco, T.T., 2008. Removal of cyanobacteria toxins from drinking water by adsorption on activated carbon fibers. Mater. Res. Ibero Am. J. 11, 371–380.

Deng, S.G., Upadhyayula, V.K.K., Smith, G.B., Mitchell, M.C., 2008. Adsorption equilibrium and kinetics of microorganisms on single-wall carbon nanotubes. IEEE Sens. J. 8, 954–962.

Di, Z.C., Ding, J., Peng, X.J., Li, Y.H., Luan, Z.K., Liang, J., 2006. Chromium adsorption by aligned carbon nanotubes supported ceria nanoparticles. Chemosphere 62, 861–865.

Di Valentin, C., Pacchioni, G., Selloni, A., 2004. Origin of the different photoactivity of N-doped anatase and rutile TiO_2. Phys. Rev. B, 70.

Dias, A.M.G.C., Hussain, A., Marcos, A.S., Roque, A.C.A., 2011. A biotechnological perspective on the application of iron oxide magnetic colloids modified with polysaccharides. Biotechnol. Adv. 29, 142–155.

Dimitrov, D.S., 2006. Interactions of antibody-conjugated nanoparticles with biological surfaces. Colloid Surf. A 282, 8–10.

Duan, S.X., Tang, R.F., Xue, Z.C., Zhang, X.X., Zhao, Y.Y., Zhang, W., et al., 2015. Effective removal of Pb(II) using magnetic $Co_{0.6}Fe_{2.4}O_4$ micro-particles as the adsorbent: synthesis and study on the kinetic and thermodynamic behaviors for its adsorption. Colloid Surf. A 469, 211–223.

Egerton, T.A., Kosa, S.A.M., Christensen, P.A., 2006. Photoelectrocatalytic disinfection of *E. coli* suspensions by iron doped TiO_2. Phys Chem Chem Phys 8, 398–406.

Einaga, H., Futamura, S., Ibusuki, T., 1999. Photocatalytic decomposition of benzene over TiO_2 in a humidified airstream. Phys. Chem. Chem. Phys. 1, 4093–4908.

Fan, F.L., Qin, Z., Bai, J., Rong, W.D., Fan, F.Y., Tian, W., et al., 2012. Rapid removal of uranium from aqueous solutions using magnetic $Fe_3O_4@SiO_2$ composite particles. J. Environ. Radioactiv. 106, 40–46.

Ficai, D., Andronescu, E., Ficai, A., Voicu, G., Vasile, B., Ionita, V., et al., 2012. Synthesis and characterization of mesoporous magnetite based nanoparticles. Curr. Nanosci. 8, 875–879.

Ficai, D., Ficai, A., Trusca, R., Vasile, B.S., Voicu, G., Guran, C., et al., 2013. Synthesis and characterization of magnetite-polysulfone micro- and nanobeads with improved chemical stability in acidic media. Curr. Nanosci. 9, 271–277.

Figueiredo, H., Quintelas, C., 2014. Tailored zeolites for the removal of metal oxyanions: overcoming intrinsic limitations of zeolites. J. Hazard. Mater. 274, 287–299.

Fornasiero, F., Park, H.G., Holt, J.K., Stadermann, M., Grigoropoulos, C.P., Noy, A., et al., 2008. Ion exclusion by sub-2-nm carbon nanotube pores. P Natl. Acad. Sci. USA 105, 17250–17255.

Freundlich, H., 1906. Uber die adsorption in lusungen. Z. Phys. Chem. 57, 385–470.

Friedrich, K.A., Henglein, F., Stimming, U., Unkauf, W., 1998. Investigation of Pt particles on gold substrates by IR spectroscopy—particle structure and catalytic activity. Colloid Surf. A 134, 193–206.

Fujishima, A., Rao, T.N., Tryk, D.A., 2000. Titanium dioxide photocatalysis. J. Photochem. Photobiol. C: Photochem. Rev. 1, 1–21.

Fujishima, A., Zhang, X.T., Tryk, D.A., 2008. TiO_2 photocatalysis and related surface phenomena. Surf. Sci. Rep. 63, 515–582.

Furno, F., Morley, K.S., Bong, B., Sharp, B.L., Arnold, P.L., Howdle, S.M., et al., 2004. Silver nanoparticle and polymeric medical device: a new approach to prevention of infection. J. Anti. Chemo. 54, 1019–1024.

Ge, F., Li, M.M., Ye, H., Zhao, B.X., 2012. Effective removal of heavy metal ions Cd^{2+}, Zn^{2+}, Pb^{2+}, Cu^{2+} from aqueous solution by polymer-modified magnetic nanoparticles. J. Hazard. Mater. 211, 366–372.

Gelover, S., Gomez, L.A., Reyes, K., Leal, M.T., 2006. A practical demonstration of water disinfection using TiO_2 films and sunlight. Water Res. 40, 3274–3280.

Ghasemzadeh, G., Momenpour, M., Omidi, F., Hosseini, M.R., Ahani, M., Barzegari, A., 2014. Applications of nanomaterials in water treatment and environmental remediation. Front. Env. Sci. Eng. 8, 471–482.

Gupta, A.K., Gupta, M., 2005. Synthesis and surface engineering of iron oxide nanoparticles for biomedical applications. Biomaterials 26, 3995–4021.
Hadi, P., Xu, M., Ning, C., Lin, C.S.K., McKay, G., 2015. A critical review on preparation, characterization and utilization of sludge-derived activated carbons for wastewater treatment. Chem. Eng. J. 260, 895–906.
Hajkova, P., Spatenka, P., Horsky, J., Horska, I., Kolouch, A., 2007. Photocatalytic Effect of TiO_2 Films on Viruses and Bacteria. Plasma Process Polym. 4, S397–S401.
Hanna, K., Kone, T., Ruby, C., 2010. Fenton-like oxidation and mineralization of phenol using synthetic Fe(II)–Fe(III) green rusts. Environ. Sci. Pollut. Res. 17, 124–134.
Hashimoto, K., Irie, H., Fujishima, A., 2005. TiO_2 photocatalysis: a historical overview and future prospects. Jpn. J. Appl. Phys. 44 (1), 8269–8285.
Hinds, B., 2012. Dramatic transport properties of carbon nanotube membranes for a robust protein channel mimetic platform. Curr. Opin. Solid. State Mater. Sci. 16, 1–9.
Ho, Y.S., 2006. Second-order kinetic model for the sorption of cadmium onto tree fern: a comparison of linear and non-linear methods. Water Res. 40, 119–125.
Ho, Y.S., McKay, G., 1999. Pseudo-second order model for sorption processes. Process Biochem. 34, 451–465.
Holt, J.K., Park, H.G., Wang, Y.M., Stadermann, M., Artyukhin, A.B., Grigoropoulos, C.P., et al., 2006. Fast mass transport through sub-2-nanometer carbon nanotubes. Science 312, 1034–1037.
Horikoshi, S., Serpone, N., 2009. Photochemistry with microwaves Catalysts and environmental applications. J. Photochem. Photobiol. C 10, 96–110.
Hu, J., Chen, G.H., Lo, I.M.C., 2005a. Removal and recovery of Cr(VI) from wastewater by maghemite nanoparticles. Water Res. 39, 4528–4536.
Hu, J., Lo, I.M.C., Chen, G.H., 2005b. Fast removal and recovery of Cr(VI) using surface-modified jacobsite (MnFe204) nanoparticles. Langmuir 21, 11173–11179.
Hu, X.Y., Yang, J., Zhang, J.D., 2011. Magnetic loading of $TiO_2/SiO_2/Fe_3O_4$ nanoparticles on electrode surface for photoelectrocatalytic degradation of diclofenac. J. Hazard. Mater. 196, 220–227.
Huang, S.H., Chen, D.H., 2009. Rapid removal of heavy metal cations and anions from aqueous solutions by an amino-functionalized magnetic nano-adsorbent. J. Hazard. Mater. 163, 174–179.
Huang, S.H., Liao, M.H., Chen, D.H., 2003. Direct binding and characterization of lipase onto magnetic nanoparticles. Biotechnol. Prog. 19, 1095–1100.
Huang, D.L., Zeng, G.M., Jiang, X.Y., Feng, C.L., Yu, H.Y., Huang, G.H., et al., 2006. Bioremediation of Pb-contaminated soil by incubating with *Phanerochaete chrysosporium* and straw. J. Hazard. Mater. 134, 268–276.
Huang, D.L., Zeng, G.M., Feng, C.L., Hu, S., Jiang, X.Y., Tang, L., et al., 2008. Degradation of lead-contaminated lignocellulosic waste by *Phanerochaete chrysosporium* and the reduction of lead toxicity. Environ. Sci. Technol. 42, 4946–4951.
Huang, D.L., Zeng, G.M., Feng, C.L., Hu, S.A., Zhao, M.H., Lai, C., et al., 2010. Mycelial growth and solid-state fermentation of lignocellulosic waste by white-rot fungus *Phanerochaete chrysosporium* under lead stress. Chemosphere 81, 1091–1097.
Humplik, T., Lee, J., O'Hern, S.C., Fellman, B.A., Baig, M.A., Hassan, S.F., et al., 2011. Nanostructured materials for water desalination. Nanotechnology, 22.
Ibanez, J.A., Litter, M.I., Pizarro, R.A., 2003. Photocatalytic bactericidal effect of TiO_2 on *Enterobacter cloacae* comparative study with other Gram (−) bacteria. J. Photoch. Photobio. A Chem. 157, 81–85.

Inoue, Y., Hoshino, M., Takahashi, H., Noguchi, T., Murata, T., Kanzaki, Y., et al., 2002. Bactericidal activity of Ag-zeolite mediated by reactive oxygen species under aerated conditions. J. Inorg. Biochem. 92, 37–42.

Irie, H., Watanabe, Y., Hashimoto, K., 2003. Nitrogen-concentration dependence on photocatalytic activity of TiO_2-xNx powders. J. Phys. Chem. B 107, 5483–5486.

Jain, P., Pradeep, T., 2005. Potential of silver nanoparticle-coated polyurethane foam as an antibacterial water filter. Biotechnol. Bioeng. 90, 59–63.

Jeong, U., Teng, X.W., Wang, Y., Yang, H., Xia, Y.N., 2007. Superparamagnetic colloids: controlled synthesis and niche applications. Adv. Mater. 19, 33–60.

Jia, Z.G., Qin, Q., Liu, J.H., Shi, H., Zhang, X.X., Hu, R.N., et al., 2015. The synthesis of hierarchical $ZnFe_2O_4$ architecture and their application for Cr(VI) adsorption removal from aqueous solution. Superlattice Microst. 82, 174–187.

Jiang, X.Y., Zeng, G.M., Huang, D.L., Chen, Y., Liu, F., Huang, G.H., et al., 2006. Remediation of pentachlorophenol-contaminated soil by composting with immobilized *Phanerochaete chrysosporium*. World J. Microbiol. Biotechnol. 22, 909–913.

Jiang, D.E., Cooper, V.R., Dai, S., 2009. Porous graphene as the ultimate membrane for gas separation. Nano Lett. 9, 4019–4024.

Joo, J., Kwon, S.G., Yu, T., Cho, M., Lee, J., Yoon, J., et al., 2005. Large-scale synthesis of TiO_2 nanorods via nonhydrolytic sol–gel ester elimination reaction and their application to photocatalytic inactivation of *E. coli*. J. Phys. Chem. B 109, 15297–15302.

Kanel, S.R., Nepal, D., Manning, B., Choi, H., 2007. Transport of surface-modified iron nanoparticle in porous media and application to arsenic(III) remediation. J. Nanopart. Res. 9, 725–735.

Kang, S., Pinault, M., Pfefferle, L.D., Elimelech, M., 2007. Single-walled carbon nanotubes exhibit strong antimicrobial activity. Langmuir 23, 8670–8673.

Kang, S., Herzberg, M., Rodrigues, D.F., Elimelech, M., 2008. Antibacterial effects of carbon nanotubes: size does matter. Langmuir 24, 6409–6413.

Khan, S.U.M., Al-Shahry, M., Ingler, W.B., 2002. Efficient photochemical water splitting by a chemically modified *n*-TiO_2. Science 297, 2243–2245.

Kim, Y., Lee, B., Yi, J., 2003. Preparation of functionalized mesostructured silica containing magnetite (MSM) for the removal of copper ions in aqueous solutions and its magnetic separation. Separ. Sci. Technol. 38, 2533–2548.

Kim, J.S., Kuk, E., Yu, K.N., Kim, J.H., Park, S.J., Lee, H.J., et al., 2007. Antimicrobial effects of silver nanoparticles. Nanomed.:Nanotechnol. Biol. Med. 3, 95–101.

Kubacka, A., Ferrer, M., Cerrada, M.L., Serrano, C., Sanchez-Chaves, M., Fernandez-Garcia, M., et al., 2009. Boosting TiO_2-anatase antimicrobial activity: polymer-oxide thin films. Appl. Catal. B-Environ. 89, 441–447.

Kumar, V.S., Nagaraja, B.M., Shashikala, V., Padmasri, A.H., Madhavendra, S.S., Raju, B.D., et al., 2004. Highly efficient Ag/C catalyst prepared by electro-chemical deposition method in controlling microorganisms in water. J. Mol. Catal. A-Chem. 223, 313–319.

Kumar, S., Nair, R.R., Pillai, P.B., Gupta, S.N., Iyengar, M.A.R., Sood, A.K., 2014. Graphene oxide-$MnFe_2O_4$ magnetic nanohybrids for efficient removal of lead and arsenic from water. ACS Appl. Mater. Interface 6, 17426–17436.

Lagergren, S., 1898. Handlingar 24, 1–39.

Lala, N.L., Ramaseshan, R., Li, B.J., Sundarrajan, S., Barhate, R.S., Liu, Y.J., et al., 2007. Fabrication of nanofibers with antimicrobial functionality used as filters: protection against bacterial contaminants. Biotechnol. Bioeng. 97, 1357–1365.

Langmuir, I., 1918. The adsorption of gases on plane surfaces of glass, mica and platinum. J. Am. Chem. Soc. 40, 1361–1403.

Lee, S.Y., Kim, H.J., Patel, R., Im, S.J., Kim, J.H., Min, B.R., 2007. Silver nanoparticles immobilized on thin film composite polyamide membrane: characterization, nanofiltration, antifouling properties. Polym. Adv. Technol. 18, 562–568.

Lee, H.S., Im, S.J., Kim, J.H., Kim, H.J., Kim, J.P., Min, B.R., 2008. Polyamide thin-film nanofiltration membranes containing TiO_2 nanoparticles. Desalination 219, 48–56.

Li, Y.H., Ding, J., Luan, Z.K., Di, Z.C., Zhu, Y.F., Xu, C.L., et al., 2003. Competitive adsorption of Pb^{2+}, Cu^{2+} and Cd^{2+} ions from aqueous solutions by multiwalled carbon nanotubes. Carbon 41, 2787–2792.

Li, L.X., Dong, J.H., Nenoff, T.M., Lee, R., 2004. Desalination by reverse osmosis using MFI zeolite membranes. J. Membrane Sci. 243, 401–404.

Li, Y.H., Di, Z.C., Ding, J., Wu, D.H., Luan, Z.K., Zhu, Y.Q., 2005. Adsorption thermodynamic, kinetic and desorption studies of Pb^{2+} on carbon nanotubes. Water Res. 39, 605–609.

Li, Q.L., Mahendra, S., Lyon, D.Y., Brunet, L., Liga, M.V., Li, D., et al., 2008a. Antimicrobial nanomaterials for water disinfection and microbial control: potential applications and implications. Water Res. 42, 4591–4602.

Li, L.X., Liu, N., McPherson, B., Lee, R., 2008b. Influence of counter ions on the reverse osmosis through MFI zeolite membranes: implications for produced water desalination. Desalination 228, 217–225.

Li, F.F., Dai, Y.Z., Gong, M., Yu, T.P., Chen, X.J., 2015. Synthesis, characterization of magnetic-sepiolite supported with TiO_2, and the photocatalytic performance over Cr(VI) and 2,4-dichlorophenol co-existed wastewater. J. Alloy Compd. 638, 435–442.

Liau, S.Y., Read, D.C., Pugh, W.J., Furr, J.R., Russell, A.D., 1997. Interaction of silver nitrate with readily identifiable groups: relationship to the antibacterial action of silver ions. Lett. Appl. Microbiol. 25, 279–283.

Liga, M.V., Bryant, E.L., Colvin, V.L., Li, Q.L., 2011. Virus inactivation by silver doped titanium dioxide nanoparticles for drinking water treatment. Water Res. 45, 535–544.

Lin, L., Lin, W., Zhu, Y.X., Zhao, B.Y., Xie, Y.C., He, Y., et al., 2005. Uniform carbon-covered titania and its photocatalytic property. J. Mol. Catal. A-Chem. 236, 46–53.

Lipnizki, F., Field, R.W., 2002. Hydrophobic pervaporation for environmental applications: process optimization and integration. Environ. Prog. 21, 265–272.

Liu, H.L., Yang, T.C.K., 2003. Photocatalytic inactivation of *Escherichia coli* and *Lactobacillus helveticus* by ZnO and TiO_2 activated with ultraviolet light. Process. Biochem. 39, 475–481.

Liu, C., Zou, B.S., Rondinone, A.J., Zhang, J., 2000a. Chemical control of superparamagnetic properties of magnesium and cobalt spinel ferrite nanoparticles through atomic level magnetic couplings. J. Am. Chem. Soc. 122, 6263–6267.

Liu, C., Zou, B., Rondinone, A.J., Zhang, Z.J., 2000b. Reverse micelle synthesis and characterization of superparamagnetic $MnFe_2O_4$ spinel ferrite nanocrystallites. J. Phys. Chem. B 104, 1141–1145.

Liu, Y., Li, J., Qiu, X., Burda, C., 2006. Novel TiO_2 nanocatalysts for wastewater purification: tapping energy from the sun. Water Sci. Technol. 54, 47–54.

Liu, J.F., Zhao, Z.S., Jiang, G.B., 2008. Coating Fe_3O_4 magnetic nanoparticles with humic acid for high efficient removal of heavy metals in water. Environ. Sci. Technol. 42, 6949–6954.

Livraghi, S., Votta, A., Paganini, M.C., Giamello, E., 2005. The nature of paramagnetic species in nitrogen doped TiO_2 active in visible light photocatalysis. Chem. Commun. 4, 498–500.

Livraghi, S., Paganini, M.C., Giamello, E., Selloni, A., Di Valentin, C., Pacchioni, G., 2006. Origin of photoactivity of nitrogen-doped titanium dioxide under visible light. J. Am. Chem. Soc. 128, 15666–15671.

Long, R.Q., Yang, R.T., 2001. Carbon nanotubes as superior sorbent for dioxin removal. J. Am. Chem. Soc. 123, 2058–2059.

Long, F., Gong, J.L., Zeng, G.M., Chen, L., Wang, X.Y., Deng, J.H., et al., 2011. Removal of phosphate from aqueous solution by magnetic Fe–Zr binary oxide. Chem. Eng. J. 171, 448–455.

Lonnen, J., Kilvington, S., Kehoe, S.C., Al-Touati, F., McGuigan, K.G., 2005. Solar and photocatalytic disinfection of protozoan, fungal and bacterial microbes in drinking water. Water Res. 39, 877–883.

Lu, C.Y., Chiu, H.S., 2006. Adsorption of zinc(II) from water with purified carbon nanotubes. Chem. Eng. Sci. 61, 1138–1145.

Lu, C.S., Chung, Y.L., Chang, K.F., 2005. Adsorption of trihalomethanes from water with carbon nanotubes. Water Res. 39, 1183–1189.

Lu, C.S., Chiu, H., Liu, C.T., 2006. Removal of zinc(II) from aqueous solution by purified carbon nanotubes: kinetics and equilibrium studies. Ind. Eng. Chem. Res. 45, 2850–2855.

Machala, J., Zboril, R., Gedanken, A., 2007. Amorphous iron(III) oxides: a review. J. Phys. Chem. B 111, 4003–4018.

Makhluf, S., Dror, R., Nitzan, Y., Abramovich, Y., Jelinek, R., Gedanken, A., 2005. Microwave-assisted synthesis of nanocrystalline MgO and its use as a bacteriocide. Adv. Funct. Mater. 15, 1708–1715.

Mauter, M.S., Elimelech, M., 2008. Environmental applications of carbon-based nanomaterials. Environ. Sci. Technol. 42, 5843–5859.

McHenry, M.E., Laughlin, D.E., 2000. Nano-scale materials development for future magnetic applications. Acta Mater. 48, 223–238.

Mitoraj, D., Janczyk, A., Strus, M., Kisch, H., Stochel, G., Heczko, P.B., et al., 2007. Visible light inactivation of bacteria and fungi by modified titanium dioxide. Photochem. Photobiol. Sci. 6, 642–648.

Morin-Crini, N., Crini, G., 2013. Environmental applications of water-insoluble beta-cyclodextrin-epichlorohydrin polymers. Prog. Polym. Sci. 38, 344–368.

Morones, J.R., Elechiguerra, J.L., Camacho, A., Holt, K., Kouri, J.B., Ramirez, J.T., et al., 2005. The bactericidal effect of silver nanoparticles. Nanotechnology 16, 2346–2353.

Mostafavi, S.T., Mehrnia, M.R., Rashidi, A.M., 2009. Preparation of nanofilter from carbon nanotubes for application in virus removal from water. Desalination 238, 271–280.

Nedelcu, I.-A., Ficai, A., Sonmez, M., Ficai, D., Oprea, O., Andronescu, E., 2014a. Silver based materials for biomedical applications. Curr. Org. Chem. 18, 173–184.

Nedelcu, I.A., Ficai, A., Sonmez, M., Ficai, D., Oprea, O., Andronescu, E., 2014b. Silver based materials for biomedical applications. Curr. Org. Chem. 18, 173–184.

Nednoor, P., Chopra, N., Gavalas, V., Bachas, L.G., Hinds, B.J., 2005. Reversible biochemical switching of ionic transport through aligned carbon nanotube membranes. Chem. Mater. 17, 3595–3599.

Nepal, D., Balasubramanian, S., Simonian, A.L., Davis, V.A., 2008. Strong antimicrobial coatings: single-walled carbon nanotubes armored with biopolymers. Nano Lett. 8, 1896–1901.

Ni, M., Leung, M.K.H., Leung, D.Y.C., Sumathy, K., 2007. A review and recent developments in photocatalytic water-splitting using TiO_2 for hydrogen production. Renew. Sust. Energ. Rev. 11, 401–425.

Nichols, W.T., Kodaira, T., Sasaki, Y., Shimizu, Y., Sasaki, T., Koshizaki, N., 2006. Zeolite LTA nanoparticles prepared by laser-induced fracture of zeolite microcrystals. J. Phys. Chem. B 110, 83–89.

Noworyta, K., Augustynski, J., 2004. Spectral photoresponses of carbon-doped TiO_2 film electrodes. Electrochem. Solid State 7, E31–E33.
Ohno, T., Mitsui, T., Matsumura, M., 2003. Photocatalytic activity of S-doped TiO_2 photocatalyst under visible light. Chem. Lett. 32, 364–365.
Oller, I., Malato, S., Sanchez-Perez, J.A., 2011. Combination of Advanced Oxidation Processes and biological treatments for wastewater decontamination—a review. Sci. Total Environ. 409, 4141–4166.
Page, K., Palgrave, R.G., Parkin, I.P., Wilson, M., Savin, S.L.P., Chadwick, A.V., 2007. Titania and silver-titania composite films on glass-potent antimicrobial coatings. J. Mater. Chem. 17, 95–104.
Pal, S., Tak, Y.K., Song, J.M., 2007. Does the antibacterial activity of silver nanoparticles depend on the shape of the nanoparticle? A study of the gram-negative bacterium *Escherichia coli*. Appl. Environ. Microb. 73, 1712–1720.
Pan, B.J., Qiu, H., Pan, B.C., Nie, G.Z., Xiao, L.L., Lv, L., et al., 2010. Highly efficient removal of heavy metals by polymer-supported nanosized hydrated Fe(III) oxides: behavior and XPS study. Water Res. 44, 815–824.
Panacek, A., Kvitek, L., Prucek, R., Kolar, M., Vecerova, R., Pizurova, N., et al., 2006. Silver colloid nanoparticles: synthesis, characterization, and their antibacterial activity. J. Phys. Chem. B 110, 16248–16253.
Pang, Y., Zeng, G.M., Tang, L., Zhang, Y., Liu, Y.Y., Lei, X.X., et al., 2011a. PEI-grafted magnetic porous powder for highly effective adsorption of heavy metal ions. Desalination 281, 278–284.
Pang, Y., Zeng, G.M., Tang, L., Zhang, Y., Liu, Y.Y., Lei, X.X., et al., 2011b. Cr(VI) reduction by *Pseudomonas aeruginosa* immobilized in a polyvinyl alcohol/ sodium alginate matrix containing multi-walled carbon nanotubes. Bioresource Technol. 102, 10733–10736.
Patrascu, J.M., Nedelcu, I.A., Sonmez, M., Ficai, D., Ficai, A., Vasile, B.S., et al., 2015. Composite scaffolds based on silver nanoparticles for biomedical applications. J. Nanomater. 2015, 1–8.
Peng, X.J., Luan, Z.K., Ding, J., Di, Z.H., Li, Y.H., Tian, B.H., 2005. Ceria nanoparticles supported on carbon nanotubes for the removal of arsenate from water. Mater. Lett. 59, 399–403.
Pham, T.D., Shrestha, R.A., Virkutyte, J., Sillanpaa, M., 2009. Recent studies in environmental applications of ultrasound. Can. J. Civil Eng. 36, 1849–1858.
Pirbazari, A.E., Saberikhah, E., Kozani, S.S.H., 2014. Fe_3O_4 wheat straw: preparation, characterization and its application for methylene blue adsorption. Water Res. Ind. 7–8, 23–37.
Qu, X.L., Alvarez, P.J.J., Li, Q.L., 2013. Applications of nanotechnology in water and wastewater treatment. Water Res. 47, 3931–3946.
Rao, G.P., Lu, C., Su, F., 2007. Sorption of divalent metal ions from aqueous solution by carbon nanotubes: a review. Sep. Purif. Technol. 58, 224–231.
Ray, P.Z., Shipley, H.J., 2015. Inorganic nano-adsorbents for the removal of heavy metals and arsenic: a review. RSC Adv. 5, 29885–29907.
Reddy, D.H.K., Lee, S.M., 2013a. Synthesis and characterization of a chitosan ligand for the removal of copper from aqueous media. J. Appl. Polym. Sci. 130, 4542–4550.
Reddy, D.H.K., Lee, S.M., 2013b. Application of magnetic chitosan composites for the removal of toxic metal and dyes from aqueous solutions. Adv. Colloid Interface 201, 68–93.
Reddy, D.H.K., Lee, S.M., 2014. Magnetic biochar composite: Facile synthesis, characterization, and application for heavy metal removal. Colloid Surf. A 454, 96–103.
Reddy, D.H.K., Yuna, Y.S., 2016. Spinel ferrite magnetic adsorbents: alternative future materials forwater purification? Coordin. Chem. Rev. 315, 90–111.

Rickerby, D.G., Morrison, M., 2007. Nanotechnology and the environment: a European perspective. Sci. Technol. Adv. Mater. 8, 19–24.
Rincon, M.E., Trujillo-Camacho, M.E., Cuentas-Gallegos, A.K., 2005. Sol–gel titanium oxides sensitized by nanometric carbon blacks: comparison with the optoelectronic and photocatalytic properties of physical mixtures. Catal. Today 107–108, 606–611.
Roco, M.C., 2003. Nanotechnology: convergence with modern biology and medicine. Curr. Opin. Biotechnol. 14, 337–346.
Sakthivel, S., Kisch, H., 2003. Daylight photocatalysis by carbon-modified titanium dioxide. Angew Chem. Int. Ed. 42, 4908–4911.
Saravanana, P., Vinodb, V.T.P., Sreedharc, B., Sashidhard, R.B., 2011. Gum kondagogu modified magnetic nano-adsorbent: an efficient protocol for removal of various toxic metal ions. Mater. Sci. Eng.: C 32, 581–586.
Savage, N., Diallo, M.S., 2005. Nanomaterials and water purification: opportunities and challenges. J. Nanopart. Res. 7, 331–342.
Schrick, B., Hydutsky, B.W., Blough, J.L., Mallouk, T.E., 2004. Delivery vehicles for zerovalent metal nanoparticles in soil and groundwater. Chem. Mater. 16, 2187–2193.
Schulte, J., Dutta, J., 2005. Nanotechnology in environmental protection and pollution. Sci. Technol. Adv. Mater. 6, 219–220.
Senapati, K.K., Borgohain, C., Phukan, P., 2012. $CoFe_2O_4$–ZnS nanocomposite: a magnetically recyclable photocatalyst. Catal. Sci. Technol. 2, 2361–2366.
Sharma, Y.C., Srivastava, V., Singh, V.K., Kaul, S.N., Weng, C.H., 2009. Nano-adsorbents for the removal of metallic pollutants from water and wastewater. Environ. Technol. 30, 583–609.
Shephard, G.S., Stockenström, S., De Villiers, D., Engelbrecht, W.J., Wessels, G.F.S., 2002. Degradation of microcystin toxins in a falling film photocatalytic reactor with immobilized titaniumdioxide catalyst. Water Res. 36, 140–146.
Shieh, K.-J., Li, M., Lee, Y.-H., Sheu, S.-D., Liu, Y.-T., Wang, Y.-C., 2006. Antibacterial performance of photocatalyst thin film fabricated by defection effect in visible light. Nanomedicine-UK 2, 121–126.
Shrivastava, S., Bera, T., Roy, A., Singh, G., Ramachandrarao, P., Dash, D., 2007. Characterization of enhanced antibacterial effects of novel silver nanoparticles. Nanotechnology 18 (22), 225103.
Singh, S., Barick, K.C., Bahadur, D., 2011. Surface engineered magnetic nanoparticles for removal of toxic metal ions and bacterial pathogens. J. Hazard. Mater. 192, 1539–1547.
Singh, K.K., Senapati, K.K., Borgohain, C., Sarma, K.C., 2015. Newly developed Fe_3O_4–Cr_2O_3 magnetic nanocomposite for photocatalytic decomposition of 4-chlorophenol in water. J. Environ. Sci.
Sint, K., Wang, B., Kral, P., 2008. Selective ion passage through functionalized graphene nanopores. J. Am. Chem. Soc. 130, 16448–16449.
Sokmen, M., Candan, F., Sumer, Z., 2001. Disinfection of *E-coli* by the Ag-TiO_2/UV system: lipidperoxidation. J. Photochem. Photobiol. A 143, 241–244.
Son, W.K., Youk, J.H., Lee, T.S., Park, W.H., 2004. Preparation of antimicrobial ultrafine cellulose acetate fibers with silver nanoparticles. Macromol. Rapid Comm. 25, 1632–1637.
Sondi, I., Salopek-Sondi, B., 2004. Silver nanoparticles as antimicrobial agent: a case study on *E-coli* as a model for Gram-negative bacteria. J. Colloid Interface Sci. 275, 177–182.
Sons, W.K., Youk, J.H., Lee, T.S., Park, W.H., 2004. Preparation of antimicrobial ultrafine cellulose acetate fibers with silver nanoparticles. Macromol. Rapid Commun. 25, 1632–1637.
Srivastava, A., Srivastava, O.N., Talapatra, S., Vajtai, R., Ajayan, P.M., 2004. Carbon nanotube filters. Nat. Mater. 3, 610–614.

Stoimenov, P.K., Klinger, R.L., Marchin, G.L., Klabunde, K.J., 2002. Metal oxide nanoparticles as bactericidal agents. Langmuir 18, 6679–6686.
Subramanian, V., Wolf, E., Kamat, P.V., 2001. Semiconductor-metal composite nanostructures. To what extent do metal nanoparticles improve the photocatalytic activity of TiO_2 films? J. Phys. Chem. B 105, 11439–11446.
Sung-Suh, H.M., Choi, J.R., Hah, H.J., Koo, S.M., Bae, Y.C., 2004. Comparison of Ag deposition effects on the photocatalytic activity of nanoparticulate TiO_2 under visible and UV light irradiation. J. Photochem. Photobiol. A 163, 37–44.
Tan, K.A., Morad, N., Teng, T.T., Norli, I., Panneerselvam, P., 2012. Removal of cationic dye by magnetic nanoparticle (Fe_3O_4) impregnated onto activated maize cob powder (MCP) and kinetic study of dye waste adsorption. APCBEE Procedia 1, 83–89.
Tian, Y., Wu, D., Jia, X.A., Yu, B.B., Zhan, S.H., 2011. Core–shell nanostructure of alpha-Fe_2O_3/Fe_3O_4: synthesis and photocatalysis for methyl orange. J. Nanomater. 2011, 1–5.
Tiraferri, A., Chen, K.L., Sethi, R., Elimelech, M., 2008. Reduced aggregation and sedimentation of zero-valent iron nanoparticles in the presence of guar gum. J. Colloid Interf. Sci. 324, 71–79.
Tiwari, D.K., Behari, J., Sen, P., 2008. Application of nanoparticles in wastewater treatment. World Appl. Sci. J. 3, 417–433.
Tu, Y.J., You, C.F., Chang, C.K., 2012. Kinetics and thermodynamics of adsorption for Cd on green manufactured nano-particles. J. Hazard. Mater. 235, 116–122.
Tu, Y.J., You, C.F., Chang, C.K., Wang, S.L., Chan, T.S., 2013. Adsorption behavior of As(III) onto a copper ferrite generated from printed circuit board industry. Chem. Eng. J. 225, 433–439.
Umebayashi, T., Yamaki, T., Itoh, H., Asai, K., 2002. Band gap narrowing of titanium dioxide by sulfur doping. Appl. Phys. Lett. 81, 454–456.
Upadhyayula, V.K.K., Deng, S.G., Mitchell, M.C., Smith, G.B., Nair, V.K., Ghoshroy, S., 2008. Adsorption kinetics of *Escherichia coli* and *Staphylococcus aureus* on single-walled carbon nanotube aggregates. Water Sci. Technol. 58, 179–184.
Upadhyayula, V.K.K., Deng, S.G., Mitchell, M.C., Smith, G.B., 2009. Application of carbon nanotube technology for removal of contaminants in drinking water: a review. Sci. Tot. Environ. 408, 1–13.
Vamathevan, V., Amal, R., Beydoun, D., Low, G., McEvoy, S., 2004. Silver metallisation of titania particles: effects on photoactivity for the oxidation of organics. Chem. Eng. J. 98, 127–139.
Vaseashta, A., Vaclavikova, M., Vaseashta, S., Gallios, G., Roy, P., Pummakarnchana, O., 2007. Nanostructures in environmental pollution detection, monitoring, and remediation. Sci. Technol. Adv. Mater. 8, 47–59.
Vimala, K., Sivudu, K.S., Mohan, Y.M., Sreedhar, B., Raju, K.M., 2009. Controlled silver nanoparticles synthesis in semi-hydrogel networks of poly(acrylamide) and carbohydrates: a rational methodology for antibacterial application. Carbohyd. Polym. 75, 463–471.
Wang, H., Lewis, J.P., 2005. Effects of dopant states on photoactivity in carbon-doped TiO_2. J. Phys. Condens. Mat. 17, L209–L213.
Wang, Y.H., Morin, G., Ona-Nguema, G., Juillot, F., Calas, G., Brown, G.E., 2011. Distinctive arsenic(V) trapping modes by magnetite nanoparticles induced by different sorption processes. Environ. Sci. Technol. 45, 7258–7266.
Wei, L.S., Yang, G., Wang, R., Ma, W., 2009. Selective adsorption and separation of chromium (VI) on the magnetic iron–nickel oxide from waste nickel liquid. J. Hazard. Mater. 164, 1159–1163.
Wei, J.B., Zhang, X.F., Liu, Q., Li, Z.S., Liu, L.H., Wang, J., 2014. Magnetic separation of uranium by $CoFe_2O_4$ hollow spheres. Chem. Eng. J. 241, 228–234.

Wick, P., Manser, P., Limbach, L.K., Dettlaff-Weglikowska, U., Krumeich, F., Roth, S., et al., 2007. The degree and kind of agglomeration affect carbon nanotube cytotoxicity. Toxicol. Lett. 168, 121–131.
Wu, P.G., Xie, R.C., Shang, J.K., 2008. Enhanced visible-light photocatalytic disinfection of bacterial spores by palladium-modified nitrogen-doped titanium oxide. J. Am. Ceram. Soc. 91, 2957–2962.
Wu, P.G., Xie, R.C., Imlay, J.A., Shang, J.K., 2009. Visible-light-induced photocatalytic inactivation of bacteria by composite photocatalysts of palladium oxide and nitrogen-doped titanium oxide. Appl. Catal. B-Environ. 88, 576–581.
Wu, Z.X., Zhao, D.Y., 2011. Ordered mesoporous materials as adsorbents. Chem. Commun. 47, 3332–3338.
Xiu, Z.M., Ma, J., Alvarez, P.J.J., 2011. Differential effect of common ligands and molecular oxygen on antimicrobial activity of silver nanoparticles versus silver ions. Environ. Sci. Technol. 45, 9003–9008.
Xiu, Z.M., Zhang, Q.B., Puppala, H.L., Colvin, V.L., Alvarez, P.J.J., 2012. Negligible particle-specific antibacterial activity of silver nanoparticles. Nano Lett. 12, 4271–4275.
Xu, P.A., Zeng, G.M., Huang, D.L., Feng, C.L., Hu, S., Zhao, M.H., et al., 2012. Use of iron oxide nanomaterials in wastewater treatment: a review. Sci. Tot. Environ. 424, 1–10.
Yamamoto, T., Yamashita, F., Tanaka, I., Matsubara, E., Muramatsu, A., 2004. Electronic states of sulfur doped TiO_2 by first principles calculations. Mater. Trans. 45, 1987–1990.
Yan, H., Pan, G., Zou, H., Li, X.L., Chen, H., 2004. Effective removal of microcystins using carbon nanotubes embedded with bacteria. Chin. Sci. Bull. 49, 1694–1698.
Yang, S.W., Gao, L., 2004. New method to prepare nitrogen-doped titanium dioxide and its photocatalytic activities irradiated by visible light. J. Am. Ceram. Soc. 87, 1803–1805.
Yantasee, W., Warner, C.L., Sangvanich, T., Addleman, R.S., Carter, T.G., Wiacek, R.J., et al., 2007. Removal of heavy metals from aqueous systems with thiol functionalized superparamagnetic nanoparticles. Environ. Sci. Technol. 41, 5114–5119.
Ye, C., Gong, Q.M., Lu, F.P., Liang, J., 2007. Adsorption of middle molecular weight toxins on carbon nanotubes. Acta Phys. Chim. Sin. 23, 1321–1324.
Yu, J.C., Yu, J.G., Ho, W.K., Jiang, Z.T., Zhang, L.Z., 2002. Effects of F-doping on the photocatalytic activity and microstructures of nanocrystalline TiO_2 powders. Chem. Mater. 14, 3808–3816.
Yu, J.C., Ho, W.K., Yu, J.G., Yip, H., Wong, P.K., Zhao, J.C., 2005. Efficient visible-light-induced photocatalytic disinfection on sulfur-doped nanocrystalline titania. Environ. Sci. Technol. 39, 1175–1179.
Yu, J.-G., Yu, L.-Y., Yang, H., Liu, Q., Chen, X.-H., Jiang, X.-Y., et al., 2015. Graphene nanosheets as novel adsorbents in adsorption, preconcentration and removal of gases, organic compounds and metal ions. Sci. Tot. Environ. 502, 70–79.
Zan, L., Fa, W., Peng, T., Gong, Z.K., 2007. Photocatalysis effect of nanometer TiO_2 and TiO_2-coated ceramic plate on Hepatitis B virus. J. Photochem. Photobiol. B 86, 165–169.
Zelmanov, G., Semiat, R., 2008. Iron(3) oxide-based nanoparticles as catalysts in advanced organic aqueous oxidation. Water Res. 42, 492–498.
Zeng, G.M., Li, X., Huang, J.H., Zhang, C., Zhou, C.F., Niu, J., et al., 2011. Micellar-enhanced ultrafiltration of cadmium and methylene blue in synthetic wastewater using SDS. J. Hazard. Mater. 185, 1304–1310.
Zhang, L.D., Fang, M., 2010. Nanomaterials in pollution trace detection and environmental improvement. Nano Today 5, 128–142.

Zhang, X.W., Zhou, M.H., Lei, L.C., 2005. Preparation of an Ag–TiO_2 photocatalyst coated on activated carbon by MOCVD. Mater. Chem. Phys. 91, 73–79.
Zhang, S.X., Zhang, Y.Y., Liu, J.S., Xu, Q., Xiao, H.Q., Wang, X.Y., et al., 2013. Thiol modified Fe_3O_4@SiO_2 as a robust, high effective, and recycling magnetic sorbent for mercury removal. Chem. Eng. J. 226, 30–38.
Zhang, L., Wu, Z., Chen, L.W., Zhang, L.J., Li, X.L., Xu, H.F., et al., 2016. Preparation of magnetic Fe_3O_4/TiO_2/Ag composite microspheres with enhanced photocatalytic activity. Solid State Sci. 52, 42–48.
Zhanga, X., Wu, J., Meng, G., Guo, X., Liua, C., Liu, Z., 2016. One-step synthesis of novel PANI–Fe_3O_4@ZnO core–shell microspheres: an efficient photocatalyst under visible lightirradiation. Appl. Surf. Sci. 336, 486.
Zhao, G.J., Stevens, S.E., 1998. Multiple parameters for the comprehensive evaluation of the susceptibility of *Escherichia coli* to the silver ion. Biometals 11, 27–32.
Zhou, L.M., Wang, Y.P., Liu, Z.R., Huang, Q.W., 2009a. Characteristics of equilibrium, kinetics studies for adsorption of Hg(II), Cu(II), and Ni(II) ions by thiourea-modified magnetic chitosan microspheres. J. Hazard. Mater. 161, 995–1002.
Zhou, Y.T., Nie, H.L., Branford-White, C., He, Z.Y., Zhu, L.M., 2009b. Removal of Cu^{2+} from aqueous solution by chitosan-coated magnetic nanoparticles modified with alpha-ketoglutaric acid. J. Colloid Interface Sci. 330, 29–37.
Zhu, Y., Ran, T.C., Li, Y.G., Guo, J.X., Li, W.X., 2006. Dependence of the cytotoxicity of multi-walled carbon nanotubes on the culture medium. Nanotechnology 17, 4668–4674.
Zodrow, K., Brunet, L., Mahendra, S., Li, D., Zhang, A., Li, Q.L., et al., 2009. Polysulfone ultrafiltration membranes impregnated with silver nanoparticles show improved biofouling resistance and virus removal. Water Res. 43, 715–723.

2

NANOTECHNOLOGY FOR WATER PURIFICATION: APPLICATIONS OF NANOTECHNOLOGY METHODS IN WASTEWATER TREATMENT

Konda Reddy Kunduru*, Michael Nazarkovsky*, Shady Farah*, Rajendra P. Pawar, Arijit Basu*, Abraham J. Domb***

**The Hebrew University of Jerusalem, Department of Medicinal Chemistry and Natural Products, Institute for Drug Research, School of Pharmacy—Faculty of Medicine, Jerusalem, Israel; **Deogiri College, Department of Chemistry, Aurangabad, Maharashtra, India*

1 Introduction

Water is the most important asset of human civilization, and potable water supply is a basic human necessity. However, we are far from meeting global demands; this problem will only increase with time (Hillie and Hlophe, 2007). Demand escalates due to population growth, global climate change, and water-quality deterioration (Ali and Aboul-Enein, 2004; Nemerow and Dasgupta, 1991; Tchobanoglous and Franklin, 1991).

Only 2.5% of the world's oceans and seas harness fresh water, FW (salts concentration of <1 g/L). However, 70% of fresh water is frozen as eternal ice. Only <1% of FW can be used for drinking. Globally, >700 million people do not have access to potable water (WHO, 2014). This problem is severe in developing nations and sub-Saharan African countries. Therefore, water treatment must be implemented in these affected places. Available technologies for water treatment are reaching their limits in providing sufficient quality to meet human and environmental needs (Qu et al., 2013a). Therefore, reuse, recycle, and repurpose are the "needs of the day."

Water Purification. http://dx.doi.org/10.1016/B978-0-12-804300-4.00002-2

Water contaminants may be organic, inorganic, and biological. Some contaminants are toxic and carcinogenic (Ali and Aboul-Enein, 2006; Ali et al., 2009; Laws, 2000) and have deleterious effects on humans and ecosystems (Ali, 2012). Some heavy metals are notorious water pollutants with high toxicity. Arsenic is one of the deadliest elements, well known since ancient times. Other heavy metals water pollutants with high toxicity are cadmium, chromium, mercury, lead, zinc, nickel, copper, and so on; they have serious toxicities (Ali, 2012). Nitrates, sulfates, phosphates, fluorides, chlorides, selenides, chromates, and oxalates show hazardous effects at high concentrations; these ions also change the taste of water. For example, high levels of fluoride in water causes fluorosis. Organic pollutants, such as pesticides, fertilizers, hydrocarbons, phenols, plasticizers, biphenyls, detergents, oils, and greases are associated with toxicities (Damià, 2005). Emerging contaminants include pharmaceuticals and personal care products (PPCPs) (Carballa et al., 2007; Ellis, 2006; Mohapatra et al., 2014). PPCPs are usually resistant against natural biodeterioration. The general origin of such compounds is household and hospital water, which contains metabolized and nonmetabolized (1) drugs, (2) drugs products, (3) additives to detergents, and (4) packaging. PPCP concentration ranges from ng/L up to μg/L, in water. Therefore, conventional water treatment provides unsatisfactory results, because treatment facilities are not equipped to remove stable low-concentrated pollutants.

Produced water containing hydrocarbons necessitates measures that are fast and simple. For instance, according to Gouma and Lee (2014), the petroliferous shale production in the USA will amount to 5 MIO barrels/d by 2017. Hydraulic fracturing (fracking) of shale or rocks are used to recover oil under high pressure. The US's mining-holes (1 MIO) produce 3.29×10^3 m^3 of produced water (frac-water) per year.

Aromatic hydrocarbons, such as benzene, toluene, ethylbenzene, and xylenes, labeled as BTEX are the most predominant components in produced water among other volatile contaminants. These volatile organic compounds can easily evaporate when water is exposed to the atmosphere. BTEX chemicals are toxic and have carcinogenic persistent contaminants. The current maximum contaminant levels (MCL, μg/liter = ppb) for BTEXs in potable water are: benzene 5; toluene 1,000; ethylbenzene 700; xylenes (total) 10,000. Since customary procedures of water purification (filtration, flocculation, and sedimentation) do not remove organic compounds at low concentrations, eliminating BTEX pollutants is challenging. Working expenses depend on various factors, but eventually are dictated by the cost of the media and means used, the cost of their arrangement, and distribution.

Microbes are responsible for several waterborne diseases. Natural organic matter (NOM) is one of the principal characteristics of water quality, and it determines the strategy of purification. Sometimes NOM is represented by total organic carbon and dissolved organic carbon (Matilainen et al., 2011). NOM is omnipresent from remains of animal or plant origins. It negates the effectiveness of certain techniques—flocculation, carbon adsorption, or filtration. Even worse, NOM produces undesirable oxidation products during chlorination treatment. NOM serves as a breeding medium for a large variety of other microorganisms stimulating growth of bacteria in an aquatic habitat. Given that NOM is various, it is not possible to remove all related objects from the environment, disfavoring water authorities. NOM pollution in the surface and ground water makes it unfit for drinking (Gledhill, 1987).

The global population is expected to reach 7.9 billion by 2020, and therefore the world may encounter severe water scarcity. Therefore, it is necessary to remove these pollutants from contaminated water to provide good health to the public (Dyson, 1996). Different methods are available for water purification (Gupta et al., 2011a,b; Gupta and Nayak, 2012; Saleh and Gupta, 2012; Saleh et al., 2011). The most important methods are screening, filtration, micro- and ultrafiltration, crystallization, sedimentation, gravity separation, flotation, precipitation, coagulation, oxidation, solvent extraction, evaporation, distillation, reverse osmosis, ion exchange, electrodialysis, electrolysis, adsorption, setting-out, centrifugal and membrane separation, fluidization, neutralization and remineralization, reduction and oxidation, and so on (Ali, 2012).

Any of the above methods can be combined depending on the type of contaminated water and prospective purpose (Fig. 2.1) (Stackelberg et al., 2004). These methods work well, but recent notorious anthropogenic pollutants (result of modern human life style) pose a challenge to purify/treat the contaminated water. Table 2.1 summarizes major limitations associated with conventional methods (Das et al., 2014). A literature survey reveals no single method sufficient to remove all pollutants from water. High operational costs prevent use of sophisticated techniques. Moreover, efficiency of the treatment plants decreases during the removal of contaminants. Thus, the existing technologies are not "100%" effective to supply potable water (Upadhyayula et al., 2009). New and improved technologies for water purification are, therefore, extremely important.

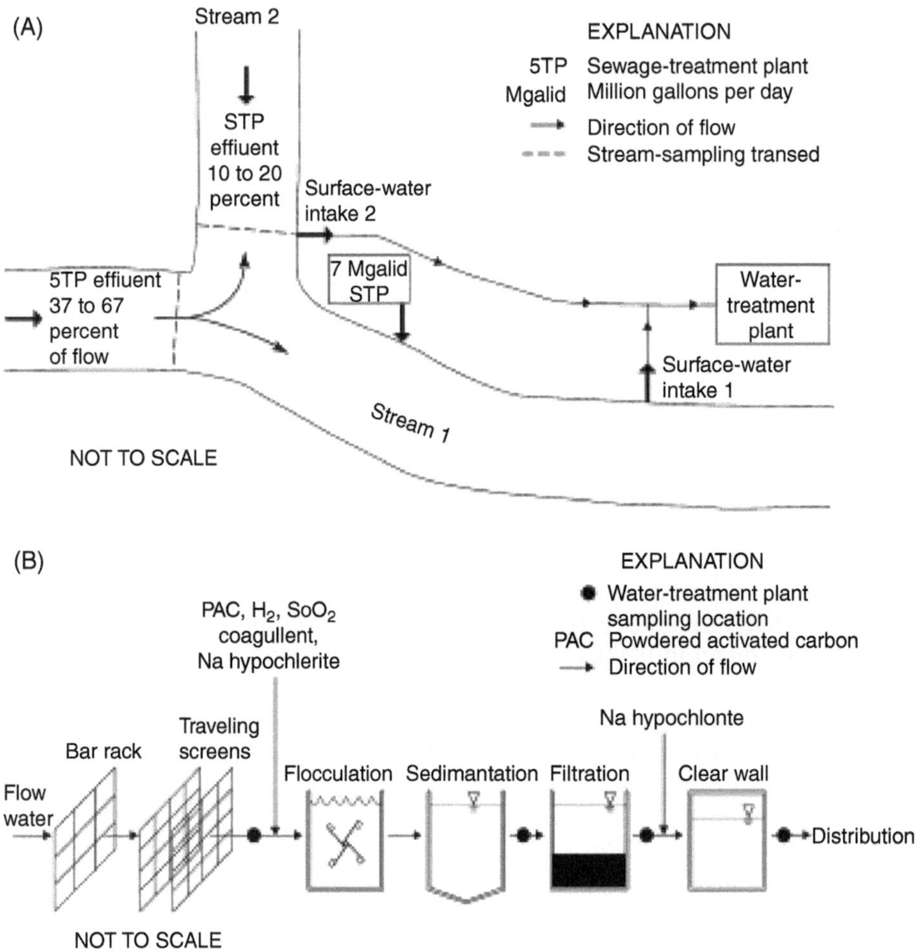

Figure 2.1. Schematic diagram showing (A) location of stream sampling sites, surface-water intakes, and drinking-water-treatment plant, and (B) physical and chemical processes used in drinking-water-treatment plant. From Stackelberg, P.E., Furlong, E.T., Meyer, M.T., Zaugg, S.D., Henderson, A.K., Reissman, D.B., 2004. Persistence of pharmaceutical compounds and other organic wastewater contaminants in a conventional drinking-water-treatment plant, Sci. Total Environ. 329, 99–113, Copyright © 2004, Elsevier.

2 Importance of Nanotechnology in Water Purification

Nanotechnology provided innovative solutions for water treatment. Nanomaterials are fabricated with features, such as high aspect ratio, reactivity, and tunable pore volume, electrostatic, hydrophilic, and hydrophobic interactions, which are useful in adsorption, catalysis, sensoring, and optoelectronics

Table 2.1 Major Limitations Associated with Conventional Water Purification Methods

Conventional Methods	Limitations
Distillation	Most contaminants remain behind and require high amounts of energy and water. Pollutants with boiling point >100°C are difficult to remove
Chemical transformation	Excess reagents are required. Product may be a low-quality mixture and cannot be released into environment. Inactive in harsh conditions. This is not highly selective method
Coagulation and flocculation	This is a complex and less-efficient method and requires alkaline additives to achieve optimum pH
Biological treatment	Microorganisms are sensitive to environmental factors and difficult to control. Intermediates damage the microbial cells. This is not cost effective. Time consuming
Ultraviolet treatment	Expensive method and inactivated by water cloudiness and turbidity. Ineffective for heavy metals and other nonliving contaminants removal
Reverse osmosis	This method removes minerals from water which is unhealthy, and the treated water will be acidic. This method cannot remove volatile organics, chemicals, chlorine, chloramines and pharmaceuticals. Requires high energy
Nanofilteration	This technique requires high energy, and pretreatment. Limited retention for salts and univalent ions. Membrane fouling will occur with limited lifetime and expensive
Ultrafiltration	This method will not remove dissolved inorganics. Requires high energy. Susceptible to particulate plugging and difficult to clean
Microfiltration	Cannot remove nitrates, fluoride, metals, sodium, volatile organics, color, and so on. Requires regular cleaning. Membrane fouling will occur. Less sensitive to microbes, especially virus
Carbon filter	Cannot remove nitrates, fluoride, metals, sodium, and so on. Clogging occurs with undissolved solids. Susceptible to mold. Requires frequent changing of filters

(Das et al., 2014). Nanotechnology-enabled processes are highly efficient, modular, and multifunctional in nature, and they provide high performance, affordable water and, wastewater treatment solutions. Materials consisting of nanoobjects are durable and instantiated by high specific surface (S_{BET}). In other words, a huge surface to volume ratio controls the interaction with pollutants and/or bacteria (Qu et al., 2013b). Nanotechnology-enabled processes for the water treatment constitute major challenges to existing methods. Nanotechnology can also be extended to the purification and utilization of unconventional water sources in an economic way.

Treating industrial wastewater with nanomaterials is also important and widespread. The remediation technologies available now are effective, but they are costly and time consuming. Nanotechnologies are advantageous in treating wastewater, since they eliminate contaminants and help in the recycling process to obtain purified water. This leads to reduction in labor, time, and expenditure to industry solving various environmental issues (Kanchi, 2014).

It should be noted that nanomaterials for purifying drinking water must be environment-friendly and nontoxic. Unsafe particles can cause severe injury to vital organs upon contact with the human body. Due to dimensional features nanoobjects may translocate to various organs, which aggravates the danger of biological damage. Thus, before introduction into the industry, toxicity performance tests must strictly be included into safety data sheets, SOPs, and other related normative documents. This chapter reviews an overview of nanotechnology applications in water and wastewater treatment.

3 Applications of Nanotechnology in Water or Wastewater Treatment

Nanomaterials are typically less than 100 nm in dimension and contain materials with novel and significantly changed physical, chemical, and biological properties (Theron et al., 2008). Materials of this scale contain novel size-dependent properties, which are different compared to their larger counterparts. Desirable nanomaterial properties, such as high surface area for adsorption and high reactivity toward photocatalysis should also have good antimicrobial properties for disinfection and also to control biofouling, should have superparamagnetism for particle separation, should contain optical and electronic properties, and should have good sensing nature to measure water quality (Qu et al., 2013a). Some nanotechnology applications for water and wastewater treatment are discussed in the following sections (Table 2.2) (Gehrke et al., 2015).

3.1 Nanoadsorption

Adsorption is a surface process wherein pollutants are adsorbed on a solid surface. Adsorption takes place in general by physical forces, but sometimes this can be attributed to weak chemical bonds (Faust and Aly, 1983). The efficiency of conventional adsorbents may be restricted by their surface area, and

Table 2.2 Overview of Different Nanomaterials in Water and Wastewater Treatment

Nanomaterial	Properties/Applications	Limitations
Nanoadsorbents	Have high specific surface and very good adsorption capacity Used to remove organic and inorganic contaminants, and bacteria	High production costs
Nanometals and nanometal oxides	Contain high specific surface area and short intraparticle diffusion distance, compressible without change in the surface area, abrasion resistant, magnetic, and photo catalytic in nature Used to remove heavy metals and radionuclides Used in media filters, slurry reactors, powders, and pellets	Less reusable
Membranes and membrane process	Highly reliable and mostly automated process. Applied in all fields of water and waste treatments	Requires high-energy source
Photocatalysis	Photocatalytic activity in UV and possibly visible light range, low human toxicity, high stability, and low cost	Reaction selectivity
Disinfection and microbial control	Strong and wide-spectrum antimicrobial activity, low toxicity to humans, ease of use	Lack of disinfection residue

the lack of selectivity (Qu et al., 2013a). Usually nanoadsorbents are used to remove inorganic and organic pollutants from water and wastewater. The unique properties of nanoadsorbents, such as small size, catalytic potential, high reactivity, large surface area, ease of separation, and large number of active sites for interaction with different contaminants make them ideal adsorbent materials for the treatment of wastewater (Ali, 2012). Carbon-based (Chowdhury and Balasubramanian, 2014; Gao et al., 2011; Jurado-Sanchez et al., 2015; Lubick, 2009; Sui et al., 2012) nanoadsorbents, metal-based (>99.5% purity) (Das et al., 2012; Zhang, 2003) nanoadsorbents, polymeric nanoadsorbents, magnetic or nonmagnetic (Arshadi et al., 2015; Chen et al., 2014; Rafiq et al., 2014) oxide composite, and zeolites are currently used nanoadsorent technologies in the treatment of water (Gehrke et al., 2015).

Adsorption efficiency of target compounds hinges on the water used. With the increasing concentrations of any substances (natural or anthropogenic) in water, the capacity for the specific compound is reduced because of competitive binding with the surface sites.

3.1.1 Carbon-Based Nanoadsorbents

3.1.1.1 Removal of Organic Contaminants

Carbon-based nanoadsorbents, such as carbon nanotubes (CNTs) are cylindrical. CNTs are explored as substitutes for activated carbon. CNTs are categorized as single-walled nanotubes and multiwalled nanotubes (MWCNTs) depending on their preparation. CNTs contain a high specific surface area with highly assessable adsorption sites. Their surface chemistry can also be modified accordingly (Yang and Xing, 2010). The hydrophobic surface of CNTs makes them form loose bundles/aggregates in aqueous medium, which reduces the active surface area. These aggregates are high-energy sites for the adsorption of organic contaminants in water (Pan et al., 2008). The reason for the adsorption of bulky organic contaminants by CNTs is the availability of larger pores in bundles and more accessible sorption sites (Ji et al., 2009). CNTs can also adsorb polar organic molecules because of diverse contaminant-CNT interactions in the form of hydrophobic effect, π–π interactions (with polycyclic aromatic hydrocarbons), hydrogen bonding (with acids, amines, alcoholic functional groups), covalent bonding, and electrostatic interactions (with positively charged organic contaminant molecules, such as antibiotics) (Chen et al., 2007; Ji et al., 2009; Lin and Xing, 2008; Yang and Xing, 2010).

3.1.1.2 Removal of Heavy Metal Ions

Surface-oxidized CNTs using hydrogen peroxide, $KMnO_4$, and nitric acid are used in the removal of Cd^{2+} from aqueous solutions (Li et al., 2003a). The oxidation of CNTs may have high adsorption capacity for metal ions with faster kinetics. The surface of oxidized CNTs contains functional groups, such as carboxylic acid, hydroxyl, and carbonyls (Fig. 2.2) (Vuković et al., 2010). These groups have good adsorbing capacity for heavy metal ions

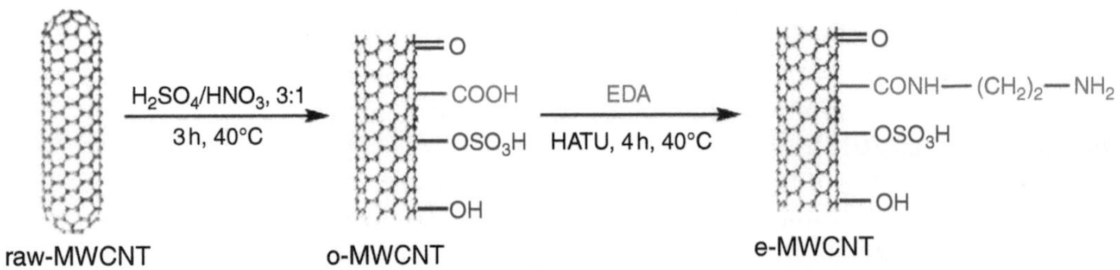

Figure 2.2. Functionalization of MWCNT for the removal of heavy metals (Vuković et al., 2010).

when the pH is above the isoelectric point of the oxidized CNT (Datsyuk et al., 2008; Li et al., 2003b; Musameh et al., 2011; Peng and Liu, 2006; Lau et al., 2015; Liu et al., 2012). Many other studies have been reported that CNTs are very good adsorbing nanomaterials for heavy metal ions, such as Cu^{2+}, Pb^{2+}, Cd^{2+}, and Zn^{2+} (Li et al., 2003a; Lu et al., 2006). A sponge made of CNTs with a dash of boron showed a very good adsorbing capacity for oil from water. These sponges are reusable once the oil is removed from them, and they are promising in the removal of oil spills for oil remediation (Hashim et al., 2012).

Even though CNTs have significant advantages, they have certain limitations even in water treatment by the adsorption method. The use of CNTs on an industrial scale for wastewater treatment plants is not expected due to high production costs (De Volder et al., 2013). Toxicity of CNTs is well known since their discovery in the early 1990s (Iijima, 1991), and many reports are available regarding health concerns (Bottini et al., 2006; Muller et al., 2006). Another limitation is that CNTs show a coagulation phenomenon with some organic contaminants and algae, and lose their nanoidentity/structure (Ali, 2012).

Regeneration of CNTs after their use as adsorbents can be achieved by altering the pH toward acidic. The regeneration will not change CNTs adsorbing capacity for metal ions. It has been reported that Zn^{+2} adsorption by CNTs decreases only 25% after the tenth regeneration, while activated carbon loses its activity almost 50% after its first regeneration (Lu et al., 2006).

In combination with graphene, aerogels of acid-treated MWCNTs exhibit improved absorbance of heavy metals from aqueous solutions (Fig. 2.3) (Sui et al., 2012). To depend upon types of MWCNTs used—pristine or acid treated (c-MWCNT)—there is a substantial difference in absorbance capacities. Aerogels containing acid-pretreated MWCNTs represent higher values of capacities in contrast to CNT-free graphene or pristine MWCNT–graphene hybrid samples. This was supposed to have been due to electrostatic interactions revealed by oxygen-based groups on the pore surface of graphene-c-MWCNT sample–carboxyl groups located on the edges of graphene sheets and walls of c-MWCNTs.

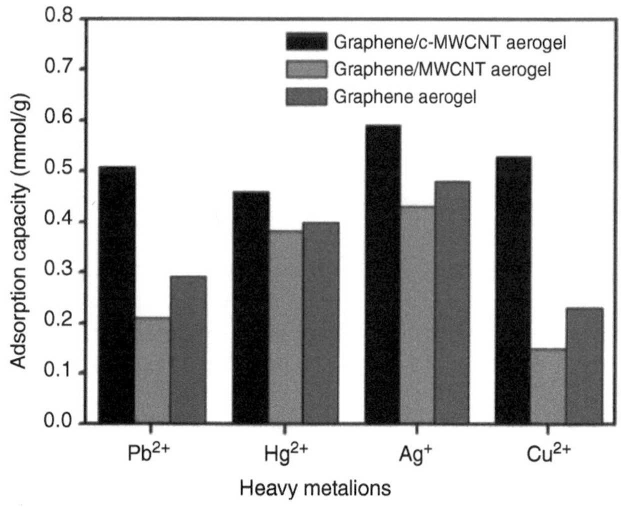

Figure 2.3. Binding capacities of graphene–MWCNTs for heavy metals (Pb^{2+}, Hg^{2+}, Ag^{+}, Cu^{2+}). From Sui, Z., Meng, Q., Zhang, X., Ma, R.Cao, B., 2012. Green synthesis of carbon nanotube-graphene hybrid aerogels and their use as versatile agents for water purification. J. Mater. Chem. 22, 8767–8771, Copyright © 2012, Royal Society of Chemistry.

3.1.1.3 Metal-Based Nanoadsorbents

Metal-based nanoadsorbents, such as iron oxide, titanium dioxide, zinc oxide, and alumina are used in heavy metal removal during water decontamination. They are effective and low-cost materials. The mechanism of action is that the oxygen in metal oxides complexed with heavy metals dissolves in contaminated water (Trivedi and Axe, 2000). As the particle size decreases, the adsorption capacity increases several fold. For example, 300 and 20 nm magnetite particles have similar degrees of adsorption for As(III). When the particle size is reduced to less than 20 nm, the specific area of the nanoparticles increases, and the adsorbing capacity for arsenic increases three times (Auffan et al., 2008). This is because of the "nanoscale effect" of the magnetite, in which surface structure creates new adsorption sites for metal ions (Auffan et al., 2009).

Magnetic nanoadsorbents (MNPs), such as maghemite (γ-Fe_2O_3), hematite (α-Fe_2O_3), and spinel ferrites ($M^{2+}Fe_2O_4$, where M^{2+}: Fe^{2+}, Cd^{2+}, Cu^{2+} Ni^{2+}, Co^{2+}, Mn^{2+}, Zn^{2+}, Mg^{2+}) are very good adsorbing materials for the collection and removal of toxic elements from contaminated water. Environmental benefits lie in their magnetic nature. They can be easily separated from reaction media by application of an external magnetic field. Ample reports are available in the literature for the use of these MNPs for the removal of a variety of elements, such as arsenic, chromium, cobalt, copper, lead, and nickel in their ionic forms (Badruddoza et al., 2013; Lei et al., 2014; Ngomsik et al., 2012; Tan et al., 2014; Tu et al., 2012).

Such magnetic oxide nanomaterials modified by functional groups—$\equiv Si(CH_2)_3NH_2/\equiv SiCH_3$ and ($\equiv Si(CH_2)_3NH_2/$ $\equiv SiC_3H_{7-n}$— are shown to act as adsorbents of undesired biopolymers from water (Melnyk and Zub, 2012).

The iron oxide surfaces are covered with −OH groups. This is because of adsorbed H_2O or structural features. Hence, the surface is not inert:

$$\text{Fe-OH} + H^+ \rightarrow FeOH^{2+} \quad (2.1)$$

$$FeOH + OH^- \rightarrow FeO^- + H_2O \quad (2.2)$$

The functionality can be varied in relation to the nature of both oxide and pH. In this respect $FeOH^+$ or $FeOH^{2+}$ ($pH < 7$) and $Fe(OH)_2^0$, $Fe(OH)_3^-$, FeO^- ($pH > 7$) are predominant on surfaces of iron oxides (Chowdhury et al., 2012). Such dependences become agents for different removals of chromium(IV) from water.

Maghemite nanotubes have been used for the removal of Cu^{2+}, Zn^{2+}, and Pb^{2+} from water (Roy and Bhattacharya, 2012).

The use of magnetite nanorods was reported for the removal of heavy metal ions, such as Fe^{2+}, Pb^{2+}, Zn^{2+}, Ni^{2+}, Cd^{2+}, and Cu^{2+} from aqueous solutions (Karami, 2013). In the same study, when compared to nanotubes, the nanorods had higher adsorption capacity for Zn^{2+} and Pb^{2+} but lower adsorption capacity for Cu^{2+}. Water soluble magnetite superparamagnetic nanoparticles had fast and selective for the adsorption of Hg^{2+} (Qi et al., 2014). The study on the role of nanohematite as an adsorbent for the removal of heavy metal ions was carried out using spiked tap water. The nanohematite particle surface contains hydroxyl groups, which allows for the adsorption of specific heavy metal ions. This model also proves that the adsorption of Pb^{2+}, Cd^{2+}, and Cu^{2+} is endothermic, and Zn^{2+} is exothermic (Shipley et al., 2013).

The application of paramagnetic ferrite-based nanoparticles ($CuFe_2O_4$) makes possible adsorption of oxidized forms of As—As(V) in the form of arsenate salts from contaminated groundwater (Tu et al., 2012). This feature aids in a rapid recovery (20 s) by force of magnetic field.

Another metal oxide, $TiO_{2,}$ was studied for the removal of arsenic. Luo et al. acquired wastewater from copper smelting industry that contained an average concentration of 3310 mg L^{-1} As(III), 24 mg L^{-1} Cu, 5 mg L^{-1} Pb, and 369 mg L^{-1} Cd. The authors successfully removed arsenic in less than 1 h to 100 mg L^{-1} from the initial concentration 3310 mg L^{-1}. They have used TiO_2 for almost 21 successive cycles after its regeneration (Luo et al., 2010).

Magnesium oxide (MgO) was noticed due to its signaled removal properties against inorganic micropollutants, such as As(V) and heavy metal ions. Flower-like nanoforms of MgO (S_{BET} = 72 m^2/g) are high-efficiency adsorbents of Cd^{2+} and Pb^{2+} with maximum capacities of 1.98 and 1.5 g/g (pH = 7), respectively. The XRD studies bear record to ion exchange mechanism between both types of metal ions and Mg^{2+}: $MgO + M^{2+}(Cd^{2+}/Pb^{2+}) = PbO/CdO + Mg^{2+}$. The growth of Mg^{2+} during the adsorption process does not, however, exceed the World Health Organization limit (450 mg/L) (Cao et al., 2012). The adsorption capacity of porous MgO nanowires toward As(V) was shown to be at 384.6 mg/g (Jia et al., 2013). In this case, the ionic exchange does not take place on the surface of MgO. Although the shape of 1-D MgO had been saved, new "plates" of magnesium arsenate were formed on the surface of nanowires. Thus, a chemical interaction occurs during As(V) adsorption without any loss of Mg^{2+} into water.

By conjugation of MgO and TiO_2 into a dual nanocomposite, the adsorbing effectiveness of metal oxides goes together with

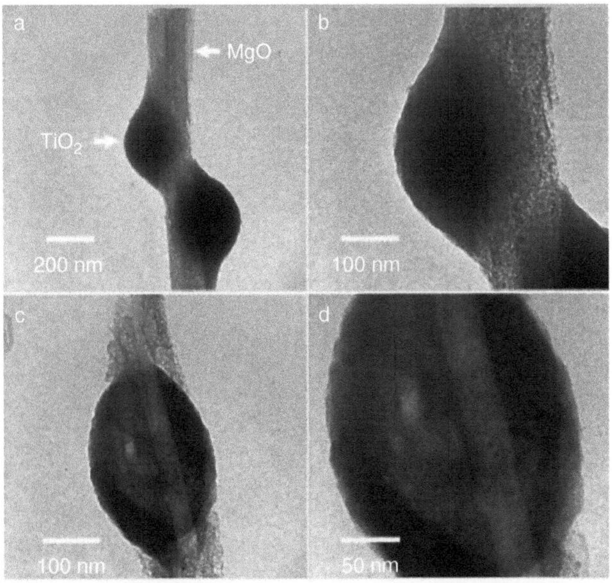

Figure 2.4. TEM images of the necklace-like MgO/TiO_2 heterojunction structures. From Jia et al. (2014), Copyright © 2013, Royal Society of Chemistry.

strongly marked photocatalytic activity (Fig. 2.4). Videlicet, As(V) removal via MgO and photocatalytic degradation of model compound (methylene blue, MB) induced by TiO_2 all complement each other despite low specific surface area (S_{BET} = 76 m^2/g). The highest capacity to adsorb As(V) was 217.8 mg/g (almost 1.8 times less than that of flower-like MgO). This is because of titanium phase introduction. Presence of arsenic did not influence the photocatalytic decomposition of MB (100% oxidation in 25 min). But, the dye reduced removal of arsenic due to possible competitive adsorption of oxidized products of MB on MgO surface.

ZnO nanoadsorbents were used to remove Zn^{2+}, Cd^{2+}, and Hg^{2+} ions from aqueous solutions in a study reported in the literature. These metal ions adsorbed onto ZnO nanoparticles at different concentrations, however, Hg^{2+} had highest adsorbing capacity. The reason may be that Hg^{2+} has the smallest hydrated ionic radii compared to other metal ions in the study (Sheela et al., 2012). Alumina nanoadsorbents can be prepared at low cost. They have high surface area and good thermal stability. They have been used for the removal of cadmium, chromium, copper, lead, and mercury metal ions (Pacheco and Rodriguez, 2001).

Metal-based nanoadsorbents are one of the highly efficient nanoadsorbents for the removal of heavy metal from water and wastewater. They have their own advantages, such as faster kinetics, high adsorption capacity, and preferable nanomaterials for heavy metal removal (Hua et al., 2012).

Long reactive iron nanoparticles (10–100 nm) as reducing materials demonstrate effectiveness as detoxicants of chlorine-containing compounds (pesticides, organic solvents, and polychlorinated biphenyls) (Fig. 2.5) (Zhang, 2003).

These contaminants, preadsorbed on zero-valent Fe, accept electrons, which provokes the reduction process with formation of ethene:

$$C_2Cl_4 + 4Fe^0 + 4H^+ \rightarrow C_2H_4 + 4Fe^{2+} + 4Cl^-$$

Monodisperse nanosilver bioconjugate particles (15 nm) synthesized using fungal mycelia (*Rhizopus oryzae*) are efficacious

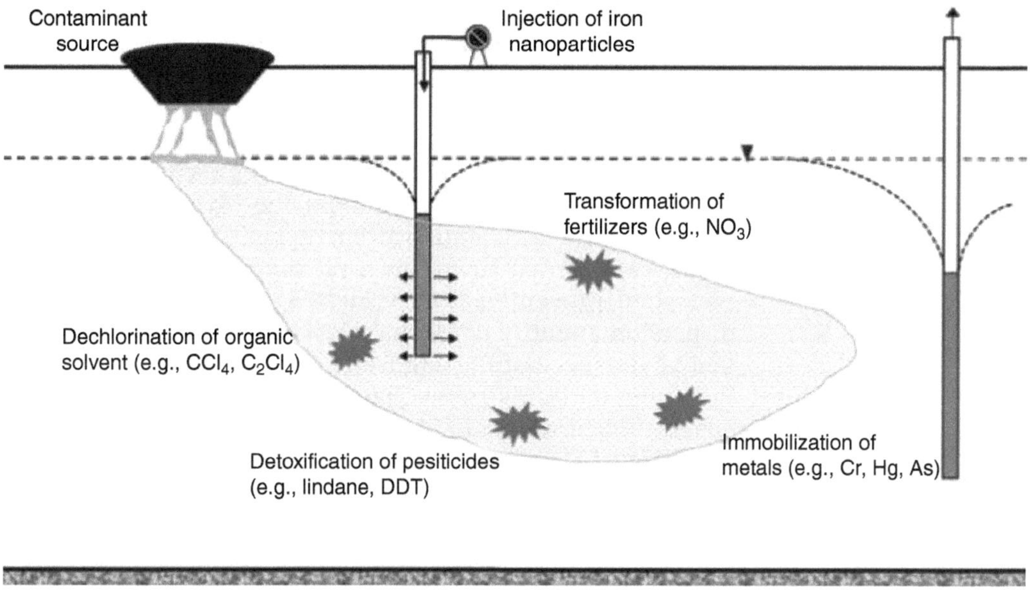

Figure 2.5. Nanoscale iron particles for in situ remediation. From Zhang, W.-X., 2003. Nanoscale iron particles for environmental remediation: an overview. J. Nanopart. Res. 5, 323–332, Copyright © 2003, Springer.

adsorbents and antimicrobal agents. The concentrations of studied pesticide (parathion and chlorpyrifos) were significantly reduced (85–99%) in contrast to gamma-hexachlorocyclohexane (γ-BHC), whereas only 16% of absorbance occurred. In addition, the amount of *Escherichia coli* (initial concentration—60 cell/mL) was not sufficient to be detected after treatment. Such a difference can be explained by means of the HSAB theory, that is, with the affinity of sulfur atoms of parathion and chlorpyrifos to Ag. Hence, hard Cl-atoms in γ-BHC cannot redound to an effective adsorption onto nanosilver particles.

Metal-based nanoadsorbents can be regenerated by changing the solution pH. Also, after several reuses and regeneration the adsorbing capacity of these nanoadsorbents is not altered. However, in some cases the adsorbing capacity is reduced after regeneration due to the formation of aggregates by van der Waals and other forces (Pan et al., 2008).

3.1.2 Polymeric Nanoadsorbents

Polymeric nanoadsorbents gained interest recently. They are used either as a system into which inorganic nanosized materials can be inserted or as a bed or template to prepare nanoparticles (Khajeh et al., 2012). The most important advantage of the polymer-inorganic nanoadsorbents is their good adsorption capacity and very good thermal stability over a wide range of pH. Further, the resistance of polymeric groups and their linkages to acid and base hydrolysis is an added advantage (Kaya et al., 2011; Khajeh et al., 2012). Some recent examples are discussed. Fe_3O_4 magnetic nanoparticles (Fe_3O_4 MNPs) modified with 3-aminopropyltriethoxysilane (APS) and copolymers of acrylic acid (AA) and crotonic acid (CA) as polymer shells (Fe_3O_4@APS@AAco-CA MNPs) were prepared. The use of a polymer shell prevented interparticle aggregation and improved dispersion stability of the nanostructures. The polymer modified MNPs successfully removed heavy metal ions, such as Cd^{2+}, Zn^{2+}, Pb^{2+}, and Cu^{2+} from aqueous solution with high maximum adsorption capacity at pH 5.5. This nanoadsorbent could be reusable in at least four cycles (Ge et al., 2012). In another study, a bimetal doped micro- and nanomultifunctional polymeric adsorbent for the removal of fluoride and arsenic(V) was developed (Fig. 2.6). The polymer was prepared by suspension

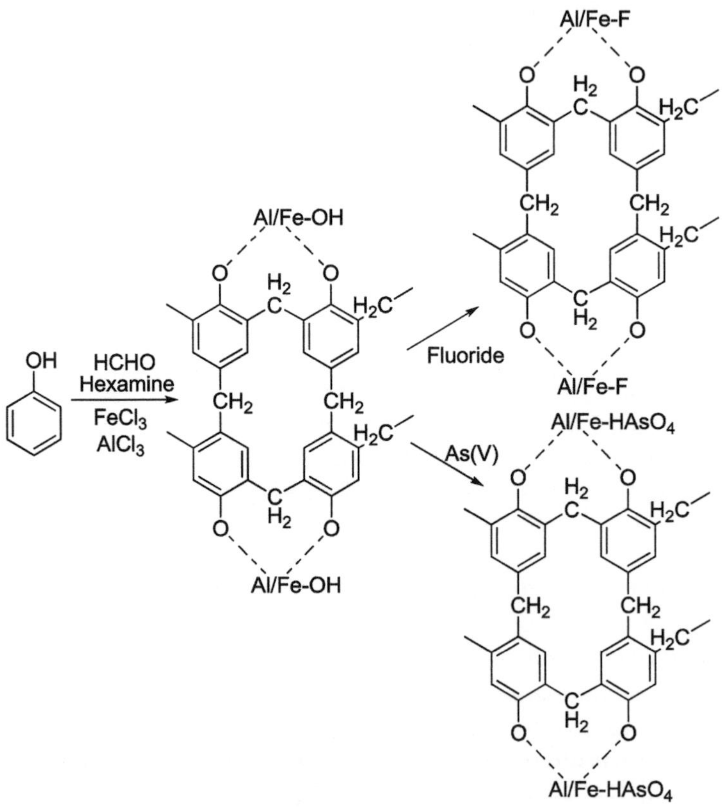

Figure 2.6. Metal-doped phenolic polymeric beads for the removal of fluoride and arsenic (Kumar et al., 2011).

polymerization. During polymerization Al and Fe salts were incorporated to obtain a bimetal-doped nanoadsorbent. The synthesized Fe-doped nanoadsorbents had very good adsorption for arsenic, compared to fluoride, whereas Al-doped nanoadsorbents showed good adsorption toward fluoride compared to arsenic (Kumar et al., 2011).

A polymeric hybrid sorbent (ZrPS-001) was fabricated for enhanced sorption of heavy metal ions, such as lead, cadmium, and zinc ions from aqueous solution by impregnating $Zr(HPO_3S)_2$ (i.e., ZrPS) nanoparticles within a porous polymeric cation exchanger D-001. The negatively charged groups bound to polymeric matrix D-001, resulting in preconcentration and permeation enhancement of target metal ions prior to sequestration. The nanoparticles are expected to sequester heavy metals through an ion-exchange process (Zhang et al., 2008).

Even though, polymeric nanoadsorbents are excellent materials with respect to their structures, pore sizes, and tunable functional groups, for the removal of heavy metals from water/wastewater the ability to make them selective for a given pollutant is a difficult task. The adsorbing capacities of these materials is rather low, and regeneration is required at high cost (Pan et al., 2008).

3.2 Membranes and Membrane Process

A membrane is a porous thin-layered material that allows water molecules to pass through it, but at simultaneously restricts the passage of bacteria, viruses, salts, and metals. Membranes use either pressure-driven forces or electrical technologies. Pressure-driven membrane technology is a perfect method for water purification to any desired quality (Kumar et al., 2014). Membrane separation processes are increasingly advanced methods for the treatment of water and wastewater. Membranes separate substances depending on pore and molecule size. It is a reliable and automated process for wastewater treatment (Gehrke et al., 2015). The challenge of membrane technology is the inherent tradeoff between membrane selectivity and permeability. This technique requires high-energy consumption due to the pressure-driven process. Fouling of membranes makes the process very complex and also reduces the life time of membranes and membrane modules (Qu et al., 2013a). The performance of the membrane system depends on the type of membrane material. Functional nanomaterial inclusion into membranes is definitely advantageous in the improvement of membrane permeability, fouling resistance, mechanical, and thermal stability.

3.2.1 Nanofiber Membranes

Electrospinning is a simple, inexpensive, and efficient technique to fabricate nanofibers. These nanofibers contain high surface area, porosity, and form nanofiber mats with complex pore structures. The physical and chemical parameters of electrospun nanofibers can be easily manipulated for different applications. Nanofiber applications have been documented in the literature. Preparation, characterization of nano fibers, and their suitability for air and water filtration applications have been presented in the literature (Ahmed et al., 2015; Balamurugan et al., 2011; Feng et al., 2013; Nasreen et al., 2013; Subramanian and Seeram, 2013). This class of membranes removes micron-sized particles from water without any significant fouling (Ramakrishna et al., 2006). Thus, nanofibers can be used in pretreatment prior to ultra filtration or reverse osmosis. Electrospinning is a widely employed technology in air treatment, not so common in wastewater treatment (Ramakrishna et al., 2006).

An electrospun nanomembrane can eliminate bacteria or viruses by size exclusion. However, the utilization of these membranes incurs difficulties because of smaller pore sizes for appropriate removal of viral agents. At such a scale, size of pores will essentially depress the onrush of water. To the contrary, a novel composite cellulose-based membrane to remove microorganisms (99.9999% removal of *E. coli*) maintaining the water stream was developed (Sato et al., 2011). Positive charged fibers trapped the negatively charged viruses. This rate of activity is in keeping with requirements specified by the National Sanitation Foundation Standard (max. 2 CFU/mL).

Poly(acrylonitrile-*co*-glycidyl methacrylate) nanofibers (100 nm in diameter) were surface functionalized with protein. Biofunctionalization made these nanofibers slightly bigger in diameter to 126 nm. The bionanohybrid electrospun nanofiber membranes undergo conformational change upon wetting during filtration and increase of pH to above the isoelectric point of the protein. This leads to the emergence of the hidden functional groups, molecules (thereby protein swelling) and nanosolids to be filtered. Moreover, swollen protein makes a higher steric hindrance, facilitating the capturing of the nanosolids, such as metal ions (Elbahri et al., 2012).

The marketed nanofiber filter, NanoCeram (Argonide Corporation, Sanford, FL, USA) is a fiber with a small diameter and large surface area (300–600 m^2/g). This fiber is scalable to kilogram through a sol–gel reaction. The resulting product is white, free-flowing made up of nanofibers (~2 nm) in diameter and ~10–100 nm length. They aggregate and can be embedded in glass and cellulose sheets, and may be used in ultra filtration. The fibers are free flowing, but gets rid of dirt, bacteria, viruses, and proteins through Columbic interactions. NanoCeram can be applied in commercial/industrial water treatment as a microbiological sampler or as a stand-alone filtration device (Karim et al., 2009). Membranes made of hydrophobic nanofiber materials might become very appropriate for separation of organic solvents, leading to higher flux efficiency (Feng et al., 2013).

3.2.2 Nanocomposite Membranes

Nanocomposite membranes are promising filtration units; they may be fabricated from mixed matrices and surface-functionalized membrane. Mixed matrix membranes use nanofillers, and most are inorganic. They are mixed to a polymeric or inorganic oxide matrix, and have substantial surface area (Gehrke et al., 2015). Hydrophilic metal oxide nanoparticles (Al_2O_3, TiO_2, and zeolite), antimicrobial nanoparticles (nano-Ag and CNTs), and (photo)catalytic nanomaterials (bimetallic nanoparticles, TiO_2) are some of the nanomaterials used for such applications (Qu et al., 2013a).

Addition of hydrophilic metal oxide nanoparticles into the membrane reduces fouling. Such addition of metal oxide nanoparticles (alumina, silica, zeolite, TiO_2, etc.) to polymeric membranes increases membrane surface hydrophilicity, water permeability, or fouling resistance (Bae and Tak, 2005; Bottino et al., 2001; Maximous et al., 2010; Pendergast and Hoek, 2011; Pendergast et al., 2010). Also, these inorganic nanoparticles enhance the mechanical and thermal stability of polymeric membranes by reducing the negative impact of compaction and heat on membrane permeability). Also, these inorganic nanoparticles enhance the mechanical and thermal stability of polymeric membranes by reducing the negative impact of compaction and heat on membrane permeability.

Nanocomposite membranes are made up of ordered mesoporous carbons as nanofillers fabricated as thin-film polymeric matrices. They are semipermeable, the top surface used in reverse osmosis. Atmospheric pressure plasma converts hydrophobic mesoporous carbons to a hydrophilic one. Small percentage of hydrophilic mesoporous carbons increases the hydrophilicity; resulting in increased pure water permeability (Kim and Deng, 2011). Thin-film nanocomposites made from polyamide and nano-NaX zeolite (40–150 nm) were reported. They were coated by interfacial polymerization using trimesoyl chloride and *m*-phenylenediamine monomers over porous polyethersulfone. This membrane shows good permeability to pure water, leaving contaminants behind the membrane (Fathizadeh et al., 2011).

Nano-Ag and CNTs prevent membrane biofouling. Inhibition of bacterial attachment or biofilm formation was observed with doping or surface grafting of nano-Ag on polymeric membranes (Mauter et al., 2011; Zodrow et al., 2009) and also inactivation of viruses (De Gusseme et al., 2011). However, authors do not mention membrane long-term efficacy against biofouling. It has been reported that direct contact of CNTs inactivate bacteria (Brady-Estévez et al., 2008). High bacterial inactivation (>90%) was observed with polyvinyl-*N*-carbazole-SWNT nanocomposite at 3 wt.% of SWNT (Ahmed et al., 2012). However, long-term filtration experiments are required to determine the impact of fouling on the antimicrobial activity of CNTs.

Photocatalytic nanoparticles inserted into membranes perform a physical separation process, and the photocatalyst contaminate degradation. Photocatalytic TiO_2 nanoparticles were grafted on metallic filters through dip coating. Combination if foul repellent and photocatalytic nanocoatings degrades unwanted substances forming a dense cake. However, such surface activation is limited to inert materials, polymeric membranes would degrade during the oxidation process (Gehrke et al., 2012).

Nanoscale metallic iron (nZVI) or zero-valent iron (ZVI) has been extensively studied for two decades for the treatment of ground water and also wastewater polluted with different organic and inorganic contaminants (Crane and Scott, 2012; Guan et al., 2015). When these nano-ZVI and noble metal supported on nano ZVI incorporated in to polymeric membrane system for the degradation of contaminants, such as chlorinated compounds, nano ZVI serves as electron donor and noble metals catalyze the degradation reaction (Qu et al., 2013a; Wu et al., 2008).

3.2.3 Thin-Film Nanocomposite Membranes

Thin-film nanocomposite (TFN) membranes are a new category of composite membranes prepared by an interfacial polymerization process. Nanoparticles are incorporated within the thin layer of the polymer to improve the properties of the interfacially polymerized layer (Lau et al., 2015). Nanomaterials, such as nanozeolates, nano-Ag, nano-TiO_2, and CNTs were incorporated as nanoparticles into active thin layers of thin film composite (Lau et al., 2015; Qu et al., 2013a).

Zeolite-polyamide nanocomposite thin films were prepared by interfacial polymerization, which results in reverse osmosis membranes with improved permeability and interfacial properties when compared to similarly formed pure polyamide thin films. In this study water permeability increased up to 80% compared to thin-film composite membrane, with the NaCl rejection largely maintained up to 94% (Jeong et al., 2007). In another study, zeolite-polyamide TFN membranes for reverse osmosis process were prepared and achieved upon usage for salt rejection (32 g/L) up to 99.4% compared to regular reverse osmosis membranes (Kurth et al., 2011; Lind et al., 2010). This technology is now commercially available; LG NanoH_2O, Inc. successfully conducted a field test of a TFN element and reported that the water flux of the TFN membrane was twice the flux of the polyamide membrane and achieved salt rejection >99.7% (Lau et al., 2015). The unique crystal structure of the zeolite molecular sieve particle is responsible for the improvement in the water flux through its internal pore structure (Lind et al., 2009). Also, hydrophilicty and negative charge of zeolite nanoparticles can cause greater affinity toward water molecules and increased repulsions of anions due to coulombic effects (Jeong et al., 2007).

In a study, incorporation of appropriate amounts of nano-TiO_2 into the thin-film composite active layer resulted in increase in the membrane rejection for salts while maintaining the permeability. In another study, a TFN nanofiltration membrane was developed by the incorporation of aminosilanized TiO_2 nanoparticles. Aminosilanized TiO_2 nanoparticles had good dispersion inside the polyamide skin layer by reducing their surface energy. At ultra-low concentration (0.005 wt.%), the functionalized TiO_2 nanoparticles improved NaCl rejection up to 54%. For higher concentrations of TiO_2 nanoparticle incorporation into thin layers, increased water flux up to twofold was observed compared to thin film composite membrane with negligible rejection loss (Rajaeian et al., 2013).

The incorporation of Ag nanoparticles into TFN membrane has also been reported. Ag nanoparticles can deactivate microorganisms during the filtration process and also reduce membrane biofouling. TFN membranes containing Ag nanoparticles in the thin layer improved water permeability and also demonstrated antibacterial effects on the growth of *Pseudomonas aeruginosa* (Kim et al., 2012).

CNTs also found application in TFN membranes. Introduction of carboxyl-functionalized MWCNTs into thin-film composite structure can improve both membrane antifouling property and chlorine resistance (Zhao et al., 2014). The better antifouling property of TFN membrane was due to the improved surface hydrophilicity and greater negative surface charge upon addition of MWCNTs. Further, this membrane was reported to exhibit better chlorine resistance when evaluated either in dynamic or immersion mode, owing to the protection of amide linkage by electron-rich MWCNTs.

3.2.4 Aquaporin-Based or Biologically Inspired Membranes

Aquaporins are pore-forming protein channels. They are ubiquitous in living cells. Under specific conditions they regulate water flux to reject most ionic molecules (Gehrke et al., 2015; Tang et al., 2013). They are selectively permeable to water; ideal material for making efficient biomimetic membranes for water purification. Aquaporins are unstable, so incorporated into small vesicles embedded in a matrix (Gehrke et al., 2015; Tang et al., 2013). Aquaporin Inside (Aquaporin A/S, Copenhagen, Denmark), first commercial membrane with embedded aquaporins. They withstand pressures up to 10 bar, and water flux rate >100 L/(hm^2). One of the example of using this product is in desalination (Gehrke et al., 2015). The experiments of (Kumar et al., 2014) show that aquaporins have exceptional water permeability. Their observations lead to postulate desalination membranes with vastly improved performance. In a study, aquaporins form self-assembled polymer vesicles forming dense hydrophobic polymer layer. It provides a nanostructured selective barrier, with significant mechanical strength (Xie et al., 2013). Recent literature concerning membrane nanotechnologies has indicted that aquaporin-based bioinspired membranes offer the best chance for revolutionary performance, but they were also seen as the furthest from commercialization (Pendergast and Hoek, 2011).

3.3 Photocatalysis

Photocatalysis is an advanced oxidation process employed in the treatment of water and wastewater. This technique is based on the oxidative elimination of micropollutants and microbial pathogens (Friedmann et al., 2010; Gehrke et al., 2015). Most organic pollutants can be degraded by heterogeneous photocatalysis (Chong et al., 2010; Fujishima et al., 2008; Gaya and Abdullah, 2008; Lazar et al., 2012). TiO_2 is a validated photocatalyst as it is readily availability, safe, and inexpensive (Qu et al., 2013a). When TiO_2 is irradiated by UV light in the range of 200–390 nm, electron-hole pair (e^--h^+)s is photoexcited. They move into the conduction (CB) and valent (VB) bands, which results in charge separation for an effective photocatalytic function depending on redox potential of a substrate. Therefore, the biodegradability of heavily decomposable substances can be increased in a pretreatment step.

$$TiO_2 + h\nu \rightarrow e^-_{(CB)} + h^+_{(VB)} \tag{2.3}$$

$$O_2 + e^-_{(CB)} \rightarrow O_2^- \tag{2.4}$$

$$h^{+}_{(VB)} + H_2O \rightarrow HO\cdot + H^{+} \quad (2.5)$$

$$\cdot OH + \cdot OH \rightarrow H_2O_2 \quad (2.6)$$

$$O_2^{-} + H_2O_2 \rightarrow \cdot OH + HO^{-} + O_2 \quad (2.7)$$

$$O_2^{-} + H^{+} \rightarrow \cdot OOH \quad (2.8)$$

$$\cdot OH + \text{Organic matter} + O_2 \rightarrow CO_2 + H_2O \quad (2.9)$$

Since, the band gap (E_g) of TiO_2 undertakes broadening (ΔE_g) within the nanorange ($d = 1 \div 12$ nm), more possible organic substrates can be involved in redox processes according to the necessary criteria of photocatalysis:

$$E_{CB} < E^0_{A/A^-}$$

$$E_{VB} > E^0_{D/D^+}$$

where E_{CB} and E_{VB}, the energy edges of conduction and valence bands, respectively; E^0_{A/A^-} and E^0_{D/D^+}, the standard potentials in respect to reduction and oxidation of an acceptor (A) or a donor (D). Another requirement to a photocatalyst is correspondence of its band gap to redox potential of $H_2O/\cdot OH$ couple ($OH^- \rightarrow \cdot OH + e^-$, $E^0 = -2.8$ V). This figure resolves the case for TiO_2 ($E_g = 3.2$ eV).

Principally, persistent compounds, such as antibiotics or other micropollutants may be eliminated through photocatalysis during polishing. UV-A radiation is ~5% of the total sunlight. This low photon efficiency limits its industrial use (Gehrke et al., 2015).

In general, activation of TiO_2 is carried out by a UV lamp, but a sunlight source or visible light lamps are also permitted. KRONO Clean 7000 (Kronos Inc., Cranbury, NJ, USA) a photocatalyst, band gap shifted to lower energy; this renders using broader spectrum in sunlight (Gehrke et al., 2015).

To enhance photocatalytic properties of titania, which consist of improving activity or red-shift for energy saving, modification techniques have been explored. For example, the consociation of nanosilica (high thermal and chemical stability) and nanotitania (n-semiconductor) gives rise to new active surface sites. Catalytic properties of silica/titania nanocomposites considerably depend on TiO_2 content and distribution. Jung and Park (2000) stated that photocatalytic activity of SiO_2/TiO_2 enhance significantly in compared to TiO_2. The highest performance was found at Ti/Si ratio = 70/30. Maximal photocatalytic activity was observed in thin composite films of SiO_2/TiO_2 at C_{SiO_2} = 5 mol.%. It is the authors' opinion that such a phenomenon deals with the size reduction of crystallites accompanied with the expansion in the number of OH-groups on the film surface (Yu et al., 2002). According to Gao and Wachs (1999), direct contact between TiO_2 and SiO_2 enables generating of new characteristics and changes in reactiveness of the surface. Such changes are provoked by the formation of Ti–O–Si bonds—their amount is directly proportional to TiO_2 concentration.

Doping of nanotitania makes possible improvement in photocatalytic properties. Inherently, dopants can be metal ions (Choi et al., 1994; Czech et al., 2015; Nazarkovsky et al., 2014), nonmetals (Takeshita et al., 2006; Zaleska et al., 2008); or other semiconductors (Hou et al., 2007; Wu, 2004; Wu and Ritchie, 2008; Zhou et al., 2008). For instance, tungsten trioxide and some fullerene derivatives, such as Fullerol and C60 encapsulated with poly(*N*-vinylpyrrolidone) and composites with TiO_2 show photocatalytic effects under visible light irradiation (Meng et al., 2011, 2013). But, they generate $^1O_2$ that contains low oxidation potential compared to TiO_2. Another study shows that to enhance the performance of TiO_2 nanoparticles, TiO_2 nanotubes, and 25–40% doping with noble metals are more efficient, reducing the e^-/h^+ pairs recombination (Lazar et al., 2012). A recent review covers effect on water purification by different operating parameters, such as TiO_2 loading, pH, temperature, dissolved oxygen, contaminant type, concentration, light wavelength, and intensity (Chong et al., 2010).

Many research groups have been investigating the combination of separation and catalytic processes using a membrane photocatalytic reactor to purify water and retain catalytic particles (Azrague et al., 2007; Gehrke et al., 2015; Ollis, 2003). When using highly efficient nanoparticles, suitable filtration system should ensure complete removal of these toxic nanoparticles. Hence, the technology is expensive and energy inefficient—use of high-pressure pumps. A good solution is that photocatalytic nanoparticle immobilization on defined materials uses suitable coating methods, such as physical or chemical vapor deposition, and wet chemical coating method. Microfilters provide both mechanical filtration and chemical decontamination. Dirt particles and larger microorganisms are eliminated by microfiltration membranes at the same time that viruses, spores, and contaminants are also degraded (Gehrke et al., 2015).

In one study, TiO_2 nanoparticles and bactericidal silver-coated metallic filters were reported. The authors used dip coating, chemical, and physical vapor deposition to produce photocatalytic thin coatings, maximum thickness 500 nm. Photocatalyst surfaces were less effective than corresponding nanoparticle suspensions, due to lesser surface area (Gehrke et al., 2012). Purific Water (Holiday, FL, USA) combined water treatment process with photocatalysis, and ceramic membrane filtration. The filtration assembly has the capacity of >4 million cubic meters per day. They were successful in degrading 1, 4-dioxane (http://www.purifics.com/).

Photocatalysis shows a substantial potential as a low-cost, environmental friendly, and sustainable water-treatment technology. Yet, there are some technical challenges for its large scale application, such as (1) catalyst optimization to improve quantum yield or to utilize visible light; (2) efficient photocatalytic reactor design and catalyst recovery/immobilization techniques; (C) better reaction selectivity (Qu et al., 2013a).

3.4 Antimicrobial Nanomaterials in Disinfection and Microbial Control

Current disinfection methods applied in the treatment of drinking water can effectively control the microbial pathogens. Research conducted in the past few decades, however, discloses a dilemma between effective disinfection and formation of harmful disinfection by-products (DBPs) (Li et al., 2008). The commonly used chemical disinfectants in the water industry are chlorine, chloramines, and ozone. They can react with other constituents in the water and generate harmful DBPs. Most are carcinogenic (Hossain et al., 2014). There were more than 600 DBPs, such as halogenated DBPs , carcinogenic nitrosamines, bromate, and so on, reported in the literature (Krasner et al., 2006). UV-disinfection processes have come out as an alternative for oxidative disinfection, since they generate fewer DBPs, while the required high dosage for certain viruses, such as adenoviruses. All these limitations urge the development of alternative methods that can enhance the robustness of disinfection while avoiding DBP formation (Li et al., 2008).

The ideal disinfectant should have the following properties (modified from Hossain et al., 2014):

1. very-broad antimicrobial activity at ambient temperature within short time;
2. cannot produce any harmful by-products during and after their use;
3. does not affect human health;
4. inexpensive and easy apply for the intended use;
5. easy to store, highly soluble in water, and must not be corrosive for any equipment or surface; and
6. amenable to safe disposal.

Materials, such as nano-Ag, nano-ZnO, nano-TiO_2, CNTs, and fullerenes exhibit antimicrobial properties without strong oxidation; they have lower tendency to form DBPs (Li et al., 2008;

Hossain et al., 2014). Brief updates regarding these nanomaterials and their antimicrobial actions are discussed in the following sections.

3.4.1 Antimicrobial Action of TiO_2 Nanoparticles

TiO_2 is a very common nanoparticle type to inactive microbes in drinking water, surface water, wastewater, and other sources (Brame et al., 2011; Dimitroula et al., 2012; Friedmann et al., 2010; Hossain et al., 2014; Li et al., 2008; Liga et al., 2011; Markowska-Szczupak et al., 2011; Matin et al., 2011; Ng et al., 2013; Pleskova et al., 2011; Simon-Deckers et al., 2009). The antibacterial mechanism of TiO_2 is due to reactive oxygen species (ROS) generation, especially hydroxyl-free radicals and peroxides formed under UV-A irradiation via oxidative and reductive pathways, respectively (Kikuchi et al., 1997; Li et al., 2008). Generated ROS destruct the cell membrane, damaging DNA and protein, releasing hazardous ions for cell malfunction, disrupting electron transfer, and hampering respiration process (Hossain et al., 2014) (Fig. 2.7).

Strong absorbance of UV-A furnish activation of TiO_2 under solar irradiation, significantly enhancing solar disinfection.

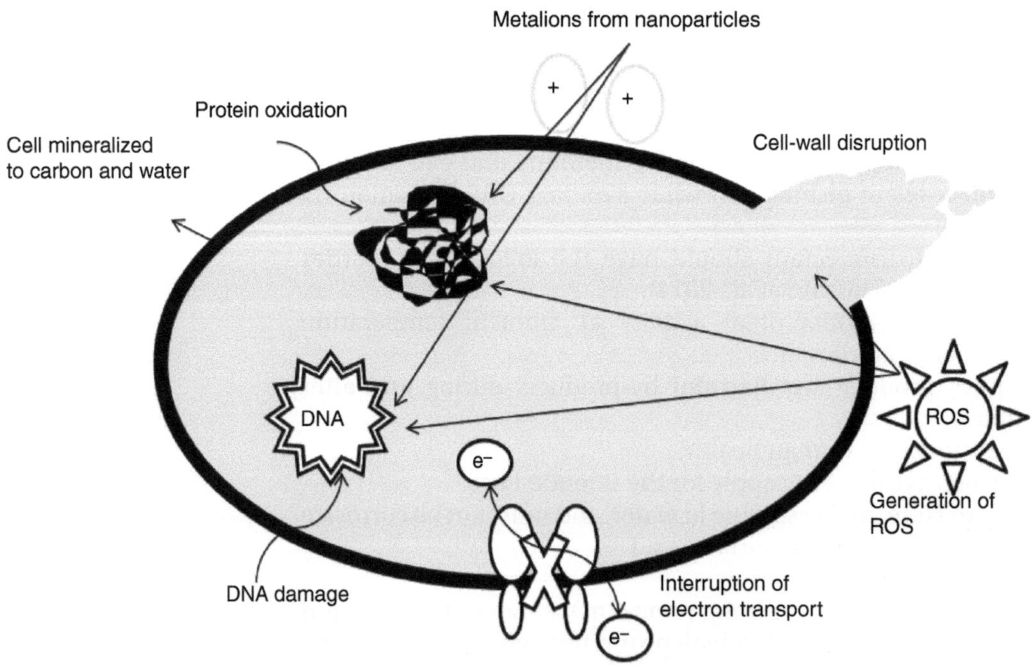

Figure 2.7. Different mechanisms of antimicrobial activities showed by nanomaterials.

However, TiO_2-based solar disinfection is a very slow process that may be a small fraction of UV-A in solar radiation. Therefore, successful research on metal or nitrogen doping to improve visible light absorbance of TiO_2 or UV-A activity is critical to the application of TiO_2 solar disinfection. Bacterial death was observed even in the dark by this nanomaterial, indicating the involvement of other unknown mechanisms (Adams et al., 2006; Li et al., 2008).

3.4.2 Antimicrobial Action of Ag Nanoparticles

Silver has been known for its antimicrobial action since ancient times. It has a wide range of industrial applications in healthcare and external medicine (Hua et al., 2012). In recent times nano-Ag has been widely used as an antimicrobial nanomaterial. They are the material of choice for water decontamination, due to the following: significant and broad antimicrobial activity, safety, and easy to fabricate. Nanosilver releases silver ions in water (Qu et al., 2013a; Xiu et al., 2011, 2012) binding to −SH groups in vital enzymes and damaging them (Liau et al., 1997; Qu et al., 2013a). Silver ions interfere with DNA replication and induce structural changes in the cell envelope (Feng et al., 2000; Qu et al., 2013a). Toxicity of nano-Ag depends rate of release of silver ions. Size, shape, coating, and crystallographic facet influences the release kinetics of silver ions. Presence of ubiquitous ligands reduces its bioavailability, and reduces its toxicity (Qu et al., 2013a; Xiu et al., 2011). In another study they show sublethal doses of silver ions favors *E. coli* growth. Suggesting these strategy could hit back if not designed properly (Qu et al., 2013a; Xiu et al., 2011). Commercial devices, such as MARATHON and Aquapure systems utilizing nano-Ag are already available. Nano-Ag was incorporated into ceramic microfilters as a barrier for pathogens, which can be employed in remote areas of developing countries (Peter-Varbanets et al., 2009).

3.4.3 Antimicrobial Action of ZnO Nanoparticles

Zinc oxide nanoparticles have been used in sunscreen lotions, coatings, and paints due to their strong UV absorption capacity and transparency to visible light (Franklin et al., 2007; Qu et al., 2013a). ZnO nanoparticles show very good antibacterial activities on a broad spectrum of bacteria. However, the antibacterial mechanism of ZnO nanoparticles is unclear, since researchers have obtained opposite results, for example, the particles size effect was established in Jones et al. (2008), but that was not consistant in Franklin et al. (2007). The reason was suggested to be the photocatalytic generation of H_2O_2 responsible for antimicrobial action of ZnO. Even though, both Zn^{+2} ion and ZnO nanoparticles show antibacterial activity. Aquatic organisms can be highly sensitive to dissolved zinc (Franklin et al., 2007). In view of its easy dissolving nature, ZnO applications in drinking water purification are limited.

3.4.4 Antimicrobial Action of Carbon Nanotubes

CNTs show antimicrobial property upon direct contact with cells (Vecitis et al., 2010). Graphene and graphite materials also exhibit similar mechanisms (Liu et al., 2012). Short, dispersed, and metallic CNTs with small diameters are more lethal (Qu et al., 2013a). CNTs effectively removes bacteria by size exclusion, and viruses by depth filtration (Brady-Estévez et al., 2008). Retained bacteria are quickly inactivated by CNTs; MWNTs directly oxidize adhering bacteria and viruses. They decontaminate water within seconds by using small intermittent voltage (Rahaman et al., 2012; Vecitis et al., 2010).

3.5 Nano Antimicrobial Polymers

Polymeric nanoparticles kill microorganisms either by releasing antibiotics, antimicrobial peptides, and antimicrobial agents or by contact-killing cationic surfaces, such as quaternary ammonium compounds, alkyl pyridiniums, or quaternary phosphonium. Different antibacterial mechanisms are reported to show how these cationic groups are able to disrupt bacterial cell membrane. The main mechanism is that the hydrophobic chains of certain lengths will penetrate and burst the bacterial membrane. It has been shown that high levels of positive charge are capable of conferring antimicrobial properties irrespective of hydrophobic chain length, perhaps by an ion exchange mechanism between the bacterial membrane and the charged surface (Beyth et al., 2015; Jain et al., 2014).

Nanopolymeric antimicrobial materials show long-term antimicrobial activity. They are nonvolatile and chemically stable. They can bind to the surface of interest and hardly permeate through biological membranes (Kenawy et al., 2007). Polycationic antimicrobials contain high surface density of active groups, which result in higher antimicrobial activity. Quaternary ammonium compounds possess a broad spectrum of antimicrobial activity against both Gram-positive and Gram-negative bacteria. Polyamines that have been reported as being highly effective antimicrobial nanoparticles are quaternary ammonium polyethylenimines, which have a broad range of bacterial targets when incorporated into various polymeric matrixes (Shvero et al., 2015).

3.5.1 Chitosan-Based Nanoparticles

Chitosan is obtained from chitin in shells. It show antibacterial activity and is a natural polymer. The nanoparticles of chitosan or its derivatives have a broad spectrum of antibacterial, antifungal, and antiviral activities (Chirkov, 2002; Li et al., 2008; No et al., 2002; Qi et al., 2004; Rabea et al., 2003). The nanoparticles of chitosan are more effective against fungi and virus compared with bacteria (Rabea et al., 2003). Among bacterial strains the antimicrobial activity of chitosan is higher for Gram-positive bacteria than for Gram-negative bacteria (Don et al., 2005; No et al., 2002). Minimum inhibitory concentrations from 18 to 5000 ppm were reported for the chitosan nanoparticles depending on the organism, pH, molecular weight, degree of polymerization, and the presence of lipids and proteins (Don et al., 2005; No et al., 2002; Rabea et al., 2003).

The main antimicrobial mechanisms of chitosan are the positively charged chitosan particles that interact with negatively charged cell membranes, causing an increase in membrane permeability and eventually rupture and leakage of intracellular components (Qi et al., 2004). The antibacterial action of chitosan derivatives containing quaternary ammonium groups, such as *N,N,N*-trimethyl chitosan, *N*-propyl-*N,N*-dimethyl chitosan, and *N*-furfuryl-*N,N*-dimethyl chitosan are stronger than those of chitosan, and they increase with decreasing pH (Ji et al., 2009; Li et al., 2008).

Nanochitosan has potential drinking water disinfection applications as an antimicrobial agent in membranes, sponges, or surface coatings of water storage tanks. Nanochitosan shows superiority over other disinfectants due to its broader spectrum of activity against bacteria, viruses, and fungi, and less toxicity toward animals and humans. However, chitosan is an effective disinfectant only at acidic pH; this is due to its solubility and the availability of charged amino groups (Li et al., 2008; No et al., 2002; Rabea et al., 2003). This limitation can be overcome by preparing water-soluble chitosan derivatives.

3.5.2 Nanopolymers-Based Water Purification Systems

N-bromohydantoin/uracil-conjugated polystyrene beads were prepared by the authors and studied controlled release of bromine in water purification systems (Farah et al., 2015a,c).

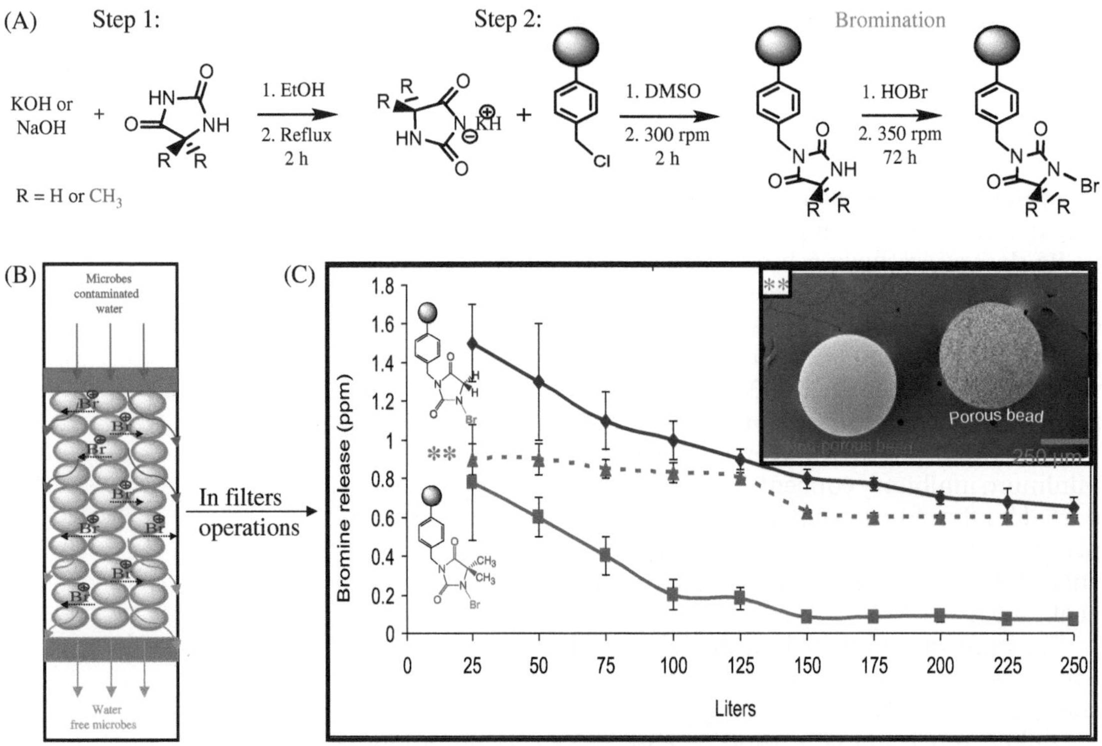

Figure 2.8. (A) *N*-bromo-hydantoin synthesis by conjugation to micronsize beads followed by bromination; (B) filters loaded with *N*-bromo-hydantoin/5,5′-dimethylhydantoin-polystyrene beads; and (C) in filters operations: hypobromous release from different beads having different nanomicro characteristics affecting oxidative halogen releasing profiles from polymeric beads/surfaces (Farah et al. 2015a).

Active bromine release from the beads into the running water and the antimicrobial efficiency were examined against *E. coli* and MS2. This resin exhibited excellent antimicrobial properties; 6- and 4-log reduction for *E. coli* and MS2, respectively, were obtained for all tested points during 250 L. Moreover, beside the effect of the *N*-haloamine structure on lasting disinfection activities, bead's nanomicro characteristics were found critical for oxidative halogen release control: rate stabilization and modulation, extension and consequently influences antimicrobial activity. This aspect is important, since the rate of halogen release influences antimicrobial activity and subsequently the material usage for different applications, Fig. 2.8B and C (Farah et al., 2015a). Bromine% of the reported materials after release study indicated the capability of antimicrobial hypobromous release for extra hundreds of liters. *N*-bromo-dimethylhydantoin with low crosslinked beads maintained their activity for 550 L. This brominated resin should be considered in filters for

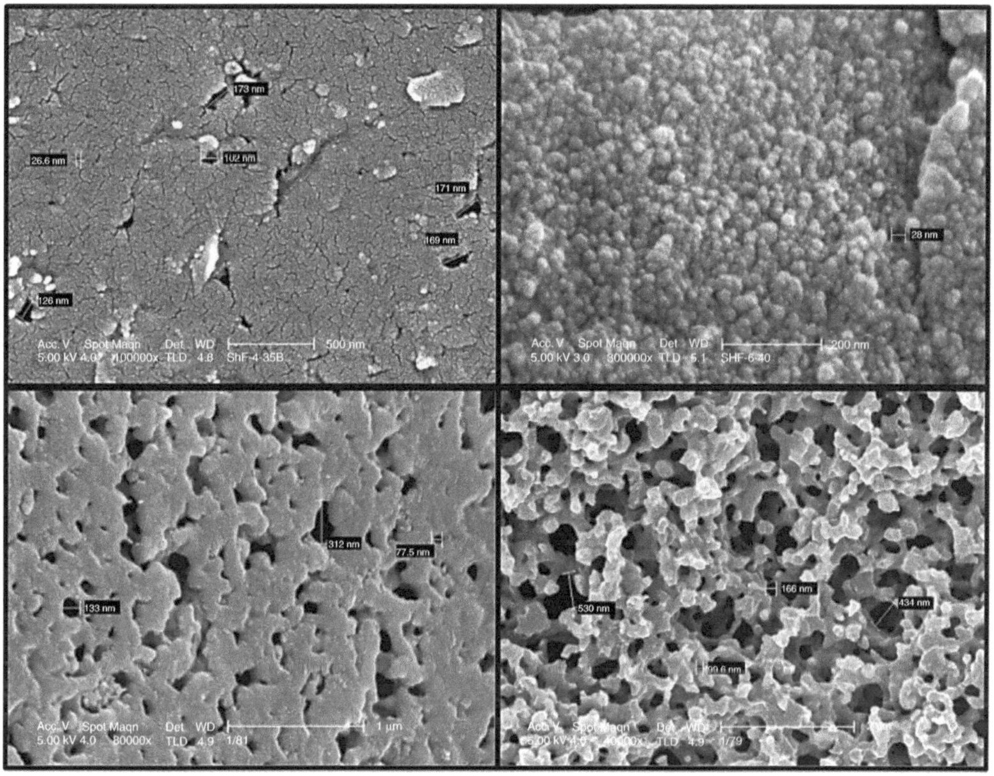

Figure 2.9. Representative photos of the nanopores onto the brominated-hydantoin beads surface or inside were found to play a critical role in oxidative halogen release control: rate stabilization and modulation, extension and consequently influences antimicrobial activity (Farah et al. 2015a).

decontamination of drinking water (Aviv et al., 2015). These materials were found to have great potential due to ability to decontaminate large volumes of contaminated water, low costs, and bromine rechargeability (Farah et al., 2015a,c) (Fig. 2.9).

Quaternary ammonium polyethylenimine nanoparticles (QA-PEI NPs) of C8 chain alkylated, and C18 modified were prepared (Fig. 2.10), and these nanoparticles were embedded into polyethylene vinyl acetate and polyethylene methacrylic acid coatings (Farah et al., 2013; Yudovin-Farber et al., 2010). These coatings were tested for their antibacterial activity against representative bacteria, *E. coli*, *P. aeruginosa*, *Staphylococcus aureus*, and heterotrophic plate count. Study was performed in both static and dynamic modes. The authors anticipated that these QA-PEI NPs will find application in water purification systems as disinfectants and for

Figure 2.10. Chemical structure of C8/C18 alkylated QA-PEI crosslinked nanoparticles.

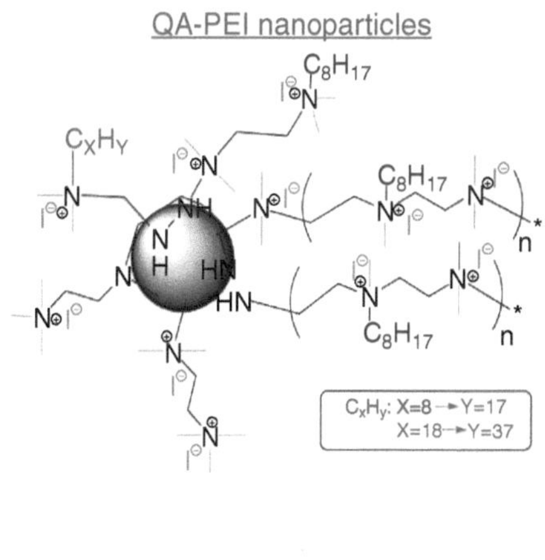

preventing the accumulation of microorganisms onto device surfaces as well as in bioadhesives and self-sterilizing surfaces (Farah et al., 2015b).

4 Regeneration of Nanoparticles

Regeneration of nanoparticles in water purification is one of the crucial aspects, since it controls the economy of water-treatment technology. pH-dependent solvents play crucial roles in regeneration of nanoparticles. They can also be achieved by applying a separation device or immobilizing nanomaterials in the treatment system. Membrane filtration is promising for the regeneration and reuse of nanomaterials due to its chemical use. Ceramic membranes are always advantageous, as they are more resistant to UV compared to polymer membranes (Qu et al., 2013a). Raw water pretreatment is very important to reduce turbidity; otherwise the suspended particles are retained by membranes, and they decrease the efficiency of the treatment. Immobilization is another technique for nanomaterials. But, current immobilization techniques have not been very successful. Development of simple and low-cost methods is required to immobilize nanoparticles without affecting their efficiency. Magnetic separation is another option for the separation of magnetic nanoparticles.

Nanomaterials coated on treatment system surfaces are released in a relatively quick and complete manner. Nanomaterials embedded into a solid matrix have slow release until they are disposed of. Identification of a nanomaterial release is a major technical hurdle for risk assessment, and it remains challenging. The details regarding different detection techniques are mentioned in the literature (Qu et al., 2013a; Silva et al., 2011; Tiede et al., 2008). According to the literature, nanoparticles can be regenerated and used for water treatment, which makes them economically viable materials. The regeneration capacity of nanoparticles may be considered an extra advantage for their popularity in wastewater treatment. However, only very few methods are available for the detection of nanomaterials in complex aqueous matrices. They are highly sophisticated, expensive, and have many limitations. Currently, rapid, sensitive, and selective nanomaterial analytical techniques are in great demand.

Management of used nanoparticles and recovered pollutants is one of the most important aspects. Everyone is aware of pollutant hazards and nanotoxicology; proper disposal must be carried out by the users. The best way is to recycle the nanoparticle and exhausted nanoparticles may be in the manufacturing of bricks, stones, and so on. Different regenerated materials may be recycled for manufacturing various commodities. The recovered organic contaminants should be treated as priority pollutants (Ali, 2012).

5 Safety, Toxicity, and Environmental Impact of Nanomaterials

Due to our current poor understanding of the fate and behavior of nanoparticles in humans and the environment, toxicity is becoming one of the urgent issues of nanotechnology. The main concerns related to nanotechnology are the hazardousness of nanoparticles and the exposure to risk (Gardner and Dhai, 2014). Biological and chemical effects on humans or the environment is the first major issue. The second is of leakage, spillage, circulation, and concentration of nanoparticles that might cause a hazard to humans or the environment (Gardner and Dhai, 2014). Properties, such as size, shape, reactivity, and so on, are making these nanomaterials very useful. The same properties can also make them harmful to the environment and toxic to humans (Hillie and Hlophe, 2007). The entry of nanoparticles into our body is possible through the skin, inhalation, ingestion, and so on. After reaching the bloodstream, they can travel to various body parts, such as the brain, heart, liver, kidneys, spleen, bone marrow, and nervous system (Ali, 2012). The toxicity of nanoparticles is due to their properties and can lead to high chemical reactivity and production of ROS. Production of ROS is possible from CNTs and metal oxides. The ROS generated will cause oxidative stress, inflammation, which results in damage to proteins, membranes, and DNA (Nel et al., 2006). There is a possibility that nanoparticles can adsorb on the body surface and alter the mechanisms of enzymes and certain proteins (Hubbs et al., 2011). Nanoparticles show their toxicity in the environment by agglomeration. Environmental risk analysis of nanoparticles is mentioned in the literature (Grieger et al., 2012). Knowledge of hazards and exposure risks of nanoparticles to the ecology is less among the scientific community, so risk assessment and management is crucial (Gardner and Dhai, 2014). The challenge is to resolve problems before nanoparticle usage starts on a large scale in water purification. There must be safety evaluation, large-scale production facilities, safe disposal of wastes, and energy efficiency. These are the major challenges that may cause delay in the large-scale application of nanotechnology in water purification (Gardner and Dhai, 2014). However, the behavior of nanoparticles inside the body is still a major question that needs to be addressed.

6 Limitations and Research Needs

Water/wastewater-treatment processes by nanotechnology show great promise in laboratory studies. Some of these technologies are marketed, and others require significant research before they can be scaled up. Their commercialization is challenging; we need to overcome many technical hurdles; make them cost effective, and safe. Research is needed before full-scale operation of nanotechnology for treating natural and wastewaters. Studies should be conducted under realistic conditions to assess the efficiency of available nanotechnology to validate the nanomaterial-enabled sensing. Another research need is to measure the long-term efficiency of available technologies, which are conducted on a laboratory scale. Commercialization of these technologies is possible only by their long-term performance in the treatment of water and wastewater. Also, adoption of an innovative technology strongly depends on the cost effectiveness and potential risks involved. At the moment, the cost of nanomaterials is very high, with few exceptions, such as nano-TiO_2, nano-scale iron oxide, and polymeric nanofibers. Cost effectiveness can be achieved by regeneration and reuse of these nanomaterials (Qu et al., 2013b). Since these materials are nanoscale, risk assessment and management is a challenge. Researchers should understand the potential hazards of these materials in the treatment of water and wastewater.

7 Conclusions

Today we need water purification technologies that provide high-quality drinking water, remove micropollutants, and intensify the industrial processes. Nanotechnology provides the opportunity; unique properties of nanoparticles are ideal candidate for developing rapid water-treatment technology. Nanoparticles may eliminate metal ions, anions, organic compounds, and microorganisms. Nanoparticle doses required for the water treatment are low, making their application relatively economical.

Different nanotechnologies are reviewed in this chapter. A few technologies are in laboratory research stage, some reached to pilot testing, and some are commercial. Among these technologies, nanoadsorbents, nanomembranes, and nanophotocatalysts are most promising. Although, these technologies have been commercialized, their potential has not been reached for large-scale use in wastewater treatment.

Risk assessment of recovered pollutants and exhausted nanoparticles still remains significantly unexplored. Therefore, ecofriendly waste management methods are required to avoid hazards and toxicities. The future of the nanoparticles in water treatment is quite progressive, but it requires collaborative efforts of academic and industrial resources to materialize a fast, economical, and feasible water-treatment technology. It will be possible by working together to solve water contamination globally.

References

Adams, L.K., Lyon, D.Y., Alvarez, P.J.J., 2006. Comparative ecotoxicity of nanoscale TiO_2, SiO_2, and ZnO water suspensions. Environ. Sci. Technol. 40, 3527–3532.

Ahmed, F., Santos, C.M., Vergara, R.A.M.V., Tria, M.C.R., Advincula, R., Rodrigues, D.F., 2012. Antimicrobial applications of electroactive PVK-SWNT nanocomposites. Environ. Sci. Technol. 46, 1804–1810.

Ahmed, F.E., Lalia, B.S., Hashaikeh, R., 2015. A review on electrospinning for membrane fabrication: challenges and applications. Desalination 356, 15–30.

Ali, I., 2012. New generation adsorbents for water treatment. Chem.Rev. 112, 5073–5091.

Ali, I., Aboul-Enein, H.Y., 2004. Chiral Pollutants: Distribution, Toxicity and Analysis by Chromatography and Capillary Electrophoresis. John Wiley & Sons, Chichester, UK.

Ali, I., Aboul-Enein, H.Y., 2006. Instrumental Methods in Metal Ions Speciation: Chromatography, Capillary Electrophoresis and Electrochemistry. Taylor & Francis Ltd, New York.

Ali, I., Aboul-Enein, H.Y., Gupta, V.K., 2009. Nano Chromatography and Capillary Electrophoresis: Pharmaceutical and Environmental Analyses. John Wiley & Sons, Hoboken, NJ.

Arshadi, M., Faraji, A.R., Amiri, M.J., 2015. Modification of aluminum-silicate nanoparticles by melamine-based dendrimer l-cysteine methyl esters for adsorptive characteristic of Hg(II) ions from the synthetic and Persian Gulf water. Chem. Eng. J. 266, 345–355.

Auffan, M., Rose, J., Proux, O., Borschneck, D., Masion, A., Chaurand, P., Hazemann, J.-L., Chaneac, C., Jolivet, J.-P., Wiesner, M.R., Van Geen, A., Bottero, J.-Y., 2008. Enhanced adsorption of arsenic onto maghemites nanoparticles: As(III) as a probe of the surface structure and heterogeneity. Langmuir 24, 3215–3222.

Auffan, M., Rose, J., Bottero, J.-Y., Lowry, G.V., Jolivet, J.-P., Wiesner, M.R., 2009. Towards a definition of inorganic nanoparticles from an environmental, health and safety perspective. Nat. Nano. 4, 634–641.

Aviv, O., Farah, S., Amir, N., Laout, N., Ratner, S., Domb, A.J., 2015. N-Bromo-dimethylhydantoin polystyrene resin for water microbial decontamination. Biomacromolecules 16, 1442–1447.

Azrague, K., Aimar, P., Benoit-Marquie, F., Maurette, M.T., 2007. A new combination of a membrane and a photocatalytic reactor for the depollution of turbid water. Appl. Catal. B 72, 197–204.

Badruddoza, A.Z.M., Shawon, Z.B.Z., Rahman, M.T., Hao, K.W., Hidajat, K., Uddin, M.S., 2013. Ionically modified magnetic nanomaterials for arsenic and chromium removal from water. Chem. Eng. J. 225, 607–615.

Bae, T.-H., Tak, T.-M., 2005. Effect of TiO_2 nanoparticles on fouling mitigation of ultrafiltration membranes for activated sludge filtration. J. Membrane Sci. 249, 1–8.

Balamurugan, R., Sundarrajan, S., Ramakrishna, S., 2011. Recent trends in nanofibrous membranes and their suitability for air and water filtrations. Membranes 1, 232.

Beyth, N., Houri-Haddad, Y., Domb, A., Khan, W., Hazan, R., 2015. Alternative antimicrobial approach: nano-antimicrobial materials. Evid. Based Complement. Alternat. Med. 2015, 16.

Bottini, M., Bruckner, S., Nika, K., Bottini, N., Bellucci, S., Magrini, A., Bergamaschi, A., Mustelin, T., 2006. Multi-walled carbon nanotubes induce T lymphocyte apoptosis. Toxicol. Lett. 160, 121–126.

Bottino, A., Capannelli, G., D'asti, V., Piaggio, P., 2001. Preparation and properties of novel organic-inorganic porous membranes. Sep. Purif. Technol. 22–23, 269–275.

Brady-Estévez, A.S., Kang, S., Elimelech, M., 2008. A single-walled-carbon-nanotube filter for removal of viral and bacterial pathogens. Small 4, 481–484.

Brame, J., Li, Q., Alvarez, P.J.J., 2011. Nanotechnology-enabled water treatment and reuse: emerging opportunities and challenges for developing countries. Trends Food Sci. Tech. 22, 618–624.

Cao, C.-Y., Qu, J., Wei, F., Liu, H., Song, W.-G., 2012. Superb adsorption capacity and mechanism of flowerlike magnesium oxide nanostructures for lead and cadmium ions. ACS Appl. Mater. Interfaces 4, 4283–4287.

Carballa, M., Omil, F., Ternes, T., Lema, J.M., 2007. Fate of pharmaceutical and personal care products (PPCPs) during anaerobic digestion of sewage sludge. Water Res. 41, 2139–2150.

Chen, W., Duan, L., Zhu, D., 2007. Adsorption of polar and nonpolar organic chemicals to carbon nanotubes. Environ. Sci. Technol. 41, 8295–8300.

Chen, L., Xin, H., Fang, Y., Zhang, C., Zhang, F., Cao, X., Zhang, C., Li, X., 2014. Application of metal oxide heterostructures in arsenic removal from contaminated water. J. Nanomater., 2.

Chirkov, S.N., 2002. The antiviral activity of chitosan (review). Appl. Biochem. Microbiol. 38, 1–8.

Choi, W., Termin, A., Hoffmann, M.R., 1994. Einflüsse von Dotierungs-Metall-Ionen auf die photokatalytische Reaktivität von TiO_2-Quantenteilchen. Angewandte Chemie 106, 1148–1149.

Chong, M.N., Jin, B., Chow, C.W.K., Saint, C., 2010. Recent developments in photocatalytic water treatment technology: a review. Water Res. 44, 2997–3027.

Chowdhury, S., Balasubramanian, R., 2014. Recent advances in the use of graphene-family nanoadsorbents for removal of toxic pollutants from wastewater. Adv. Colloid Interface Sci. 204, 35–56.

Chowdhury, S.R., Yanful, E.K., Pratt, A.R., 2012. Chemical states in XPS and Raman analysis during removal of Cr(VI) from contaminated water by mixed maghemite–magnetite nanoparticles. J. Hazard. Mater. 235-236, 246–256.

Crane, R.A., Scott, T.B., 2012. Nanoscale zero-valent iron: future prospects for an emerging water treatment technology. J. Hazard. Mater. 211-212, 112–125.

Czech, B., Nazarkovsky, M.A., Gun'ko, V.M., 2015. Photocatalytic activity of SnO_2-doped SiO_2@TiO_2 nanocomposites. In: Petkov, P. (Ed.), Nanoscience Advances in CBRN Agents Detection, Information, Energy Security. Springer, The Netherlands, pp. 255–264.

Damià, B., 2005. Emerging Organic Pollutants in Waste Waters and Sludge. Springer, New York.

Das, S.K., Khan, M.M.R., Guha, A.K., Das, A.R., Mandal, A.B., 2012. Silver-nano biohybride material: synthesis, characterization and application in water purification. Bioresour. Technol. 124, 495–499.

Das, R., Ali, M.E., Hamid, S.B.A., Ramakrishna, S., Chowdhury, Z.Z., 2014. Carbon nanotube membranes for water purification: a bright future in water desalination. Desalination 336, 97–109.

Datsyuk, V., Kalyva, M., Papagelis, K., Parthenios, J., Tasis, D., Siokou, A., Kallitsis, I., Galiotis, C., 2008. Chemical oxidation of multiwalled carbon nanotubes. Carbon 46, 833–840.

De Gusseme, B., Hennebel, T., Christiaens, E., Saveyn, H., Verbeken, K., Fitts, J.P., Boon, N., Verstraete, W., 2011. Virus disinfection in water by biogenic silver immobilized in polyvinylidene fluoride membranes. Water Res. 45, 1856–1864.

De Volder, M.F., Tawfick, S.H., Baughman, R.H., Hart, J., 2013. Carbon nanotubes: present and future commercial applications. Science 339, 535–539.

Dimitroula, H., Daskalaki, V.M., Frontistis, Z., Kondarides, D.I., Panagiotopoulou, P., Xekoukoulotakis, N.P., Mantzavinos, D., 2012. Solar photocatalysis for the abatement of emerging micro-contaminants in wastewater: synthesis, characterization and testing of various TiO_2 samples. Appl. Catal. B 117–118, 283–291.

Don, T.-M., Chen, C.-C., Lee, C.-K., Cheng, W.-Y., Cheng, L.-P., 2005. Preparation and antibacterial test of chitosan/PAA/PEGDA bi-layer composite membranes. J. Biomater. Sci. Polym. Ed. 16, 1503–1519.

Dyson, T., 1996. Population and Food: Global Trends and Future Prospects. Routledge, London.

Elbahri, M., Homaeigohar, S., Abdelaziz, R., Dai, T., Khalil, R., Zillohu, A.U., 2012. smart metal–polymer bionanocomposites as omnidirectional plasmonic black absorber formed by nanofluid filtration. Adv. Funct. Mater. 22, 4771–4777.

Ellis, J.B., 2006. Pharmaceutical and personal care products (PPCPs) in urban receiving waters. Environ. Pollut. 144, 184–189.

Farah, S., Khan, W., Farber, I., Kesler-Shvero, D., Beyth, N., Weiss, E.I., Domb, A.J., 2013. Crosslinked QA-PEI nanoparticles: synthesis reproducibility, chemical modifications, and stability study. Polym. Adv. Technol. 24, 446–452.

Farah, S., Aviv, O., Daif, M., Kunduru, K.R., Laout, N., Ratnor, S., Beyth, N., Domb, A.J., 2015a. *N*-bromohydantoin grafted polystyrene beads: synthesis and nano-micro beads characteristics for achieving controlled release of active oxidative bromine and extended microbial inactivation efficiency. J. Polym. Sci. A 54, 596–610.

Farah, S., Aviv, O., Laout, N., Ratner, S., Beyth, N., Domb, A.J., 2015b. Quaternary ammonium polyethylenimine nanoparticles for treating bacterial contaminated water. Colloids Surf. B Biointerfaces 128, 614–619.

Farah, S., Aviv, O., Laout, N., Ratner, S., Domb, A.J., 2015c. Antimicrobial *N*-brominated hydantoin and uracil grafted polystyrene beads. J. Control. Release 216, 18–29.

Fathizadeh, M., Aroujalian, A., Raisi, A., 2011. Effect of added NaX nano-zeolite into polyamide as a top thin layer of membrane on water flux and salt rejection in a reverse osmosis process. J. Membr. Sci. 375, 88–95.

Faust, S.D., Aly, O.M., 1983. Chemistry of Water Treatment. Butterworth, MA, Stoneham.

Feng, Q.L., Wu, J., Chen, G.Q., Cui, F.Z., Kim, T.N., Kim, J.O., 2000. A mechanistic study of the antibacterial effect of silver ions on *Escherichia coli* and *Staphylococcus aureus*. J. Biomed. Mater. Res. 52, 662–668.

Feng, C., Khulbe, K.C., Matsuura, T., Tabe, S., Ismail, A.F., 2013. Preparation and characterization of electro-spun nanofiber membranes and their possible applications in water treatment. Sep. Purif. Technol. 102, 118–135.

Franklin, N.M., Rogers, N.J., Apte, S.C., Batley, G.E., Gadd, G.E., Casey, P.S., 2007. Comparative toxicity of nanoparticulate ZnO, Bulk ZnO, and $ZnCl_2$ to a freshwater microalga (*Pseudokirchneriella subcapitata*): the importance of particle solubility. Environ. Sci. Technol. 41, 8484–8490.

Friedmann, D., Mendive, C., Bahnemann, D., 2010. TiO_2 for water treatment: parameters affecting the kinetics and mechanisms of photocatalysis. Appl. Catal. B 99, 398–406.

Fujishima, A., Zhang, X., Tryk, D.A., 2008. TiO_2 photocatalysis and related surface phenomena. Surf. Sci. Rep. 63, 515–582.

Gao, X., Wachs, I.E., 1999. Titania-silica as catalysts: molecular structural characteristics and physico-chemical properties. Catal. Today 51, 233–254.

Gao, W., Majumder, M., Alemany, L.B., Narayanan, T.N., Ibarra, M.A., Pradhan, B.K., Ajayan, P.M., 2011. Engineered graphite oxide materials for application in water purification. ACS Appl. Mater. Interfaces 3, 1821–1826.

Gardner, J., Dhai, A., 2014. Nanotechnology and water: ethical and regulatory considerations. In: Mishra, A.K. (Ed.), Application of Nanotechnology in Water Research. John Wiley & Sons, MA, pp. 1–20.

Gaya, U.I., Abdullah, A.H., 2008. Heterogeneous photocatalytic degradation of organic contaminants over titanium dioxide: a review of fundamentals, progress and problems. J. Photochem. Photobiol. C 9, 1–12.

Ge, F., Le, M.-M., Ye, H., Zhao, B.X., 2012. Effective removal of heavy metal ions Cd^{2+}, Zn^{2+}, Pb^{2+}, Cu^{2+} from aqueous solution by polymer-modified magnetic nanoparticles. J. Hazard. Mater. 211–212, 366–372.

Gehrke, I., Keuter, V., Groπ, F., 2012. Development of nanocomposite membranes with photocatalytic surfaces. J. Nanosci. Nanotechnol. 12, 9163–9168.

Gehrke, I., Geiser, A., Somborn-Schulz, A., 2015. Innovations in nanotechnology for water treatment. Nanotech. Sci. Appl. 8, 1–17.

Gledhill, W.E., 1987. Microbial toxicity and degradation test methodology: an industrial perspective. Toxicity Assess. 2, 89–96.

Gouma, P.I., Lee, J., 2014. Photocatalytic nanomats clean up produced water from fracking. Transl. Mater. Res. 1, 025002.

Grieger, K.D., Linkov, I., Hansen, S.F., Baun, A., 2012. Environmental risk analysis for nanomaterials: review and evaluation of frameworks. Nanotoxicology 6, 196–212.

Guan, X., Sun, Y., Qin, H., Li, J., Lo, I.M.C., He, D., Dong, H., 2015. The limitations of applying zero-valent iron technology in contaminants sequestration and the corresponding countermeasures: the development in zero-valent iron technology in the last two decades (1994–2014). Water Res. 75, 224–248.

Gupta, V.K., Nayak, A., 2012. Cadmium removal and recovery from aqueous solutions by novel adsorbents prepared from orange peel and Fe_2O_3 nanoparticles. Chem. Eng. J. 180, 81–90.

Gupta, V.K., Agarwal, S., Saleh, T.A., 2011a. Chromium removal by combining the magnetic properties of iron oxide with adsorption properties of carbon nanotubes. Water Res. 45, 2207–2212.

Gupta, V.K., Agarwal, S., Saleh, T.A., 2011b. Synthesis and characterization of alumina-coated carbon nanotubes and their application for lead removal. J. Hazard. Mater. 185, 17–23.

Hashim, D.P., Narayanan, N.T., Romo-Herrera, J.M., Cullen, D.A., Hahm, M.G., Lezzi, P., Suttle, J.R., Kelkhoff, D., Munoz-Sandoval, E., Ganguli, S., Roy, A.K., Smith, D.J., Vajtai, R., Sumpter, B.G., Meunier, V., Terrones, H., Terrones, M., Ajayan, P.M., 2012. Covalently bonded three-dimensional carbon nanotube solids via boron induced nanojunctions. Sci. Rep. 2, 363.

Hillie, T., Hlophe, M., 2007. Nanotechnology and the challenge of clean water. Nat. Nano. 2, 663–664.

Hossain, F., Perales-Perez, O.J., Hwang, S., Roman, F., 2014. Antimicrobial nanomaterials as water disinfectant: applications, limitations and future perspectives. Sci. Total Environ. 466–467, 1047–1059.

Hou, L.-R., Yuan, C.-Z., Peng, Y., 2007. Synthesis and photocatalytic property of SnO_2/TiO_2 nanotubes composites. J. Hazard. Mater. 139, 310–315.
Hua, M., Zhang, S., Pan, B., Zhang, W., Lv, L., Zhang, Q., 2012. Heavy metal removal from water/wastewater by nanosized metal oxides: a review. J. Hazard. Mater. 211–212, 317–331.
Hubbs, A.F., Mercer, R.R., Benkovic, S.A., Harkema, J., Sriram, K., Schwegler-Berry, D., Goravanahally, M.P., Nurkiewicz, T.R., Castranova, V., Sargent, L.M., 2011. Nanotoxicology—pathologist's perspective. Toxicol. Pathol. 39, 301–324.
Iijima, S., 1991. Helical microtubules of graphitic carbon. Nature 354, 56–58.
Jain, A., Duvvuri, L.S., Farah, S., Beyth, N., Domb, A.J., Khan, W., 2014. Antimicrobial polymers. Adv. Healthc. Mater. 3, 1969–1985.
Jeong, B.-H., Hoek, E.M.V., Yan, Y., Subramani, A., Huang, X., Hurwitz, G., Ghosh, A.K., Jawor, A., 2007. Interfacial polymerization of thin film nanocomposites: a new concept for reverse osmosis membranes. J. Membr. Sci. 294, 1–7.
Ji, L., Chen, W., Duan, L., Zhu, D., 2009. Mechanisms for strong adsorption of tetracycline to carbon nanotubes: a comparative study using activated carbon and graphite as adsorbents. Environ. Sci. Technol. 43, 2322–2327.
Jia, Y., Luo, T., Yu, X.-Y., Sun, B., Liu, J.-H., Huang, X.-J., 2013. A facile template free solution approach for the synthesis of dypingite nanowires and subsequent decomposition to nanoporous MgO nanowires with excellent arsenate adsorption properties. RSC Adv. 3, 5430–5437.
Jones, N., Ray, B., Ranjit, K.T., Manna, A.C., 2008. Antimicrobial activity of ZnO nanoparticles suspensions on a broad spectrum of microorganisms. FEMS Microb. Lett. 279, 71–76.
Jung, K.Y., Park, S.B., 2000. Enhanced photoactivity of silica-embedded titania particles prepared by sol-gel process for the decomposition of trichloroethylene. Appl. Catal. B 25, 249–256.
Jurado-Sanchez, B., Sattayasamitsathit, S., Gao, W., Santos, L., Fedorak, Y., Singh, V.V., Orozco, J., Galarnyk, M., Wang, J., 2015. Self-propelled activated carbon janus micromotors for efficient water purification. Small 11, 499–506.
Kanchi, S., 2014. Nanotechnology for water treatment. J. Environ. Anal. Chem. 1, e102.
Karami, H., 2013. Heavy metal removal from water by magnetite nanorods. Chem. Eng. J. 219, 209–216.
Karim, M.R., Rhodes, E.R., Brinkman, N., Wymer, L., Fout, G.S., 2009. New electropositive filter for concentrating enteroviruses and noroviruses from large volumes of water. Appl. Environ. Microbiol. 75, 2393–2399.
Kaya, I.G.B., Duranoglu, D., Beker, U., Senkal, B.F., 2011. Development of polymeric and polymer-based hybrid adsorbents for chromium removal from aqueous solution. CLEAN—Soil, Air Water 39, 980–988.
Kenawy, E.-R., Worley, S.D., Broughton, R., 2007. The chemistry and applications of antimicrobial polymers. A state-of-the-art review. Biomacromolecules 8, 1359–1384.
Khajeh, M., Laurent, S., Dastafkan, K., 2012. Nanoadsorbents: classification, preparation, and applications (with emphasis on aqueous media). Chem. Rev. 113, 7728–7768.
Kikuchi, Y., Sunada, K., Iyoda, T., Hashimoto, K., Fujishima, A., 1997. Photocatalytic bactericidal effect of TiO_2 thin films: dynamic view of the active oxygen species responsible for the effect. J. Photochem. Photobiol. A Chem. 106, 51–56.
Kim, E.-S., Deng, B., 2011. Fabrication of polyamide thin-film nano-composite (PA-TFN) membrane with hydrophilized ordered mesoporous carbon (H-OMC) for water purifications. J. Membr. Sci. 375, 46–54.
Kim, E.-S., Hwang, G., Gamal El-Din, M., Liu, Y., 2012. Development of nanosilver and multi-walled carbon nanotubes thin-film nanocomposite membrane for enhanced water treatment. J. Membr. Sci. 394–395, 37–48.

Krasner, S.W., Weinberg, H.S., Richardson, S.D., Pastor, S.J., Chinn, R., Sclimenti, M.J., Onstad, G.D., Thruston, A.D., 2006. Occurrence of a new generation of disinfection byproduct. Environ. Sci. Technol. 40, 7175–7185.

Kumar, V., Talreja, N., Deva, D., Sankararamakrishnan, N., Sharma, A., Verma, N., 2011. Development of bi-metal doped micro- and nano multi-functional polymeric adsorbents for the removal of fluoride and arsenic(V) from wastewater. Desalination 282, 27–38.

Kumar, S., Ahlawat, W., Bhanjana, G., Heydarifard, S., Nazhad, M.M., Dilbaghi, N., 2014. Nanotechnology-based water treatment strategies. J. Nanosci. Nanotechnol. 14, 1838–1858.

Kurth, C.J., Burk, R., Green, J., 2011. Utilizing nanotechnology to enhance RO membrane performance for seawater desalination. In: IDA World Congress—Perth Convention and Exhibition Centre (PCEC). 4–9 September, 2011, Perth, Western Australia.

Lau, W.J., Gray, S., Matsuura, T., Emadzadeh, D., Paul Chen, J., Ismail, A.F., 2015. A review on polyamide thin film nanocomposite (TFN) membranes: history, applications, challenges and approaches. Water Res. 80, 306–324.

Laws, E.A., 2000. Aquatic Pollution: An Introductory Text, third ed. John Wiley & Sons, New York.

Lazar, M., Varghese, S., Nair, S., 2012. Photocatalytic water treatment by titanium dioxide: recent updates. Catalysts 2, 572.

Lei, Y., Chen, F., Luo, Y., Zhang, L., 2014. Three-dimensional magnetic graphene oxide foam/Fe_3O_4 nanocomposite as an efficient absorbent for Cr(VI) removal. J. Mater. Sci. 49, 4236–4245.

Li, Y.-H., Ding, J., Luan, Z., Di, Z., Zhu, Y., Xu, C., Wu, D., Wei, B., 2003a. Competitive adsorption of Pb^{2+}, Cu^{2+} and Cd^{2+} ions from aqueous solutions by multiwalled carbon nanotubes. Carbon 41, 2787–2792.

Li, Y.-H., Wang, S., Luan, Z., Ding, J., Xu, C., Wu, D., 2003b. Adsorption of cadmium(II) from aqueous solution by surface oxidized carbon nanotubes. Carbon 41, 1057–1062.

Li, Q., Mahendra, S., Lyon, D.Y., Brunet, L., Liga, M.V., Li, D., Alvarez, P.J.J., 2008. Antimicrobial nanomaterials for water disinfection and microbial control: potential applications and implications. Water Res. 42, 4591–4602.

Liau, S.Y., Read, D.C., Pugh, W.J., Furr, J.R., Russell, A.D., 1997. Interaction of silver nitrate with readily identifiable groups: relationship to the antibacterialaction of silver ions. Lett. Appl. Microbiol. 25, 279–283.

Liga, M.V., Bryant, E.L., Colvin, V.L., Li, Q., 2011. Virus inactivation by silver doped titanium dioxide nanoparticles for drinking water treatment. Water Res. 45, 535–544.

Lin, D., Xing, B., 2008. Adsorption of phenolic compounds by carbon nanotubes: role of aromaticity and substitution of hydroxyl groups. Environ. Sci. Technol. 42, 7254–7259.

Lind, M.L., Ghosh, A.K., Jawor, A., Huang, X., Hou, W., Yang, Y., Hoek, E.M.V., 2009. Influence of zeolite crystal size on zeolite-polyamide thin film nanocomposite membranes. Langmuir 25, 10139–10145.

Lind, M.L., Eumine Suk, D., Nguyen, T.-V., Hoek, E.M.V., 2010. Tailoring the structure of thin film nanocomposite membranes to achieve seawater RO membrane performance. Environ. Sci. Technol. 44, 8230–8235.

Liu, L., Liu, J., Sun, D.D., 2012. Graphene oxide enwrapped Ag3PO4 composite: towards a highly efficient and stable visible-light-induced photocatalyst for water purification. Catal. Sci. Technol. 2, 2525–2532.

Lu, C., Chiu, H., Liu, C., 2006. Removal of zinc(II) from aqueous solution by purified carbon nanotubes. Kinetics and equilibrium studies. Ind. Eng. Chem. Res. 45, 2850–2855.

Lubick, N., 2009. Cap and degrade: a reactive nanomaterial barrier also serves as a cleanup tool. Environ. Sci. Technol. 43, 235–1235.
Luo, T., Cui, J., Hu, S., Huang, Y., Jing, C., 2010. Arsenic removal and recovery from copper smelting wastewater using TiO_2. Environ. Sci. Technol. 44, 9094–9098.
Markowska-Szczupak, A., Ulfig, K., Morawski, A.W., 2011. The application of titanium dioxide for deactivation of bioparticulates: an overview. Catal. Today 169, 249–257.
Matilainen, A., Gjessing, E.T., Lahtinen, T., Hed, L., Bhatnagar, A., Sillanpää, M., 2011. An overview of the methods used in the characterisation of natural organic matter (NOM) in relation to drinking water treatment. Chemosphere 83, 1431–1442.
Matin, A., Khan, Z., Zaidi, S.M.J., Boyce, M.C., 2011. Biofouling in reverse osmosis membranes for seawater desalination: phenomena and prevention. Desalination 281, 1–16.
Mauter, M.S., Wang, Y., Okemgbo, K.C., Osuji, C.O., Giannelis, E.P., Elimelech, M., 2011. Antifouling ultrafiltration membranes via post-fabrication grafting of biocidal nanomaterials. ACS Appl. Mater. Interfaces 3, 2861–2868.
Maximous, N., Nakhla, G., Wong, K., Wan, W., 2010. Optimization of Al_2O_3/PES membranes for wastewater filtration. Sep. Purif. Technol. 73, 294–301.
Melnyk, I.V., Zub, Y.L., 2012. Preparation and characterization of magnetic nanoparticles with bifunctional surface layer $Si(CH_2)_3NH_2/SiCH_3$ (or SiC_3H_{7-n}). Micropor. Mesopor. Mat. 154, 196–199.
Meng, Z.-D., Zhu, L., Choi, J.-G., Park, C.-Y., Oh, W.-C., 2011. Preparation, characterization and photocatalytic behavior of WO_3-fullerene/TiO_2 catalysts under visible light. Nanoscale Res. Lett. 6, 1–11.
Meng, Z.-D., Zhu, L., Ye, S., Sun, Q., Ullah, K., Cho, K.-Y., Oh, W.-C., 2013. Fullerene modification CdSe/TiO_2 and modification of photocatalytic activity under visible light. Nanoscale Res. Lett. 8, 1–10.
Mohapatra, D.P., Brar, S.K., Tyagi, R.D., Picard, P., Surampalli, R.Y., 2014. Analysis and advanced oxidation treatment of a persistent pharmaceutical compound in wastewater and wastewater sludge-carbamazepine. Sci. Total Environ., 58–75.
Muller, J., Huaux, F., Lison, D., 2006. Respiratory toxicity of carbon nanotubes: how worried should we be? Carbon 44, 1048–1056.
Musameh, M., Hickey, M., Kyratzis, I., 2011. Carbon nanotube-based extraction and electrochemical detection of heavy metals. Res. Chem. Intermed. 37, 675–689.
Nasreen, S.a.a.N., Sundarrajan, S., Nizar, S.a.S., Balamurugan, R.Ramakrishna, S., 2013. Advancement in electrospun nanofibrous membranes modification and their application in water treatment. Membranes 3, 266–284.
Nazarkovsky, M.A., Gun'ko, V.M., Wójcik, G., Czech, B., Sobieszek, A., Skubiszewska-Zięba, J., Janusz, W., Skwarek, E., 2014. Band-gap change and photocatalytic activity of silica/titania composites associated with incorporation of CuO and NiO chemistry. Phys. Technol. Surf. 5, 421–437.
Nel, A., Xia, T., Mädler, L., Li, N., 2006. Toxic potential of materials at the nanolevel. Science 311, 622–627.
Nemerow, N., Dasgupta, A., 1991. Industrial and Hazardous Waste Treatment. Van Nostrand Reinhold, New York.
Ng, L.Y., Mohammad, A.W., Leo, C.P., Hilal, N., 2013. Polymeric membranes incorporated with metal/metal oxide nanoparticles: a comprehensive review. Desalination 308, 15–33.
Ngomsik, A.-F., Bee, A., Talbot, D., Cote, G., 2012. Magnetic solid-liquid extraction of Eu(III), La(III), Ni(II) and Co(II) with maghemite nanoparticles. Sep. Purif. Technol. 86, 1–8.

No, H.K., Young Park, N., Ho Lee, S., Meyers, S.P., 2002. Antibacterial activity of chitosans and chitosan oligomers with different molecular weights. Int. J. Food Microbiol. 74, 65–72.

Ollis, D.F., 2003. Integrating photocatalysis and membrane technologies for water treatment. Ann. N. Y. Acad. Sci. 984, 65–84.

Pacheco, S., Rodriguez, R., 2001. Adsorption properties of metal ions using alumina nano-particles in aqueous and alcoholic solutions. J. Sol-Gel Sci. Technol. 20, 263–273.

Pan, B., Lin, D., Mashayekhi, H., Xing, B., 2008. Adsorption and hysteresis of bisphenol A and 17O ± -ethinyl estradiol on carbon nanomaterials. Environ. Sci. Technol. 42, 5480–5485.

Pendergast, M.M., Hoek, E.M.V., 2011. A review of water treatment membrane nanotechnologies. Energy Environ. Sci. 4, 1946–1971.

Pendergast, M.T.M., Nygaard, J.M., Ghosh, A.K., Hoek, E.M.V., 2010. Using nanocomposite materials technology to understand and control reverse osmosis membrane compaction. Desalination 261, 255–263.

Peng, Y., Liu, H., 2006. Effects of oxidation by hydrogen peroxide on the structures of multiwalled carbon nanotubes. Ind. Eng. Chem. Res. 45, 6483–6488.

Peter-Varbanets, M., Zurbrugg, C., Swartz, C., Pronk, W., 2009. Decentralized systems for potable water and the potential of membrane technology. Water Res. 43, 245–265.

Pleskova, S.N., Golubeva, I.S., Verevkin, Y.K., Pershin, E.A., Burenina, V.N., Korolichin, V.V., 2011. Photoinduced bactericidal activity of TiO_2 films. Appl. Biochem. Microbiol. 47, 23–26.

Qi, L., Xu, Z., Jiang, X., Hu, C., Zou, X., 2004. Preparation and antibacterial activity of chitosan nanoparticles. Carbohydr. Res. 339, 2693–2700.

Qi, X., Li, N., Xu, Q., Chen, D., Li, H., Lu, J., 2014. Water-soluble Fe_3O_4 superparamagnetic nanocomposites for the removal of low concentration mercury(II) ions from water. RSC Adv. 4, 47643–47648.

Qu, X., Alvarez, P.J.J., Li, Q., 2013a. Applications of nanotechnology in water and wastewater treatment. Water Res. 47, 3931–3946.

Qu, X., Brame, J., Li, Q., Alvarez, P.J.J., 2013b. Nanotechnology for a safe and sustainable water supply: enabling integrated water treatment and reuse. Acc. Chem. Res. 46, 834–843.

Rabea, E.I., Badawy, M.E.T., Stevens, C.V., Smagghe, G., Steurbaut, W., 2003. Chitosan as antimicrobial agent: applications and mode of action. Biomacromolecules 4, 1457–1465.

Rafiq, Z., Nazir, R., Shah, M.R., Ali, S., 2014. Utilization of magnesium and zinc oxide nano-adsorbents as potential materials for treatment of copper electroplating industry wastewater. J. Environ. Chem. Eng. 2, 642–651.

Rahaman, M.S., Vecitis, C.D., Elimelech, M., 2012. electrochemical carbon-nanotube filter performance toward virus removal and inactivation in the presence of natural organic matter. Environ. Sci. Technol. 46, 1556–1564.

Rajaeian, B., Rahimpour, A., Tade, M.O., Liu, S., 2013. Fabrication and characterization of polyamide thin film nanocomposite (TFN) nanofiltration membrane impregnated with TiO_2 nanoparticles. Desalination 313, 176–188.

Ramakrishna, S., Fujihara, K., Teo, W.-E., Yong, T., Ma, Z., Ramaseshan, R., 2006. Electrospun nanofibers: solving global issues. Mater. Today 9, 40–50.

Roy, A., Bhattacharya, J., 2012. Removal of Cu(II), Zn(II) and Pb(II) from water using microwave-assisted synthesized maghemite nanotubes. Chem. Eng. J. 211–212, 493–500.

Saleh, T., Gupta, V., 2012. Column with CNT/magnesium oxide composite for lead(II) removal from water. Environ. Sci. Pollut. Res. 19, 1224–1228.

Saleh, T.A., Agarwal, S., Gupta, V.K., 2011. Synthesis of MWCNT/MnO_2 and their application for simultaneous oxidation of arsenite and sorption of arsenate. Appl. Catal. B 106, 46–53.
Sato, A., Wang, R., Ma, H., Hsiao, B.S., Chu, B., 2011. Novel nanofibrous scaffolds for water filtration with bacteria and virus removal capability. J. Electron. Microsc. 60, 201–209.
Sheela, T., Nayaka, Y.A., Viswanatha, R., Basavanna, S., Venkatesha, T.G., 2012. Kinetics and thermodynamics studies on the adsorption of Zn(II), Cd(II) and Hg(II) from aqueous solution using zinc oxide nanoparticles. Powder Technol. 217, 163–170.
Shipley, H., Engates, K., Grover, V., 2013. Removal of Pb(II), Cd(II), Cu(II), and Zn(II) by hematite nanoparticles: effect of sorbent concentration, pH, temperature, and exhaustion. Environ. Sci. Pollut. Res. 20, 1727–1736.
Shvero, D.K., Zatlsman, N., Hazan, R., Weiss, E.I., Beyth, N., 2015. Characterisation of the antibacterial effect of polyethyleneimine nanoparticles in relation to particle distribution in resin composite. J. Dent. 43, 287–294.
Silva, B.F.D., Perez, S., Gardinalli, P., Singhal, R.K., Mozeto, A.A., Barcelo, D., 2011. Analytical chemistry of metallic nanoparticles in natural environments. Trends Anal. Chem. 30, 528–540.
Simon-Deckers, A., Loo, S., Mayne-L'hermite, M., Herlin-Boime, N., Menguy, N., Reynaud, C., Gouget, B., Carriere, M., 2009. Size-, composition- and shape-dependent toxicological impact of metal oxide nanoparticles and carbon nanotubes toward bacteria. Environ. Sci. Technol. 43, 8423–8429.
Stackelberg, P.E., Furlong, E.T., Meyer, M.T., Zaugg, S.D., Henderson, A.K., Reissman, D.B., 2004. Persistence of pharmaceutical compounds and other organic wastewater contaminants in a conventional drinking-water-treatment plant. Sci.Total Environ. 329, 99–113.
Subramanian, S., Seeram, R., 2013. New directions in nanofiltration applications—are nanofibers the right materials as membranes in desalination? Desalination 308, 198–208.
Sui, Z., Meng, Q., Zhang, X., Ma, R., Cao, B., 2012. Green synthesis of carbon nanotube-graphene hybrid aerogels and their use as versatile agents for water purification. J. Mater. Chem. 22, 8767–8771.
Takeshita, K., Yamakata, A., Ishibashi, T.-A., Onishi, H., Nishijima, K., Ohno, T., 2006. Transient IR absorption study of charge carriers photogenerated in sulfur-doped TiO_2. J. Photochem. Photobiol. A 177, 269–275.
Tan, L., Xu, J., Xue, X., Lou, Z., Zhu, J., Baig, S.A., Xu, X., 2014. Multifunctional nanocomposite Fe_3O_4@SiO_2-mPD/SP for selective removal of Pb(II) and Cr(VI) from aqueous solutions. RSC Adv. 4, 45920–45929.
Tang, C.Y., Zhao, Y., Wang, R., Helix-Nielsen, C., Fane, A.G., 2013. Desalination by biomimetic aquaporin membranes: review of status and prospects. Desalination 308, 34–40.
Tchobanoglous, G., Franklin, L.B., 1991. Wastewater engineering: treatment, disposal and reuse. McGraw Hill Inc., New York.
Theron, J., Walker, J.A., Cloete, T.E., 2008. Nanotechnology and water treatment: applications and emerging opportunities. Crit. Rev. Microbiol. 34, 43–69.
Tiede, K., Boxall, A.B.A., Tear, S.P., Lewis, J., David, H., Hassellov, M., 2008. Detection and characterization of engineered nanoparticles in food and the environment. Food Addit. Contam. Part A 25, 795–821.
Trivedi, P., Axe, L., 2000. Modeling Cd and Zn sorption to hydrous metal oxides. Environ. Sci. Technol. 34, 2215–2223.
Tu, Y.-J., You, C.-F., Chang, C.-K., Wang, S.-L., Chan, T.-S., 2012. Arsenate adsorption from water using a novel fabricated copper ferrite. Chem. Eng. J. 198–199, 440–448.

Upadhyayula, V.K.K., Deng, S., Mitchell, M.C., Smith, G.B., 2009. Application of carbon nanotube technology for removal of contaminants in drinking water: a review. Sci. Total Environ. 408, 1–13.
Vecitis, C.D., Zodrow, K.R., Kang, S., Elimelech, M., 2010. Electronic-structure-dependent bacterial cytotoxicity of single-walled carbon nanotubes. ACS Nano 4, 5471–5479.
Vuković, G.D., Marinković, A.D., Cˇolić, M., Ristić, M.Đ., Aleksić, R., Perić-Grujić, A.A., Uskoković, P.S., 2010. Removal of cadmium from aqueous solutions by oxidized and ethylenediamine-functionalized multi-walled carbon nanotubes. Chem. Eng. J. 157, 238–248.
WHO, 2014. Progress on Drinking Water and Sanitation. 2014 Update. Available from: http://apps.who.int/iris/bitstream/10665/112727/1/9789241507240_eng.pdf
Wu, C.-H., 2004. Comparison of azo dye degradation efficiency using UV/single semiconductor and UV/coupled semiconductor systems. Chemosphere 57, 601–608.
Wu, L., Ritchie, S.M.C., 2008. Enhanced dechlorination of trichloroethylene by membrane-supported Pd-coated iron nanoparticles. Environ. Prog. 27, 218–224.
Wu, Z., Dong, F., Zhao, W., Guo, S., 2008. Visible light induced electron transfer process over nitrogen doped TiO_2 nanocrystals prepared by oxidation of titanium nitride. J. Hazard. Mater. 157, 57–63.
Xie, W., He, F., Wang, B., Chung, T.-S., Jeyaseelan, K., Armugam, A., Tong, Y.W., 2013. An aquaporin-based vesicle-embedded polymeric membrane for low energy water filtration. J. Mater. Chem. A 1, 7592–7600.
Xiu, Z.-M., Ma, J., Alvarez, P.J.J., 2011. Differential effect of common ligands and molecular oxygen on antimicrobial activity of silver nanoparticles versus silver ions. Environ. Sci. Technol. 45, 9003–9008.
Xiu, Z.-M., Zhang, Q.-B., Puppala, H.L., Colvin, V.L., Alvarez, P.J.J., 2012. Negligible particle-specific antibacterial activity of silver nanoparticles. Nano Lett. 12, 4271–4275.
Yang, K., Xing, B., 2010. Adsorption of organic compounds by carbon nanomaterials in aqueous phase: polanyi theory and its application. Chem. Rev. 110, 5989–6008.
Yu, J., Yu, J., Zhao, X., 2002. The effect of SiO_2 addition on the grain size and photocatalytic activity of TiO_2 thin films. J. Sol-Gel Sci. Technol. 24, 95–103.
Yudovin-Farber, I., Beyth, N., Weiss, E., Domb, A., 2010. Antibacterial effect of composite resins containing quaternary ammonium polyethyleneimine nanoparticles. J. Nanopart. Res. 12, 591–603.
Zaleska, A., Sobczak, J.W., Grabowska, E., Hupka, J., 2008. Preparation and photocatalytic activity of boron-modified TiO_2 under UV and visible light. Appl. Catal. B 78, 92–100.
Zhang, W.-X., 2003. Nanoscale iron particles for environmental remediation: an overview. J. Nanopart. Res. 5, 323–332.
Zhang, Q., Pan, B., Pan, B., Zhang, W., Jia, K., Zhang, Q., 2008. Selective sorption of lead, cadmium and zinc ions by a polymeric cation exchanger containing nano-$Zr(HPO_3S)_2$. Environ. Sci. Technol. 42, 4140–4145.
Zhao, H., Qiu, S., Wu, L., Zhang, L., Chen, H., Gao, C., 2014. Improving the performance of polyamide reverse osmosis membrane by incorporation of modified multi-walled carbon nanotubes. J. Membr. Sci. 450, 249–256.
Zhou, M., Yu, J., Liu, S., Zhai, P., Jiang, L., 2008. Effects of calcination temperatures on photocatalytic activity of SnO_2/TiO_2 composite films prepared by an EPD method. J. Hazard. Mater. 154, 1141–1148.
Zodrow, K., Brunet, L.N., Mahendra, S., Li, D., Zhang, A., Li, Q., Alvarez, P.J.J., 2009. Polysulfone ultrafiltration membranes impregnated with silver nanoparticles show improved biofouling resistance and virus removal. Water Res. 43, 715–723.

3

NANOTECHNOLOGY FOR DRINKING WATER PURIFICATION

Ratna G. Suthar, Bin Gao
University of Florida, Department of Agricultural and Biological Engineering, Gainesville, FL, United States

1 Introduction

Water is an essential but often overlooked nutrient in the human diet. In fact, the average adult human requires the consumption of approximately 2 L (8 glasses) of water daily to maintain the body's respiratory balance. However, no arbitrary source can be safely used to meet this requirement. Although two-thirds of the Earth's surface is water, only 0.3% is freshwater consumable by the 7.3 billion humans that populate the Earth. The remaining 99.7% is comprised of saltwater, glacial bodies, and water vapor unusable by humans.

Potential water scarcity is being recognized as a threat to human activity. Water resources continue to diminish due to overuse, waste, and pollution. The current water treatment capacities for securing clean water will be unable to meet the needs of the growing population (Fig. 3.1) (Hu and Apblett, 2014). An estimated 663 million people lack access to improved drinking water sources whereas 946 million people do not have water to fulfill their basic sanitation necessities (WHO and UNICEF, 2015).

Clean drinking water is becoming an increasingly limited resource in many parts of the world and as a consequence water purification technologies are gaining major attention worldwide. Sustainable development of water treatment technologies requires innovative techniques to provide safe drinking water in a cost and energy efficient manner. Nanotechnology in particular, offers the potential of long term solutions to increase energy efficiency and lower costs, through the adaptation of advanced filtration materials that enable greater water quality and reuse.

This chapter highlights the recent developments of nanoscale materials and processes for treatment of surface water, groundwater that are plagued by contaminants, such as toxic metals, organic and inorganic compounds, bacteria, and viruses. After the

Water Purification. http://dx.doi.org/10.1016/B978-0-12-804300-4.00003-4

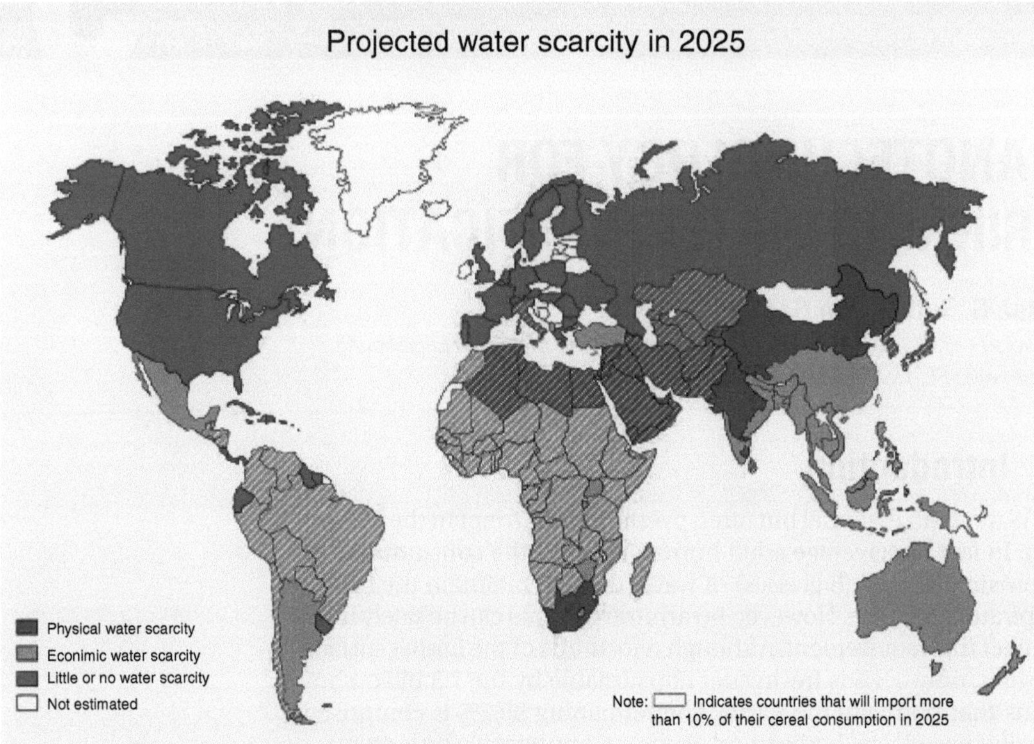

Figure 3.1. Projected global water scarcity in 2025. An area is considered water scarce when a large population does not have access to safe and affordable water to satisfy their needs for drinking, washing or livelihoods for a significant period of time. IWMI, 2000.

introduction of water sources and pollutants, the chapter provides a brief overview of current water purification processes, which will aid the reader in understanding and evaluating nanomaterials compared against current technologies. Application possibilities are limitless; however, this chapter will primarily focus on nanoadsorbents and nanoenhanced membrane technology as examples of applications of nanotechnology in drinking water purification.

1.1 Sources of Drinking Water

Where does *your* drinking water come from? For most urban residents, drinking water comes from two major sources: groundwater and surface water. The high salt concentration in seawater makes it unsuitable for most human applications, and removal of the salt is costly. This leaves only about 0.3% of the water on the Earth as liquid freshwater. The vast majority (approximately 98.5%) of the Earth's freshwater is held under the surface of the earth in different geological formations is known as *groundwater* (also called subsurface water). Large ground water bodies are

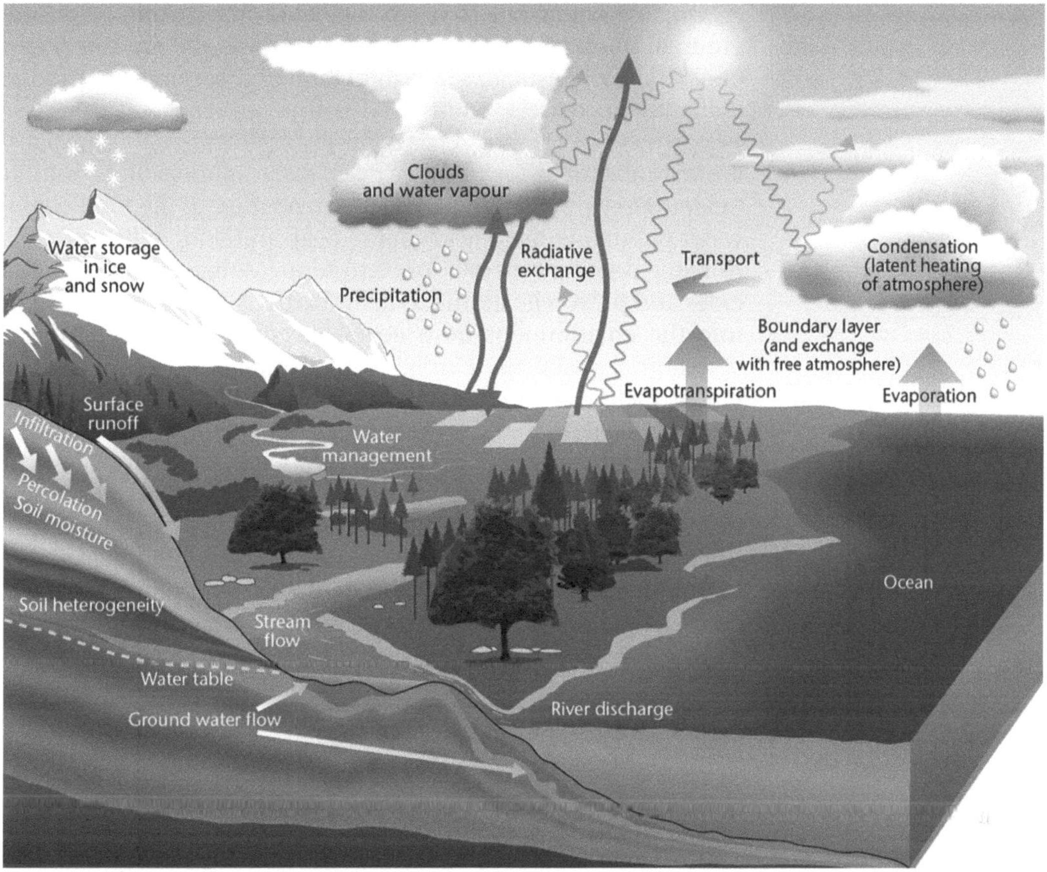

Figure 3.2. The hydrologic cycle (National Water Institute).

called "aquifers." The water enters this formation via percolation from precipitation and direct seepage from surface water sources. The remaining freshwater sources, including rivers and lakes, are collectively known as *surface water*. It is automatically replenished through precipitation and naturally lost by way of evaporation and discharge to oceans through the hydrologic cycle (Fig. 3.2). The amount of water that we use from groundwater and surface water varies by region. Generally, groundwater is cleaner and requires less treatment. Surface water, however, is easier to obtain.

1.2 Common Contaminants in Drinking Water

The Environmental Protection Agency (EPA) categorizes drinking water contaminants into the following categories: organic, inorganic, disinfectants, disinfectant by-products, microorganisms, and radionuclides. Contaminants may be naturally occurring or result from human activity.

Organic chemicals that are regulated in drinking water are a group of human-made chemical compounds and are components of a variety of pesticides and industrial and commercial products, including degreasers, paints, and petroleum distillates (Minnesota Department of Health, 2005a). For example, phenolic compounds are manufactured for plastics, dyes, drugs, antioxidants, and pesticides. They pose serious danger when entering the food chain as water pollutants, as they affect the taste and odor of fish, as well as drinking water at very low concentration (Mohan et al., 2014). Pesticides and polycyclic aromatic hydrocarbons (PAHs) are introduced into the environment from economically large-scale production agricultural practices (Fig. 3.3). Important pesticide remediation targets include organophosphorus, organochlorine, carbamate, triazine, and chlorophenoxy acid compounds (Mohan et al., 2014).

Inorganic contaminants are heavy metals, salts, and other compounds that do not contain carbon. These chemicals contaminate water supplies as a result of human activity; although, many are naturally occurring in certain geographic areas. Additionally, fluoride is added to water to supplement dental care, although if the water contains too much fluoride, it can result in adverse effects, such as

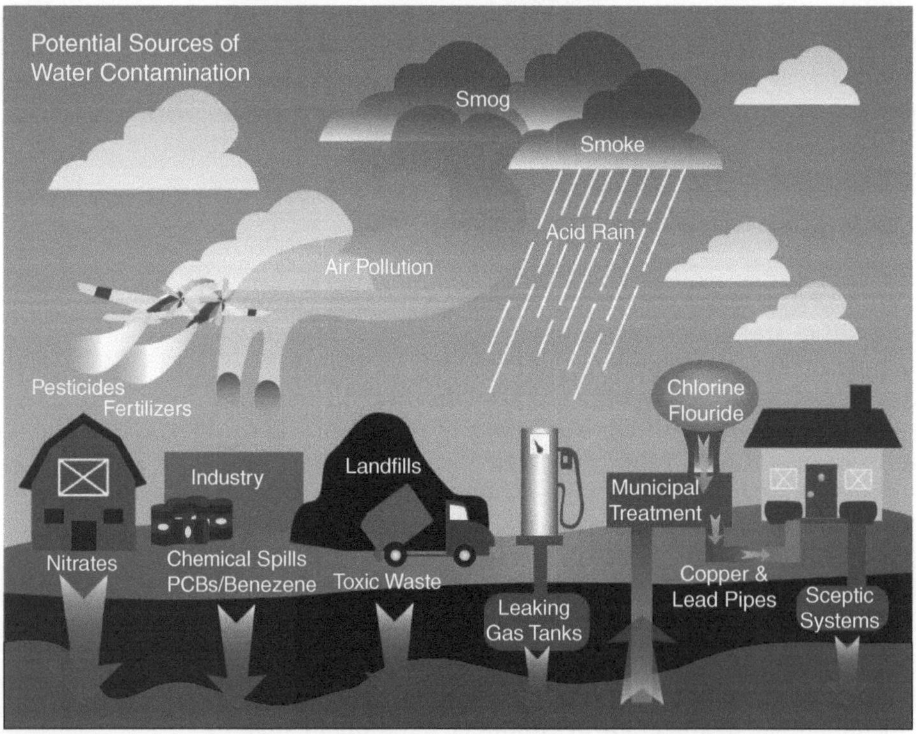

Figure 3.3. Drinking water contaminants may be naturally occurring as well as synthetic. Graphic created by Audrey Lawrence.

mottling of the teeth (Minnesota Department of Health, 2005b). Discharged heavy metals present a serious threat to human health and natural waters (FAO, 1999). In this century, heavy metals appear among the main pollutants as they pose serious health threats even when present at very low concentrations (Davydova, 1998). Eventual build-up of heavy metals in human tissues occurs over time and leads to damage, that eventually cause harmful physiological effects. Lead and copper are heavy metals that are rarely found in source waters. Usually, these contaminants enter the water through the corrosion of materials in the distribution system, including household plumbing (Minnesota Department of Health, 2005b).

To protect drinking water from disease-causing microorganisms, or pathogens, a disinfectant agent, such as chlorine is added. However, standard disinfection practices can be complicated because certain microbial pathogens, such as *Cryptosporidium*, are highly resistant to standard disinfection measures (EPA, 2013). Furthermore, such disinfectants themselves react with other contaminants in the water to form disinfectant by-products, such as trihalomethanes and haloacetic acids, which may pose health risks (EPA, 2013).

Microorganisms are biological contaminants in water. They are also referred to as microbes or microbiological contaminants. Examples of biological or microbial contaminants include bacteria, viruses, protozoan, and parasites. Origins may include either directly or indirectly, by sewage, by other wastes, or by human or animal excrement (FAO, 1999).

Radioactive contaminants are chemical elements with an unbalanced number of protons and neutrons resulting in unstable atoms that may emit ionizing radiation. Examples of radiological contaminants include cesium, plutonium, and uranium (EPA, 2009). Also termed radionuclides, radiological contaminants may enter water from soils, from the disposal and storage of radioactive wastes, or from mining practices of phosphorus or uranium. While these elements are usually naturally occurring, contamination of drinking water from radionuclides may be primarily from man-made sources (Mohan et al., 2014). Some people who drink water containing alpha emitters, radium, or uranium in excess of the MCLs over many years may have an increased risk of getting cancer (Minnesota Department of Health, 2005b).

The EPA sets the maximum contaminant level (MCL), which is the maximum level of a contaminant present in drinking water at which no anticipated adverse effect on the health of persons would be expected upon consumption, and which allows an adequate margin of safety. The EPA-defined MCLs for common contaminants in drinking water including, organics, inorganic, disinfectants, disinfectant by-products, microorganisms, and radionuclides, are listed in Table 3.1.

Table 3.1 EPA-Defined MCLs for Common Contaminants in Drinking Water

Drinking Water Contaminants				
Contaminant	MCLG[a] (mg/L)[2]	MCL or TT[a] (mg/L)[b]	Potential Health Effects from Long-Term Exposure Above the MCL (Unless Specified as Short-Term)	Sources of Contaminant in Drinking Water
Microorganisms				
Cryptosporidium	Zero	TT[c]	Gastrointestinal illness (such as diarrhea, vomiting, and cramps)	Human and animal fecal waste
Giardia lamblia	Zero	TT[c]	Gastrointestinal illness (such as diarrhea, vomiting, and cramps)	Human and animal fecal waste
Heterotrophic plate count (HPC)	n/a	TT[c]	HPC has no health effects; it is an analytic method used to measure the variety of bacteria that are common in water. The lower the concentration of bacteria in drinking water, the better maintained the water system is	HPC measures a range of bacteria that are naturally present in the environment
Legionella	Zero	TT[c]	Legionnaire's disease, a type of pneumonia	Found naturally in water; multiplies in heating systems
Total coliforms (including fecal coliform and *E. coli)*	Zero	5.0%[d]	Not a health threat in itself; it is used to indicate whether other potentially harmful bacteria may be [e]	Coliforms are naturally present in the environment; as well as feces; fecal coliforms and *E. coli* only come from human and animal fecal waste
Turbidity	n/a	TT[c]	Turbidity is a measure of the cloudiness of water. It is used to indicate water quality and filtration effectiveness (such as whether disease-causing organisms are present). Higher turbidity levels are often associated with higher levels of disease-causing microorganisms, such as viruses, parasites, and some bacteria. These organisms can cause symptoms, such as nausea, cramps, diarrhea, and associated headaches	Soil runoff
Viruses (enteric)	zero	TT[c]	Gastrointestinal illness (such as diarrhea, vomiting, and cramps)	Human and animal fecal waste
Disinfection By-Products				
Bromate	zero	0.010	Increased risk of cancer	By-product of drinking water disinfection
Chlorite	0.8	1.0	Anemia; infants and young children: nervous system effects	By-product of drinking water disinfection
Haloacetic acids (HAA5)	n/a[f]	0.060[g]	Increased risk of cancer	By-product of drinking water disinfection

Total trihalomethanes (TTHMs)	>n/a[f]	>0.080[g]	Liver, kidney, or central nervous system problems; increased risk of cancer	By-product of drinking water disinfection
Disinfectants				
Chloramines (as CL_2)	MRDLG = 4[1]	MRDL = 4.0[a]	Eye/nose irritation; stomach discomfort, anemia	Water additive used to control microbes
Chlorine (as Cl_2)	MRDLG = 4[1]	MRDL = 4.0[a]	Eye/nose irritation; stomach discomfort	Water additive used to control microbes
Chlorine dioxide (as ClO_2)	MRDLG = 0.8[1]	MRDL = 0.8[a]	Anemia; infants and young children: nervous system effects	Water additive used to control microbes
Inorganic Chemicals				
Antimony	0.006	0.006	Increase in blood cholesterol; decrease in blood sugar	Discharge from petroleum refineries; fire retardants; ceramics; electronics; solder
Arsenic	0	0.010 as of 01/23/06	Skin damage or problems with circulatory systems, and may have increased risk of getting cancer	Erosion of natural deposits; runoff from orchards, runoff from glass and electronics production wastes
Asbestos (fiber > 10 μm)	7 million fibers per liter (MFL)	7 MFL	Increased risk of developing benign intestinal polyps	Decay of asbestos cement in water mains; erosion of natural deposits
Barium	2	2	Increase in blood pressure	Discharge of drilling wastes; discharge from metal refineries; erosion of natural deposits
Beryllium	0.004	0.004	Intestinal lesions	Discharge from metal refineries and coal-burning factories; discharge from electrical, aerospace, and defense industries
Cadmium	0.005	0.005	Kidney damage	Corrosion of galvanized pipes; erosion of natural deposits; discharge from metal refineries; runoff from waste batteries and paints
Chromium (total)	0.1	0.1	Allergic dermatitis	Discharge from steel and pulp mills; erosion of natural deposits
Copper	1.3	TT[g]; action level = 1.3	Short-term exposure: gastrointestinal distress Long-term exposure: liver or kidney damage People with Wilson's disease should consult their personal doctor if the amount of copper in their water exceeds the action level	Corrosion of household plumbing systems; erosion of natural deposits
Cyanide (as free cyanide)	0.2	0.2	Nerve damage or thyroid problems	Discharge from steel/metal factories; discharge from plastic and fertilizer factories

(*Continued*)

Table 3.1 EPA-defined MCLs for common contaminants in drinking water (*cont.*)

Drinking Water Contaminants				
Contaminant	**MCLG[a] (mg/L)[2]**	**MCL or TT[a] (mg/L)[b]**	**Potential Health Effects from Long-Term Exposure Above the MCL (Unless Specified as Short-Term)**	**Sources of Contaminant in Drinking Water**
Microorganisms				
Fluoride	4.0	4.0	Bone disease (pain and tenderness of the bones); children may get mottled teeth	Water additive which promotes strong teeth; erosion of natural deposits; discharge from fertilizer and aluminum factories
Lead	Zero	TT[g]; action level = 0.015	Infants and children: delays in physical or mental development; children could show slight deficits in attention span and learning abilities Adults: kidney problems; high blood pressure	Corrosion of household plumbing systems; erosion of natural deposits
Mercury (inorganic)	0.002	0.002	Kidney damage	Erosion of natural deposits; discharge from refineries and factories; runoff from landfills and croplands
Nitrate (measured as nitrogen)	10	10	Infants below the age of 6 months who drink water containing nitrate in excess of the MCL could become seriously ill and, if untreated, may die. Symptoms include shortness of breath and blue-baby syndrome	Runoff from fertilizer use; leaking from septic tanks, sewage; erosion of natural deposits
Nitrite (measured as nitrogen)	1	1	Infants below the age of 6 months who drink water, containing nitrite in excess of the MCL could become seriously ill and, if untreated, may die. Symptoms include shortness of breath and blue-baby syndrome	Runoff from fertilizer use; leaking from septic tanks, sewage; erosion of natural deposits
Selenium	0.05	0.05	Hair or fingernail loss; numbness in fingers or toes; circulatory problems	Discharge from petroleum refineries; erosion of natural deposits; discharge from mines
Thallium	0.0005	0.002	Hair loss; changes in blood; kidney, intestine, or liver problems	Leaching from ore-processing sites; discharge from electronics, glass, and drug factories

Organic Chemicals				
Acrylamide	Zero	TT[h]	Nervous system or blood problems; increased risk of cancer	Added to water during sewage/wastewater treatment
Alachlor	Zero	0.002	Eye, liver, kidney or spleen problems; anemia; increased risk of cancer	Runoff from herbicide used on row crops
Atrazine	0.003	0.003	Cardiovascular system or reproductive problems	Runoff from herbicide used on row crops
Benzene	Zero	0.005	Anemia; decrease in blood platelets; increased risk of cancer	Discharge from factories; leaching from gas storage tanks and landfills
Benzo(*a*)pyrene (PAHs)	Zero	0.0002	Reproductive difficulties; increased risk of cancer	Leaching from linings of water storage tanks and distribution lines
Carbofuran	0.04	0.04	Problems with blood, nervous system, or reproductive system	Leaching of soil fumigant used on rice and alfalfa
Carbon tetrachloride	Zero	0.005	Liver problems; increased risk of cancer	Discharge from chemical plants and other industrial activities
Chlordane	Zero	0.002	Liver or nervous system problems; increased risk of cancer	Residue of banned termiticide
Chlorobenzene	0.1	0.1	Liver or kidney problems	Discharge from chemical and agricultural chemical factories
2,4-D	0.07	0.07	Kidney, liver, or adrenal gland problems	Runoff from herbicide used on row crops
Dalapon	0.2	0.2	Minor kidney changes	Runoff from herbicide used on rights of way
l,2-Dibromo- 3-chloropropane (DBCP)	Zero	0.0002	Reproductive difficulties; increased risk of cancer	Runoff/leaching from soil fumigant used on soybeans, cotton, pineapples, and orchards
o-Dichlorobenzene	0.6	0.6	Liver, kidney, or circulatory system problems	Discharge from industrial chemical factories
p-Dichlorobenzene	0.075	0.075	Anemia; liver, kidney or spleen damage; changes in blood	Discharge from industrial chemical factories
1,2-Dichloroethane	Zero	0.005	Increased risk of cancer	Discharge from industrial chemical factories
1,1-Dichloroethylene	0.007	0.007	Liver problems	Discharge from industrial chemical factories
cis-1,2-Dichloroethylene	0.07	0.07	Liver problems	Discharge from industrial chemical factories
trans-1,2-Dichloroethylene	0.1	0.1	Liver problems	Discharge from industrial chemical factories
Dichloromethane	Zero	0.005	Liver problems; increased risk of cancer	Discharge from drug and chemical factories
1,2-Dichloropropane	Zero	0.005	Increased risk of cancer	Discharge from industrial chemical factories

(*Continued*)

Table 3.1 EPA-defined MCLs for common contaminants in drinking water (*cont.*)

Drinking Water Contaminants				
Contaminant	**MCLG[a] (mg/L)[2]**	**MCL or TT[a] (mg/L)[b]**	**Potential Health Effects from Long-Term Exposure Above the MCL (Unless Specified as Short-Term)**	**Sources of Contaminant in Drinking Water**
Microorganisms				
Di(2-ethylhexyl) adipate	0.4	0.4	Weight loss, liver problems, or possible reproductive difficulties	Discharge from chemical factories
Di(2-ethylhexyl) phthalate	Zero	0.006	Reproductive difficulties; liver problems; increased risk of cancer	Discharge from rubber and chemical factories
Dinoseb	0.007	0.007	Reproductive difficulties	Runoff from herbicide used on soybeans and vegetables
Dioxin (2,3,7,8-TCDD)	Zero	0.00000003	Reproductive difficulties; increased risk of cancer	Emissions from waste incineration and other combustion; discharge from chemical factories
Diquat	0.02	0.02	Cataracts	Runoff from herbicide use
Endothall	0.1	0.1	Stomach and intestinal problems	Runoff from herbicide use
Endrin	0.002	0.002	Liver problems	Residue of banned insecticide
Epichlorohydrin	Zero	TT[h]	Increased cancer risk, and over a long period of time, stomach problems	Discharge from industrial chemical factories; an impurity of some water treatment chemicals
Ethylbenzene	0.7	0.7	Liver or kidneys problems	Discharge from petroleum refineries
Ethylene dibromide	Zero	0.00005	Problems with liver, stomach, reproductive system, or kidneys; increased risk of cancer	Discharge from petroleum refineries
Glyphosate	0.7	0.7	Kidney problems; reproductive difficulties	Runoff from herbicide use
Heptachlor	Zero	0.0004	Liver damage; increased risk of cancer	Residue of banned termiticide
Heptachlor epoxide	Zero	0.0002	Liver damage; increased risk of cancer	Breakdown of heptachlor
Hexachloroben zene	Zero	0.001	Liver or kidney problems; reproductive difficulties; increased risk of cancer	Discharge from metal refineries and agricultural chemical factories
Hexachlorocycl opentadiene	0.05	0.05	Kidney or stomach problems	Discharge from chemical factories
Lindane	0.0002	0.0002	Liver or kidney problems	Runoff/leaching from insecticide used on cattle, lumber, gardens

Methoxychlor	0.04	0.04	Reproductive difficulties	Runoff/leaching from insecticide used on fruits, vegetables, alfalfa, livestock
Oxamyl (Vydate)	0.2	0.2	Slight nervous system effects	Runoff/leaching from insecticide used on apples, potatoes, and tomatoes
Polychlorinated biphenyls (PCBs)	Zero	0.0005	Skin changes; thymus gland problems; immune deficiencies; reproductive or nervous system difficulties; increased risk of cancer	Runoff from landfills; discharge of waste chemicals
Pentachlorophenol	Zero	0.001	Liver or kidney problems; increased cancer risk	Discharge from wood preserving factories
Picloram	0.5	0.5	Liver problems	Herbicide runoff
Simazine	0.004	0.004	Problems with blood	Herbicide runoff
Styrene	0.1	0.1	Liver, kidney, or circulatory system problems	Discharge from rubber and plastic factories; leaching from landfills
Tetrachloroethylene	Zero	0.005	Liver problems; increased risk of cancer	Discharge from factories and dry cleaners
Toluene	1	1	Nervous system, kidney, or liver problems	Discharge from petroleum factories
Toxaphene	Zero	0.003	Kidney, liver, or thyroid problems; increased risk of cancer	Runoff/leaching from insecticide used on cotton and cattle
2,4,5-TP (Silvex)	0.05	0.05	Liver problems	Residue of banned herbicide
12 4-Trichlorobenzene	0.07	0.07	Changes in adrenal glands	Discharge from textile finishing factories
111-Trichloroethane	0.20	0.2	Liver, nervous system, or circulatory problems	Discharge from metal degreasing sites and other factories
1,1,2-Trichloroethane	0.003	0.005	Liver, kidney, or immune system problems	Discharge from industrial chemical factories
Trichloroethylene	Zero	0.005	Liver problems; increased risk of cancer	Discharge from metal degreasing sites and other factories
Vinyl chloride	Zero	0.002	Increased risk of cancer	Leaching from PVC pipes; discharge from plastic factories
Xylenes (total)	10	10	Nervous system damage	Discharge from petroleum factories; discharge from chemical factories
Radionuclides				
Alpha particles	None[g]—zero	15 picocuries per liter (PCi/L)	Increased risk of cancer	Erosion of natural deposits of certain minerals that are radioactive and may emit a form of radiation known as alpha radiation

(*Continued*)

Table 3.1 EPA-defined MCLs for common contaminants in drinking water (*cont.*)

Drinking Water Contaminants				
Contaminant	**MCLG[a] (mg/L)[2]**	**MCL or TT[a] (mg/L)[b]**	**Potential Health Effects from Long-Term Exposure Above the MCL (Unless Specified as Short-Term)**	**Sources of Contaminant in Drinking Water**
Microorganisms				
Beta particles and photon emitters	None[g] —zero	4 millirems per year	Increased risk of cancer	Decay of natural and man-made deposits of certain minerals that are radioactive and may emit forms of radiation known as photons and beta radiation
Radium 226 and Radium 228 (combined)	None[g]—zero	5 PCi/L	Increased risk of cancer	Erosion of natural deposits
Uranium	Zero	30 UG/Las of 12/08/03	Increased risk of cancer, kidney toxicity	Erosion of natural deposits

United States Environmental Protection Agency (Notes)

[a] Definitions:

- Maximum Contaminant Level Goal (MCLG): The level of a contaminant in drinking water below, which there is no known or expected risk to health. MCLGs allow for a margin of safety and are nonenforceable public health goals.
- MCL: The highest level of a contaminant that is allowed in drinking water. MCLs are set as close to MCLGs as feasible using the best available treatment technology and taking cost into consideration. MCLs are enforceable standards.
- Maximum Residual Disinfectant Level Goal (MRDLG): The level of a drinking water disinfectant below, which there is no known or expected risk to health. MRDLGs do not reflect the benefits of the use of disinfectants to control microbial contaminants.
- Treatment Technique (TT): A required process intended to reduce the level of a contaminant in drinking water.
- Maximum Residual Disinfectant Level (MRDL): The highest level of a disinfectant allowed in drinking water. There is convincing evidence that addition of a disinfectant is necessary for control of microbial contaminants.

[b] Units are in milligrams per liter (mg/L) unless otherwise noted. mg/L are equivalent to parts per million (ppm).

[c] EPA's surface water treatment rules require systems using surface water or ground water under the direct influence of surface water to

- disinfect their water,
- filter their water, or
- meet criteria for avoiding filtration so that the following contaminants are controlled at the following levels:
  - *Cryptosporidium:* Unfiltered systems are required to include *Cryptosporidium* in their existing watershed control provisions
  - *Giardia lamblia:* 99.9% removal/inactivation.
  - Viruses: 99.99% removal/inactivation.
  - *Legionella:* No limit, but EPA believes that if *Giardia* and viruses are removed/inactivated, according to the treatment techniques in the Surface Water Treatment Rule, *Legionella* will also be controlled.
  - Turbidity: For systems that use conventional or direct filtration, at no time can turbidity (cloudiness of water) go higher than 1 Nephelometric Turbidity Unit (NTU), and samples for turbidity must be less than or equal to 0.3 NTUs in at least 95% of the samples in any month. Systems that use filtration other than the conventional or direct filtration must follow state limits, which must include turbidity at no time exceeding 5 NTUs.
  - Heterotrophic Plate Count (HPC): No more than 500 bacterial colonies per milliliter.

- Long Term 1 Enhanced Surface Water Treatment: Surface water systems or groundwater under the direct influence (GWUDI) systems serving fewer than 10,000 people must comply with the applicable Long Term 1 Enhanced Surface Water Treatment Rule provisions (such as turbidity standards, individual filter monitoring, *Cryptosporidium* removal requirements, updated watershed control requirements for unfiltered systems).
- Long Term 2 Enhanced Surface Water Treatment Rule: This rule applies to all surface water systems or ground water systems under the direct influence of surface water. The rule targets additional *Cryptosporidium* treatment requirements for higher risk systems and includes provisions to reduce risks from uncovered finished water storage facilities and to ensure that the systems maintain microbial protection as they take steps to reduce the formation of disinfection by-products.
- Filter Backwash Recycling: The Filter Backwash Recycling rule requires systems that recycle to return specific recycle flows through all processes of the system's existing conventional or direct filtration system or at an alternate location approved by the state.

[d] No more than 5.0% samples total coliform-positive (TC-positive) in a month. (For water systems that collect fewer than 40 routine samples per month, no more than 1 sample can be total coliform-positive per month.) Every sample that has total coliform must be analyzed for either fecal coliforms or *E. coli* if two consecutive TC-positive samples, and one is also positive for *E.coli* fecal coliforms, system has an acute MCL violation.

[e] Fecal coliform and *E. coli* are bacteria whose presence indicates that the water may be contaminated with human or animal wastes. Disease-causing microbes (pathogens) in these wastes can cause diarrhea, cramps, nausea, headaches, or other symptoms. These pathogens may pose a special health risk for infants, young children, and people with severely compromised immune systems.

[f] Although there is no collective MCLG for this contaminant group, there are individual MCLGs for some of the individual contaminants:

- Trihalomethanes: bromodichloromethane (zero); bromoform (zero); dibromochloromethane (0.06 mg/L): chloroform (0.07 mg/L).
- Haloacetic acids: dichloroacetic acid (zero); trichloroacetic acid (0.02 mg/L); monochloroacetic acid (0.07 mg/L). Bromoacetic acid and dibromoacetic acid are regulated with this group but have no MCLGs.

[g] Lead and copper are regulated by a treatment technique that requires systems to control the corrosiveness of their water. If more than 10% of tap water samples exceed the action level, water systems must take additional steps. For copper, the action level is 1.3 mg/L, and for lead is 0.015 mg/L.

[h] Each water system must certify, in writing, to the state (using third-party or manufacturer's certification) that when acrylamide and epichlorohydrin are used to treat water, the combination (or product) of dose and monomer level does not exceed the levels specified, as follows:

- Acrylamide = 0.05% dosed at 1 mg/L (or equivalent)
- Epichlorohydrin = 0.01% dosed at 20 mg/L (or equivalent)

Modified from EPA 816-F-09-004.

2 Overview of the Water Purification Process

Raw, untreated, water undergoes various treatment processes before it reaches our tap. In the beginning of the 20th century, many of the water treatment methods used today were already established (Hazen, 1914). These methods include conventional drinking water treatment processes, consisting of coagulation, flocculation, sedimentation, filtration, and chlorine disinfection (Fig. 3.4) (Fuller, 1933). The aforementioned processes are chemically and operationally intensive; requiring large systems, infrastructure, and engineering expertise, which make them burdensome, inefficient, time consuming, and costly (Kumar et al., 2014).

Water is often pretreated with chlorine: chlorine disinfects by killing *biological* contaminants, such as bacteria, protozoa, and viruses that may cause illness in humans. Chlorine also prevents the growth of algae at the treatment plant that may hinder treatment of water and cause taste and odor problems. Enough chlorine is added to maintain a residual of 1 part per million or residual chlorine in water after filtration in order to prevent regrowth of microorganisms in the distribution systems pipes and home plumbing (Safe Water Drinking Foundation). Next, in the coagulation process aluminum sulfate (alum) is added to the water with the objective of destabilizing natural fine *physical* particulate matter suspended in water. These impurities found in water suspensions consist of charged colloids ranging in size from 5 nm to 1 μm and particulates greater than 0.5 mm (EPA, 1999). Most of

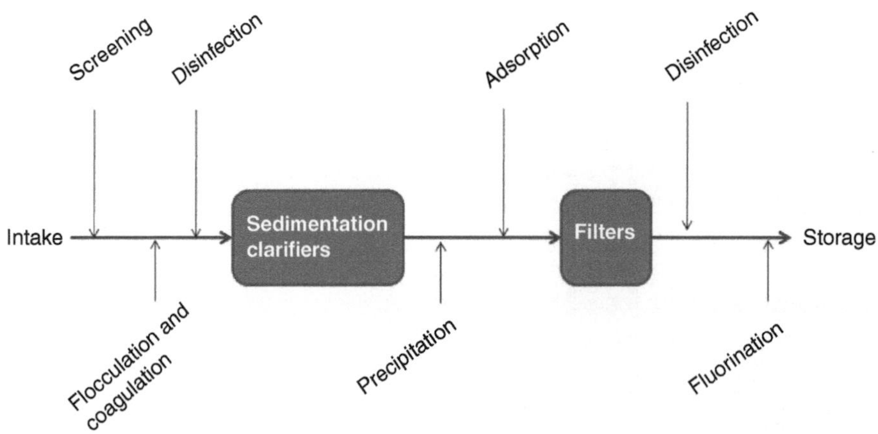

Figure 3.4. Water purification process.

these particles will not settle by gravity due to their charge and because they are less dense than water. The addition of alum causes these particles and colloids to clump together to form heavier particles called flocculocides (floc) which settle in water. Flocculation is the aggregation of destabilized particles into larger masses that are easier to remove from water than the original particles. After flocculation, the water and floc moves slowly through large basins known as sedimentation or settling basins. Coagulation typically occurs in less than 10 s, whereas flocculation occurs over a period of 20–45 min (Crittenden et al., 2012). Floc along with other materials, such as clay and silt based turbidity and natural organic matter are removed in the sedimentation process. These impurities include microbial contaminants, toxic metals, synthetic organic chemicals, iron, manganese, and humic substances. Humic substances come from soil and are produced within natural water and sediments by chemical and biological processes, such as the decay of vegetation (McCarthy and Zachara, 1989). Removal of humic substances from drinking water is desirable since they form disinfection by-products when chlorine is added to the water. The last step in purifying the water is accomplished by filtration through a bed of sand and gravel. As water filters through the sand, the remaining particles of suspended matter are trapped in the sand bed.

Fluoride is added to the filtered water at each of the plants to reduce tooth decay. Treatment plants maintain fluoride levels between 0.7 and 1.2 part per million in the treated water (American Dental Association, 2005). Finally, lime (or calcium oxide) is added to the treated water before the water is added to the distribution system. This raises the pH of the water to 8 standard units, reducing the ability of the water to corrode water mains and home plumbing materials, such as copper, lead, and brass (Fig. 3.4).

Existing technologies for water purification have many limitations in their time consuming and costly manner; nanotechnology contributes a framework for offering rapid detection and removal of contaminants by providing water treatment at low cost (Doroodmand et al., 2012; Shannon et al., 2008). Further, the chemicals used for the intensive treatment of water involves chlorine, ammonia, hydrochloric acid, ozone, alum, permanganate, ferric salts, corrosion control chemicals, coagulants, filtration aids, ion exchange resins, and residuals can add to the contamination freshwater sources (Shannon et al., 2008).

The water industry is required to produce drinking water of high quality. There is a clear need for the development of cost-effective and stable materials and methods to address the challenges of providing the potable water in adequate amounts.

3 Applications of Nanotechnology in Water Purification

Nanomaterials are materials that are modified or synthesized in the nanoscale (1–100 nm) using a bottom-up approach and can exhibit properties that are not observed in the macro- or microscale. Nanotechnology has been promising in applications across different fields but recent applications for water treatment has emerged as a fast-developing, promising area. They are promising solutions for water supplies contaminated by organic and inorganic compounds, heavy metal ions, and microorganisms (Theron et al., 2008).

Recent advances strongly suggest that many of the current problems involving water quality can be addressed and potentially resolved using nanoadsorbents, nanoenhanced membranes, nanometal oxides, and nano photo catalysts. Table 3.2 shows a summary of relevant literature. Among them, three categories show most promising application at the full scale in the near future based on their stages in research and development, commercial availability and cost of nanomaterials involved, and compatibility with the existing infrastructure: nanoadsorbents, nanotechnology enabled membranes, and nanophotocatalysts. In this chapter, we will discuss the first two, nanoadsorbents and nanoenhanced membranes. They both have commercial products, although they have not been applied in large-scale water treatment.

There is a significant need for advanced water purification technologies, particularly to ensure a high quality of drinking water, eliminate water pollutants, and intensify industrial production processes by the use nanotechnology enabled purification processes. Nanoengineered materials, such as nanoadsorbents, nanometal oxides, nanomembranes, and nanophotocatalysts, offer the potential for novel water technologies that can be easily integrated to customer-specific applications (Gehrke et al., 2015). Most of them can be compatible with existing treatment technologies and can be integrated simply in conventional modules. Retention and reuse of nanomaterials is a key aspect of nanotechnology enabled device design due to both cost and public health concerns (Qu et al., 2013a).

One of the most important advantages of nanomaterials when compared with conventional water technologies is their ability to integrate various properties, resulting in multifunctional systems, such as nanocomposite membranes that enable both particle retention and elimination of contaminants (Gehrke et al., 2015).

Table 3.2 Survey of Representative Materials Advancing the use of Nanotechnology in Water Purification

Overview of Nanotechnology-Enabled Water Purification Materials		
Nanostructured membrane		
CNT	Bacteria and viruses	Srivastava et al. (2004)
Nanoreactive membrane		
Nanosilver impregnate membranes	Bacteria	Zodrow et al. (2009)
Ag/TiO_2 nanofiber membrane	Bacteria	Liu et al. (2012)
Alumina membrane formed from A-alumoxane	Synthetic dyes	DeFriend et al. (2003)
Alumina membranes functionalized with poly(styrene sulfonate) or poly(allylamine hydrochloride)	Divalent cations	Stanton et al. (2003)
Silica and cellulose-based membranes functionalized with amino acid homo-polymers	Metal ions	Bhattacharyya et al. (1998); Ritchie et al. (2001)
Alumina or polymeric membranes with gold nanoparticles	4-Nitrophenol	Dotzauer et al. (2006)
Polymer-impregnated ceramic TiO_2 filters	PAHs	Arkas et al. (2005)
TiO_2/Al_2O_3 composite membranes	Direct Black 168 dye	Zhang et al. (2006)
TiO_2 photocatalytic nanofiltration membrane	Methyl orange, azo-dye	Romanos et al. (2012); Athanasekou et al. (2012)
Polymer-impregnated ceramic alumina and silicon–carbon filters	Trihalogen methanes, PAHs, pesticide	Allabashi et al. (2007)
Nanoparticle/nanomaterial		
Nanocrystalline zeolites	Toluene, nitrogen dioxide	Song et al. (2004, 2005)
Synthetic zeolite nanoparticles	Humic acid	Orha et al. (2011)
Graphitized CNT	1,2-Dichlorobenzene	Peng et al. (2003)
CeO_2–CNTs	Metal ions	Di et al. (2006); Peng et al. (2005)
CNT functionalized with polymer	*p*-Nitrophenol	Salipira et al. (2007)
Carbon nanoparticles conjugated with polyethylenimine	Metal ions	Khaydarov et al. (2010)
CNT hydrite	Arsenate	Poinern et al. (2010)

(*Continued*)

Table 3.2 Survey of Representative Materials Advancing the use of Nanotechnology in Water Purification (*cont.*)

Overview of Nanotechnology-Enabled Water Purification Materials		
CNT sheets	Metal ions	Tofighy and Mohammadi (2011)
CNT functionalized with Fe	Benzene, toluene, dimethylbenzene, metal ions	Jin et al. (2007); Lu and Liu (2006); Lu et al. (2005)
SWCNTs and MWCNTs	Trihalomethanes (THMs), metal ions, roxarsone, ethylbenzene, herbicides, cyanobacterial toxins	Lu et al. (2005, 2006b); Li et al. (2003); Hu et al. (2012); Bina et al. (2011); Cai et al. (2005); Zhou et al. (2006b, 2006a, 2007); Yan et al. (2006, 2008); Albuquerque Júnior et al. (2008)
Mesoporous aluminosilicate spheres with nanosized Fe_3O_4	Mercury	Guo et al. (2008a)
Silica-coated Fe_3O_4	Metal ions	Huang and Hu (2008)
Fe_3O_4 particles encapsulated in thiol containing polymers	Metal ions	Shin and Jang (2007)
Fe_3O_4 nanoparticles functionalized with dithiocarbamate groups	Mercury	Girginova et al. (2010); Figueira et al. (2011)
Fe_4O_3 magnetic nanoparticles coated with humic acid	Metal ions	Liu et al. (2008)
Zero-valent iron nanoparticles (nZVI)	Polychlorinated biphenyls (PCBs), inorganic ions, chlorinated organic compounds heavy metal ions	He and Zhao (2005); Choe et al. (2000); Cao et al. (2005); Yang and Lee (2005); Zhang et al. (1998); Cheng et al. (2007); Ponder et al. (2000); Kanel et al. (2005, 2006); Li et al. (2006); Xu and Zhao (2007); Üzüm et al. (2009); Boparai et al. (2011)
Nano photocatalyst materials		
Nanocrystalline TiO_2	Metal ions	Pena et al. (2005)
Nitrogen (N)-doped TiO_2	Azo dyes	Liu et al. (2005)
Fe(III)-doped TiO_2	Phenol	Nahar et al. (2006)
Supported TiO_2 nanoparticles	Aromatic pollutants	López-Muñoz et al. (2005)
Silver-doped titanium dioxide nanoparticles	Bacteria	Liga et al. (2011)
Manganese-doped ZnO NPs	Methylene blue	Ullah and Dutta (2008)
Nanotubes Bi_2O_3	Chromium ions	Qin et al. (2012)

Table 3.2 Survey of Representative Materials Advancing the use of Nanotechnology in Water Purification (*cont.*)

Overview of Nanotechnology-Enabled Water Purification Materials		
Bi_2O_3 and Au/Bi_2O_3 nanorods	Orange II dye	Anandan et al. (2010)
CeO_2	Dyes	Zhai et al. (2007); Ji et al. (2009b); Borker and Salker (2007)
Nanocomposite plasmonic photocatalyst Ag–AgCl/CeO_2	Methyl orange	Wang et al. (2011)
Nano WO_3	*E. coli*	Gondal et al. (2009)
Photocatalyst CdS coated with CdS nanoparticles	Dyes and phenolic compounds	Yang et al. (2009)
ZnS nanoporous nanoparticles	Eosin	He and Zhao (2005)

Modified from: Mohmood et al., 2013.

Further, nanomaterials enable higher processing and efficiency due to their unique characteristics, such as a high reaction rate.

3.1 Adsorption

Adsorption is a surface phenomenon where molecules of a substance (adsorbate) adhere on to the surface of a solid (adsorbent). Various factors affecting adsorption are temperature, the nature of the adsorbate and adsorbent, the presence of other impurities, particle size, contact time, and chemical environment (Kumar et al., 2014). Surface area and pore size are significant factors that determine the number and availability of adsorption sites. Typically, there is an inverse relationship between surface area and pore size; the smaller the pores size, the greater the surface area. Additionally, depending on the size of the adsorbate and pores size, adsorption may be limited by steric effects of the adsorbate (Hu and Apblett, 2014). Adsorption is commonly employed as a polishing step to remove organic and inorganic contaminants in water and wastewater treatment processes. Conventional adsorbents limit efficiency with less surface area or active sites, the lack of selectivity, and the adsorption kinetics (Qu et al., 2013a).

Activated carbon is the most common commercial adsorbent used in water treatment. Typically made from carbon-rich organic materials (coconut husks, wood, peat, coal, and synthetic sources) that are heated to high temperatures (>700°C), it is ultimately pyrolized in an oxygen-deprived environment (Hu and Apblett, 2014). In water treatment, it is used in one of two forms: granular-activated carbon (GAC) and powder-activated carbon (PAC). PAC (particle size: 20–50 μm) is usually released directly to water sources at various stages of the water treatment process and is removed in sedimentation or filtration stages. GAC (particle size: 0.5–3 mm) is typically employed as a fixed bed between filtration and disinfection.

Activated carbon has been long used as a wide-spectrum adsorbent. However, regeneration problems also exist. This makes activated carbon expensive for treating wastewater, therefore utilizing in developing areas is problematic. Low-cost, locally available materials with comparable adsorption capacities are needed (Mohan et al., 2014).

Nanomaterials offer significant improvement due to their advantageous properties that include small size, high surface area, and abundant sorption sites, short intraparticle diffusion distance, and tunable pore size and surface chemistry along with ease of regeneration and reuse (Qu et al., 2013a).

3.1.1 Carbon Nanomaterials

3.1.1.1 Carbon Nanotubes

Carbon nanotubes (CNTs) are an allotrope of carbon in the form of cylindrical macromolecules. The available surface areas for adsorption on individual CNTs are their external surfaces (Yang and Xing, 2010). In an aqueous environment, CNT structures aggregate due to the hydrophobicity, decreasing the effective surface area. On the other hand, CNT aggregates contain large interstitial spaces, which are high-energy adsorption sites for organic contaminants (Pan et al., 2008). Although activated carbon and CNT possess comparable surface area, a significant amount of surface area on activated carbon is attributed to micropores, which are inaccessible to bulky organic molecules, such as antibiotics and pharmaceuticals (Ji et al., 2009a; Pan and Xing, 2008). Thus, CNTs have much higher adsorption capacity for some bulky organic molecules because of their larger pores in aggregates and more accessible sorption sites (Qu et al., 2013a; Tian et al., 2013a).

Several studies show that CNTs are better adsorbents than activated carbon for heavy metals (Cu^{+2}, Pb^{+2}, Cd^{+2}, and Zn^{+2}) and the adsorption kinetics are fast on CNTs due to highly accessible adsorption sites and the short intraparticle diffusion distance (Li et al., 2003; Lu et al., 2006a). Functionalized CNTs combined with natural sands can also be used to effectively remove heavy metals, such as Pb^{2+} and Cu^{2+} (Tian et al., 2012). The surface functional groups (carboxyl, hydroxyl, and phenol) of CNTs are the key adsorption sites for metal ions, mainly through electrostatic attraction and chemical bonding (Rao et al., 2007).

A major drawback of activated carbon is its low affinity for low molecular weight polar organic compounds (Yang and Xing, 2010). CNTs strongly adsorb these polar organic compounds due diverse contaminant-CNT interactions including the hydrophobic effect, pi–pi interactions, hydrogen bonding, covalent bonding, and electrostatic interactions (Yang and Xing, 2010). The pi electron rich CNT surface allows pi–pi interactions with organic molecules with C=C bonds or benzene rings, such as PAHs and aromatic polar compounds (Chen et al., 2007; Lin and Xing, 2008; Pan et al., 2008). The —COOH, —OH, $—NH_2$ functional groups inherent to organic compounds could also form hydrogen bonds with the graphitic CNT surface that donates electrons (Yang et al., 2008). Such electrostatic attraction facilitates the adsorption of positively charged organic chemicals, such as antibiotics at the appropriate pH (Ji et al., 2009a; Tian et al., 2013b).

Overall, CNTs may not substitute activated carbon as a wide-spectrum adsorbent; since CNT surface chemistry can be altered to target specific contaminants, they may aid in polishing steps to remove recalcitrant compounds or in preconcentration of trace organic contaminants for analytical purposes (Qu et al., 2013a). Such applications would require small quantity of materials and therefore are less sensitive to the material cost.

Regeneration of adsorbents is key in evaluating the cost-effectiveness. Adsorption of metal ions on CNTs can be easily reversed by reducing the solution pH. Moreover, it has been shown that adsorption capacity remains relatively stable after regeneration. Lu et al. (2006a) reported that Zn^{+2} adsorption capacity of single-walled nanotube and multiwalled nanotubes decreased less than 25% after 10 regeneration and reuse cycles, while, that of activated carbon was reduced by more than 50% after one regeneration.

3.1.2 Metal-Based Nanoadsorbents

Metal oxides, such as iron oxide, titanium dioxide, and alumina are effective, low cost adsorbents for heavy metals and radionuclides (Qu et al., 2013a). The sorption is mainly controlled by a complexation mechanism between dissolved heavy metal contaminants and the oxygen in metal oxides (Koeppenkastrop and De Carlo, 1993). It is a two-step process encompassing a fast adsorption of metal ions on the external surface, followed by the rate-limiting intraparticle diffusion along the micropore walls (Trivedi and Axe, 2000). The nanoscale version of these metal oxides has higher adsorption capacity and faster kinetics because of the higher specific surface area, shorter intraparticle diffusion distance and larger number of surface reaction sites (Qu et al., 2013a). For example, as the particle size of nanomagnetite decreased from 300 to 11 nm, its arsenic adsorption capacity increased by more than 100 times (Yean et al., 2005). Much of this observed increase in adsorption was attributed to the increase in specific surface area as the 300 and 20 nm magnetite particles have similar surface area normalized arsenic adsorption capacity (Auffan et al., 2008, 2009). However, the specific surface area normalized adsorption capacity increased when 11 nm magnetite nanoparticles absorbed three times more arsenic than 20 nm magnetite, suggested a "nanoscale effect". The "nanoscale effect" was attributed to the change of surface structure which creates new adsorption sites (Auffan et al., 2009).

Iron oxide nanoparticles, particularly nanomaghemite and nanomagnetite, are considered superparamagnetic. Magnetism allows ease of separation and can be further taken advantage of as the core material in a core–shell nanoshell structure (Qu et al., 2013a). Surface modification achieved by the attachment of such inorganic shells or/and organic molecules not only stabilizes the nanoparticles by preventing oxidation, but also provides specific functionalities that can be selective for ion uptake (Mohmood et al., 2013).

Porous pellets can be formed from metal oxide nanocrystals without significantly compromising their surface area when moderate pressure is applied (Lucas et al., 2001). Thus, they can be applied in forms of fine powders and porous pellets, which are the likely forms to be used in industry (Qu et al., 2013a).

Metal based nanomaterials have shown great potential to outcompete activated carbon in the arena of heavy metal removal, such as arsenic, lead, mercury, copper, cadmium, chromium, and nickel (Sharma et al., 2009). Although a good overall adsorbent for many organic and inorganic contaminants, activated carbon has limited capacity for arsenic, especially for As(V) which has attracted much attention (Daus et al., 2004). Nanomagnetite and TiO_2 have shown arsenic adsorption superior to standard activated carbon (Mayo et al., 2007; Deliyanni et al., 2003). Such metal oxide nanoadsorbents also can be impregnated into the framework of activated carbon or other porous materials to achieve simultaneous removal of arsenic and organic cocontaminants, which favors point-of-use applications (Hristovski et al., 2009a, 2009b).

Like CNTs, metal oxide nanoadsorbents can be easily regenerated by changing solution pH to desorb contaminants (Sharma et al., 2009). In many cases, the adsorption capacity of metal oxide nanoadsorbents is consistent after several regeneration and reuse cycles (Hu et al., 2006). Above all, metal-based nanoadsorbents can be produced at relatively low cost (Qu et al., 2013a). The high adsorption capacity, low cost, and regeneration ease make metal-based nanoadsorbent a technologically and economically advantageous selection.

3.1.3 Nanocomposites as Innovative Adsorbents

Nanocomposites are defined as a matrix to which nanoscale materials have been added, where one of the dimensions of at less of one of the material is less than 100 nm, to improve a particular property of the composite material. Porous supporting materials can be impregnated with a variety of nanomaterials; resulting in stable nanocomposites that combine the advantages of both nanomaterials performing synergistically. Nanoparticles are often modified with a stabilizing matrix, such as polymers, to prevent aggregation and increase usability; however, in some cases these surface-bound stabilizers coat the reactive particle surface and may thereby inhibit the sorption and degradation of target compounds (Carpenter et al., 2015). Nonetheless, many nanocomposites have been developed that have great potential to be used as innovative adsorbents for drinking water purification. This section provides some examples of nanocomposites utilizing natural biopolymers, synthetic polymers, and carbon materials as the scaffolding.

3.1.3.1 Natural Polymer Nanocomposites

Natural polymers (i.e., biopolymers), such as cellulose, lignin, hemicellulose, chitosan, and so on, have been used as supporting materials for various type nanoparticles to form novel biopolymer-based nanocomposites with special properties. Among the biopolymers, cellulose and liglin are most commonly used biopolymers for composite materials.

Cellulose is an abundant, inexpensive and renewable resource. When it is interwoven, nanocellulose can form highly porous and mechanically strong bulk materials, such as nanocellulose papers, films, or aerogels (Cai et al., 2012; Sehaqui et al., 2010; Taniguchi and Okamura, 1998). While the paper packaging, automotive, personal care, construction, and textiles industries have recognized cellulose nanomaterials' potential, cellulose nanomaterials have great potential for application in water treatment technologies.

Cellulose nanomaterials find applications in water purification in a solitary manner, as well as by serving as a stabilizing scaffold for reactive nanoparticles. Specifically, iron oxide nanoparticles aggregation is a key concern when employing them as adsorbents for arsenic removal (Yu et al., 2013). Nata et al. (2011) designed a simple method for growing aminated iron oxide particles onto a cellulose nanofibril matrix which exhibited significantly higher arsenic removal [36.49 mg/g As(V)] than any previously published iron oxide-based adsorbents.

The most widely used semiconductor photocatalyst in water treatment, TiO_2, due to its low cost, chemical stability, and abundance has been utilized to create a nanocomposite with cellulose nanofibrils. Hybrid Au/TiO_2- and Ag/TiO_2-cellulose nanofibril composites that removed methylene blue, a model for organic contaminants, by 75 and 70%, respectively, through both adsorption and photocatalytic degradation (Snyder et al., 2013). Overall, addition of the metallic nanoparticles increased the mechanical strength of the cellulose nanofibril films, making them more robust and recyclable (Carpenter et al., 2015).

Next to cellulose, lignin is the second most abundant biopolymer, and the main source of aromatic structures on earth (Duval and Lawoko, 2014). As a phenolic macromolecule its complex structure varies considerably depending on the plant species and the isolation process (Duval and Lawoko, 2014). Lignin possesses two types of adsorption sites (carboxyl and phenol groups) that participate in the sorption mechanism; the phenolic groups have been shown to have a higher affinity toward metal ions (Guo et al., 2008b). Therefore, lignin-based nanocomposites present good potential to adsorb various chemical compounds, particularly cations. The excellent sorptive properties of the composite materials may come from both the lignin matrix and the nanoparticles within the porous structures.

Biopolymers-based nanocomposite membranes using propionated lignin and cellulose triacetate (CTA) have been prepared by Nevárez et al. (2011) for water purification applications. The study utilized three types of lignin, namely, kraft (KL), organosolv (OL), and hydrolytic (HL) which were propionated to increase compatibility with CTA.

For determining membrane performance, groundwater containing high concentrations of fluoride, arsenic, calcium, sodium, and magnesium, was filtrated. Membrane rejection was 15–35% for anions present, and 12–42% for monovalent cations and 27–54% for divalent cations (Nevárez et al., 2011). Arsenic and fluoride removal by membranes was found to be affected by ionic and organic matter present in groundwater. Nanocomposites with raw lignins (MKL and MOL) improved their permeability properties with propionation (MPKL and MPOL), except for arsenic, which was slightly less rejected (Nevárez et al., 2011).

For all membranes obtained, the homogeneity was clearly improved with addition of chemically modified lignin as the associated groups improved their incorporation into the matrix due to a correspondingly higher solubility (Nevárez et al., 2011). The propionated kraft lignin-reinforced membrane showed superior mechanical properties with the highest Young's modulus, as well as tensile strength. Surface roughness of the nanocomposite

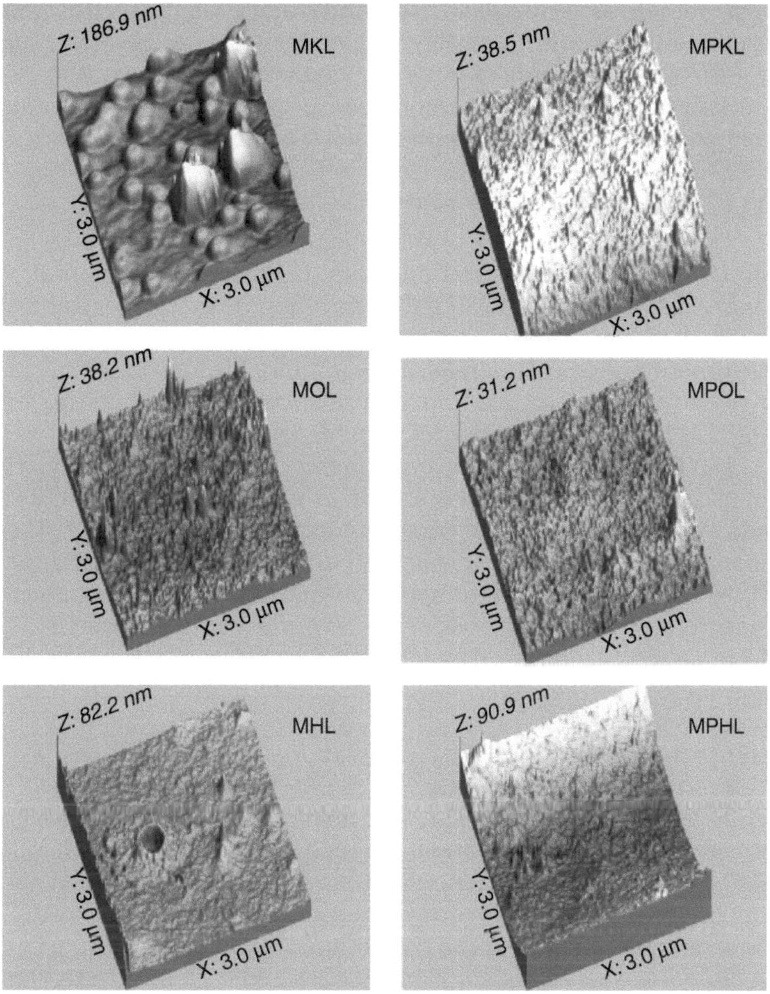

Figure 3.5. Topographic images obtained by AFM for raw and propionated lignin nanocomposites. Reprinted with permission from Nevárez et al., 2011.

membranes was also studied using atomic force microscopy (AFM) (Fig. 3.5). It is obvious from the images that significant smoothing occurred in the nanocomposite membranes after propionation (Thakur et al., 2014).

Nanobiopolymers, such as cellulosic nanofibrillar structures attract a great deal of attention as components in nanocomposites due to their abundance, their renewable and environmentally friendly nature, and their outstanding mechanical properties. One of the major challenges has been to find efficient ways to extract biopolymers, such as cellulosic fibrils and lignins from different source materials, including wood, agricultural residues, and forest residues (Hubbe et al., 2008). A second major challenge is frequent incompatibility with a variety of composite materials. Last, the water-swellable nature of some biopolymers, such as cellulose, especially in noncrystalline regions, can also be a concern in various composite materials and applications (Hubbe et al., 2008).

3.1.3.2 Synthesized Polymer Nanocomposites

Due to their excellent physicochemical properties and highly porous nature, synthesized polymers have been used as matrix material for nanocomposites. For example, polymeric membranes have been synthesized and common membrane type for water treatment due to their straightforward pore forming mechanism, flexibility, ease of installation, and relatively low costs compared to other inorganic membranes (Pendergast and Hoek, 2011a).

Nanoparticles have also been incorporated inside a polymer matrix to make functional membranes with specific capability to adsorb heavy metals from water. For example, Daraei et al. (2012) incorporated polyaniline/iron(II,III) oxide (PANI/Fe_3O_4) nanoparticles inside polyethersulfone (PES) matrix using a phase inversion method for the purpose of removing the contaminants Cu (II) from aqueous solution. The efficient dispersion of nanoparticles was observed at relatively low loading (0.1%), providing more accessible active sites for sorption of copper ions. Regeneration of this membrane by EDTA (ethylendiamine tetraacetate) was possible, thus allowing for reuse.

In another study, Tetala and Stamatialis (2013) developed a copper removal membrane by incorporating chitosan beads (20–40 μm) inside ethylene vinyl alcohol (EVAL) matrix. While the particle size is significantly larger than the nanoscale, a similar strategy was used for the membrane fabrication, resulting in a membrane with fast adsorption kinetics and high adsorption capacity (225.7 mg/g) for Cu (II) ions (Yin and Deng, 2015).

Similarly, hydrous manganese dioxide (HMO) and Fe_3O_4 nanoparticles (60 nm) were embedded into the PES and polyvinyl chloride (PVC) matrices, respectively, for the purpose of removing lead from aqueous solution (Gholami et al., 2014; Jamshidi Gohari et al., 2013a). Also, removal of As (III) was achieved in a study utilizing Fe–Mn binary oxide (FMBO) to prepare a PES nanocomposite membrane (Jamshidi Gohari et al., 2013b).

These studies demonstrated the possibility of preparing nanocomposite membranes containing selective adsorbents for purification of water. However, the relatively high water permeability and therefore short contact time implies the sorption could be limited by kinetics, in addition to adsorption capacity, which may hinder practical application (Yin and Deng, 2015).

3.1.3.3 Carbon Nanocomposites

Carbon materials, such as activated carbon and biochar are often used as a matrix for preparations of various types of carbon-based nanocomposties for water purifications. Biochar, highly porous pyrolyzed biomass, alone and modified with nanomaterials have been shown in recent literature to have legitimate applications in water filtration (Mohan et al., 2014). While activated carbon is ideal for removing contaminants from water, it is costly to make. Biochar, on the other hand, requires less investment and can be utilized sustainably. Biochar may be used for removing contaminants from water while also being loaded with nutrients for subsequent use as a soil amendment, providing long-term sorption capacity and a sustainable fertilizer (Bernd et al., 2013). Biochar could replace coal-, coconut shell-, and wood-based activated carbons as a low cost adsorbent for contaminants and pathogens (Mohan et al., 2014). Biochars exhibit potential for water purification by facilitating adsorption of hydrocarbons, other organics, and some inorganic metal ions (Hale et al., 2012; Mohan et al., 2012).

Incorporating engineered nanoparticles into the biochar structure could improve biochar functioning in not only water treatment applications but also soil fertility enhancement and carbon sequestration (Inyang et al., 2014).

Nanocomposites containing MgO were made by pyrolyzing (600°C, N_2) $MgCl_2$ with the following biomass feedstocks: sugar beet tailings, sugarcane bagasse, cottonwoods, pine woods, and peanut shells (Zhang et al., 2012a). These nanocomposites were used to remove phosphate and nitrate from water. MgO particles were well dispersed over the biochar surface. Sugar beet tailing composites averaged an adsorption capacity of 835 mg/g of phosphate. Peanut shell composites prevailed with a nitrate adsorption capacity of 94 mg/g. Aforementioned high nitrate and phosphate capacities may be due to surface area and porosity enrichment by introducing MgO (Zhang et al., 2012a).

Hybrid multiwalled carbon nanotube (MWCNT)-coated biochars were made by dipping biomass into various concentrations of CNT solutions prior to slow pyrolysis (Inyang et al., 2014). Untreated hickory and bagasse biochars and CNT–biochar composites were characterized for use in methylene blue adsorption. CNT addition significantly enhanced the hickory and bagasse char thermal stabilities, surface areas and pore volumes (Inyang et al., 2014). Electrostatic attraction dominated methylene blue sorption and diffusion controlled the adsorption rate (Inyang et al., 2014).

Mubarak et al. (2013) compared functionalized CNTs and magnetic biochar from empty fruit branches for Zn^{2+} the adsorption capacity. Maximum Zn^{2+} adsorption capacities were 1.05 and 1.18 mg/g for functionalized CNT and magnetic biochar, respectively (Mubarak et al., 2013).

A biochar/AlOOH nanoflake nanocomposite was fabricated from $AlCl_3$-pretreated biomass through slow pyrolysis in N_2 at 600°C for 1 h (Zhang and Gao, 2013). This was a highly effective adsorbent to remove arsenic, methylene blue, and phosphate. The Langmuir capacity of methylene blue adsorption on the biochar/AlOOH (~85,000 mg/kg) was 10 times higher than that of the corresponding biochar without aluminum treatment (8,000 mg/kg) (Zhang et al., 2012b). The Langmuir adsorption capacity of phosphate and arsenic on the biochar/AlOOH was ~135,000 and ~17,410 mg/kg, respectively (Zhang and Gao, 2013).

Adsorbent magnetization is an emerging area of water treatment to overcome the filtration problems of nonmagnetic adsorbents (Mohan et al., 2014). The advantage of magnetic adsorbents is that they can easily be recovered using low strength external magnetic fields from water containing contaminants, such as suspended solids, oil, and grease. Impurities may cause adsorbent fouling requiring frequent separation/regeneration. Magnetic separation simplifies isolation and washing, followed by redispersion.

Zhang et al. (2013) reported a γ-Fe_2O_3 modified biochar with a As(V) sorption capacity of 3147 mg/kg. A magnetic (saturation magnetization of 69.2 emu/g) biochar, with colloidal or nanosized γ-Fe_2O_3 particles embedded in the porous matrix, fabricated via $FeCl_3$-treated cottonwood pyrolysis at 600°C in N2 environment for 1 h (Zhang et al., 2013). A large quantity of γ-Fe_2O_3 particles with sizes ranging hundreds of nanometers to several micrometers grew within the porous biochar.

Carbon adsorbents, such as activated carbon and biochar are versatile materials that have been used to treat ground water, drinking water, synthetic industrial wastewater, and actual wastewater. It is very difficult to directly compare adsorption capacities due to a lack of consistency in the literature data (Mohan et al., 2014). Cost studies are not analyzed in the literature; however, the cost of biomass (including its transport to pyrolysis site) is the largest single contributor (Mohan et al., 2014).

3.2 Membranes

The basic goal of water treatment is to remove contaminants from water. Membranes facilitate this by providing a physical barrier preventing the transport of contaminant matter. Microscopic openings in the membrane allow water molecules to pass, but not compounds that are larger than the opening.

As the key component of water treatment and reuse, membranes provide a high level of automation, require less land and chemical use, and the modular configuration allows flexible design (Qu et al., 2013b).

Membrane filtration can be functionalized either as dead-end filtration or cross-flow filtration. In dead-end filtration, the feed water flows perpendicular to the membrane surface where all solids will amass onto the membrane surface during filtration, and removed via backwashing. In cross-flow filtration, the feed water is parallel to the membrane surface. Since the majority of the solid contaminants will pass with the water instead of collecting on the membrane surface, the system can function at higher flux.

The performance of membrane systems is largely decided by the membrane material. Most membranes are made from synthetic organic polymers, which can be either hydrophobic or hydrophilic. Hydrophilic polymer membranes include cellulose and its derivatives, polyacrylonitrile, hydrophilized polysulfone, hydrophilized polyethylene, and so on; hydrophobic membranes include polytetrafluroethylene (PTFE), polyethylene (PE), polyvinylidene fluoride (PVDF), and many more (Hu and Apblett, 2014).

A fundamental challenge of membrane technology is the inherent tradeoff between membrane selectivity and permeability (Qu et al., 2013a). High energy consumption is a significant barrier to the vast applications of pressure driven membrane processes. Membrane fouling also adds to the energy consumption along with adding complexity to the design and operation process. Additionally, it reduces the longevity of membranes and membrane units. The development of membranes with high permeability and rejection, and good antifouling property is much needed for water purification under the context of energy efficiency and cost effectiveness (Yin and Deng, 2015).

Three membrane nanotechnologies that have shown promise in overcoming such challenges will be discussed: aligned CNTs, biomimetic membranes, and thin film nanocomposite (TFNC) membranes.

3.2.1 CNT and Biomimetic Membranes

Both aligned CNT and biomimetic membranes utilize the principle of nanochannels that allow permeation of water molecules in a single file with exceptional rates (Qu et al., 2013b). These channels in CNTs are the nanotubes themselves, and in biomimetic membranes these channels are considered aquaporins. The selectivity for water is achieved in CNT by diameter and chemistry of the nanotube opening, and in the case of aquaporins via the unique hourglass shape, size, and chemical structure (Lawton et al., 2001; Mauter and Elimelech, 2008).

CNT-membranes provide near frictionless water flow through them with the retention of a broad spectrum of water pollutants. Aligned CNTs have been shown both experimentally and theoretically to provide water permeation much faster than theoretically calculated, due to the atomic smoothness of the nanosized channel, and the one dimensional single-file ordering of water molecules while passing through the nanotubes (Holt et al., 2006; Hummer et al., 2001).

Incorporation of vertically aligned CNTs and aquaporins into membrane matrices, even at a low percentage could provide flux exceeding that of current commercial seawater reverse osmosis (RO) membranes (Kumar et al., 2007; Pendergast and Hoek, 2011b). The higher flux may not significantly improve the energy consumption of seawater desalination by RO but it can greatly reduce the footprint and cost of wastewater treatment by RO, whose theoretical energy consumption is much lower (Elimelech and Phillip, 2011).

The inner hollow cavity of CNTs provides a great possibility for desalinating water. However, the high rejection for salt and small molecules is challenging for aligned CNT membranes due to the lack of CNTs with uniformly subnanometer diameter (Qu et al., 2013b). Functional group gating at the nanotube opening has been proposed to enhance the selectivity of aligned CNT membranes (Mauter and Elimelech, 2008). Grafting bulky functional groups at the tube opening could physically exclude salts. However, steric exclusion will significantly reduce membrane permeability (Nednoor et al., 2005). Thus, at the current stage, aligned CNT membranes are not capable of desalination. To achieve reliable salt rejection, the CNT diameter must be uniformly smaller than 0.8 nm (Hinds, 2012).

Many biological membranes are highly selective and permeable. Aquaporins are protein channels that regulate flux of water through cell membranes. Their high selectivity and water permeability makes their use in polymeric membranes an attractive approach to improve membrane performance. Aquaporin-Z from *Escherichia coli* has been incorporated into amphiphilic triblock-polymer vesicles, which added water permeability at least an order of magnitude over the original vesicles with full rejection to glucose, glycerol, salt, and urea (Kumar et al., 2007). Limited success was achieved in a study by Kaufman et al. (2010) coating aquaporin incorporated lipid bilayers on commercial nanofiltration membranes.

A key barrier for both aquaporin and aligned CNT membranes is the scale-up of the nanomaterial production and membrane fabrication. Large-scale production and purification of aquaporins are very challenging. The first hurdle is the complicated methods for the synthesis of CNTs with uniform pore size and distribution (Schnorr and Swager, 2011). To date, chemical vapor deposition (CVD) is the most common way to make aligned nanotubes. The current CVD method is often unable to synthesize size-controlled pores with uniform distribution. Bottom up approach could be used for synthesizing well-controlled CNT-chirality with desired pore diameters and structural integrity (Omachi et al., 2013). However, the method is expensive and time consuming, making them unsuitable for industrial production. A continuous high-yield CVD prototype has been designed for producing vertically aligned CNT, paving the way for large-scale production (Guzmán de Villoria et al., 2011). A high-yield CVD process and post manufactured alignment using a magnetic field for production of aligned CNT membranes (Guzmán de Villoria et al., 2011; Mauter et al., 2010).

The second complication might be the reduction of pore size and diameter of the CNT membranes. Smaller pores with better desalination properties can be possible with only single-walled carbon nanotubes (SWCNTs) and double-walled carbon nanotubes (DWCNTs) grown on epoxy or mixed composite matrices (Das et al., 2014). However, both S/DWCNTs are expensive and hence the large-scale and sustainable synthesis of such membranes is not commercially feasible (Das et al., 2014).

3.2.2 Thin Film Nanocomposites

The focus of TFN membranes development is incorporating nanomaterials into the active layer of thin film composite (TFC) membranes by doping in casting solutions or surface modification. Nanomaterials that have been researched for such applications include nanozeolites, nano-Ag, nano-TiO_2, and CNTs. Effect of nanoparticles on membrane permeability and selectivity depends on the type, size, and amount of nanoparticles added.

3.2.2.1 Zeolites and Ag

Nanozeolites are the most frequently used in TFN and have shown significant enhancement in membrane permeability (Qu et al., 2013a). Addition of nanozeolites leads to more permeable, negatively charged, and thicker active layer (Lind et al., 2009). A study by Jeong et al. (2007) reported water permeability increased up to 80% over the TFC membrane, with salt rejection largely maintained (93.9 ± 0.3%). TFN membranes doped with 250 nm nanozeolites at 0.2 wt.% achieved higher permeability and better salt rejection (>99.4%) than commercial RO membranes (Lind et al., 2010). It was hypothesized that the small, hydrophilic pores of nanozeolites create superior paths for water. However, water permeability increased even with pore-filled zeolites, although less than the pore-open ones, which could be attributed to defects at the zeolite–polymer interface (Qu et al., 2013a). Utilizing antimicrobial agents such as Ag^+ with nanozeolites carriers imparts an antifouling property to the membrane (Lind et al., 2009). The zeolite TFN technology has reached the early stage of commercialization: QuantumFlux, a seawater TFN RO membrane, is now commercially available.

3.2.2.2 TiO_2

In has been shown that the incorporation of nano-TiO_2 (up to 5 wt.%) into the TFC active layer increases the membrane rejection slightly while maintaining the permeability (Lee et al., 2008). Upon UV irradiation, TiO_2 has the capacity to degrade organic contaminants and inactivate microorganisms. This aids in decreasing of organic and biological fouling, as well as removing contaminants that are unrestrained by the membrane. However, the close adjacency between the photocatalyst and the membrane may also lead to detrimental effects on polymeric membrane materials, which needs to be addressed for long-term efficacy (Chin et al., 2006).

Due to their antimicrobial properties, unaligned CNTs also find their place in TFN membranes. Tiraferri et al. (2011) covalently bonded single-walled nanotubes to a TFC membrane surface. This approach is advantageous as it uses relatively small amounts of the nanomaterial, minimizing disturbance in the active layer. The resulting TFN membrane exhibited moderate antibacterial properties (60% inactivation of bacteria attached on the membrane surface in 1 h contact time), possibly reducing or delaying membrane biofouling.

Incorporating of functional nanomaterials into membranes offers the opportunity to improve the critical factors of membrane permeability, fouling resistance, mechanical, and thermal stability, as well as creating new functions in contaminant degradation and self-cleaning (Qu et al., 2013a).

4 Concluding Remarks

Table 3.3 summarizes the nanotechnology comparisons made in this chapter.

Table 3.3 Current and Potential Applications of Nanotechnology in Drinking Water Treatment Nanotechnologies Discussed in this Chapter

Applications	Nanoalternatives	Desirable Nanomaterial Properties	Enabled Technologies
Adsorption	CNT	High specific surface area, highly assessable adsorption sites, diverse contaminant-CNT interactions, tunable surface chemistry, easy reuse	Contaminant preconcentration/ detection, adsorption of recalcitrant contaminants
	Nanoscale metal oxide	High specific surface area, short intraparticle diffusion distance, more adsorption sites, compressible without significant surface area reduction, easy reuse, some are superparamagnetic	Adsorptive media filters, slurry reactors
	Nanofibers with core–shell structure	Tailored shell surface chemistry for selective adsorption, reactive core for degradation, short internal diffusion distance	Reactive nanoadsorbents
	Nanocomposites	Immobilized, synergistic effects of both materials	Nanoadsorbents, selective adsorbents
	Nano modified Biochar	High surface area, sustainably produced from agricultural/environmental residues, can house a variety of nanoparticles to target contaminants	Adsorption media filters
Membranes and membrane processes	Nanozeolites	Molecular sieve, hydrophilicity	High permeability TFNC membranes
	Nano-Ag	Strong and wide-spectrum antimicrobial activity, low toxicity to humans	Antibiofouling membranes
	CNTs	Antimicrobial activity (unaligned CNTs)	Antibiofouling membranes
		Small diameter, atomic smoothness of inner surface, tunable opening chemistry, high mechanical and chemical stability	Aligned carbon nanotube membranes
	Aquaporin	High permeability and selectivity	Aquaporin membranes
	Nano-TiO_2	Photocatalytic activity, hydrophilicity, high chemical stability	Reactive membranes, high performance TFNC membranes

Modified from: Qu et al., 2013a.

Nanotechnology for water drinking water treatment is gaining momentum through a global perspective. The unique properties of nanomaterials and their emergence with current water treatment technologies present great opportunities to revolutionize water treatment. Although many nanotechnology strategies highlighted in this review are still in the laboratory stage, some have made their way to pilot testing or even large scale commercialization.

However, there are still potential drawbacks. Major practical challenges are the cost of nanostructured materials along with the difficulty in scaling up nanobased treatment processes for commercial use. In addition, health and safety issues around the use of nanomaterials have to be addressed in the domestic water industry, particularly with respect to the direct application of nanoparticles into the receiving natural bodies of water. Materials functionalized with nanoparticles incorporated or deposited on their surface have risk potential, since nanoparticles might leach into the environment where they can accumulate for long periods of time. As a result nanomaterials can be embedded into various platforms (e.g., nanocomposites as discussed in this chapter) to avoid leaching into the system. However, current immobilization techniques require further optimizations to mitigate the loss in efficiency. Research is also needed to stabilize, as well as distribute nanoparticles throughout the hosting platforms without significantly impacting overall performance. Nevertheless, nanoengineered materials offer great potential for water innovations in the coming decades, in particular for decentralized treatment systems, point-of-use devices, and heavily degradable contaminants.

The performance of various nanotechnologies in treating natural waters needs to be tested. Future studies must be done under more realistic conditions to assess the applicability and efficiency of different nanotechnology enabled strategies. Although nanotechnology enabled water treatment processes have shown great promise in laboratory studies, their readiness for commercializationis not uniform. Some are already on the market, while others require significant research before they can be considered for full-scale applications. Their future development and commercialization face a variety of challenges including technical hurdles, cost-effectiveness, and potential environmental and human risk. Additionally, the longevity of these nanotechnologies is largely unknown as most lab studies were conducted for relatively short intervals. If integration of nanotechnologies is a future goal, then studies that address the long-term performance of water and wastewater treatment nanotechnology strategies is imperative. Direct comparison of nanotechnology-enabled systems with current technologies would be challenging but fruitful.

Cohesion between nanotechnologies, current drinking water purification processes, and infrastructure is lacking but essential. Most developed countries already have sustained treatment plants, and distribution systems with the expectation that these outmoded technologies will remain for decades to come. Thus, there must be a way to implement nanotechnology with minimal changes to existing infrastructure in the near future.

It is anticipated that the growing body of research and development on this topic that will lead to nanomaterials playing a vital role in future drinking water treatment.

References

Albuquerque Júnior, E.C.D., Méndez, M.O.A., Coutinho, A.D.R., Franco, T.T., 2008. Removal of cyanobacteria toxins from drinking water by adsorption on activated carbon fibers. Mater. Res. 11, 371–380.

Allabashi, R., Arkas, M., Hörmann, G., Tsiourvas, D., 2007. Removal of some organic pollutants in water employing ceramic membranes impregnated with cross-linked silylated dendritic and cyclodextrin polymers. Water Res. 41, 476–486.

American Dental Association. 2005. Fluoride Fact Sheet. http://www.ada.org/~/media/ADA/Member%20Center/FIles/fluoridation_facts.ashx

Anandan, S., Lee, G.-J., Chen, P.-K., Fan, C., Wu, J.J., 2010. Removal of orange II dye in water by visible light assisted photocatalytic ozonation using Bi_2O_3 and Au/Bi_2O_3 nanorods. Ind. Eng. Chem. Res. 49, 9729–9737.

Arkas, M., Eleades, L., Paleos, C.M., Tsiourvas, D., 2005. Alkylated hyperbranched polymers as molecular nanosponges for the purification of water from polycyclic aromatic hydrocarbons. J. Appl. Polym. Sci. 97, 2299–2305.

Athanasekou, C.P., Romanos, G.E., Katsaros, F.K., Kordatos, K., Likodimos, V., Falaras, P., 2012. Very efficient composite titania membranes in hybrid ultrafiltration/photocatalysis water treatment processes. J. Memb. Sci. 392–393, 192–203.

Auffan, M., Rose, J., Proux, O., Borschneck, D., Masion, A., Chaurand, P., Hazemann, J.-L., Chaneac, C., Jolivet, J.-P., Wiesner, M.R., Van Geen, A., Bottero, J.-Y., 2008. Enhanced adsorption of arsenic onto maghemites nanoparticles: As(III) as a probe of the surface structure and heterogeneity. Langmuir 24, 3215–3222.

Auffan, M., Rose, J., Bottero, J.-Y., Lowry, G.V., Jolivet, J.-P., Wiesner, M.R., 2009. Towards a definition of inorganic nanoparticles from an environmental, health and safety perspective. Nat. Nanotechnol. 4, 634–641.

Bernd, M., Steffen, W., Karsten, A., Lübken, M., 2013. EGU general assembly, potential dual use of biochar for wastewater treatment and soil amelioration. Geophysics Research Abstract, Vienna, Austria.

Bhattacharyya, D., Hestekin, J.A., Brushaber, P., Cullen, L., Bachas, L.G., Sikdar, S.K., 1998. Novel poly-glutamic acid functionalized microfiltration membranes for sorption of heavy metals at high capacity. J. Memb. Sci. 141, 121–135.

Bina, B., Pourzamani, H., Rashidi, A., Amin, M.M., 2011. Ethylbenzene removal by carbon nanotubes from aqueous solution. J. Environ. Pub. Health 2012.

Boparai, H.K., Joseph, M., O'carroll, D.M., 2011. Kinetics and thermodynamics of cadmium ion removal by adsorption onto nano zerovalent iron particles. J. Hazard. Mater. 186, 458–465.

Borker, P., Salker, A.V., 2007. Solar assisted photocatalytic degradation of naphthol blue black dye using Ce1−xMnxO_2. Mater. Chem. Phys. 103, 366–370.
Cai, Y.-Q., Cai, Y.-E., Mou, S.-F., Lu, Y.-Q., 2005. Multi-walled carbon nanotubes as a solid-phase extraction adsorbent for the determination of chlorophenols in environmental water samples. J. Chromatogr. A 1081, 245–247.
Cai, J., Liu, S., Feng, J., Kimura, S., Wada, M., Kuga, S., Zhang, L., 2012. Cellulose–silica nanocomposite aerogels by in situ formation of silica in cellulose gel. Angew. Chem. 124, 2118–2121.
Cao, J., Elliott, D., Zhang, W.-X., 2005. Perchlorate reduction by nanoscale iron particles. J. Nanopart. Res. 7, 499–506.
Carpenter, A.W., de Lannoy, C.-F., Wiesner, M.R., 2015. Cellulose nanomaterials in water treatment technologies. Environ. Sci. Technol. 49, 5277–5287.
Chen, W., Duan, L., Zhu, D., 2007. Adsorption of polar and nonpolar organic chemicals to carbon nanotubes. Environ. Sci. Technol. 41, 8295–8300.
Cheng, R., Wang, J.-L., Zhang, W.-X., 2007. Comparison of reductive dechlorination of *p*-chlorophenol using Fe^0 and nanosized Fe^0. J. Hazard. Mater. 144, 334–339.
Chin, S.S., Chiang, K., Fane, A.G., 2006. The stability of polymeric membranes in a TiO_2 photocatalysis process. J. Membr. Sci. 275, 202–211.
Choe, S., Chang, Y.-Y., Hwang, K.-Y., Khim, J., 2000. Kinetics of reductive denitrification by nanoscale zero-valent iron. Chemosphere 41, 1307–1311.
Crittenden, J.C., Trussell, R.R., Hand, D.W., Howe, K.J., Tchobanoglous, G., 2012. MWH's Water Treatment: Principles and Design. John Wiley & Sons, Hoboken, New Jersey.
Daraei, P., Madaeni, S.S., Ghaemi, N., Salehi, E., Khadivi, M.A., Moradian, R., Astinchap, B., 2012. Novel polyethersulfone nanocomposite membrane prepared by PANI/Fe_3O_4 nanoparticles with enhanced performance for Cu(II) removal from water. J. Membr. Sci. 415–416, 250–259.
Das, R., Ali, M.E., Hamid, S.B.A., Ramakrishna, S., Chowdhury, Z.Z., 2014. Carbon nanotube membranes for water purification: a bright future in water desalination. Desalination 336, 97–109.
Daus, B., Wennrich, R., Weiss, H., 2004. Sorption materials for arsenic removal from water: a comparative study. Water Res. 38, 2948–2954.
Davydova, S.L., 1998. Heavy metals as main pollutants of the next century. Crit. Rev. Anal. Chem. 28, 377–381.
Defriend, K.A., Wiesner, M.R., Barron, A.R., 2003. Alumina and aluminate ultrafiltration membranes derived from alumina nanoparticles. J. Membr. Sci. 224, 11–28.
Deliyanni, E.A., Bakoyannakis, D.N., Zouboulis, A.I., Matis, K.A., 2003. Sorption of As(V) ions by akaganéite-type nanocrystals. Chemosphere 50, 155–163.
Di, Z.-C., Ding, J., Peng, X.-J., Li, Y.-H., Luan, Z.-K., Liang, J., 2006. Chromium adsorption by aligned carbon nanotubes supported ceria nanoparticles. Chemosphere 62, 861–865.
Doroodmand, M.M., Tahvildar, Z., Sheikhi, M.H., 2012. Multi-walled carbon nanotubes/polyacrylonitrile composite as novel semi-permeable filter for water treatment process. Sci. Adv. Mater. 4, 1085–1095.
Dotzauer, D.M., Dai, J., Sun, L., Bruening, M.L., 2006. Catalytic membranes prepared using layer-by-layer adsorption of polyelectrolyte/metal nanoparticle films in porous supports. Nano Lett. 6, 2268–2272.
Duval, A., Lawoko, M., 2014. A review on lignin-based polymeric, micro- and nano-structured materials. React. Funct. Polym. 85, 78–96.
Elimelech, M., Phillip, W.A., 2011. The future of seawater desalination: energy, technology, and the environment. Science 333, 712–717.
EPA, 1999. Particles Contributing to Turbidity. EPA Guidance Manual Turbidity Provisions.

EPA. 2009. Drinking Water Contaminants—Standards and Regulations. New York. https://www.epa.gov/dwstandardsregulations
EPA, 2013. Basic Information about Disinfection Byproducts in Drinking Water: Total Trihalomethanes, Haloacetic Acids, Bromate, and Chlorite.
FAO, 1999. Potential pollutants, their sources and their impacts. In: Sciortino, J.A., Ravikumar, R. (Eds.), Fishery Harbour Manual on the Prevention of Pollution. Bay of Bengal Programme for Fisheries Management (BOBP/MAG/22), Madras, India.
Figueira, P., Lopes, C.B., Daniel-da-Silva, A.L., Pereira, E., Duarte, A.C., Trindade, T., 2011. Removal of mercury (II) by dithiocarbamate surface functionalized magnetite particles: application to synthetic and natural spiked waters. Water Res. 45, 5773–5784.
Fuller, G.W., 1933. Progress in water purification. J. Am. Water Works Assoc. 25 (11), 1566–1576.
Gehrke, I., Geiser, A., Somborn-Schulz, A., 2015. Innovations in nanotechnology for water treatment. Nanotechnol., Sci. Appl. 8, 1–17.
Gholami, A., Moghadassi, A., Hosseini, S., Shabani, S., Gholami, F., 2014. Preparation and characterization of polyvinyl chloride based nanocomposite nanofiltration-membrane modified by iron oxide nanoparticles for lead removal from water. J. Ind. Eng. Chem. 20, 1517–1522.
Girginova, P.I., Daniel-da-Silva, A.L., Lopes, C.B., Figueira, P., Otero, M., Amaral, V.S., Pereira, E., Trindade, T., 2010. Silica coated magnetite particles for magnetic removal of Hg^{2+} from water. J. Colloid Interface Sci. 345, 234–240.
Gondal, M.A., Dastageer, M.A., Khalil, A., 2009. Synthesis of nano-WO_3 and its catalytic activity for enhanced antimicrobial process for water purification using laser induced photo-catalysis. Catal. Commun. 11, 214–219.
Guo, L., Li, J., Zhang, L., Li, J., Li, Y., Yu, C., Shi, J., Ruan, M., Feng, J., 2008a. A facile route to synthesize magnetic particles within hollow mesoporous spheres and their performance as separable Hg^{2+} adsorbents. J. Mater. Chem. 18, 2733–2738.
Guo, X., Zhang, S., Shan, X.-Q., 2008b. Adsorption of metal ions on lignin. J. Hazard. Mater. 151, 134–142.
Guzmán de Villoria, R., Hart, A.J., Wardle, B.L., 2011. Continuous high-yield production of vertically aligned carbon nanotubes on 2D and 3D substrates. ACS Nano 5, 4850–4857.
Hale, S.E., Lehmann, J., Rutherford, D., Zimmerman, A.R., Bachmann, R.T., Shitumbanuma, V., O'Toole, A., Sundqvist, K.L., Arp, H.P.H., Cornelissen, G., 2012. Quantifying the total and bioavailable polycyclic aromatic hydrocarbons and dioxins in biochars. Environ. Sci. Technol. 46, 2830–2838.
Hazen, A., 1914. Clean Water and How to get it. J. Wiley & Sons, Inc., New York.
He, F., Zhao, D., 2005. Preparation and characterization of a new class of starch-stabilized bimetallic nanoparticles for degradation of chlorinated hydrocarbons in water. Environ. Sci. Technol. 39, 3314–3320.
Hinds, B., 2012. Dramatic transport properties of carbon nanotube membranes for a robust protein channel mimetic platform. Curr. Opin. Solid State Mater. Sci. 16, 1–9.
Holt, J.K., Park, H.G., Wang, Y., Stadermann, M., Artyukhin, A.B., Grigoropoulos, C.P., Noy, A., Bakajin, O., 2006. Fast mass transport through sub-2-nanometer carbon nanotubes. Science 312, 1034–1037.
Hristovski, K.D., Nguyen, H., Westerhoff, P.K., 2009a. Removal of arsenate and 17α-ethinyl estradiol (EE2) by iron (hydr)oxide modified activated carbon fibers. J. Environ. Sci. Health Part A 44, 354–361.
Hristovski, K.D., Westerhoff, P.K., Möller, T., Sylvester, P., 2009b. Effect of synthesis conditions on nano-iron (hydr)oxide impregnated granulated activated carbon. Chem. Eng. J. 146, 237–243.

Hu, A., Apblett, A., 2014. Nanotechnology for Water Treatment and Purification. Springer, Switzerland.
Hu, J, Chen, G, Lo, I, 2006. Selective removal of heavy metals from industrial wastewater using maghemite nanoparticle: performance and mechanisms. J. Environ. Eng. 132, 709–715.
Hu, J., Tong, Z., Hu, Z., Chen, G., Chen, T., 2012. Adsorption of roxarsone from aqueous solution by multi-walled carbon nanotubes. J. Colloid Interface Sci. 377, 355–361.
Huang, C., Hu, B., 2008. Silica-coated magnetic nanoparticles modified with γ-mercaptopropyltrimethoxysilane for fast and selective solid phase extraction of trace amounts of Cd, Cu, Hg, and Pb in environmental and biological samples prior to their determination by inductively coupled plasma mass spectrometry. Spectrochim. Acta B 63, 437–444.
Hubbe, M.A., Rojas, O.J., Lucia, L.A., Sain, M., 2008. Cellulosic nanocomposites: a review. BioResources 3, 929–980.
Hummer, G., Rasaiah, J.C., Noworyta, J.P., 2001. Water conduction through the hydrophobic channel of a carbon nanotube. Nature 414, 188–190.
Inyang, M., Gao, B., Zimmerman, A., Zhang, M., Chen, H., 2014. Synthesis, characterization, and dye sorption ability of carbon nanotube–biochar nanocomposites. Chem. Eng. J. 236, 39–46.
IWMI, 2000. World water supply and demand in 2025. In: Rijsberman, F.R. (Ed.), World Water Scenario Analyses. World Water Council, Marseille, France.
Jamshidi Gohari, R., Lau, W.J., Matsuura, T., Halakoo, E., Ismail, A.F., 2013a. Adsorptive removal of Pb(II) from aqueous solution by novel PES/HMO ultrafiltration mixed matrix membrane. Sep. Purif. Technol. 120, 59–68.
Jamshidi Gohari, R., Lau, W.J., Matsuura, T., Ismail, A.F., 2013b. Fabrication and characterization of novel PES/Fe–Mn binary oxide UF mixed matrix membrane for adsorptive removal of As(III) from contaminated water solution. Sep. Purif. Technol. 118, 64–72.
Jeong, B.-H., Hoek, E.M.V., Yan, Y., Subramani, A., Huang, X., Hurwitz, G., Ghosh, A.K., Jawor, A., 2007. Interfacial polymerization of thin film nanocomposites: a new concept for reverse osmosis membranes. J. Memb. Sci. 294, 1–7.
Ji, L., Chen, W., Duan, L., Zhu, D., 2009a. Mechanisms for strong adsorption of tetracycline to carbon nanotubes: a comparative study using activated carbon and graphite as adsorbents. Environ. Sci. Technol. 43, 2322–2327.
Ji, P., Zhang, J., Chen, F., Anpo, M., 2009b. Study of adsorption and degradation of acid orange 7 on the surface of CeO_2 under visible light irradiation. Appl. Catal. B Environ. 85, 148–154.
Jin, J., Li, R., Wang, H., Chen, H., Liang, K., Ma, J., 2007. Magnetic Fe nanoparticle functionalized water-soluble multi-walled carbon nanotubules towards the preparation of sorbent for aromatic compounds removal. Chem. Commun. 28 (4), 386–388.
Kanel, S.R., Manning, B., Charlet, L., Choi, H., 2005. Removal of Arsenic(III) from groundwater by nanoscale zero-valent iron. Environ. Sci. Technol. 39, 1291–1298.
Kanel, S.R., Grenèche, J.-M., Choi, H., 2006. Arsenic(V) removal from groundwater using nano scale zero-valent iron as a colloidal reactive barrier material. Environ. Sci. Technol. 40, 2045–2050.
Kaufman, Y., Berman, A., Freger, V., 2010. Supported lipid bilayer membranes for water purification by reverse osmosis. Langmuir 26, 7388–7395.
Khaydarov, R.A., Khaydarov, R.R., Gapurova, O., 2010. Water purification from metal ions using carbon nanoparticle-conjugated polymer nanocomposites. Water Res. 44, 1927–1933.

Koeppenkastrop, D., De Carlo, E.H., 1993. Uptake of rare earth elements from solution by metal oxides. Environ. Sci. Technol. 27, 1796–1802.
Kumar, M., Grzelakowski, M., Zilles, J., Clark, M., Meier, W., 2007. Highly permeable polymeric membranes based on the incorporation of the functional water channel protein Aquaporin Z. Proc. Natl. Acad. Sci. 104, 20719–20724.
Kumar, S., Ahlawat, W., Bhanjana, G., Heydarifard, S., Nazhad, M.M., Dilbaghi, N., 2014. Nanotechnology-based water treatment strategies. J. Nanosci. Nanotechnol. 14, 1838–1858.
Lawton, R.O., Nair, U.S., Pielke, R.A., Welch, R.M., 2001. Climatic impact of tropical lowland deforestation on nearby montane cloud forests. Science 294, 584–587.
Lee, H.S., Im, S.J., Kim, J.H., Kim, H.J., Kim, J.P., Min, B.R., 2008. Polyamide thin-film nanofiltration membranes containing TiO_2 nanoparticles. Desalination 219, 48–56.
Li, Y.-H., Ding, J., Luan, Z., Di, Z., Zhu, Y., Xu, C., Wu, D., Wei, B., 2003. Competitive adsorption of Pb^{2+}, Cu^{2+} and Cd^{2+} ions from aqueous solutions by multiwalled carbon nanotubes. Carbon 41, 2787–2792.
Li, X.-Q., Elliott, D.W., Zhang, W.-X., 2006. Zero-valent iron nanoparticles for abatement of environmental pollutants: materials and engineering aspects. Crit. Rev. Solid State Mater. Sci. 31, 111–122.
Liga, M.V., Bryant, E.L., Colvin, V.L., Li, Q., 2011. Virus inactivation by silver doped titanium dioxide nanoparticles for drinking water treatment. Water Res. 45, 535–544.
Lin, D., Xing, B., 2008. Adsorption of phenolic compounds by carbon nanotubes: role of aromaticity and substitution of hydroxyl groups. Environ. Sci. Technol. 42, 7254–7259.
Lind, M.L., Jeong, B.-H., Subramani, A., Huang, X., Hoek, E.M.V., 2009. Effect of mobile cation on zeolite-polyamide thin film nanocomposite membranes. J. Mater. Res. 24, 1624–1631.
Lind, M.L., Eumine Suk, D., Nguyen, T.-V., Hoek, E.M.V., 2010. Tailoring the structure of thin film nanocomposite membranes to achieve seawater ro membrane performance. Environ. Sci. Technol. 44, 8230–8235.
Liu, Y., Majetich, S.A., Tilton, R.D., Sholl, D.S., Lowry, G.V., 2005. TCE dechlorination rates, pathways, and efficiency of nanoscale iron particles with different properties. Environ. Sci. Technol. 39, 1338–1345.
Liu, J.-F., Zhao, Z.-S., Jiang, G.-B., 2008. Coating Fe_3O_4 magnetic nanoparticles with humic acid for high efficient removal of heavy metals in water. Environ. Sci. Technol. 42, 6949–6954.
Liu, L., Liu, Z., Bai, H., Sun, D.D., 2012. Concurrent filtration and solar photocatalytic disinfection/degradation using high-performance Ag/TiO_2 nanofiber membrane. Water Res. 46, 1101–1112.
López-Muñoz, M.-J., Grieken, R.V., Aguado, J., Marugán, J., 2005. Role of the support on the activity of silica-supported TiO_2 photocatalysts: structure of the TiO_2/SBA-15 photocatalysts. Catal. Today 101, 307–314.
Lu, C., Liu, C., 2006. Removal of nickel(II) from aqueous solution by carbon nanotubes. J. Chem. Technol. Biotechnol. 81, 1932–1940.
Lu, C., Chung, Y.-L., Chang, K.-F., 2005. Adsorption of trihalomethanes from water with carbon nanotubes. Water Res. 39, 1183–1189.
Lu, C., Chiu, H., Liu, C., 2006a. Removal of zinc(II) from aqueous solution by purified carbon nanotubes: kinetics and equilibrium studies. Ind. Eng. Chem. Res. 45, 2850–2855.
Lu, C., Chung, Y.-L., Chang, K.-F., 2006b. Adsorption thermodynamic and kinetic studies of trihalomethanes on multiwalled carbon nanotubes. J. Hazard. Mater. 138, 304–310.

Lucas, E., Decker, S., Khaleel, A., Seitz, A., Fultz, S., Ponce, A., Li, W., Carnes, C., Klabunde, K.J., 2001. Nanocrystalline metal oxides as unique chemical reagents/sorbents. Chem. Eur. J. 7, 2505–2510.

Mauter, M.S., Elimelech, M., 2008. Environmental applications of carbon-based nanomaterials. Environ. Sci. Technol. 42, 5843–5859.

Mauter, M.S., Elimelech, M., Osuji, C.O., 2010. Nanocomposites of vertically aligned single-walled carbon nanotubes by magnetic alignment and polymerization of a lyotropic precursor. ACS Nano 4, 6651–6658.

Mayo, J.T., Yavuz, C., Yean, S., Cong, L., Shipley, H., Yu, W., Falkner, J., Kan, A., Tomson, M., Colvin, V.L., 2007. The effect of nanocrystalline magnetite size on arsenic removal. Sci. Technol. Adv. Mater. 8, 71–75.

Mccarthy, J.F., Zachara, J.M., 1989. Subsurface transport of contaminants. Environ. Sci. Technol. 23, 496–502.

Minnesota Department of Health. 2005a. Drinking Water Standards for Contaminants Volatile and Synthetic Organic Chemicals. http://www.health.state.mn.us/divs/eh/water/factsheet/com/voc_soc.html

Minnesota Department of Health. 2005b. Drinking Water Standards for Contaminants: Microbiological, Radiological, and Inorganic Contaminants. http://www.health.state.mn.us/divs/eh/water/factsheet/com/ioc.html

Mohan, D., Sharma, R., Singh, V.K., Steele, P., Pittman, C.U., 2012. Fluoride removal from water using bio-char, a green waste, low-cost adsorbent: equilibrium uptake and sorption dynamics modeling. Indust. Eng. Chem. Res. 51, 900–914.

Mohan, D., Sarswat, A., Ok, Y.S., Pittman, C.U., 2014. Organic and inorganic contaminants removal from water with biochar, a renewable, low cost and sustainable adsorbent—a critical review. Biores. Technol. 160, 191–202.

Mohmood, I., Lopes, C., Lopes, I., Ahmad, I., Duarte, A., Pereira, E., 2013. Nanoscale materials and their use in water contaminants removal—a review. Environ. Sci. Pollut. Res. 20, 1239–1260.

Mubarak, N., Alicia, R., Abdullah, E., Sahu, J., Haslija, A.A., Tan, J., 2013. Statistical optimization and kinetic studies on removal of Zn^{2+} using functionalized carbon nanotubes and magnetic biochar. J. Environ. Chem. Eng. 1, 486–495.

Nahar, M.S., Hasegawa, K., Kagaya, S., 2006. Photocatalytic degradation of phenol by visible light-responsive iron-doped TiO_2 and spontaneous sedimentation of the TiO_2 particles. Chemosphere 65, 1976–1982.

Nata, I.F., Sureshkumar, M., Lee, C.-K., 2011. One-pot preparation of amine-rich magnetite/bacterial cellulose nanocomposite and its application for arsenate removal. RSC Adv. 1, 625–631.

Nednoor, P., Chopra, N., Gavalas, V., Bachas, L.G., Hinds, B.J., 2005. Reversible biochemical switching of ionic transport through aligned carbon nanotube membranes. Chem. Mater. 17, 3595–3599.

Nevárez, L.M., Casarrubias, L.B., Canto, O.S., Celzard, A., Fierro, V., Gómez, R.I., Sánchez, G.G., 2011. Biopolymers-based nanocomposites: membranes from propionated lignin and cellulose for water purification. Carbohydr. Polym. 86, 732–741.

Omachi, H., Nakayama, T., Takahashi, E., Segawa, Y., Itami, K., 2013. Initiation of carbon nanotube growth by well-defined carbon nanorings. Nat. Chem. 5, 572–576.

Orha, C., Pop, A., Lazau, C., Grozescu, I., Tiponut, V., Manea, F., 2011. Structural characterization and the sorption properties of the natural and synthetic zeolite. J. Optoelectron. Adv. Mater. 13, 544–549.

Pan, B., Xing, B., 2008. Adsorption mechanisms of organic chemicals on carbon nanotubes. Environ. Sci. Technol. 42, 9005–9013.

Pan, B., Lin, D., Mashayekhi, H., Xing, B., 2008. Adsorption and hysteresis of bisphenol A and 17α-ethinyl estradiol on carbon nanomaterials. Environ. Sci. Technol. 42, 5480–5485.
Pena, M.E., Korfiatis, G.P., Patel, M., Lippincott, L., Meng, X., 2005. Adsorption of As(V) and As(III) by nanocrystalline titanium dioxide. Water Res. 39, 2327–2337.
Pendergast, M.M., Hoek, E.M., 2011a. A review of water treatment membrane nanotechnologies. Energy Environ. Sci. 4, 1946–1971.
Pendergast, M.M., Hoek, E.M.V., 2011b. A review of water treatment membrane nanotechnologies. Energy Environ. Sci. 4, 1946–1971.
Peng, X., Li, Y., Luan, Z., Di, Z., Wang, H., Tian, B., Jia, Z., 2003. Adsorption of 1,2-dichlorobenzene from water to carbon nanotubes. Chem. Phys. Lett. 376, 154–158.
Peng, X., Luan, Z., Ding, J., Di, Z., Li, Y., Tian, B., 2005. Ceria nanoparticles supported on carbon nanotubes for the removal of arsenate from water. Mater. Lett. 59, 399–403.
Poinern, G.E.J., Parsonage, D., Issa, T.B., Ghosh, M., Paling, E., Singh, P., 2010. Preparation, characterization and As (V) adsorption behaviour of CNT-ferrihydrite composites. Int. J. Eng. Sci. Technol. 2.
Ponder, S.M., Darab, J.G., Mallouk, T.E., 2000. Remediation of Cr(VI) and Pb(II) aqueous solutions using supported. Nanoscale zero-valent iron. Environ. Sci. Technol. 34, 2564–2569.
Qin, F., Li, G., Wang, R., Wu, J., Sun, H., Chen, R., 2012. Template-free fabrication of Bi_2O_3 and $(BiO)_2CO_3$ nanotubes and their application in water treatment. Chem.-A Eur. J. 18, 16491–16497.
Qu, X., Alvarez, P.J.J., Li, Q., 2013a. Applications of nanotechnology in water and wastewater treatment. Water Res. 47, 3931–3946.
Qu, X., Brame, J., Li, Q., Alvarez, P.J.J., 2013b. Nanotechnology for a safe and sustainable water supply: enabling integrated water treatment and reuse. Acc. Chem. Res. 46, 834–843.
Rao, G.P., Lu, C., Su, F., 2007. Sorption of divalent metal ions from aqueous solution by carbon nanotubes: a review. Sep. Purif. Technol. 58, 224–231.
Ritchie, S.M.C., Kissick, K.E., Bachas, L.G., Sikdar, S.K., Parikh, C., Bhattacharyya, D., 2001. Polycysteine and other polyamino acid functionalized microfiltration membranes for heavy metal capture. Environ. Sci. Technol. 35, 3252–3258.
Romanos, G.E., Athanasekou, C.P., Katsaros, F.K., Kanellopoulos, N.K., Dionysiou, D.D., Likodimos, V., Falaras, P., 2012. Double-side active TiO_2-modified nanofiltration membranes in continuous flow photocatalytic reactors for effective water purification. J. Hazard. Mater. 211–212, 304–316.
Safe Water Drinking Foundation. Water Quality Tests. https://www.safewater.org/PDFS/communitywatertestkit/Water_Quality_Tests.pdf
Salipira, K.L., Mamba, B.B., Krause, R.W., Malefetse, T.J., Durbach, S.H., 2007. Carbon nanotubes and cyclodextrin polymers for removing organic pollutants from water. Environ. Chem. Lett. 5, 13–17.
Schnorr, J.M., Swager, T.M., 2011. Emerging applications of carbon nanotubes. Chem. Mater. 23, 646–657.
Sehaqui, H., Liu, A., Zhou, Q., Berglund, L.A., 2010. Fast preparation procedure for large flat cellulose and cellulose/inorganic nanopaper structures. Biomacromolecules 11, 2195–2198.
Shannon, M.A., Bohn, P.W., Elimelech, M., Georgiadis, J.G., Marinas, B.J., Mayes, A.M., 2008. Science and technology for water purification in the coming decades. Nature 452, 301–310.

Sharma, Y.C., Srivastava, V., Singh, V.K., Kaul, S.N., Weng, C.H., 2009. Nano-adsorbents for the removal of metallic pollutants from water and wastewater. Environ. Technol. 30, 583–609.
Shin, S., Jang, J., 2007. Thiol containing polymer encapsulated magnetic nanoparticles as reusable and efficiently separable adsorbent for heavy metal ions. Chem. Commun., 4230–4232.
Snyder, A., Bo, Z., Moon, R., Rochet, J., Stanciu, L., 2013. Reusable photocatalytic titanium dioxide–cellulose nanofiber films. J Colloid Interface Sci. 399, 92–98.
Song, W., Justice, R.E., Jones, C.A., Grassian, V.H., Larsen, S.C., 2004. Synthesis, characterization, and adsorption properties of nanocrystalline ZSM-5. Langmuir 20, 8301–8306.
Song, W., Li, G., Grassian, V.H., Larsen, S.C., 2005. Development of improved materials for environmental applications: nanocrystalline nay zeolites. Environ. Sci. Technol. 39, 1214–1220.
Srivastava, A., Srivastava, O.N., Talapatra, S., Vajtai, R., Ajayan, P.M., 2004. Carbon nanotube filters. Nat. Mater. 3, 610–614.
Stanton, B.W., Harris, J.J., Miller, M.D., Bruening, M.L., 2003. Ultrathin, multilayered polyelectrolyte films as nanofiltration membranes. Langmuir 19, 7038–7042.
Taniguchi, T., Okamura, K., 1998. New films produced from microfibrillated natural fibres. Polym. Int. 47, 291–294.
Tetala, K.K.R., Stamatialis, D.F., 2013. Mixed matrix membranes for efficient adsorption of copper ions from aqueous solutions. Sep. Purif. Technol. 104, 214–220.
Thakur, V.K., Thakur, M.K., Raghavan, P., Kessler, M.R., 2014. Progress in green polymer composites from lignin for multifunctional applications: a review. ACS Sustain. Chem. Eng. 2, 1072–1092.
Theron, J., Walker, J., Cloete, T., 2008. Nanotechnology and water treatment: applications and emerging opportunities. Crit. Rev. Microbiol. 34, 43–69.
Tian, Y., Gao, B., Morales, V.L., Wu, L., Wang, Y., Muñoz-Carpena, R., Cao, C., Huang, Q., Yang, L., 2012. Methods of using carbon nanotubes as filter media to remove aqueous heavy metals. Chem. Eng. J. 210, 557–563.
Tian, Y., Gao, B., Chen, H., Wang, Y., Li, H., 2013a. Interactions between carbon nanotubes and sulfonamide antibiotics in aqueous solutions under various physicochemical conditions. J. Environ. Sci. Health A 48, 1136–1144.
Tian, Y., Gao, B., Morales, V.L., Chen, H., Wang, Y., Li, H., 2013b. Removal of sulfamethoxazole and sulfapyridine by carbon nanotubes in fixed-bed columns. Chemosphere 90, 2597–2605.
Tiraferri, A., Vecitis, C.D., Elimelech, M., 2011. Covalent binding of single-walled carbon nanotubes to polyamide membranes for antimicrobial surface properties. ACS Appl. Mater. Interfaces 3, 2869–2877.
Tofighy, M.A., Mohammadi, T., 2011. Adsorption of divalent heavy metal ions from water using carbon nanotube sheets. J. Hazard. Mater. 185, 140–147.
Trivedi, P., Axe, L., 2000. Modeling Cd and Zn sorption to hydrous metal oxides. Environ. Sci. Technol. 34, 2215–2223.
Ullah, R., Dutta, J., 2008. Photocatalytic degradation of organic dyes with manganese-doped ZnO nanoparticles. J. Hazard. Mater. 156, 194–200.
Üzüm, Ç., Shahwan, T., Eroğlu, A.E., Hallam, K.R., Scott, T.B., Lieberwirth, I., 2009. Synthesis and characterization of kaolinite-supported zero-valent iron nanoparticles and their application for the removal of aqueous Cu^{2+} and Co^{2+} ions. Appl. Clay Sci. 43, 172–181.
Wang, H., Yang, L., Yu, H., Peng, F., 2011. A highly efficient and stable visible-light plasmonic photocatalyst Ag–AgCl/CeO_2. World J. Nano Sci. Eng. 1, 129.

WHO and UNICEF. 2015. Progress on drinking water and sanitation Update. http://files.unicef.org/publications/files/Progress_on_Sanitation_and_Drinking_Water_2015_Update_.pdf.
Xu, Y., Zhao, D., 2007. Reductive immobilization of chromate in water and soil using stabilized iron nanoparticles. Water Res. 41, 2101–2108.
Yan, H., Gong, A., He, H., Zhou, J., Wei, Y., Lv, L., 2006. Adsorption of microcystins by carbon nanotubes. Chemosphere 62, 142–148.
Yan, X.M., Shi, B.Y., Lu, J.J., Feng, C.H., Wang, D.S., Tang, H.X., 2008. Adsorption and desorption of atrazine on carbon nanotubes. J. Colloid Interface Sci. 321, 30–38.
Yang, G.C.C., Lee, H.-L., 2005. Chemical reduction of nitrate by nanosized iron: kinetics and pathways. Water Res. 39, 884–894.
Yang, K., Xing, B., 2010. Adsorption of organic compounds by carbon nanomaterials in aqueous phase: polanyi theory and its application. Chem. Rev. 110, 5989–6008.
Yang, K., Wu, W., Jing, Q., Zhu, L., 2008. Aqueous adsorption of aniline, phenol, and their substitutes by multi-walled carbon nanotubes. Environ. Sci. Technol. 42, 7931–7936.
Yang, Y.-H., Ren, N., Zhang, Y.-H., Tang, Y., 2009. Nanosized cadmium sulfide in polyelectrolyte protected mesoporous sphere: a stable and regeneratable photocatalyst for visible-light-induced removal of organic pollutants. J. Photochem. Photobiol. A Chem. 201, 111–120.
Yean, S., Cong, L., Yavuz, C.T., Mayo, J.T., Yu, W.W., Kan, A.T., Colvin, V.L., Tomson, M.B., 2005. Effect of magnetite particle size on adsorption and desorption of arsenite and arsenate. J. Mater. Res. 20, 3255–3264.
Yin, J., Deng, B., 2015. Polymer-matrix nanocomposite membranes for water treatment. J. Memb. Sci. 479, 256–275.
Yu, X., Tong, S., Ge, M., Zuo, J., Cao, C., Song, W., 2013. One-step synthesis of magnetic composites of cellulose@ iron oxide nanoparticles for arsenic removal. J. Mater. Chem. A 1, 959–965.
Zhai, Y., Zhang, S., Pang, H., 2007. Preparation, characterization and photocatalytic activity of CeO_2 nanocrystalline using ammonium bicarbonate as precipitant. Mater. Lett. 61, 1863–1866.
Zhang, M., Gao, B., 2013. Removal of arsenic, methylene blue, and phosphate by biochar/AlOOH nanocomposite. Chem. Eng. J. 226, 286–292.
Zhang, W.-X., Wang, C.-B., Lien, H.-L., 1998. Treatment of chlorinated organic contaminants with nanoscale bimetallic particles. Catal. Today 40, 387–395.
Zhang, H., Quan, X., Chen, S., Zhao, H., Zhao, Y., 2006. Fabrication of photocatalytic membrane and evaluation its efficiency in removal of organic pollutants from water. Sep. Purif. Technol. 50, 147–155.
Zhang, M., Gao, B., Yao, Y., Xue, Y., Inyang, M., 2012a. Synthesis of porous MgO-biochar nanocomposites for removal of phosphate and nitrate from aqueous solutions. Chem. Eng. J. 210, 26–32.
Zhang, M., Gao, B., Yao, Y., Xue, Y., Inyang, M., 2012b. Synthesis, characterization, and environmental implications of graphene-coated biochar. Sci. Tot. Environ. 435, 567–572.
Zhang, M., Gao, B., Varnoosfaderani, S., Hebard, A., Yao, Y., Inyang, M., 2013. Preparation and characterization of a novel magnetic biochar for arsenic removal. Biores. Technol. 130, 457–462.
Zhou, Q., Xiao, J., Wang, W., 2006a. Using multi-walled carbon nanotubes as solid phase extraction adsorbents to determine dichlorodiphenyltrichloroethane and its metabolites at trace level in water samples by high performance liquid chromatography with UV detection. J. Chromatogr. A 1125, 152–158.

Zhou, Q., Xiao, J., Wang, W., Liu, G., Shi, Q., Wang, J., 2006b. Determination of atrazine and simazine in environmental water samples using multiwalled carbon nanotubes as the adsorbents for preconcentration prior to high performance liquid chromatography with diode array detector. Talanta 68, 1309–1315.

Zhou, Q., Xiao, J., Wang, W., 2007. Comparison of multiwalled carbon nanotubes and a conventional absorbent on the enrichment of sulfonylurea herbicides in water samples. Anal. Sci. 23, 189–192.

Zodrow, K., Brunet, L., Mahendra, S., Li, D., Zhang, A., Li, Q., Alvarez, P.J.J., 2009. Polysulfone ultrafiltration membranes impregnated with silver nanoparticles show improved biofouling resistance and virus removal. Water Res. 43, 715–723.

4

APPLICATION OF NANOTECHNOLOGY IN DRINKING WATER PURIFICATION

Alberto Figoli*, Mir Saeed Seyed Dorraji, Ali Reza Amani-Ghadim†**

**Institute on Membrane Technology (ITM-CNR), Rende, Italy; **University of Zanjan, Applied Chemistry Research Laboratory, Department of Chemistry, Faculty of Science, Zanjan, Iran; †Azarbaijan Shahid Madani University, Applied Chemistry Research Laboratory, Department of Chemistry, Faculty of Science, Tabriz, Iran*

1 Introduction

Safe and reliable water supply is an important challenge for the human society. Demand for clean water sources is increasing owing to the rapid industrialization of developing countries, population growth, and long-term droughts and it is expected that the problems with accessing clean water and sanitation will be worse in the next decades. On the other hand, waterborne diseases are reported in both developing and developed countries. In developing countries, it is estimated that 50% of the population does not have access to clean water. Accordingly, in developing countries, deficiency of clean water producing system is the main problem. However, in countries where conventional treatment and distribution systems adequately spread, new waterborne problems are emerging. Dangerous contaminants, such as heavy metals, organic contaminants, and anions are introduced into water supplies via agricultural and industrial activities. New organic contaminants including pesticides, hormones, drugs, various aromatic compound, and so on could not be removed by chemical, biological, and photolytic processes occurring in environment. The conventional treatment processes are not able to remove them while most of them are carcinogenic and cause the formation of dangerous disinfection by-products (DBPs) even at very low concentrations. Moreover, some microbial pathogens are resistant to conventional disinfectants, such as chlorination and

Water Purification. http://dx.doi.org/10.1016/B978-0-12-804300-4.00004-6

ozonation. The existence of pollutants not only impacts health and sanitation but also reduces food safety. It is clear that availability of clean water is essential for all aspects of food production including preparation, distribution, and consumption.

With the increasing waterborne diseases caused by different organic, inorganic, and microorganisms, the innovation of new and efficient treatment processes is crucial. Nanotechnology exhibits great promise for developing new processes and enabling and improving technologies for removal of the mentioned types of pollutants. Significant research has been carried out on utilizing nanotechnology-enabled water-treatment processes and many of them exhibit improved performance over conventional technologies. However, the potentials and limitations of nanotechnology as a perfect treatment process must be discussed.

In this chapter, we introduce some nanotechnology-enabled water-treatment processes in which their efficacy is extensively investigated in laboratory and large-scale practical treatment scales. Mostly, commercialized nanotechnology-enabled technologies utilize silver nanoparticles, zero-valent iron (ZVI) nanoparticles, nanostructured photocatalysts, membrane-based nanotechnology, and nanoadsorbents, all of which are applied to resolve waterborne problems.

2 Antimicrobial Nanotechnology for Water Disinfection

Chemical-based disinfection agents (chlorine, chlorine dioxide, ozone) and ultraviolet (UV) radiations are the disinfection methods currently used in water treatment to overcome waterborne microbiological contamination. Both the chemical disinfectants and UV possess wide microbicidal action and can effectively control microbial pathogens including bacteria, algae, fungi, yeast, virus, and mold spores (Nameni and Economy, 2014). However, considerable researches in the past few decades have revealed a challenge between effective disinfection of chemical disinfectants and formation of disinfectant by products (DBPs) (Grellier et al., 2015; Matilainen and Sillanpää, 2010). Chlorine and chlorinated disinfection agents may react with naturally present organic matter, as well as iodide and bromide ions, to produce a range of DBPs. Various types of DBPs, such as trihalomethanes (THMs), haloacetic acids, chlorite, halonitromethanes, haloacetonitriles, haloamides, halofuranones, iodo-THMs, nitrosamines, and others have been reported in the literature (Singer, 2006; Krasner et al., 2006; Matilainen and Sillanpää, 2010; Hua and Reckhow, 2007). The formation of DBPs in ozonation process, such as formaldehyde and bromate (Matilainen and Sillanpää, 2010) and in UV irradiation, such as nitrogenous DBPs is proven (Zhao et al., 2008; Kolkman et al., 2015). Longer exposure time for effectiveness is other disadvantage for disinfection by UV irradiation. Moreover, some pathogens are resistant to conventional disinfectants which require extremely high disinfectant dosage, leading to more DBPs formation (Li et al., 2008). Even developed countries which utilize advanced water purification infrastructures report disease outbreaks from waterborne infections (Reynolds, 2007). Accordingly, innovative disinfection technologies not only must enhance the reliability of disinfection in elimination of the emerging pathogens, but also avoid DBPs formation.

Recently, researchers have displayed the microbicidal efficacy of several nanomaterials including nanometallic particles, nanophotocatalysts, and carbon-based nanomaterials (Li et al., 2008). The mentioned nanomaterials are weak oxidation agents. Accordingly, the harmful DBPs formation is not expected in disinfection by them.

Nanometallic particles, especially silver (Ag) display antimicrobial properties, which referred to their oligodynamic effect (Mahmoudi and Serpooshan, 2012). Among nanometallic particles, silver nanoparticles (AgNPs) are widely studied and commercialized because of (1) possessing wide microbicidal action, (2) low toxicity in comparison with other nanometallic particles, (3) ease of incorporation on a wide range of low-cost materials for application in filters and (4) historical application of silver compounds and silver ion as disinfectant (Yu et al., 2013; Praveena and Aris, 2015, Butkus et al., 2003). Fig. 4.1 shows the impact of AgNPs on *Escherichia coli* cells which was recorded by scanning electron microscope and transmission electron microscope (Loo et al., 2015).

Several bactericidal mechanisms of AgNPs have been proposed, however, discussion still exists concerning the exact mechanism. The bactericidal efficacy of the AgNPs is ascribed to oligodynamic effect of the dissolved silver ions and special physicochemical properties of AgNPs that lead to the destruction of bacteria (Li et al., 2008). The following mechanisms have been proposed (Kittler et al., 2010; Morones et al., 2005; Loo et al., 2015):

1. AgNPs release Ag^+ ions in low concentration and silver ions taken up by the cell. Ag^+ ions instantaneously react with thiol groups of enzymes leading to the production of reactive oxygen species (ROS), such as superoxide anion radical and hydrogen peroxide. Simultaneously, Ag^+ ions prevent ATP synthesis and the electron transport in cellular respiratory chain.
2. AgNPs alter the membrane surface properties by adhesing to the outer surface of membrane. Moreover, it is reported that AgNPs react with hydrogen peroxide and form hydroxyl radicals leading to destroy the cell membrane and cell death.

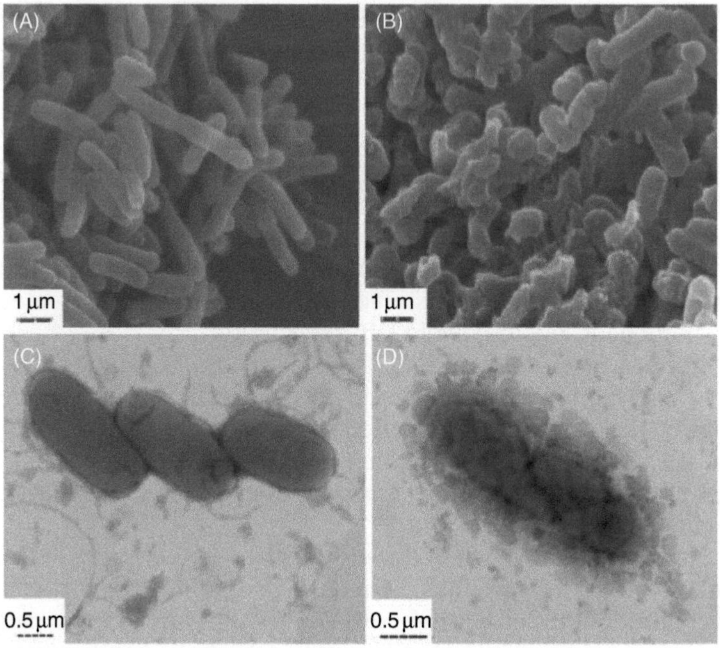

Figure 4.1. Impact of AgNPs on *E. coli* cells observed by SEM (A, B) and TEM (C, D). (A, C) Cell structure of native *E. coli* cells; (B, D) cell structure of *E. coli* cells treated with 50 μg/ml AgNPs. [Reprinted from (Li, W.-R., Xie, X.-B., Shi, Q.-S., Zeng, H.-Y., Ou-Yang, Y.-S., Chen, Y.-B., 2010. Antibacterial activity and mechanism of silver nanoparticles on Escherichiacoli. Appl. Microbiol. Biotechnol. 85, 1115–1122.), copyright (2009), with kind permission from Springer Science and Business Media]

3. AgNPs could degrade lipopolysaccharide molecules of membrane and create "pits." In this case, membrane permeability is increased and AgNPs damage DNA. Fig. 4.2 summarizes the bactericidal mechanisms of AgNPs decorated on cryogels (Loo et al., 2015). However, the microbicidal mechanism is still only partially understood, silver nanoparticles are a promising alternative for conventional disinfection processes.

It is clear that AgNPs could not be utilized directly in water disinfection. For practical applications, AgNPs must be incorporated into inert supports. On the other hand, AgNPs are mainly designed for point-of-use (POU) applications which include water purification filters. Thus, the selection of appropriate support is an important factor in design of POU filters (Praveena and Aris, 2015). The systems containing AgNPs incorporated into support must exhibit wide-range microbicidal property in long period of time. It means that, silver desorption rate of filters is an important parameter in designing of filters based on AgNPs and must be determined

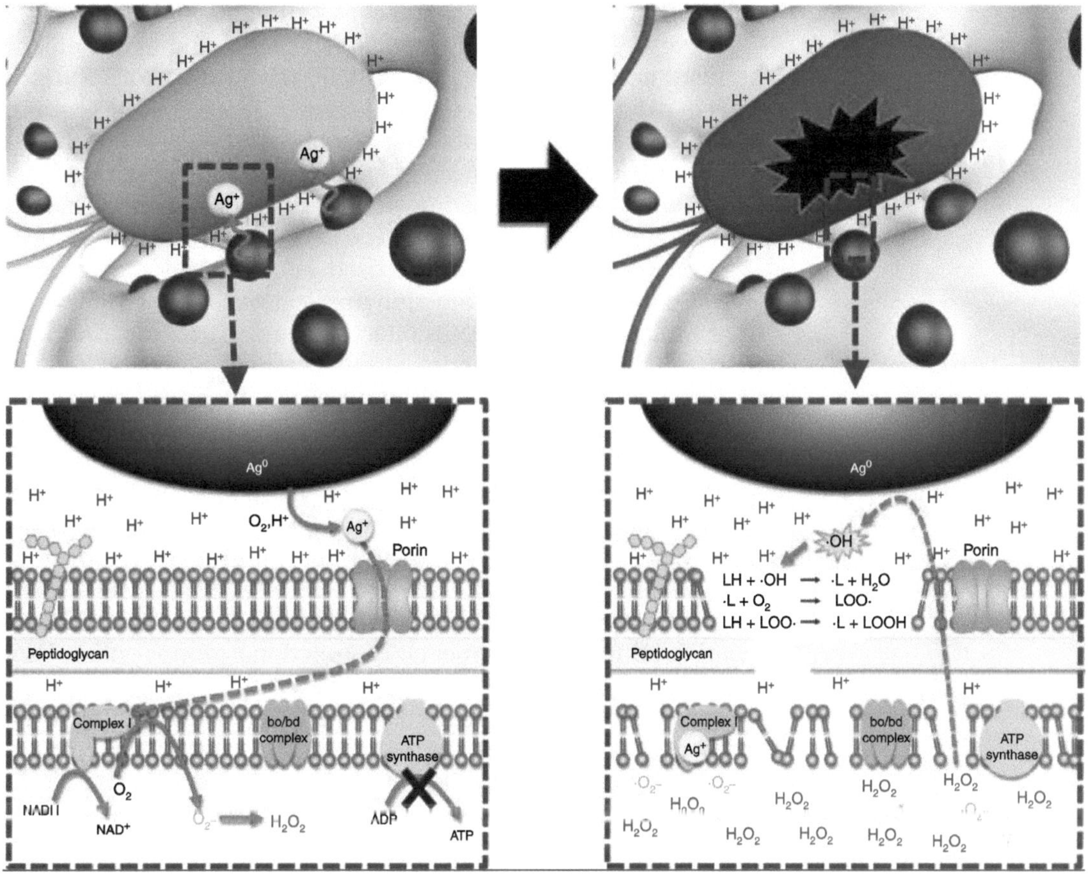

Figure 4.2. Summarizing proposed bactericidal mechanism of Ag-NPs. Ag^+ is taken up by the bacteria and disturbs ATP synthesis and the electron transport chain. The latter mechanism is the indirect formation of ROS. The hydroxyl radicals, generated via decomposition of H_2O_2, can initiate a peroxidative sequence that inevitably results in cell death by destroying the cell membrane (·L and LOO· are a carbon-centered radical and a peroxyl radical, respectively). [Reprinted from (Loo, S.-L., Krantz, W.B., Fane, A.G., Gao, Y., Lim, T.-T., Hu, X., 2015. Bactericidalmechanisms revealed for rapid water disinfection by superabsorbent cryogelsdecorated with silver nanoparticles. Environ. Sci. Technol. 49, 2310–2318.), copyright (2015), with kind permission from American Chemical Society]

before practical application. Silver attrition in AgNPs-based filters decrease silver content available for microbicidal efficacy and subsequently efficient disinfection of filter in a long period of time is decreased. Moreover, silver concentration in the disinfected water must be below the permissible limits. The US Environmental Protection Agency's (EPA) maximum contamination level for silver in water is 100 ppb (EPA, 2011). The cost of AgNPs-based filters is another important criteria in POU design. The common supports involved in silver nanoparticles' incorporation for microbicidal application are ceramic (Rayner et al., 2013; Ren and Smith, 2013), cellulose (Dankovich and Gray, 2011), granulated-activated carbon and activated carbon fibers (ACF) (Karumuri et al., 2013; Le Pape et al., 2002), polyurethane (Jain and Pradeep, 2005), and inorganic fiberglass substrate (Nangmenyi et al., 2009).

Investigation on using silver nanoparticles incorporated in ceramics and their disinfection performance has been conducted by several researchers. Ceramics coated with silver nanoparticles have been widely studied and used because they are inexpensive and efficient in disinfection applications. The various types of ceramic-decorated AgNPs filters have been manufactured and reported that their bactericidal efficacy is acceptable (Kallman et al., 2011). However, the filtration durability and release of silver is mainly affected by silver loading amount on the ceramics, water-quality parameters, and ceramic-pore structure (Oyanedel-Craver and Smith, 2008). Commonly, pot-shaped, cylindrical ceramic and ceramic membranes decorated with AgNPs are placed in container as POU filter (Kim and Van Der Bruggen, 2010).

Polysulphone, polyethersulphone, and cellulose acetate membranes are the common polymers in water-treatment applications owing to stiffness of polymer chain, functional groups polarity, and low-fouling affinity in polymeric membranes (Lufrano et al., 2000). Studies have been reported about the incorporation of AgNPs in membrane (Kim and Van Der Bruggen, 2010). AgNPs incorporated in membrane to prevent the microbial growth and biofilm formation on the membrane used in water-treatment processes, such as reverse osmosis. Biofilm formation on membrane causes to be fouling which decreases membrane permeability, permeate quality, and increases energy cost of water treatment (Zodrow et al., 2009). The main challenge in incorporating the AgNPs in membrane is weak dispersion of the nanoparticles in polymer matrix (Yin and Deng, 2015). However, it has been reported that decorating silver nanoparticles in polyethersulphones leads to bactericidal activity to it with silver release lower than the WHO and the USEPA drinking water standard values (Praveena and Aris, 2015).

Granular-activated carbon (GAC) and fiber-activated carbon (FAC) have been widely studied as a support for AgNPs due to high specific surface area, pore size distribution, and no biological pollution. At the moment, the commercial POU and point of entry filter are manufactured and utilized in water purification and it is believed that the removal efficiency of *E. coli* is 100% and silver release below WHO and USEPA standards. In addition to the microbicidal property of AgNPs, silver nanoparticles could impel the formation of ROS on ACF which inhibits from bacteria growth (Park and Jang, 2003).

The efficacy of AgNPs–ACF filters in disinfection depends on the activation method of ACF or GAC, silver content, and the specific surface area of the activated carbon (Praveena and Aris, 2015). However, it has been reported that active carbon substrates have lower the bactericidal impact of AgNPs. Moreover, because AgNPs–FAC POU filters are not able to eliminate bacterial growth completely, it is believed that the AgNPs decorated on active carbon substrates exhibit bacteriostatic efficiency which stops bacteria reproducing without killing them (Le Pape et al., 2002).

AgNPs-based disinfection process is a promising method which could resolve the water disinfection problems in developing and developed countries. However, more investigations are needed to overcome the AgNPs disadvantages. In this way, in addition to discussed supports, the efficiency of new advanced and engineered supports including fiber glass, superabsorbent cryogels, engineered ceramics, chitosan silica beads, and some polymers were investigated.

3 Zero-Valent Iron Nanoparticles

Nanoparticles are an important class of materials that are highly interested for removal of contaminants which could not be removed easily by conventional water-treatment techniques owing to their small size (1–100 nm) and enhanced reactivity. Nanoparticles possess important function in the removal of waterborne pathogens, toxic metals and anions, and chlorinated organics, which are introduced into the water supplies from both natural and human activities source. Compared with more-conventional water-treatment approaches, the utilization of nanoparticles provides a more rapid, cost-effective removal process (Savage and Diallo, 2005).

Iron is one of the most abundant metals on earth, making up about 5% of the earth's crust. Due to ZVI unique features, such as nontoxic, easy to produce, cost effective, strong reductive potential (E_0 = −0.44 V vs SHE), simple reduction process equipment is considered by researchers for decontamination application (Fu et al., 2014).

3.1 Zero-Valent Iron Nanoparticles Structure and Synthesis

Zero-valent iron nanoparticles (ZVI-NPs) possess higher reactivity and higher surface area than ZVI. Moreover, the unique capabilities including sufficient mobility within porous media and sufficient reactive durability provide high interest for both in situ and ex situ applications in remediation of contaminated groundwater. ZVI-NPs can be easily injected to the contaminated zone within slurry under pressure and/or by gravity and remove contaminants effectively (Zhang, 2003). Alternatively, nanoparticles can be immobilized on solid supports, such as polymers, mesoporous silica, bentonite, kaolinite chitosan, multiwalled carbon nanotubes (MWCNTs) which not only avoid agglomeration of nanoparticles and provide an easy operation but also maintain the excellent reduction ability of ZVI-NPs (Crane and Scott, 2012).

ZVI-NPs could be synthesized via top–down approach which involves breakdown or restructure of bulk material to the nanoscale mechanically. A good sample for top–down approach is ball-milling technique, in which stainless steel balls are utilized as the grinding media to fragment the starting iron materials into pieces with dimensions less than 100 nm (Li et al., 2009).

Bottom–up methods are the other synthesis approach that involve constructing a nanomaterial from basic building blocks, such as atoms or molecules with both physical and chemical methods. The synthesis of ZVI-NPs via borohydride reduction of ferrous salts in the presence of starch as a stabilizer under nitrogen atmosphere is the most widely used bottom-up method which leads to prepare ZVI-NPs in the range of 10–100 nm in diameter (Glavee et al., 1995).

Currently, ZVI-NPs are produced in commercial-scale production via reduction of goethite (α-FeOOH) and/or hematite (α-Fe_2O_3) by hydrogen gas at high temperature or by electrolytic reduction of ferrous iron (Liu et al., 2012a; Nurmi et al., 2005). Typically, ZVI-NPs have a typical core–shell structure (Fig. 4.3). The core consists of ZVI-NPs whereas the shell consists of mixed valent Fe(II) and Fe(III) oxides as a result of metallic iron oxidation. The oxide film of borohydride-made ZVI-NPs exhibits a smooth amorphous oxide and the H_2-reduced ZVI-NPs contain crystalline oxide structures (Nurmi et al., 2005; Yan et al., 2010a,b). The synthesis process, particle size, and storage conditions are the determining parameters in the structure of the oxide phase (Wang et al., 2010).

3.2 Target Contaminants and Removal Mechanism

In the surface of ZVI-NPs, various phenomena occur including iron corrosion, particle stabilization, contaminant adsorption, and redox transformations. Chlorinated organic compounds, such as trichloroethylene (TCE), tetrachloroethylene, polychlorinated biphenyls, and many organochlorine pesticides are the main highly toxic environmental problems due to their very-low

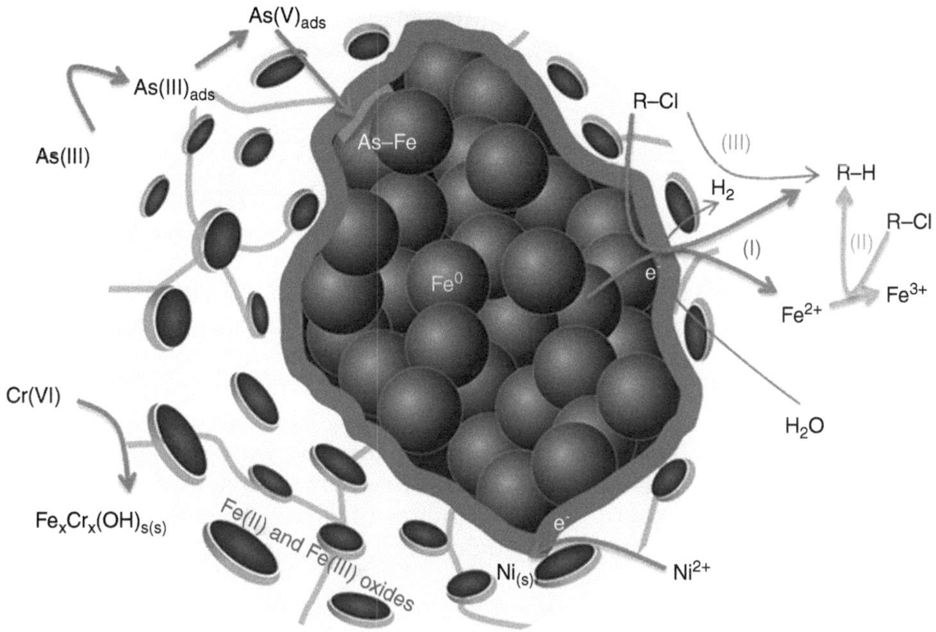

Figure 4.3. Core–shell structure of ZVI-NPs and its role in contaminant removal.

biodegradability. ZVI-NPs have been successfully employed for dehalogination of 1,4-chlorinatedethane, TCE, and carbon tetrachloride using ZVI.

In addition, ZVI-NPs have successfully degraded a large number of chlorinated organic compounds, including TCE (Tseng et al., 2011), vinyl chloride, (Cho and Choi, 2010; Wei et al., 2012), carbon tetrachloride (Zhang X1 et al., 2011). So, experimental results are promising for the development of remediation technologies to treat the contaminated sites with chlorinated organic compounds. In the removal of chlorinated organics, ZVI-NPs act as a potent and cost-effective electron donor from the iron surface to chlorinated hydrocarbons.

The half-reaction of chlorinated organic compounds has the standard reduction potential ranging from +0.5 to +1.5 V versus SHE [Eq. (4.1)] (Matheson and Tratnyek, 1994). By considering the ZVI-NPs standard reduction potential [Eq. (4.2)], it is clear that strong thermodynamic driving force for chlorinated organic compounds reduction is by ZVI-NPs.

$$R-Cl+2e^-+H^+ \rightarrow RH+Cl^- \qquad E^0=0.5-1.5\ \text{V} \qquad (4.1)$$

$$Fe^{2+}+2e^- \rightarrow Fe^0 \qquad E^0=-0.44\ \text{V} \qquad (4.2)$$

Beside the mentioned mechanism, the reduction of water molecules [Eq. (4.3)] is featured because ZVI-NPs are used in underground contaminant plume which contains water.

$$2H_2O + 2e^- \rightarrow H_{2(g)} + 2OH^- \qquad E^0 = -0.82\ \text{V} \tag{4.3}$$

So, corrosion products including ferrous iron and hydrogen are possible reducing agents relative to contaminants, such as chlorinated organic compounds (Matheson and Tratnyek, 1994). Fig. 4.3 is a simplified representation of the three possible mechanisms involved in the chlorinated organic compounds degradation by ZVI-NPs. Pathway I involves direct electron transfer form Fe(0) to the adsorbed halocarbon (RCl) resulting in dechlorination and production of ferrous. Pathway III shows that H_2 from the anaerobic corrosion of Fe(II) might react with RCl. It seems that hydrogenation plays a minor role in most systems because ZVI-NPs are covered with precipitates of oxides (or carbonates and sulfides) resulted from iron corrosion, unless an effective catalyst, such as a noble metal is present at the surface of ZVI-NPs (bimetallic ZVI-NPs particles). Now the question is: how the electrons transform from ZVI-NPs to chlorinated organics in spite of the oxide layer between them. P. Tratnyek and R. Johnson presented the heuristic model consisting of three mechanisms including (Timothy and Donald, 1999):

1. Electron could directly transform from ZVI-NPs to chlorinated organics in a corrosion pit, or similar defect in the oxide film.
2. The oxide film could act as a semiconductor which mediates electron transformation.
3. The oxide film could act as a coordinating surface containing sites of Fe(II), which are complex and can reduce chlorinated organics (Fig. 4.4).

This mechanism implies that ZVI-NPs is a multifunctional remediation agent because ZVI core contains a relatively rich electron source for reductive remediation and the oxide shell

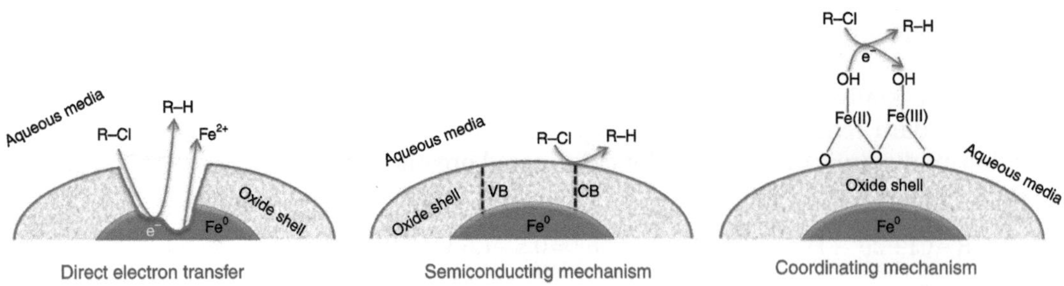

Figure 4.4. Removal mechanism of chlorinated hydrocarbons by ZVI-NPs.

provides adsorption sites and the initial surface of precipitation and surface mineralization. Subsequently, the capability of ZVI-NPs as an emerging environmental remediation agent could be utilized to remove a variety of inorganic contaminants, including nitrate (Sohn et al., 2006; Yoshino and Kawase, 2013), phosphate (Yoshino and Kawase, 2013), Cr(VI) (Cao and Zhang, 2006; Schrick et al., 2004), As(III) and As(V) (Kanel et al., 2005; Dudu et al., 2015), heavy metals (Shokes and Möller, 1999), and radionuclides (Crane and Scott, 2014).

The inorganic cations and anions occur in the earth's crust and could be introduced to groundwater by dissolving their salts through natural processes or human activities, such as landfill leachate, sewage, leachate from mining, or industrial spills (Hashim et al., 2011). The efficacy of ZVI-NPs in the removal of chromium (Mitra et al., 2011), nickel (Calabrò et al., 2012), lead (Shi et al., 2013), copper (Xiao et al., 2011), and zinc (Shi et al., 2013) has been proven experimentally. Briefly, in the case of removal of heavy metals, species with a more positive reduction potential than iron, such as Cu(II) and Ag(I), are predominantly removed in their reduced states by receiving the electron. For other species with a lower reduction potential than iron, adsorption on the iron oxide shell is an important mechanism (Yan et al., 2013). Cr(VI) removal by ZVI-NPs has been extensively studied due to high toxicity and abundance. ZVI-NPs were successfully applied in groundwater remediation. The removal mechanism involves instantaneous adsorption of Cr(VI) on the surface of ZVI-NPs oxide shell and electron transfer due to positive reduction potential of Cr(VI) than iron occurs. The result of this reaction is the reduction of Cr(VI) to Cr^{3+} with the oxidation of ZVI to Fe^{3+}, followed by the precipitation of sparingly soluble mixed Cr(III) and Fe(III) hydroxides [Eqs. (4.4) and (4.5)] (Fu et al., 2014). Cr(III) is less mobile and less toxic than Cr(VI) and is as an essential trace element in humans.

$$Cr_2O^{2-}_{7(aq)} + 2Fe^0_{(s)} + 14H^+_{(aq)} \rightarrow 2Cr^{3+}_{(aq)} + 2Fe^{3+}_{(aq)} + 7H_2O_{(l)} \quad (4.4)$$

$$Cr^{3+}_{(aq)} + Fe^{3+}_{(aq)} + 6OH^-_{(aq)} \rightarrow Cr(OH)_{3(s)} + Fe(OH)_{3(s)} \quad (4.5)$$

Similarly, aqueous U(VI) species (UO_2^{2+}) removed by coupled reduction and coprecipitation of UO_2 precipitates by various iron-bearing materials (ZVI, green rust, magnetite) (Crane and Scott, 2014).

Arsenic is a very toxic contaminant existing in groundwater mostly in the oxidation states of As(III) (arsenite) and As(V) (arsenate). The arsenic level in drinking water is to be less than 10 μg/L (Fu et al., 2014). It is found that the ZVI-NPs are very effective in the removal of arsenic; however, the removal mechanism is complicated. Experimental evidences are presented that all adsorption, reduction, surface precipitation, and coprecipitation with various iron corrosion products act in the removal mechanism. As(VI) can be reduced by ZVI-NPs to either As^0 or (III). The remaining As(VI) is adsorbed onto oxide shell of ZVI-NPs. The formed As(III), is either adsorbed or coprecipitated at ZVI-NPs surface. Moreover, some As(III) is also oxidized to As(VI) by hydroxyl radicals or by iron oxides via formation of iron oxide As(III) surface complexes. Arsenic species on iron nanoparticles adsorb on ZVI-NPs by forming inner-sphere complexes with the (hydr)oxide shell or diffuse through the oxide layer to form an As–Fe intermetallic phase adjacent to the Fe(0) core. The reported results imply dual reduction–oxidation function of ZVI-NPs owing to its core–shell composition (Yan et al., 2010a,b, 2012).

It is mentioned that As(III) oxidation could be performed by hydroxyl radicals. The formation of the oxidant species in the reductive media provided with ZVI-NPs imparts the oxidative aspect of ZVI-NPs technology. In fact, in the presence of oxygen, ROS including hydrogen peroxide (H_2O_2), hydroxy (·OH) and peroxyl ($HO_2^{\cdot}$) radicals are produced via a two-electron reduction of dissolved oxygen by Fe^0 results in hydrogen peroxide formation. Then, hydrogen react with ferrous ion (Fenton reaction) and produces hydroxyl radicals. In neutral conditions, ferrous ion may react with dissolved oxygen directly to produce superoxide anion radical (O_2^-) (Keenan and Sedlak, 2008; Joo et al., 2005). Because of small size, large surface area, and subsequently high corrosion rate of ZVI-NPs, these are emerging heterogeneous Fenton reagents. The combination of dual redox behavior of ZVI-NPs results in increasing the degradation of persistent organic pollutants, such as azo dyes (Xu and Wang, 2011).

Nitrate is one of the serious contaminants of groundwater supplies which both developed and developing countries are facing. It naturally occurs in the nitrogen cycle that is the stable form of nitrogen for oxygenated systems. Due to its high solubility in water, nitrate is a widespread groundwater contaminant and introduced to water supplies from chemical fertilizers, pesticides, animal-feeding operations, petroleum products, and waste contamination through storm and urban runoff (Bhatnagar and Sillanpää, 2011). Because of the toxic effects of nitrate on humans including induction of "blue-baby syndrome" and the potential formation of carcinogenic nitrosamines, WHO (2004) recommended a maximum concentration of nitrate–nitrogen (NO_3^-–N) 10 mg/L in drinking water.

ZVI-NPs has a great efficacy to effective denitrification by its reduction. It has been shown that nitrate is mostly transformed

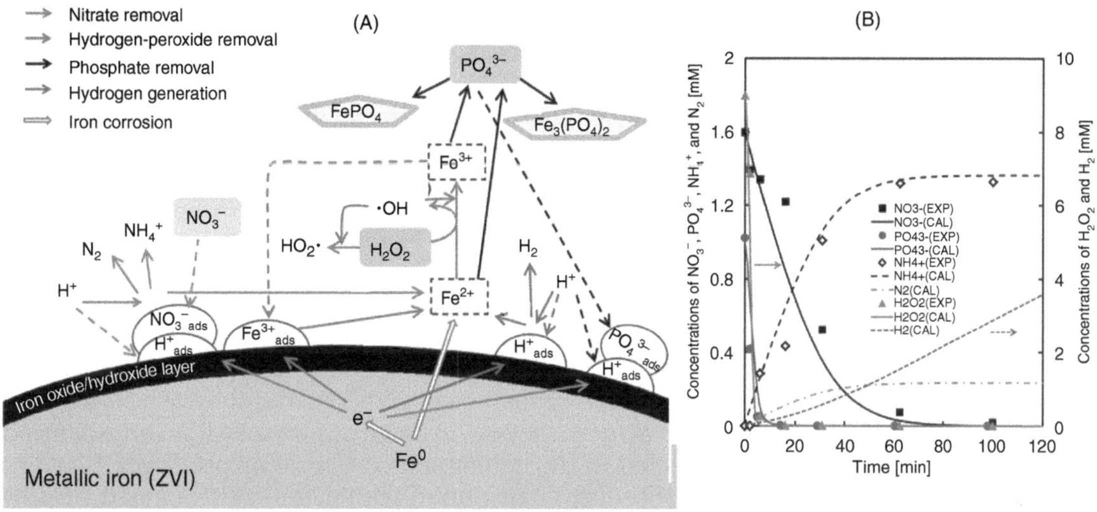

Figure 4.5. (A) Simultaneous mechanisms for the removal of nitrate, hydrogen peroxide, and phosphate by zero valent iron and (B) efficiency of ZVI in the removal of nitrate and phosphate. [Reprinted from (Yoshino, H., Kawase, Y., 2013. Kinetic modeling and simulation of zero-valent ironwastewater treatment process: simultaneous reduction of nitrate, hydrogenperoxide, and phosphate in semiconductor acidic wastewater. Ind. Eng.Chem. Res. 52, 17829–17840.), copyright (2013), with kind permission from American Chemical Society]

into ammonia. Nitrate could be absorbed onto the ZVI-NPs surface, and then reduction and desorption followed in consecutive steps (Ahn et al., 2000; Yoshino and Kawase, 2013).

Yoshino and Kawase developed a kinetic model for simultaneous removal of nitrate and phosphate from real-semiconductor acidic wastewater by ZVI with considering the redox dual behaviour of ZVI. Their kinetic model was based on assuming the following steps which are presented in Fig. 4.5A: (1) corrosion of ZVI; (2) competitive adsorption of proton, nitrate, ferric ion, and phosphate at the active site on the iron; (3) degradation of adsorbed nitrate; (4) reduction of adsorbed proton or generation of hydrogen gas; (5) recycling of ferric ion at the iron surface; and (6) decomposition of hydrogen peroxide. The developed model could adequatly simulate the removal of nitrate and phosphate confirming the supposed mechanisim in Fig. 4.5B.

It has been proven that ZVI-NPs possess high efficacy in the wide-range contaminants species removal from simple synthetic solutions. However, less work has been performed for investigation of ZVI-NPs performance chemically for complex and/or real cases. Surprisingly, limited number of field investigations have reported an unexpected need to provide supplementary treatment processes because of poor ZVI-NPs efficiency. We mean that more studies for practical application are needed to examine the reversible nature of remediation in complex and/or natural water using nZVI.

The foundation of treatment technology utilizing ZVI-NPs was pioneered in the 1990s. Gillham and O'Hannesin (1994) reported that ZVI was able to reduce a group of halogenated aliphatic compounds in groundwater which led to the development of iron permeable-reactive barriers (PRBs). A permeable reactive barrier (PRB) is an engineered zone including vertical trenches filled by ZVI. PRBs are placed in the flowpath of an underground contaminant zone and treat contaminated groundwater. All geologic, hydraulic, and chemical parameters must be determined to implement to perform an effective site evaluation. Due the unique properties of ZVI-NPs, diminutive size, and enhanced reactivity, they could be incorporated within a suspension and injected to the contaminated plume by pressurized feed or gravity effectively transported within the contaminated sites by the flow of groundwater.

3.3 Permeable Reactive Barrier Technologies

PRB systems filled with ZVI and ZVI-NPs could remove chlorinated organics (by reductive dechlorination) metals, metalloids, and radionuclides may be removed via reductive precipitation, surface adsorption, complexation, or coprecipitation with the Fe oxyhydroxides which we will discuss (Cundy et al., 2008; Crane and Scott, 2012). Fig. 4.6 shows a schematic of PRBs technology with zero ZVI-NPs which is designed for mobile or immobile contamination (Tratnyek and Johnson, 2006).

The significant loss of porosity and permeability within PRBs systems has been reported due to the formation of mineral precipitates (such as aragonite, calcite, and siderite) derived from reactions with local groundwater. Also, iron corrosion products, such as goethite, mackinawite, and green rusts are other challenges (Liang et al., 2005; Morrison et al., 2006; Henderson and Demond, 2007; Wilkin et al., 2005). However, formation of secondary iron-corrosion byproducts, such as iron sulfides and ferrihydrite within ZVI-NP PRBs may enhance their removal efficiency of various contaminants by mentioned mechanisms (Henderson and Demond, 2007).

In laboratory research works, operational parameters including the ZVI-NPs size and amount, pH, contact time, water composition (various anions and cations existence), pollutant concentration, and so on. are widely investigated and optimized. In practical application including injection of ZVI-NPs in contaminated underground zones, various operational parameters must be considered

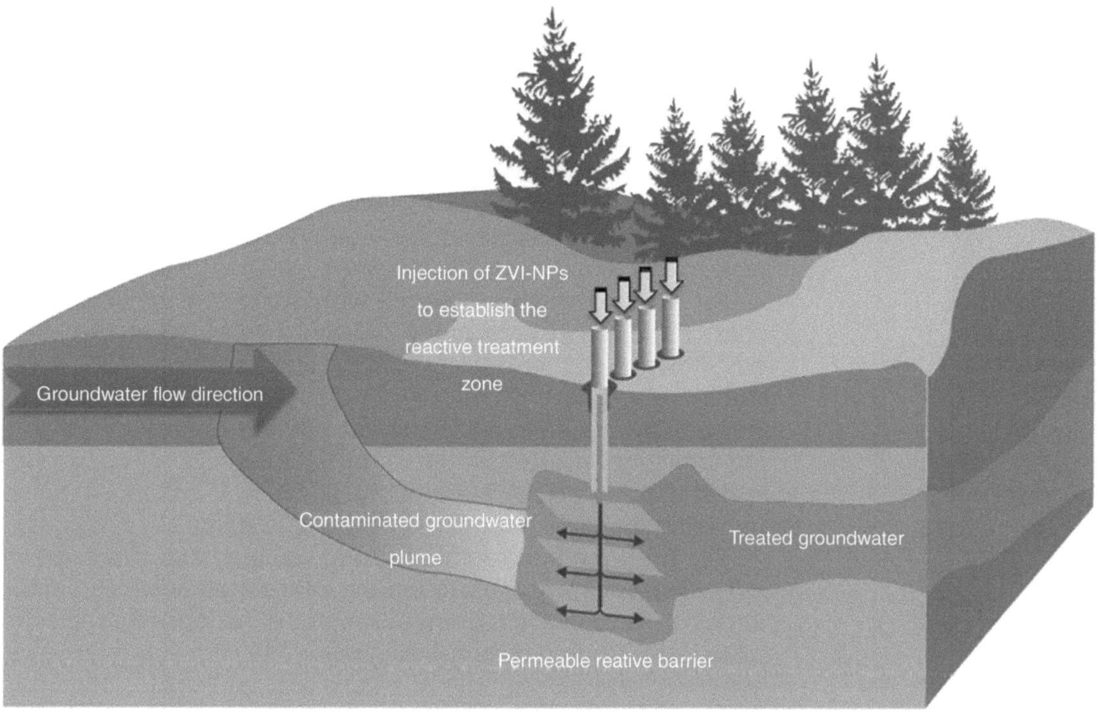

Figure 4.6. Permeable reactive barrier utilizing ZVI-NPs injection technology for treating contaminated zones.

in ZVI-NPs deployment because the structure, geochemistry, and hydrogeology of each polluted site is unique. These parameters affecting ZVI-NPs performance are the particle size range mobility, reactivity and longevity, injection strategy, and the material's ecotoxicity (Crane and Scott, 2012).

3.4 Reactivity and Longevity of ZVI-NPs

It is logical that by lowering the ZVI size, its performance for decontamination is improved because of the higher surface area for reaction. In other words, the smaller ZVI-NPs would facilitate the performance for contaminant removal. Although smaller ZVI-NPs increase the removal performance, it reduces longevity. So, for optimal ZVI performance, the ZVI-NPs must be optimized taking into account reactivity and longevity. Other effective factor in insite application of ZVI-NPs is their low mobility which is caused by: (1) ZVI-NPs aggregation followed by gelation, (2) the formation of massive corrosion product of ZVI-NPs, and (3) attachment of ZVI-NPs to mineral surfaces and carbonaceous materials or via microbial removal (Kanel et al., 2008). Several solutions have been proposed to enhance the stability and mobility of ZVI-NPs, including modification of the ZVI-NPs surface properties with polyelectrolytes or nonionic surfactants, immobilization on solid support, and dispersing the particles in oil–water emulsions. Modification of surface is implemented by surface coating of ZVI-NPs which improves the particle stability although reduces the reactivity. Polyelectrolyte specially anionic polymers are desirable because negative surface charges reduce the affinity of ZVI-NPs for the natural soil media (Petosa et al., 2010).

For zones which are polluted with dense nonaqueous liquid phase (DNAPL), dispersing the ZVI-NPs in oil–water suspensions was shown that is very effective (Berge and Ramsburg, 2009; Berge and Ramsburg, 2010). Typically, food-grade vegetable oil and surfactants have been used to prepare the emulsion for pilot and full-scale projects. In addition to the enhanced stability, it was revealed that the microbial reduction of TCE was enhanced due to providing the carbon source (Quinn et al., 2005).

The third method to enhance the stability and dispersion of ZVI-NPs is their immobilization on solid supports which enable the ZVI-NPs to use in continuous systems. Various types of solid materials including carbon-based materials, silica, clays, zeolites, bentonite, and montmorillonite have been utilized for preparing the composite by ZVI-NPs but none of them has been evaluated in real contaminated sites (Yan et al., 2013). As mentioned earlier, a certainty to increase the chemical reactivity of ZVI-NPs, a thin layer of nobler metals including Pd, Pt, Ag, Ni, Cu, and so on is deposited on iron which leads to a considerable increase of ZVI-NPs. The prepared bimetallics was found to be very effective. In bimetallic electrochemical couples, ZVI-NPs behave as a sacrificial anode which protects the noble metal. In bimetallic ZVI-NPs, both electron transfer through noble metal to contaminant and reduction via produced hydrogen act for decontamination. Among different bimetallic systems, it was reported that ZVI-Pd nanoparticles are extremely effective for chlorinated organic compounds removal and effectively reduce the decontamination of toxic intermediates (Huang et al., 2013).

Admittedly, the higher efficiency and rate of decontamination in bimetallic systems is due to the higher corrosion rate of Fe^0. Accordingly, the reactive longevity will be poor. The toxicity of noble metals is another problem. It seems that the bimetallic systems are desirable for insite application where short-time decontamination is concerned. Until 2012, no insite application utilizing bimetallic systems was reported in Europe (Mueller et al., 2012) while they are common in United States (Crane and Scott, 2012).

It is obvious that ZVI-NPs possess a high potential in decontamination of important and toxic water supply pollutant from both technical and economical views. However, the following challenges must be more investigated: (1) high reactivity with maintaining the longevity, (2) high mobility of particles and predicting the mobility of particles, and (3) the impact of heterogeneous nature of groundwater systems on insite applications and scaling up the laboratory findings to insite implementation.

4 Advanced Oxidation Processes with Nanostructured Photocatalysts

Advanced oxidation processes (AOPs) are defined as water and wastewater-treatment processes which involve the generation of hydroxyl radicals to effect water purification (Glaze et al., 1987). In water-treatment application, AOPs have been mainly targeted to remove and degrade taste and odor-causing compounds, natural organic matters, pesticides, hormones, and pharmaceuticals which are introduced into water supplies from agricultural and industrial activities in trace amount. As mentioned earlier, the existence of organics is not only directly toxic for health but also cause formation of DBPs. Degradation of residual toxicity in water resulting from organic contaminants, such as DBPs and disinfection are also an emerging application of AOPs. The main features of AOPs are the production of nonselective and powerful oxidizing agent hydroxyl radicals (E^0 = 2.8 eV) which react very rapidly with most inorganic and organic pollutants in drinking water. Compared to other advanced treatment processes used to treat chemical and biological persistent pathogens, AOPs possess a number of advantages in water which are listed as follows (Linden and Mohseni, 2014):

1. rapid decontamination for most contaminants
2. degradation of pollutant, not adsorping it
3. no generation of solid residuals
4. capability of virtual mineralization of the contaminant
5. no selectivity

However, some disadvantages including production of toxic byproducts, dependence on treated water quality, and capital intensive are challenging. In AOPs, hydroxyl radicals are commonly generated by homogeneous and heterogeneous processes listed in the following (Malato et al., 2013):

- Ozonation at high pH

  The included key reactions:

$$O_3 + OH^- \rightarrow O_2 + HO_2^- \tag{4.6}$$

$$O_3 + HO_2^- \rightarrow O_2 + HO^{\cdot} + O_2^- \tag{4.7}$$

- Catalytic ozonation in the presence of transition metal ions
- Ozone and hydrogen peroxide (O_3/H_2O_2)
- Fenton (Fe^{2+}/H_2O_2) and Fenton-like reactions (Fe^{3+}/H_2O_2)
  The included key reaction:

$$Fe^{2+} + H_2O_2 \rightarrow Fe^{3+} + HO^{\cdot} + HO^{-} \tag{4.8}$$

- UV irraduiation/H_2O_2
  The included key reaction:

$$H_2O_2 + h\upsilon \rightarrow 2HO^{\cdot} \tag{4.9}$$

4.1 Nanostructured Photocatalysts

In addition to the homogenous process, the heterogeneous process has been utilized for hydroxyl radical and reactive oxygen species generation. In a typical photocatalytic process, a semiconductor, such as TiO_2, ZnO, CdS, and so on, called nanostructured photocatalyst, is irradiated by light with sufficient energy leads to reactive radical generations via initiating a series of reactions (Amani-Ghadim et al., 2015). It is interesting that the photocatalytic properties of TiO_2, as first investigated photocatalyst, was first investigated by Akira Fujishima and applied for water splitting (Fujishima and Honda, 1972). A number of important advantages are exhibited including (1) ambient operating temperature and pressure, (2) complete mineralization of pollutants and their intermediates into CO_2 and other inorganics, and (3) low operating costs in decontamination with nanostructured photocatalyst. These important advantages attract researchers to extend photocatalyst feasible applications in water treatment (Malato et al., 2009). AOPs with nanostructured photocatalyst are basically designed and utilized for organic degradation. In addition to organic degradation, photocatalytic processes are the emerging disinfection methods for various types of pathogens (Dalrymple et al., 2010).

4.2 Mechanism of Photocatalytic Process

The photocatalytic processes is initiated with irradiation of photocatalyst suspension by photons having energy equal to or greater than photocatalyst band-gap energy and leads to excite the electron from valence band (VB) of nanostructured to its conduction band (CB) in femtoseconds and generate the electrons (e^-) and holes (h^+) (called charge carriers) in conduction and VBs, respectively (Mahyar and Amani-Ghadim, 2011). The band-gap energy for TiO_2 with anatase phase and ZnO nanostructures are 3.2 and 3.3 eV, respectively which are related to UV-light wavelength range (Amani-Ghadim and Dorraji, 2015).

- Excitation:

$$\text{Photocatalyst} + h\nu \rightarrow h^+ + e^- \quad (4.10)$$

The generated charge carriers could recombine and emit light or heat:

- Recombination:

$$h^+ + e^- \rightarrow \text{heat} \quad (4.11)$$

Also, charge carriers may migrate to the catalyst surface. In these conditions, electron-hole recombination will be immediately supressed after photon excitation and they survive enough long to participate in redox reactions (Furube et al., 2001). At the photocatalyst surface, the generated holes are trapped by adsorbed water molecules (onto photocatalyst support) and/or adsorbed hydroxyl ions lead to generate the hydroxyl radical:

- Trapping:

$$H_2O_{ads} + h^+ \rightarrow {}^{\cdot}OH_{ads} + H^+ \quad (4.12)$$

$$OH^-_{ads} + h^+ \rightarrow {}^{\cdot}OH_{ads} \quad (4.13)$$

Moreover, the generated holes could be trapped by electron donors, especially organic pollutants, which cause to direct oxidation of organic pollutants. In the presence of dissolved oxygen molecules, known as electron scavengers, generated electrons are trapped by the oxygen.

$$O_{2(ads)} + e^- \rightarrow O_2^{\cdot-} \quad (4.14)$$

The result of Eq. (4.14) is the formation of superoxides radical ($O_2^{\cdot-}$) which can be further protonated to produce ROS, such as peroxyl radical and hydrogen peroxide:

$$O_2^{\cdot-} + H^+ \rightarrow HO_2^{\cdot} \quad (4.15)$$

$$O_2^{\cdot-} + 2H^+ \rightarrow H_2O_2 \quad (4.16)$$

The formation of various types of radicals provides very strong oxidative media and effective degradation of the contaminants.

- Radical attack:

$${}^{\cdot}OH_{ads} \text{ or } O^{\cdot-}_{2(ads)} + R \rightarrow \text{intermediate} \quad (4.17)$$

$${}^{\cdot}OH_{ads} \text{ or } O^{\cdot-}_{2(ads)} + \text{intermediate} \rightarrow \text{final degradation products} \quad (4.18)$$

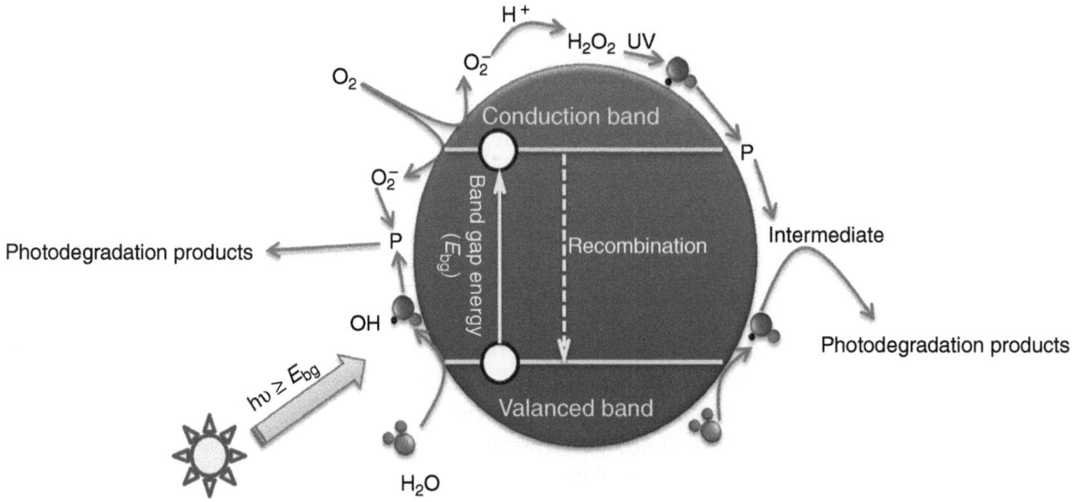

Figure 4.7. Mechanism of photoinduced electron-hole pair formation in a semiconductor and degradation of water pollutant (P).

Fig. 4.7 summarizes the mentioned mechanism. It should be considered that all of these reactions occurred in the presence of both oxygen and water molecules. In the absence of water molecules, the highly oxidative hydroxyl radicals could not be generated. Accordingly, the photoexcited electron recombines with the VB hole with simultaneous dissipation of heat energy or generated holes could react with some simple organic compounds (e.g., oxalate and formic acid) (Byrne and Eggins, 1998).

There is a great interest to prepare the photocatalyst in nanodimensions. Nanostructured photocatalysts in various morphologies, such as nanosphere (zero dimensional), nanofibers, tubes, and rods (one dimensional) and nanosheets (two dimensional) display a large surface area-to-volume (Nakata and Fujishima, 2012). This important feature of nanostructured photocatalysts elevate the oxidation capability and the efficient charge separation and trapping at the physical surface (Nagaveni et al., 2004a,b; Siddiquey et al., 2008).

4.3 Challenges in the Development of Photocatalytic Water-Treatment Process

Although the conventional photocatalytic processes, mostly utilizing TiO_2 and ZnO, are implemented in ambient operational conditions, the main challenge is the cost efficacy arises by utilizing the UV illumination due to the relatively large band gap of titanium dioxide and zinc oxide. Accordingly, the necessity of UV irradiation will be accompanied by high operating costs and it is very attractive to employ plentiful solar and/or visible irradiation. Furthermore, the relatively rapid e^- and h^+ recombination waste the energy and reduce quantum yield. Quantum yield is defined as the number of molecules being decomposed per photon absorbed (Daneshvar et al., 2007). Accordingly, the main drawback of common photocatalysts, especially TiO_2, is the relatively large band-gap which only allows the absorption of UV light (<5% of solar energy).

The nanostructured photocatalysts are commonly utilized in suspension state in slurry reactors caused to introduce the pollution of water with nanostructures. In slurry application, the necessity of additional separation of catalyst step, such as sedimentation, crossflow filtration and membrane filtration for avoiding the loss of catalyst is another challenge (Chong et al., 2010).

4.3.1 Modification of Nanostructured Photocatalyst for Absorption of Sun and Visible Light

To design and develop the photoresponse nanostructured photocatalyst for visible spectrum, considerable researches have been reported. These attempts improve the photocatalytic activity through coupling different semiconductor photocatalysts, modifying the surface of the semiconductor with metals and nonmetal-ion doping, formation of composite with CNTs and coupling of photocatalyst with photosensitizers (Di Paola et al., 2012; Khodam et al., 2015).

CNTs facilitate the charge transfer due to their unique electronic properties associated with their special one-dimensional structure. CNTs coupling with photocatalyst increase the longevity of the generated charge carriers via transfer of electrons from the CB of photocatalyst to the CB of the CNTs (Di Paola et al., 2012). It is also recommended that electron transfer from CNTs to the CB of TiO_2, acting as photosensitizer, results in enhanced generation of super oxide anion radicals. The photogenerated electrons and holes initiate the formation of hydroxyl radicals which are responsible for the degradation of the contaminants (Wang et al., 2005). The proposed mechanism justifies the activity of CNTs–TiO_2 composite under visible-light irradiation.

The doping of photocatalysts with transitional and rare earth metal ions have been widely conducted to enhance the photocatalytic activity. In this case, the photocatalyst is excited by visible light due to incorporation of metal ions into the photocatalyst lattice leading to the creation of impure energy levels in the photocatalyst band gap (Malato et al., 2009). Doping of TiO_2 with nonmetals, especially nitrogen, is one of the common methods to activate the nanostructured photocatalysts under visible-light irradiation which was proposed by Sato in 1986 (Sato, 1986). Upto date, various nonmetals, such as nitrogen, fluorine, carbon, and so on are doped in TiO_2 to activate by visible light. They mainly enhance the photocatalytic activity and the mechanism is not exactly known. However, two common mechanisms about the photoactivity of the TiO_2 doped with nonmetals are:

1. N-TiO_2 could absorb the visible light because the energy of N 2p state close with O 2p states in TiO_2 doped with nitrogen, subsequently the band gap is narrowed (Asahi et al., 2001).
2. TiO_2 oxygen site is substituted by nitrogen atom forming isolated impure energy levels above the VB illumination with visible light only excites electrons in the impurity energy level (Irie et al., 2003).

Coupling of TiO_2 photocatalyst with photosensitizers is one of the most-studied methods to photoactivate the wide band-gap photocatalysts by visible irradiation widely used for energy conversion in solar cells. In this system, the photosensitizer typically is the dye preadsorbed onto the TiO_2 surface and excited by visible-light irradiation. The dye sensitization is initiated by the photoexcitation of dye molecules. The process is followed by electron transfer from the excited dye to CB of TiO_2 resulting in the generation of cation radical. When suitable electron donors are available, the oxidized dye is subsequently regenerated. Otherwise, the cation radical could be decomposed itself. The electron existing in the CB of the TiO_2, would be transferred to the electron acceptor, mainly oxygen, to produce ROS, such as superoxide anion radical. Various dyes including ruthenium bipyridine derivatives, metal porphyrins, and metal-free organic dyes have been tested for the sensitization of TiO_2 (Zhang et al., 2014a). Other photosensitizers, such as catechol, EDTA, and phenolic compound are evaluated for photsensetization TiO_2 by visible light and it is revealed that they act in other mechanism named ligand-to-metal charge transfer (LMCT) mechanism. In the LMCT sensitization, the electron is directly photoexcited from the ground state adsorbate (without involving the excited state of the adsorbate) to CB of TiO_2 (Fig. 4.8A) (Zhang et al., 2014a).

Semiconductor coupling with other semiconductor is a widely studied method used to utilize visible light and improve photocatalytic efficiency. In this method, a large band-gap semiconductor is coupled with a small band-gap semiconductor with a more-negative CB level. CB electrons could be transferred from the small band-gap semiconductor to the large band-gap semiconductor leading to a wide electron-hole separation achievement (Fig. 4.8B) (Malato et al., 2009).

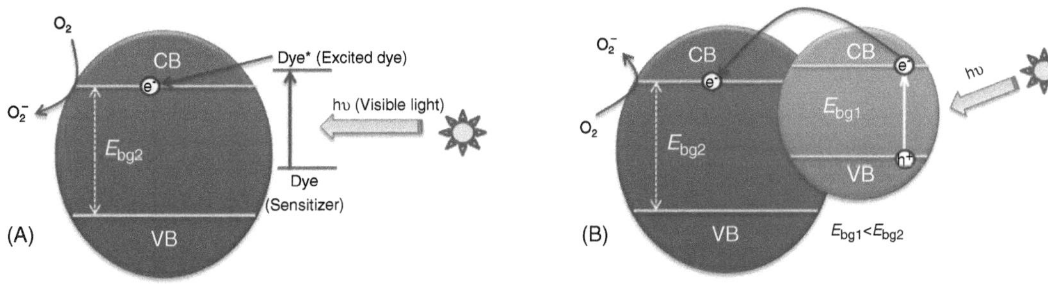

Figure 4.8. The excitation mechanism in (A) photocatalysis with photosensitizers and (B) semiconductor-semiconductor photocatalyst.

4.3.2 Nanostructured Photocatalyst Immobilization onto Inert Supports

As mentioned, the nanostructured photocatalyst is generally used in the form of powder suspended in slurry reactor. This type of application requires photocatalyst separating and recovering step at the end of photocatalytic-treatment operation. This application challenge is mainly resolved via immobilizing the photocatalyst on an inert support. Numerous inert supports, such as glass, quartz, concrete or ceramics, nonwoven paper, and cement are utilized for immobilizing the TiO_2 and TiO_2 supported on glass (as a self-cleaning glass) and nonwoven paper are commercially available (Khataee et al., 2010, 2011; Malato et al., 2009). However, the immobilization of photocatalyst on the supports suffers from the low degradation rate. The immobilization of photocatalyst reduces the number of active sites and prevents the effective photon penetration. Accordingly, the supported photocatalysts have an operational difficulty in large-scale treatment systems.

4.4 Photocatalytic Reactors

The design of suitable reactor is an important step in the practical application of photocatalyst in water-treatment plants on a large scale. From the point of photocatalyst state, two main types of reactors are (1) the slurry photoreactors which operate with suspended photocatalyst and (2) fixed-supports reactors which employ the supported photocatalysts. The slurry reactors require a separation unit for the recovery of photocatalyst particles. The surface area of catalyst per unit volume and light distribution are the main factors affecting the reactor efficacy. However, other factors, such as reactor-field configuration (series or parallel), treatment-system mode (batch or continuous), flow rate, pressure drop, pretreatment, catalyst- and oxidant-loading method, pH control must be considered. The annular slurry photoreactor, cascade photoreactor, downflow contactor reactor, and so on are samples of designed reactors (Chong et al., 2010). In the case of photoreactors which are designed to apply with solar light, the reactor is equipped with several solar collectors (Malato et al., 2009).

5 Membranes for Water Purification Based on Nanotechnology

Owing to its low cost and high efficiency, membrane technology has dominated water purification technologies (Madaeni et al., 2015). Membrane-based water purification processes are now among the most important and versatile technologies for conventional drinking-water production, wastewater treatment, ultrapure water production, desalination, and water reuse. Nanotechnology promises to dramatically enhance water purification membrane technology (Hoek et al., 2014). This section presents a review of six emerging nanotechnology-based membrane intended for use in water purification.

5.1 Nanostructured Photocatalytic Membranes

Based on the ISO definition, membranes with pores in the range of 1–100 nm would clearly be considered as nanostructured materials. Hence, ultrafiltration and most nanofiltration membranes that reject particles of about 0.2–200 kDa corresponding to a nanoscale diameter of around 1–100 nm should be classified as nanostructured membranes (Bhadra and Mitra, 2013; Muellera et al., 2012; Madaeni et al., 2015). A promising method for solving problems concerning separation of the nanophotocatalyst as well as products and by-products of photodegradation from the solution is application of nanostructured photocatalytic membranes (Mozia, 2010). Photocatalytic membranes have shown great potential for use in water purification since they combine the physical separation of membrane filtration, and the organic degradation and antibacterial property achieved by nanophotocatalyst in a single unit (Leong et al., 2014). There are two different configurations of photocatalytic membrane reactor: membranes with immobilized nanophotocatalyst and a slurry reactor coupled with a membrane filtration unit. In the first configuration, nanophotocatalysts immobilized on the membrane surface or within the membrane pores become part of the membrane structure (Liu, 2013). A variety of materials including polymers (such as polyamide, polyvinylidenefluoride, polyethersulfone, polyurethane, polyethylene terephthalate, polyester, polyacrylonitrile, and polytetrafluoroethylene) ceramics (e.g., silica, alumina, and zeolite) have been used as supports for the preparation of photocatalytic membranes. The ceramic membranes are more stable than polymeric membranes under UV irradiation with superior resistance against chemical attacks by free radicals generated in photocatalytic reactions (Pendergast and Hoek, 2011). Nanophotocatalysts can be either coated onto the polymeric membrane surface (Rahimpour et al., 2011; Barni et al., 1995) or dispersed in the polymer solution for membrane casting (Alaoui et al., 2009; Molinari et al., 2004). In order to prepare photocatalytic ceramic membranes, sol dip-coating procedure is widely used (Ma et al., 2009; Choi et al., 2007; Leong et al., 2014). In addition to photocatalytic composite membranes, preparation of pure nanostructured photocatalytic membranes has also been reported (Hu et al., 2011; Liao et al., 2012; Albu et al., 2007).

In nanostructured photocatalytic membranes, the photocatalysis, disinfection, separation, and antifouling functions are simultaneous (Choi et al., 2009). Antibactericidal effect of nanostructured photocatalytic membranes was greatly investigated. Most of the photocatalytic membranes showed good disinfection performance. The antibacterial effect was due to the presence of ROS which are the photocatalytic reaction products under UV irradiation (Rahimpour et al., 2011; Damodar et al., 2009; Leong et al., 2014). Removal of 4 log of *E. coli* in 1.5 h of UV irradiation by a TiO_2–Al_2O_3 membrane prepared by sol dip-coating procedure has been reported (Choi et al., 2007). Damodar et al. (2009) found that the elimination of almost 100% of *E. coli* by TiO_2/PVDF membrane could be achieved after 1 min of UV irradiation.

Application of photocatalytic membranes is a promising method to inactivate viruses and other microorganisms in drinking water. Zheng et al. reported that removal of more than 5 log of bacteriophage f2 on average after 24 h by a slurry reactor containing TiO_2 nanoparticles coupled with a membrane-filtration unit under UV irradiation. Bacteriophage f2 in feed tank was 5.22 log. The f2 was primarily inactivated by hydroxyl radicals during the photocatalysis process, and the membrane served mainly as a barrier (Zheng et al., 2015). Natural organic matter is one of the concerning pollutants in the water purification system because it forms DBPs when in contact with chlorine in the disinfection process (Leong et al., 2014). According to Zhang et al. (2008), pure TiO_2 nanowire membranes prepared by hydrothermal method could remove nearly 100% humic acid and 93.6% total organic carbon in a continuous operation mode.

5.2 Electrospun Nanofiber Membranes

Electrospinning is a simple, efficient, and inexpensive way to make ultra fine fibers using various materials (e.g., polymers, ceramics, or even metals) (Li and Xia, 2004; Qu et al., 2013).

In early days, electrospinning technique was mainly used to produce polymeric nanofibers.

More than 100 polymers, both synthetic and natural, have been successfully electrospun into nanofibers, mostly from polymer solutions since any polymers may be electrospun into nanofibers, provided that the polymer molecular weight is sufficiently high and the solvent can be evaporated in time during the jet transit period over a distance between the spinneret and the collector (Feng et al., 2013). A schematic diagram to interpret electrospinning of polymer nanofibers is illustrated in Fig. 4.9.

In 2002, Shao and coworkers first fabricated silica nanofibers by using the electrospinning technique (Shao et al., 2002). Afterward, a large number of inorganic nanofibers have been prepared by the electrospinning way and the calcination process (Li et al., 2006; Dai et al., 2011; Lu and Wang, 2014). The heavy metals are quite often mixed with water resources and are also well distributed in the entire environment. Nanofibers could play a major role in the removal of these pollutants (Nabeela Nasreen et al., 2013). Sang et al. (2008) used nanofiber membrane prepared from chloridized polyvinyl chloride by electrospinning process for the removal of divalent metal cations (Cd^{2+} and Pb^{2+}) from the simulated groundwater. Pant et al. (2014) reported that silver-doped fly ash/polyurathene spider-web-like electrospun nanofibers could be used in adsorptive removal of arsenic, destructive removal of

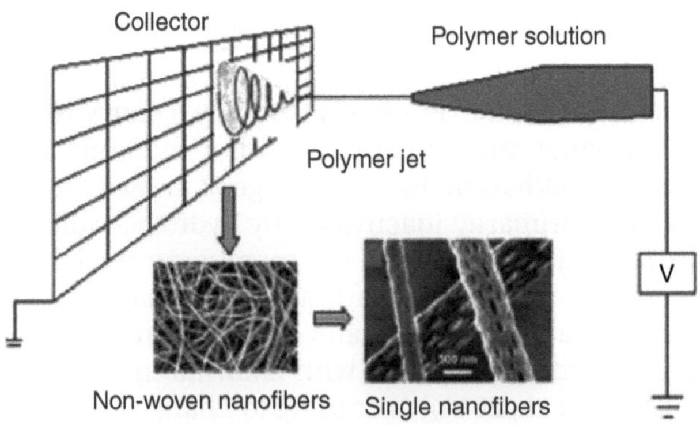

Figure 4.9. Schematic of electro-spinning of nanofiber. [Reprinted from (Huang, Z.-M., Zhang, Y.-Z., Kotaki, M., Ramakrishna, S., 2003. A reviewon polymer nanofibers by electrospinning and their applications innanocomposites. Compos. Sci. Technol. 63, 2223–2253.), copyright (2003), with kind permission from Elsevier]

microorganisms, and filtration removal of particulate impurities through a single membrane using continuous filtration process.

Taha et al. (2012) reported the preparation of amine-functionalized cellulose acetate/silica composite nanofiber. The amine-functionalized nanofibers, due to the electrostatic interaction and chelation process, enables adsorption of chromium(VI). Nanofibers are presently investigated as potential membranes for desalination due to their improved flux performance (Nabeela Nasreen et al., 2013; Feng et al., 2013). Feng et al. (2008) tried to use electrospun nanofibers for desalination by membrane distillation. Polyvinylidene fluoride nanofiber membrane could produce potable water with NaCl concentration less than 280 ppm from a saline water containing 6 wt.% NaCl. Prince et al. (2012) reported the application of poly(vinylidene fluoride)-clay nanocomposite nanofiber membranes with greater than 99.95% salt rejection in a contact membrane distillation process. There is another application for electrospun nanofibers in water purification, which is to use, such as filter for the removal of nanoparticles, including waterborne bacteria, for example, *E. coli* in drinking water. For example, a nanofibrous microfiltration membrane system, fabricated from surface-functionalized polyacrylonitrile electrospun nanofibers, was demonstrated for the effective removal of bacteria (e.g., *E. coli*) and bacteriophage (e.g., *MS2*) from contaminated water. For membrane modification, dualvinyl and trivinyl monomers were polymerized on the surface of the polyacrylonitrile electrospun nanofibers, which also improved the membrane mechanical properties due to the formation of a crosslinkage (Ma et al., 2014). The electrospun nanofibers of inorganic semiconductors could be applied in the field of photocatalysis. Also, the electrospun polymeric nanofibers with large surface area and porous structure provide a platform to support semiconductor nanostructures for photocatalytic applications (Lu and Wang, 2014). The photocatalytic performance of the inorganic electrospun nanofibers was significantly relying on the morphology, size, structure, crystal form, and chemical composition (Li et al., 2012). Zhang et al. (2009) investigated the photocatalytic efficiencies of TiO_2 continuous and porous nanofiber membranes and suggested that this membrane has potential applications for environmental purifications. Sundaramurthy et al. (2012) reported the photocatalytic properties of the electrospun Fe_2O_3 nanofibers under the visible light. Liu et al. (2012b) synthesized TiO_2 nanofibers (Fig. 4.10) using electrospinning with the addition of Ag, which indicates photocatalytic activity under solar light.

Liu et al. (2007) reported the preparation of TiO_2/SnO_2 composite nanofibers through a side-by-side dual spinneret

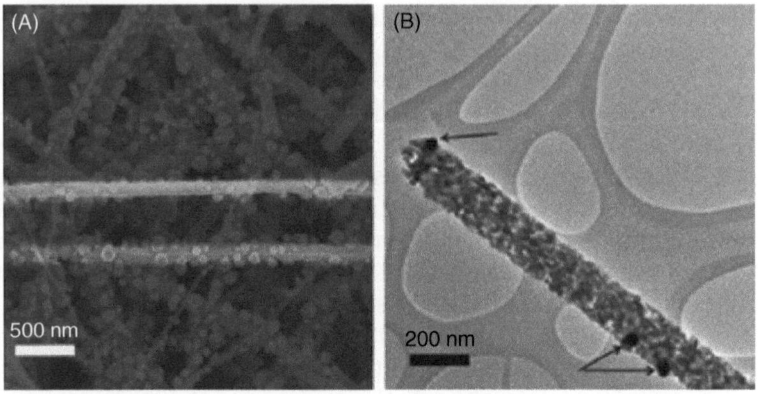

Figure 4.10. (A) SEM and (B) TEM images of electrospun TiO_2 nanofibers with Ag. [Reprinted from (Liu, L., Liu, Z., Bai, H., Sun, D.D., 2012b. Concurrent filtration and solarphotocatalytic. disinfection/degradation using high-performance Ag/TiO_2 nanofiber membrane. Water Res. 46, 1101–1112.), with Permission from Elsevier]

electrospinning process. The photocatalytic activity of as-spun TiO_2/SnO_2 composite nanofibers was superior to the pure TiO_2 nanofibers, because of increased charge separation of the photogenerated electrons. It is well known that SnO_2 has a band-gap energy of 3.8 eV and a low-CB level than TiO_2, thus the photogenerated electrons will flow from TiO_2 to SnO_2, while the holes will move from SnO_2 to TiO_2, resulting in an efficient separation of electron/hole pairs and a high quantum efficiency (Lu and Wang, 2014; Daghrir et al., 2013). Preparation of Ag/AgCl-coated electrospun polyacrylonitrile nanofibers by electrospinning process followed by an electroless plating procedure has been reported (Lei et al., 2011). The as-prepared Ag/AgCl/polyacrylonitrile nanofibers exhibited an excellent photocatalytic performance under visible-light irradiation. The composite nanofibers also indicated a long-term stability and recycling property, which could be used for practical applications.

5.3 Carbon Nanotube Membranes

CNTs are composed of cylindrical graphite sheets rolled up in a tube-like structure with the appearance of latticework fence (Thostenson et al., 2001).

CNTs are attracting advanced membrane technologies for water desalination since they provide low-energy solution for water treatment. CNT membranes provide near frictionless water flow through them with the retention of a broad range of water pollutants. The inner hollow cavity of CNTs provide a great possibility for desalinating water. The high aspect ratios, smooth hydrophobic walls, and inner-pore diameter of CNTs allow ultraefficient transport of water molecules (Das et al., 2014). Carbon nanotube-based membranes may also have longer lifetime than conventional membrane materials due to the excellent mechanical properties that CNTs exhibit (Pendergast and Hoek, 2011). Carbon nanotube-based membranes performance often rely on their processing and fabrication procedures. Currently, two types of carbon nanotube membranes are available: (1) vertical aligned and (2) mixed matrix carbon nanotube membranes (Ahn et al., 2012). Using molecular dynamic simulation, the desalination potential of a virtual-aligned carbon nanotube membrane was estimated according to the size of the nanotubes (Corry, 2011). The results indicated that desalination capacity decreased as pore size increased. Based on the molecular dynamic simulation results, the inner diameter of nanotubes should be 0.6 nm, if the desalination capacity of a virtual aligned CNT membrane is to correspond with to that of a reverse osmosis membrane (Corry, 2008, 2011). Currently, CNTs can be produced with an inner diameter of 1–2 nm. Although a narrower carbon nanotube pore size is essential for achieving enhanced rejection efficiency desalination, it may not be practical with current technology. It seems that virtual aligned CNT membranes will have salt-rejection performances similar to those of commercialized nanofiltration membranes. Ultrafine CNTs (inner diameter ~0.6 nm) should be employed and their surface charge should be elevated to enhance desalination capacity to the level of a reverse osmosis membrane (Ahn et al., 2012). The potential of CNT filters in the removal of microbial contaminants, namely, *E. coli* (2–5 μm), *Staphylococcus aureus* (~1 μm), and the *poliovirus* (~25 nm) from drinking water was reported by Srivastava et al. (2004).

Shawky et al. (2011) prepared MWCNT/aromatic polyamide nanocomposite membrane. The nanocomposite membrane synthesized with 15 mg/g MWCNT in a 10% aromatic polyamide casting solution rejected NaCl and humic acid by factors of 3.17 and 1.67, respectively, relative to the polyamide membrane without MWCNT. However, there may be a reduction in longer term membrane performance due to adsorptive fouling since the addition of MWCNT increased membrane hydrophobicity. Though carbon nanotube-based membranes have shown great promises to serve as an excellent aid in desalination and water purification applications, however, there are numerous challenges involved with each step of membrane making starting from growth of CNTs to membrane performance evaluation and scale up (Kar et al., 2012). The safety of the carbon nanotube membranes is also an important issue to be resolved. Concerns will be raised about the environmental and health impacts of CNTs that may be released from CNT-based membranes during operation (Lam et al., 2006).

5.4 Graphene Membranes

One of the attractive new additions into the carbon family is graphene, the one-atom thick sheets of carbon (Geim and Novoselov, 2007; Gupta et al., 2012). Since Geim and Novoselov were awarded the Nobel Prize in Physics in 2010, researches on graphene have developed rapidly in both academic and industrial applications (Sengupta et al., 2011).

Graphene can be defined as one-atom thick two-dimensional sheets, consisting of sp^2-bonded carbon atoms arranged in a hexagonal, honeycomb lattice (Mahmoud et al., 2015). On the other hand, graphene oxide is regarded as a single two-dimensional asymmetrical carbon sheet with one-atom layer thick and has abundant oxygen containing functional groups (e.g., hydroxyl, carboxyl, carbonyl, and epoxy groups) on the edge and basal plane (Gonçalves et al., 2010). Nanopores could be introduced into the structure of graphene with the unsaturated carbon atoms at the pore edge passivated by chemical functional groups (Hegab and Zou, 2015). Forming nanopores of varying sizes in a monolayer graphene can be achieved by electron-beam exposure, ultraviolet-induced oxidative etching, diblock copolymer templating, and helium ion-beam drilling (Mahmoud et al., 2015). A number of studies have shown that membranes based on graphene with artificial nanopores have great promise for nanofiltration and reverse osmosis-based seawater desalination (Cohen-Tanugi and Grossman, 2014; Cohen-Tanugi and Grossman, 2012). In the membrane fabrication, graphene-based materials could potentially be applied to (2) freestanding membranes, (2) on the surface of membranes, or (3) through incorporating casted membranes. The first procedure utilizes graphene-based materials as a separating layer directly, the second employs them for membrane-surface modification and in the third procedure, graphene-based materials are incorporated into a polymer matrix (Hegab and Zou, 2015).

Incorporating graphene-based materials in the polymer matrix is a promising method, in applying these materials in the desalination membranes, rather than surface coating (Stankovich et al., 2006). Recently, graphene-based membranes open the door for ideal next-generation membranes as a cost effective and sustainable alternative to the long-existing thin-film composite membranes for water-purification applications. This can be attributed to the unique properties of graphene-based materials including distinctive structural characteristics (Ruan et al., 2012), high mechanical strength (Leenaerts et al., 2008) and negligible thickness (Lu and Huang, 2009). The advancement in molecular simulation of graphene family opens the door for their potential contribution in developing novel membrane desalination technologies. For example, the efficiency of NaCl rejection by nanoporous graphene membrane was simulated using a molecular dynamics modeling (Cohen-Tanugi and Grossman, 2012). While the pores in the structure of graphene must exceed a critical size in order to permeate water molecules; they must also be narrower than a maximum diameter in order to effectively hinder the passage of salt ions. The results of this research for permeate salinity suggest that this maximum diameter is around 5.5 Å, that is, the majority of salt ions approaching the pore entrance are able to pass through the membrane beyond this diameter. This study concluded that one layer of nanoporous graphene can successfully remove salt from water with rates ranging from 10 to 100 $L/cm^2/day/MPa$, which is two to three times greater than can be achieved using the RO diffusion membrane. This work highlights the promise of atomically thin, periodic nanostructures, such as graphene for water desalination (Cohen-Tanugi and Grossman, 2012).

Introducing charged species to nanopores in the structure of graphene by oxidation and other chemical groups can prove beneficial in desalination (Loh et al., 2010). Evidently, graphene nanopores subjected to negatively charged functional groups exhibit excellent rejection of salt ions (Han et al., 2013). When dealing with seawater, charged nanopores in graphene proved advantageous in the desalination process using electro dialysis, as the nanopores had a much lower energy requirement compared to current reverse osmosis membranes or multistage flash distillation (Oren, 2008; Mahmoud et al., 2015). The removal of bacteria and other organisms from water is an important process, not only for water purification but also biofouling control in membrane filtration. The graphene-based membranes indicate superior antifouling properties against common waterborne bacteria because of antibacterial activity of graphene-based materials. The antibacterial activity of graphene-based materials is attributed to cellular membrane stress induced by sharp edges of graphene nanosheets, which may result in physical damages on cell membranes, leading to a loss of bacterial membrane integrity (Mahmoud et al., 2015). Sun et al. (2015) reported an antibiofouling membrane by fabricating graphene oxide–silver nanoparticles composite onto cellulose acetate membrane. The presence of graphene oxide–silver nanoparticles composite on the membrane exhibited a strong antibacterial activity, leading to an inactivation of 86% *E. coli* after contacting with the membrane for 2 h. This study may have great potential in developing high-performance antibiofouling membrane for membrane-separation processes. Preparation of graphene oxide-coated polyamide membranes has been reported by Choi et al. (2013) for reverse osmosis desalination application.

The prepared membrane indicated an inherent improvement in the membrane's resistance to foulants and chlorine ions. Integrating graphene oxide sheets with polymers was found to increase the hydrophilicity of the sheets, a much-desired property, while reducing the roughness of surface properties on the sheets. This, in turn, leads to improved antifouling properties. The strong chemical nature of graphene oxide allowed it to act as a protective layer for the membrane during chlorination, preserving the high degree of salt rejection (Choi et al., 2013; Mahmoud et al., 2015). Poly(vinylidene fluoride) membranes containing different concentrations of graphene oxide as an additive were prepared via a phase inversion method by Xia and Ni (2015). The results showed that graphene oxide greatly improved the hydrophilic character and permeability of the membrane due to oxygen-containing functional groups. Regarding parameters, such as pure water flux, hydrophilicity, and antifouling properties, the optimum concentration of graphene oxide was 0.5 wt.% in the casting solution. The results also revealed that the poly(vinylidene fluoride)/graphene-oxide membrane better removed natural organic matter especially relatively-low molecular weight components.

5.5 Zeolite Membranes

Zeolites are naturally occurring aluminosilicate minerals with highly uniform subnanometre and nanometre scale crystalline structures (Kumakiri et al., 2000). A few common zeolite materials employed in membranes include MFI-type, sodalite (SOD), and Linde Type A. Zeolite ZSM-5 (MFI) is the most commonly applied zeolite in membranes (Dyer, 1988). Typical zeolite membranes are amorphous silicate, aluminosilicate, or aluminophosphate crystalline structures formed via hydrothermal synthesis, in situ layer-by-layer crystallization and dry-gel conversion in the presence of a template-water vapor (Kazemimoghadam, 2010; Kumakiri et al., 2000).

In 2001, molecular dynamics simulations indicated that zeolite-based membranes previously applied for gas separations may be applicable for aqueous osmotic separations as well. Since then, nanozeolite membranes have been studied extensively for reverse osmosis desalination (Bhadra and Mitra, 2014). Zeolite crystals consist of a three-dimensional framework. This structure contains cavities that allow for the movement and containment of ions and water molecules. Pore size and framework density are the primary factors of concern when considering zeolites for water separations (Pendergast and Hoek, 2011).

Separations in zeolitic materials occur primarily through (1) molecular sieving, (2) competitive adsorption, or (3) ion exchange. Li et al. (2004) reported the application of MFI-type zeolite membranes (thickness: 3 mm) for reverse osmosis desalination (with 0.1 M NaCl feed solution at 2.07 MPa). Water flux was 0.112 $kg/m^2/h$ with 76.7% Na rejection. Nanozeolites are also the most frequently used dopants in thin-film nanocomposite membranes and have exhibited potential in enhancing membrane permeability (Qu et al., 2013). Lind et al. (2010) found that thin-film nanocomposite membranes doped with 250 nm nanozeolites at 0.2 wt.% achieved moderately higher permeability and better salt rejection (>99.4%) than commercial reverse osmosis membranes. Nanozeolites were also used as carriers for antimicrobial agents, which revealed antifouling property to the membrane (Qu et al., 2013).

The zeolite thin-film nanocomposite technology has reached the early stage of commercialization. QuantumFlux, seawater thin-film nanocomposite reverse osmosis membrane, is now commercially available (www.nanoH_2O.com).

5.6 Nanocomposite Membranes

A variety of nanomaterials have been used in the structure of membrane for successful fabrication of nanocomposite membrane. Reduced fouling via enhanced hydrophilicity is a goal of many nanocomposite studies. More details about performance of nanoenhanced membranes by CNTs, graphene, zeolite, and photocatalytic nanoparticles in water purification were presented in the previous sections. Here, we intend to provide an insight into nanoenhanced membranes by other nanomaterials (such as nanoclay, nanosilver, and nanosized metal oxides) and their applications in water purification.

Clay minerals are traditionally classified as a group of hydrated phyllosilicates, and may also be considered as hydroxides of silicon, aluminum, magnesium, or other metal ions, either natural or synthetic (Bergaya and Lagaly, 2013). Clay can be characterized as two-dimensional nanomaterial with anisotropic sheet-like morphology. Typically, clay can be divided into two broad groups: cationic clay and anionic clay. The cationic clay has negatively charged alumino silicate layers balanced by interlayer cations, which is mostly prepared from the nature minerals, while the anionic clay has positively charged metal hydroxide layers with an interlayer region containing anions, which is generally synthetic (Dawson and Oreffo, 2013; Vaccari, 1998; Dong et al., 2015). Cationic and anionic clays will possibly be very promising nanofiller materials for preparation of nanocomposite membranes with improved antifouling properties and separation performance by taking advantages of their charged properties and hydrophilic nature. For example, Dong et al. used Na-montmorillonite, the cationic clay and Mg/Al-layered double hydroxide, the anionic clay, as the inorganic nanofiller material to fabricate thin-film nanocomposite membranes for desalination (Dong et al., 2015). The salt rejection of Na-montmorillonite-polyamide thin-film nanocomposite membranes was all around 99.0% with different clay loading. At the 0.10 wt.% layered double hydroxide loading, the nanocomposite membrane showed the best salt rejection (99.3%). The results of membrane separation performance revealed that improved permeability and salt rejection have been obtained through the incorporation of nanoclays compared with pristine polyamide membrane. The antibacterial nature of silver nanoparticles brought them into the field of nanocomposite membranes. It was demonstrated that the adding of silver nanoparticles into the membrane matrix not only makes the membranes bactericidal, but also results in a significant improvement in virus removal (Madaeni et al., 2015).

Incorporation of antimicrobial nanomaterials, such as silver into membranes also offers an innovative potential solution to biofouling control (Li et al., 2008). Zodrow et al. (2009) found that silver nanoparticles incorporated into polysulfone ultrafiltration membranes exhibited antimicrobial properties toward a variety of bacteria, including *E. coli* K12 and *Pseudomonas mendocina* KR1, and the MS2 bacteriophage. Nanosilver incorporation also increased membrane hydrophilicity, reducing the potential for other types of membrane fouling. The influence of incorporation of nanoparticles, such as Fe_3O_4 (Huang et al., 2012), Al_2O_3 (Yan et al., 2009), SiO_2 (Jin et al., 2012), ZrO_2 (Zheng et al., 2011), and so on on the performance of nanocomposite membranes has been investigated. The results indicated the best performance in terms of pure water flux, rejection, and fouling resistance.

6 Nanoadsorbent for Water Purification

Nanomaterials are typically defined as materials smaller than 100 nm in at least one dimension. At this scale, materials often possess novel size-dependent properties different from their large counterparts, many of which have been explored for applications in water treatment (Qu et al., 2013). A very important one is that most of the atoms that have high chemical activity and adsorption capacity to many metal ions are on the surface of the nanomaterials (Kalfa et al., 2009; Liu et al., 2005). The surface atoms are unsaturated and are thus subject to combination with other element ions by static electricity (Hang et al., 2002). Therefore, nanomaterials can strongly adsorb many substances including trace metals (Cui et al., 2006) and polar organic compounds (Khajeh et al., 2013). Nanoadsorbents offer significant improvement with their extremely high specific surface area and associated sorption sites, short intraparticle diffusion distance, and tunable pore size and surface chemistry (Qu et al., 2013).

The potential of nanomaterials as nanoadsorbents and their many advances afforded in the water purification has been also outlined (Khajeh et al., 2013; Kyzas and Matis, 2015). More details about developments in nanoadsorbents and their performance in water treatment are presented in the following sections.

6.1 Metal-Based Nanoadsorbents

Metal-based nanomaterials, such as metallic nanoparticles, metallic oxide nanoparticles, and nanostructured mixed oxides have been explored to remove a variety of heavy metals, such as arsenic, lead, mercury, copper, cadmium, chromium, nickel, and have shown great potential to outcompete activated carbon (Sharma et al., 2009). These nanomaterials have higher adsorption capacity and faster kinetics because of the higher specific surface area, shorter intraparticle diffusion distance, and larger number of surface reaction sites (Qu et al., 2013). For instance, as the particle size of nanomagnetite decreased from 300 to 11 nm, its arsenic adsorption capacity increased more than 100 times (Yean et al., 2005). For example, a nanomaterial for fluoride removal from drinking water was prepared by coating Fe–Ti bimetallic oxide on magnetic Fe_3O_4 particles by coprecipitation. The adsorbent was a nanosized composite with a core–shell structure comprising a magnetic Fe_3O_4 core of 10–20 nm in diameter and an amorphous adsorbent shell of several nanometer thicknesses (Zhang et al., 2014b).

6.2 Carbon-Based Nanoadsorbents

Adsorption of water contaminants on CNTs is widely studied and extensively reviewed (Upadhyayula et al., 2009; Rao et al., 2007). CNTs as nanoadsorbent media, able to remove wide range of contaminant heavy metals, such as Cr^{3+} (Di et al., 2006); Pb^{2+}; Cu^{2+} and Cd^{2+} (Li et al., 2003); metalloids, such as arsenic compounds (Peng et al., 2005); organics, such as polycyclic aromatic organic compounds (Gotovac et al., 2007); and a range of biological contaminants (Upadhyayula et al., 2009), including bacteria (Akasaka and Watari, 2009), viruses (Mostafavi et al., 2009), natural organic matters (Hyung and Kim, 2008), and cyanobacterial toxins (Albuquerque et al., 2008).

An important consideration for CNTs is that, by introducing different functional groups on their surface, properties including polarity and hydrogen-bonding potential can be altered, which in turn can affect sorption selectivity and capacity (ALOthman et al., 2012; Khajeh et al., 2013).

6.3 Clay-Based Nanoadsorbents

Clays and their modified forms have received widespread attention for their use as adsorbents of metal ions and organic micropollutants from aqueous solution because they are readily available and much cheaper than activated carbon, zeolites, and membranes (de Paiva et al., 2008; Unuabonah and Taubert, 2014). Recently, layered double hydroxides also known as hydrotalcite-like compounds or anionic clays attracted attention in the removal of uranium(VI) (Zhang et al., 2012), lead (Liang et al., 2010), Cr(VI) (Guo et al., 2011), fluoride hardness (Lv et al., 2007), sulfate (Noori Sepehra et al., 2014), and so on.

6.4 Polymeric Nanoadsorbents

Dendrimers are tailored adsorbents that are capable of removing both organic and heavy metals. Their interior shells can be hydrophobic for sorption of organic compounds while the exterior branches can be tailored (e.g., hydroxyl or amine terminated) for adsorption of heavy metals. The sorption can be based on complexation, electrostatic interactions, hydrophobic effect, and hydrogen bonding (Crooks et al., 2001; Qu et al., 2013). For example, polyamidoamine dendrimers which have amide groups and functional nitrogen arranged in regular "branched upon branched" patterns are promising candidate materials for use in water purification (Zhou et al., 2015). Ottaviani et al. (2000) found that polyamidoamine dendrimers have a high adsorption capacity for U(VI), which indicated that dendrimers are the potential adsorbents for the separation and recovery of uranium from aqueous solution. Zhou et al. (2015) also reported that the growth of polyamidoamine dendrimers on the surface of palygorskite, a natural clay, as successful strategy for efficient removal of Pb^{2+} ions from water. Polymer-inorganic nanoscale hybrids also could be counted as desirable adsorbents due to good adsorption capacity and good chemical stability in a wide pH range (Khajeh et al., 2013).

7 Conclusions

The new organic contaminants including pesticides, hormones, drugs, various aromatic compound, heavy metals, resistant microbial pathogens, and so on are not removable by conventional treatment process. Accordingly, the appearance of these contaminants arising from expanding industrial activities requires new technologies for removing them from water supplies. The new-developed treatment processes based on the nanotechnology have been extensively studied in literature and the results are promising in preventing the spread of waterborne diseases. In this regard, the silver nanoparticles are effective in disinfection of water without leaving the harmful DBPs. However, the potential ecotoxicity of silver nanoparticles and increasing microbial resistance to them are the considerable problems in utilizing silver nanoparticles in large-scale practical applications. ZVI-NPs are the powerful reducing agent that have been successfully utilized to dechlorinate the chlorinated organic compounds and immobilize toxic heavy metals, such as hexavalent chromium, uranium, and arsenic. Pilot-scale fields are extensively implemented to treat the contaminated aquifers by injection of nanoscale ZVI and constructing permeable reactive barriers. The uncertainties about the groundwater remediation, hydrogeochemical and biological aspects of in situ application of nanoscale ZVI require further studies.

Various water contaminants including natural organic matters, persistent organic pollutants, and resistant microbial pathogens are effectively removed by photocatalytic processes under UV and solar irradiation. The main advantage of photocatalytic process using semiconductors is the mineralization of the organics. The low quantum yields, number of destructed molecules per a photon, as a result of high electron hole recombination is one of the main challenges to use the photocatalytic processes in large-scale applications. The other challenges that must be investigated are: (1) design of new photocatalysts or improvement of common photocatalyst for higher quantum yields and exciting by solar irradiation, (2) design and modeling photocatalytic reactors to collect maximum solar energy, and (3) optimization of the operational conditions. As discussed, membranes based on nanotechnology have a great potential for removal of undesired constituents from drinking water. Also, the performance of nanomaterials as nanoadsorbents was explained in removal of pollutants from drinking water. Even though nanotechnology possesses a great potential to enhance the water quality used in different industrial, agricultural, sanitary, and drinking consumptions, toxicity and environmental risks of nanoscale materials must be considered. In other words, the employment of nanoscale materials in water-treatment processes may cause to release them into environment and introduce a new class of pollutants that threaten the public health. Unfortunately, the number of studies in this area is very low and a very large volume of studies should be done about the environmental risks of nanoscale materials.

References

Ahn, S.C., Oh, S.-Y., Cha, D.K., 2008. Enhanced reduction of nitrate by zero-valent iron at elevated temperatures. J. Hazard. Mater. 156, 17–22.

Ahn, C.H., Baek, Y., Lee, C., Kim, S.O., Kim, S., Lee, S., Kim, S.H., Bae, S.S., Park, J., Yoon, J., 2012. Carbon nanotube-based membranes: fabrication and application to desalination. J. Ind. Eng. Chem. 18, 1551–1559.

Akasaka, T., Watari, F., 2009. Capture of bacteria by flexible carbon nanotubes. Acta Biomater. 5, 607–612.

Alaoui, O.T., Nguyen, Q.T., Mbareck, C., Rhlalou, T., 2009. Elaboration and (vinylidene fluoride)–anatase TiO_2 composite membranes in degradation of dyes. Appl. Catal. A 358, 13–20.

Albu, S.P., Ghicov, A., Macak, J.M., Hahn, R., Schmuki, P., 2007. Self-organized, freestanding TiO_2 nanotube membrane for flow-through photocatalytic applications. Nano Lett. 7, 1286–1289.

Albuquerque, Jr., E.C., Mendez, M.O.A., Coutinho, A.D.R., Franco, T.T., 2008. Removal of cyanobacterial toxins from drinking water by adsorption on activated carbon fibers. Mater. Res. 11, 370–380.

ALOthman, Z.A., Habila, M., Yilmaz, E., Soylak, M., 2012. Solid phase extraction of Cd(II), Pb(II), Zn(II) and Ni(II) from food samples using multiwalled carbon nanotubes impregnated with 4-(2-thiazolylazo)resorcinol. Microchim. Acta 177, 397–403.

Amani-Ghadim, A.R., Dorraji, M.S.S., 2015. Modeling of photocatalyatic process on synthesized ZnO nanoparticles: kinetic model development and artificial neural networks. Appl. Catal. B, 539–546.

Amani-Ghadim, A.R., Alizadeh, S., Khodam, F., Rezvani, Z., 2015. Synthesis of rod-like α-FeOOH nanoparticles and its photocatalytic activity in degradation of an azo dye: empirical kinetic model development. J. Mol. Catal. A, 60–68.

Asahi, R., Morikawa, T., Ohwaki, T., Aoki, K., Taga, Y., 2001. Visible-light photocatalysis in nitrogen-doped titanium oxides. Science 293, 269–271.

Barni, B., Cavicchioli, A., Riva, E., Zanoni, L., Bignoli, F., Bellobono, I.R., Gianturco, F., De Giorgi, A., Muntau, H., Montanarella, L., Facchetti, S., Castellano, L., 1995. Laboratory-scale photodegradation of phenol in aqueous solution by photocatalytic membranes immobilizing titanium dioxides. Chemosphere 30, 1847–1860.

Bergaya, F., Lagaly, G., 2013. General introduction: clays, clay minerals, and clay science. In: Faïza, B., Gerhard, L. (Eds.), Developments in Clay Science. Elsevier, Boston, pp. 1–19.

Berge, N.D., Ramsburg, C.A., 2009. Oil-in-water emulsions for encapsulated delivery of reactive iron particles. Environ. Sci. Technol. 43, 5060–5066.

Berge, N.D., Ramsburg, C.A., 2010. Iron-mediated trichloroethene reduction within nonaqueous phase liquid. J. Contam. Hydrol. 118, 105–116.

Bhadra, M., Mitra, S., 2013. Nanostructured membranes in analytical chemistry. TrAC-Trend Anal. Chem. 45, 248–263.

Bhadra, M., Mitra, S., 2014. Advances in nanostructured membranes for water desalination. In: Street, A., Sustich, R., Duncan, J., Savage, N. (Eds.), Nanotechnology Applications for Clean Water. Elsevier, Boston, pp. 109–122.

Bhatnagar, A., Sillanpää, M., 2011. A review of emerging adsorbents for nitrate removal from water. Chem. Eng. J. 168, 493–504.

Butkus, M.A., Edling, L., Labare, M.P., 2003. The efficacy of silver as a bactericidal agent: advantages, limitations and considerations for future use. J. Water Supply: Res. Technol. -AQUA 52, 407–416.

Byrne, J.A., Eggins, B.R., 1998. Photoelectrochemistry of oxalate on particulate TiO_2 electrodes. J. Electroanal. Chem. 457, 61–72.

Calabrò, P.S., Moraci, N., Suraci, P., 2012. Estimate of the optimum weight ratio in zero-valent iron/pumice granular mixtures used in permeable reactive barriers for the remediation of nickel contaminated groundwater. J. Hazard. Mater. 207-208, 111–116.

Cao, J., Zhang, W.-X., 2006. Stabilization of chromium ore processing residue (COPR) with nanoscale iron particles. J. Hazard. Mater. 132, 213–219.

Cho, Y., Choi, S.-I., 2010. Degradation of PCE, TCE and 1,1,1-TCA by nanosized FePd bimetallic particles under various experimental conditions. Chemosphere 81, 940–945.

Choi, H., Stathatos, E., Dionysiou, D.D., 2007. Photocatalytic TiO_2 films and membranes for the development of efficient wastewater treatment and reuse systems. Desalination 202, 199–206.

Choi, H., Al-Abed, S.R., Dionysiod, D.D., 2009. Nanostructured titanium oxide film- and membrane-based photocatalysis for water treatment. In: Savage, N., Diallo, M., Duncan, J., Street, A., Sustich, R. (Eds.), Nanotechnology Applications for Clean Water. William Andrew, p. 44, 39-46.

Choi, W., Choi, J., Bang, J., Lee, J.-H., 2013. Layer-by-layer assembly of graphene oxide nanosheets on polyamide membranes for durable reverse-osmosis applications. ACS Appl. Mater. Interfaces 5, 12510–12519.

Chong, M.N., Jin, B., Chow, C.W.K., Saint, C., 2010. Recent developments in photocatalytic water treatment technology: a review. Water Res. 44, 2997–3027.

Cohen-Tanugi, D., Grossman, J.C., 2012. Water desalination across nanoporous graphene. Nano Lett. 12, 3602–3608.

Cohen-Tanugi, D., Grossman, J.C., 2014. Water permeability of nanoporous graphene at realistic pressures for reverse osmosis desalination. J. Chem. Phys. 141, 074704.

Corry, B., 2008. Designing carbon nanotube membranes for efficient water desalination. J. Phys. Chem. B 112, 1427–1434.

Corry, B., 2011. Water and ion transport through functionalised carbon nanotubes: implications for desalination technology. Energ. Environ. Sci. 4, 751–759.

Crane, R.A., Scott, T.B., 2012. Nanoscale zero-valent iron: future prospects for an emerging water treatment technology. J. Hazard. Mater. 211-212, 112–125.

Crane, R., Scott, T., 2014. The removal of uranium onto carbon-supported nanoscale zero-valent iron particles. J. Nanopart. Res. 16, 1–13.

Crooks, R.M., Zhao, M.Q., Sun, L., Chechik, V., Yeung, L.K., 2001. Dendrimer-encapsulated metal nanoparticles: synthesis, characterization, and applications to catalysis. Accounts Chem. Res. 34, 181–190.

Cui, Y., Chang, X., Zhai, Y., Zhu, X., Zheng, H., Lian, N., 2006. ICP-AES determination of trace elements after preconcentrated with p-dimethylaminobenzaldehyde-modified nanometer SiO_2 from sample solution. Microchem. J. 83, 35–41.

Cundy, A.B., Hopkinson, L., Whitby, R.L.D., 2008. Use of iron-based technologies in contaminated land and groundwater remediation: a review. Sci. Total Environ. 400, 42–51.

Daghrir, R., Drogui, P., Robert, D., 2013. Modified TiO_2 for environmental photocatalytic applications: a review. Ind. Eng. Chem. Res. 52, 3581–3599.

Dai, Y.Q., Liu, W.Y., Formo, E, Sun, Y.M., Xia, Y.N., 2011. Ceramic nanofibers fabricated by electrospinning and their applications in catalysis, environmental science, and energy technology. Polym. Adv. Technol. 22, 326–338.

Dalrymple, O.K., Stefanakos, E., Trotz, M.A., Goswami, D.Y., 2010. A review of the mechanisms and modeling of photocatalytic disinfection. Appl. Catal. B 98, 27–38.

Damodar, R.A., You, S.-J., Chou, H.-H., 2009. Study the self cleaning, antibacterial and photocatalytic properties of TiO_2 entrapped PVDF membranes. J. Hazard. Mater. 172, 1321–1328.

Daneshvar, N., Aber, S., Seyed Dorraji, M.S., Khataee, A.R., Rasoulifard, M.H., 2007. Photocatalytic degradation of the insecticide diazinon in the presence of prepared nanocrystalline ZnO powders under irradiation of UV-C light. Sep. Purif. Technol. 58, 91–98.

Dankovich, T.A., Gray, D.G., 2011. Bactericidal paper impregnated with silver nanoparticles for point-of-use water treatment. Environ. Sci. Technol. 45, 1992–1998.

Das, R., Ali, M.E., Bee Abd Hamid, S., Ramakrishna, S., Zaman Chowdhury, Z., 2014. Carbon nanotube membranes for water purification: a bright future in water desalination. Desalination 336, 97–109.

Dawson, J.I., Oreffo, R.O.C., 2013. Clay: new opportunities for tissue regeneration and biomaterial design. Adv. Mater. 25, 4069–4086.

de Paiva, L.B., Morales, A.R., Valenzuela Diaz, F.R., 2008. Organoclays: properties, preparation and applications. Appl. Clay Sci. 42, 8–24.

Di, Z.C., Ding, J., Peng, X.J., Li, Y.H., Luan, Z.K., Liang, J., 2006. Chromium adsorption by aligned carbon nanotubes supported ceria particles. Chemosphere 62, 861–865.

Di Paola, A., García-López, E., Marcì, G., Palmisano, L., 2012. A survey of photocatalytic materials for environmental remediation. J. Hazard. Mater. 211-212, 3–29.

Dong, H., Wu, L., Zhang, L., Chan, H., Gao, C., 2015. Removal of As(V), Cr(III) and Cr(VI) from aqueous environments by poly(acrylonitril-co-acrylamidopropyl-trimethyl ammonium chloride)-based hydrogels. J. Environmen. Manage. 494, 92–103.

Dudu, T.E., Sahiner, M., Alpaslan, D., Demirci, S., Aktas, N., 2015. Removal of As(V), Cr(III) and Cr(VI) from aqueous environments by poly(acrylonitril-co-acrylamidopropyl-trimethyl ammonium chloride)-based hydrogels. J. Environ. Manage. 161, 243–251.

Dyer, A., 1988. An Introduction to Zeolite Molecular Sieves. John Wiley & Sons Ltd, UK.

EPA. 2011. 2011 Edition of the Drinking Water Standards and Health Advisories, EPA 820-R-11-002). Washington, DC: Office of Water US Environmental Protection Agency (EPA). Available from: http://water.epa.gov/action/advisories/drinking/upload/dwstandards2011.pdf

Feng, C., Khulbe, K.C., Matsuura, T., Gopal, R., Kaur, S., Ramakrishna, S., Khayet, M., 2008. Production of drinking water from saline water by air-gap membrane distillation using polyvinylidene fluoride nanofiber membrane. J. Membr. Sci. 311, 1–6.

Feng, C., Khulbe, K.C., Matsuura, T., Tabe, S., Ismail, A.F., 2013. Preparation and characterization of electro-spun nanofiber membranes and their possible applications in water treatment. Sep. Purif. Technol. 102, 118–135.

Fu, F., Dionysiou, D.D., Liu, H., 2014. The use of zero-valent iron for groundwater remediation and wastewater treatment: a review. J. Hazard. Mater. 267, 194–205.

Fujishima, A., Honda, K., 1972. Electrochemical photolysis of water at a semiconductor electrode. Nature 238, 37–38.

Furube, A., Asahi, T., Masuhara, H., Yamashita, H., Anpo, M., 2001. Direct observation of a picosecond charge separation process in photoexcited platinum-loaded TiO_2 particles by femtosecond diffuse reflectance spectroscopy. Chem. Phys. Lett. 336, 424–430.

Geim, A.K., Novoselov, K.S., 2007. The rise of graphene. Nat. Mater. 6, 183–191.

Gillham, R.W., O'Hannesin, S.F., 1994. Enhanced degradation of halogenated aliphatics by zero-valent iron. Ground Water 32, 958–967.

Glavee, G.N., Klabunde, K.J., Sorensen, C.M., Hadjipanayis, G.C., 1995. Chemistry of borohydride reduction of iron(II) and iron(III) ions in aqueous and nonaqueous media. formation of nanoscale Fe, FeB, and Fe2B powders. Inorg. Chem. 34, 28–35.

Glaze, W.H., Kang, J.-W., Chapin, D.H., 1987. The chemistry of water treatment processes involving ozone, hydrogen peroxide and ultraviolet radiation. Ozone Sci. Eng. 9, 335–352.
Gonçalves, G., Marques, P.A.A.P., Barros-Timmons, A., Bdkin, I., Singh, M.K., Emami, N., 2010. Graphene oxide modified with PMMA via ATRP as a reinforcement filler. J. Mater. Chem. 20 (44), 9927–9934.
Gotovac, S., Yang, C.M., Hattori, Y., Takahashi, K., Kanoh, H., Kaneko, K., 2007. Adsorption of polyaromatic hydrocarbons on single walled carbon nanotubes of different functionalities and diameters. J. Colloid Interface Sci. 314, 18–24.
Grellier, J., Rushton, L., Briggs, D.J., Nieuwenhuijsen, M.J., 2015. Assessing the human health impacts of exposure to disinfection by-products—a critical review of concepts and methods. Environ. Int. 78, 61–81.
Guo, X., Zhang, F., Peng, Q., Xu, S., Lei, X., Evans, D.G., Duan, X., 2011. Layered double hydroxide/eggshell membrane: an inorganic biocomposite membrane as an efficient adsorbent for Cr(VI) removal. Chem. Eng. J. 166, 81–87.
Gupta, S.S., Sreeprasad, T.S., Maliyekkal, S.M., Das, S.K., Pradeep, T., 2012. Graphene from sugar and its application in water purification. ACS Appl. Mater. Interfaces 4, 4156–4163.
Han, Y., Xu, Z., Gao, C., 2013. Ultrathin graphene nanofiltration membrane for water purification. Adv. Funct. Mater. 23, 3693–3700.
Hang, Y., Quin, Y., Jiang, Z., Hu, B., 2002. Direct analysis of trace rare earth elements by fluorination assisted ETV-ICP-AES with slurry sampling through nano-sized TiO_2 separation/preconcentration. Anal. Sci. 18, 843–846.
Hashim, M.A., Mukhopadhyay, S., Sahu, J.N., Sengupta, B., 2011. Remediation technologies for heavy metal contaminated groundwater. J. Environ. Manage. 92, 2355–2388.
Hegab, H.M., Zou, L., 2015. Graphene oxide-assisted membranes: fabrication and potential applications in desalination and water purification. J. Membr. Sci. 484, 95–106.
Henderson, A.D., Demond, A.H., 2007. Long-term performance of zero-valent iron permeable reactive barriers: a critical review. Environ. Eng. Sci. 24, 401–423.
Hoek, E.M.V., Pendergast, M.T.M., Ghosh, A.K., 2014. Nanotechnology-based membranes for water purification. In: Street, A., Sustich, R., Duncan, J., Savage, N. (Eds.), Nanotechnology Applications for Clean Water. William Andrew, Boston, pp. 133–154.
Hu, A., Zhang, X., Oakes, K.D., Peng, P., Zhou, Y.N., Servos, M.R., 2011. Hydrothermal growth of free standing TiO_2 nanowire membranes for photocatalytic degradation of pharmaceuticals. J. Hazard. Mater. 189, 278–285.
Hua, G., Reckhow, D.A., 2007. Characterization of disinfection byproduct precursors based on hydrophobicity and molecular size. Environ. Sci. Technol. 41, 3309–3315.
Huang, Z.Q., Zheng, F., Zhang, Z., Xu, H.T., Zhou, K.M., 2012. The performance of the PVDF-Fe_3O_4 ultrafiltration membrane and the effect of a parallel magnetic field used during the membrane formation. Desalination 292, 64–72.
Huang, Q., Liu, W., Peng, P.A., Huang, W., 2013. Reductive dechlorination of tetrachlorobisphenol A by Pd/Fe bimetallic catalysts. J. Hazard. Mater. 262, 634–641.
Hyung, H., Kim, J.H., 2008. Natural organic matter (NOM) adsorption to multi walled carbon nanotubes: effect on NOM characteristics and water quality parameters. Environ. Sci. Technol. 42, 4416–4421.
Irie, H., Watanabe, Y., Hashimoto, K., 2003. Nitrogen-concentration dependence on photocatalytic activity of TiO_2 xNx powders. J. Phys. Chem. B 107, 5483–5486.

Jain, P., Pradeep, T., 2005. Potential of silver nanoparticle-coated polyurethane foam as an antibacterial water filter. Biotechnol. Bioeng. 90, 59–63.

Jin, L., Shi, W., Yu, S., Yi, X., Sun, N., Ma, C., 2012. Preparation and characterization of a novel PA-SiO_2 nanofiltration membrane for raw water treatment. Desalination 298, 34–41.

Joo, S.H., Feitz, A.J., Sedlak, D.L., Waite, T.D., 2005. Quantification of the oxidizing capacity of nanoparticulate zero-valent iron. Environ. Sci. Technol. 39, 1263–1268.

Kalfa, O.M., Yalcinkaya, O., Turker, A.R., 2009. Synthesis of nano B_2O_3/TiO_2 composite material as a new solid phase extractor and its application to preconcentration and separation of cadmium. J. Hazard. Mater. 166, 455–461.

Kallman, E., Oyanedel-Craver, V., Smith, J., 2011. Ceramic filters impregnated with silver nanoparticles for point-of-use water treatment in rural Guatemala. J. Environ. Eng. 137, 407–415.

Kanel, S.R., Manning, B., Charlet, L., Choi, H., 2005. Removal of arsenic(III) from groundwater by nanoscale zero-valent iron. Environ. Sci. Technol. 39, 1291–1298.

Kanel, S.R., Goswami, R.R., Clement, T.P., Barnett, M.O., Zhao, D., 2008. Two dimensional transport characteristics of surface stabilized zero-valent iron nanoparticles in porous media. Environ. Sci. Technol. 42, 896–900.

Kar, S., Bindal, R.C., Tewari, P.K., 2012. Carbon nanotube membranes for desalination and water purification: challenges and opportunities. Nano Today 7, 385–389.

Karumuri, A.K., Oswal, D.P., Hostetler, H.A., Mukhopadhyay, S.M., 2013. Silver nanoparticles attached to porous carbon substrates: robust materials for chemical-free water disinfection. Mater. Lett. 109, 83–87.

Kazemimoghadam, M., 2010. New nanopore zeolite membranes for water treatment. Desalination 251, 176–180.

Keenan, C.R., Sedlak, D.L., 2008. Factors affecting the yield of oxidants from the reaction of nanoparticulate zero-valent iron and oxygen. Environ. Sci. Technol 42, 1262–1267.

Khajeh, M., Laurent, S., Dastafkan, K., 2013. Nanoadsorbents: classification, preparation, and applications (with emphasis on aqueous media). Chem. Rev. 113, 7728–7768.

Khataee, A.R., Fathinia, M., Aber, S., 2010. Kinetic modeling of liquid phase photocatalysis on supported TiO_2 nanoparticles in a rectangular flat-plate photoreactor. Ind. Eng. Chem. Res. 49, 12358–12364.

Khataee, A.R., Amani-Ghadim, A.R., Rastegar Farajzade, M., Valinazhad Ourang, O., 2011. Photocatalytic activity of nanostructured TiO_2-modified white cement. J. Exp. Nanosci. 6, 138–148.

Khodam, F., Rezvani, Z., Amani-Ghadim, A.R., 2015. Fabrication of a novel ZnO/MMO/CNT nanohybrid derived from multi-cationic layered double hydroxide for photocatalytic degradation of azo dye under visible light. RSC Adv. 5, 19675–19685.

Kim, J., Van Der Bruggen, B., 2010. The use of nanoparticles in polymeric and ceramic membrane structures: review of manufacturing procedures and performance improvement for water treatment. Environ. Pollut. 158, 2335–2349.

Kittler, S., Greulich, C., Diendorf, J., Köller, M., Epple, M., 2010. Toxicity of silver nanoparticles increases during storage because of slow dissolution under release of silver ions. Chem. Mater. 22, 4548–4554.

Kolkman, A., Martijn, B.J., Vughs, D., Baken, K.A., Van Wezel, A.P., 2015. Tracing nitrogenous disinfection byproducts after medium pressure UV water treatment by stable isotope labeling and high resolution mass spectrometry. Environ. Sci. Technol. 49, 4458–4465.

Krasner, S.W., Weinberg, H.S., Richardson, S.D., Pastor, S.J., Chinn, R., Sclimenti, M.J., Onstad, G.D., Thruston, A.D., 2006. Occurrence of a new generation of disinfection byproducts. Environ. Sci. Technol. 40, 7175–7185.

Kumakiri, I., Yamaguchi, T., Nakao, S., 2000. Application of a zeolite: a membrane to reverse osmosis process. J. Chem. Eng. Jpn. 33, 333–336.

Kyzas, G.Z., Matis, K.A., 2015. Nanoadsorbents for pollutant's removal: a review. J. Mol. Liq. 203, 159–168.

Lam, C., James, J.T., McCluskey, R., Arepalli, S., Hunter, R.L., 2006. A review of carbon nanotube toxicity and assessment of potential occupational and environmental health risks. CRC Cr. Rev. Toxicol. 36, 189–217.

Le Pape, H., Solano-Serena, F., Contini, P., Devillers, C., Maftah, A., Leprat, P., 2002. Evaluation of the anti-microbial properties of an activated carbon fibre supporting silver using a dynamic method. Carbon 40, 2947–2954.

Leenaerts, O., Partoens, B., Peeters, F.M., 2008. Graphene: a perfect nanoballoon. Appl. Phys. Lett 93, 193107.

Lei, J.Y., Wang, W., Song, M.X., Dong, B., Li, Z.Y., Wang, C., Li, L.J., 2011. Ag/AgCl coated polyacrylonitrile nanofiber membranes: synthesis and photocatalytic properties. React. Funct. Polym. 71, 1071–1076.

Leong, S., Razmjou, A., Wang, K., Hapgood, K., Zhang, X., Wang, H., 2014. TiO_2 based photocatalytic membranes: a review. J. Membr. Sci. 472, 167–184.

Li, D., Xia, Y.N., 2004. Electrospinning of nanofibers: reinventing the wheel? Adv. Mater., 16.

Li, Y.H., Ding, J., Luan, Z.K., Di, Z.C., Zhu, Y.F., Xu, C.L., Wu, D.H., Wei, B.Q., 2003. Competitive adsorption of Pb^{2+}, Cu^{2+} and Cd^{2+} ions from aqueous solutions by multiwalled carbon nanotubes. Carbon 41, 2787–2792.

Li, L., Dong, J., Nenoff, T., Lee, R., 2004. Desalination by reverse osmosis using MFI zeolite membranes. J. Membr. Sci. 243, 401–404.

Li, D., McCann, J.T., Xia, Y.N., 2006. Electrospinning: a simple and versatile technique for producing ceramic nanofibers and nanotubes. J. Am. Ceram. Soc. 89, 1861–1869.

Li, Q., Mahendra, S., Lyon, D.Y., Brunet, L., Liga, M.V., Li, D., Alvarez, P.J.J., 2008. Antimicrobial nanomaterials for water disinfection and microbial control: potential applications and implications. Water Res. 42, 4591–4602.

Li, S., Yan, W., Zhang, W.-X., 2009. Solvent-free production of nanoscale zero-valent iron (nZVI) with precision milling. Green Chem. 11, 1618–1626.

Li, S.D., Dong, Y.H., Guo, M.R., 2012. Photocatalytic activity of different morphologies TiO_2 nanofibers. Appl. Surf. Sci. 258, 8015–8019.

Liang, L., Moline, G.R., Kamolpornwijit, W., West, O.R., 2005. Influence of hydrogeochemical processes on zero-valent iron reactive barrier performance: a field investigation. J. Contam. Hydrol. 78, 291–312.

Liang, X., Hou, W., Xu, Y., Sun, G., Wang, L., Sun, Y., Qin, X., 2010. Sorption of lead ion by layered double hydroxide intercalated with diethylenetriaminepentaacetic acid. Colloid Surf. A 366, 50–57.

Liao, J., Lin, S., Pan, N., Li, S., Cao, X., Cao, Y., 2012. Fabrication and photocatalytic properties of free-standing TiO_2 nanotube membranes with through-hole morphology. Mater. Charact. 66, 24–29.

Lind, M.L., Suk, D.E., Nguyen, T.V., Hoek, E.M.V., 2010. Tailoring the structure of thin film nanocomposite membranes to achieve seawater RD membrane performance. Environ. Sci. Technol. 44, 8230–8235.

Linden, K.G., Mohseni, M., 2014. Advanced oxidation processes: applications in drinking water treatment. In: AHUJA, S. (Ed.), Comprehensive Water Quality and Purification. Elsevier, Waltham.

Liu, C., 2013. Advances in membrane technologies for drinking water purification. In: Ahuja, S. (Ed.), Comprehensive Water Quality and Purification. Elsevier, pp. 7–97.

Liu, Y., Liang, P., Guo, L., 2005. Nanometer titanium dioxide immobilized on silica gel as sorbent for preconcentration of metal ions prior to their determination by inductively coupled plasma atomic emission spectrometry. Talanta 68, 25–30.

Liu, Z.Y., Sun, D.D., Guo, P., Leckie, J.O., 2007. An efficient bicomponent TiO_2/SnO_2 nanofiber photocatalyst fabricated by electrospinning with a side-by-side dual spinneret method. Nano Lett. 7, 1081–1085.

Liu, H.B., Chen, T.H., Chang, D.Y., Chen, D., Liu, Y., He, H.P., Yuan, P., Frost, R., 2012a. Nitrate reduction over nanoscale zero-valent iron prepared by hydrogen reduction of goethite. Mater. Chem. Phys. 133, 205–211.

Liu, L., Liu, Z., Bai, H., Sun, D.D., 2012b. Concurrent filtration and solar photocatalytic. disinfection/degradation using high-performance Ag/TiO_2 nanofiber membrane. Water Res. 46, 1101–1112.

Loh, K.P., Bao, Q., Ang, P.K., Yang, J., 2010. The chemistry of graphene. J. Mater. Chem. 20, 2277–2289.

Loo, S.-L., Krantz, W.B., Fane, A.G., Gao, Y., Lim, T.-T., Hu, X., 2015. Bactericidal mechanisms revealed for rapid water disinfection by superabsorbent cryogels decorated with silver nanoparticles. Environ. Sci. Technol. 49, 2310–2318.

Lu, Q., Huang, R., 2009. Nonlinear mechanics of single-atomic layer graphene sheets. Int. J. Appl. Mech. 1, 443–467.

Lu, X., Wang, C., 2014. Electrospun nanofiber-based photocatalysts. In: Ding, B., Yu, J. (Eds.), Electrospun nanofibers for energy and environmental applications. Springer, Berlin, pp. 371–401.

Lufrano, F., Squadrito, G., Patti, A., Passalacqua, E., 2000. Sulfonated polysulfone as promising membranes for polymer electrolyte fuel cells. J. Appl. Polym. Sci. 77, 1250–1256.

Lv, L., He, J., Wei, M., Evans, D.G., Zhou, Z., 2007. Treatment of high fluoride concentration water by MgAl–CO_3 layered double hydroxides: kinetic and equilibrium studies. Water Res. 41, 1534–1542.

Ma, N., Fan, X., Quan, X., Zhang, Y., 2009. Ag–TiO_2/HAP/Al_2O_3 bioceramic composite membrane: fabrication, characterization and bactericidal activity. J. Membr. Sci. 336, 109–117.

Ma, H., Hsiao, B.S., Chu, B., 2014. Functionalized electrospun nanofibrous microfiltration membranes for removal of bacteria and viruses. J. Membr. Sci. 452, 446–452.

Madaeni, S.S., Ghaemi, N., Rajabi, H., 2015. Advances in polymeric membranes for water treatment. In: Basile, A., Cassano, A., Rastogi, N. (Eds.), Advances in Membrane Technologies for Water Treatment. Woodhead Publishing, Oxford, pp. 3–41.

Mahmoud, K.A., Mansoor, B., Mansour, A., Khraisheh, M., 2015. Functional graphene nanosheets: the next generation membranes for water desalination. Desalination 356, 208–225.

Mahmoudi, M., Serpooshan, V., 2012. Silver-coated engineered magnetic nanoparticles are promising for the success in the fight against antibacterial resistance threat. ACS Nano 6, 2656–2664.

Mahyar, A., Amani-Ghadim, A.R., 2011. Influence of solvent type on the characteristics and photocatalytic activity of TiO nanoparticles prepared by the sol–gel method. Micro Nano Lett. 6, 244–248.

Malato, S., Fernández-Ibáñez, P., Maldonado, M.I., Blanco, J., Gernjak, W., 2009. Decontamination and disinfection of water by solar photocatalysis: recent overview and trends. Catal. Today 147, 1–59.

Malato, S., Fernández-Ibáñez, P., Maldonado, M.I., Oller, I., 2013. Solar photocatalytic processes: water decontamination and disinfection. In: Suib, S.L. (Ed.), New and Future Developments in Catalysis. Elsevier, Amsterdam.

Matheson, L.J., Tratnyek, P.G., 1994. Reductive dehalogenation of chlorinated methanes by iron metal. Environ. Sci. Technol. 28, 2045–2053.

Matilainen, A., Sillanpää, M., 2010. Removal of natural organic matter from drinking water by advanced oxidation processes. Chemosphere 80, 351–365.

Mitra, P., Sarkar, D., Chakrabarti, S., Dutta, B.K., 2011. Reduction of hexa-valent chromium with zero-valent iron: batch kinetic studies and rate model. Chem. Eng. J. 171, 54–60.

Molinari, R., Pirillo, F., Falco, M., Loddo, V., Palmisano, L., 2004. Photocatalytic degradation of dyes by using a membrane reactor. Chem. Eng. Process. Process Intensif. 43, 1103–1114.

Morones, J.R., Elechiguerra, J.L., Camacho, A., Holt, K., Kouri, J.B., Ramírez, J.T., Yacaman, M.J., 2005. The bactericidal effect of silver nanoparticles. Nanotechnology 16, 2346.

Morrison, S.J., Mushovic, P.S., Niesen, P.L., 2006. Early breakthrough of molybdenum and uranium in a permeable reactive barrier. Environ. Sci. Technol. 40, 2018–2024.

Mostafavi, S.T., Mehrnia, M.R., Rashidi, A.M., 2009. Preparation of nanofilter from carbon nanotubes for application in virus removal from water. Desalination 238, 271–280.

Mozia, S., 2010. Photocatalytic membrane reactors (PMRs) in water and wastewater treatment: a review. Sep. Purif. Technol. 73, 71–91.

Mueller, N., Braun, J., Bruns, J., Cˇerník, M., Rissing, P., Rickerby, D., Nowack, B., 2012. Application of nanoscale zero valent iron (NZVI) for groundwater remediation in Europe. Environ. Sci. Pollut. Res. 19, 550–558.

Muellera, N.C., Bruggen, B., Keuterc, V., Luis, P., Melin, T., Pronk, W., 2012. Nanofiltration and nanostructured membranes—should they be considered nanotechnology or not? J. Hazard. Mater. 211-212, 275–280.

Nabeela Nasreen, S.A.A., Sundarrajan, S., Syed Nizar, S.A., Balamurugan, R., Ramakrishna, S., 2013. Advancement in electrospun nanofibrous membranes modification and their application in water treatment. Membranes 3, 266–284.

Nagaveni, K., Sivalingam, G., Hegde, M.S., Madras, G., 2004a. Photocatalytic degradation of organic compounds over combustion-synthesized nano-TiO_2. Environ. Sci. Technol. 38, 1600–1604.

Nagaveni, K., Sivalingam, G., Hegde, M.S., Madras, G., 2004b. Solar photocatalytic degradation of dyes: high activity of combustion synthesized nano TiO_2. Appl. Catal. B 48, 83–93.

Nakata, K., Fujishima, A., 2012. TiO_2 photocatalysis: design and applications. J. Photochem. Photobiol. C 13, 169–189.

Nameni, G., Economy, J., 2014. Nanometallic particles for oligodynamic microbial disinfection. In: Street, A., Sustich, R., Duncan, J., Savage, N. (Eds.), Nanotechnology Applications for Clean Water (second ed.). William Andrew Publishing, Oxford, (Chapter 18).

Nangmenyi, G., Yue, Z., Mehrabi, S., Mintz, E., Economy, J., 2009. Synthesis and characterization of silver-nanoparticle-impregnated fiberglass and utility in water disinfection. Nanotechnology 20, 495705.

Noori Sepehra, M., Yetilmezsoyb, K., Marofic, S., Zarrabia, M., Ghaffarid, H.R., Fingase, M., Foroughi, M., 2014. Synthesis of nanosheet layered double hydroxides at lower pH: optimization of hardness and sulfate removal from drinking water samples. J. Taiwan Ins. Chem. Eng. 45, 2786–2800.

Nurmi, J.T., Tratnyek, P.G., Sarathy, V., Baer, D.R., Amonette, J.E., Pecher, K., Wang, C., Linehan, J.C., Matson, D.W., Penn, R.L., Driessen, M.D., 2005. Characterization and properties of metallic iron nanoparticles: spectroscopy, electrochemistry, and kinetics. Environ. Sci. Technol. 39, 1221–1230.

Oren, Y., 2008. Capacitive deionization (CDI) for desalination and water treatment—past, present and future (a review). Desalination 228, 10–29.

Ottaviani, M.F., Favuzza, P., Bigazzi, M., Turro, N.J., Jockusch, S., Tomalia, D.A., 2000. A TEM and EPR investigation of the competitive binding of uranyl ions to starburst dendrimers and liposomes: potential use of dendrimers as uranyl ion sponges. Langmuir 16, 7368–7372.

Oyanedel-Craver, V.A., Smith, J.A., 2008. Sustainable colloidal-silver-impregnated ceramic filter for point-of-use water treatment. Environ. Sci. Technol. 42, 927–933.

Pant, H.R., Kim, H.J., Joshi, M.K., Pant, B., Park, C.H., Kim, J., Hui, I., K.S. Kim, C.S., 2014. One-step fabrication of multifunctional composite polyurethane spider-web-like nanofibrous membrane for water purification. J. Hazard. Mater. 264, 25–33.

Park, S.-J., Jang, Y.-S., 2003. Preparation and characterization of activated carbon fibers supported with silver metal for antibacterial behavior. J. Colloid Interface Sci. 261, 238–243.

Pendergast, M.T.M., Hoek, E.M.V., 2011. A review of water treatment membrane nanotechnologies. Energy Environ. Sci. 4, 1946–1971.

Peng, X., Luan, Z., Ding, J., Di, Z., Li, Y.H., Tian, B., 2005. Ceria nanoparticles supported nanotubes for removal of arsenate in water. Mater. Lett. 59, 399–403.

Petosa, A.R., Jaisi, D.P., Quevedo, I.R., Elimelech, M., Tufenkji, N., 2010. Aggregation and deposition of engineered nanomaterials in aquatic environments: role of physicochemical interactions. Environ. Sci. Technol. 44, 6532–6549.

Praveena, S., Aris, A., 2015. Application of low-cost materials coated with silver nanoparticle as water filter in *Escherichia coli* removal. Water Qual. Expo. Health, 1–9.

Prince, J.A., Singh, G., Rana, D., Matsuura, T., Anbharasi, V., Shanmugasundaram, T.S., 2012. Preparation and characterization of highly hydrophobic poly(vinylidene fluoride)—clay nanocomposite nanofiber membranes (PVDF–clay NNMs) for desalination using direct contact membrane distillation. J. Membr. Sci. 397-398, 80–86.

Qu, X., Alvarez, P.J.J., Li, Q., 2013. Applications of nanotechnology in water and wastewater treatment. Water Res. 47, 3931–3946.

Quinn, J., Geiger, C., Clausen, C., Brooks, K., Coon, C., O'Hara, S., Krug, T., Major, D., Yoon, W.-S., Gavaskar, A., Holdsworth, T., 2005. Field demonstration of dnapl dehalogenation using emulsified zero-valent iron. Environ. Sci. Technol. 39, 1309–1318.

Rahimpour, A., Jahanshahi, M., Rajaeian, B., Rahimnejad, M., 2011. TiO_2 entrapped nano-composite PVDF/SPES membranes: preparation, characterization, antifouling and antibacterial properties. Desalination 278, 343–353.

Rao, G.P., Lu, C., Su, F., 2007. Sorption of divalent metal ions from aqueous solutions by carbon nanotubes: a review. Sep. Purif. Technol. 58, 224–231.

Rayner, J., Zhang, H., Schubert, J., Lennon, P., Lantagne, D., Oyanedel-Craver, V., 2013. Laboratory investigation into the effect of silver application on the bacterial removal efficacy of filter material for use on locally produced ceramic water filters for household drinking water treatment. ACS Sustainable Chem. Eng. 1, 737–745.

Ren, D., Smith, J.A., 2013. Retention and transport of silver nanoparticles in a ceramic porous medium used for point-of-use water treatment. Environ. Sci. Technol. 47, 3825–3832.

Reynolds, K.A., 2007. Water quality monitoring: lessons from the developing world. Water Conditioning Purif. 39, 66–68.

Ruan, M., Hu, Y., Guo, Z., Dong, R., Palmer, J., Hankinson, J., Berger, C., de Heer, W.A., 2012. Epitaxial graphene on silicon carbide: introduction to structured graphene. MRS Bull. 37, 1138–1147.

Sang, Y., Li, F., Gu, Q., Liang, C., Chen, J., 2008. Heavy metal-contaminated ground water treatment by a novel nanofiber membrane. Desalination 223, 349–360.

Sato, S., 1986. Photocatalytic activity of NOx-doped TiO_2 in the visible light region. Chem. Phys. Lett. 123, 126–128.

Savage, N., Diallo, M., 2005. Nanomaterials and water purification: opportunities and challenges. J. Nanopart. Res. 7, 331–342.

Schrick, B., Hydutsky, B.W., Blough, J.L., Mallouk, T.E., 2004. Delivery vehicles for zerovalent metal nanoparticles in soil and groundwater. Chem. Mater. 16, 2187–2193.

Sengupta, R., Bhattacharya, M., Bandyopadhyay, S., Bhowmick, A.K., 2011. A review on the mechanical and electrical properties of graphite and modified graphite reinforced polymer composites. Prog. Polym. Sci. 36, 638–670.

Shao, C.L., Kim, H., Gong, J., Lee, D., 2002. A novel method for making silica nanofibres by using electrospun fibres of polyvinylalcohol/silica composite as precursor. Nanotechnology 13, 635–637.

Sharma, Y.C., Srivastava, V., Singh, V.K., Kaul, S.N., Weng, C.H., 2009. Nano-adsorbents for the removal of metallic pollutants from water and wastewater. Environ. Technol. 30, 583–609.

Shawky, H.A., Chae, S.-R., Lin, S., Wiesner, M.R., 2011. Synthesis and characterization of a carbon nanotube/polymer nanocomposite membrane for water treatment. Desalination 272, 46–50.

Shi, J., Yi, S., He, H., Long, C., Li, A., 2013. Preparation of nanoscale zero-valent iron supported on chelating resin with nitrogen donor atoms for simultaneous reduction of $Pb2^+$ and and $NO3^-$. Chem. Eng. J. 230, 166–171.

Shokes, T.E., Möller, G., 1999. Removal of dissolved heavy metals from acid rock drainage using iron metal. Environ. Sci. Technol. 33, 282–287.

Siddiquey, I.A., Furusawa, T., Sato, M., Honda, K., Suzuki, N., 2008. Control of the photocatalytic activity of TiO_2 nanoparticles by silica coating with polydiethoxysiloxane. Dyes Pigments 76, 754–759.

Singer, P.C., 2006. DPBs in drinking water: additional scientific and policy considerations for public health protection. J. Am. Water Works Assess. 98, 73–79.

Sohn, K., Kang, S.W., Ahn, S., Woo, M., Yang, S.-K., 2006. Fe(0) nanoparticles for nitrate reduction: stability, reactivity, and transformation. Environ. Sci. Technol. 40, 5514–5519.

Srivastava, A., Srivastava, O.N., Talapatra, S., Vajtai, R., Ajayan, P.M., 2004. Carbon nanotube filters. Nat. Mater. 3, 610–614.

Stankovich, S., Dikin, D.A., Dommett, G.H., Kohlhaas, K.M., Zimney, E.J., Stach, E.A., Piner, R.D., Nguyen, S.T., Ruoff, R.S., 2006. Graphene-based composite materials. Nature 442, 282–286.

Sun, X.-F., Qin, J., Xia, P.-F., Guo, B.-B., Yang, C.-M., Song, C., Wang, S.-G., 2015. Graphene oxide-silver nanoparticle membrane for biofouling control and water purification. Chem. Eng. J. 281, 53–59.

Sundaramurthy, J., Kumar, P.S., Kalaivani, M., Thavasi, V., Mhaisalkar, S.G., Ramakrishna, S., 2012. Superior photocatalytic behaviour of novel 1D nanobraid and nanoporous Fe_2O_3 structures. RSC Adv. 2, 8201–8208.

Taha, A.A., Wu, Y.N., Wang, H., Li, F., 2012. Preparation and application of functionalized cellulose acetate/silica composite nanofibrous membrane via electrospinning for Cr(VI) ion removal from aqueous solution. J. Environ. Manage. 112, 10–16.
Thostenson, E.T., Ren, Z., Chou, T.W., 2001. Advances in the science and technology of carbon nanotubes and their composites: a review. Compos. Sci. Technol. 61, 1899–1912.
Timothy, J.G., Donald, L.S., 1999. Kinetics and mechanisms of reactions at the mineral-water interface: an overview. American Chemical Society, Washington DC, pp. 2–11.
Tratnyek, P.G., Johnson, R.L., 2006. Nanotechnologies for environmental cleanup. Nano Today 1, 44–48.
Tseng, H.-H., Su, J.-G., Liang, C., 2011. Synthesis of granular activated carbon/zero valent iron composites for simultaneous adsorption/dechlorination of trichloroethylene. J. Hazard. Mater. 192, 500–506.
Unuabonah, E.I., Taubert, A., 2014. Clay–polymer nanocomposites (CPNs): adsorbents of the future for water treatment. Appl. Clay Sci. 99, 83–92.
Upadhyayula, V.K.K., Deng, S., Mitchell, M.C., Smith, G.B., 2009. Application of carbon nanotube technology for removal of contaminants in drinking water: a review. Sci. Total Environ. 408, 1–13.
Vaccari, A., 1998. Preparation and catalytic properties of cationic and anionic clays. Catal. Today 41, 53–71.
Wang, W., Serp, P., Kalck, P., Faria, J.L., 2005. Visible light photodegradation of phenol on MWNT-TiO_2 composite catalysts prepared by a modified sol–gel method. J. Mol. Catal. A 235, 194–199.
Wang, Q., Lee, S., Choi, H., 2010. Aging study on the structure of Fe0-nanoparticles: stabilization, characterization, and reactivity. J. Phys. Chem. C 114, 2027–2033.
Wei, Y.-T., Wu, S.-C., Yang, S.-W., Che, C.-H., Lien, H.-L., Huang, D.-H., 2012. Biodegradable surfactant stabilized nanoscale zero-valent iron for in situ treatment of vinyl chloride and 1,2-dichloroethane. J. Hazard. Mater. 211-212, 373–380.
WHO, 2004. Guidelines for Drinking-Water Quality. Geneva: World Health Organization. Available from: http://www.who.int/water_sanitation_health/dwq/GDWQ2004web.pdf
Wilkin, R.T., Su, C., Ford, R.G., Paul, C.J., 2005. Chromium-removal processes during groundwater remediation by a zerovalent iron permeable reactive barrier. Environ. Sci. Technol. 39, 4599–4605.
Xia, S., Ni, M., 2015. Preparation of poly(vinylidene fluoride) membranes with graphene oxide addition for natural organic matter removal. J. Membr. Sci. 473, 54–62.
Xiao, S., Ma, H., Shen, M., Wang, S., Huang, Q., Shi, X., 2011. Excellent copper(II) removal using zero-valent iron nanoparticle-immobilized hybrid electrospun polymer nanofibrous mats. Colloid Surf. A 381, 48–54.
Xu, L., Wang, J., 2011. A heterogeneous Fenton-like system with nanoparticulate zero-valent iron for removal of 4-chloro-3-methyl phenol. J. Hazard. Mater. 186, 256–264.
Yan, L., Hong, S., Li, M.L., Li, Y.S., 2009. Application of the Al_2O_3-PVDF nanocomposite tubular ultrafiltration (UF) membrane for oily wastewater treatment and its antifouling research. Sep. Purif. Technol. 66, 347–352.
Yan, W., Herzing, A.A., Kiely, C.J., Zhang, W.-X., 2010a. Nanoscale zero-valent iron (nZVI): aspects of the core-shell structure and reactions with inorganic species in water. J. Contam. Hydrol. 118.
Yan, W., Ramos, M.a.V., Koel, B.E., Zhang, W.-X., 2010b. Multi-tiered distributions of arsenic in iron nanoparticles: observation of dual redox functionality enabled by a core–shell structure. Chem. Commun. 46, 6995–6997.

Yan, W., Vasic, R., Frenkel, A.I., Koel, B.E., 2012. Intraparticle reduction of arsenite (As(III)) by nanoscale zerovalent iron (nZVI) investigated with in situ X-ray absorption spectroscopy. Environ. Sci. Technol. 46, 7018–7026.

Yan, W., Lien, H.-L., Koel, B.E., Zhang, W.-X., 2013. Iron nanoparticles for environmental clean-up: recent developments and future outlook. Env. Sci. Process. Impact. 15, 63–77.

Yean, S., Cong, L., Yavuz, C.T., Mayo, J.T., Yu, W.W., Kan, A.T., Colvin, V.L., Tomson, M.B., 2005. Effect of magnetite particle size on adsorption and desorption of arsenite and arsenate. J. Mater. Res. 20, 3255–3264.

Yin, J., Deng, B., 2015. Polymer-matrix nanocomposite membranes for water treatment. J. Membr. Sci. 479, 256–275.

Yoshino, H., Kawase, Y., 2013. Kinetic modeling and simulation of zero-valent iron wastewater treatment process: simultaneous reduction of nitrate, hydrogen peroxide, and phosphate in semiconductor acidic wastewater. Ind. Eng. Chem. Res. 52, 17829–17840.

Yu, S.-J., Yin, Y.-G., Liu, J.-F., 2013. Silver nanoparticles in the environment. Env. Sci. Process. Impact. 15, 78–92.

Zhang, W.-X., 2003. Nanoscale iron particles for environmental remediation: an overview. J. Nanopart. Res. 5, 323–332.

Zhang, X., Du, A.J., Lee, P., Sun, D.D., Leckie, J.O., 2008. TiO_2 nanowire membrane for concurrent filtration and photocatalytic oxidation of humic acid in water. J. Membr. Sci. 313, 44–51.

Zhang, X., Xu, S., Han, G., 2009. Fabrication and photocatalytic activity of TiO_2 nanofiber membrane. Mater. Lett. 63, 1761–1763.

Zhang, X., Ji, L., Wang, J., Li, R., Liu, Q., Zhang, M., Liu, L., 2012. Removal of uranium(VI) from aqueous solutions by magnetic Mg–Al layered double hydroxide intercalated with citrate: kinetic and thermodynamic investigation. Colloid Surf. A 414, 220–227.

Zhang, G., Kim, G., Choi, W., 2014a. Visible light driven photocatalysis mediated via ligand-to-metal charge transfer (LMCT): an alternative approach to solar activation of titania. Energy Environ. Sci. 7, 954–960.

Zhang, C., Chen, L., Wang, T.J., Su, C.L., Jin, Y., 2014b. Synthesis and properties of a magnetic core-shell composite nano-adsorbent for fluoride removal from drinking water. Appl. Surf. Sci. 317, 552–559.

Zhang X1, D.B., Guo, J., Wang, Y., Lan, Y., 2011. Ligand-assisted degradation of carbon tetrachloride by microscale zero-valent iron. J. Environ. Manage. 92, 1328–1333.

Zhao, Y.-Y., Boyd, J.M., Woodbeck, M., Andrews, R.C., Qin, F., Hrudey, S.E., Li, X.-F., 2008. Formation of N-nitrosamines from eleven disinfection treatments of seven different surface waters. Environ. Sci. Technol. 42, 4857–4862.

Zheng, Y.M., Zou, S.W., Nadeeshani Nanayakkara, K.G., Matsuura, T., Paul Chen, J., 2011. Adsorptive removal of arsenic from aqueous solution by a PVDF/zirconia blend flat sheet membrane. J. Membr. Sci. 374, 1–11.

Zheng, X., Wang, Q., Chen, L., Wang, J., Cheng, R., 2015. Photocatalytic membrane reactor (PMR) for virus removal in water: performance and mechanisms. Chem. Eng. J. 277, 124–129.

Zhou, S., Xue, A., Zhang, Y., Li, M., Li, K., Zhao, Y., Xing, W., 2015. Novel polyamidoamine dendrimer-functionalized palygorskite adsorbents with high adsorption capacity for $Pb^{2+}$and reactive dyes. Appl. Clay Sci. 107, 220–229.

Zodrow, K., Brunet, L., Mahendra, S., Li, D., Zhang, A., Li, Q., Alvarez, P.J.J., 2009. Polysulfone ultrafiltration membranes impregnated with silver nanoparticles show improved biofouling resistance and virus removal. Water Res. 43, 715–723.

5

NANOTECHNOLOGY-BASED FILTERS FOR COST-EFFECTIVE DRINKING WATER PURIFICATION IN DEVELOPING COUNTRIES

Maggy NB Momba*, Lerato Baloyi*, Lizzy Mpenyana-Monyatsi*, Ilunga Kamika**

**Tshwane University of Technology, Department of Environmental, Water and Earth Sciences, Faculty of Science, Pretoria, South Africa; **University of South Africa, Department of Environmental Science, College of Agriculture and Environmental Sciences, Florida, South Africa*

1 Introduction

Water is a vital component of our planet and life. However, only 1% of the world's freshwater is accessible for extraction and use (Peter-Varbanets et al., 2009). Since the beginning of this century, water scarcity in both quantity and quality has become one of the major problems, and the perspectives on climate change, rapid population growth, fast industrialization and urbanization, increased living standards, and poor water-use strategies have aggravated water scarcity at the global scale. Although access to safe and sufficient water for human consumption has long been a priority in many water policies worldwide (European water policies, UNESCO, WHO, South African Water Acts, etc.), the alarming increase in pollution of this valuable resource around the globe, especially in rural sub-Saharan countries, constitutes a public health concern. It has been estimated that the waterborne diseases cause the death of approximately 20 million people every year; in addition, nonfatal infections of more than 200 million people each year (Leonard et al., 2003). Children are the most affected with about 5000–6000 children dying every day from water-related illnesses, such as diarrheal diseases (Ashbolt, 2004; Hutton et al., 2007). In the 1990s, the death toll of children due to diarrheal diseases was greater than the sum of people killed in conflicts since the Second World War (Smith, 2009). This situation is more acute in rural

Water Purification. http://dx.doi.org/10.1016/B978-0-12-804300-4.00005-8

communities in developing countries as safe freshwater supplies are limited and the use of unconventional water sources, such as stormwater, contaminated freshwater, and brackish water is high (Amin et al., 2014). This is due to the fact that existing water treatment systems and extensive distribution systems associated with huge centralized schemes are no longer sustainable (Weber, 2002). Decentralized water purification systems should be considered in rural areas where centralized systems are not feasible.

To ensure the quality of water sources and improve human health, considerable effort has been put into scientific research, technological development, monitoring, and environmental protection, as well as consumer information and public awareness (Corcoran et al., 2010). Considering the growing demand for potable water and the shortage of water supplies, water reuse will soon become imperative in densely populated areas. Therefore, developing innovative, efficient, cost-effective, and sustainable strategies for water treatment to acceptable quality is a major technological challenge. Nanotechnology has been reported to play a major role in ameliorating or mitigating many water purification and quality issues (Savage and Diallo, 2005; Bottero et al., 2006). This chapter highlights the current trends of improved drinking water systems in developing countries and the relevance of nanotechnology in purifying water to an acceptable quality, which is safe for human consumption.

2 Trends of Improved Drinking Water Systems in Developing Countries

2.1 Concept of Clean and Safe Drinking Water

The concept of clean and safe drinking water as being essential to public health is one of the oldest theories demonstrated in 350 BC, when Hippocrates recommended the process of boiling water to inactivate "impurities" (Montgomery and Elimelech, 2007). The understanding of this concept by the developed world, such as Europe and North America and the advent of community water service have been one of the greatest public health advances of the 20th century in these parts of the world. This has resulted in a drastic reduction of waterborne diseases and protection of public health in developed countries. Yet, in the 21st century, there is evidence that millions of people in the developing world still have no access to safe water supply, although it is a basic necessity. Consumption of unsafe water and lack of sanitation have generated a vicious cycle that has hindered economic and social development and growth of developing countries. This is typical for hundreds of millions of African, Asian, and Latin American families who are paying every day in lost income for their lack of access to improved drinking water and sanitation services (WHO/UNICEF, 2005a). Inadequacies in safe water supply have been proven not only to affect public health adversely, both directly and indirectly, but also to prevent good sanitation and hygiene (Hunter et al., 2010).

Given the imperative role that clean drinking water plays in the everyday lives of people, the world leaders have advocated for improved drinking water systems in developing countries to enhance public health and subsequently the social and economic development. Contrary to unprotected water sources, the term "improved access" refers to households that obtain water from sources that are superior to traditional sources (WHO/UNICEF, 2005b). An index for global source drinking water quality has been developed by the World Health Organization (WHO) and the United Nations Children's Fund (WHO/UNICEF), Joint Monitoring Programme for Water-supply and Sanitation (JMP), which monitors progress on the Millennium Development Goal (MDG) water supply target (WHO/UNICEF, 2010) (Fig. 5.1).

Based on the aforementioned classification, global monitoring of access to improved water supply has been carried out in response to international development policies, while also generating knowledge that updates the development and facilitates the implementation of these policies (Fukuda-Parr, 2008; Meier et al., 2014).

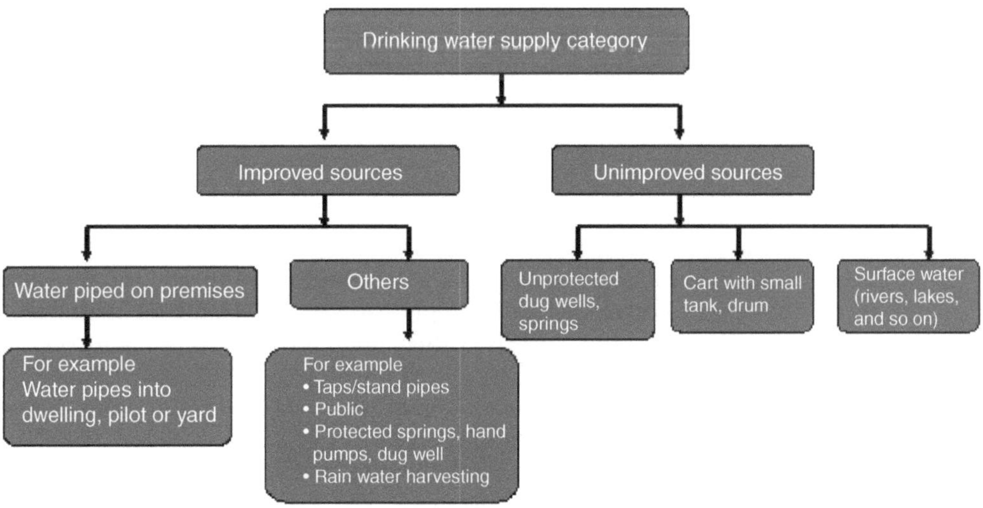

Figure 5.1. Drinking water source categories available to developing countries. Adapted from WHO/UNICEF JMP 2010

2.2 Timeline of Progress Made in Providing Access to Improved Drinking Water Supply in Developing Countries

Access to water supply and sanitation has received considerable attention in the United Nations (UN) policies and in its collaborative programs with member states for many decades, by raising awareness about water, stimulating debate, and focusing on the dangers that arise from access to unimproved water supply. Since the 1930s, international monitoring of drinking water supply and sanitation has been ongoing and its major aim is the improvement of people's health. In the past, this monitoring was carried out by the League of Nations Health Organization; subsequently by the WHO, and today jointly by the WHO and the UNICEF through their Joint Monitoring Programme (JMP) (Bartram et al., 2014). Although the JMP reports every 2-years on progress against MDG indicators for both water supply and sanitation, for the purpose of this chapter, the main focus is only on the access to improved water supply.

The first global survey on access to water supply was conducted in 1962 and covered 75 developing countries. This survey mainly dealt with urban water supply conditions only (Dieterich and Hernderson, 1963). In 1970, a second global survey extended its coverage on community water supply conditions in both urban and rural areas (Pineo and Subrahmanyam, 1975). Based on this latter survey, a revised global target (also called a middecade review) was approved in 1975 by the 25th World Health Assembly (WHA) for the Second United Nations Development Decade (WHO, 1976). The middecade review covered a 5-year period between 1971 and 1975 and focused only on a total of 67 developing countries, excluding the population of China. During its conference on Human Settlements (Habitat) in 1976, the United Nations set 1990 as the target date for community water supply and sanitation to all in both urban and rural areas. At its 35th session in November 1980, the United Nations General Assembly launched the decade and by its resolution 35/18, member states made their commitments to substantially improve the standards and levels of access to drinking water supply and sanitation service by the year 1990 (WHO, 1981). In 2000, the United Nations as part of its MDGs set a key target (MDG Target 7c), which aims at reducing by half the proportion of people without sustainable access to safe drinking water and basic sanitation by 2015 (WHO/UNICEF, 2000). In 2003, the WHO launched the Household Water Treatment (HWT) and Safe Storage Network. This Network consisted of a consortium of nearly 100 organizations working throughout developing nations. This Network aims at promoting collaboration, generating research, and exploring measures to scale up pilot projects (WHO/UNICEF, 2005a; Montgomery and Elimelech, 2007). The WHO declared 2005–2015 the decade of water and the goal is to establish the framework to eventually provide full access to water supply and sanitation for all people. On 28 July 2010, the UN General Assembly adopted Resolution 64/292, which recognizes safe and clean drinking water and sanitation as a human right, essential to the full enjoyment of life and all other human rights (UN, 2010). To accomplish this resolution, the UN member states and international organizations were then called to offer funding, technology, and other resources to assist poorer developing countries to increase their efforts in providing clean, accessible and affordable drinking water, and sanitation services for everyone; and since 2012 water has been playing a prominent role in the Rio + 20 process and discussions around the Post-2015 Development Agenda (WHO/UNICEF, 2005a) [UN-Water Decade Programme on Capacity Development (UNW-DPC 2015)].

Tremendous efforts have been undertaken over the years and resulted in major achievements on access to safe drinking water in developing countries from 1970 to 2010 as can be seen in Table 5.1. The MDG target of halving by 2015 the proportion of the population without sustainable access to safe drinking was met in 2010, 5-years ahead of schedule (UN, 2010). Despite the progress, however, much work still remains to be done in developing countries, especially in rural areas. Throughout the 4 decades of great work by global member states, the use of unimproved drinking-water sources still persists and varies both among the regions and within the countries. However, the disparity in water supply services between urban and rural communities is the common denominator in developing countries (Table 5.2) and it has become a matter of great concern. The percentage of the urban population served with improved drinking water supply by far exceeds that of the rural population. Among the three developing regions, Africa represents the region with least access to improved water supply. The percentage of the total water supply in Africa increased only from 29% to 66% from 1970 to 2010. Of this percentage, the access to improved water supply in urban areas was found to increase fivefold in 1970, threefold in 1975 and almost twofold in 1990, compared to those of the rural population during the same periods. While no drastic increase in access to improved water supply was noted between 1990 and 2010, the percentages of urban water supply were almost twofold those of rural water supply (Table 5.1). These figures clearly show that for many decades, a large number of people in rural areas of Africa do not have

Table 5.1 Percentages of Access to Improved Drinking Water in Urban and Rural Regions of Global Developing Countries and Three Continents—Africa, Asia, and Latin America/The Caribbean—Between 1970 and 2010

Regions	1970 (%)	1975 (%)	1990 (%)	2000 (%)	2010 (%)
Africa					
Urban area	67	65	84	85	85
Rural area	13	21	44	47	54
Total water supply	21	29	57	62	66
Asia					
Urban area	50	68	94	93	96
Rural area	9	19	67	75	82
Total water supply	17	29	76	81	89
Latin America and the Caribbean					
Urban area	61	76	92	93	98
Rural area	24	30	56	62	81
Total water supply	54	57	82	85	94
Global Water Supply					
Urban area	67	75	93	94	95
Rural area	14	22	59	69	79
Total water supply	29	35	70	79	86

WHO (1976); WHO/UNICEF (2000, 2012); AMCOW (2012)

access to clean water. This means that rural populations still relied on unimproved water sources for the past 4 decades. By 2010, the WHO/UNICEF pointed out a total of 10 countries with the largest population without access to an improved drinking water source. These countries are part of Asia (China: 119 million, India: 97 million, Indonesia: 43 million, Bangladesh: 28 million) and sub-Saharan Africa (Nigeria: 66 million, Ethiopia: 46 million, Democratic Republic of Congo: 36 million, United Republic of Tanzania: 21 million, Sudan: 18 million, and Kenya: 17 million).

Table 5.2 Progress in Population Served (in Millions) With Improved Water Supply in Three Regions in Developing Countries Versus Global Supply Between 1970 and 2010

	1970*			1975*			1990			2000			2010		
Region	**TP**	**PS**	**PU**	**TP**	**PS**	**PU**	**TP**	**PS**	**PU**	**IP**	**PS**	**PU**	**TP**	**PS**	**PU**
Africa															
Urban area	NA	0.013	NA	NA	0.014	NA	197	166	31	297	253	44	409	348	61
Rural area	NA	0.012	NA	NA	0.019	NA	418	183	235	487	231	256	613	331	282
Total	NA	0.024	NA	NA	0.032	NA	615	349	266	784	484	300	1022	679	343
Asia															
Urban area	NA	0.076	NA	NA	1.2	NA	1029	972	57	1352	1254	98	1665	1598	67
Rural area	NA	0.063	NA	NA	1.4	NA	2151	1433	718	2331	1736	595	2339	1918	421
Total	NA	1.30	NA -	NA	2.7	NA	3180	2405	775	3683	2990	693	4004	3516	488
Latin America and the Caribbean															
Urban area	NA	1.1	NA	NA	1.4	NA	313	287	26	391	362	29	496	486	10
Rural area	NA	0.028	NA	NA	0.021	NA	128	72	56	128	79	49	94	76	18
Total	NA	1.4	NA	NA	1.0	NA	441	359	82	519	441	78	590	562	28
Global															
Urban area	NA	308	NA	NA	390	NA	2292	2179	113	2845	2672	173	2570	2432	138
Rural area	NA	169	NA	NA	248	NA	2974	1961	1013	3210	2284	926	3046	2325	721
Total	NA	477	NA	NA	638	NA	5 266	4 140	1 126	6055	4956	1099	5616	4757	859

TP, Total population; PS, population served; PU, population unserved; (*) only 67 countries included in this survey; NA, information not available.
WHO (1976); WHO (2012)

Politicians in developing countries have been accused of being responsible for not making available improved drinking water supply in rural areas. This is due to the fact that the elite who hold power are urban-based and their policies, in spite of the rhetoric, have clearly favored the urban areas, where their power centers normally are (Biswas, 1983). Furthermore, inadequate planning, insufficient budget, incomplete execution of plans, limited appreciation of the problems of rural people and a lack of understanding and emphasis by donor countries and agencies have contributed to this situation in rural areas (Biswas, 1983). As a result, millions continue to perish from preventable illnesses every year in developing countries.

2.3 Impact of the Lack of Improved Drinking Water Supply on Public Health in Developing Countries

The lack of improved domestic water supply plays a substantial role in the transmission of waterborne disease and water-washed diseases. In its report on global water supply and sanitation assessment, the WHO/UNICEF (2000) pointed out those nearly 4 billion cases of diarrhea each year cause 2.2 million deaths, mostly among children under the age of 5. This is equivalent to one child dying every 15 s, or 20 jumbo jets crashing every day. The report highlighted that these deaths represent almost 15% of all child deaths under the age of 5 in developing countries. This rate can be reduced on average by between one-quarter and one-third by introducing water, sanitation, and hygiene interventions. Not only diarrheal diseases deteriorate the health of communities without access to improved drinking water supply, intestinal worms also infect about 10% of the population of the developing world. It is also estimated that 6 million people are blind from trachoma and the population at risk from this disease is approximately 500 million. Two hundred million people in the world are infected with schistosomiasis, of whom 20 million suffer severe consequences (WHO/UNICEF, 2000).

Furthermore, in 2004, the WHO/UNICEF report revealed that infants and young children in sub-Saharan Africa and South Asia continue to be victims of developing world failure to make safe drinking water and basic sanitation services available to impoverished people. Women and children are the main victims as they endure the indignity, shame, and sickness that result from a lack of cleanliness. An updated progress report on drinking water by the WHO/UNICEF (2014) revealed that globally 748 million people still lack access to an improved drinking water source. The poor and marginalized people in developing countries are among these peoples and almost a quarter of them (173 million) rely on untreated surface water, and over 90% live in rural areas. This report clearly indicates that if current trends continue, there will still be 547 million people without an improved drinking water supply in 2015. Unless people have access to safe drinking water and basic sanitation, efforts to prevent death from diarrhea or to reduce the burden of diseases, such as ascariasis, dracunculiasis, hookworm infection, schistosomiasis, and trachoma are doomed to failure. In light of this, there is an obvious need to assess, prioritize, and manage in a systematic manner the safety of water for the millions of people lacking access to improved basic services in developing countries.

It is time for developing countries to develop low-cost technologies with the maximum utilization of local resources in terms of manpower and materials that can go a long way in reducing investment costs, thus permitting a greater number of people to be served. If only one barrier is possible to eliminate waterborne diseases, it has to be disinfection, unless evidence exists that chemical contaminants are more harmful than the risk from ingestion of microbial pathogens. Two criteria should therefore be considered for the selection of technologies: (1) the technology should have disinfection capability to remove microbial contaminants; and (2) the technology should be cost-effective and applicable for domestic use, especially in rural communities. Nanotechnology has great potential for providing access to safe drinking water that complies with international standards. Nanotechnology applications in drinking water treatment including bimetallic nanoparticles have been explored as alternative water disinfection tools.

3 Nanotechnology for Safe Drinking Water

For decades now, the access of freshwater is becoming more and more difficult due to the fast growing of population expected to reach 9 billion by 2050 (Smith, 2009). Based on the aforementioned, a novel innovative water technology is urgently needed in order to improve water quality and access and ensure supply of safe drinking water to all, especially to rural communities who are mostly forgotten. In 2005, the Allianz AG and the Organization for Economic Co-operation and Development (OECD) (2005) highlighted the potential impact of nanotechnology for water treatment in developing countries and the National Research Council (NRC) appointed a committee to develop a research strategy for environmental, health, and safety aspects of engineered nanomaterials (NMs). This nanotechnology shows promising potential for significant improvements of advanced materials and manufacturing techniques with a vast range of applications, which are critical for the future competitiveness of national industries (Miyazaki and Islam, 2007; Masoka et al., 2012).

As promising as it can be, the core of success for nanotechnology has been based on the manipulation and production of matter at the atomic and molecular scale, while retaining the optical properties, high surface area to mass ratio and quantum effects at the nanoscale (Schutte and Focke, 2007). Furthermore, matters involved in this technology namely NMs have different strength, conductivity, color, and reactivity, and therefore can be "tuned" to build faster, lighter, and stronger devices (Schutte and Focke, 2007). It is therefore not surprising to see the use of engineered NMs in commercially available products increasing almost exponentially from 54 in 2005 and possibly to 3400 by 2020, containing mainly silver materials (Fig. 5.2) (Wwics, 2011). The ability to alter these materials' physical and chemical properties to have an enhanced specific surface area, smaller size, and high chemical reactivity has enabled NMs to be manufactured and marketed in a wide variety of sectors, such as electronics, cosmetics, pharmaceuticals, information technology, agriculture, food, medical, water treatment, and environmental protection.

In the interests of finding solutions to social, economic, and environmental challenges, as well as scientific and technological innovations, developing countries have realized the need to address

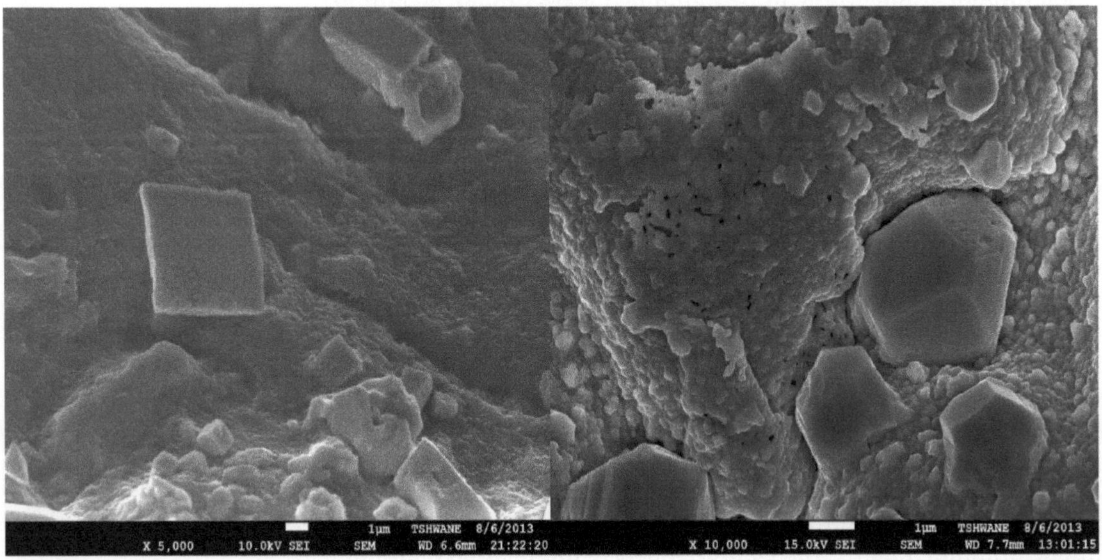

Figure 5.2. Scanning electron microscope (SEM) image of silver nanoparticles embedded in polymers.

their pressing problems, for example, water crisis, and therefore are taking a lead in the development, usage, and marketing of the nanotechnology products (Schutte and Focke, 2007; Singer et al., 2005). This is due to the fact that the available water resource is approximately 1% freshwater and 97% salt water, and the drinkable resources are still labeled as a scarce resource (Hinrichsen and Tacio, 2002; Fenwick, 2006; Schwarzenbach et al., 2006).

Water scarcity due to droughts, a sharp increase in demand or overexploitation and poor water quality leads to food shortages. With the human population expected to increase by 34% in respect of the present situation by the year 2050, there will therefore be an increase in global demand for food, water, and energy (Gutiérrez et al., 2012). Hence, nanotechnology can be used to improve the productivity and efficiency of crop and livestock production, remove or minimize pollutants, and purify water, reduce drug loading and efficacy in medicine, while increasing the safety, bioavailability, nutritional value, and affordability to meet with future population growth trends. Furthermore, nanotechnology is set to make a considerable impact in the water sector as it provides the possibility of water disinfection at the point of use (Smith, 2009). The advantages and disadvantages of relevant nanotechnologies used in water sectors are evaluated later.

3.1 Nanosorbents

Nanosorbents are widely used as separation media in water purification to remove inorganic and organic pollutants from contaminated water (Savage and Diallo, 2005). Nanoparticles used as nanosorbents are attractive because of their large surface area and high affinity toward target compounds compared to bulk counterparts (Gautam et al., 2013). With an aim to understand these particles, several studies have been carried out investigating the unique properties of nanoparticles and their affinity to metal ions and anions (Rajan, 2011; Bhattacharya et al., 2013; Amin et al., 2014). In their study, Li et al. (2003) reported that carbon nanotubes have 3–4 times more sorption capacities than powdered activated carbon with a sorption capacity of 97.08 mg/g for Pb^{2+}, 24.49 mg/g for Cu^{2+} and 10.86 mg/g for Cd^{2+} at pH 5.0 and at room temperature. Furthermore, chitosan nanoparticles have also been reported as having very high sorption capacities for lead (398 mg Pb/g) due to their high affinity for phosphate. Apart from chemical pollutants, several nanosorbents have been studied for their microbial removal capacity (Brady-Estévez et al., 2008; De La Rosa-Gomez et al., 2008; Kang et al., 2009). Based on the aforementioned, nanosorbents have been used in the various water treatment industries and effective nanosorbents are classified in the following three main categories:

1. Inorganic materials (e.g., zeolites), in the form of silver zeolite, have been reported as antibacterial agents against coliform microorganisms from wastewater in a column system (De La Rosa-Gomez et al., 2008).
2. Carbon-based adsorbents (e.g., carbon nanotubes) have demonstrated antimicrobial properties against microorganisms, including bacteria, such as *Escherichia coli* (Kang et al., 2009) and *Salmonella* spp. (Arias and Murray, 2009), and viruses, such as the MS2 bacteriophage (Brady-Estévez et al., 2008).
3. Organic polymers (e.g., dendrimers), such as silver-based complexes and nanocomposites (Fig. 5.3) have been reported to be effective antimicrobial agents in vitro against *Staphylococcus aureus, Pseudomonas aeruginosa,* and *E. coli* (Balogh et al., 2001).

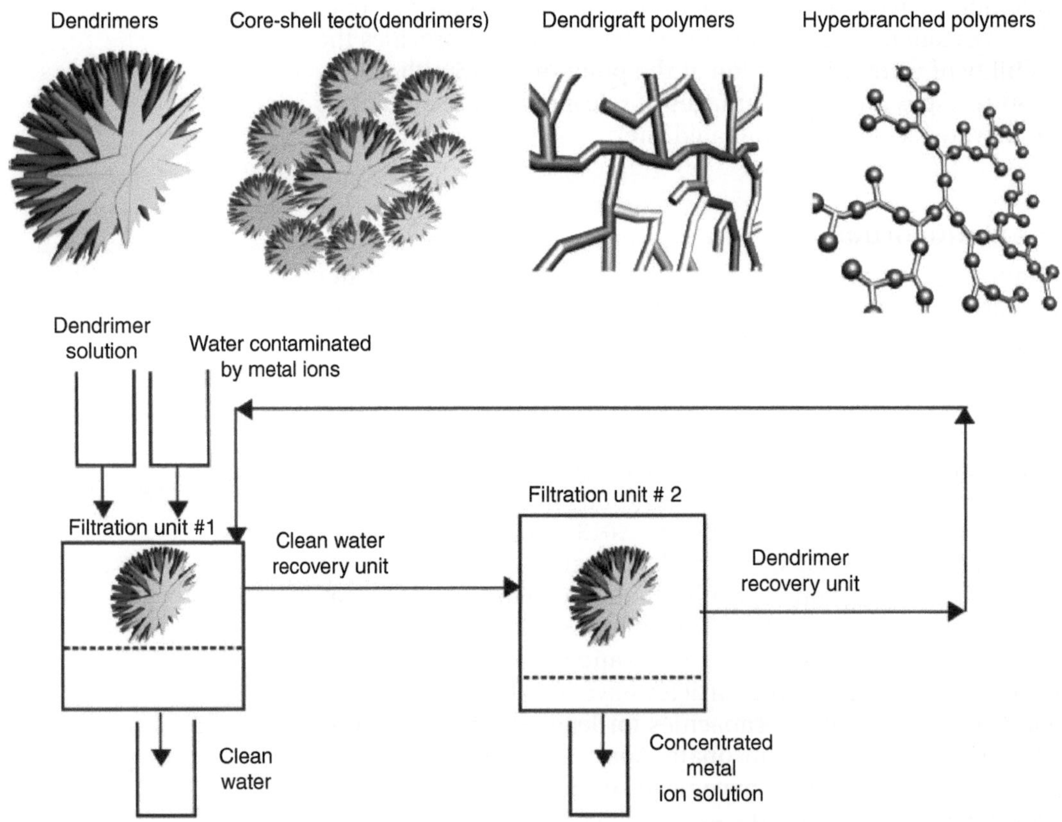

Figure 5.3. Recovery of metal ions from aqueous solutions by dendrimer-enhanced filtration (Theron et al., 2008, Copyright 2005 American Chemical Society).

3.1.1 Carbon-Based Nanosorbents

Carbon-based nanosorbents also called carbon nanotubes or carbonaceous nanomaterials are widely used for the treatment of drinking water and are central to many nanotechnology projects (Amin et al., 2014). These nanosorbents can be as small as 1 nm in diameter, with their length reaching several microns. Since their discovery, carbon-based nanosorbents have attracted considerable attention due to their outstanding structural, mechanical, and chemical properties (Popov, 2004). They are stronger than any other known material in the universe and can be SWCNT or MWCNT (Fig. 5.4). Carbon-based nanosorbents are also some of the most selective nanosorbents and can be used as smart nanophase extraction agents to remove organic pollutants, such as drug molecules from aqueous solution (Rajan, 2011). The selectivity and sorption capacity of carbon nanosorbents depends whether they are single- or multiwalled (Savage and Diallo, 2005). Multiwalled carbon nanosorbents have been reported to have better sorbent capacity for volatile organic compounds and water-soluble dyes (Li et al., 2003). In their study, Mangun et al. (2001) synthesized nanoporous carbon fibers and found that these have much higher organic sorption equilibrium constants than granular activated carbon. Furthermore, Peng et al. (2003) also evaluated the organic sorption capacity of

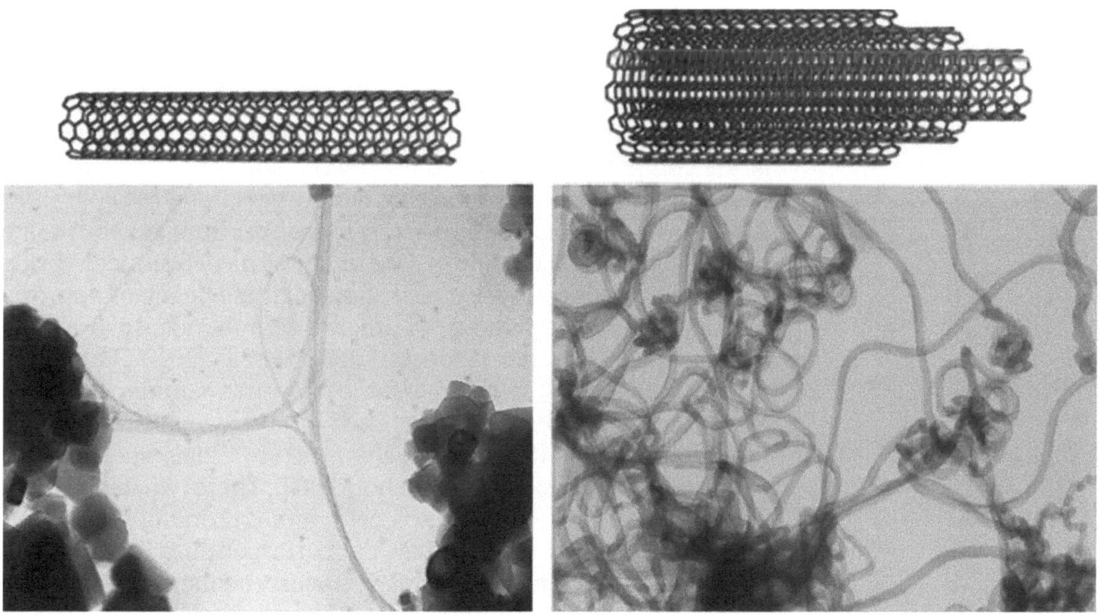

Figure 5.4. Molecular representations of single-walled (left) and multiwalled (right) carbon nanotubes with typical transmission electron micrographs below Donaldson et al. (2006)

carbon nanotubes and revealed that the latter takes only 40 min. to adsorb 1.2 dichlorobenzene and reaches equilibrium with a maximum sorption capacity of 30.8 mg/g. Moreover, the carbon nanotubes' potential for sidewall functionalization and surface modification placed them in an attractive position as one of the best support phases for water treatment (Wang et al., 2005, 2006). It should be mentioned that even though studies have shown potential efficiency of carbon nanotube as adsorbents for a variety of pollutants in drinking water, their high cost and poor solubility constantly hamper their practical application (Theron et al., 2008).

3.1.2 Polymeric Nanosorbents

Advances in nanotechnology have resulted in the synthesis of polymeric nanosorbents with chemically and structurally different properties. Polymeric nanosorbents, such as dendrimers have received much attention due to both their fast rate adsorption and their selective sensitivity (Pradeep and Anshup, 2009). Dendrimers are three-dimensional globular nanoparticles with a core, interior branch cells, and terminal branch cells (Diallo et al., 2009). These polymers display several critical physicochemical properties that make them attractive as separation and reaction media for water purification and can remove organic and inorganic pollutants, as well as microbes (Tully and Frechet, 2001). During their application, organic pollutants are adsorbed by the interior hydrophobic shells, while inorganic pollutants, such as metals are adsorbed at the exterior branches by the tailored (Gehrke et al., 2015). Dendrimers are basic building blocks that enable specific nanostructures to be built to meet the existing needs and solve the evolving problems. To tackle complex issues, hybrid polymeric nanosorbents have been developed, such as incorporating dendrimers in an ultrafiltration device to improve its removal ability for copper from aqueous solution (Diallo et al., 2005). Furthermore, chitosan-dendrimer hybrids have been produced which also show high efficiency of approximately 99% for the removal of anionic compounds, such as dye from textile wastewater (Sadeghi-Kiakhani et al., 2013). Poly (propylene imine) (PPI) dendrimers and polyamidoamine (PAMAM) dendrimers have been reported as the most widely studied and useful dendrimers due to their chemical structures and physical characteristics (Pradeep and Anshup, 2009). These two dendrimers are comprised of a core, branch cells, and a terminal group, and PPI is considerably smaller than the corresponding PAMAM. Beside the high efficiency of polymeric nanosorbents, especially their hybrid counterparts, in removing pollutants from drinking water, their commercialization has encountered problems due to the cost involved during the complex multistage synthesis. So far, there are only very few companies in the People's Republic of China that have attempted to venture into the commercialization of these precious nanosorbents (Gehrke et al., 2015).

3.2 Nanocatalyst and Magnetic Nanoparticles

A catalyst is a substance that promotes the chemical reactions of other materials without becoming permanently involved in the reaction (Hillie et al., 2006). Nanocatalysts and magnetic nanoparticles coated with different compounds, for instance natural chitosan, have been used to remove pollutants from the aqueous environment and groundwater due to the presence of selective affinity for diverse contaminating substances (Fig. 5.4) (Singer et al., 2005; Hillie et al., 2006). Nanocatalysts have been found to have a very good degradation and reduction potential, are more reactive, remove a wider range of environmental contaminants, easier to inject and achieve deeper subsurface penetration (Hillie et al., 2006; Lukhele et al., 2010). Furthermore, additions of chemicals are not required, it is easy to operate and maintain and it is selective on the substances to be removed (Zhang, 2013). They, however, require supporting substrates for their application in water treatment (Lukhele et al., 2010). It has been reported that the coating of magnetic nanoparticles with citric acid allows metal ions to bind and produces a high affinity, therefore enabling the removal of heavy metals from soil and water (Fig. 5.5) (Singer et al., 2005; Tang and Lo, 2013). Furthermore, these materials are recycled and reused, hence reducing treatment cost, and eliminating energy-related cost (Singer et al., 2005; Hillie et al., 2006). The sorption capacity of magnetic NMs is higher,

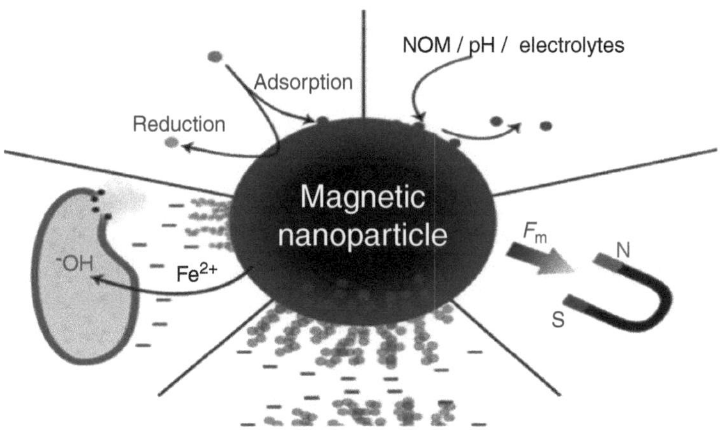

Figure 5.5. Contaminant removal mechanisms of magnetic nanoparticles.
Adapted from Tang and Lo, 2013

therefore they are considered to be effective adsorbents at low pH, irreversible sorption providing an efficient storage sink for collection of waste (Hillie et al., 2006). Zero-valent metal oxides especially bimetals have been classified as one of the metallic nanoparticles with strong catalytic properties (Cheng and Wu, 2000). These nanocatalysts mostly catalyze the reduction of inorganic compounds (metal ions) but can also catalyze organic compounds (chlorinated organic pollutants). In the past decade, titanium dioxide (TiO_2) has emerged as the prominent nanocatalyst for water treatment for its well-known advanced oxidative and reductive process and is being used mainly for the photo-degradation of organic pollutants, such as phenolic compounds (Lazar et al., 2012).

One of the most important applications in the field of nanotechnology is the development of nanosensors to detect and identify pollutants and pathogens in polluted waters. Appropriate treatment methods, that is, nanotechnology, are then used to progressively remove dissolved salts, degrade water contaminants, capture particles, and pathogenic microbes to achieve cleaner water and treat wastewater. Through the innovation of such technology a real-time detection of the presence of contaminants and pathogens is possible without laboratory testing of samples and time-consuming techniques. Sensors using nanometer-scale detectors will enhance and improve health, maintain safe food, and water supply while, allowing for use of otherwise unusable water resources (Hillie et al., 2006; Qu et al., 2013). Multiwalled carbon nanotubes and single-walled carbon nanotubes have been functionalized with cerium dioxide, hydroxyl, phosphate functional groups to remove heavy metals and organics from water (Lukhele et al., 2010). Nanomagnets have also been developed for the removal of heavy metals. Other NMs that are bioactive, such as silver and magnesium oxide, can kill bacteria and might be used in place of chlorine to disinfect water (Lukhele et al., 2010).

3.3 Nanofiltration Membranes

A breakthrough in drinking water production was made by the development of composite membranes, called nanofiltration (NF) membranes, also known as polymeric permeable membranes, for the removal of pollutants by providing the possibility of refining, improving water, speed, and accuracy (Fig. 5.6) (Van Der Bruggen and Vandecasteele, 2003; Mousavi and Rezaei, 2011). These nanostructured filtration membranes with a pore size of around 0.001 μm are pressure-driven membrane filtration processes and are able to remove organic matter, viruses, cysts, and salts (Schutte, 2006). Nevertheless, membrane fouling is a major drawback in this filtration process and poses a serious problem that challenges the viability of the use of membrane. However, there are not many other problems

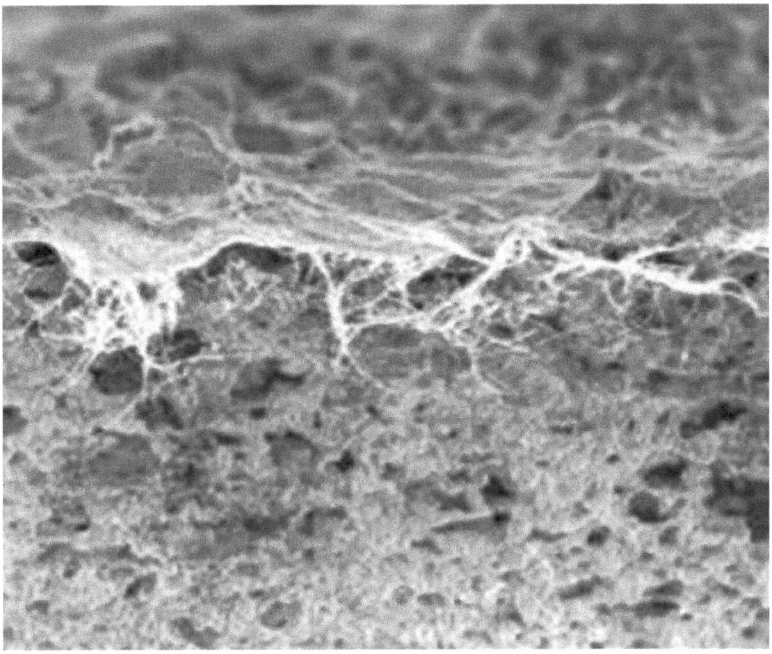

Figure 5.6. Plasma-treated nanofilters: cross-sectional image of the membrane. Adapted from image courtesy of (Yang et al., 2013)

linked to these techniques except the cost factor, which is also an issue. This new technique is a pressure-driven membrane process used between ultrafiltration membranes and reverse osmosis. NF membranes are capable of filtering both bacterial and viral pathogens (Van Der Bruggen et al., 2003; Singer et al., 2005; Grimshaw et al., 2009), hard, natural organic material (NOM), micropollutants, such as pesticides and VOCs, salt, nitrates, and arsenic can be simultaneously or partly removed (Van Der Bruggen et al., 2001; Van Der Bruggen and Vandecasteele, 2003). These membranes are cheaper to operate, energy saving, have a fast flow rate, durable, easy to clean, reusable, and heat-resistant when compared to reverse osmosis membranes, making them to be the latest innovation in the membrane technology with great potential in the desalination of brackish waters (Hillie et al., 2006; Lukhele et al., 2010). Research has been conducted by the Stephen and Nancy Grand Water Research Institute of Israel on the treatment and conversion of salt water to freshwater for drinking and irrigation purposes using reverse osmosis (Hillie et al., 2006). NF membranes are able to retain nutrients present in water that are essential and required for normal functioning of the body. Uniformly vertically aligned and tightly packed carbon nanotubes coated with silicon wafer are able to remove almost all kinds of water contaminates.

Since South Africa and many sub-Saharan countries are seen as water-scarce countries, NF possibilities to provide water through the desalination of saline water will be a lifesaving situation. NF products are currently developed by the University of Stellenbosch and the University of North-West in South Africa (Hillie et al., 2006). Also, these technologies are proposed to be an alternative for reverse osmosis and can be used for desalination (Hillie et al., 2006). Nanotechnology has lessened some of the labor-intensive and chemical methods required to treat water. As a result, more people are expected to access clean drinkable water in African countries, and more countries are now warming to the idea of water technology.

In a study done to calculate investment requirements (Van Der Bruggen et al., 2001) it was indicated that the final price for the treatment of groundwater is realistic, showing that NF is a valuable option for groundwater treatment. Water quality and quantity affect peoples' livelihoods and also their health status and this should be taken into consideration. Theron et al. (2008) reported the following water filtration membranes produced from NMs: (1) nanostructured membranes from NMs, such as carbon nanotubes, nanoparticles, and dendrimers; and (2) nanoreactive membranes from metal nanoparticles and other NMs.

3.3.1 Nanocomposite Membranes

The nanocomposite membranes consist of mixed matrix membranes and surface-functionalized membranes, such as nanofillers associated to metallic oxide nanoparticles, such as Al_2O_3 or TiO_2 (Pendergast and Hoek, 2011). The metal oxide NPs assist in improving the stability (mechanic or thermal) and the permeability of the membranes. Mostly, the resistance of these membranes to fouling, as well as the antimicrobial properties is increased by the addition of bimetallic nanoparticles (e.g., FeCuO) or metal oxide nanoparticles (e.g., TiO_2). Recently, in their study, Kim and Deng (2011) have reported a new polyamide thin-film nanocomposite (PA-TFN) membrane with hydrophilized ordered mesoporous carbon (H-OMC) for water treatment. However, one of the main challenges of these techniques is membrane clogging (Bae et al., 2006). To avoid it, membranes are functionalized by coating them with organic-oxidizing substances, such as metal oxide NPs, and bimetallic NPs (Pendergast and Hoek, 2011).

3.3.2 Nanofiber Membranes

Nanofiber membranes, also called electrospun nanofibrous membranes, have attracted increasing attention in the field of water treatment as they possess several unique attributes, such as a high porosity and an interconnected open pore structure, a higher surface area to volume ratio, and submicron pore sizes (Huang et al., 2013). Beside their effectiveness in air filtration, these membranes have been successfully used for the filtration of water (Yoon et al., 2009). However, NF membranes have mostly been avoided because of their poor mechanical strength, weak bonding between fibers, nonwoven and highly porous structure (Huang et al., 2013). Membrane processes using different types of membranes are becoming increasingly popular for the production of drinking water from seawater, brackish water, wastewater, surface water, and groundwater (Ventresque et al., 2000). Table 5.3 illustrates a number of nanostructured and nanoreactive membranes used in the water filtration technology.

Table 5.3 Some Nanostructured and Nanoreactive Membranes Used in Water Filtration Technology (Theron et al., 2010)

Membrane	Pollutants	References
Carbon nanotubes	Bacteria and viruses	Srivastava et al. (2004)
Nanocapillary array membranes (NCA)	*	Chatterjee et al. (2005)
Alumina membrane formed from A-alumoxane	Synthetic dyes	DeFriend et al. (2003)
Alumina membranes functionalized with poly(styrene sulfonate) or poly(allylamine hydrochloride)	Divalent cations	Stanton et al. (2003)
Silica and cellulose-based membranes functionalized with amino acid homopolymers	Metal ions	Bhattacharyya et al. (1998); Ritchie et al. (1999); Ritchie et al. (2001)
Polycarbonate track-etched membranes functionalized with amino acid homopolymers	Metal ions	Hollman and Bhattacharyya (2004)
Pt/Fe laden cellulose acetate film	Trichloroethylene (TCE)	Meyer et al. (2004)
Zero-valent Fe-laden cellulose acetate membrane	TCE	Wu et al. (2005)
Ni/Fe or Pd/Fe laden polyacrylic acid/polyether sulfone composite membranes	TCE	Xu et al. (2005)
Ni/Fe laden cellulose acetate membrane	TCE	Wu and Ritchie (2006)
Polymer-impregnated ceramic alumina and silicon-carbon filters	Trihalogen methanes, PAHs, pesticide	Allabashi et al. (2007)
Polymer-impregnated ceramic TiO_2 filters	Polycyclic aromatic hydrocarbons (PAHs)	Arkas et al. (2006)
Alumina or polymeric membranes with gold nanoparticles	4-nitrophenol	Dotzauer et al. (2006)

3.4 Advantages and Disadvantage of the Techniques

Nowadays, the need for safe drinking water is increasing fast due to the rapid population growth, climate change leading to severe droughts and contamination of water bodies. This situation is more severe as approximately 748 million people still lack access to improved drinking water sources (WHO/UNICEF, 2014). In the past decade, researchers have identified potential benefits of nanotechnology (Table 5.4) in many fields including the water sector and stated that nanotechnology will bring innovative answers to most of our existing scientific questions (Savage and Diallo, 2005; Theron et al., 2008; Fleischer and Grunwald, 2008). Hence, nanotechnology is going to play an important role in addressing fundamental issues, such as health, energy, and water (Grimshaw et al., 2014). Water purification using NF technology or through adsorption and catalytic degradation processes was made possible by the advances achieved and mysteries revealed in the quantum world. The impact of nanotechnology on the development of tools and techniques for water treatment will be more pronounced in the near future. As lack of safe drinking water in many developing countries hampers the advancement and the social security of their communities, nanotechnology in water recycling, seawater desalination, and water remediation is more efficient and cost-effective and is expected to provide the solution (El Saliby et al., 2008).

Table 5.4 Some Potential Applications of Nanotechnology in Water Treatment (Amin et al., 2014)

Application	Examples of Nanoparticles	Some Novel Properties
Adsorption	Carbon nanotubes/nanofibers and metal oxide nanoparticles	High specific surface area and accessible adsorption sites, selective and more adsorption sites, short intraparticle diffusion distance, tunable surface chemistry, easy reuse, and so forth
Disinfection	Nanosilver/titanium dioxide and carbon nanotubes	Strong antimicrobial activity, low toxicity and cost, high chemical stability ease of use, and so on.
Photocatalysis	Nano-TiO_2 and fullerene derivates	Photocatalytic activity in solar spectrum, low human toxicity, high stability and selectivity, low cost, and so on.
Membranes	Nano-Ag/TiO_2/zeolites/magnetite and carbon nanotubes	Strong antimicrobial activity, hydrophilicity low toxicity to humans, high mechanical and chemical stability, high permeability and selectivity, photocatalytic activity, and so on.

In South Africa, for example, cost-effective NF membranes have been used for treating contaminated water at a point of consumption in rural areas (Smith, 2009). Nanotechnology is considered to be tool ensuring sustainable development of rural communities in developing and developed worlds and this is in relation to its capacity to produce safe drinking water from seawater, contaminated water, and wastewater (Theron et al., 2008). Nevertheless, nanotechnology applications show two apparent challenges, namely (1) the loss of nanoparticles' antimicrobial or antiviral activities; and (2) bacteria may become resistant to these novel nanoparticles. Furthermore, it is also widely accepted that the same properties which render nanoparticles novel for a wide range of applications also represent a risk associated to their toxicity to human and environment as their inherent physicochemical properties allow them to penetrate into any type of cells. Moreover, the understanding of these risks at an early stage and the practical approach toward the management of such risks becomes critical. The environmental fate, transport, and toxicity of any material are critical issues for water purification. Not much is known about the environmental fate, transport, and toxicity of NMs (Colvin, 2003; Lecoanet et al., 2004). No systematic investigations of the hydrolytic, oxidative, photochemical, and biological stability of NMs (e.g., dendrimers, carbonaceous nanoparticles, metal oxides) in natural and engineered environmental systems have been conducted.

4 The Governmental and Institutional Involvement Across Africa

A necessary condition to the development of adapted and appropriate technologies is the presence of sufficient research and development investments, whether in the public or private sector (Cozzens et al., 2013). Lack of support for nanotechnology science and applications in developing countries could force these countries to either abandon nanotechnology options or import (transfer) existing technologies from other countries. There is a lack of emphasis on societal issues and risk assessment principles yet economies and public sectors have heavily invested in nanotechnology.

Meanwhile (Saidi, 2009), emphasizes the point on strengthening worldwide capacities in science, technology, and innovation to provide prospects for human development of which nanotechnology is expected to play an instrumental role. It is paramount that the technology developed can be transferred to the country that needs it the most; however, there should also be an appropriate and effective approach, such as technological adaptations, technical capability, infrastructure, and market potential to prevent unforeseen failures of the technology (Hillie and Hlophe, 2007).

The involvement of the private sector is tricky because their main concern is whether there is sufficient infrastructure, basic capital, and sufficient economic incentives to invest in nanotechnology. The uncertainties of investment and technologies make it almost impossible for domestic companies to invest in long-term research and development programs and lack of access to capital (Cozzens et al., 2013). However, African countries, such as Egypt, South Africa, Botswana, Nigeria, Morocco, Tunisia, Kenya, Sudan, Algeria, Ethiopia, and Zimbabwe have made investments and local research teams are involved in nanotechnology.

In South Africa, nanotechnology and its applications have been the most successful as compared to other African countries. This successful introduction of nanotechnology was due to the national government funding motivating stakeholders, research institutions, and academic institutions to carry out more research on nanoparticles' benefits and the risks associated with the usage. Moreover, the Department of Science and Technology (DST) launched a 10-year Innovation Plan for Nanotechnology and that had given rise to other bodies. The South African approach to nanotechnology shows how the country's policy is influenced, albeit timidly, by national interests (e.g., risk assessment, occupational health, and strategic social focus) (Musee et al., 2011).

In 2002, the South African Nanotechnology Initiative (SANi) was launched, followed by National Nanotechnology Strategy (the Strategy) that was published in 2006, with the aim to support and promote nanotechnology research and development (R&D), and develop human capacity in this field. Later on, National Innovation Centers (NICs) were developed containing two centers: (1) National Centre for Nano-Structured Materials (NCNSM) at the Council for Scientific and Industrial Research (CSIR) and (2) Mining Technology (Mintek). They serve as national multiuser research facilities. Both NICs are financially supported by the DST and have established partnerships with private companies, for example, SASOL, ECO-Structure International, Biomass Corp., and De Beers. Since 2007, CSIR at the Natural Resources and the Environment (NRE)/DST has been collaborating with academic institutions, the Water Research Commission, and international academic institutions to assess the environmental risk and management associated with NMs (Cele et al., 2009).

Apart from South Africa, Maghreb countries, such as Egypt, Morocco, and Algeria have shown high interest in nanotechnology. In Egypt, the Nanotechnology Research Centre has been funded by the Information Technology Industry Development Agency and the Science and Technological Development Fund in partnership with IBM, and was launched in 2009. This project has been initiated to probe the possibility of developing devices for enhancing health provision, and water purification, mostly targeting the rural areas. While the UN Department of Economic and Social Affairs (DESA), for example, is engaged in providing support to the construction of high-tech R&D and industrial parks in Ghana and Senegal, Egypt and Kenya already have techno parks and Ethiopia is currently developing a similar project (Masoka et al., 2012; Musee et al., 2011). In Morocco, projects on the implementation of an international laboratory for molecular chemistry, creation of a Euro-Mediterranean competence pool in microtechnology and nanotechnology, purification and preservation of Moroccan water resources, and urban waste treatment have already been started (Bouoiyour, 2006). In the year 2006, National Initiative for Nanosciences and Nanotechnologies was launched, and an industrial high-tech park was built in Rabat (Technopolis) (Masoka et al., 2012).

5 Case Studies on Cost-Effective Drinking Water Purification Systems for Developing Countries Particularly sub-Saharan Countries

This study was conducted using groundwater sources to assess the performance of the cation resin-silver nanoparticle filter systems in inhibiting bacterial regrowth and biofilm formation in drinking water distribution systems using a laboratory-scale unit (not published). The effectiveness of cation resin-silver nanoparticle filter systems was compared to that of chlorine for a period of 30 days. Plastic-based pipe materials (polyvinyl chloride-PVC) and metallic-based pipe materials (galvanized steel-GS), which are generally used in drinking water distribution systems were used for the purpose of this study. The disinfection process was carried out using c. 2.5 mg/L free chlorine and c. 0.030 mg/L free silver ions in a cation resin-silver nanoparticle filter system. Non-disinfected water was added as a control water system. The evaluation of the process relied on coliforms and heterotrophic plate count bacteria. Scanning electron microscopy (SEM) was utilized to view the regrowth and formation of biofilm on PVC and GS piping material.

5.1 Laboratory-Scale Test Unit

The filter system was made of a PVC column, with a diameter of 2 cm and a length of 30 cm. Each filter was packed with 40 g of silver nanoparticle-coated resins as illustrated in Fig. 5.7. The silver nanoparticle-coated resins were synthesized and characterized according to the method previously described by Mpenyana-Monyatsi et al. (2012) and Mthombeni et al. (2012). The biofilm study was performed by using a modified Pedersen device described by Momba et al. (1998, 1999, 2000) and Momba and Kaleni (2002). Three laboratory-scale systems (Fig. 5.7) were used during the study period (raw intake water, chlorinated water and resin-silver nanoparticle disinfected water). The system for the raw intake water and chlorinated water consisted of a 1 × 50 L^2 polyethylene drum, which was attached to a peristaltic pump, a Pedersen device (26× 10× 3 cm^3) and a tap, while the system for the resin-silver nanoparticle water consisted of a 1 L × 50 L polyethylene drum, which was attached to a peristaltic pump, a resin-silver nanoparticle filter system, a Pedersen device (26× 10× 3 cm^3) and a tap. All components were connected using latex tubing (8 mm diameter, 4 m length). Groundwater sample was continuously pumped into the system. The flow rate of all the water samples through the system was 300 mL/h.

PVC and GS materials are generally used for drinking water distribution systems and were therefore considered in order to study bacterial regrowth and biofilm formation. Twenty coupons of each

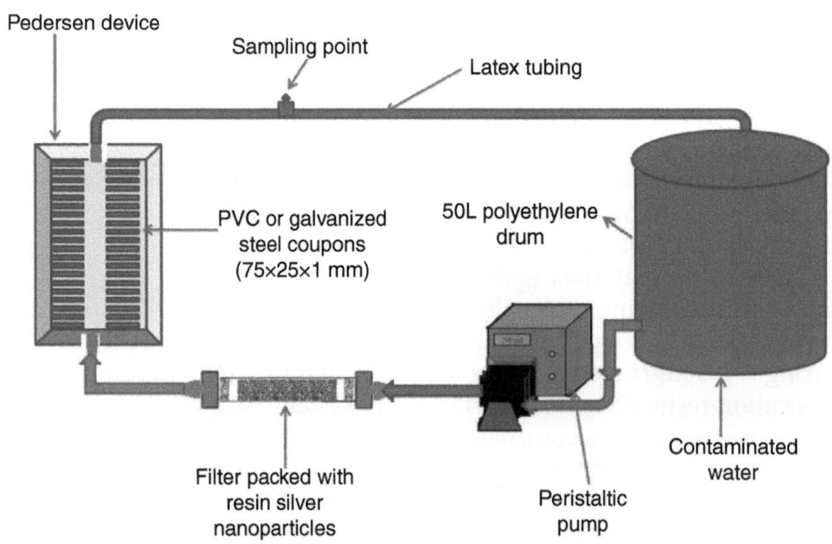

Figure 5.7. Schematic diagram of laboratory-scale unit for biofilm formation.

piping material (PVC and GS with the size of a microscope slide of 75 mm x 25 mm x 1 mm) were vertically inserted into the Pedersen device. Prior to use, all three systems were thoroughly washed with sterile distilled water. The coupons were cleaned in a detergent solution, rinsed first in tap water and then disinfected with chorine at 5 mg/L for 24 h. Thereafter, the coupon was dechlorinated with sodium thiosulphate(± 17.5 mg/L) and flushed with sterile distilled water to remove any residues (Momba et al., 1998).

5.2 Treatment of Groundwater

5.2.1 Disinfection of Water Source

Chlorine and resin-silver nanoparticle filter systems were used to disinfect the groundwater collected from the Delmas borehole in Mpumalanga Province, South Africa. Prior to the circulation of the groundwater through the chlorinated system, the water was subjected to a chlorination process. The chlorinated water was produced by adding 2.5 mg/L free chlorine into a 50 L drum that contained the groundwater water sample. For the resin-silver nanoparticle system, the raw intake groundwater from the 50 L drum was pumped continuously into the resin-silver nanoparticle filter system for disinfection before entering the Pedersen device as depicted in Fig. 5.7.

5.2.2 Physicochemical and Microbiological Analyses

The concentration of residual free chlorine was measured by *N*, *N*-diethyl-*p*-phenylenediamine (DPD powder pillows, Hach) using the chlorine colorimeter C401 (Eutech instrument), and the residual silver ion was measured by atomic absorption spectrophotometry (SpectrAA 220FS), according to the standard method (Apha, 1998). The temperature, pH and turbidity of the water samples were also measured using the standard method (Apha, 1998). Nitrates (NO_3^-), sulfates (SO_4^{2-}), and total organic carbon (TOC) concentrations were determined by using the Spectroquant Nova 400 manual water analyzer (Merck, South Africa) and the photometric test kits (Merck, South Africa). Table 5.5 summarizes the physicochemical parameters.

For the microbiological analysis of the quality of groundwater in the nondisinfected water system, water samples were collected in sterile glass bottles, while for the treated water systems, the collection was done using sterile bottles, which contained 1 mL of 15% sodium thiosulfate to neutralize the disinfection process. Total coliforms, fecal coliforms and presumptive *E.coli* were detected by using membrane filtration techniques incorporating chromocult coliform and mFC agar (Merck, South Africa). The plates were incubated at 37°C for 24 h. Heterotrophic plate count (HPC)

Table 5.5 Physicochemical and Microbiological Characteristics of Tested Water Prior to Treatment During the Study

Parameters	Units	Raw Water	Chlorinated Water	Ag Nanoresin	Sans-241 (2011)
pH		7.38 ± 0.05	7.21 ± 0.10	7.36 ± 0.09	5–9.5
Temperature	°C	24 ± 0.14	24 ± 0.11	24 ± 0.12	≤25
Turbidity	NTU	3.28 ± 0.11	3.26 ± 0.15	3.28 ± 0.10	≤1
SO_4^{2-}	mg/L	26.36 ± 1.18	29.05 ± 0.91	26.36 ± 1.18	≤250
NO_3^{2-}	mg/L as N	1.04 ± 1.02	1.06 ± 1.04	1.04 ± 1.03	≤10
TOC	mg/L	2.30 ± 0.18	2.36 ± 0.16	2.30 ± 0.18	≤10
Free chlorine	mg/L	0	2.5 ± 0.02	N/A	0.5
Silver ions	mg/L	N/A	N/A	0.03	—
Total coliforms	Per 100mL	1.4×10^4	1.4×10^4	1.4×10^4	≤10
Fecal coliform	Per 100mL	2.8×0^3	2.8×10^3	2.8×10^3	—
E.coli	Per 100mL	3.0×10^2	3.0×10^2	3.0×10^2	—

bacteria were enumerated by using a standard spread plate procedure incorporating nutrient agar (Merck), according to the Standard Methods for Examination of Water and Wastewater (Apha, 1998). Analyses were carried out in triplicate.

5.3 Biofilm Formation

Attached viable counts—For each type of piping material (PVC and GS: 75× 25× 1 mm³), 2 coupons were removed from the Pedersen devices after 24, 48, and 72 h from all 3 systems. Thereafter, the coupons were removed once a week for a period of three weeks with the exception of the fourth week where coupons were removed twice. The removed coupons were subsequently immersed in 50 mL sterile centrifuge tubes with 30 mL sterile saline water. The tubes were then vortexed for 2 min using a vortex mixer to detach the biofilms from the coupon surface. The resultant suspensions were then serially diluted in a sterile saline solution (0.9% w/v NaCl). For the enumeration of attached heterotrophic plate count bacteria, 100 μL of diluted biofilm suspension were then placed on spread plates in triplicate onto nutrient agar plates and finally incubated at 35°C for 48 h. The following equation was used to calculate the number of HPC bacteria (cfu/cm²):

$$\text{HPC}\,(\text{cfu}/\text{cm}^2) = N \times D/\text{Surface area of slides} \qquad (5.1)$$

Where N, average number of colonies and D, dilution factor (Momba and Kaleni, 2002).

Visualization of biofilm formation—One coupon of each material (25 mm × 25 mm × 1 mm) was removed from the Pedersen devices after 24,336 and 720 h from all 3 systems. The removed coupons were then rinsed with sterile MilliQ water for 30 s to remove any unattached cells. Subsequently, the coupons were fixed and treated sequentially with 20 mL of 2.5% glutaraldehyde (30 min); 0.15 M phosphate-buffer (3 min × 15 min); 50% ethanol (1 min × 15 min); 70% ethanol (1 min × 15 min); 90% ethanol (1 min × 15 min); and 100% ethanol (3 min × 15 min). Thereafter, the coupons were dried for 3 h using supercritical carbon dioxide. Finally, the dried samples were coated with gold plasma so as to make them visible under the scanning electron microscope (SEM) (JEOL JSM 840).

The results in Table 5.5 indicate that the physicochemical characteristics of the intake groundwater complied with the limits set by SANS (South African National Standard) 241, with the exception of turbidity levels that were above the recommended limit (≤1 NTU) (WHO, 2006; Sans-241, 2011). It is well known that high levels of turbidity enhance the survival of microorganisms in water and render the disinfection process ineffective (WHO, 2006). Although the temperature of the intake water was within the SANS 241/WHO recommended limits, it has been affirmed that temperatures above 15°C contribute to the formation of biofilm in drinking water systems (Power and Nagy, 1999; Zacheus et al., 2000). It is evident that both turbidity and temperature could influence bacterial regrowth and biofilm formation in treated water systems.

Results of this study also revealed that the microbiological quality of the Delmas borehole water did not comply with the recommended limits (Table 5.5). The average counts for heterotrophic culturable bacteria, total coliforms, fecal coliforms, and *E.coli* by far exceeded the limits that were set for water intended for human consumption (Table 5.5). Such water sources may increase the risks of infectious disease transmission if they are utilized for domestic use without prior effective treatment. These results corroborated the findings of earlier studies that reported an outbreak of waterborne diseases within the Delmas community who totally depended on boreholes as their unique source of water supply in 2005. Typhoid and diarrheal diseases resulted in 49 deaths within this community due to the consumption of water from these water sources (Masinga, 2005; Nel et al., 2009).

Although an appropriate concentration of free chlorine (2.5 mg/L) was used to treat the Delmas borehole water during the study period, the treated water that contained free chlorine residual of 1.7 mg/L could not produce safe drinking water within the first 30 min of contact. Heterotrophic bacteria and coliform bacteria including *E.coli* were still present in chlorinated water even in the presence of the free chlorine residual (0.7 mg/L). Virto et al. (2005) discovered those two Gram-positive organisms, *Listeria monocytogenes* and *Bacillus subtilis*, and two Gram-negative organisms, *Yersinia enterocolitica* and *E. coli* were more resistant to chlorine concentrations of between 0.3 mg/L and 0.7 mg/L and between 0.8 mg/L and 1 mg/L. During the present study, the potential for bacterial regrowth was observed between day 2 and day 30, and was accompanied by an increase in the bacterial counts from 2 log to 6 log cfu/mL in the chlorinated water system (Fig. 5.8B). This confirms the report by Momba (1997) and Williams and Braun-Howland (2003) that microorganisms can regrow in water even in the presence of residual chlorine. Contrary to the chlorination process, the bacterial regrowth in water treated with resin-silver nanoparticle filter system was only observed between day 21 and day 30 with a bacterial count increase of up to 3 log cfu/mL (Fig. 5.8C). This system provided a longer silver residual of 0.03–0.01 mg/L that inhibited the bacterial regrowth, and this residual concentration was by far less than the limit set by the WHO, which is 0.1 mg/L (WHO, 2006). Similarly, Lin et al. (1996) achieved a complete inhibition of *Legionella pneumophila* from hospital hot water recirculating systems with residuals of silver ion between 0.04 mg/L and 0.08 mg/L for a period of 5 days. The bacteriostatic properties of the silver ion have been reported by previous investigators, who found that this chemical disinfectant inactivates bacteria by disrupting the disulfide bonds of proteins in their cell membrane or by inhibiting DNA replication (Oyanedel-Craver and Smith, 2008; Sharma et al., 2009). The results of the present study suggest that the bacterial regrowth in a resin-silver nanoparticle filter system was due to the depletion of free residual disinfectants. The maintenance of a sufficient and stable silver ion residual may therefore be considered to be a prerequisite in preserving the integrity of this system. However, it is important to note that the concentration of silver ions in drinking water must be within the recommended limit.

Although the formation of biofilms on PVC and GS coupons was observed in all the water systems, the attached viable bacterial counts were noticeable on the surfaces of both piping materials exposed to untreated and chlorinated water systems within the first day of their exposure (Fig. 5.9A–B), with an average of 0.5 log

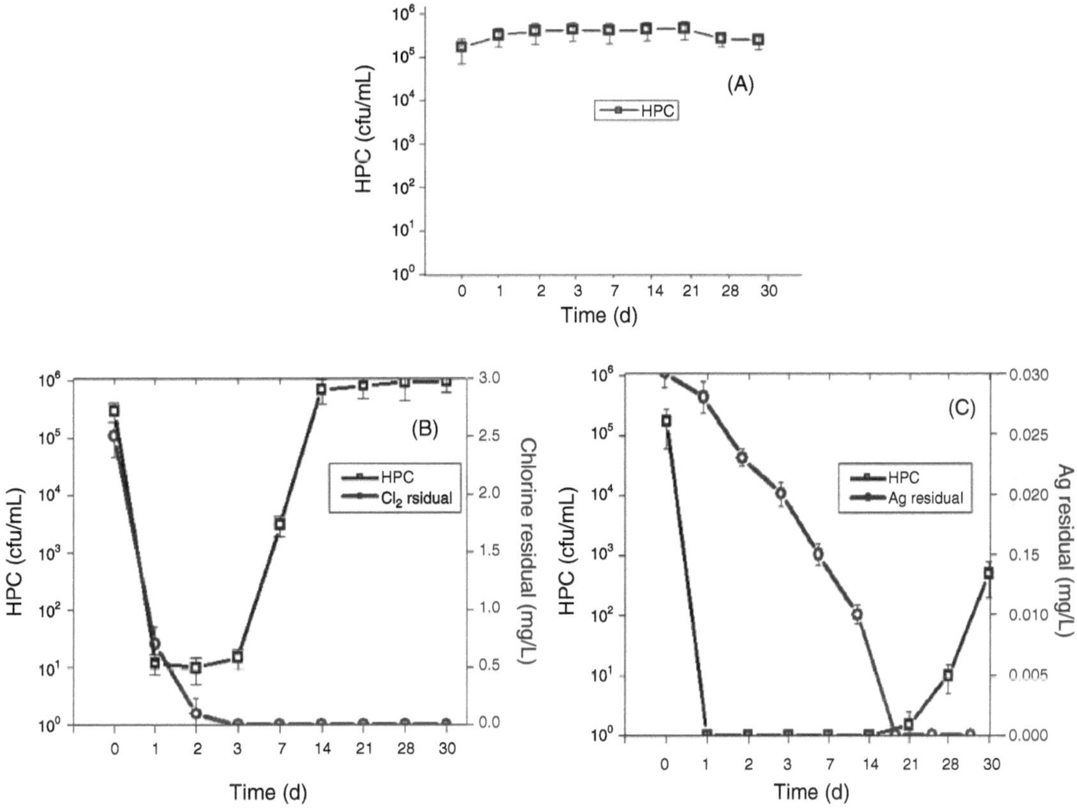

Figure 5.8. Growth and regrowth of heterotrophic plate count in (A) untreated groundwater system (control); (B) chlorinated groundwater system; and (C) resin–silver nanoparticle filter system during the experimental study

and 1 log cfu/cm^2 of HPC colonizing the PVC and GS coupons inserted in the untreated system, respectively. Between day 1 and day 30, the viable bacterial counts on these piping materials in both systems progressed to a maximum average of 4.8 log and 6 log cfu/cm^2 HPC for PVC and GS coupons, respectively. There was a significant difference ($p < 0.05$) in the viable counts attached to both coupons throughout the trial. Similar observations were also reported by Momba and Kaleni (2002) in terms of bacterial adhesion on the piping materials exposed to untreated and chlorinated water systems. Various researchers also observed the formation of biofilm on plastic and GS surfaces (Lehtola et al., 2004; Ilhan-Sungur et al., 2007; Dog˘ruöz et al., 2009). Although, in the present study, the formation of biofilm on PVC and GS was quite similar, the results indicated that the HPC values were slightly higher on GS than on PVC pipes. Tsvetanova (2006) also reported higher bacterial counts in steel pipes than in plastic pipes. The attachment of the

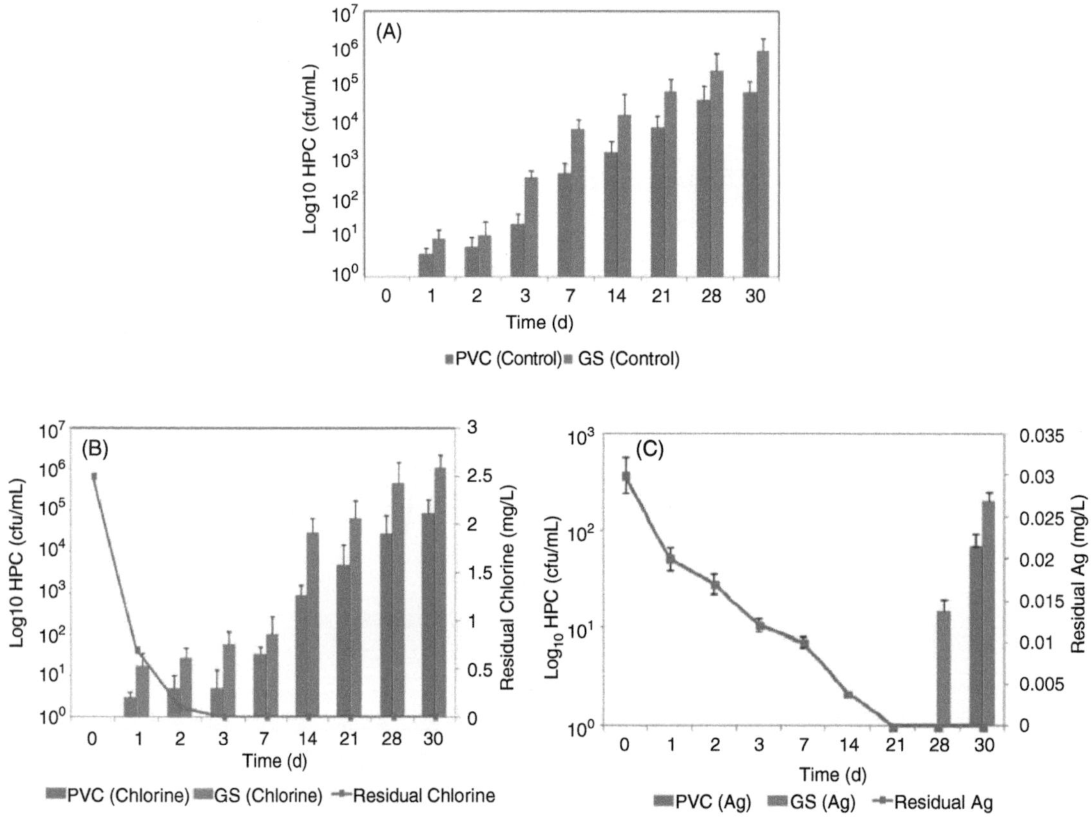

Figure 5.9. Average counts of viable bacteria attached to the surface of PVC and GS coupons exposed to (A) untreated groundwater system (control); (B) chlorinated groundwater system; and (C) resin–silver nanoparticle filter system

heterotrophic plate count bacteria to the surface of test piping materials exposed to chlorinated water is supported by negative weak correlations of free chlorine residual concentrations with bacterial numbers ($r = -233$ for PVC and $r = -245$ for GS).

With the resin-silver nanoparticle filter system, the results indicate that this system inhibited biofilm formation due to the antibacterial activity of the silver ions, even though this activity decreases over time (Fig. 5.9C). A complete inhibition of biofilm formation was achieved in both PVC and GS piping materials for a period of 27 days with a residual silver ion ranging between 0.030 mg/L and 0.010 mg/L. Findings by Liu et al. (1998) indicated that a small concentration of silver ion can be used to inhibit bacterial counts in water systems. Negative correlations were found between silver ion residual concentrations and attached bacterial numbers under the silver nanoparticles resin filter treatment ($r = -0.326$ for PVC and $r = -0.351$ for GS). Compared to the chlorine treatment, these negative correlations were found to be much stronger and more significant (at $p < 0.05$) in the resin-silver nanoparticle filter groundwater system.

The SEM clearly exhibited the inhibition and the presence of biofilms on the surfaces of both piping materials using different water systems from day 1 to day 30 (Fig. 5.10A–F). SEM assessment of uncontaminated GS and PVC pipes demonstrated a clear difference between the two piping materials, where the GS pipe appeared to have a rough surface and PVC appeared to have a smooth surface (Fig. 5.10A,D). However, after 30 days, large amounts of biofilms had been deposited at the surface of both piping materials in untreated and chlorinated water systems. Moderately different concentrations of microorganisms with various morphologies were clearly observed depending on the type of pipe material. A relatively smaller amount of microorganisms was observed on the PVC pipes, which seemed to have smooth surfaces, while the rough surfaces of the GS pipes were covered with large numbers of microorganisms. According to Chang et al. (2003) pipes made of rough surface materials, such as GS have a greater biofilm regrowth capability than those that have smooth surfaces, such as PVC material. Yu et al. (2010) also reported that the surface structure of piping materials appears to be the main reason for biofilm formation on metal pipes. However, the crystalline structures of silver nanoparticles were observed on the surface of GS piping material between 1 and 7 days, while in PVC they were observed even after 7 days. Biofilm formation on both pipes was observed after 30 days with lower numbers of attached cells on GS piping, while PVC piping material also exhibited a lower number of cells with silver nanoparticle crystal structures. Results of this experimental study revealed the effectiveness of the resin-silver nanoparticle filter system in controlling the growth of attached heterotrophic plate count bacteria in a laboratory-scale unit.

6 Conclusions

Considering the worldwide growing population and industrialization, the demand for potable water and the shortage of water supplies is becoming more and more an alarming situation, especially in developing countries. While access to safe drinking is a human right, approximately 748 million people do not yet enjoy this right, worldwide. With many advantages on its side and the potential benefits of its convergence with other available treatment technologies, nanotechnology is showing great promise in developing countries as a feasible means of treating water to potable standards. Even though many nanotechnologies are still at the experimental

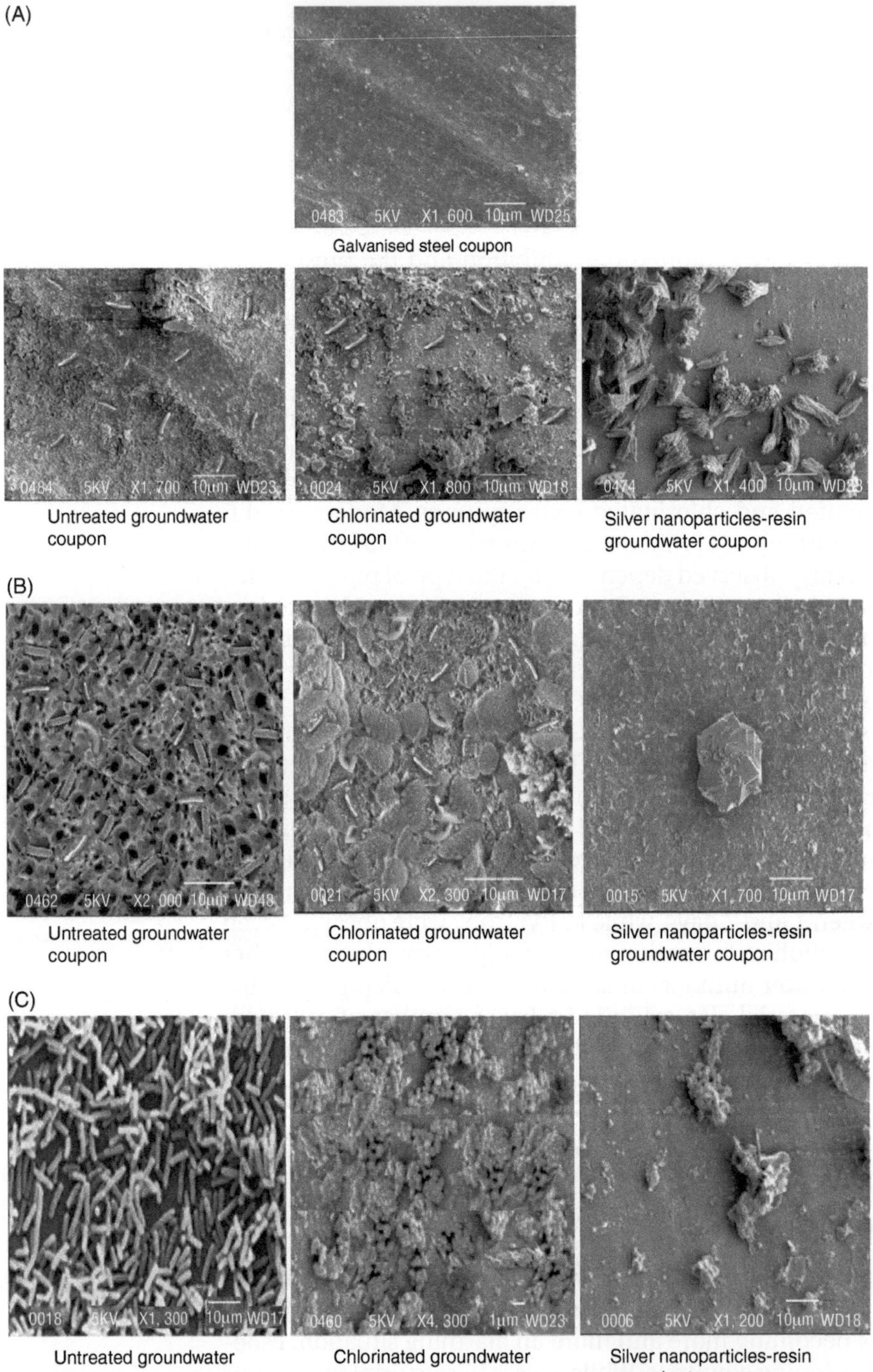

Figure 5.10. (A) SEM images depicting microbiological features of GS coupons before and after an exposure period of 1 day to untreated groundwater, chlorinated groundwater, and a resin–silver nanoparticle filter system. (B) SEM images depicting microbiological features of GS coupons after an exposure period of 7 days to untreated groundwater, chlorinated groundwater, and a resin–silver nanoparticle filter system. (C) SEM images depicting microbiological features of GS coupons after an exposure period of 30 days to untreated groundwater, chlorinated groundwater, and a resin–silver nanoparticle filter system.

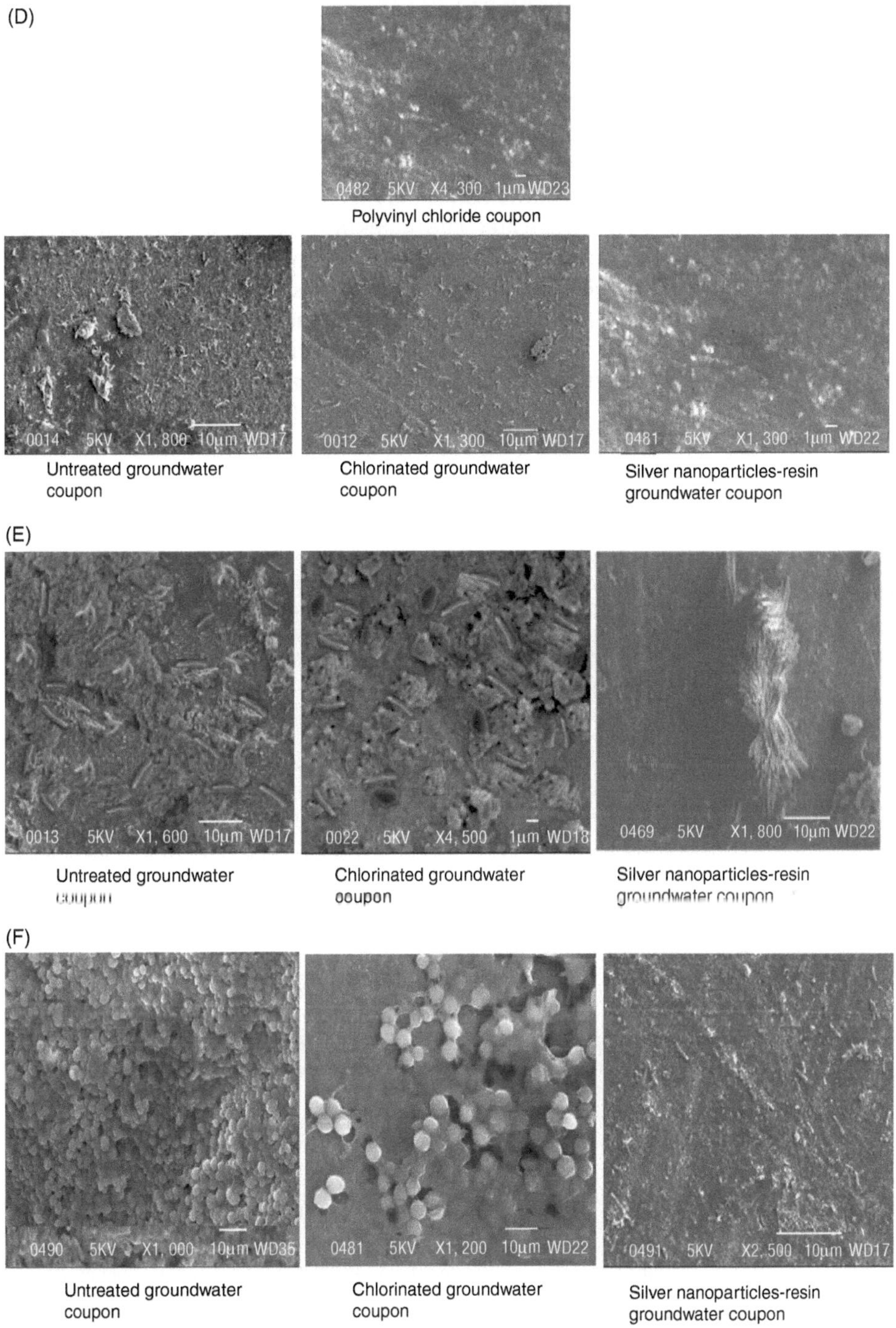

Figure 5.10. (***Continued***) (D) SEM images depicting microbiological features of PVC coupons before and after an exposure period of 1 day to untreated groundwater, chlorinated groundwater, and a resin–silver nanoparticle filter system. (E) SEM images depicting microbiological features of PVC coupons before and after an exposure period of 7 days to untreated groundwater, chlorinated groundwater, and a resin-silver nanoparticle filter system. (F) SEM images depicting microbiological features of PVC coupons before and after an exposure period of 30 days to untreated groundwater, chlorinated groundwater, and a resin-silver nanoparticle filter system.

stage in laboratories, it is widely recognized that these techniques with their ability to manipulate atoms and molecules at nanometer resolutions could play a significant role in resolving the issues related to water shortages and water quality. Therefore, African countries should endeavor to exploit the use of nanotechnology by establishing a symbiotic relationship with scientific, industrial, and institutional sectors since this emerging technology will make a major contribution toward finding creative solutions to address future challenges associated with the scarcity of safe drinking water.

References

Allabashi, R., Arkas, M., Hörmann, G., Tsiourvas, D., 2007. Removal of some organic pollutants in water employing ceramic membranes impregnated with cross-linked silylated dendritic and cyclodextrin polymers. Water Res. 41, 476–486.

AMCOW (2012), A snapshot of drinking water and sanitation in Africa—2012 update, 2012, Publication prepared for the African Ministers' Council on Water as a contribution to the Fourth Africa Water Week, Cairo, 14–15 May 2012; African Ministers' Council on Water (AMCOW), FCT-Abuja, Nigeria.

Amin, M., Alazba, A., Manzoor, U., 2014. A Review of Removal of pollutants from water/wastewater using different types of nanomaterials. Adv. Mater. Sci. Eng. 2014, 1–24.

Apha, 1998. Standard methods for the examination of water and wastewater, eighteenth ed. American Public Health Association, American Water Works Association, Water Environment Federation, Washington, DC.

Arias, C.A., Murray, B.E., 2009. Antibiotic-resistant bugs in the 21st century—a clinical super-challenge. N. Engl. J. Med. 360, 439–443.

Arkas, M., Allabashi, R., Tsiourvas, D., Mattausch, E.-M., Perfler, R., 2006. Organic/inorganic hybrid filters based on dendritic and cyclodextrin "nanosponges" for the removal of organic pollutants from water. Environ. Sci. Technol. 40, 2771–2777.

Ashbolt, N.J., 2004. Risk analysis of drinking water microbial contamination versus disinfection by-products (DBPs). Toxicology 198, 255–262.

Bae, T.-H., Kim, I.-C, Tak, T.-M., 2006. Preparation and characterization of fouling-resistant TiO_2 self-assembled nanocomposite membranes. J. Membr. Sci. 275, 1–5.

Balogh, L., Swanson, D.R., Tomalia, D.A., Hagnauer, G.L., Mcmanus, A.T., 2001. Dendrimer-silver complexes and nanocomposites as antimicrobial agents. Nano Lett. 1, 18–21.

Bartram, J., Brocklehurst, C., Fisher, M.B., Luyendijk, R., Hossain, R., Wardlaw, T., Gordon, B., 2014. Global monitoring of water supply and sanitation: history, methods and future challenges. Int. J. Environ. Res. Public Health. 11, 8137–8165.

Bhattacharyya, D., Hestekin, J.A., Brushaber, P., Cullen, L., Bachas, L.G., Sikdar, S.K., 1998. Novel poly-glutamic acid functionalized microfiltration membranes for sorption of heavy metals at high capacity. J. Membr. Sci. 141, 121–135.

Bhattacharya, S., Saha, I., Mukhopadhyay, A., Chattopadhyay, D., Chand, U., 2013. Role of nanotechnology in water treatment and purification: potential applications and implications. Int. J. Chem. Sci. Technol. 3, 59–64.

Biswas, A.K., 1983. Major water problems facing the world. Int. J. Water Res. D 1, 1–14.

Bottero, J.Y., Rose, J., Wiesner, M.R., 2006. Nanotechnologies: tools for sustainability in a new wave of water treatment processes. Integr. Environ. Assess Manag. 2, 391–395.

Bouoiyour, J., 2006. Migration, Diaspora et Développement Humain in Cinquantenaire de l'indépendance du Maroc. Rapport sur le Développement Humain, Royaume du Maroc 2005, 457–483.
Brady-Estévez, A.S., Kang, S., Elimelech, M., 2008. A single-walled-carbon-nanotube filter for removal of viral and bacterial pathogens. Small 4, 481–484.
Cele, L.M., Ray, S.S, Coville, N.J., 2009. Guest editorial: nanoscience and nanotechnology in South Africa. S. Afr. J. Sci. 105, 242–1242.
Chang, Y.C., Le Puil, M., Biggerstaff, J., Randall, A.A., Schulte, A., Taylor, J.S., 2003. Direct estimation of biofilm density on different pipe material coupons using a specific DNA-probe. Mol. cell. probe. 17, 237–243.
Chatterjee, A.N., Cannon, D.M., Gatimu, E.N., Sweedler, J.V., Aluru, N.R., Bohn, P.W., 2005. Modeling and simulation of ionic currents in three-dimensional microfluidic devices with nanofluidic interconnects. J. Nanopart. Res. 7, 507–516.
Cheng, S.-F., Wu, S.-C., 2000. The enhancement methods for the degradation of TCE by zero-valent metals. Chemosphere 41, 1263–1270.
Colvin, V.L., 2003. The potential environmental impact of engineered nanomaterials. Nat. Biotechnol. 21, 1166–1170.
Corcoran, E., Nellemann, C., Baker, E., Bos, R., Osborn, D., Savelli, H., 2010. Sick water? The central role of wastewater management in sustainable development: a rapid response assessment. UNEP, UN-HABITAT, GRID-Arendal, Birkeland Trykkeri AS, Norway.
Cozzens, S., Cortes, R., Soumonni, O., Woodson, T., 2013. Nanotechnology and the millennium development goals: water, energy, and agri-food. J. Nanopart. Res. 15, 2001.
DeFriend, K.A., Wiesner, M.R., Barron, A.R., 2003. Alumina and aluminate ultrafiltration membranes derived from alumina nanoparticles. J. Membr Sci. 224 (1–2), 11–28.
De La Rosa-Gomez, I., Olguin, M., Alcantara, D., 2008. Antibacterial behavior of silver-modified clinoptilolite–heulandite rich tuff on coliform microorganisms from wastewater in a column system. J. Environ. Manage. 88, 853–863.
Diallo, M.S., Christie, S., Swaminathan, P., Johnson, J.H., Goddard, W.A., 2005. Dendrimer enhanced ultrafiltration. 1. Recovery of Cu (II) from aqueous solutions using PAMAM dendrimers with ethylene diamine core and terminal NH_2 groups. Environ. Sci. Technol. 39, 1366–1377.
Diallo, M.S., Duncan, J.S., Savage, N., Street, A., Sustich, R., 2009. Nanotechnology Solutions for Improving Water Quality. In: Nora, S., Mamadou, D., Jeremiah, D., Anita, S., Richard Sustich a2-Nora Savage, M.D.J.D.A.S., Richard, S. (Eds.), Nanotechnology Applications for Clean Water. William Andrew Publishing, Boston, (Chapter 38).
Dieterich, B.H, Hernderson, J.M., 1963. Urban water supply conditions and needs in seventy-five developing countries. WHO, Geneva.
Doğruöz, N., Göksay, D., Ilhan-Sungur, E., Cotuk, A., 2009. Pioneer colonizer microorganisms in biofilm formation on galvanized steel in a simulated recirculating cooling-water system. J. Basic Microb. 49, S5–S12.
Donaldson, K., Aitken, R., Tran, L., Stone, V., Duffin, R., Forrest, G., Alexander, A., 2006. Carbon nanotubes: a review of their properties in relation to pulmonary toxicology and workplace safety. Toxicol. Sci. 92 (1), 5–22.
Dotzauer, D.M., Dai, J., Su, L., Bruening, M.L., 2006. Catalytic membranes prepared using layer–by–layer adsorption of polyelectrolyte/metal nanoparticle films in porous supports. Nano Lett. 6, 2268–2272.
El Saliby, I., Shon, H., Kandasamy, J, Vigneswaran, S., 2008. Nanotechnology for Wastewater Treatment: in brief. Encyclopedia of Life Support System.(EOLSS). Developed under the Auspices of the UNESCO, Eolss Publishers, UK.
Fenwick, A., 2006. Waterborne infectious diseases—could they be consigned to history? Science 313, 1077–1081.

Fleischer, T., Grunwald, A., 2008. Making nanotechnology developments sustainable. A role for technology assessment? J. Cleaner Prod. 16, 889–898.
Fukuda-Parr, S., 2008. Are the MDGs priority in development strategies and aid programmes? Only few are! Working Paper, International Poverty Centre.
Gautam, P., Madathil, D., Nair, A.B., 2013. Nanotechnology in waste water treatment: a review. Int. J. Chem. Tech. Res. 5, 2303–2308.
Gehrke, I., Geiser, A., Somborn-Schulz, A., 2015. Innovations in nanotechnology for water treatment. Nanotechnol. Sci. Appl. 8, 1–17.
Grimshaw, D.J., Gudza, L.D., Stilgoe, J., 2009. How Can Nanotechnologies Fulfill the Needs of Developing Countries? In: Nora, S., Mamadou, D., Jeremiah, D., Anita, S., Richard Susticha2-Nora Savage, M.D.J.D.A.S., Richard, S. (Eds.), Nanotechnology Applications for Clean Water. William Andrew Publishing, Boston, (Chapter 35).
Grimshaw, D.J., Gudza, L.D., Stilgoe, J., 2014. How can nanotechnologies fulfil the needs of developing countries? In: Diallo, M., Duncan, J., Savage, N., Street, A., Sustich, R.C. (Eds.), Nanotechnology applications: Solutions for improving water quality. William Andrew Publishing, Norwich, NY.
Gutiérrez, F.J., Mussons, M.L., Gatón, P., Rojo, R., 2012. Nanotechnology and food industry. INTECH Open Access Publisher.
Hillie, T., Hlophe, M., 2007. Nanotechnology and the challenge of clean water. Nat. Nanotechnol. 2, 663–664.
Hillie, T., Munasinghe, M., Hlope, M, Deraniyagala, Y., 2006. Overview and comparison of conventional water treatment technologies and nano-based treatment technologies. Workshop on Nanotechnology, Water and Development, 10–12 Chennai, India. Meridian-Institute.
Hinrichsen, D., Tacio, H., 2002. The coming freshwater crisis is already here. Finding the Source: The Linkages Between Population and Water, Woodrow Wilson International Center for Scholars. ESCP Publication, Spring, Washington, DC.
Hollman, A.M., Bhattacharyya, D., 2004. Pore assembled multilayers of charged polypeptides in microporous membranes for ion separation. Langmuir 20, 5418–5424.
Huang, Y.H., Peddi, P.K., Zeng, H., Tang, C.L., Teng, X., 2013. Pilot-scale demonstration of the hybrid zero-valent iron process for treating flue-gas-desulfurization wastewater: part II. Water Sci. Technol. 67, 239–246.
Hunter, P.R., Macdonald, A.M, Carter, R.C., 2010. Water supply and health. PLoS Med. 7, 1350.
Hutton, G., Haller, L, Bartram, J., 2007. Economic and health effects of increasing coverage of low cost household drinking-water supply and sanitation interventions to countries off-track to meet MDG target 10. OMS.
Ilhan-Sungur, E., Cansever, N., Cotuk, A., 2007. Microbial corrosion of galvanized steel by a freshwater strain of sulphate reducing bacteria (*Desulfovibrio* sp.). Corros. Sci. 49, 1097–1109.
Kang, S., Mauter, M.S., Elimelech, M., 2009. Microbial cytotoxicity of carbon-based nanomaterials: implications for river water and wastewater effluent. Environ.l Sci. Technol. 43, 2648–2653.
Kim, E.-S., Deng, B., 2011. Fabrication of polyamide thin-film nano-composite (PA-TFN) membrane with hydrophilized ordered mesoporous carbon (H-OMC) for water purifications. J. Membr. Sci. 375, 46–54.
Lazar, M.A., Varghese, S., Nair, S.S., 2012. Photocatalytic water treatment by titanium dioxide: recent updates. Catalysts 2, 572–601.
Lecoanet, H.F., Bottero, J.-Y., Wiesner, M.R., 2004. Laboratory assessment of the mobility of nanomaterials in porous media. Environ. Sci. Technol. 38, 5164–5169.

Lehtola, M.J., Miettinen, I.T., Keinänen, M.M., Kekki, T.K., Laine, O., Hirvonen, A., Vartiainen, T., Martikainen, P.J., 2004. Microbiology, chemistry and biofilm development in a pilot drinking water distribution system with copper and plastic pipes. Water Res. 38, 3769–3779.
Leonard, P., Hearty, S., Brennan, J., Dunne, L., Quinn, J., Chakraborty, T., O'kennedy, R., 2003. Advances in biosensors for detection of pathogens in food and water. Enzyme Microb. Technol. 32, 3–13.
Li, W.-X., Stampfl, C., Scheffler, M., 2003. Why is a noble metal catalytically active? The role of the O-Ag interaction in the function of silver as an oxidation catalyst. Phys. Rev. Lett. 90, 256102.
Lin, Y.-S.E., Vidic, R.D., Stout, J.E., Victor, L.Y., 1996. Individual and combined effects of copper and silver ions on inactivation of *Legionella pneumophila*. Water Res. 30, 1905–1913.
Liu, Z., Stout, J.E., Boldin, M., Rugh, J., Diven, W.F., Victor, L.Y., 1998. Intermittent use of copper-silver ionization for *Legionella* control in water distribution systems: a potential option in buildings housing individuals at low risk of infection. Clin. Infect. Dis. 26, 138–140.
Lukhele, L., Mamba, B., Momba, M., Krause, R., 2010. Water disinfection using novel cyclodextrin polyurethanes containing silver nanoparticles supported on carbon nanotubes. J. Appl. Sci. 10, 65–70.
Mangun, C.L., Yue, Z., Economy, J., Maloney, S., Kemme, P., Cropek, D., 2001. Adsorption of organic contaminants from water using tailored ACFs. Chem. Mater 13, 2356–2360.
Masinga, S., 2005. Is Government underestimating deaths in Delmas typhoid and diarrhoea outbreak. TAC Electr. Newsl. Sunday September 18 2005, http://www.tac.org.za/newsletter/2005/ns18_09_2005.htm. Cited 10 January 2015.
Masoka, X., Utembe, W., Sekobe, G., Gulumian, M., 2012. Nanotechnology research & occupational health and safety in Africa. Afr. Newsl. 22, 56.
Meier, B.M., Kestenbaum, J.G., Kayser, G.L., Amjad, U.Q., Bartram, J., 2014. Examining the practice of developing human rights indicators to facilitate accountability for the human right to water and sanitation. J. Hum. Rights Pract. 6, 159–181.
Meyer, D.E., Wood, K., Bachas, L.G., Bhattacharyya, D., 2004. Degradation of chlorinated organics by membrane-immobilized nanosized metals. Environ. Prog. 23, 232–242.
Miyazaki, K., Islam, N., 2007. Nanotechnology systems of innovation—an analysis of industry and academia research activities. Technovation 27, 661–675.
Momba, M., 1997. The Impact of Disinfection Processes on Biofilm Formation in Potable Water Distribution Systems. Universiteit van Pretoria, Pretoria.
Momba, M.N., Kaleni, P., 2002. Regrowth and survival of indicator microorganisms on the surfaces of household containers used for the storage of drinking water in rural communities of South Africa. Water Res. 36, 3023–3028.
Momba, M.N., Cloete, T.E., Venter, S.N., Kfir, R., 1998. Evaluation of the impact of disinfection processes on the formation of biofilms in potable surface water distribution systems. Water Sci. Technol. 38, 283–289.
Momba, M.N., Cloete, T.E., Venter, S.N., Kfir, R., 1999. Examination of the behaviour of *Escherichia coli* in biofilms established in laboratory-scale units receiving chlorinated and chloraminated water. Water Res. 33, 2937–2940.
Momba, M., Kfir, R., Venter, S.N, Cloete, T.E., 2000. Overview of biofilm formation in distribution systems and its impact on the deterioration of water quality.
Montgomery, M.A., Elimelech, M., 2007. Water and sanitation in developing countries: including health in the equation. Environ. Sci. Technol. 41, 17–24.
Mousavi, S.R., Rezaei, M., 2011. Nanotechnology in agriculture and food production. J. Appl. Environ. Biol. Sci. 1, 414–419.

Mpenyana-Monyatsi, L., Mthombeni, N.H., Onyango, M.S., Momba, M.N., 2012. Cost-effective filter materials coated with silver nanoparticles for the removal of pathogenic bacteria in groundwater. Int. J. Environ. Res. Public Health 9, 244–271.
Mthombeni, N.H., Mpenyana-Monyatsi, L., Onyango, M.S., Momba, M.N., 2012. Breakthrough analysis for water disinfection using silver nanoparticles coated resin beads in fixed-bed column. J. Hazard. Mater. 217–218, 133–140.
Musee, N., Thwala, M., Nota, N., 2011. The antibacterial effects of engineered nanomaterials: implications for wastewater treatment plants. J. Environ. Monitor. 13, 1164–1183.
Nel, J., Xu, Y., Batelaan, O., Brendonck, L., 2009. Benefit and implementation of groundwater protection zoning in South Africa. Water Resour. Manage. 23, 2895–2911.
Oyanedel-Craver, V.A., Smith, J.A., 2008. Sustainable colloidal-silver-impregnated ceramic filter for point-of-use water treatment. Environ. Sci. Technol. 42, 927–933.
Pendergast, M.M., Hoek, E.M.V., 2011. A review of water treatment membrane nanotechnologies. Energy Environ. Sci. 4, 1946–1971.
Peng, X., Li, Y., Luan, Z., Di, Z., Wang, H., Tian, B., Jia, Z., 2003. Adsorption of 1, 2-dichlorobenzene from water to carbon nanotubes. Chem. Phys. Lett. 376, 154–158.
Peter-Varbanets, M., Zurbrügg, C., Swartz, C., Pronk, W., 2009. Decentralized systems for potable water and the potential of membrane technology. Water Res. 43, 245–265.
Pineo, C.S, Subrahmanyam, D.V., 1975. Community water supply and excreta disposal situation in the developing countries. A commentary.
Popov, V.N., 2004. Carbon nanotubes: properties and application. Mater. Sci. Eng. R 43, 61–102.
Power, K.N., Nagy, L.A., 1999. Relationship between bacterial regrowth and some physical and chemical parameters within Sydney's drinking water distribution system. Water Res. 33, 741–750.
Pradeep, T., Anshup, 2009. Noble metal nanoparticles for water purification: a critical review. Thin Solid Films 517, 6441–6478.
Qu, X., Alvarez, P.J.J, Li, Q., 2013. Applications of nanotechnology in water and wastewater treatment. Water Res. 47, 3931–3946.
Rajan, C., 2011. Nanotechnology in groundwater remediation. Int. J. Environ. Sci. D 2, 182–187.
Ritchie, S.M.C., Bachas, L.G., Olin, T., Sikdar, S.K., Bhattacharyya, D., 1999. Surface modification of silica- and cellulose-based microfiltration membranes with functional polyamino acids for heavy metal sorption. Langmuir 15, 6346–6357.
Ritchie, S.M.C., Kissick, K.E., Bachas, L.G., Sikdar, S.K., Parikh, C., Bhattacharyya, D., 2001. Polycysteine and other polyamino acid functionalized microfiltration membranes for heavy metal capture. Environ. Sci. Technol. 35, 3252–3258.
Sadeghi-Kiakhani, M., Arami, M., Gharanjig, K., 2013. Dye removal from colored-textile wastewater using chitosan-PPI dendrimer hybrid as a biopolymer: optimization, kinetic, and isotherm studies. J. Appl. Polym. Sci. 127, 2607–2619.
Saidi, T., 2009. The Expectations and Challenges in the Development of Nanotechnology In Sub-Saharan Africa. Maastricht Graduate School of Govrnance, Maastricht, Netherlands.
Sans-241, 2011. South African National Standard: Drinking Water.
Savage, N., Diallo, M., 2005. Nanomaterials and water purification: opportunities and challenges. J. Nanopart. Res. 7, 331–342.
Schutte, F., 2006. Handbook for the operation of water treatment works. WRC report no. TT 265/06, Water Research Commission. Pretoria.

Schutte, C., Focke, W., 2007. Evaluation of nanotechnology for application in water and wastewater treatment and related aspects in South Africa. University of Pretoria, Pretoria.
Schwarzenbach, R.P., Escher, B.I., Fenner, K., Hofstetter, T.B., Johnson, C.A., Von Gunten, U., Wehrli, B., 2006. The challenge of micropollutants in aquatic systems. Science 313, 1072–1077.
Sharma, V.K., Yngard, R.A., Lin, Y., 2009. Silver nanoparticles: green synthesis and their antimicrobial activities. Adv. Colloid Interface Sci. 145, 83–96.
Singer, P.A., Salamanca-Buentello, F., Daar, A.S., 2005. Harnessing nanotechnology to improve global equity. Issues Sci.Technol. 21 (4), 57–64.
Smith, L., 2009. Municipal compliance with water services policy: A challenge for water security. Development Planning Division, Development Bank of Southern Africa, Midrand.
Srivastava, A., Srivastava, O.N., Talapatra, S., Vajtai, R., Ajayan, P.M., 2004. Carbon nanotube filters. Nat. Mater. 3 (9), 610–614.
Stanton, B.W., Harris, J.J., Miller, M.D., Bruening, M.L., 2003. Ultrathin, multilayered polyelectrolyte films as nanofiltration membranes. Langmuir 19, 7038–7042.
Tang, S.C.N., Lo, I.M.C., 2013. Magnetic nanoparticles: essential factors for sustainable environmental applications. Water Res. 47 (8), 2613–2632.
Theron, J., Walker, J., Cloete, T., 2008. Nanotechnology and water treatment: applications and emerging opportunities. Crit. Rev. Microbiol. 34, 43–69.
Theron, J., Walker, J., Cloete, T., Cloete, T., De Kwaadsteniet, M., Botes, M., Lopez-Romero, J., 2010. Nanotechnology in water treatment applications. Caister Academic Press, Norfolk.
Tsvetanova, Z., 2006. Study of biofilm formation on different pipe materials in a model of drinking water distribution system and its impact on microbiological water quality. In: Chirila, E. (Ed.), Chemicals as Intentional and Accidental Global Environmental Threats. Springer, Netherlands, pp. 463–468.
Tully, D.C., Frechet, J.M.J., 2001. Dendrimers at surfaces and interfaces: Chemistry and applications. J. Chem. Commun. 14, 1229–1239.
UN, 2010. Resolution A/RES/64/292,;1; The human right to water and sanitation (28 July 2010). General Assembly, sixty-fourth session.
Van Der Bruggen, B., Vandecasteele, C., 2003. Removal of pollutants from surface water and groundwater by nanofiltration: overview of possible applications in the drinking water industry. Environ. Pollut. 122, 435–445.
Van Der Bruggen, B., Everaert, K., Wilms, D., Vandecasteele, C., 2001. Application of nanofiltration for removal of pesticides, nitrate and hardness from ground water: rejection properties and economic evaluation. J. Membr. Sci. 193, 239–248.
Van Der Bruggen, B., Vandecasteele, C., Van Gestel, T., Doyen, W., Leysen, R., 2003. A review of pressure-driven membrane processes in wastewater treatment and drinking water production. Environ. Prog. 22, 46–56.
Ventresque, C., Gisclon, V., Bablon, G., Chagneau, G., 2000. An outstanding feat of modern technology: the Mery-sur-Oise nanofiltration treatment plant (340,000 m 3/d). Desalination 131, 1–16.
Virto, R., Manas, P., Alvarez, I., Condon, S., Raso, J., 2005. Membrane damage and microbial inactivation by chlorine in the absence and presence of a chlorine-demanding substrate. Appl. Environ. Microbiol. 71, 5022–5028.
Wang, Y., Iqbal, Z., Mitra, S., 2005. Microwave-induced rapid chemical functionalization of single-walled carbon nanotubes. Carbon 43, 1015–1020.
Wang, Y., Iqbal, Z., Mitra, S., 2006. Rapidly functionalized, water-dispersed carbon nanotubes at high concentration. J. Am. Chem. Soc. 128, 95–99.
Weber, J.W., 2002. Distributed optimal technology networks: a concept and strategy for potable water sustainability. Water Sci. Technol. 46, 241–246.

WHO, 1976. Community water supply and wastewater disposal. World Health Organization Chronicle 30, 329–334.
WHO, 1981. Drinking-water and sanitation, 1981–1990: a way to health, a WHO contribution to the International Drinking Water Supply and Sanitation Decade. World Health Organization, Geneva.
WHO, 2006. Guidelines for drinking-water quality, third ed. Disponibile all'indirizzo: http://www.who.int/water.sanitation_health/dwq X, 924156251
WHO, 2012. Global costs and benefits of drinking-water supply and sanitation interventions to reach the MDG target and universal coverage. World Health Organization, Geneva.
WHO/UNICEF, 2000. Global water supply and sanitation assessment 2000 report, OMS.
WHO/UNICEF, 2005a. Water for Life, Making it Happen. Geneva: WHO/UNICEF Joint Monitoring Programme for Water Supply and Sanitation. Available from: http://www.wssinfo.org/pdf/JMP_05_text.pdf
WHO/UNICEf, 2005b. WHO 2004 Meeting the MDG Drinking Water and Sanitation Target: A Mid-Term Assessment of Progress. UNICEF/WHO, Geneva, Switzerland.
WHO/UNICEF, 2010. Progress on Sanitation and Drinking Water: 2010 Update. France.
WHO/UNICEF, 2012. Progress on Drinking Water and Sanitation: 2012 Update. World Health Organization, New York, United States of America.
WHO/UNICEF, 2014. Progress on Drinking-Water and Sanitation–2014 Update. World Health Organization, Geneva, Switzerland.
Williams, M.M., Braun-Howland, E.B., 2003. Growth of *Escherichia coli* in model distribution system biofilms exposed to hypochlorous acid or monochloramine. Appl. Environ. Microbiol. 69, 5463–5471.
Wu, L., Ritchie, S.M.C., 2006. Removal of trichloroethylene from water by cellulose acetate supported bimetallic Ni/Fe nanoparticles. Chemosphere 63, 285–292.
Wu, L., Shamsuzzoha, M., Ritchie, S.M.C., 2005. Preparation of cellulose acetate supported zero-valent iron nanoparticles for the dechlorination of trichloroethylene in water. J. Nanopart. Res. 7, 469–476.
WWICS, 2011. A Nanotechnology Consumer Products Inventory Project on Emerging Nanotechnologies. Woodrow Wilson International centre for Scholars.
Xu, J., Dozier, A., Bhattacharyya, D., 2005. Synthesis of nanoscale bimetallic particles in polyelectrolyte membrane matrix for reductive transformation of halogenated organic compounds. J. Nanopart. Res. 7, 449–461.
Yang, H.Y., Han, Z.J., Yu, S.F., Pey, K.L., Ostrikov, K., Karnik, R., 2013. Carbon nanotube membranes with ultrahigh specific adsorption capacity for water desalination and purification. Nature Communications. 2220.
Yoon, K., Hsiao, B.S., Chu, B., 2009. Formation of functional polyethersulfone electrospun membrane for water purification by mixed solvent and oxidation processes. Polymer 50, 2893–2899.
Yu, J., Kim, D., Lee, T., 2010. Microbial diversity in biofilms on water distribution pipes of different materials. Water Sci. and Techn. 61, 163–171.
Zacheus, O.M., Iivanainen, E.K., Nissinen, T.K., Lehtola, M.J., Martikainen, P.J., 2000. Bacterial biofilm formation on polyvinyl chloride, polyethylene and stainless steel exposed to ozonated water. Water Res. 34, 63–70.
Zhang, H., 2013. Application of Silver Nanoparticles in Drinking Water Purification. Doctor in Philosophy, University of Rhode Island, Rhode Island.

6

DEVELOPMENT OF MAGNETIC NANOPARTICLES FOR FLUORIDE AND ORGANIC MATTER REMOVAL FROM DRINKING WATER

Ashraf F. Ali*, Sahar M. Atwa,†, Emad M. El-Giar****

**National Research Centre, Inorganic Chemistry Department, Dokki, Cairo, Egypt; **University of Louisiana at Monroe, School of Sciences, Monroe, LA, United States; †Cairo University, Department of Chemistry, Faculty of Science, Giza, Egypt*

1 Introduction

Safe drinking water is an essential component for food, health, and wellness of humans. Without water, a person would die in a few days. The quality of drinking water is as important as its quantity. According to the World Health Organization (WHO), any water intended for drinking should be clean and free of toxic contaminants and pathogenic microorganisms (Galal-Gorchev, 1993). However, the available clean and safe drinking water is increasingly shrinking due to the global population growth, increased industrial activities, and climate changes.

Almost all sources of drinking water (tap water, lakes, rivers, groundwater, etc.) are heavily polluted with industrial and/or environmental toxins. Moreover, many waterborne pathogens in drinking water lead to several global endemic diseases, such as diarrhea, dysentery, typhoid, gastroenteritis, cholera, and jaundice.

The major drinking water contaminants include pesticides, halogenated and nonhalogenated organic compounds including volatile organic compounds, BETX fuel compounds (benzene, toluene, ethylbenzene, and xylene), heavy metal ions, turbidity, inorganic salts, pathogens (bacteria, viruses, fungi, protozoans), and disinfection by-products (Howd, 2002).

Several traditional treatment techniques have been used to remove the different contaminants and pathogens from drinking

Water Purification. http://dx.doi.org/10.1016/B978-0-12-804300-4.00006-X

water. These techniques include reverse osmosis (RO), nanofiltration (NF), activated sludge, activated carbon (AC), lime softening, and coagulation–flocculation. Each of these methods has its own merits. However, none of them is capable of removing all toxins and pathogens from drinking water.

The emergence of new contaminants and pathogens along with the new strict water-quality standards and regulations mandated the search for new reliable, cost effective and stable materials, and analytical technologies to address the new challenges and replace the conventional water treatment methods. Due to their unique physicochemical properties and large surface to volume ratio, nanomaterials are very efficient separation media for water purification. The past two decades have witnessed an increased interest in the use of nanomaterials, especially nanoparticles (NPs), in drinking water purification (Iseli et al., 2009).

This chapter highlights the recent advances in developing and use of magnetic NP materials (metal and metal oxide NPs, as well as their nanocomposites) in the treatment of drinking water with emphasis on the elimination of fluoride ions and organic matter (OM). The chapter also covers the properties, advantages, and the challenges associated with the use of nanomaterials in drinking water purification systems.

2 Drinking Water Purification: Importance and Challenges

Water is an essential commodity for the life on earth and access to sufficient quantities of safe drinking water is a major problem worldwide especially in the developing countries. The annual number of deaths worldwide due to water-related diseases, mainly amebiasis and diarrhea caused by bacteria, is more than 3.4 million people. About 5000–6000 children die everyday due to the consumption of contaminated drinking water. The annual number of deaths due to lack of safe drinking water and poor water treatment is more than the combined number of people who die in wars, due to terrorist acts, and weapons of mass destruction. Most of these deaths occur in Africa and Asia (Day, 2011).

Due to global climate changes, world population growth, increased industrial waste, the excessive use of pesticides and herbicides, ecological degradation, and current strict health-related regulations, the available water resources are very limited and the purification of drinking water is a not only a big challenge but also expensive.

Several organizations and federal agencies, such as the WHO, the Environmental Protection Agency (EPA), the United States Environmental Protection Agency (USEPA), the American Water Workers Association (AWWA), the European Union (EU), and many others have been working on funding research and developing standards, guidelines, acts, and regulations to determine and remove the various contaminants in drinking water (Benner, 2004).

The technology to remove contaminants from drinking water must be highly efficient in removing the contaminants at the molecular level, economically feasible, environmentally friendly, and can be easily applied in developing countries.

3 Conventional Water Purification Methods

The current conventional technologies used for removing contaminants from drinking water include the following methods where each method has its own merits and limitations.

3.1 Filtration

This involves the flow of drinking water through a filter (cartridge). Different types of filters, such as ceramic, sediment, biosand, fabric, AC, and rapid-rate filters are used. Filtration methods are usually combined with another water-treatment method, such as deaeration, ozonation, and chlorination. Depending on the type of filter used, filtration is partially effective in removing turbidity (e.g., sand, clay), some bacteria (e.g., *Fecal coliform, Escherichia coli*), metal ions, and some OM.

In general, all filtration methods are not 100% effective in removing all water contaminants. Some filters are good in removing dissolved ions while others are better in removing OM or bacteria. However, filtration methods are normally good for household use but relatively expensive and require regular cleaning and/or replacement of the filters, as well as education and training.

3.2 Desalination

Desalination is a relatively old but expensive method to produce drinking water from salty waters. The development of new equipment and technology constantly reduces the cost and makes the process cost-competitive with the other water-treatment methods. More than 12,500 desalination plants exist throughout the world and produce about 40 million cubic meters of drinking water per day (WHO, 2007). The quality of the desalinated water depends on the quality of the crude water source as well as the desalination technology used.

Desalination is usually conducted in one of two major technologies: membrane-based desalination (either as electrical-driven or as pressure-driven technique) and thermal desalination (distillation).

Depending on pore diameter of the membrane (Fig. 6.1) and the amount of the applied pressure (Table 6.1), the pressure-driven membrane methods have been categorized into four classes. These are microfiltration (MF), NF, ultrafiltration (UF), and RO. However, according to the EPA, RO is considered the best available technique for the removal of a wide range of contaminants from water including inorganic ions, radionuclides, volatile organics, pharmaceuticals, and emerging contaminants (Kaushik, 2012).

While the pressure-driven desalination methods extract water from the dissolved salts, electrical-driven methods [such as electrodialysis (ED), electrodeionization, and capacitive deionization] extract the dissolved salts and retain the pure water where a DC voltage is applied to opposite ends of a membrane stack and the dissolved ions are attracted to the opposite membrane terminals leaving pure water in the chamber.

Both polymer-based membranes (usually made from synthetic polymers) and inorganic-based membranes (thin-film composites, ceramics, or metallic membranes) are commonly used in membrane filtration (Ritchie, 2003). Inorganic-based membranes are usually more stable and chemically resistant but

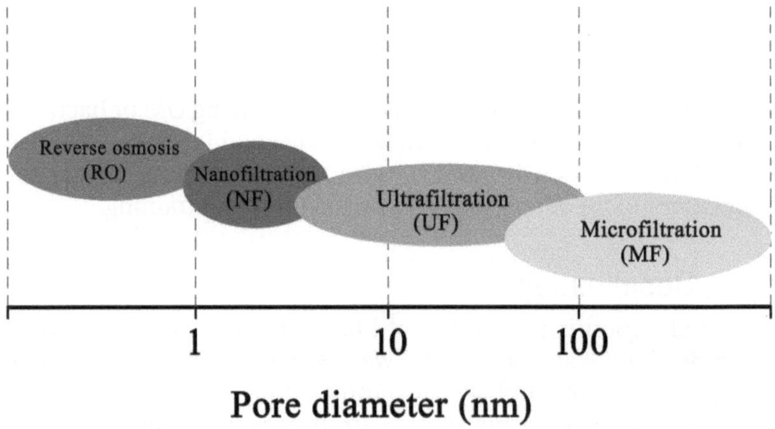

Figure 6.1. Range of nominal membrane pore sizes in the different membrane filtration technologies. From Perry, R. H., Green, D. W., 1997. Perry's Chemical Engineers' Handbook, seventh ed. McGraw-Hill, New York.

Table 6.1 Characteristics of Applications of Pressure-Driven Membrane-Based Desalination Techniques

Technique	Appl. Pressure (kPa)	Min. Particle Size Removed	Application (Type, Average Removal Efficiency %)
Microfiltration (MF)	30–500	0.1–3 μm	• Particle/turbidity removal (>99) • Bacteria/protozoa removal (>99.99)
Ultrafiltration (UF)	30–500	0.01–0.1 μm	• Particle/turbidity removal (>99) • Bacteria/protozoa removal (>99.999) • TOC removal (<20) • Virus removal/(partial credit only)
Nanofiltration (NF)	500–1000	200–400 Da	• Turbidity removal (>99) • Color removal (>98) • TOC removal (DBP control) (>95) • Hardness removal (softening) (>90) • SOC removal (≥500 Da) (0–100) • Sulfate removal (>97) • Virus removal (>95)
Reverse osmosis (RO)	1000–5000	50–200 Da	• Desalination (>99) • Color and DOC removal (>97) • Radionuclide removal (not including Rn) (>97) • Nitrate removal (85–95) • Pesticide/SOC removal (0–100) • Virus removal (>95) • As, Cd, Cr, Pb, F^- removal (40–>98)

DBP, Disinfection-by-products; DOC, dissolved organic matter; SOC, synthetic organic contaminant; TOC, total organic carbon.
Source: Younos, T., Tulou, K. E, 2005. Overview of desalination techniques. J. Contemp. Water Res. Educ. 132, 3–10.

more expensive than polymer-based membranes. Very recently, organic-inorganic hybrid membranes and nanocomposites have been used in desalination. Although effective in removing most of the water contaminants, membrane-based methods are still expensive and require regular membrane testing (using electronic devices) and replacement when deemed necessary.

Distillation is the oldest method of water purification. It involves evaporating contaminated water and condensing the fresh water into a storage tank. The process requires a source of energy to boil the water. To ensure a good-quality drinking water, the distillation process may be repeated several times depending on the quality of the contaminated water. Distillation is effective in eliminating all salts (both soluble and insoluble), nitrates, sulfates, heavy metals, pathogens, and several biological compounds. However, the process is relatively slow and can consume up to 5 L of water to produce 1 L of distilled water. In addition, distillation is not suitable for removing organic contaminants which normally have boiling points lower than that of water. Accordingly, when present in the crude water, compounds, such as volatile organic compounds, pesticides, herbicides, fertilizers, and chlorine solutions can evaporate and condense back with the distilled water (Bharadwaj et al., 2008).

3.3 Ultraviolet Irradiation

Ultraviolet (UV) irradiation is an effective method for killing microorganisms (mainly bacteria and viruses) that cause waterborne diseases in drinking water. In this method, drinking water in transparent plastic tanks is exposed to a dose of UV light (usually above 30,000 mw/cm^2) from a UV lamp. Depending on the UV light intensity, this method can kill most of the microorganisms in the drinking water. As a main advantage, UV irradiation does not involve adding any additives to the drinking water and, hence, it does not alter the taste or odor of the water. Also the UV light does not produce any harmful by-products and the method requires minimum maintenance. However, UV irradiation alone is insufficient to produce safe drinking water. The UV radiation cannot remove chemical contaminants or volatile organics. In addition, certain ions (e.g., iron, sulfites), turbidity, and OM in water may absorb the UV radiation and result in lowering the efficiency of the radiation. Accordingly, UV irradiation is usually paired with a filtration method, such as RO or AC filtration.

3.4 Chemical Treatment

As the name suggests, certain chemicals (e.g., coagulants, flocculants, and disinfectants) are added to water to remove toxins and pathogens. As shown below, some of these chemicals affect the taste and/or the odor of the purified water.

3.4.1 Chemical Oxidation (Chemical Disinfection)

This method involves the addition of an oxidant (disinfectant) to the drinking water to destroy the pathogens and microbes, such as bacteria, viruses, and coliforms. The oxidant, either directly added (such as potassium permanganate, ozone, chlorine, sodium hypochlorite, chlorine dioxide) or produced on site (e.g., sodium hypochlorite), kills the pathogens within 30 min through changing the pH of the water (Groeber, 1996). The concentration of the disinfectant should be calculated based on the quality of water treated.

Although the method is very effective and can remove almost all pathogens, the use of disinfectants usually produces undesirable taste, odors, and sometimes very harmful and carcinogenic by-products (e.g., if chlorine is used). In addition, the presence of colloids and solid suspended particles in the treated water can lower the efficiency of the process as some of the pathogens can be embedded in these solid particles. Moreover, the oxidants are not effective in removing heavy metals or turbidity. Accordingly, this method is usually combined with another method, such as filtration or coagulation for complete removal of both pathogens and chemical contaminants.

3.4.2 Advanced Oxidation Processes

This oxidation method is very effective in destroying organic contaminants (such as phenols, substituted phenols, chloroalkenes, and pharmaceuticals) from water. The method was first introduced in the mid-1980s when Glaze et al. used O_3 and O_3/H_2O_2 systems to oxidize trichloroethylene and tetrachloroethylene in groundwater (Glaze and Kang, 1988). Since then, this method has been in use for treatment of drinking water, wastewater, groundwater, and soil. The method is also used to eliminate the taste and odor compounds from aqueous media.

The advanced oxidation process (AOP) involves the use of a powerful advanced oxidation system to generate hydroxide radicals ($OH^\bullet$) in situ to oxidize the OM to the harmless products carbon dioxide and water. An advanced oxidation system usually consist of a strong oxidant (e.g., O_3), a catalyst (usually a transition metal ion, such as Fe^{2+}), and a source of radiation (e.g., UV, ultrasound, electron beam). Some of the typical AOP systems are H_2O_2/UV, O_3/UV, O_3/ultrasound, O_3/H_2O_2, O_3/H_2O_2/UV, Fenton's reagent (Fe^{2+}/H_2O_2), O_2/TiO_2/UV, and electron beam irradiation.

When applied correctly, the AOP is very effective and can reduce the concentration of the organic contaminants from several hundreds of parts per million (ppm) to sub-parts per billion (sub-ppb) (Sundstrom et al., 2013). However, this method is relatively expensive and requires special safety in the use of the reactive oxidants and radiation sources.

3.4.3 Coagulation–Flocculation Process

Coagulation and flocculation are two successive processes used to remove suspended solid, turbidity, and pathogens from drinking water, as well as wastewater (Carvalho Bongiovani et al., 2014).

Coagulation takes place first, and then flocculation follows. Most of the suspended particles in water are negatively charged. Coagulants are chemical reagents that are added to neutralize the surface charge of the suspended particles and hence allow the particles to grow into larger particles called flocs. For a successful removal of contaminants, the coagulation step has to be completed before flocculation starts. Accordingly, the selection of the coagulants and flocculants requires a good knowledge about the nature as well as the characteristics of the suspended particles and microbes present in water under treatment. For an effective removal of contaminants, the coagulation and flocculation technique is usually combined with filtration or chemical oxidation.

3.4.3.1 Coagulation

In this step, the suspended particles are neutralized (destabilized) through the addition of a suitable coagulant with a charge opposite to that of the particles. This allows the particles to collide more often, stick together, and form microflocs. For an efficient coagulation step, the coagulant should be rapidly mixed with the water for about 3 min to allow for a homogeneous dispersion of the coagulant in water. The most-common coagulant reagents are ferrous sulfate, ferric sulfate, ferric chloride, sodium aluminate, alum, and some naturally derived compounds (Gimbel and Malzer, 2000). Coagulation is an effective method to remove several contaminates from drinking water and wastewater. However, for some particles, such as F^- ions, this method does not lower the concentration to the desirable (permissible) level.

3.4.3.2 Flocculation

Flocculation is the step that follows coagulation. It is a slow gentle mixing step that involves the addition of a flocculant to increase the size of the microflocs to become visible suspended particles (macroflocs) that can be easily removed by filtration (Kawamura, 1976). Coagulant aids, high molecular weight polymers, may be added in this step to help in binding the microflocs and hence enhance the formation of the macroflocs. The flocculation step is usually done either in a single chamber or a series of chambers and the whole process takes about 20–60 min. This step requires careful control of the mixing rate to avoid shearing or tearing apart the floc.

3.5 Electrocoagulation–Flotation

Electrocoagulation–flotation (ECF) is an alternative to the conventional coagulation–flocculation method. ECF is an electrochemical process in which the contaminants are removed through the application of a DC voltage between two metal electrodes (an anode and a cathode) immersed in a tank that has the contaminated water. The anode is usually an iron, steel, or aluminum plate while the cathode can be either of the same anode material or an inert metal. The coagulant (metal ions) is generated through the electrolytic oxidation (dissolution) of the anode material. Depending on the type of anode used, the dissolution of the anode releases metal ions (Fe^{2+}, Fe^{3+}, Al^{3+}, etc.) in the tank that react with particles in the polluted water and finally precipitate as hydroxide complexes in the bottom of the tank. These hydroxides grow to form agglomerates which fall in the bottom of the tank, then be removed through filtration. However, in a typical ECF apparatus, the agglomerates float to the top of the tank through adhering to the hydrogen gas bubbles that are formed from the reduction of water at the cathode. The floated particulates can then be skimmed from the surface of the tank.

ECF is rapidly growing, especially in the wastewater, industrial water, and groundwater treatment, because of its low cost and ability to treat a large volume of polluted water. However, the efficiency of an ECF reactor depends on several factors including the type and volume of water under treatment, pH, current density, electrolysis time, and type as well as dimensions of the electrodes used.

4 Nanoparticles in Water Purification

Nanomaterials are a class of materials that have at least one dimension in the nm range (typically 1–100 nm) (Fig. 6.2). They are considered a bridge between bulk (massive) and atomic (molecular) materials. Nanomaterials take different forms and shapes including NPs, nanowires, dendrimers, nanopowders, nanofibers, nanotubes, nanoflakes, nanoplates, nanogranules, nanohelixes, nanorings, nanoclays, nanomembranes, nanobelts, nanosheets, nanocages, nanocomposites, nanocapsules, and nanoflowers (some examples are given in Fig. 6.3). Some nanomaterials exist naturally (e.g., dust, smoke, soot), while the rest are synthesized for specific applications. Among the synthetic nanomaterials, fumed silica and carbon black have been extensively used in many applications in the automotive (rubber) and sporting equipment industries.

NPs are nanomaterials with all three dimensions in the nm range. They have several special physical and biochemical properties, such as high surface area, large surface to volume ratio, easy surface modification and separation, increased reaction kinetics,

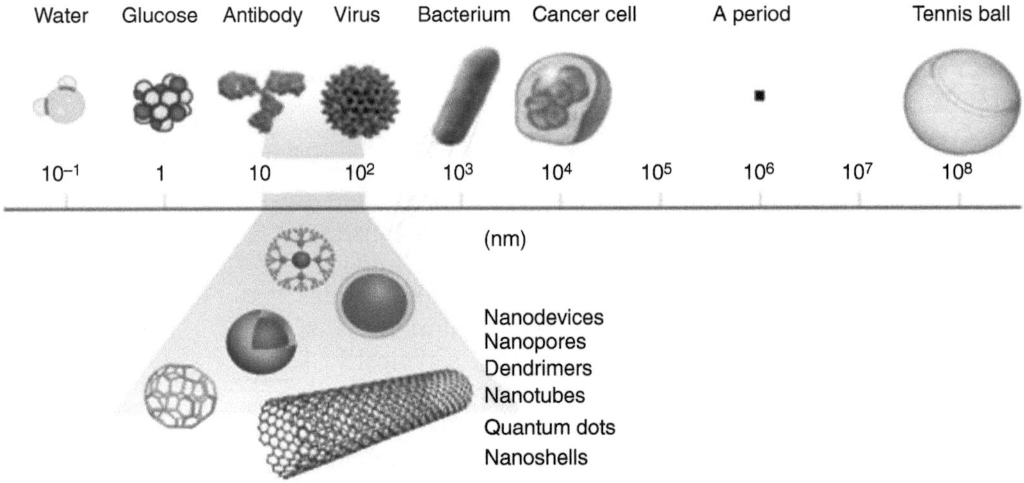

Figure 6.2. Size comparison of nanoparticle with other larger-sized materials. From Amin, M. T., Alazba, A. A., Manzoor, U., 2014. A review of removal of pollutants from water/wastewater using different types of nanomaterials. Adv. Mater. Sci. Eng., 1–25.

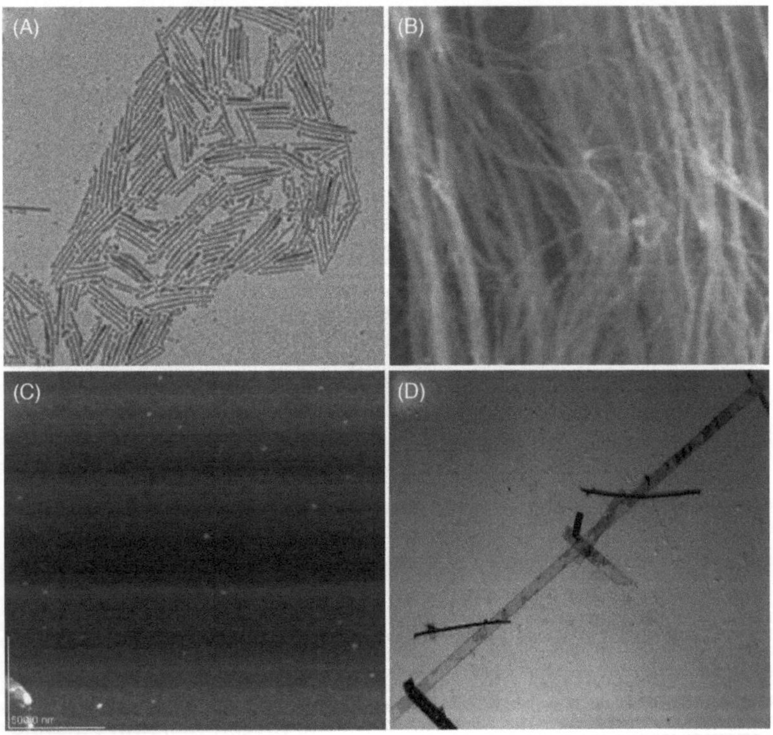

Figure 6.3. Examples of different types of nanomaterials, crystals, tubes, particles, and belts. (A) TEM image of TiO_2 nanocrystals, (B) SEM image of aligned CNTs, (C) AFM image of ZnO nanoparticles, and (D) optical microscope image of Sn-doped ZnO nanobelts. From Amin, M. T., Alazba, A. A., Manzoor, U., 2014. A review of removal of pollutants from water/wastewater using different types of nanomaterials. Adv. Mater. Sci. Eng. 1–25.

modifiable surface functionalization, excellent optical and electrical properties, unique adsorption properties, disinfectant capabilities, and versatility for carrying various therapeutic payloads. Due to their special properties, the last two decades have witnessed an increased interest in the synthesis and utilization of NPs and their nanocomposites for many applications in several fields including the environmental analysis, biology, agriculture, food, drinking water and wastewater treatment, electroanalysis, microfluidics, optical imaging and optical biosensors, cosmetics and dermal products, pharmaceuticals, medical, and biomedical fields.

Magnetic nanoparticles (MNPs) are a class of NPs that can be manipulated using magnetic field gradient. MNPs have the characteristics of NPs. In addition, MNPs have magnetic properties and can be easily separated under external magnetic field. The magnetic properties of MNPs depend on several factors including the size, shape, and morphology of the particles, the chemical composition, the extent of imperfection of the crystal lattice of the particles, and the interaction forces among the particles as well as with the surrounding matrix.

MNPs have been synthesized for use in many technological applications including molecular and magnetic resonance imaging (MRI) for medical diagnosis and therapy, hyperthermia, gene therapy and targeted drug delivery, neural tissue applications, microfluidic sensors, electrochemical sensors, environmental analysis including water treatment, biological analysis, catalysis, magnetic recording and data storage media, biomedical analysis, and ferrofluids for audio speakers (Tran et al., 2010).

The use of nanomaterials in the treatment of drinking water, groundwater, and wastewater is fast growing and anticipated to lead to high-performance purification systems that can reduce the concentration of toxic metal ions to the ppb level, kill the pathogens, and provide high-quality fresh water that meets all quality specifications. The biggest advantage of using NPs in water purification is the large surface-to-volume ratio which enables effective adsorption of many chemical and biological substances, and hence effective removal of the contaminants at very low concentrations, and greatly reduces the waste. In addition, the use of NPs allows for new reactions at the atomic level to take place that do not usually take place in conventional-size purifiers.

Due to their low toxicity, excellent magnetic properties, and large adsorption capacity toward many substances, MNPs are used as adsorbents and nanocatalysts in water treatment. When injected into a source of contaminated water, with their tiny sizes and high dispersibility, MNPs can easily penetrate deeper and provide a much faster and better detection, quantification, and treatment than conventional water treatment methods. In this regard, regular and modified MNPs have been used to remove almost all types of contaminants, from OM, inorganic salts, heavy metals, and dyes to bacteria and viruses, in drinking water and wastewater.

4.1 The Most Common Nanoparticles Used in Water Purification

As mentioned earlier, NPs have a wide range of applications in many fields including water purification. However, in this section, the attention is given to some of the NPs that are commonly used as adsorbents to filter out the contaminants, either in the pure state, modified, or included in a composite, in drinking water as well as wastewater treatment.

4.1.1 Iron-Based Nanomaterials

MNPs of iron oxide/hydroxide exist naturally in the earth's crust in several stable forms including magnetite (Fe_3O_4), hematite (α-Fe_2O_3), maghemite (γ-Fe_2O_3), goethite, ferrates, hydrous ferric oxide, and granular ferric hydroxide. Among these, the oxides Fe_3O_4, α-Fe_2O_3, and γ-Fe_2O_3 are the most-common iron-based MNPs used as adsorbents in water treatment (Dave and Chopda, 2014). In addition, nanoscale zero-valent iron (NZVI) has emerged and became commercially available for the treatment of water especially for the remediation of groundwater and wastewater.

When used in water treatment, iron oxides NPs are capable of treating large volumes of treatment water, very convenient for magnetic separation, and do not produce toxic by-products. In addition, as adsorbents, iron oxides MNPs are capable of reacting with (binding to) different functional groups and ligands via different methods to make the particles specific for the removal of specific contaminants from water.

4.1.1.1 Magnetite (Fe_3O_4)

Magnetite NPs have unique superparamagnetic characteristics and excellent adsorption properties that enable the particles to be used in many applications including water treatment. Pristine and modified magnetite NPs were used to remove a variety of contaminants from drinking water, groundwater, and wastewater such as heavy metals, natural OM (NOM), radionuclides, Pb^{2+}, As^{3+}, $As^{5+,}$ phosphorous, and nitrogen.

Modified magnetite NPs, such as mixed magnetite–hematite, SiO_2–Fe_3O_4, Pd–Fe_3O_4, FeS–Fe_3O_4, and polymer-coated magnetite (e.g., chitosan (CTS)–polypyrrole–magnetite, CTS–Al_2O_3–magnetite, humic acid–magnetite) MNPs have been used for the removal of a variety of water pollutants (e.g., La^{3+}, Zn^{2+}, As^{5+}, Pb^{2+}, Cr^{6+}, Cr^{3+}, Cd^{2+}, Hg^{2+}, dyes, NOM, and halogenated organic compounds) with higher adsorption capacity than the pure MNPs (Ahmed et al., 2013). Furthermore, in most of these studies, the results also showed that the modified magnetite MNPs are more stable and easier to separate at the end of the process than the pure magnetite NPs. In addition, ligand-modified magnetite NPs based on iron oxide attached to a powerful polydentate organic ligand such as EDTA or maltol were used to remove several heavy metals (Cd^{2+}, Pb^{2+}, and Cu^{2+}) from contaminated water (Ahmed et al., 2013).

4.1.1.2 Maghemite (γ-Fe_2O_3)

Maghemite (γ-Fe_2O_3) is another iron oxide that has a dual advantage when used in water treatment applications. It acts as both a photocatalyst and magnetic nanomaterial which makes it very effective in the degradation and removal of contaminants. Pristine, surface modified, and nanocomposites-based maghemite MNPs have been used as adsorbents to remove several contaminants from different water samples (drinking water, wastewater, groundwater, acid mine drainage, and aqueous solutions) with remarkable adsorption efficiency that approached 100% for some contaminants. Among the contaminants removed using maghemite MNPs are Cs^+, Se^{4+}, heavy metal ions, NO_2^-, NO_3^-, NOM, and several dyes [e.g., rose bengal, methylene blue (MB), methyl orange (MO), brilliant cresyl blue, thionine, Janus green B, and Congo red (CR)] (Cheng et al., 2012).

Several nanocomposites containing maghemite MNPs have been synthesized and used for water treatment. Thin-film nanocomposites based on maghemite MNPs, with different concentrations, in porous polysulfone supports have been synthesized as a membrane for groundwater purification using RO (Al-Hobaib et al., 2015). The results showed a 98% NaCl rejection with a membrane containing 0.3% γ-Fe_2O_3. Maghemite MNPs intercalated into an alumino-silicate matrix was used as a photocatalyst for the catalytic reduction and removal of Cr^{6+} from aqueous solutions. The study involved the use of potentiometry to monitor the change in the concentration of Cr^{6+} in the solutions.

A composite of maghemite/magnetite/carbon aerogel nanostructure was synthesized and used for the removal of the organic dye Rhodamine B from aqueous solutions with an efficiency exceeding 98%. The composite showed a great potential as an adsorbent for the removal of organic dyes from wastewater. Another composite based on CTS/maghemite NPs was used to effectively remove MO from aqueous solutions with a 90% removal in less than 20 min (Obeid et al., 2013).

4.1.1.3 Nanoscale Zero-Valent Iron

NZVI is another iron-based NP material that is used in the remediation and treatment of all types of water (drinking water, groundwater, wastewater, industrial water) to remove a variety of contaminants including organic compounds (e.g., chloromethanes, chlorobenzenes, trihalomethanes), dyes, pesticides, radionuclides (e.g., uranium), nitrates, selenium, heavy metals (e.g., As, V, Mn, Mo, Ag, Hg, Ni, Zn), bacteria, and viruses. NZVI acts as a reducing agent and an adsorbent at the same time where it reduces the OM to harmless carbon compounds and causes the heavy metals to agglomerate and deposit. To increase their surface area and improve their reactivity, the NZVI NPs are also modified with other substance, such as palladium, AC, $Ca(OH)_2$, and polymers (e.g., polyacrylic acid) (Li et al., 2015).

Pristine and modified NZVI NPs can be directly added to the contaminated water for in situ treatment or immobilized in membranes for ex situ. Both regular and modified NZVI NPs have shown a great stability over a wide pH range and temperature for up two months of use. Compared to the regular NZVI, modified NZVI is more efficient and faster in removing a wide range of contaminants. Both types of NZVI NPs can be regenerated for reuse. However, the efficiency of the NPs in both types of NZVI is decreased by the presence of competing anions. The NZVI particles may also corrode over time and become less effective (Kanel et al., 2005).

4.1.2 Manganese Dioxide

Naked MnO_2 NPs have been used as an adsorbent in water purification to remove several contaminants (including metal ions and dyes) from water environments and aqueous solutions. δ-MnO_2 NPs were used to remove Pb^{2+}, Zn^{2+}, and MB (Wang et al., 2014). However, hybrid, composite-incorporated, and functionalized MnO_2 NPs have more applications and showed better adsorption capacity in water treatment than the naked particles. In one study, MnO_2 and Fe MNPs were used as adsorption medium to remove organic and inorganic contaminants from water with more than 90% of the compounds completely removed in about 1 h. In addition, polymer-modified MnO_2 NPs were used to effectively remove the heavy metal ions Pb^{2+}, Cu^{2+}, Cd^{2+}, and Ni^{2+} from water (Zhu and Li, 2015).

4.1.3 Silver Nanoparticles

Among all nanomaterials, silver nanoparticles (AgNPs) represent one of the major nanoproducts. Due to their excellent antimicrobial properties, AgNPs are extensively used in the manufacture of several products including water disinfectants, consumer products (oral care, supplements, cosmetics/skin care, and wound dressing gels), and biomedical devices. A recent study showed that a 20 ppm colloidal solution of AgNPs is enough to provide a 100% cure rate for malaria and help boost the immune system. As applied in water treatment, either in the pure state, modified with functional groups, or incorporated in nanocomposites, AgNPs have demonstrated excellent capabilities to inactivate microbes and remove several contaminants including heavy metals. AgNPs in the pure state were used to remove several contaminates from drinking water and wastewater including Cr^{6+} and bacteria. Ag/Bi_2O_3, Ag/ZnO, and ZnO/Ag/CdO nanocomposites were prepared for use as photocatalytic adsorbents in water treatment.

A CTS –Ag–Cu–CNT multifunctional nanocomposite was used for the removal of Cu^{2+}, Cd^{2+}, and Pb^{2+} ions from an aqueous solution in about 10 min with nearly 100% efficiency. $Ag/CoFe_2O_4$/graphene-oxide (GO) nanocomposite was synthesized for the removal of bacteria and Pb^{2+} ions from contaminated water. In addition, poly(Na acrylate)-based cryogels decorated with AgNPs were synthesized for use for point-of-use disinfection of drinking water (Loo et al., 2013).

4.1.4 Aluminum-Based Nanomaterials

Pure and modified aluminum-based NPs (Al, Al_2O_3, $Al(OH)_3$, and AlOOH) have been used as effective adsorbents to filter out several contaminates including OM, dyes, fluoride ions, and heavy metals from drinking water and wastewater (Iorio et al., 2008). Pure Al_2O_3 NPs were used to remove F^- ions from aqueous solutions. The performance of the F^- ion adsorption was studied as a function of several experimental factors including contact time, pH, initial F^- concentration, temperature, and influence of interfering anions (e.g., PO_4^{3-}, SO_4^{2-}, and CO_3^{2-}). The results showed that the presence of the interfering anions affects the F^- ion adsorption capacity and the maximum F^- removal occurs at pH 6.15.

In another investigation, the performance of Al_2O_3 NPs modified with three different compounds (humic acid, an extract of walnut shell, 1,5-diphenylcarbazone) to remove heavy metals (Cu^{2+}, Cd^{2+}, and Ni^{2+}) was studied. The results showed that the modified Al_2O_3 NPs are better adsorbents than the pristine ones. In addition, Al_2O_3 modified with 2,5-diphenylcarbazone showed the maximum adsorption capacity for the heavy metals

(Mahdavi et al., 2015). γ-Aluminum oxide hydroxide (AlOOH) NPs supported on CTS shell have been used to remove F^- ion from drinking water. The results showed a rapid adsorption of the F^- ion with more than 80% adsorption within the first 20 min while equilibrium was achieved within 1 h of treatment (Wan et al., 2015). Aluminum-modified hematite, zeolitic tuff, and calcite adsorbents were used and compared to remove F^- ions from drinking water. The highest fluoride ion sorption capacity was found for aluminum-modified zeolite.

Sulfate-doped Fe_3O_4/Al_2O_3 MNPs were developed for F^- ion removal from drinking water (Chai et al., 2013). The kinetic data showed that the adsorption of F^- ions was rapid in the beginning of the treatment followed by a slower process with ~85–90% of the F^- ion removed in the first 20 min of the treatment while 100% removal was achieved in about 8 h.

4.1.5 Titanium Dioxide

The past four decades have witnessed the emergence of TiO_2 as an excellent photocatalyst and adsorbent in numerous industries and applications (Leong et al., 2014). With its combined oxidative and adsorptive properties, pristine TiO_2 NPs have been used to remove several metals from water with a great efficiency. The metals removed included arsenic (As^{3+} and As^{4+}), lead, cadmium, copper, zinc, and nickel. Because of its photoelectrocatalytic and photocatalytic disinfection power, TiO_2 has also been used as an adsorbent for environmental purification. Pure TiO_2 has been used to filter out several contaminants including bacteria; waterborne OM, viruses, and emerging pollutants such as pharmaceutical and personal care products (Zheng et al., 2013). TiO_2 has also been involved in several composite NPs (e.g., Au/TiO_2, Fe_3O_4/TiO_2, TiO_2–graphene/Fe_3O_4) for the photocatalytic removal of various water pollutants including toxic organic compounds, such as azo dyes, phenols, endocrine-disrupting chemicals, and bacteria.

4.1.6 Zirconium-Based Materials

Due to their chemical stability and nontoxicity, zirconium-based NPs (mainly Zr and ZrO_2) either in the pristine state or included in composite membranes have been used to effectively remove several contaminants from drinking water and wastewater including fluoride ions, organic dyes, and heavy metals (Cui et al., 2012).

4.1.7 *Carbon-Based Materials*

Carbon nanomaterials have many excellent properties such as low cytotoxicity, chemical inertness, high biocompatibility, fast flow rate, high permeability, great thermal stability, and excellent physical, mechanical, and electronic properties. Due to these unique properties, carbon-based nanomaterials have been used as adsorbents to remove various inorganic, as well as organic pollutants and radionuclides from drinking water, groundwater, and wastewater (Smith and Rodrigues, 2015). In these applications, carbon nanomaterials have been use in a variety of forms including membranes and nanomesh either in the pure state or supported with other materials. With the advances in the water treatment technologies and instrumentation, it is anticipated that carbon-based nanomaterials will be less expensive, more durable, and easier to clean and reuse. Examples of the C-based nanomaterials are AC, carbon fibers, graphite, carbon nanoparticles (CNPs), graphene oxide (GO), and carbon nanotubes (CNTs). In this section, the focus is on the carbon nanomaterials that are extensively used in water purification.

4.1.7.1 Carbon Nanoparticles

CNPs in the pure state, oxidized (O-CNPs), and modified have been used as adsorbents in water treatment and CNP-based composites have been used for the removal of several heavy metals from aqueous solutions. Pure CNPs were used in the removal of N-nitrosamines, NOM, and Ni^{2+}. O-CNPs were used as a recyclable adsorbent for efficient removal of Cu^{2+} ions, dyes and N-doped magnetic CNPs were used for the removal of Cr^{3+} from aqueous solutions (Shin et al., 2011). A nitrogen-doped CNPs coated on the surface of stainless steel wire was investigated as an adsorbent fiber for aromatic polycyclic, hydrocarbons, some acid esters, and UV filters in water.

4.1.7.2 Graphene Oxide

GO has become one of the main adsorbents in water purification especially for wastewater treatment. Many studies have been reported on the water treatment applications of GO in several states, such as pure state, surface modified, or incorporated in a hybrid composite or a membrane. Pristine GO was used as an adsorbent to remove several pollutants from aqueous solutions and wastewater samples (Lee et al., 2013).

Magnetic GO showed a high removal efficiency toward water pollutants. However, it is difficult to separate the GO particles from water after the treatment. To solve this issue but retain their great removal efficiency, magnetic GO NPs have been incorporated in hybrid nanocomposites and membranes for the removal of a variety of water pollutants. In this regard, a magnetic composite gel composed of GO NPs and poly(vinyl alcohol) has been synthesized. The gel showed a great efficiency in the adsorption of MB and methyl violet dyes. In another study, magnetic GO NPs were immobilized in a composite containing calcium alginate. The composite was successful in the adsorption of MB (Li et al., 2013). GO nanosheets supported on sepiolite composites were also used to remove uranium from aqueous solutions.

4.1.7.3 Carbon Nanotubes

Among all carbon-based nanosorbents used in drinking water and wastewater treatment, CNTs received special attention due to their unique and tunable properties, exceptional capabilities, and effectiveness in removing many of the chemical as well as biological contaminants (Liu et al., 2013). For their use in water treatment, CNTs have been prepared in the form of membranes. Several synthesis methods have been proposed for making these membranes. However, the most popular synthesis method is through coating a silicon wafer with a metal NP catalyst that allows the CNTs to grow vertically in the form of a membrane with pores in the nm range. To increase the stability of the membrane, the space between the CNTs can be filled with a ceramic material.

The CNT membranes are very effective and are able to filter out almost all kinds of water contaminants including fluoride ions, several heavy metals, metalloids, NOM, turbidity, bacteria, viruses, and cyanobacterial toxins (Das et al., 2014). These membranes are also durable, and easy to clean and reuse. However, CNT membranes are still relatively expensive. It is anticipated that, with the advances in research and technology, the CNT membranes to be much less expensive.

4.1.8 Zeolites

Zeolites are a family of porous crystalline aluminosilicates that are widely used as adsorbents or ion-exchange media in the drinking water, groundwater, and wastewater treatments. Zeolites exist naturally but can also be synthesized in the laboratory. Natural and modified zeolites are usually used to remove ammonium ions and metal contaminants, such as iron, manganese, calcium, and heavy metals (e.g., arsenic, lead, copper, mercury, silver, chromium, zinc, cadmium) (Vatin et al., 2014). In addition, modified zeolites are also used to remove viruses, bacteria, and organic contaminants (Bai et al., 2010). The efficiency of zeolites in removing or filtering out contaminants depends on several factors including the composition of the zeolite material and the nature of water under treatment (e.g., its pH value, types of contaminants present and their concentrations). Because zeolites are naturally available, zeolites-based devices are relatively inexpensive but require occasional regeneration using an acid solution.

4.1.9 Magnesium-Based Materials Oxide

Pure and modified Mg-based NPs, such as MgO, $Mg(OH)_2$, and activated MgO have been used as adsorbents for the removal of several contaminants from drinking water and groundwater including fluoride ions, phosphates, nitrogen species, OM, heavy metals, suspended solids, and bacteria. The results of several studies showed that MgO NPs are very efficient in the removal of the pollutants especially the fluoride ions. Modified MgO was used to remove fluoride ions from groundwater in presence of several other coexisting ions with a high defluoridation capacity. Phosphates and bicarbonates are the two most competitors for fluoride removal (Maliyekkal et al., 2010).

4.2 Synthesis Methods of Magnetic Nanoparticles

The successful applications of the MNPs depend on both the synthesis method and the surface modification of the NPs. The past two decades have witnessed a great effort devoted to the synthesis of MNPs due to their unique properties and potential applications in many fields. Several methods and protocols have been reported in the literature for the synthesis of MNPs. These methods can be generally classified into three main routes: physical, chemical, and biological methods (Xu et al., 2014). Table 6.2 summarizes the advantages and disadvantages of each individual synthesis method (Xu et al., 2014).

Examples of the physical methods are gas-phase deposition and electron-beam lithography. These methods are relatively inexpensive, simple, and produce pure NPs. However, in the physical synthesis methods, it is difficult to control the size of the NPs and usually produce NPs with a wide size distribution range. In addition, these methods are not always suitable for the synthesis of all types of NPs.

The chemical methods are widely used because they are straightforward and produce high yield of the NPs. Although some methods require drastic experimental conditions such as high temperatures and/or pressures, the size of the NPs can be controlled by adjusting the reaction parameters (Xu et al., 2014).

Table 6.2 Different Synthesis Methods of Magnetic Nanoparticles (MNPs)

Synthesis Methods			
General	**Specific**	**Advantages**	**Disadvantages**
(i) Physical methods	Gas-phase deposition	• Easy to perform	• Difficult to control particle size
	Electron beam lithography	• Good control of interparticle spacing	• Expensive • Requires complex machining
(ii) Chemical methods	Sol–gel synthesis	• Inexpensive • Excellent control of particle size and morphology	• Low wear resistance • High permeability • Weak bonding
	Oxidation	• Uniform size • Narrow particle size distribution	• Small-sized ferrite colloids
	Chemical coprecipitation	• Simple and reproducible • Efficient	• Not appropriate for the synthesis of highly pure and accurate stoichiometric phases
	Hydrothermal reactions	• Easy control of particle size and shape	• Requires high temperature and pressure
	Flow injection	• Good reproducibility • High mixing homogeneity along with excellent control of the process	• Requires special mixing methods of reagents in a capillary reactor
	Electrochemical	• Easy control of particle size	• Lack of reproducibility
	Aerosol/vapor phase	• High yield	• High temperature
	Sonochemical decomposition	• Narrow particle size distribution	• Synthesis mechanism not fully understood
	Supercritical fluid	• No organic solvents involved • Efficient control of particle size	• Extremely high (critical) temperature and pressure
	Using nanoreactors	• Precise control of particle size	• Complex reaction conditions
(iii) Microbial methods	Microbial incubation	• Low cost • Good reproducibility • High yield	• Time-consuming process

Source: Xu, J., Sun, J., Wang, Y., Sheng, J., Wang, F., Sun, M., 2014. Application of iron magnetic nanoparticles in protein immobilization. Molecules 19, 11465–11486.

Several investigations have proven that the particle size and dispersity of the NPs can be tailored by controlling the experimental parameters of the chemical synthesis method such as temperature, the nature and the concentration ratio of the reactants, ionic strength, the pH of the medium, the mixing rate, etc. (Clement et al., 2015). In general, the particle size increases with increasing temperature and/or reaction time.

The microbial (biological) approach for the synthesis of MNPs includes the use of different organisms such as bacteria, yeast, fungi, plant and plant extracts, and algae. These methods have several merits including high yield, low cost, and good reproducibility. However, biological methods involve a fermentation step which is a time-consuming process (Pawar et al., 2013).

In this section and due to the limited space, the focus is on the most popular and commonly used methods to synthesize MNPs.

4.2.1 Coprecipitation

This is the most conventional and most popular method used to synthesize many MNPs, such as magnetite (Fe_3O_4), maghemite (γ-Fe_2O_3), ZnS, Ni-doped ZnO, cobalt ferrite ($CoFe_2O_4$), magnesium ferrite ($MgFe_2O_4$), Ga-substituted cobalt ferrite, and Nd-doped strontium ferrites (Chen et al., 2013). The MNPs produced by the co-deposition method can be either used directly or incorporated in a composite. As an example for the method, in the coprecipitation synthesis of Fe_3O_4 MNPs, a mixture of Fe^{3+} and Fe^{2+} ions in a suitable molar ratio is mixed in a basic solution [e.g., NaOH, KOH, NH_4OH, $(C_2H_5)_4NOH$) in an inert atmosphere] (Mascolo et al., 2013). The size and shape of the MNPs produced from this method depends on the experimental conditions, such as the pH of the medium, the ionic strength, the molar ratio of the reactants, the type of the salts and solvent used, the rate of mixing, the digestion time, the type of the base used, the reaction temperature, and the presence or absence of a magnetic field (Chen et al., 2013). The control of the experimental conditions allows the production of uniform MNPs with a narrow particle size distribution.

4.2.2 Thermal Decomposition (Thermolysis)

Thermal decomposition is the second most commonly used method to synthesize MNPs. The method was used to prepare several MNPs including pure and mixed metals (e.g., Fe, Co, Co/Fe), metal oxides (e.g., magnetite), metal sulfides (e.g., MnS, PbS), and nanocomposites (e.g., Pt-based NP composites) (Liu et al., 2011). In addition, Park et al. developed a general thermal decomposition synthesis procedure based on the use of metal–surfactant complexes to produce uniform monodispersed nanocrystals of several transition metals (Cu, Ni, Cr, Fe, Pd), metal oxides (MnO, NiO, $MnFe_2O_4$, Fe_3O_4, γ-Fe_2O_3, and $CoFe_2O_4$), and metal sulfides (MnS, CdS, ZnS, PbS).

Thermal decomposition involves the pyrolysis of a low-stability organic complex (precursor) of the substance (metal or oxide) of interest. For example, for the synthesis of iron metal and iron oxide NPs, complexes such as $Fe(CO)_5$, ferrocene, iron(III) acetoacetate, $[Fe(CON_2H_4)_6](NO_3)_3$ are used as precursors (Amara and Margel, 2013). The synthesis can be performed in one step or two steps depending on the product of interest. Thermal decomposition offers a good control over the properties of the MNPs (such as the particle size, size distribution, shape, crystallinity, and magnetic properties) through the control of the synthesis conditions. Another advantages for the method when used to prepare pure metal NPs from the thermal decomposition of metal carbonyl precursors ($M_x(CO)_y$) is that the carbonyl compound contains the metal atoms in the zero-valent oxidation state, thus, no extra reducing agent is necessary.

Microwave-induced thermal decomposition of precursors is another simple, rapid, cost-effective, and energy-saving modification of the thermal decomposition method. This method has been used for the synthesis of stable metal and metal oxide NPs using carbonyl precursors in ionic liquids. The method was also used to synthesize metal oxide powders (e.g., Cr_2O_3) through the solid-state decomposition of a suitable precursor.

4.2.3 Hydrothermal Synthesis

Supercritical hydrothermal synthesis is a relatively simple and environmentally friendly method for the formation or growth of crystals in aqueous solutions in a sealed container under high temperature (in the range of 130–250°C) and high pressure (0.3–4 MPa) water conditions (Kim et al., 2012). As the reaction is done in aqueous medium, this method is also known as supercritical water hydrothermal synthesis. An advantage of this method is that the starting substances are usually insoluble in water under normal temperatures and pressures.

With the proper control of the experimental conditions, hydrothermal synthesis can grow crystals and produce well-crystalized superparamagnetic NPs. The method has been used for the synthesis of several metal oxide NPs including single oxides, such as CuO, $AlOOH/Al_2O_3$, Fe_2O_3, NiO, ZrO_2, Co_3O_4, mixed oxides, such as ceria–zirconia (Kim et al., 2012), and complex oxides, such as $K_2O{\cdot}6TiO_2$. The method is also suitable for the preparation of nanomaterials with unusual shapes, such as α-Fe_2O_3 nanocubes and hollow sea urchin-like α-Fe_2O_3 nanostructures (Zhang et al., 2009).

4.2.4 Reverse Micelles and Microemulsion Technology

Recently, there has been an increased interest in the synthesis of NPs within "nanoreactors," such as a micelle. Reverse micelles or water-in-oil (w/o) microemulsions are nanosized water droplets dispersed in a continuous oil medium. The micelles are stabilized by the addition of a surfactant whose molecules accumulate at the w/o interface. Within a reverse micelle, the NPs are synthesized only through the materials introduced there and without substance supply from outside. Owing to this situation, the reverse micelles technology offers good control and modulation of the size of the NPs. In general, the size and degree of dispersity of the NPs depend on the solvent and surfactant, the concentrations of the reagents, the addition of an electrolyte, and the molar ratio of water to surfactant concentrations (Eastoe et al., 2006).

Microemulsion systems were successfully used as reaction media to synthesize several nanomaterials including superparamagnetic iron oxide and $MnFe_2O_4$, CdS, Co, Cu, and Ag_2S NPs (Siong et al., 2014). NPs synthesized using the reverse micelles method have been used in different applications including catalysis, food analyses, nanofluids, and water treatment (Siong et al., 2014).

4.2.5 Sonochemical Synthesis

In sonochemical synthesis, a high intensity ultrasonic irradiation is used to induce the formation of particles with small size and high surface area. The method has many merits including simplicity, controlled reaction conditions, fast reaction rate, uniform NP shape, and small-size distribution. Due to these advantages, several studies have been reported on the use of the sonochemical process to prepare various nanomaterials such as metals, metal oxides, metal carbides, and nanocomposites.

Among the nanomaterials synthesized using the sonochemical route are Ag, hematite (Fe_2O_3), magnetite (Fe_3O_4), Fe_3O_4/ZrO_2, TiO_2, PdO/silica, CdTe, HgS, Eu-doped ZnO, Co-ferrite ($CoFe_2O_4$), titanium carbide, GO-wrapped gold, and γ-Al_2O_3-doped ZnO nanocomposites. The method has also been used to synthesize several new nanomaterials with unique properties (e.g., unusual magnetic properties) (Sivasankaran and Kishor Kumar, 2015).

4.3 Surface Modification (Functionalization) of Magnetic Nanoparticles

Bare MNPs are generally highly reactive under ambient conditions and hence are susceptible to air oxidation and degradation if exposed to certain environments, such as strong acidic solutions.

In addition, due to their large surface area, high surface energy, and hence strong magnetic interactions, bare MNPs tend to form aggregates in aqueous media. This aggregation limits the dispersion of the MNPs, lowers the surface energy, and hence decreases their strength and reactivity. Accordingly, for their applications in various potential fields, MNPs should be stabilized through surface modification (functionalization) (Piquemal et al., 2013). In their applications in water treatment, the surface modification of the MNPs is very critical for the stability and monodispersity of the particles in the different types of water and also for the affinity of the MNPs toward the different contaminants (Ling et al., 2010).

Generally, surface modification of bare MNPs is done through coating the NPs with an inorganic material (e.g., silica, gold, silver). The coating minimizes the interaction forces (van der Waals forces) among the particles, increases the dispersity of the particles, and minimizes the interaction between the particles and the system environment (Ma and Liu, 2007). In addition to improving the stability of the MNPs, the coatings also provide active sites for covalent bonding of certain ligands and functional groups with the NPs. These functional groups increase the solubility and selectivity of the MNPs in aqueous solutions.

Different types of materials and compounds, both natural and synthetic, are used in the surface modification process of the MNPs. These materials are introduced either during or after the synthesis of the particles. The most commonly used materials in surface modification of MNPs are: (1) naturals dispersants, such as starch, gelatin, CTS, ethyl cellulose, albumin, liposomes, and dextran; (2) surfactants and chelating agents such as carboxylic acids (e.g., lauric acid, acrylic acid, oleic acid, alkylphosphonic acids, and alkylsulphonic acids); (3) polymers such as polypyrrole, polyethylene glycol, polyvinyl alcohol, polythiophene, polydopamine, polyvinylpyrrolidone, poly(lactic-*co*-glycolic acid), and poly(ethylene-*co*-vinyl acetate); and (4) inorganic oxides and metals such as silica, gold, and silver.

Several techniques are currently used to introduce functional groups to MNPs. Among these, silanization is the most popular and widely employed one because it has the following merits: easy to perform, highly stable under several experimental conditions (e.g., high acidity, pH change, and change in electrolyte concentration), inert to redox reactions, can be performed in aqueous as well as nonaqueous media, and shows low toxicity (Mulvaney et al., 2000).

4.4 Characterization of Magnetic Nanoparticles

As mentioned earlier, the physical and chemical properties of the MNPs control the applications of the particles. Accordingly, comprehensive structural, optical, electrical, electrochemical, and thermal characterization of the MNPs from the early stages of synthesis to their applications and aging is required and is a routine part of any study that involves the synthesis and/or the application of the NPs. Surface characterization of the NPs is also required to understand the health, safety, and environmental impact issues related to their use especially for their applications in medicine and water treatment.

The most common analysis techniques that are currently used to detect, measure, and characterize the NPs are transmission electron microscopy, scanning electron microscopy, scanning transmission electron microscopy, X-ray photoelectron spectroscopy, X-ray diffraction, Auger electron spectroscopy, Fourier transform infrared spectroscopy, vibrating sample magnetometry, diffraction and differential scanning calorimetry, thermogravimetric analysis, time of flight secondary ion mass spectrometry, low-energy ion scattering, UV/Vis spectroscopy, nuclear magnetic resonance, and scanning probe microscopy including atomic force microscopy and scanning tunneling microscopy (Baer, 2011). Usually, any investigation involves several of these analysis techniques depending on several factors including the starting materials, experimental conditions, physical state, size and shape of the NPs, and the required application of the NPs produced. Each of these techniques provides specific information and has its own merits as well as level of accuracy (Baer, 2011).

4.5 Technologies Currently Used for Fluoride Ion Removal

Traces of fluoride salts are normally present in all types of natural water. The concentration of fluoride ion (F^-) varies from 0.5 mg/L in rivers and lakes to about 1.0 mg/L in seawater. In groundwater, the concentration of F^- ions can be as high as 10.0 mg/L depending on the chemical nature of the rocks as well as the type of fluoride mineral present. F^- ion may be an essential ion for animals and humans. However, the concentration of the ion must be carefully monitored and strictly controlled. For humans, fluoride ion is added to drinking water at very low level (up to 1.0 mg/L) to reduce tooth decay and dental cavities. On the other hand, F^- ion is a carcinogen and all salts of fluorine are toxic (UNICEF, 2015). So, higher concentrations of F^- ion (typically above 2.0 mg/L) are harmful and can induce dental fluorosis (also known as mottled enamel), especially for children under nine (Jagtap et al., 2012). The maximum permissible level of F^- ion content in drinking water set by the WHO is 1.5 mg/L. Many studies showed that long-term ingestion of high-fluoride (3.0 mg/L or above) drinking water can lead to many diseases including skeletal fluorosis, brain damage, thyroid disorder, infertility, hypertension, chronic kidney diseases, and bladder cancer (Ringe, 2004). Accordingly, it is extremely necessary to develop cost-effective and high-performance technologies to remove excess F^- ions from drinking water.

This section covers the general methods for the removal of F^- ions from water sources with emphasis on drinking water.

4.5.1 Use of Alternative Water Sources/Resources

If the concentration of fluoride ion in the drinking water supply of a community is constantly higher than the permissible levels, there are two general options to reduce and control the fluoride content. The first option is the search for and/or making alternative water supplies with acceptable fluoride concentration (Renuka and Pushpanjali, 2013). This option includes: (1) drilling for new wills that have water with lower fluoride levels, (2) mixing high fluoride water with low fluoride water to bring the fluoride concentration within the permissible range, (3) rainwater harvesting where the rain is collected and stored using simple rain catchment systems, such as rooftop and land surface catchment systems, (4) water transport from a distant source of safe water with low fluoride content, and (5) use of two parallel (dual) water sources where a source of low fluoride level is used for drinking as well as cooking and another source with high fluoride content for all other purposes.

The above techniques have some advantages but they are only acceptable as a short-term solution. However, for long-term solutions, these techniques have many disadvantages including the high cost of the infrastructure and equipment, long time to plan, implement and run, frequent monitoring and maintenance, and require comprehensive educational programs to increase of the community awareness of the problem and the solutions. In addition, in many areas throughout the world, most of these methods may not be feasible to implement.

4.5.2 Use of Defluoridation Techniques

The second option to produce safe drinking water with permissible fluoride content is the use of defluoridation techniques. Defluoridation is the process of lowering the fluoride ion concentration in drinking water to the optimal range. It is the most practical and efficient method for fluoride removal. The defluoridation process can be done either at the source (treatment on a large scale at the community level) or at the point of use (treatment on a small scale at the household level). The selection of a specific defluoridation method, especially in developing countries, is controlled by several factors including the cost, capacity, and simplicity of the method; availability of the chemicals; and acceptance of the taste and odor of the purified water.

Various defluoridation methods have been employed. These methods can be broadly divided into four main categories: (1) chemical precipitation methods, (2) ion-exchange methods, (3) adsorption methods, and (4) other methods, which include ED, Donnan dialysis, RO, and electrochemical methods (Renuka and Pushpanjali, 2013).

In general, these methods are effective in removing the fluoride ion. However, some of these techniques are not convenient for use on a large scale because of various unfavorable factors such as the high costs of installation, operation, and maintenance; complexity of the treatment; and generation of toxic by-products. Among these techniques, adsorption is the most commonly and widely used one because of its simplicity, high efficiency, low cost, and versatility.

4.5.2.1 Chemical Precipitation Methods

The precipitation defluoridation methods involve the addition of suitable chemicals (coagulants and coagulant aids) to react with the fluoride in the contaminated water and form insoluble salts (precipitates) that can be, then, removed from water. Fluoride can also be removed by coprecipitation or adsorption on the surface of the precipitate. The most common chemicals used in the precipitation techniques are calcium salts (calcite, lime, and $CaCl_2$) either used alone or with aluminum or magnesium salts along with coagulant aids. The major disadvantage of the precipitation techniques is that the added chemicals usually change the pH, taste, or odor of the treated water, and in some cases, the water will be unsuitable for drinking.

A very famous example of the precipitation technique is the Nalgonda technique. This technique involves the addition of the following chemicals in sequence: aluminum salt, lime, bleaching powder, followed by rapid mixing, then coagulation, sedimentation, filtration, and disinfection. The amount of the aluminum salt added depends on the alkalinity and concentration of fluoride in the water under treatment. The common aluminum slats used in the process are aluminum sulfate (alum) or aluminum chloride. The salt is selected based on the level of sulfate and chloride concentrations in the water under treatment to avoid exceeding levels of these anions in water above the permissible levels.

The Nalgonda technique is very versatile and has been successfully used for water purification at the individual as well as community levels in several developing countries including India and Tanzania.

Recently, a new chemical technique, known as contact precipitation, was introduced for the removal of fluoride ion from water (Dahi, 1997). The technique involves the addition of calcium and phosphate compounds to the raw water to precipitate the fluoride ions as CaF_2 and/or fluorapatite. The process involves the addition of bone charcoal to catalyze the precipitation. The technique is cost-effective, highly efficient, and was used at the community level in some of the developing countries.

4.5.2.2 Ion-Exchange Methods

In ion-exchange defluoridation, raw water is passed through a bed containing a cationic or an anion inorganic exchange resin where the fluoride ions are exchanged with the ions in the resin (Jamhour, 2005). Some of the common commercial anion resins (usually in the form of hydroxide or chloride) used in fluoride removal are ResinTech SIR-900, ResinTech SBG2, Tulson A-27, Amberlite XE, and Lewatit MIH-59 (Ingle et al., 2014). The fluoride exchange capacity of the resins depends on the pH of the raw water as well as the ratio of the fluoride concentration to the total anions concentration. Each resin has its own maximum fluoride exchange capacity. Some of the resins may introduce taste and odor to the treated water due to the resin degradation products. It is also recommended for some of the resins to be first washed with hot water and cycled several times through exhaustion and regeneration. The method is effective in removing 90–95% of the fluoride ion (Ingle et al., 2014). However, the resin is expensive, the method produces a large volume of the waste, and the efficiency of the fluoride removal is affected by the presence of other ions in the raw water.

4.5.2.3 Adsorption Methods

In the adsorption process, the water contaminants (the adsorbates) attach to the surface of the adsorption medium (the adsorbent) through physical (physisorption) and chemical (chemisorption) interactions. Due to their small size and high surface area, NPs have high affinity to absorb many contaminants and hence are considered the best adsorption media for air and water purification and remediation of wastewater as well as groundwater.

A wide range of materials has been used as adsorption media for the defluoridation of water (Verma et al., 2014). These materials include aluminum-based materials (e.g., activated alumina,

alumina and amorphous aluminum hydroxide), calcium-based materials (e.g., calcite, limestone, calcium phosphate), iron-based materials (e.g., hydrous ferric oxide, granular ferric hydroxide), carbon-based materials (e.g., activated charcoal, CNTs, multiwalled CNTs), natural materials (e.g., clays, tamarind seeds, zeolite F-9), biomaterials (e.g., chitin, CTS derivatives, algal and fungal biomasses), and nanomaterials.

The efficiency of fluoride removal varies from one adsorbent to another. Natural adsorbents such as serpentine, bauxite, kaolinite, and clays showed low fluoride uptake capacities. On the other hand, processed and synthetic adsorbents have shown higher fluoride uptake capacities. Among these materials, bone char, activated alumina, calcined clays, and nanomaterials have been successfully used in the defluoridation process. In this section, some applications of these materials for the removal of fluoride ions from different water samples are given.

4.5.2.3.1 Adsorption Using Aluminum-Based Materials Various studies have been conducted on the defluoridation of different types of water using aluminum-based adsorbents (e.g., activated alumina, aluminum hydroxide, and aluminum oxide hydroxide). Most of the studies focused on the use of activated alumina (Al_2O_3) either alone or modified (impregnated) with a transition metal ion compound [e.g., $La(OH)_3$, CuO, CaO, MnO_2, $La^{3+} + Y^{3+}$] to enhance the efficiency of the process (Kumar et al., 2011). Alumina forms complexes with fluoride. Based on the published results, the removal capacity of the F^- depends mainly on the pH of the crude water. Other factors that affect the fluoride uptake include temperature and zeta potential (Bahena et al., 2002).

Activated alumina is considered the best adsorbent for removing fluoride ion from drinking water worldwide. Domestic defluoridation units based on activated alumina are currently used in several developed as well as developing countries. However, this technology is relatively expensive and its efficiency is affected by several factors including the specific grade of activated alumina, particle size, and the nature of water under treatment (e.g., the pH, the presence of other ions in the treated water such as bicarbonates and sulfates). In addition, activated alumina requires frequent activation and its adsorption efficiency decreases with the frequent use and regeneration (Renuka and Pushpanjali, 2013).

4.5.2.3.2 Bone Char Defluoridation Bone char, activated alumina, and RO systems are the three methods recommended by the EPA and Water Quality Association for the removal of fluoride from drinking water. Bone char was used in the past in the USA and is currently used in several developing countries (including Thailand, Sri Lanka, and some African countries) for water defluoridation. Bone char consists of ground animal bones that have been charred to remove all OM. Bone char mainly consists of calcium carbonate, calcium phosphate, and AC. Fluoride removal takes place through the replacement of the carbonate in the char with the fluoride ion.

Based on several investigations, a number of factors affect the uptake capacity of the bone char in the removal of fluoride from raw water. Among these factors are: (1) the method of preparation and quality of the char; the results showed that the use of low-grade bone char adds bad taste and odor to the treated water, (2) the presence of other interfering ions, especially arsenic ions, in the raw water decreases the fluoride uptake capacity, and (3) the amount of acid added to neutralize the excess base used in the regeneration step as bone char dissolves in acids. In addition, the efficiency of the defluoridation process depends on the amount of char used, temperature, pH of raw water, contact time, and concentration of fluoride in the raw water. The results showed a maximum defluoridation efficiency of 62–66% when the process was conducted under the optimum conditions (45°C, pH = 7.0–7.5, and duration time = 9 h). The efficiency was further enhanced through the addition of brushite and calcium hydroxide to the raw water in a pretreatment step to precipitate out the fluoride (Larsen et al., 1993).

To improve the fluoride adsorption properties of the bone char, very recently, Rojas-Mayorga et al. (2015) described a new route for the synthesis of bone chars using a CO_2 atmosphere at different pyrolysis temperatures. The authors used the new bon chars as adsorbent for fluoride removal from water. The authors reported that the chars formed at 700°C showed the best adsorption properties. In another study, the same research group synthesized new surface-modified bone chars via metallic doping using aluminum and iron salts and used them as an adsorbent for fluoride adsorption from drinking water. The results showed that the surface modification of the bon char with aluminum sulfate enhanced the fluoride adsorption properties up to 600% relative to the unmodified bon char.

4.5.2.3.3 Adsorption Using Nanomaterials Due to their unique properties and large surface area, nanomaterials have been used as nanosorbents to remove fluoride ion from all types of water. Several studies were conducted on the use of several nanomaterials in defluoridation including CNTs, metal and metal oxide NPs, surface-modified NPs, and nanocomposites (Narayan, 2010). In this section, some examples of these studies are given.

Pristine mesoporous γ-Al_2O_3, MgO-loaded Al_2O_3, ZrO_2-loaded Al_2O_3, and CaO-loaded Al_2O_3 materials were synthesized using a simple and cost-effective method (Dayananda et al., 2015). The four systems were used as adsorbents for the removal of fluoride from water samples. The results showed that the modification of Al_2O_3 with the metal oxide NPs greatly enhanced the fluoride ion adsorption capacity of Al_2O_3 as well as the adsorption kinetics. In the case of MgO/Al_2O_3 adsorbent, the adsorption capacity increased from 56 to 90% while a 15 wt.% ZrO_2 loading increased the adsorption capacity of Al_2O_3 from 28 to 52%.

Adeno et al. (2014) used aluminum oxide hydroxide (AlOOH) NPs to remove F^- from water samples. The authors studied the effect of several parameters (initial pH, fluoride concentration, contact time, and adsorbent dose) on the adsorption capacity. The results showed that the removal efficiency of F^- increases as the adsorbent dose increases and also as the pH increases from pH 3 to 8. The maximum F^- removal occurred at pH 7 with an adsorbent dose of 1.6 g/L and an initial F^- concentration of 20 mg/L.

Li et al. (2003a) used multiwalled CNTs, aligned CNTs and alumina-supported CNTs (Al_2O_3/CNTs) (Li et al., 2003b) for the removal of F^- from water. The authors studied the effect of several parameters (pH, temperature, adsorbent concentration) on the fluoride adsorption capacities of the nanomaterials. The highest adsorption capacity was obtained at 25°C in the pH range 6–9 for Al_2O_3/CNTs containing 30% Al_2O_3 loading.

He and Chen (2014) synthesized zirconium nanoparticles (ZrNPs) for use as an adsorbent for the defluoridation of water. The authors studied the effect of interfering some anions and NOM on the F^- ion adsorption capacity of the ZrNPs. The ZrNPs showed a high adsorption capacity for fluoride in the pH range 3–10 with the maximum capacity obtained at pH 4. The presence of PO_4^{3-}, NO_4^{3-}, and NOM did not significantly inhibit the fluoride removal while both HCO_3^- and SiO_3^{2-} retarded the removal.

Riahi et al. (2015) prepared Fe_3O_4 superparamagnetic NPs modified with ZrO_2 and used them as nanoadsorbent in the removal of excess fluoride from aqueous solutions. The authors investigated the effect of initial F^- concentration, mass ratio of ZrO_2 to Fe_3O_4, pH, adsorption time, and coexistence of other anions on the adsorption capacity of the ZrO_2/Fe_3O_4 NPs. The results showed that the F^- maximum adsorption capacity was achieved at pH 2.5. The authors tested the new sorbent for the removal of F^- from real samples and concluded that ZrO_2/Fe_3O_4 NPs could be excellent adsorbents for treatment of fluoride-contaminated water samples.

Chai et al. (2013) synthesized sulfate-doped Fe_3O_4/Al_2O_3 NPs and used them as a novel adsorbent for the F^- removal from drinking water. The authors evaluated the F^- removal efficiency of the NPs under different experimental conditions. The kinetic data showed that the F^- adsorption on the NPs was rapid in the beginning followed by a slower rate with 90% F^- removal achieved in the first 20 min of treatment while the equilibrium was achieved in about 8 h. The NPs showed a good performance in the pH 4–10 range with a 90 and 70% efficiency with initial F^- concentration of 10 and 50 mg/L, respectively. The results also indicated that, with the exception of PO_4^{3-} ions, the coexistence of other anions (NO_3^-, Cl^-, and SO_4^{2-}) did not evidently inhibit the defluoridation process by the Fe_3O_4/Al_2O_3 NPs.

Jayarathna et al. (2015) synthesized γ-Fe_2O_3 NPs and used them as adsorbent for the removal of F^- from groundwater. The results showed a rapid adsorption with 95 ± 3% occurred in the first 15 min of treatment. In addition, the higher F^- removal capacities were observed at low pH values.

Using inexpensive starting materials, Rout et al. (2015) synthesized an iron oxide-based nanocomposite (IBNC), a titania-based nanocomposite (TBNC), and a micro carbon fiber (MCF) for use as adsorbents for fluoride uptake from aqueous solutions. The three adsorbents showed great performances with 97, 92, and 88% fluoride removal at an optimum time of 60 min for IBNC, TBNC, and MCF, respectively.

4.5.2.4 Other Methods

Methods, such as RO, ED, and electrochemical methods have been used for the defluoridation of drinking water. Although effective in removing fluoride salts from water, these methods are used only on a small scale due to certain experimental procedures that limit their usage on a large scale.

4.5.2.4.1 Reverse Osmosis and Electrodialysis Pressure-driven membrane filtration systems include MF, UF, NF, and RO. As shown in Fig. 6.1 and Table 6.1, the membranes differ in the pore diameter and also in the amount of the applied pressure. Accordingly, the four systems differ in the particles removed (rejected) as shown in Fig. 6.4.

RO is a process in which dissolved inorganic solids are removed from drinking water and aqueous solutions by forcing the feed water through a semipermeable membrane. In RO, a hydraulic pressure higher than the osmotic pressure, applied using a high-pressure pump, is exerted on one side of a semipermeable membrane which forces the purified water across the membrane leaving the solid pollutants behind (Kaushik, 2012). Several factors affect the performance of any RO system and hence control the

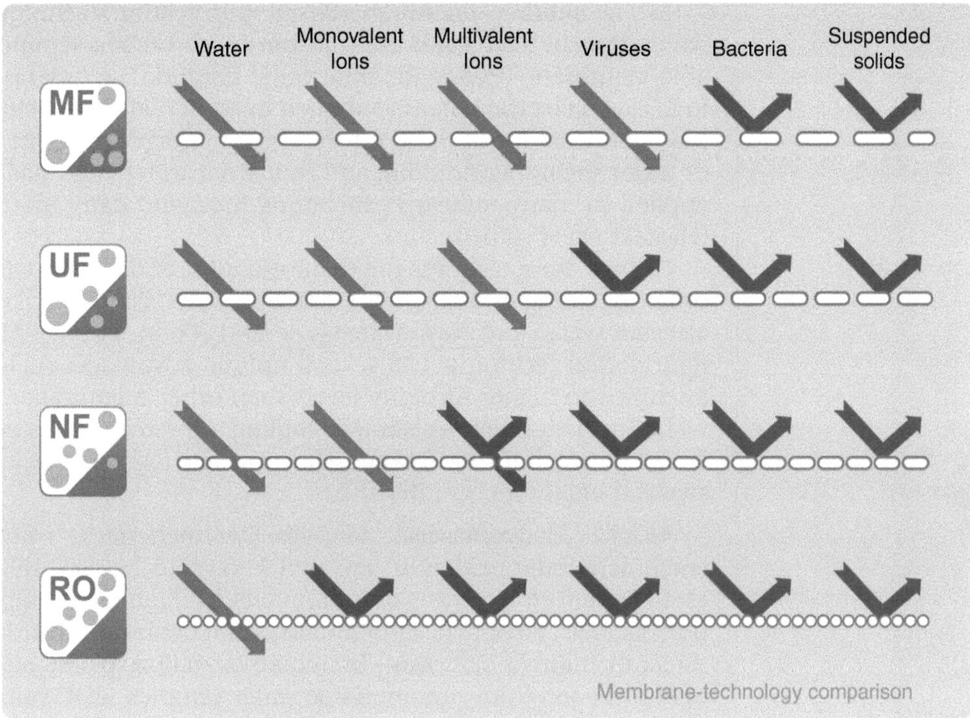

Figure 6.4. A comparison between the filtration capacities of the different membrane technologies used in water purification. Courtesy of Koch Membrane Systems (KMS), Inc. (http://www.kochmembrane.com), with permission.

quality of the purified water. These factors include the temperature, applied pressure, the type of membrane used, and the properties (type, size, pH, and concentration) of the total dissolved solids in the feed water.

Several studies have been reported on the use of RO for the removal of fluoride salts from different type of water including drinking water, seawater, groundwater, and brackish water with rejection efficiency between 45 and 90% depending on the nature of the feed water and the other experimental parameters (Nasr et al., 2013). RO is very convenient and environmentally friendly technique that does not use or produce chemicals of any type, requires a minimum amount of power, and removes all contaminants that cause water to smell, taste, or have color. However, the method uses an enormous amount of water which requires a long treatment time and also produces a large amount of wastewater. In addition, the membrane is sensitive to the temperature and pH and the RO units are also subject to clogging and fouling by the solid contaminants.

ED is another membrane-based purification technique in which ions in feed water are transported through a semipermeable membrane leaving the pure water behind. The driving force in ED is an electric potential applied in an ED cell. The technique has been used for decades for the desalination of different types of water including drinking and industrial waters and has been applied in many industries including food and dairy industries (Hestekin et al., 2010).

ED has been used for the defluoridation of different types of water including drinking water, brackish water, geothermal water, artesian wells, and wastewater (Lee and Moon, 2014). This defluoridation technique has several unique advantages including tolerance to higher turbidity levels than other membrane-based techniques and high resistant to fouling. Moreover, ED is not affected by pH in the 0–10 range. However, the technique is expensive and highly energy intensive.

4.5.2.4.2 Electrochemical Methods Electrochemical methods (such as anodic oxidation) are well known to be very selective and efficient in removing contaminants from aqueous solutions (Pulkka et al., 2014). The contaminants can be destroyed either by direct or indirect oxidation. In anodic oxidation for the removal of fluoride ion from contaminated water samples, a DC voltage is applied between a suitable sacrificial anode (usually Al metal or alloy) and a cathode (usually a stainless steel plate) in an electrochemical cell (a tank) containing the contaminated water (Renuka and Pushpanjali, 2013). The Al anode dissolves to produce Al^{3+} which precipitates as $Al(OH)_3$. Fluoride ions adsorb to the surface of $Al(OH)_3$ and are removed from the tank through filtration. The current density applied in the process is calculated depending on the volume of the water under treatment and the concentration of fluoride ion in it. This method is operator-friendly, does not require the addition of chemicals, occupies less space, consumes less electrical energy, produces a small quantity of sludge, and the operating parameters can be easily modified. However, the sacrificial anode needs to be replaced periodically.

5 Technologies Currently Used for Organic Matter Removal

Natural, synthetic, biodegradable, and dissolved OM in surface water and industrial wastewater cause several serious problems and diseases. In addition, the presence and degradation of organic materials in drinking water affect the quality as well as the physical properties of the water such as the color, odor, and taste.

Moreover, NOM indirectly adds more pollutants to the water and reduce the efficiency of most of the conventional water treatment methods in two different ways: (1) it reacts with most of the disinfectants (e.g., chlorine, ozone) to produce other harmful or undesired disinfection by-products, and (2) acts as a source of nutrition for the growth of bacteria in potable water-distribution systems (Osterhus et al., 2007).

The major sources of OM in drinking water are domestic, agricultural, forestry, and industrial activities. Among these sources, industrial wastewater is the most dangerous especially in developing countries where the untreated industrial effluents are discharged directly into surface water bodies (Koukal et al., 2004).

OM that may exist in groundwater, wastewater, and drinking water are very diverse and include the following classes: alcohols, high molecular weight hydrocarbons (e.g., gasoline), phenol and substituted phenols (e.g., chlorophenols, bisphenol A), dyes, detergents, oils, greases, plasticizers, aromatic solvents (e.g., benzene, toluene, nitrobenzene, carbon tetrachloride), chlorinated aromatics (chlorobenzene, chloronaphthalene), pharmaceuticals, and pesticides, as well as herbicides (e.g., organochlorine, organosulfur, and nitrogen-based compounds, such as atrazine, simazine, lindane, triazophos, quinalphos, carbofuran, and monochrotofos) (Egeberg et al., 1999).

Due to their unique catalytic activity, several studies have been reported on the use of MNPs (such as zero-valent iron, copper, gold, and silver) to successfully filter out the OM from water. In these reports, the organic pollutants were removed from water through adsorption on the surface of the MNPs. To prevent the MNPs from going in the purified water after treatment, the MNPs are usually incorporated or loaded in a support such as AC, activated alumina, or a polymeric membrane (Manimegalai et al., 2014). In this section, the removal efficiencies of some of the OM are discussed.

5.1 Pesticides

Pesticides get into surface and drinking water through several sources including household, agricultural, and industrial activities. Due to the excellent adsorption capacity of bare and modified MNPs, recently, several studies on the removal of pesticides using MNPs have been reported. Most of these studies focus on the adsorption efficiency of atrazine.

In a very recent study, Ouali et al. (2015) synthesized Fe_2O_3/palygorskite MNPs and used them to study the adsorption removal of the fungicide fenarimol from aqueous samples. The authors assessed the effect of several experimental parameters (e.g., the adsorbent mass, reaction time, initial pesticide concentration, and desorption stability) on the adsorption efficiency of fenarimol on the surface of the Fe_2O_3/palygorskite MNPs relative to two other adsorbents (sifted palygorskite and purified palygorskite). The adsorption of fenarimol on the surface of Fe_2O_3/palygorskite MNPs was the highest with an adsorption rate of 70%.

Lindane (γ-hexachlorocyclohexane) is one of the world's very common and extensively used insecticides especially in the 1940–90s. Derbalah et al. (2015) reported the adsorption of lindane on the surface of mesoporous alumina nanoparticles (MA–NPs). The influences of pH, temperature, contact time, and adsorbate–adsorbent concentration were studied. The results showed that the maximum adsorption capacity of lindane is 25.54 mg/g at 20°C and the MA–NPs exhibited an uptake efficiency of lindane higher than 80%. Applying the NPs in real water samples from tap and lake water sources contaminated with lindane showed that the MA–NPs can be a good adsorbent for water treatment. In another study, Elliott et al. (2005, 2009) proposed a technique for removing lindane from groundwater using NZVI NPs synthesized by mixing sodium borohydride and ferrous sulfate heptahydrate solutions. The results showed that more than 95% of lindane was removed within 48 h of the treatment of the water with the NZVI.

Satapanajaru et al. (2008) examined the effectiveness of NZVI NPs to dechlorinate atrazine in contaminated water and soil samples under different experimental conditions. The results revealed that NZVI was excellent in the destruction of atrazine. When Pd was added as a catalyst, the NZVI/Pd particles showed higher kinetic destruction rate in both contaminated water and soil. In another study, Bezbaruah et al. (2009) used NZVI NPs for the remediation of the pesticides alachlor and atrazine in water. The results showed that 92–96% of alachlor was removed within 72 h of the treatment. On the other hand, no degradation was observed for atrazine.

Saifuddin et al. (2011) prepared a crosslinked composite based on Ag NPs embedded in CTS for use as an adsorbent in point-of-use drinking water filtration system for household to remove pesticides in water. The authors made an online filter using a column of the composite and used it to study the adsorption and release behaviors of atrazine under different experimental conditions. The results showed that the CTS–AgNPs composite removed atrazine with a high adsorption capacity.

Pradeep and Bootharaju (2015) proposed a convenient and cost-effective method for the removal of the pesticides chlorpyrifos and malathions from groundwater using an adsorption column containing activated alumina loaded with silver or gold NPs. The authors also studied the degradation mechanism of chlorpyrifos in drinking water by noble MNPs and concluded that the rate of degradation of chlorpyrifos by AgNPs is higher than its degradation by gold NPs.

Rakhshan and Pakizeh (2015) synthesized a thin film polyamide nanocomposite membrane containing silica NPs modified with oleic acid and used it as an adsorbent for the removal of three triazines (namely, atrazine, propazine, and prometryn) from water. The results revealed that the addition of the modified SiO_2 particles to the polyamide membrane decreased the pore size and surface roughness of the membrane and improved its rejection efficiency for the pesticides. Moreover, the authors studied the effect of molecular weight, hydrophobicity, and dipole moment of the triazines on the rejection efficiency. The results showed that prometryn, the triazine with the largest molecular weight, had the highest rejection rate (99%) while propazine showed the lowest rejection rate (84–90%) due to its high dipole moment.

Manimegalai et al. (2014) synthesized AgNPs supported on a cellulose acetate membrane and used it to mineralize the pesticides chlorpyrifos and malathion. The authors studied the effect of the concentration of the AgNPs and the pesticides on the mineralization time. The results showed that the AgNPs are effective in the mineralization process and does not differentiate between the two pesticides. In addition, the results also showed that the higher the concentration of the AgNPs, the shorter the time required for the complete removal of the pesticides.

5.2 Hydrocarbons and Halocarbons

Aromatic hydrocarbons, such as BTEX, are found naturally in petroleum products, such as crude oil, gasoline, and diesel fuel. They are also extensively used as solvents in the synthesis of many products including paints, coatings, oils, resins, rubber, leather, consumer products, plastics, pesticides, etc. Produced water (water produced by oil and gas wells during the recovery process) contains large quantities of BTEX (Ranck et al., 2005). In addition to BTEX, contaminated groundwater contains oxygenated additives such as methyl-t-butyl ether (MTBE) which is believed to be present in almost all surface and groundwaters under urban areas (Klinger et al., 2002). The removal of these dangerous contaminates from water requires the use of efficient and cost-effective treatment and purification methods. Nanomaterials have been used as an effective adsorption media for the removal of halocarbons and hydrocarbons, including BTEX and MTBE, from different water sources and aqueous solutions (Gupta and Kulkarni, 2011).

Gupta and Kulkarni (2011) prepared a nanocomposite foam based on poly(dimethylsiloxane) (PDMS) incorporated with gold NPs and used it to remove oil spills from water samples. The results showed that the foam has a high swelling ability against BTEX. The foam is also an effective adsorbent for sulfur-containing contaminants such as thioanisole. The foam is very stable in harsh chemical environments, easy to regenerate, and can be a good adsorbent for wastewater treatment.

Torabian et al. (2014) synthesized NPs based on Fe_3O_4 MNPs modified with 3-mercaptopropyltrimethoxysiline and grafted with allyl glycidyl ether. The modified MNPs were used as an adsorbent to remove polyaromatic hydrocarbons (anthracene and pyrene) in contaminated water samples. The authors studied the effect of several experimental parameters on the adsorption capacity of the MNPs. The results showed that the modified MNPs have a great adsorption capacity toward polyaromatic hydrocarbons, the adsorption equilibrium was attained in 10 min, and the maximum adsorption capacity was achieved at pH = 7.0 and 20°C.

Russo et al. (2015) synthesized NZVI NPs supported on natural zeolite and used them for the catalytic degradation and removal of benzene and MTBE from aqueous solutions. The degradation experiments were conducted in a fixed bed (where the NPs were placed in a vertical glass column) at pH 3–4 at room temperature. The results showed ~80% degradation for benzene and a 52% for MTBE.

Saien and Shahrezaei (2012) used TiO_2 NPs as a photocatalyst in presence of UV light for the removal of organic pollutants (both aliphatic and aromatic compounds) from a real petroleum refinery wastewater sample. At the optimum experimental conditions, a maximum reduction in chemical oxygen demand of more than 78% was achieved in about 2 h. In addition, a GC/MS analysis showed that several petroleum compounds in the samples were degraded with high efficiencies.

Orlov et al. (2006) prepared TiO_2 NPs modified with AuNPs and used them for the photocatalytic degradation of MTBE and 4-chlorophenol in aqueous solutions. The results showed that, compared to the unmodified TiO_2 NPs, the Au/TiO_2 NPs were much more effective in the removal of both contaminants with an increase in the degradation rate by 50 and 100% for the removal of 4-chlorophenol and MTBE, respectively. In another study, Nikazar et al. (2014) used TiO_2 NPs and solar UV light (UV/TiO_2) for the photocatalytic removal of MTBE from aqueous solutions. The authors studied the effect of pH and the concentration of MTBE as well as the TiO_2 NPs on the efficiency of the removal process. The results showed an 88% removal in 2 h using 50 ppm of MTBE and 2 g/L of TiO_2.

5.3 Organic Dyes

Synthetic organic dyes are one of the major and most-toxic pollutants in water. The most common sources of dyes are the following industries: pickling, paper and pulp, photography, cosmetic, printing, textile, plastics, rubber, and leather (Sharma et al., 2011). Several studies have been reported on the use of nanoadsorbents for the removal of dyes from water samples. In this section some examples are given.

MO is one of the very common water-soluble azo dyes (commonly known as a pH indicator) that is extensively used in several industries including the textile, paper, printing, and food industries and mostly discharged in industrial wastewater (Malviya et al., 2015). Other studies have been reported on the removal of MO from aqueous solutions using NPs and nanocomposites as the adsorption medium (Girginov et al., 2012). Tajizadegan et al. (2015) investigated the adsorption of MO from an aqueous solution on the surface of ZnO–Al_2O_3 nanocomposite under different experimental conditions. The results showed that the adsorption occurred spontaneously, and under optimal conditions, the percent removal of MO was 97% at an equilibrium time of 30 min.

CR is another water-soluble acid azo dye that can be used as a pH indicator. This dye is extensively used in the textile industry. Several studies have been reported on the removal (decolorization) of CR from aqueous solutions through adsorption on the surface of different nanomaterials (e.g., Ag, ZnO, TiO_2, and Fe_2O_3) (Chelkar et al., 2014). In one of these studies, Madrakian et al. (2014) synthesized a γ-Fe_2O_3/TiO_2 nanocomposite and used it to investigate the removal efficiency of CR (as an example of an anionic dye) and Janus Green B (as an example of a cationic dye) from water samples. The authors studied the ability of the nanocomposite to adsorb and degrade the dyes under various sunlight radiations. The results showed that the photocatalytic activity of the nanocomposite improved under irradiation and the nanocomposite can be used for wastewater treatment.

In another study, Wang and Wang (2007) synthesized a series of biopolymer CTS/montmorillonite (MMT) nanocomposites and used them as adsorbents for the removal of CR. The authors studied the effect of several experimental parameters (initial pH, CTS/MMT molar ratio, temperature) on the adsorption capacity of the nanocomposites. In addition, the authors compared the performance of the CTS/MMT biopolymer nanocomposites with the performance of the separate CTS and MMT polymer nanocomposites. The results showed that the adsorption capacity of the CTS/MMT nanocomposite was higher than the mean adsorption value for the separate polymer nanocomposites.

Pal and Deb (2014) used AC coated with either AgNPs or AuNPs to remove CR from aqueous solutions. The authors investigated the effect of various experimental parameters on the adsorption efficiency. The results showed that the percent removal of the dye increases with increasing the contact time and the equilibrium time was independent of the initial dye concentration. The results also indicated that the adsorption efficiency of the AgNPs- and AuNPs-coated AC is about 88% which makes these NPs suitable adsorbents for the removal of CR from water.

Afkhami and Moosavi (2010) evaluated the adsorption and desorption of CR on the surface of maghemite NPs (γ-Fe_2O_3) under several experimental conditions. The results revealed that the maghemite NPs have an adsorption capacity to CR that is much higher than that recorded with many other adsorbents which makes γ-Fe_2O_3 a good adsorbent for the removal of CR from wastewater. The maximum adsorption occurred at pH 5.9.

MB is a cationic dye that is also found in industrial wastewater. Several investigations have been reported on the use of nanomaterials for the removal of MB from aqueous solutions and wastewater (Middea et al., 2015). In one of these studies, Dutta et al. (2015) synthesized mesoporous SiO_2 NPs (MSN) decorated with SnO_2 quantum dots and used it to investigate the removal of MB from industrial wastewater. The authors studied the effect of various variables (e.g., the amount of adsorbents, pH, temperature, dye concentration, contact time, salt concentration) on the removal efficiency of MB. The results showed that the SnO_2/MSN NPs adsorb almost 100% of MB within 5 min at room temperature. Silica nanosheets, derived from vermiculite as an adsorption medium, were used to study the adsorption properties of MB in aqueous solutions (Zhao et al., 2008). The effect of the several experimental conditions (initial pH, temperature, dye concentration, and contact time) on the adsorption efficiency was studied. The results revealed that increasing the pH and temperature decreased the adsorption while increasing the dye concentration increased the adsorption.

Moussavi and Mahmoudi (2009) synthesized MgO NPs and used them for the removal of azo and anthraquinone reactive dyes (blue 19 and red 198) from industrial wastewater under different experimental conditions. Both dyes showed a good adsorption capacity with a maximum percent removal of more than 98% achieved at the following optimum experimental conditions: adsorbent dose = 0.2 g, pH = 8.0, dye concentration = 50–300 mg/L, and 50 min contact time.

6 Nanomaterial-Based Commercially Available Water Treatment Devices

The last decade has witnessed the appearance of several commercial filtration systems devices based on NP adsorbents for individual, household, and community use for purification of drinking water (GDNP, 2006). These devices are simple, portable, inexpensive, but very effective and able to filter out almost all water contaminants including viruses, bacteria, turbidity, inorganic, and OMs. Some of these devices were designed specifically for use in developing countries.

In one of these commercially available devices, alumina nanofibers electrochemically deposited on glass filter substrates were used to make cartridge filters (Yarrington et al., 2011). These filters are cheap, easy to use, and have the advantage of having high electropositive charge which allows the filters to strongly adsorb many negatively charged contaminants including natural organic colloids at a fast rate with high efficiency.

In another filtration device based on nanofibers, the filtration medium consists of several nanofiber layers where each layer is added to remove certain contaminates (e.g., dirt, OM, colloidal particles). The nanofiber materials used in the filtration layers include a variety of materials such as alumina, cellulose, ceramics, resins, and polymers (Koslow, 2003).

A device designed to treat water in fish tanks involves the use of light nanoporous ceramic material hosting aerobic bacteria within its pores (TARA, 2010). The bacteria convert different pollutants into nontoxic substances. The ceramic material can also be modified to remove other pollutants such as heavy metals, biological contaminants, and phosphates (GDNP, 2006).

Among the current examples of devices that use the latest innovations in nanotechnology are the PureMadi (2015), the LifeStraw (Vestergaard.com, 2015b), and the Koch membrane [Koch Membrane Systems (KMS), 2015a] devices.

PureMadi is a not-for-profit organization led by James Smith at the University of Virginia (U.Va.) working with several other international institutes and organizations to provide sustainable

Figure 6.5. University of Virginia's PureMadi project for drinking-water purification in South Africa. (A) A coating of Ag or Cu NPs is applied to each PureMadi filter, (B) some of the PureMadi filters ready to go, (C) a woman flow-testing a batch of PureMadi filters, and (D) MadiDrop disks. Courtesy of PureMadi (http://www.puremadi.org/), with permission.

solutions for global water and health problems. PureMadi also includes students and faculty members from several schools and departments at U.Va. PureMadi developed two innovative filters (PureMadi and MadiDrop) for point-of-use drinking water treatment in developing countries (Fig. 6.5). Both filters are porous ceramic devices made of local clay, sawdust, and water. Once prepared and fired, a PureMadi filter is coated with a dilute solution of Ag or Cu NPs while a MadiDrop tablet is impregnated with the same NPs. The NPs act as disinfectants for waterborne pathogens (Abebe Lydia et al., 2014). Ceramic filtration is one of the common and popular methods for household water purifications. Ceramic water filters are the products of the work of many researchers, organizations, and institutions over many years (Abebe Lydia et al., 2014).

A PureMadi filter is a flowerpot-like filter that removes more than 99.9% of the pathogens in water, produces 1–3 L of safe and clean water per hour, and can serve a family of six for 2–5 years. The MadiDrop tablet can be used either alone or in conjunction with the PureMadi filter. The MadiDrop tables are usually dropped into a storage container that has the untreated water. Over time, the tablet releases Ag^{+} or Cu^{2+} ions that kill the pathogens in water. The MadiDrop filter is smaller, cheaper, lighter, and easier to transport than the PureMadi filter (Ehdaie et al., 2014). However, due to the way it works, the MadiDrop tablet is not as effective as the PureMadi filter in removing the sediment in water that causes the color and taste. Nevertheless, both systems have been used in South Africa and Guatemala with great success and expected to expand to many other developing countries (Abebe Lydia et al., 2014).

Vestergaard introduced the LifeStraw devices as a breakthrough solution for purification of any surface water source worldwide especially in developing countries. The portable versions of the LifeStraw devices have been deployed following almost every natural disaster since 2005. In addition, the family versions of the LifeStraw are they key source of sustainable clean and safe drinking water for more than 4.5 million people in Kenya (Vestergaard.com, 2015a). Fig. 6.6 shows some examples of the instant microbiological LifeStraw point-of-use water purifiers for personal, family, community, educational, workplace, health facility, and institutional settings. These chemical-free plastic devices use hollow fiber membranes as the adsorbent and remove over 99.99% of bacteria, viruses, and protozoan parasites. In addition to the pathogens, the LifeStraw devices are good in reducing turbidity and do not require electrical power or batteries to operate.

Innovations and miniaturization in membrane-based filtration devices are not limited to drinking water applications but also go to other types of water including wastewater. Koch Membrane Systems (KMS), Inc., another world-class developer and manufacturer of innovative membrane filtration systems, introduced several systems in different shapes for treatment of drinking water and industrial wastewater [Koch Membrane Systems (KMS), 2015b].

Figure 6.6. Different models of Vestergaard's LifeStraw tools currently used in Kenya for water purification. (A) and (B) Two different models of LifeStraw, (C) LifeStraw Family 1.0, (D) LifeStraw Family 2.0, (E) LifeStraw Community, and (F) LifeStraw Go bottle. Courtesy of Vestergaard.com (http://www.vestergaard.com/), with permission. From Vestergaard.com. 2015b. LifeStraw. Available from: http://www.vestergaard.com/our-products/lifestraw; Inhabitat.com, 2015. Gallery: design for the other 90%: LifeStraw water purifier. Available from: http://inhabitat.com/design-for-the-other-90-lifestraw/

Fig. 6.7 shows two of the durable and mechanically cleanable UF membrane-based tubular devices (FEG PLUS and ULTR-COR) for treatment of difficult industrial wastewater. The devices are robust open-channel configurations that have high tolerance for pH, as well as temperatures and can handle extremely high suspended solids loads [Koch Membrane Systems (KMS), 2015a].

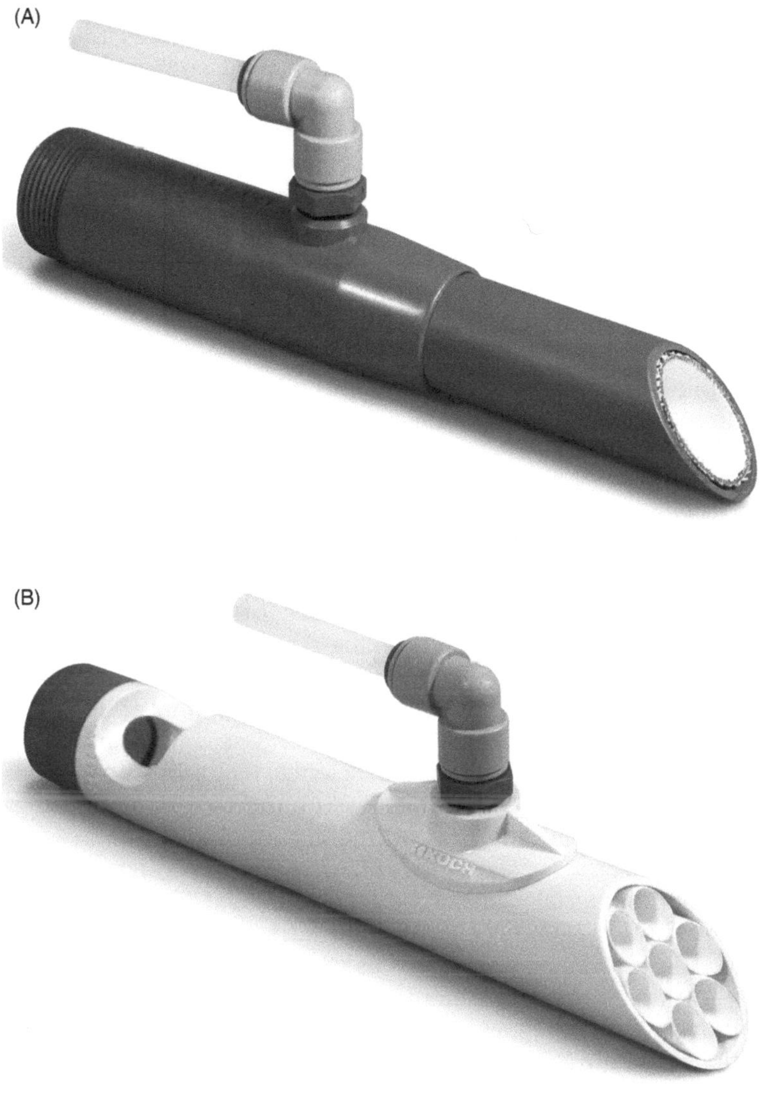

Figure 6.7. Two of the virtually uncloggable ABCOR tubular membranes used for treatment of difficult industrial wastewater. (A) FEG PLUS tubular membrane and (B) ULTRA-COR. Courtesy of Koch Membrane Systems (KMS), Inc. (http://www.kochmembrane.com), with permission.

7 Challenges and Prospectives

Today the world is facing drinking water and wastewater problems that require immediate long-term solutions to secure safe (clean and free of toxins, pollutants, and pathogens) water at the lowest possible cost. Several water pollutants including emerging contaminants, such as pharmaceuticals and personal care products (PPCPs), are currently detected at very low concentrations in natural and drinking water (Mu et al., 2014). Although they may not be of imminent threat to human health, the complete removal of these contaminants is highly desirable. This requires advanced effective techniques and high throughput analytical instrumentations to screen, monitor, detect, and completely remove these contaminants in trace levels in a feasible and economic way. The need for advanced instrumentations will be one of top priorities in the purification of drinking water in the future especially with the new standards and regulations that lowers the maximum permissible levels of certain contaminants to the sub-ppb range.

Industrial and international bodies and organizations, such as the Organization for Economic Co-operation and Development (OECD), the EU, International Organization for Standardization (ISO), Responsible Care, The European Chemical Industry Council (CEFIC), and the European Technology Platform for Sustainable Chemistry (SusChem ETP) are working on monitoring the safety and environmental performance of nanomaterials not only in their applications in drinking-water purification but also in the rest of their applications (Dunn et al., 2014; Bencko and Ungvary, 1994). In 2007, the OECD established the Working Party on Nanotechnology (WPN) to examine, assess, and advise on emerging policy-relevant issues in science, technology, and innovation related to the responsible development of nanotechnology (OECD, 2015). In addition, these organizations are currently working on setting a set of standards and methods for testing the safety of the nanomaterials that can be used globally (CEFIC, 2015).

As shown earlier, because of their attractive physicochemical properties, MNPs have emerged as great versatile tools for many applications including removing different contaminants and pathogens from all types of water. However, the biggest challenge in the use of the NPs in water treatment is the difficulty of removing the NPs after the completion of the purification process which requires expensive postpurification process. In addition, there are only few studies on the toxicity and potential harmful health and environmental impact of the NPs. It is very important to understand the interaction of the NPs with human and living animals' organs (Singh, 2015).

Other challenges with the MNPs, at least for the design and synthesis of some reactive NPs are: (1) the full control of the experimental synthesis and modification parameters to achieve NPs with a narrow size distribution, high dispersibility, and homogeneous composition especially in large-scale syntheses,

(2) understanding the properties and behavior of the NPs as they are synthesized, used, and evolved (or age) in a certain environment (Karakoti et al., 2012), and (3) integration of some of these novel and reactive nanomaterials into existing water purification devices (Savage and Diallo, 2005).

With the increasing use of the MNPs in many fields, it is anticipated that surface functionalization and modification of MNPs to introduce additional affinity and selectivity to the NPs will attract more attention in the field of nanotechnology. In addition, the research and development in this area will also focus on the toxicity and degradability of the bare as well as the surface modified MNPs. These challenges mandate the use of advanced nanotechnology techniques for water purification. It is anticipated that the use of nanomaterials in water treatment will be one of the top topics for years with an ultimate goal of delivering high-quality drinking water that is free from current and emerging pollutants and meets all quality specifications from nonfresh water sources using highly efficient, cost-effective, and environmentally acceptable techniques and technologies (OECD, 2011). One of the long-term goals for the use of nanomaterials-based water purification devices is the development of smart filtration membranes with embedded sensors to control the performance and selectivity of the membrane (Savage and Diallo, 2005).

References

Abebe Lydia, S., Smith James, A., Narkiewicz, S., Oyanedel-Craver, V., Conaway, M., Singo, A., Amidou, S., Mojapelo, P., Brant, J., Dillingham, R., 2014. Ceramic water filters impregnated with silver nanoparticles as a point-of-use water-treatment intervention for HIV-positive individuals in Limpopo Province, South Africa: a pilot study of technological performance and human health benefits. J. Water Health 12, 288–300.

Adeno, F., Mulugeta, E., Zewge, F., Chebude, Y., 2014. Adsorptive removal of fluoride from water using nanoscale aluminium oxide hydroxide (AlOOH). Bull. Chem. Soc. Ethiop. 28, 215–227.

Afkhami, A., Moosavi, R., 2010. Adsorptive removal of Congo red, a carcinogenic textile dye, from aqueous solutions by maghemite nanoparticles. J. Hazard. Mater. 174, 398–403.

Ahmed, M.A., Ali, S.M., El-Dek, S.I., Galal, A., 2013. Magnetite-hematite nanoparticles prepared by green methods for heavy metal ions removal from water. Mater. Sci. Eng. B 178, 744–751.

Al-Hobaib, A.S., Al-Sheetan, K.M., El Mir, L., 2015. Effect of iron oxide nanoparticles on the performance of polyamide membrane for ground water purification. Mater. Sci. Semicond. Process. 42, 107–110.

Amara, D., Margel, S., 2013. Synthesis and characterization of elemental iron and iron oxide nano/microcomposite particles by thermal decomposition of ferrocene. Nanotechnol. Rev. 2, 333–357.

Baer, D.R., 2011. Surface characterization of nanoparticles: critical needs and significant challenges. J. Surf. Anal. 17, 163–169.

Bahena, J.L.R., Cabrera, A.R., Valdivieso, A.L., Urbina, R.H., 2002. Fluoride adsorption onto α-Al_2O_3 and its effect on the zeta potential at the alumina-aqueous electrolyte interface. Sep. Sci. Technol. 37, 1973–1987.

Bai, Y., Sun, Q., Xing, R., Wen, D., Tang, X., 2010. Removal of pyridine and quinoline by bio-zeolite composed of mixed degrading bacteria and modified zeolite. J. Hazard. Mater. 181, 916–922.

Bencko, V., Ungvary, G., 1994. Risk assessment of chemicals—a central European perspective. Cent. Eur. J. Public Health 2, 70–72.

Benner, T.C., 2004. Brief survey of EPA standard-setting and health assessment. Environ. Sci. Technol. 38, 3457–3464.

Bezbaruah, A.N., Thompson, J.M., Chisholm, B.J., 2009. Remediation of alachlor and atrazine contaminated water with zero-valent iron nanoparticles. J. Environ. Sci. Health Part B 44, 518–524.

Bharadwaj, R., Singh, D., Mahapatra, A., 2008. Seawater desalination technologies. Int. J. Nucl. Desalin. 3, 151–159.

Carvalho Bongiovani, M., Camacho, F.P., Nishi, L., Ferri Coldebella, P., Cardoso Valverde, K., Vieira, A.M.S., Bergamasco, R., 2014. Improvement of the coagulation/flocculation process using a combination of *Moringa Oleifera* Lam with anionic polymer in water treatment. Environ. Technol. 35, 2227–2236.

Chai, L., Wang, Y., Zhao, N., Yang, W., You, X., 2013. Sulfate-doped Fe_3O_4/Al_2O_3 nanoparticles as a novel adsorbent for fluoride removal from drinking water. Water Res. 47, 4040–4049.

Chelkar, P., Shetty, M., Packiyam, E.J.E., Bhat, P.R., Jayadev, K., Rao, V., 2014. Decolorization of dye Congo red by *Aspergillus niger* silver nanoparticles. J. Microbiol. Biotechnol. Res. 4, 36–39.

Chen, F., Xie, S., Zhang, J., Liu, R., 2013. Synthesis of spherical Fe_3O_4 magnetic nanoparticles by co-precipitation in choline chloride/urea deep eutectic solvent (DES). Mater. Lett. 112, 177–179.

Cheng, Z., Tan, A.L.K., Tao, Y., Shan, D., Ting, K.E., Yin, X.J., 2012. Synthesis and characterization of iron oxide nanoparticles and applications in the removal of heavy metals from industrial wastewater. Int. J. Photoenergy 608298, 5.

Clement, S., Deng, W., Drozdowicz-Tomsia, K., Liu, D., Zachreson, C., Goldys, E.M., 2015. Bright, water-soluble CeF_3 photo-, cathodo-, and X-ray luminescent nanoparticles. J. Nanopart. Res. 17, 1–9.

Cui, H., Li, Q., Gao, S., Shang, J.K., 2012. Strong adsorption of arsenic species by amorphous zirconium oxide nanoparticles. J. Ind. Eng. Chem. 18, 1418–1427.

Dahi, E., 1997. Development of the contact precipitation method for appropriate defluoridation of water. In: 2nd International Workshop on Fluorosis and Defluoridation of Water, The International Society for Fluoride Research, November 19–25, Nazreth, Ethiopia, pp. 128–137.

Das, R., Ali, M.E., Bee Abd hamid, S., Ramakrishna, S., Chowdhury, Z.Z., 2014. Carbon nanotube membranes for water purification: a bright future in water desalination. Desalination 336, 97–109.

Dave, P.N., Chopda, L.V., 2014. Application of iron oxide nanomaterials for the removal of heavy metals. J. Nanotechnol., 1–14.

Day, W.W, 2011. World Water Day 2011. Available from: http://www.un.org/waterforlifedecade/swm_cities_zaragoza_2010/pdf/01_water_quality_and_sanitation.pdf

Dayananda, D., Sarva, V.R., Prasad, S.V., Arunachalam, J., Ghosh, N.N., 2015. A simple aqueous solution based chemical methodology for preparation of mesoporous alumina: efficient adsorbent for defluoridation of water. Part. Sci. Technol. 33, 8–16.

Derbalah, A., El-Safty, S.A., Shenashen, M.A., Abdel Ghany, N.A., 2015. Mesoporous alumina nanoparticles as host tunnel-like pores for removal and recovery of insecticides from environmental samples. ChemPlusChem 80, 1119–1126.

Dunn, G., Bakker, K., Harris, L., 2014. Drinking water quality guidelines across Canadian provinces and territories: jurisdictional variation in the context of decentralized water governance. Int. J. Environ. Res. Public Health 11, 4634–4651.

Dutta, D., Thakur, D., Bahadur, D., 2015. SnO_2 quantum dots-decorated silica nanoparticles for fast removal of cationic dye (methylene blue) from wastewater. Chem. Eng. J. 281, 482–490.

Eastoe, J., Hollamby, M.J., Hudson, L., 2006. Recent advances in nanoparticle synthesis with reversed micelles. Adv. Colloid Interface Sci. 128-130, 5–15.

Egeberg, P.K., Gjessing, E.T., Ratnaweera, H. (Eds.), 1999. Typing of Natural Organic Matter in Water. Proceedings of a Workshop, Kristiansand, Norway, 3–6 June 1998. Environ. Int. 25 (2/3).

Ehdaie, B., Krause, C., Smith, J.A., 2014. Porous ceramic tablet embedded with silver nanopatches for low-cost point-of-use water purification. Environ. Sci. Technol. 48, 13901–13908.

Elliott, D.W., Spear, S.T., Zhang, W.X., 2005. Novel products from the degradation of lindane by nanoscale zero valent iron. Abstracts of Papers, 229th ACS National Meeting, IEC-167, San Diego, CA, United States, March 13–17.

Elliott, D.W., Lien, H.-L., Zhang, W.-X., 2009. Degradation of lindane by zero valent iron nanoparticles. J. Environ. Eng. 135, 317–324.

The European Chemical Industry Council (CEFIC), 2015. Q&A on nanomaterials and nanotechnologies. Available from: http://www.cefic.org/Documents/PolicyCentre/Nanomaterials/Cefic%20Nano%20QAs%20for%20public%203%20October%202012_FINAL.pdf.

Galal-Gorchev, H., 1993. WHO guidelines for drinking-water quality. Water Supply 11, 1 16.

Global Dialogue on Nanotechnology and the Poor: Opportunities and Risks (GDNP), 2006. Overview and Comparison of Conventional and Nano-Based Water Treatment Technologies. Meridian Institute, Chennai, India, Available from: http://www.merid.org/~/media/Files/Projects/nano-waterworkshop/watertechpaper.ashx

Gimbel, R., Malzer, H.-J., 2000. Potable water treatment. Biotechnology (second ed.) 11c, 371–412.

Girginov, C., Stefchev, P., Vitanov, P., Dikov, H., 2012. Silver doped TiO_2 photocatalyst for methyl orange degradation. J. Eng. Sci. Technol. Rev. 5, 14–17.

Glaze, W.H., Kang, J.W., 1988. Advanced oxidation processes for treating groundwater contaminated with TCE and PCE: laboratory studies. J. Am. Water Works Assoc. 80, 57–63.

Groeber, M., 1996. Chemical oxidation treatment. EPA Environ. Eng. Source, 261–270.

Gupta, R., Kulkarni, G.U., 2011. Removal of organic compounds from water by using a gold nanoparticle-poly(dimethylsiloxane) nanocomposite foam. ChemSusChem 4, 737–743.

He, J., Chen, J.P., 2014. A zirconium-based nanoparticle: essential factors for sustainable application in treatment of fluoride containing water. J. Colloid Interface Sci. 416, 227–234.

Hestekin, J., Ho, T., Potts, T., 2010. Electrodialysis in the food industry. Membr. Technol. 3, 75–104.

Howd, R.A., 2002. Can we protect everybody from drinking water contaminants? Int. J. Toxicol. 21, 389–395.

Ingle, N.A., Dubey, H.V., Kaur, N., Sharma, I., 2014. Defluoridation techniques: which one to choose. J. Health Res. Rev. 1, 1–4.

Iorio, M., Pan, B., Capasso, R., Xing, B., 2008. Interaction of polymerin with aluminum oxide nanoparticles for potential water remediation. Preprints of Extended Abstracts, ACS National Meeting of the American Chemical Society, Division of Environmental Chemistry, 48, pp. 386–389.

Iseli, A., Kwen, H.-D., Rajagopalan, S., 2009. Nanomaterials for environmental remediation. Nanoscale Mater. Chem. (second ed.), 649–679.

Jagtap, S., Yenkie, M.K., Labhsetwar, N., Rayalu, S., 2012. Fluoride in drinking water and defluoridation of water. Chem. Rev. 112, 2454–2466.

Jamhour, R.M.A.Q., 2005. New inorganic ion-exchange material for the selective removal of fluoride from potable water using ion-selective electrode. Am. J. Environ. Sci. 1, 1–4.

Jayarathna, L., Bandara, A., Ng, W.J., Weerasooriya, R., 2015. Fluoride adsorption on γ-Fe_2O_3 nanoparticles. J. Environ. Health Sci. Eng. 13 (54), 10.

Kanel, S.R., Manning, B., Charlet, L., Choi, H., 2005. Removal of arsenic(III) from groundwater by nanoscale zero-valent iron. Environ. Sci. Technol. 39, 1291–1298.

Karakoti, A.S., Munusamy, P., Hostetler, K., Kodali, V., Kuchibhatla, S., Orr, G., Pounds, J.G., Teeguarden, J.G., Thrall, B.D., Baer, D.R., 2012. Preparation and characterization challenges to understanding environmental and biological impacts of ceria nanoparticles. Surf. Interface Anal. 44, 882–889.

Kaushik, M.R., 2012. Reverse osmosis—a unique technique of water purification. J. Biol. Chem. Res. 29, 9, 73-81.

Kawamura, S., 1976. Considerations on improving flocculation. J. Am. Water Works Assoc. 68, 328–336.

Kim, J.-R., Lee, K.-Y., Suh, M.-J., Ihm, S.-K., 2012. Ceria-zirconia mixed oxide prepared by continuous hydrothermal synthesis in supercritical water as catalyst support. Catal. Today 185, 25–34.

Klinger, J., Sacher, F., Brauch, H.-J., 2002. Methyl tertiary-butyl ether (MTBE). A problem for water suppliers? GWF Wasser/Abwasser 143, 166–171.

Koch Membrane Systems (KMS), Inc. 2015a. Available from: http://www.kochmembrane.com/PDFs/Brochures/konsolidator-brochure.aspx

Koch Membrane Systems (KMS), Inc. 2015b. Membrane technologies: targeted technology makes the difference. Koch Membrane Systems (KMS), Inc. Available from: http://www.kochmembrane.com/Learning-Center/Technologies.aspx

Koslow, E. E. 2003. Microporous filter media, filtration systems containing same, and methods of making and using it. PCT Int. Appl., WO 2003064013 A1 20030807.

Koukal, B., Dominik, J., Vignati, D., Arpagaus, P., Santiago, S., Ouddane, B., Benaabidate, L., 2004. Assessment of water quality and toxicity of polluted Rivers Fez and Sebou in the region of Fez (Morocco). Environ. Pollut. 131, 163–172.

Kumar, E., Bhatnagar, A., Kumar, U., Sillanpaeae, M., 2011. Defluoridation from aqueous solutions by nano-alumina: characterization and sorption studies. J. Hazard. Mater. 186, 1042–1049.

Larsen, M.J., Pearce, E.I., Jensen, S.J., 1993. Defluoridation of water at high pH with use of brushite, calcium hydroxide, and bone char. J. Dent. Res. 72, 1519–1525.

Lee, H.-J., Moon, S.-H., 2014. Electrodialysis desalination. Desalination, 287–326.

Lee, J., Chae, H.-R., Won, Y.J., Lee, K., Lee, C.-H., Lee, H.H., Kim, I.-C., Lee, J.-M., 2013. Graphene oxide nanoplatelets composite membrane with hydrophilic and antifouling properties for wastewater treatment. J. Membr. Sci. 448, 223–230.

Leong, S., Razmjou, A., Wang, K., Hapgood, K., Zhang, X., Wang, H., 2014. TiO_2 based photocatalytic membranes: a review. J. Membr. Sci. 472, 167–184.

Li, Y.-H., Wang, S., Zhang, X., Wei, J., Xu, C., Luan, Z., Wu, D., 2003a. Adsorption of fluoride from water by aligned carbon nanotubes. Mater. Res. Bull. 38, 469–476.

Li, Y.H., Wang, S., Zhang, X., Wei, J., Xu, C., Luan, Z., Wu, D., Wei, B., 2003b. Removal of fluoride from water by carbon nanotube supported alumina. Environ. Technol. 24, 391–398.

Li, Y., Du, Q., Liu, T., Sun, J., Wang, Y., Wu, S., Wang, Z., Xia, Y., Xia, L., 2013. Methylene blue adsorption on graphene oxide/calcium alginate composites. Carbohydr. Polym. 95, 501–507.

Li, Y., Cheng, W., Sheng, G., Li, J., Dong, H., Chen, Y., Zhu, L., 2015. Synergetic effect of a pillared bentonite support on Se(VI) removal by nanoscale zero valent iron. Appl. Catal. B 174–175, 329–335.

Ling, M.M., Wang, K.Y., Chung, T.-S., 2010. Highly water-soluble magnetic nanoparticles as novel draw solutes in forward osmosis for water reuse. Ind. Eng. Chem. Res. 49, 5869–5876.

Liu, Y., Li, D., Sun, S., 2011. Pt-based composite nanoparticles for magnetic, catalytic, and biomedical applications. J. Mater. Chem. 21, 12579–12587.

Liu, X., Wang, M., Zhang, S., Pan, B., 2013. Application potential of carbon nanotubes in water treatment: a review. J. Environ. Sci. 25, 1263–1280.

Loo, S.-L., Fane, A.G., Lim, T.-T., Krantz, W.B., Liang, Y.-N., Liu, X., Hu, X., 2013. Superabsorbent cryogels decorated with silver nanoparticles as a novel water technology for point-of-use disinfection. Environ. Sci. Technol. 47, 9363–9371.

Ma, Z., Liu, H., 2007. Synthesis and surface modification of magnetic particles for application in biotechnology and biomedicine. China Particuol. 5, 1–10.

Madrakian, T., Afkhami, A., Haryani, R., Ahmadi, M., 2014. Synthesis of γ-Fe_2O_3/TiO_2 nanocomposite and its application in removal of dyes from water samples by adsorption and degradation processes. RSC Adv. 4, 44841–44847.

Mahdavi, S., Jalali, M., Afkhami, A., 2015. Heavy metals removal from aqueous solutions by Al_2O_3 nanoparticles modified with natural and chemical modifiers. Clean Technol. Environ. Policy 17, 85–102.

Maliyekkal, S.M., Anshup, Antony, K.R., Pradeep, T., 2010. High yield combustion synthesis of nanomagnesia and its application for fluoride removal. Sci. Total Environ. 408, 2273–2282.

Malviya, A., Jaspal, D., Sharma, P., Dubey, A., 2015. Isothermal mathematical modeling for decolorizing water—a comparative approach. Sustain. Environ. Res. 25, 53–58.

Manimegalai, G., Shanthakumar, S., Sharma, C., 2014. Silver nanoparticles: synthesis and application in mineralization of pesticides using membrane support. Int. Nano Lett. 4, 1–5.

Mascolo, M.C., Pei, Y., Ring, T.A., 2013. Room temperature co-precipitation synthesis of magnetite nanoparticles in a large pH window with different bases. Materials 6, 5549–5567.

Middea, A., Spinelli, L.S., Souza, F.G., Neumann, R., Gomes, O.D.F.M., Fernandes, T.L.A.P., de Lima, L.C., Barthem, V.M.T.S., De Carvalho, F.V., 2015. Synthesis and characterization of magnetic palygorskite nanoparticles and their application on methylene blue remotion from water. Appl. Surf. Sci. 346, 232–239.

Moussavi, G., Mahmoudi, M., 2009. Removal of azo and anthraquinone reactive dyes from industrial wastewaters using MgO nanoparticles. J. Hazard. Mater. 168, 806–812.
Mu, R., Shi, H., Adams, C., Ma, Y., Eichholz, T., Timmons, T, 2014. Occurrence screening of PPCPs and their removal study in the Missouri drinking water system. Abstracts, 49th Midwest Regional Meeting of the American Chemical Society, MWRM-173, Columbia, MO, United States, November 12–15.
Mulvaney, P., Liz-Marzan, L.M., Giersig, M., Ung, T., 2000. Silica encapsulation of quantum dots and metal clusters. J. Mater. Chem. 10, 1259–1270.
Narayan, R., 2010. Use of nanomaterials in water purification. Mater. Today 13, 44–46.
Nasr, A.B., Charcosset, C., Amar, R.B., Walha, K., 2013. Defluoridation of water by nanofiltration. J. Fluorine Chem. 150, 92–97.
Nikazar, M., Gorji, L.M., Shojae, S., Keynejad, K., Haghighaty, A.H., Jalili, F., Mirzahosseini, A.R.H., 2014. Removal of methyl tertiary-butyl ether (MTBE) from aqueous solution using sunlight and nano TiO_2. Energ. Source. 36, 2305–2311.
Obeid, L., Bee, A., Talbot, D., Jaafar, S.B., Dupuis, V., Abramson, S., Cabuil, V., Welschbillig, M., 2013. Chitosan/maghemite composite: a magsorbent for the adsorption of methyl orange. J. Colloid Interface Sci. 410, 52–58.
Organization for Economic Co-operation and Development (OECD), 2011. Fostering nanotechnology to address global challenges: water. Available from: http://www.oecd.org/sti/nano/47601818.pdf
Organization for Economic Co-operation and Development (OECD), 2015. Science and technology policy: nanotechnology. Available from: http://www.oecd.org/sti/nano/
Orlov, A., Chan, M.S., Jefferson, D.A., Zhou, D., Lynch, R.J., Lambert, R.M., 2006. Photocatalytic degradation of water-soluble organic pollutants on TiO_2 modified with gold nanoparticles. Environ. Technol. 27, 747–752.
Osterhus, S., Azrague, K., Leiknes, T., Odegaard, H., 2007. Membrane filtration for particles removal after ozonation-biofiltration. Water Sci. Technol. 56, 101–108.
Ouali, A., Belaroui, L.S., Bengueddach, A., Galindo, A.L., Pena, A., 2015. Fe_2O_3-palygorskite nanoparticles, efficient adsorbates for pesticide removal. Appl. Clay Sci. 115, 67–75.
Pal, J., Deb, M.K., 2014. Efficient adsorption of Congo red dye from aqueous solution using green synthesized coinage nanoparticles coated activated carbon beads. Appl. Nanosci. 4, 967–978.
Pawar, S.H., Ghosh, S.J., Karande, V.A., Patil, R.N., 2013. Recent advances in microbial synthesis of nanoparticles and their applications. Recent Adv. Microbiol. 1, 1–32.
Piquemal, J.-Y., Ait Atmane, K., Viau, G., Ibrahim, M., Decorse, P., Lamouri, A., Chehimi, M.M., 2013. Preparation, surface modification and functionalization of anisotropic ferromagnetic metal nanoparticles: towards multifunctional materials. Appl. Surf. Chem. Nanomater., 205–239.
Pradeep, T., Bootharaju, M.S., 2015. Noble metal nanosystems for drinking water purification: from nanoparticles to clusters. Aquananotechnology, 511–532.
Pulkka, S., Martikainen, M., Bhatnagar, A., Sillanpaa, M., 2014. Electrochemical methods for the removal of anionic contaminants from water—a review. Sep. Purif. Technol. 132, 252–271.
PureMadi, 2015. Available from: http://www.puremadi.org/
Rakhshan, N., Pakizeh, M., 2015. Removal of triazines from water using a novel OA modified SiO_2/PA/PSf nanocomposite membrane. Sep. Purif. Technol. 147, 245–256.

Ranck, J.M., Bowman, R.S., Weeber, J.L., Katz, L.E., Sullivan, E.J., 2005. BTEX removal from produced water using surfactant-modified zeolite. J. Environ. Eng. 131, 434–442.
Renuka, P., Pushpanjali, K., 2013. Review on defluoridation techniques of water. Int. J. Eng. Sci. 2, 86–94.
Riahi, F., Bagherzadeh, M., Hadizadeh, Z., 2015. Modification of Fe_3O_4 superparamagnetic nanoparticles with zirconium oxide; preparation, characterization and its application toward fluoride removal. RSC Adv. 5, 72058–72068.
Ringe, J.D., 2004. Fluoride and bone health. In: Holick, M.F., Dawson-Hughes, B. (Eds.), Nutrition and Bone Health. Humana Press, Inc., Totowa, NJ, pp. 345–362.
Ritchie, S.M.C., 2003. Polymer grafted membranes. Membr. Sci. Technol. Ser. 8, 299–327.
Rojas-Mayorga, C.K., Silvestre-Albero, J., Aguayo-Villarreal, I.A., Mendoza-Castillo, D.I., Bonilla-Petriciolet, A., 2015. A new synthesis route for bone chars using CO_2 atmosphere and their application as fluoride adsorbents. Micropor. Mesopor. Mater. 209, 38–44.
Rout, T.K., Verma, R., Dennis, R.V., Banerjee, S., 2015. Study the removal of fluoride from aqueous medium by using nano-composites. J. Encaps. Adsorp. Sci. 5, 38–52.
Russo, A.V., Lobo, D.N.D., Jacobo, S.E., 2015. Removal of MTBE in columns filled with modified natural zeolites. Procedia Mater. Sci. 8, 375–382.
Saien, J., Shahrezaei, F., 2012. Organic pollutants removal from petroleum refinery wastewater with nanotitania photocatalyst and UV light emission. Int. J. Photoenergy 703074, 5.
Saifuddin, N., Nian, C.Y., Zhan, L.W., Ning, K.X., 2011. Chitosan-silver nanoparticles composite as point-of-use drinking water filtration system for household to remove pesticides in water. Asian J. Biochem. 6, 142–159.
Satapanajaru, T., Anurakpongsatorn, P., Pengthamkeerati, P., Boparai, H., 2008. Remediation of atrazine-contaminated soil and water by nano zero valent iron. Water, Air, Soil Pollut. 192, 349–359.
Savage, N., Diallo, M.S., 2005. Nanomaterials and water purification: opportunities and challenges. J. Nanopart. Res. 7, 331–342.
Sharma, P., Kaur, H., Sharma, M., Sahore, V., 2011. A review on applicability of naturally available adsorbents for the removal of hazardous dyes from aqueous waste. Environ. Monit. Assess. 183, 151–195.
Shin, K.-Y., Hong, J.-Y., jang, J., 2011. Heavy metal ion adsorption behavior in nitrogen-doped magnetic carbon nanoparticles: isotherms and kinetic study. J. Hazard. Mater. 190, 36–44.
Singh, D., 2015. Carbon nanoparticles: occupational exposure and genotoxic effects. Res. Rev. J. Toxicol. 5, 9–26.
Siong, C.T., Daik, R., Abdul Hamid, M.A., 2014. Thermally conductive of nanofluid from surfactant doped polyaniline nanoparticle and deep eutectic ionic liquid. AIP Conf. Proc. 1614, 381–385.
Sivasankaran, S., Kishor Kumar, M.J., 2015. A novel sonochemical synthesis of nano-size silicon nitride and titanium carbide. Ceram. Int. 41, 11301–11305.
Smith, S.C., Rodrigues, D.F., 2015. Carbon-based nanomaterials for removal of chemical and biological contaminants from water: a review of mechanisms and applications. Carbon 91, 122–143.
Society for Technology and Action for Rural Advancement (TARA), 2010. Access to safe water for the bottom of pyramid: strategies for disseminating technology benefits secondary research report. New Delhi, India. Available from: http://r4d.dfid.gov.uk/PDF/Outputs/water/Secondary_Research_Report.pdf

Sundstrom, G., Coulter, B., Hall, C., Knapp, A, 2013. Advanced oxidation process technology for ultrapure water in the semiconductor industry. Official Proceedings—74th International Water Conference, pp. 1051–1063.

Tajizadegan, H., Torabi, O., Heidary, A., Golabgir, M.H., Jamshidi, A., 2015. Study of methyl orange adsorption properties on ZnO–Al_2O_3 nanocomposite adsorbent particles. Desalin. Water Treat. 57, 12324–12334.

Torabian, A., Panahi Homayoun, A., Nabi Bid Hendi, G.R., Mehrdadi, N., 2014. Synthesis, modification and graft polymerization of magnetic nano particles for PAH removal in contaminated water. J. Environ. Health Sci. Eng. 12, 105.

Tran, D.L., Le, V.H., Pham, H.L., Hoang, T.M.N., Nguyen, T.Q., Luong, T.T., Ha, P.T., Nguyen, X.P., 2010. Biomedical and environmental applications of magnetic nanoparticles. Adv. Nat. Sci. Nanosci. Nanotechnol. 1, 1–5.

UNICEF. 2015. Fluoride in water: an overview. Available from: http://www.unicef.org/wash/files/fluoride.pdf

Vatin, N.I., Chechevichkin, V.N., Chechevichkin, A.V., Shilova, Y., Yakunin, L.A., 2014. Application of natural zeolites for aquatic and air medium purification. Appl. Mech. Mater. 587-589, 565–572.

Verma, N., Verma, H., Meena, R.C., 2014. Removal of fluoride from drinking water: an overview. Int. J. Pharm. Res. Bio-Sci. 3, 328–351.

Vestergaard.com, 2015a. About Us. Available from: http://www.vestergaard.com/about-us

Vestergaard.com, 2015b. LifeStraw. Available from: http://www.vestergaard.com/our-products/lifestraw

Wan, Z., Chen, W., Liu, C., Liu, Y., Dong, C., 2015. Preparation and characterization of γ-AlOOH@CS magnetic nanoparticle as a novel adsorbent for removing fluoride from drinking water. J. Colloid Interface Sci. 443, 115–124.

Wang, L., Wang, A., 2007. Adsorption characteristics of Congo red onto the chitosan/montmorillonite nanocomposite. J. Hazard. Mater. 147, 979–985.

Wang, M., Pang, P., Koopal, L.K., Qiu, G., Wang, Y., Liu, F., 2014. One-step synthesis of δ-MnO_2 nanoparticles using ascorbic acid and their scavenging properties to Pb(II), Zn(II) and methylene blue. Mater. Chem. Phys. 148, 1149–1156.

The World Health Organization (WHO), 2007. Desalination for safe water supply: guidance for the health and environmental aspects applicable to desalination, Geneva. Available from: http://www.who.int/water_sanitation_health/gdwqrevision/desalination.pdf

Xu, J., Sun, J., Wang, Y., Sheng, J., Wang, F., Sun, M., 2014. Application of iron magnetic nanoparticles in protein immobilization. Molecules 19, 11465–11486.

Yarrington, C.D., Son, S.F., Foley, T.J., Obrey, S.J., Pacheco, A.N., 2011. Nano aluminum energetics: the effect of synthesis method on morphology and combustion performance. Propellants Explos. Pyrotech. 36, 551–557.

Zhang, F., Yang, H., Xie, X., Li, L., Zhang, L., Yu, J., Zhao, H., Liu, B., 2009. Controlled synthesis and gas-sensing properties of hollow sea urchin-like α-Fe_2O_3 nanostructures and α-Fe_2O_3 nanocubes. Sens. Actuators B 141, 381–389.

Zhao, M., Tang, Z., Liu, P., 2008. Removal of methylene blue from aqueous solution with silica nano-sheets derived from vermiculite. J. Hazard. Mater. 158, 43–51.

Zheng, X., Chen, D., Wang, Z., Lei, Y., Cheng, R., 2013. Nano-TiO_2 membrane adsorption reactor (MAR) for virus removal in drinking water. Chem. Eng. J. 230, 180–187.

Zhu, Q., Li, Z., 2015. Hydrogel-supported nanosized hydrous manganese dioxide: synthesis, characterization, and adsorption behavior study for Pb^{2+}, Cu^{2+}, Cd^{2+} and Ni^{2+} removal from water. Chem. Eng. J. 281, 69–80.

7

MICROORGANISMS: NEW TRENDS IN ENVIRONMENT-FRIENDLY AND ENERGY-SAVING WATER PURIFICATION

Mariana Oana Mihaela Fufă*,, Roxana Cristina Popescu*,†, Alexandru Mihai Grumezescu*, Alina Maria Holban*,‡**

**Department of Science and Engineering of Oxide Materials and Nanomaterials, Faculty of Applied Chemistry and Materials Science, University Politehnica of Bucharest, Bucharest, Romania; **National Institute for Laser, Plasma and Radiation Physics, Lasers Department, Magurele, Bucharest, Romania; † "Horia Hulubei", National Institute of Physics and Nuclear Engineering, Department of Life and Environmental Physics, Magurele, Bucharest, Romania; ‡University of Bucharest, Department of Botanic-Microbiology, Faculty of Biology, Bucharest, Romania*

1 Introduction

Drinking water contamination represents a current concern of the scientific community, if we consider that various types of pollutants (including liquid and gaseous suspended agents or chemical and biological agents) are incriminated in different stages of contamination process, including water resources and distribution systems pollution (Kostyla et al., 2015; Lendowski et al., 2015; Völker et al., 2016; Wu et al., 2016b; Zhou et al., 2016), but also reservoir and bottled container pollution (Felipe-Sotelo et al., 2015; Guadayol et al., 2016; Khan et al., 2016; Mata et al., 2015). When it comes to purification of polluted water, tremendous resources (including scientific, technological, and financial) have been straightened toward this particular field, given the healthcare conditions related to contaminated water consumption, like oral healthcare-related conditions (Arakeri et al., 2014; Bretzler and Johnson, 2015; Craig et al., 2015), gastrointestinal disorders (Mahanta et al., 2016;

Water Purification. http://dx.doi.org/10.1016/B978-0-12-804300-4.00007-1

Prevost et al., 2016; Sorensen et al., 2015), cardiovascular diseases (Butts et al., 2015; Talukder et al., 2016), neurological disorders (de Joode et al., 2016; Gerke et al., 2016), respiratory complications (Cancilla et al., 2016; Kruse et al., 2016), metal-associated skin conditions (Bhowmik et al., 2015; Brahman et al., 2016), and region-related carcinogenic potential (Bulka et al., 2016; Wang et al., 2016a; West et al., 2016; Zhang et al., 2016a).

The tremendous progress lately reported in nanotechnology enabled the successful development, assessment, and utilization of novel micro- and nanostructured materials, with promising experimental results regarding the efficient and targeted removal of water contaminants. Therefore, inorganic or organic and composite low-dimensional structured materials revealed impressive potential in current water purification strategies, including microfiltration (Amaral et al., 2016; Metcalfe et al., 2016; Ragab et al., 2016), ultrafiltration (Kenari and Barbeau, 2016; Sheng et al., 2016; Wu et al., 2016a), nanofiltration (Lai et al., 2016; Song et al., 2016; Zhao et al., 2015), and reverse osmosis (Ben-Sasson et al., 2016; Chen et al., 2016; Weinrich et al., 2016).

Given the alarming phenomenon of water contamination and the incessant interest toward efficient low-cost water purification procedures, the worldwide scientific community turned its attention toward the revaluation of ecological strategies to remove water contaminants. Therefore, by considering the beneficial acknowledged role of several types of microorganisms to degrade contaminant agents or to interfere with the cellular cycle of various pathogens, a current approach in prevention and treatment of pollution includes the revaluation or interaction with biocellular mechanisms. Herein, the environment-friendly removal of water pollutants first requires the thorough understanding of the main pollutants and contaminants and also the mechanistic ability of microorganisms to interact with various materials and substances.

The biological diversity and specific activity of aquatic microorganisms enabled a summarized classification of their ecological-related functions, namely: (1) production function—given their particular metabolism, aquatic microorganisms participate, and mediate vital biogeochemical reactions, including the composition and decomposition processes of organic molecules, nutrient, and energy cycling; (2) regulation function—thanks to their beneficial interactions with various inorganic and organic structures, aquatic microorganisms play a significant role during stabilizing the variability of natural processes and resource flow and the elimination of toxin and toxin-related metabolic products; (3) organization function—aquatic microorganisms significantly contribute to the physical and biological structuring of system, being a fundamental support within the self-organizing rules of water ecosystem (Brachet et al., 2015; Ramachandra, 2005).

Within the following paragraphs, we will briefly discuss about the relevance and ecological role of aquatic microorganisms, including herein the general remarks and properties and the current trends in water purification strategy. Also, relevant data regarding the novel nanotechnology-guided strategies for water pollutants removal will be pointed out.

2 Microorganisms in Water Purification

Bacteria are ubiquitous prokaryotic organisms, able to survive in planktonic and adhered forms within every surrounding environment, being capable of living and reproducing in extreme conditions. Among the aquatic ecosystem, bacterial species possess the most diversified biological variety and the most versatile metabolic activity, being capable of both autotrophic and heterotrophic nutrition.

Bacterial autotrophs are able to provide their nourishment and energy requirements either by light absorption (photoautotrophs) or inorganic molecules oxidation (chemoautotrophs). Photoautotrophic bacteria specifically possess photosynthetic pigments within their cellular membrane and use solar energy to fix inorganic vital elements (such as carbon, nitrogen, phosphorous, and sulfur) and to produce oxygen. Within this process, also known as anoxygenic photosynthesis, bacterial species specifically use hydrogen, ammonia, ferrous ions, or sulfide as hydrogen source to reduce inorganic structures (in comparison to the conventional photosynthesis process from green plants, where water acts as a hydrogen donor). Relevant types of photoautotrophic bacteria (but not limited to) related to water self-purification include the families: *Chlorobiaceae*—green sulfur bacteria with chlorophyll *a*, *c*, *d*, or *e* pigment (Naeher et al., 2016; Wilson et al., 2004), *Chromatiaceae*—purple sulfur bacteria with bacteriopurpurin pigment (Martínez-Alonso et al., 2005; Naeher et al., 2016), *Cyanothece*—nitrogen-fixing bacteria with chlorophyll *a*, phycocyanin, and phycoerythrin pigment (Dechatiwongse et al., 2014; Mota et al., 2015), *Ectothiorhodospiraceae*—purple sulfur bacteria with bacteriopurpurin pigment (Islam et al., 2016), *Nostocaceae*—nitrogen-fixing bacteria (Kutovaya and Watson, 2014; Magnuson et al., 2011), but also *Prochlorococcus* spp. with chlorophyll *b* pigment (Hynes et al., 2015; Mitbavkar et al., 2012), *Synechococcus* spp.—nitrogen-fixing bacteria with chlorophyll *a*, phycocyanin, allophycocyanin, allophycocyanin

B, and phycoerythrin pigment (Liu et al., 2016a; Tryggvesson et al., 2015), and *Synechocystis* spp.—nitrogen-fixing bacteria (Li et al., 2016a; Young et al., 2011). Thanks to their specific metabolic activity, chemoautotrophic bacteria are capable to produce organic compounds from inorganic sources, by revaluing the energy resulted from oxidation of inorganic molecules. Nitrifying bacteria possess specific enzymes involved during the nitrification process, namely ammonia monooxygenase (*amo*) and nitrite oxidoreductase (*nxr*), which enable energy production by oxidation of ammonia to hydroxylamine, respectively nitrite to nitrate. Relevant examples of nitrifying bacteria involved in water ecosystem include the *Nitrosomonas* spp. (Perez-Garcia et al., 2014; Wahman and Speitel, 2015) and *Nitrosococcus* spp. (Calvó and Garcia-Gil, 2004; Nishigaya et al., 2016) *amo*-containing bacteria, but also the *Nitrobacter* spp. (Huang et al., 2010; Liang et al., 2015), *Nitrospina* spp. (Spieck et al., 2014; Vekeman et al., 2013), and *Nitrospira* spp. (Cébron and Garnier, 2005; Daims et al., 2016) *nxr*-containing bacterial species. Sulfur-reducing and sulfate-reducing bacteria derive their energy by oxidizing hydrogen sulfide to elemental sulfur and sulfate to intermediary inorganic sulfur compounds, respectively. *Arcobacter* spp. (Laishram et al., 2016; Salas-Massó et al., 2016), *Beggiatoa* (Aranda et al., 2015; MacGregor et al., 2013), and *Thiobacillus* spp. (Lestari et al., 2016; Mohamed et al., 2014) bacterial subfamilies are representative examples for this particular type of reducing bacteria. *Ferrobacillus* spp. (Lei et al., 2016), *Gallionella* spp. (de Vet et al., 2011), *Geobacter* spp. (Si et al., 2015), and *Shewanella* spp. (Wang et al., 2016b) are relevant bacterial genera pertaining to the iron bacteria chemoautotrophs, which derive the required energy by oxidation of ferrous ions into ferric form.

Heterotrophic bacteria play a crucial role during the decomposition of organic matter and the recycling of nutrients and minerals in aquatic systems, since they particularly use preformed organic substances to form inorganic carbon and to generate energy. Most of the bacterial species within the aquatic ecosystem are bacterial heterotrophs, which are secondary or tertiary producers (due to the fact that they require external source for organic nutrition) with tremendous metabolic activity to degrade the resulted biomass. Depending on the nourishment process, three types of heterotrophic bacteria are involved in the self-protection process of aquatic environment, namely saprophytic, symbiotic, and parasitic bacteria (Soares Gomes et al., 2014). Saprophytic bacteria derive their energy and nutritional requirements from various forms of dead organic matter, by breaking down the complex organic substances into simple soluble forms thanks to the secreted exogenous enzymes. The specific activity of such bacterial species can be performed either in aerobic conditions (complete decay process) or anaerobic conditions (incomplete fermentation and putrefaction processes). Relevant representatives of bacterial saprophytes refer to nitrogen-fixing *Actinobacteria* spp. (Ghai et al., 2014) and *Azotobacter* spp. (Chug et al., 2016), sulfate-reducing *Desulfatiferula* spp. (Hakil et al., 2014) and *Desulfovibrio* spp. (Hadj-Saïd et al., 2015; Truong et al., 2013), *Leptospira* spp. (Benacer et al., 2013), and iron-reducing *Magnetospirillum* spp. (Li et al., 2015; Suzuki et al., 2007). Symbiotic bacteria particularly coexist in association with other organisms, such relation being beneficial for both organisms: the heterotrophic bacteria derive the required nutrition and energy from the host organism, while the host's biological activity is supported by bacterial mechanisms. Within the aquatic ecosystem, some particular bacteria can live together with a wide variety of organisms, including algae (Nakajima et al., 2015; Ramanan et al., 2015, 2016), arthropods (Anderson et al., 2008; Zhang et al., 2016b), cnidarians (Chimetto et al., 2008; Dirren et al., 2014; Stabili et al., 2011), molluscs (Bayat et al., 2016; Pales Espinosa et al., 2013; Martins et al., 2014), and vertebrates (Prayitno and Sarwan, 2015). Parasitic bacteria specifically occur onto or inside living eukaryotic complex structures (plants and animals) and derive their organic requirements from their hosts by secreting toxic enzymes which affects the structural and functional characteristics of the host cells.

The polyphyletic group of algae is represented by both autotrophic and heterotrophic unicellular up to multicellular organisms that are spread worldwide within the aquatic environment and particularly possess within their structure the chloroplasts (specialized organelles with a rich amount of chlorophyll *a* and various accessory pigments, which enable differential response under light irradiation). Given their enrich content in photosynthetic pigments, the specific Calvin cycle involved during the fabrication of organic molecules from carbon dioxide and their adaptive ultrastructure and function (which enabled their distribution in various regions of the aquatic ecosystem), algae represent the major and primary source for organic products within aquatic environments. *Chlorophyta* spp. (green algae with chlorophyll *a* and *b*), *Cryptophyta* spp. (brown, red or blue–green algae with type *a* and *c* chlorophylls and phycobiliproteic accessory pigments), *Dinophyta* spp. (golden–brown protists with high content of chlorophyll *a* and *c* and peridinin, dinoxanthin, and diadinoxanthin specific auxiliary xanthophylls), *Euglenophyta* spp. (green flagellates with chlorophyll *a* and *b*), *Glaucophyta* spp. (rare freshwater green algae with type a chlorophyll and accessory phycobiliproteins), *Haptophyta* spp. (unicellular motile algae with chlorophylls *a* and *c* and auxiliary fucoxanthin pigment), *Heterokontophyta* spp. (golden–brown or brownish–green unicellular flagellates with fucoxanthin and type *a* and *c* chlorophylls), and *Rhodophyta* spp. (red algae with type *a* chlorophyll and rich amounts of phycoerythrin, phycocyanin, and allophycocyanin supplementary pigments) are the representative algal phyla within the aquatic ecosystem (De Clerck et al., 2012; Hernández-Becerril, 2014; McCourt, 2016; Salmaso and Tolotti, 2009).

Within the wide diversity of algae polyphyletic group there is also included the phytoplankton—photoautotrophic microscopic organisms suspended within the superficial layers of water that produce organic nutrient molecules from water dissolved carbon dioxide. Given their reduced dimensions, phytoplankton is a primary producer for carbon-based nutrients, but their essential role is related to the aquatic nitrogen assimilation (around 70% from the global production). Thanks to their enrich enzymatic activity, phytoplankton are capable to transform a wide variety of macronutrients into simpler nutrient forms, including herein phosphate and silicic acid, but also the abundant N-based compounds: nitrate, nitrite, ammonium, molecular N_2, dissolved organic N (urea, amino acids, humic substances, purines, and vitamins) and particulate N. The diversity and function of phytoplankton are essential aspects related to the maintenance and proper activity of aquatic ecosystems, since they play a significant role during the recycling of vital inorganic nutrients and represent the main food source for zooplankton and some aquatic vertebrate (Qing et al., 2016; Laplace-Treyture and Feret, 2016).

Among the aquatic environments, fungi are uncommon microorganisms, but their functional role within the natural cycling of essential inorganic elements is crucial. Fungi exclusively possess heterotrophic nutrition, meaning that these microorganisms derive their organic food and energy by secreting specific enzymes in their vicinity. The secreted digestive exoenzymes enable the decomposition of near-by complex organic compounds into simple inorganic molecules that can be easily absorbed by fungi. The reduced abundance of fungi within the aquatic ecosystem may be assigned to the active competition for organic resources between them and other microorganisms, especially bacteria. The most abundant and important fungal species related to water self-purification are the hyphomycetes, also known as Ingold (hyphomycetes) or moulds, which play a significant role during the decay of leaves and other organic matter (Bärlocher, 2016; Chauvet et al., 2016). Relevant examples of aquatic hyphomycetes include *Anguillospora* spp. (Sati and Singh, 2014), *Canalisporium* spp. (Sri-indrasutdhi et al., 2010), *Candelabrum* spp. (Kodsueb et al., 2016), *Delitschia* spp. (Sun et al., 2011), *Dendrospora* spp. (Descals and Webster, 1980), *Helicodendron* spp. (Abdullah et al., 2000; Goss, 1985), *Lemonniera* spp. (Campbell et al., 2006; Baschien et al., 2013), *Tricladium* spp., *Triscelophorus* spp. (Duarte et al., 2015).

Protozoa are unicellular eukaryotic microorganisms, with non-photosynthetic autotrophic (osmotrophy—absorption of soluble nutrients from surrounding environment by osmosis process) and heterotrophic (phagotrophy—active or passive engulfing of solid nutrients within the specific vacuoles) nutrition. Protozoa play a significant role in maintaining the trophic structure of aquatic ecosystems, since the phagocytosis-based nutrition mainly involves feeding on other microorganisms—including bacteria, algae, or other protists (Shanan et al., 2015; Xu and Xu, 2016). This specific activity of protozoa differentiate them from other aquatic organisms involved in water self-purification, but is also a controversial behavior, given the fact that protozoa can act as vector for pathogenic microorganisms. *Ciliophora* spp. (ciliate protozoa), *Mastigophora* spp. (flagellate protozoa), and *Sarcodina* spp. (pseudopod amoebae) are the representative types of protozoa phyla with significant relevance for water ecosystem (Finlay and Esteban, 2013; Horan, 2003).

3 Wastewater Conversion

Significant progress regarding the novel technological, industrial, and evolutionary strategies has been reported during the last centuries. Even if such progress has a beneficial impact, in terms of increased quality of human lives, this also has a negative side, which concerns the quality of global environment. Until recent times, the potential adversely indirect effects of global development on surrounding media have not represented a real concern. Better late than never, the worldwide community became aware of these effects and their long-term-related impact on Earth existence and survival. Therefore, the current trend in day-to-day activities involves a rather environmentally friendly approach, which consists in redirecting impressive resources (including scientific and financial, technological, and industrial) toward new concepts and novel strategies that will guarantee the mitigation or/and elimination of environmental-harmful anthropogenic effects.

Wastewater is the generic term assigned to any water that specifically contains a wide variety of contaminant agents, including plant nutrients and pathogenic microorganisms, metallic, and organic pollutants, but also microcontaminants (small molecular compounds resulted from synthetic products) (Hespanhol, 1997). Wastewater is the negative result of human influence against the aquatic environment, which can originate from various daily activities, including agricultural, commercial, domestic, and industrial. Depending on their origin, anthropogenic-derived wastewater may be classified as follows: (1) domestic wastewater (excreta-derived black water and domestic washing-derived grey water), (2) agricultural wastewater (pesticides, vegetables, and animal wastes can negatively affect the quality of nearby water sources), and (3) industrial wastewater (metallurgy and nuclear industry, chemical and pharmaceutical industry, and food industry are few industrial domains which are responsible for wastewater production).

Given the complex variety of wastewater (depending on its origin), systematic, and efficient treatments to remove the concerned water pollutants are required. Therefore, wastewater treatment requires complex stages to remove the contaminants: (1) preliminary treatment—the resulted effluent is purged by debris, grit, scum, oil, and grease; (2) primary treatment—solid objects and suspended solid particles are removed by mechanical processes (such as screening and sedimentation); (3) secondary treatment—the biological-assisted removal of dissolved and colloid organic compounds (within this process, various microorganisms that derive their organic food from complex organic molecules are involved); (4) tertiary treatment—supplementary stage that provides the efficient removal of remaining inorganic compounds and pathogenic microorganisms (Butler et al., 1995; von Sperling, 2007).

As described previously, native aquatic microorganisms possess tremendous potential to revalue the inorganic and organic complex substances from their surrounding environment, being of a crucial importance within the aquatic ecosystem (but not limited to). By considering their particular nourishment pathways and the versatility of wastewater (in terms of complex inorganic and organic nutrients), the international community turned its attention toward the revaluation of microorganisms as potential candidates for water pollutants removal and wastewater conversion (Table 7.1).

Table 7.1 Types of Microorganisms and Their Impact in the Purification of Various Waste Waters by Different Approaches

Type of Wastewater	Microorganisms	Resulted Products	References
Agro-industrial wastewater	*Debaryomyces etchellsii* yeast	Lipid production	Arous et al. (2016)
Butanol fermentation wastewater	*Trichosporon dermatis* yeast	Microbial oil	Peng et al. (2013)
Cassava flour wastewater	Lactic acid bacteria	Biohydrogen production	Gomes et al. (2016)
Domestic wastewater	Algae	Biofuel production	Xin et al. (2016)
Domestic wastewater	*Chlorella vulgaris* microalgae *Enterobacter asburiae, Klebsiella* spp. or *Raoultella ornithinolytica* bacteria	Nutrient recovery	
Domestic wastewater/ industrial wastewater	Purple phototrophic bacteria	Nutrient recovery	Hülsen et al. (2016)
Molasses and olive mill wastewaters	*Saccharomyces cerevisiae* MAK-1	Bioethanol production	Sarris et al. (2014)
Municipal wastewater	Algae	Nutrient recovery	Shayan et al. (2016)
Municipal wastewater	Microalgae	Biofuel production	Novoveská et al. (2016)
Residual wastewaters	Mixed microbial cultures	Bio-polymers (polyhydroxyalkanoates)	
Sea food processing wastewater	*Stenotrophomonas* bacterial strains	Power generation	Jayashree et al. (2016)
Synthetic wastewater	UMN266 microalgal strain	Bioethanol production	Wang et al. (2015)

4 Nanoparticle Production by Bacteria

The main use of nanoparticles in water purification and environmental safety is given by their ability to remove heavy metal ions. Madrakian et al. (2015) developed maghemite nanoparticles modified with mercaptoethylamino-based homopolymers for the removal of Ag, Hg, Pb, and Cd ions in water solutions. Also based on iron oxide nanoparticles, the system developed by Ge et al. (2012) uses Fe_3O_4/3-aminopropyltriethoxysilane/acrylic acid/ crotonic acid for the removal of Cd, Zn, Pb, and Cu in aqueous solutions. Gupta et al. (2016) used CuO nanoparticles as adsorbents for Cr ions. Razzaz et al. (2016) obtained a membrane-like system based on chitosan and TiO_2 nanoparticles to act as adsorbents for Cu and Pb.

Nanoparticles have been proved to work as excellent sorbent systems with improved properties, due to their enhanced reactivity.

Jiang et al. (2015) obtained Fe_3O_4–chitosan–polyaniline (PANI) microparticles using the coprecipitation method to be used as magnetic adsorbents for aromatic compounds. The functioning principle of the system is based on the π–π interactions that are establishing between the polyaniline and the aromatic compound.

Also based on magnetite nanoparticles, the system developed by Yu et al. (2015) can be used for the removal of organic solvents, like diesel oil, from wastewater.

More recently, Li et al. (2016b) obtained magnetite-reduced graphene aerogels for their use in water decontamination and purification. They proved efficiency for oils, organic solvents, arsenite ions, and dyes sorption capability.

Also, many nanoparticles are used in water purification, due to their natural antimicrobial properties, like: Ag (Franci et al., 2015; El-Zahry et al., 2015; Thuptimdang et al., 2015), ZnO (Chandrappa and Venkatesha, 2012; Lee et al., 2014; Ng et al., 2013), Cu and CuO (Ramyadevi et al., 2012; Shankar and Rhim, 2014), MnO_2 (Wang et al., 2016), MgO (Zhen-Xing and Bin-Feng, 2014; Leung et al., 2014), Se (Bartůněk et al., 2015; Tran and Webster, 2011), and TiO_2 (Kubacka et al., 2014; Park et al., 2012) nanoparticles.

The importance of microorganism decontamination of water is crucial for human health, as water is a vital compound. The incidence of death due to contaminated water consumption is extremely high: 1.8 million deaths due to diarrheal diseases. This phenomenon is encountered mostly in developing countries worldwide, where people have no access to safe drinkable water or sanitation. Thus, finding simple and reusable methods for water purification is imperative (Anshup, 2009).

Bindhu et al. (2016) obtained MgO nanoparticles using a conventional wet chemical method and their antibacterial properties were evaluated against both Gram negative (*Pseudomonas aeruginosa*) and Gram positive (*Staphylococcus aureus*) strains, which are usually encountered in water. The effectiveness of the nanoparticle treatment was higher in case of *S. aureus*, probably due to the lower complexity of the membrane structure and the higher surface available for contact with the oxide nanoparticles.

Fly ash or activated charcoal is commonly used in water purification, as being classified as one of the biological processes used in this type of application, due to its high surface area and enhanced adsorption properties. Joshi et al. (2015) modified fly ash particles with AgNPs and IONPs using a hydrothermal approach. The composite material's antibacterial efficiency was proved against *Escherichia coli*, the bactericidal efficiency increasing under UV light exposure, due to the high photocatalytic effect of the resulted particles. Moreover, the fly ash–AgNPs–IONPs particles showed adsorption ability for Pb(II) ions in aqueous solutions.

However, these nanoparticles are usually embedded in different types of filtering membranes used in water purification. Table 7.2 presents some current examples from the scientific literature regarding the use of different types of nanoparticles in the fabrication of composite membranes.

Biogenic nanoparticles have recently grown in interest due to their properties like good polydispersity and biocompatibility, because of the natural functionalizing that occurs especially on intracellular-synthesized nanoparticles. Nanoparticles synthesized by means of bacteria and yeast are considered green approaches, as they do not use corrosive and toxic precursors and substances and are more environmentally friendly. The metal salts are reduced by means of intracellular or extracellular agents (terpenoids, flavones, alkaloids, pigments, proteins, or amino acids) that occur naturally in these microorganisms.

The use of nanoparticles in water decontamination also offers other advantages like the complete removal of the contaminant, even at low concentrations, again due to the enhanced reactivity of nanoparticles and their high surface area. Thus, lower waste is generated after the purification process (Anshup, 2009). Furgal et al. (2015) used MnO_2 nanoparticles produced by *Pseudomonas putida* strain when exposed to different micropollutants at different concentrations relevant to those from the polluted environments. These particles can be used for the removal of different residual organic matter in wastewater, like estrone and 17-α-ethinyl estradiol steroid hormones. The effectiveness of this system is at ppm level.

Regarding the use of biologically synthesized nanoparticles, one major property for which they are used in the fabrication of water filtering membranes is given by their antimicrobial and antiadherence properties. Thus, all biogenic nanoparticles are tested for their antibiofouling properties. For example, Sonkusre and Cameotra (2015) obtained *Bacillus licheniformis*—synthesized selenium nanoparticles that inhibited the adherence of *S. aureus* bacteria. *Stenotrophomonas maltophilia* was also used for the reduction of selenite to selenium nanoparticles.

Table 7.2 Examples of Nanoparticles Used in the Fabrication of Composite Filtering Membranes for Water Purification

Type of Nanoparticle	System Description	Type of filtering/ Membrane Pore Size	Evaluation	References
AgNPs	NH_2-polyethersulfone-AgNPs	—	In vitro antibacterial evaluation against *E.coli* strains	Haider et al. (2016)
	AgNPs dispersed in polysulfone/ AgNPs synthesized in situ in polysulfone	Antibiofoulingnanofiltration and microfiltration	In vitro silver ions release and measurement using ICP-MS; in vitro antibacterial activity and antiadherence potential against *E. coli* and *Pseudomonas aeruginosa* strains	Andrade et al. (2015)
	Nitrocellulose membrane impregnated with AgNPs of different provenience: commercial and Biosynthesized (extracellular from *Aspergillus niger*, *Cryptococcus laurentii*, and *Rhodotorula glutinis*	Microfiltration	In vitro antibacterial evaluation against *E. coli*, *Enterococcus faecalis* and *Pseudomonas aeruginosa*;	Fernández et al. (2016)
	Polyamide thin films incorporated with AgNPs via arc plasma deposition onto polysulfone membranes	Reverse osmosis filtration	Water flux determination and NaCl salt rejection evaluation were done using a lab-scaled cross flow filtration system; in vitro antibacterial efficiency against *E. coli*, *P. aeruginosa*, and *S. aureus* strains; in vitro silver ion release determination using ICP-MS	Park et al. (2016)
	AgNPs/poly(vinylidene fluoride) (PVDF)membrane fabrication using phase inversion induced by immersion precipitation technique	Reverse osmosis filtration	In vitro silver ion release using ICP-MS; organic antifouling property measurement using the water flux recovery after surface and inner fouling; in vitro antibacterial potential against *E. coli* strains; in vitro antibiofouling property evaluation	Li et al. (2013)
	Polyamide-AgNPs biosynthesized using *L. fermentum* strain	Nanofiltration	Filtration performance (water permeation and salt rejection) evaluation using a dead-end filtration cell; in vitro silver release using ICP-MS; in vitro antibacterial activity against *P. aeruginosa* and *E. coli*	Liu et al. (2015)

	Cellulose acetate- graphene oxide-AgNPs membrane	—	In vitro continuous filtration test; in vitro antibacterial activity against *E. coli* strains after 2 h from contact	Sun et al. (2015)
TiO_2 NPs	Nylon-6 nanofiber membranes-TiO_2 NPs	Microfiltration	Photodegradation test using methylene blue; in vivo toxicity test of chlorophenols using Daphia magna models	An et al. (2014)
	Polyether sulfone (PES)/polyvinyldiene fluoride (PVDF) hydrophobic/PVDF hydrophilic membranes- TiO_2 nanoparticles synthesized at the surface of the membranes via hydrolysis method	—	In vitro permeation testing, organic antifouling properties; photodegradation test of methylene blue and noninflammatory drugs (diclofenac and ibuprofen)	Fischer et al. (2015)
	Nanofiber membranes- TiO_2 nanoparticles	—	In vitro antibacterial activity against *S. aureus*, humic aids biofouling testing, water flux testing;	Daels et al. (2015)
CuNPs/CuO NPs	Thin film composite commercial membrane- Cu/CuO NPs	Reverse osmosis filtration	In vitro copper dissolution testing; in vitro antibacterial efficiency against *E. coli*	Ben-Sasson et al. (2016)
	Thin film composite- CuNPs obtained using spray- and spin-assisted layer-by-layer (SSLbL)	Reverse osmosis filtration	Membrane water permeation flux testing, salt rejection capability, antibiofouling property	Ma et al. (2016)
Si-based NPs	PVDF membrane- SiO_2- poly(methyl meth acrylate)-*block*-poly(2-dimethylaminoethyl methacrylate) copolymer (PMMA-*b*-PDMAEMA) grafted	—	Hydrophobicity, permeability and antifouling properties testing	Zhu et al. (2016)
	Polyamide membrane- MCM-48 capped nanoparticles via interfacial polymerization	—	Water flux evaluation; salt rejection testing; durability testing	Liu et al. (2016b)
	Zeolite nanoparticles- polysulfone membrane	—	Water flux evaluation; sorption capability for lead and nickel ions	Yurekli (2016)
Mg-based NPs	Polyethersulfone membrane-$Mg(OH)_2$ nanoparticles synthesized in situ	—	Hydrophobicity evaluation, permeation flux, retention rate, antifouling properties measurements	

Ashajyothi et al. (2016) has recently used biogenic synthesized nanoparticles as antifungal and antimycelium agents: silver, copper, zinc oxide, respectively gold nanoparticles, obtained using an extracellular method from *Enterococcus faecalis* nonpathogenic strain. Except from AuNPs, all of the nanoparticles exhibited high antifungal activity.

E. coli were also used for the synthesis of AgNPs, capped with L-cysteine (Perni et al., 2014). These proved antibacterial activities against *E. coli* and *S. aureus* strains. Another example of biogenically synthesized silver nanoparticles is given by scientists who employed a method using *Pseudomonas putida* Gram-negative bacteria.

Fernández et al. (2016) obtained nitrocellulose-based membranes for water filtration, impregnated with silver nanoparticles, from different sources: commercial AgNPs and biosynthesized AgNPs. The biosynthesized silver nanoparticles were obtained using an extracellular method from *Aspergillus niger, Cryptococcus laurentii,* and *Rhodotorula glutinis.* The as-obtained samples were then evaluated for their antibacterial potential for relevant bacteria strains (*E. coli, Enterococcus faecalis,* and *Pseudomonas aeruginosa*). The most promising results were proven for the sample containing the AgNPs synthesized via *Rhodotorula glutinis* culture, the explanation being given by the polysaccharide functionalizing of these nanoparticles.

Bio-Ag^0 or biogenic silver nanoparticles are synthesized in the presence of *Lactobacillus fermentum* strains and have grown interest due to their antibiofouling and antimicrobial properties. There are many studies that describe the obtaining and testing of water filtration membranes incorporating bio-Ag^0.

Liu et al. (2015) obtained polyamide membranes for nanofiltration grafted with AgNPs obtained biosynthesized using a *L. fermentum* strain. These membranes were compared with polyamide membranes embedded with chemically synthesized nanoparticles, by means of morphology, hydrophobicity, flux properties, silver release, and antibacterial properties. Both membranes showed good flux properties (by means of water permeation and salt rejection capabilities), however, the membrane containing biosynthesized nanoparticles proved to be more stable depending on the lack of AgNPs release. Also, this system proved to be more efficient by means of its antibacterial activity.

De Gusseme et al. (2011) also obtained PVDF membranes for microfiltration, embedded with AgNPs biosynthesized using *L. fermentum.* These systems proved antiviral properties against UZ1 bacteriophages under continuous flux. However, the filtering potential of the membrane was limited, proving applicability to small scales.

Other example is given by the obtaining of polyethersulfone (PES)-based membranes incorporated with bio-Ag^0 (Zhang et al., 2012). These proved hydrophobic characters increased permeate flux, antibacterial, and antiadherence properties against *E. coli* and *P. aeruginosa* strains.

5 Conclusions and Perspectives

Given the complex activity of various microorganisms (both nonpathogenic and pathogenic strains) and the current trend toward the novel ecofriendly technological approach, the scientific community turned its attention to the revaluation of biocellular mechanisms, to use natural approaches. Thereby, according to a plethora of recent research studies, the successful removal of water pollutants or/and the conversion of wastewater (resulting thus purified water and nontoxic biogenic products) can be performed by using various microorganisms. The multitudinous healthcare conditions associated with consumption of contaminated water and the adverse effects of wastewater against global ecosystem are major concerns of modern world. Therefore, significant resources (especially scientific and economical) have been directed toward this particular domain. Given the promising experimental results reported in ecological purification of water and conversion of wastewater, microorganisms seems to represent ideal candidates for novel next-generation strategies related to global management of water sources.

References

Abdullah, S.K., Cano, J., Descals, E., Guarro, J., 2000. The aero-aquatic *Helicodendron microsporum* n. sp. from Mallorca, Spain. Mycol. Res. 104 (3), 375–377.

Amaral, P., Partlan, E., Li, M., Lapolli, F., Thompson Mefford, O., Karanfil, T., Ladner, D.A., 2016. Superfine powdered activated carbon (S-PAC) coatings on microfiltration membranes: effects of milling time on contaminant removal and flux. Water Res. 100, 429–438.

An, S., Lee, M.W., Joshi, B.N., Jo, A., Jung, J., Yoon, S.S., 2014. Water purification and toxicity control of chlorophenols by 3D nanofiber membranes decorated with photocatalytic titania nanoparticles. Ceram. Int. 40 (2), 3305–3313.

Anderson, L., Halary, S., Lechaire, J.P., Boudier, T., Frébourg, G., Marco, S., Zbinden, M., Gaill, F., 2008. Tomography of bacteria–mineral associations within the deep-sea hydrothermal vent shrimp *Rimicaris exoculata*. CR Chim. 11 (3), 268–280.

Andrade, P.F., de Faria, A.F., Oliveira, S.R., Arruda, M.A.Z., Gonçalves, M. do C., 2015. Improved antibacterial activity of nanofiltration polysulfone membranes modified with silver nanoparticles. Water Res. 81, 333–342.

Anshup, Pradeep, T., 2009. Noble metal nanoparticles for water purification: a critical review. Thin Solid Films 517 (24), 6441–6478.

Arakeri, G., Hunasgi, S., Colbert, S., Merkx, M.A.W., Brennan, P.A., 2014. Role of drinking water copper in pathogenesis of oral submucous fibrosis: a prospective case control study. Brit. J. Oral Maxillofac. Surg. 52 (6), 507–512.

Aranda, C.P., Valenzuela, C., Matamala, Y., Godoy, F.A., Aranda, N., 2015. Sulphur-cycling bacteria and ciliated protozoans in a Beggiatoaceae mat covering organically enriched sediments beneath a salmon farm in a southern Chilean fjord. Mar. Poll. Bull. 100 (1), 270–278.

Arous, F., Frikha, F., Triantaphyllidou, I.- E., Aggelis, G., Nasri, M., Mechichi, T., 2016. Potential utilization of agro-industrial wastewaters for lipid production by the oleaginous yeast *Debaryomyces etchellsii*. J. Clean. Prod. 133, 899–909.

Ashajyothi, C., Prabhurajeshwar, C., Handral, H.K., Chandrakanth, K.R., 2016. Investigation of antifungal and anti-mycelium activities using biogenic nanoparticles: an ecofriendly approach, Environmental Nanotechnology. Monitor. Manage. 5, 81–87.

Bärlocher, F., 2016. Aquatic hyphomycetes in a changing environment. Fungal Ecol. 19, 14–27.

Bartůněk, V., Junková, J., Šuman, J., Kolářová, K., Rimpelová, S., Ulbrich, P., Sofer, Z., 2015. Preparation of amorphous antimicrobial selenium nanoparticles stabilized by odor suppressing surfactant polysorbate 20. Mater. Lett. 152, 207–209.

Baschien, C., Tsui, C.K.M., Gulis, V., Szewzyk, U., Marvanová, L., 2013. The molecular phylogeny of aquatic hyphomycetes with affinity to the *Leotiomycetes*. Fungal Biol. 117 (9), 660–672.

Bayat, Z., Hassanshahian, M., Hesni, M.A., 2016. Study the symbiotic crude oil-degrading bacteria in the mussel *Mactra stultorum* collected from the Persian Gulf. Marine Poll. Bull. 105 (1), 120–124.

Benacer, D., Woh, P.Y., Zain, S.N.M., Amran, F., Thong, K.L., 2013. Pathogenic and saprophytic Leptospira species in water and soils from selected urban sites in Peninsular Malaysia. Microbes Environ. 28 (1), 135–140.

Ben-Sasson, M., Lu, X., Nejati, S., Jaramillo, H., Elimelech, M., 2016. In situ surface functionalization of reverse osmosis membranes with biocidal copper nanoparticles. Desalination 388, 1–8.

Bhowmik, A.K., Alamdar, A., Katsoyiannis, I., Shen, H., Ali, N., Ali, S.M., Bokhari, H., Schäfer, R.B., Eqani, S.A.M.A.S., 2015. Mapping human health risks from exposure to trace metal contamination of drinking water sources in Pakistan. Sci. Tot. Environ. 538, 306–316.

Bindhu, M.R., Umadevi, M., Micheal, K.M., Arasu, M.V., Al-Dhabi, N.A., 2016. Structural, morphological and optical properties of MgO nanoparticles for antibacterial applications. Mater. Lett. 166, 19–22.

Brachet, C., Magnier, J., Valensuela, D., Petit, K., Fribourg-Blanc, B., Bernex, N., Scoullos, M., Tarlock, D., 2015. The handbook for management and restoration of aquatic ecosystems in river and lake basins (provided by INBO, in partnership with GWP and IOWater).

Brahman, K.D., Kazi, T.G., Afridi, H.I., Baig, J.A., Arain, S.S., Talpur, F.N., Kazi, A.G., Ali, J., Panhwar, A.H., Arain, M.B., 2016. Exposure of children to arsenic in drinking water in the Tharparkar region of Sindh, Pakistan. Sci. Tot. Environ. 544, 653–660.

Bretzler, A., Johnson, C.A., 2015. The geogenic contamination handbook: addressing arsenic and fluoride in drinking water. Appl. Geochem. 63, 642–646.

Bulka, C.M., Jones, R.M., Turyk, M.A., Stayner, L.T., Argos, M., 2016. Arsenic in drinking water and prostate cancer in Illinois counties: an ecologic study. Environ. Res. 148, 450–456.

Butler, A., Carty, G., Crowe, M., Flanagan, P., Lambert, M., 1995. Wastewater treatment manual (provided by EPA), pp. 11–18.

Butts, C.D., Bloom, M.S., Neamtiu, I.A., Surdu, S., Pop, C., Anastasiu, D., Fitzgerald, E.F., Gurzau, E.S., 2015. A pilot study of low-moderate drinking water arsenic contamination and chronic diseases among reproductive age women in Timiş County, Romania. Environ. Toxicol. Pharmacol. 40 (3), 1001–1004.

Calvó, L., Garcia-Gil, L.J., 2004. Use of amoB as a new molecular marker for ammonia-oxidizing bacteria. J. Microbiol. Methods 57 (1), 69078.

Campbell, J., Shearer, C., Marvanová, L., 2006. Evolutionary relationships among aquatic anamorphs and teleomorphs: *Lemonniera*, *Margaritispora*, and *Goniopila*. Mycol. Res. 110 (9), 1025–1033.

Cancilla, J.C., Aroca-Santos, R., Wierzchoś, K., Torrecilla, J.S., 2016. Hazardous aromatic VOC quantification through spectroscopic analysis and intelligent modeling to assess drinking water quality. Chemometr. Intell. Lab. Syst. 156, 102–107.

Cébron, A., Garnier, J., 2005. *Nitrobacter* and *Nitrospira* genera as representatives of nitrite-oxidizing bacteria: detection, quantification and growth along the lower Seine River (France). Water Res. 39 (20), 4979–4992.

Chandrappa, K.G., Venkatesha, T.V., 2012. Electrochemical synthesis and photocatalytic property of zinc oxide nanoparticles. Nano-Micro Lett. 4 (1), 14–24.

Chauvet, E., Cornut, J., Sridhar, K.R., Selosse, M.A., Bärlocher, F., 2016. Beyond the water column: aquatic hyphomycetes outside their preferred habitat. Fungal Ecol. 19, 112–127.

Chen, D., Zhao, X., Li, F., Zhang, X., 2016. Rejection of nuclides and silicon from boron-containing radioactive waste water using reverse osmosis. Sep. Purif. Technol. 163, 92–99.

Chimetto, L.A., Brocchi, M., Thompson, C.C., Martins, R.C.R., Ramos, H.R., Thompson, F.L., 2008. Vibrios dominate as culturable nitrogen-fixing bacteria of the Brazilian coral *Mussismilia hispida*. Syst. Appl. Microbiol. 31 (4), 312–319.

Chug, R., Gour, V.S., Mathur, S., Kothari, S.L., 2016. Optimization of Extracellular Polymeric Substances production using *Azotobacter beijreinckii* and *Bacillus subtilis* and its application in chromium (VI) removal. Biores. Technol. 214, 604–608.

Craig, L., Lutz, A., Berry, K.A., Yang, W., 2015. Recommendations for fluoride limits in drinking water based on estimated daily fluoride intake in the Upper East Region, Ghana. Sci. Tot. Environ. 532, 127–137.

Daels, N., Radoicic, M., Radetic, M., De Clerck, K., Van Hulle, S.W.H., 2015. Electrospun nanofiber membranes functionalised with TiO_2 nanoparticles: evaluation of humic acid and bacterial removal from polluted water. Sep. Purif. Technol. 149, 488–494.

Daims, H., Lücker, S., Wagner, M., 2016. A new perspective on microbes formerly known as nitrite-oxidizing bacteria. Trends Microbiol. 24 (9), 699–712.

De Clerck, O., Bogaert, K.A., Leliaert, F., 2012. Diversity and evolution of algae: primary endosymbiosis. Adv. Bot. Res. 64, 55–86.

De Gusseme, B., Hennebel, T., Christiaens, E., Saveyn, H., Verbeken, K., Fitts, J.P., Boon, N., Verstraete, W., 2011. Virus disinfection in water by biogenic silver immobilized in polyvinylidene fluoride membranes. Water Res. 45 (4), 1856–1864.

de Joode, B.W., Barbeau, B., Bouchard, M.F., Mora, A.M., Skytt, A., Córdoba, L., Quesada, R., Lundh, T., Lindh, C.H., Mergler, D., 2016. Manganese concentrations in drinking water from villages near banana plantations with aerial mancozeb spraying in Costa Rica: Results from the Infants' Environmental Health Study (ISA). Environ. Pollut. 215, 247–257.

de Vet, W.W.J.M., Dinkla, I.J.T., Rietveld, L.C., van Loosdrecht, M.C.M., 2011. Biological iron oxidation by *Gallionella* spp. in drinking water production under fully aerated conditions. Water Res. 45 (17), 5389–5398.

Dechatiwongse, P., Srisamai, S., Maitland, G., Hellgardt, K., 2014. Effects of light and temperature on the photoautotrophic growth and photoinhibition of nitrogen-fixing cyanobacterium *Cyanothece* sp. ATCC 51142. Algal Res. 5, 103–111.

Descals, E., Webster, J., 1980. Taxonomic studies on aquatic hyphomycetes II. The *Dendrospora* aggregate. Trans. Br. Mycol. Soc. 74 (1), 135–158.

Dirren, S., Salcher, M.M., Blom, J.F., Schweikert, M., Posch, T., 2014. Ménage-à-trois: the amoeba *Nuclearia* sp. from Lake Zurich with its ecto- and endosymbiotic bacteria. Protist 165 (5), 745–758.

Duarte, S., Batista, D., Bärlocher, F., Cássio, F., Pascoal, C., 2015. Some new DNA barcodes of aquatic hyphomycete species. Mycoscience 56 (1), 102–108.

El-Zahry, M.R., Mahmoud, A., Refaat, I.H., Mohamed, H.A., Bohlmann, H., Lendl, B., 2015. Antibacterial effect of various shapes of silver nanoparticles monitored by SERS. Talanta 138, 183–189.

Felipe-Sotelo, M., Henshall-Bell, E.R., Evans, N.D.M., Read, D., 2015. Comparison of the chemical composition of British and Continental European bottled waters by multivariate analysis. J. Food Composit. Anal. 39, 33–42.

Fernández, J.G., Almeida, C.A., Fernández-Baldo, M.A., Felici, E., Raba, J., Sanz, M.I., 2016. Development of nitrocellulose membrane filters impregnated with different biosynthesized silver nanoparticles applied to water purification. Talanta 146, 237–243.

Finlay, B.J., Esteban, G.F., 2013. Protozoa. In: Levin, S.A. (Ed.), Encyclopedia of Biodiversity. second ed. Academic Press, UK, pp. 286–297.

Fischer, K., Grimm, M., Meyers, J., Dietrich, C., Gläser, R., Schulze, A., 2015. Photoactive microfiltration membranes via directed synthesis of TiO_2 nanoparticles on the polymer surface for removal of drugs from water. J. Membr. Sci. 478, 49–57.

Franci, G., Falanga, A., Galdiero, S., Palomba, L., Rai, M., Morelli, G., Galdiero, M., 2015. Silver nanoparticles as potential antibacterial agents. Molecules 20 (5), 8856–8874.

Furgal, K.M., Meyer, R.L., Bester, K., 2015. Removing selected steroid hormones, biocides and pharmaceuticals from water by means of biogenic manganese oxide nanoparticles in situ at ppb levels. Chemosphere 136, 321–326.

Ge, F., Li, M.-M., Ye, H., Zhao, B.-X., 2012. Effective removal of heavy metal ions Cd^{2+}, Zn^{2+}, Pb^{2+}, Cu^{2+} from aqueous solution by polymer-modified magnetic nanoparticles. J. Hazard. Mater. 211–212, 366–372.

Gerke, T.L., Little, B.J., Maynard, J.B., 2016. Manganese deposition in drinking water distribution systems. Sci. Tot. Environ. 541, 184–193.

Ghai, R., Megumi Mizuno, C., Picazo, A., Camacho, A., Rodriguez-Valera, F., 2014. Key roles for freshwater Actinobacteria revealed by deep metagenomic sequencing. Mol. Energy 23, 6073–6090.

Gomes, S.D., Fuess, L.T., Mañunga, T., Gomes, P.C.F. de L., Zaiat, M., 2016. Bacteriocins of lactic acid bacteria as a hindering factor for biohydrogen production from cassava flour wastewater in a continuous multiple tube reactor. Int. J. Hydrogen Energy 41 (19), 8120–8131.

Goss, R.D., 1985. The anamorph genus *Helicodendron*. Trans. Br. Mycol. Soc. 84 (3), 423–435.

Guadayol, M., Cortina, M., Guadayol, J.M., Caixach, J., 2016. Determination of dimethyl selenide and dimethyl sulphide compounds causing off-flavours in bottled mineral waters. Water Res. 92, 149–155.

Gupta, V.K., Chandra, R., Tyagi, I., Verma, M., 2016. Removal of hexavalent chromium ions using CuO nanoparticles for water purification applications. J. Colloid Interface Sci. 478, 54–62.

Hadj-Saïd, J., Pandelia, M.E., Léger, C., Fourmond, V., Dementin, S., 2015. The carbon monoxide dehydrogenase from *Desulfovibrio vulgaris*. Biochim. Biophys. Acta (BBA)—Bioenergetics 1847 (12), 1574–1583.

Haider, M.S., Shao, G.N., Imran, S.M., Park, S.S., Abbas, N., Tahir, M.S., Hussain, M., Bae, W., Kim, H.T., 2016. Aminated polyethersulfone-silver nanoparticles (AgNPs-APES) composite membranes with controlled silver ion release for antibacterial and water treatment applications. Mater. Sci. Eng.: C 62, 732–745.

Hakil, F., Amin-Ali, O., Hirschler-Réa, A., Mollex, D., Grossi, V., Duran, R., Matheron, R., Cravo-Laureau, C., 2014. *Desulfatiferula berrensis* sp. nov., a n-alkene-degrading sulfate-reducing bacterium isolated from estuarine sediments. Int. J. Syst. Evol. Microbiol. 64 (2), 540–544.

Hernández-Becerril, D.U., 2014. Biodiversity of marine planktonic algae (Cyanobacteria, Prasinophyceae, Euglenophyta, Chrysophyceae, Dictyochophyceae, Eustigmatophyceae, Parmophyceae, Raphidophyceae, Bacillariophyta, Cryptophyta, Haptophyta, Dinoflagellata) in Mexico. Rev. Mex. Biodivers. 85 (Suppl. 1), 44–53.

Hespanhol, I., 1997. Wastewater as a resource. In: Helmer, R., Hespanhol, I. (Eds.), Water Pollution Control—A Guide to the Use of Water Quality Management Principles (provided by WHO).

Horan, N., 2003. Protozoa. In: Mara, D., Horan, N. (Eds.), Handbook of Water and Wastewater Microbiology. Academic Press, UK, pp. 69–76.

Huang, Z., Gedalanga, P.G., Asvapathanagul, P., Olson, B.H., 2010. Influence of physicochemical and operational parameters on *Nitrobacter* and *Nitrospira* communities in an aerobic activated sludge bioreactor. Water Res. 44 (15), 4351–4358.

Hülsen, T., Barry, E.M., Lu, Y., Puyol, D., Keller, J., Batstone, D.J., 2016. Domestic wastewater treatment with purple phototrophic bacteria using a novel continuous photo anaerobic membrane bioreactor. Water Res. 100, 486–495.

Hynes, A.M., Blythe, B.J., Binder, B.J., 2015. An individual-based model for the analysis of *Prochlorococcus* diel cycle behavior. Ecol. Model. 301, 1–15.

Islam, M.S., Zhang, Y., McPhedran, K.N., Liu, Y., El-Din, M.G., 2016. Mechanistic investigation of industrial wastewater naphthenic acids removal using granular activated carbon (GAC) biofilm based processes. Sci. Tot. Environ. 541, 238–246.

Jayashree, C., Tamilarasan, K., Rajkumar, M., Arulazhagan, P., Yogalakshmi, K.N., Srikanth, M., Rajesh Banu, J., 2016. Treatment of seafood processing wastewater using upflow microbial fuel cell for power generation and identification of bacterial community in anodic biofilm. J. Environ. Manag. 180, 351–358.

Jiang, X., Cheng, J., Zhou, H., Li, F., Wu, W., Ding, K., 2015. Polyaniline-coated chitosan-functionalized magnetic nanoparticles: preparation for the extraction and analysis of endocrine-disrupting phenols in environmental water and juice samples. Talanta 141, 239–246.

Joshi, M.K., Pant, H.R., Liao, N., Kim, J.H., Kim, H.J., Park, C.H., Kim, C.S., 2015. In-situ deposition of silver–iron oxide nanoparticles on the surface of fly ash for water purification. J. Colloid Interface Sci. 453, 159–168.

Kenari, S.L.D., Barbeau, B., 2016. Understanding ultrafiltration fouling of ceramic and polymeric membranes caused by oxidized iron and manganese in water treatment. J. Membr. Sci. 516, 1–12.

Khan, M.R., Wabaidur, S.M., Alothman, Z.A., Busquets, R., Naushad, M., 2016. Method for the fast determination of bromate, nitrate and nitrite by ultra performance liquid chromatography–mass spectrometry and their monitoring in Saudi Arabian drinking water with chemometric data treatment. Talanta 152, 513–520.

Kodsueb, R., Lumyong, S., McKenzie, E.H.C., Bahkali, A.H., Hyde, K.D., 2016. Relationships between terrestrial and freshwater lignicolous fungi. Fungal Ecol. 19, 155–168.

Kostyla, C., Bain, R., Cronk, R., Bartram, J., 2015. Seasonal variation of fecal contamination in drinking water sources in developing countries: a systematic review. Sci. Tot. Environ. 514, 333–343.

Kruse, E.B., Wehner, A., Wisplinghoff, H., 2016. Prevalence and distribution of *Legionella* spp. in potable water systems in Germany, risk factors associated with contamination, and effectiveness of thermal disinfection. Am. J. Infect. Control 44 (4), 470–474.

Kubacka, A., Diez, M.S., Rojo, D., Bargiela, R., Ciordia, S., Zapico, I., Albar, J.P., Barbas, C., Martins dos Santos, V.A.P., Fernández-García, Marcos, Ferrer, Manuel, 2014. Understanding the antimicrobial mechanism of TiO_2-based nanocomposite films in a pathogenic bacterium. Scient. Rep.s 4134, 4.

Kutovaya, O.A., Watson, S.B., 2014. Development and application of a molecular assay to detect and monitor geosmin-producing cyanobacteria and actinomycetes in the Great Lakes. J. Great Lakes Res. 40 (2), 404–414.

Lai, G.S., Lau, W.J., Goh, P.S., Ismail, A.F., Yusof, N., Tan, Y.H., 2016. Graphene oxide incorporated thin film nanocomposite nanofiltration membrane for enhanced salt removal performance. Desalination 387, 14–24.

Laishram, M., Rathlavath, S., Lekshmi, M., Kumar, S., Nayak, B.B., 2016. Isolation and characterization of *Arcobacter* spp. from fresh seafood and the aquatic environment. Int. J. Food Microbiol. 232 (2), 87–89.

Laplace-Treyture, C., Feret, T., 2016. Performance of the Phytoplankton Index for Lakes (IPLAC): a multimetric phytoplankton index to assess the ecological status of water bodies in France. Ecol. Indic. 69, 686–698.

Lee, J.-H., Kim, Y.-G., Cho, M.H., Lee, J., 2014. ZnO nanoparticles inhibit Pseudomonas aeruginosa biofilm formation and virulence factor production. Microbiol. Res. 169 (12), 888–896.

Lei, Y., Zhang, G., Ai, C., Zhuang, S., 2016. Bioleaching of sphalerite by the native mesophilic iron-oxidizing bacteria from a lead-zinc tailing. Procedia Environ. Sci. 31, 554–559.

Lendowski, L., Färber, H., Holy, A., Darius, A., Ehrich, B., Wippermann, C., Küfner, B., Exner, M., 2015. Accidental contamination of a German town's drinking water with sodium hydroxide. Int. J. Hyg. Environ. Health 218 (3), 366–369.

Lestari, R.A.S., Sediawan, E.B., Syamsiah, S., Sarto, Teixeira, J.A., 2016. Hydrogen sulfide removal from biogas using a salak fruit seeds packed bed reactor with sulfur oxidizing bacteria as biofilm. J. Environ. Chem. Eng. 4 (2), 2370–2377.

Leung, Y.H., Ng, A.M., Xu, X., Shen, Z., Gethings, L.A., Wong, M.T., Chan, C.M., Guo, M.Y., Ng, Y.H., Djurišić, A.B., Lee, P.K., Chan, W.K., Yu, L.H., Phillips, D.L., Ma, A.P., Leung, F.C., 2014. Mechanisms of antibacterial activity of MgO: non-ROS mediated toxicity of MgO nanoparticles towards *Escherichia coli*. Small 10 (6), 1171–1183.

Li, X., Pang, R., Li, J., Sun, X., Shen, J., Han, W., Wang, L., 2013. In situ formation of Ag nanoparticles in PVDF ultrafiltration membrane to mitigate organic and bacterial fouling. Desalination 324, 48–56.

Li, K., Chen, C., Chen, C., Wang, Y., Wei, Z., Pan, W., Song, T., 2015. Magnetosomes extracted from *Magnetospirillum magneticum* strain AMB-1 showed enhanced peroxidase-like activity under visible-light irradiation. Enzyme Microb. Technol. 72, 72–78.

Li, T., Zhang, Y., Shi, M., Pei, G., Chen, L., Zhang, W., 2016a. A putative magnesium transporter Slr1216 involved in sodium tolerance in *Cyanobacterium synechocystis* sp. PCC 6803. Algal Res. 17, 202–210.

Li, Yong, Zhang, Ruofang, Tian, Xike, Yang, Chao, Zhou, Zhaoxin, 2016b. Facile synthesis of Fe_3O_4 nanoparticles decorated on 3D graphene aerogels as broad-spectrum sorbents for water treatment. Appl. Surf. Sci. 369, 11–18.

Liang, Y., Li, D., Zhang, X., Zeng, H., Yang, Z., Cui, S., Zhang, J., 2015. Stability and nitrite-oxidizing bacteria community structure in different high-rate CANON reactors. Biores. Technol. 175, 189–194.

Liu, S., Fang, F., Wu, J., Zhang, K., 2015. The anti-biofouling properties of thin-film composite nanofiltration membranes grafted with biogenic silver nanoparticles. Desalination 375, 121–128.

Liu, Y., Alessi, D.S., Owttrim, G.W., Kenney, J.P.L., Zhou, Q., Lalonde, S.V., Konhauser, K.O., 2016a. Cell surface acid-base properties of the *Cyanobacterium synechococcus*: influences of nitrogen source, growth phase and N:P ratios. Geochim. Cosmochim. Acta 187, 179–184.

Liu, L., Zhu, G., Liu, Z., Gao, C., 2016b. Effect of MCM-48 nanoparticles on the performance of thin film nanocomposite membranes for reverse osmosis application. Desalination 394, 72–82.

Ma, W., Soroush, A., Luong, T.V.A., Brennan, G., Rahaman, M.S., Asadishad, B., Tufenkji, N., 2016. Spray- and spin-assisted layer-by-layer assembly of copper nanoparticles on thin-film composite reverse osmosis membrane for biofouling mitigation. Water Res. 99, 188–199.

MacGregor, B.J., Biddle, J.F., Harbort, C., Matthysse, A.G., Teske, A., 2013. Sulfide oxidation, nitrate respiration, carbon acquisition, and electron transport pathways suggested by the draft genome of a single orange Guaymas Basin Beggiatoa (Cand. *Maribeggiatoa*) sp. filament. Mar. Genom. 11, 53–65.

Madrakian, T., Afkhami, A., Zadpour, B., Ahmadi, M., 2015. New synthetic mercaptoethylamino homopolymer-modified maghemite nanoparticles for effective removal of some heavy metal ions from aqueous solution. J. Ind. Eng. Chem. 21, 1160–1166.

Magnuson, A., Krassen, H., Stensjö, K., Ho, F.M., Styring, S., 2011. Modeling Photosystem I with the alternative reaction center protein PsaB2 in the nitrogen fixing cyanobacterium *Nostoc punctiforme*. Biochim. Biophys. Acta (BBA). 1807 (9), 1152–1161.

Mahanta, R., Chowdhury, J., Nath, H.K., 2016. Health costs of arsenic contamination of drinking water in Assam, India. Econ. Anal. Policy 49, 30–42.

Martínez-Alonso, M., Van Bleijswijk, J., Gaju, N., Muyzer, G., 2005. Diversity of anoxygenic phototrophic sulfur bacteria in the microbial mats of the Ebro Delta: a combined morphological and molecular approach. FEMS Microbiol. Ecol. 52 (3), 339–350.

Martins, E., Figueras, A., Novoa, B., Serrão Santos, R., Moreira, R., Bettencourt, R., 2014. Comparative study of immune responses in the deep-sea hydrothermal vent mussel *Bathymodiolus azoricus* and the shallow-water mussel *Mytilus galloprovincialis* challenged with Vibrio bacteria. Fish Shellfish Immunol. 40 (2), 485–499.

Mata, A.T., Ferreira, J.P., Oliveira, B.R., Batoréu, M.C., BarretoCrespo, M.T., Pereira, V.J., Bronze, M.R., 2015. Bottled water: analysis of mycotoxins by LC–MS/MS. Food Chem. 176, 455–464.

McCourt, R., 2016. Archaeplastida: diversification of red algae and the green plant lineage. Encycl. Evolution. Biol. 2016, 101–106.

Metcalfe, D., Jarvis, P., Rockey, C., Judd, S., 2016. Pre-treatment of surface waters for ceramic microfiltration. Sep. Purif. Technol. 162, 173–180.

Mitbavkar, S., Rajaneesh, K.M., Anil, A.C., Sundar, D., 2012. Picophytoplankton community in a tropical estuary: detection of *Prochlorococcus*-like populations. Estuar. Coast. Shelf Sci. 107, 159–164.

Mohamed, A.A., Eweda, W.E.E., Heggo, A.M., Hassan, E.A., 2014. Effect of dual inoculation with arbuscular mycorrhizal fungi and sulphur-oxidising bacteria on onion (*Allium cepa* L.) and maize (*Zea mays* L.) grown in sandy soil under green house conditions. Ann. Agric. Sci. 59 (1), 109–118.

Mota, R., Pereira, S.B., Meazzini, M., Fernandes, R., Santos, A., Evans, C.A., De Philippis, R., Wright, P.C., Tamagnini, P., 2015. Effects of heavy metals on *Cyanothece* sp. CCY 0110 growth, extracellular polymeric substances (EPS) production, ultrastructure and protein profiles. J. Proteom. 120, 75–94.

Naeher, S., Suga, H., Ogawa, N.O., Takano, Y., Schubert, C.J., Grice, K., Ohkouchi, N., 2016. Distributions and compound-specific isotopic signatures of sedimentary chlorines reflect the composition of photoautotrophic communities and their carbon and nitrogen sources in Swiss lakes and the Black Sea. Chem. Geol. 441, 138–147.

Nakajima, T., Fujikawa, Y., Matsubara, T., Karita, M., Sano, A., 2015. Differentiation of a free-living alga into forms with ecto- and endosymbiotic associations with heterotrophic organisms in a 5-year microcosm culture. Biosystems 131, 9–21.

Ng, Y.H., Leung, Y.H., Liu, F.Z., Ng, A.M.C., Gao, M.H., Chan, C.M.N., Djurišić, A.B., Leung, F.C.C., Chan, W.K., 2013. Antibacterial activity of ZnO nanoparticles under ambient illumination—the effect of nanoparticle properties. Thin Solid Films 542, 368–372.

Nishigaya, Y., Fujimoto, Z., Yamazaki, T., 2016. Optimized inhibition assays reveal different inhibitory responses of hydroxylamine oxidoreductases from beta- and gamma-proteobacterial ammonium-oxidizing bacteria. Biochem. Biophys. Res. Commun. 476 (3), 127–133.

Novoveská, L., Zapata, A.K.M., Zabolotney, J.B., Atwood, M.C., Sundstrom, E.R., 2016. Optimizing microalgae cultivation and wastewater treatment in large-scale offshore photobioreactors. Algal Res. 18, 86–94.

Pales Espinosa, E., Tanguy, A., Le Panse, S., Lallier, F., Allam, B., Boutet, I., 2013. Endosymbiotic bacteria in the bivalve *Loripes lacteus*: localization, characterization and aspects of symbiont regulation. J. Exp. Mar. Biol. Ecol. 448, 327–336.

Park, S., Lee, S., Kim, B., Lee, S., Lee, J., Sim, S., Gu, M., Yi, J., Lee, J., 2012. Toxic effects of titanium dioxide nanoparticles on microbial activity and metabolic flux. Biotechnol. Bioprocess Eng. 17, 276–282.

Park, S.-H., Kim, S.H., Park, S.-J., Ryoo, S., Woo, K., Lee, J.S., Kim, T.-S., Park, H.-D., Park, H., Park, Y.-I., Cho, J., Lee, J.-H., 2016. Direct incorporation of silver nanoparticles onto thin-film composite membranes via arc plasma deposition for enhanced antibacterial and permeation performance. J. Membr. Sci. 513, 226–235.

Peng, W.-f., Huang, C., Chen, X.-f., Xiong, L., Chen, X.-d., Chen, Y., Ma, L.-l., 2013. Microbial conversion of wastewater from butanol fermentation to microbial oil by oleaginous yeast *Trichosporon dermatis*. Renew. Energy 55, 31–34.

Perez-Garcia, O., Villas-Boas, S.G., Swift, S., Chandran, K., Singhal, N., 2014. Clarifying the regulation of NO/N_2O production in *Nitrosomonas europaea* during anoxic–oxic transition via flux balance analysis of a metabolic network model. Water Res. 60, 267–277.

Perni, Stefano, Hakala, Veera, Prokopovich, Polina, 2014. Biogenic synthesis of antimicrobial silver nanoparticles capped with L-cysteine. Colloid. Surf. A 460, 219–224.

Prayitno, S.B., Sarwan, S., 2015. The diversity of gut bacteria associated with milkfish (*Chanos Chanos Forsksal*) from Northern Coast of Central Java, Indonesia. Procedia Environ. Sci. 23, 375–384.

Prevost, B., Goulet, M., Lucas, F.S., Joyeux, M., Moulin, L., Wurtzer, S., 2016. Viral persistence in surface and drinking water: suitability of PCR pre-treatment with intercalating dyes. Water Res. 91, 68–76.

Qing, W., Yue, H., Shuqun, L., Sen, P., Huabing, Z., 2016. Microbial mechanisms of using enhanced ecological floating beds for eutrophic water improvement. Biores. Technol. 211, 451–456.

Ragab, D., Gomaa, H.G., Sabouni, R., Salem, M., Ren, M., Zhu, J., 2016. Micropollutants removal from water using microfiltration membrane modified with ZIF-8 metal organic frameworks (MOFs). Chem. Eng. J. 300, 273–279.

Ramachandra, T.V., 2005. Aquatic ecosystems: conservation, restoration and management. In: Ramachandra, T.V., Ahalya, N., Murthy, C.R. (Eds.), Aquatic Ecosystems: Conservation, Restoration and Management. Capital Publishing Company, India.

Ramanan, R., Kang, Z., Kim, B.H., Cho, D.H., Jin, L., Oh, H.M., Kim, H.S., 2015. Phycosphere bacterial diversity in green algae reveals an apparent similarity across habitats. Algal Res. 8, 140–144.

Ramanan, R., Kim, B.H., Cho, D.H., Oh, H.M., Kim, H.S., 2016. Algae–bacteria interactions: evolution, ecology and emerging applications. Biotechnol. Adv. 34 (1), 14–29.

Ramyadevi, J., Jeyasubramanian, K., Marikani, A., Rajakumar, G., Rahuman, A.A., 2012. Synthesis and antimicrobial activity of copper nanoparticles. Mater. Lett. 71, 114–116.

Razzaz, A., Ghorban, S., Hosayni, L., Irani, M., Aliabadi, M., 2016. Chitosan nanofibers functionalized by TiO_2 nanoparticles for the removal of heavy metal ions. J. Taiwan Inst. Chem. Eng. 58, 333–343.

Salas-Massó, N., Andree, K.B., Furones, M.D., Figueras, M.J., 2016. Enhanced recovery of *Arcobacter* spp. using NaCl in culture media and re-assessment of the traits of *Arcobacter marinus* and *Arcobacter halophilus* isolated from marine water and shellfish. Sci. Tot. Environ. 566–567, 1355–1361.

Salmaso, N., Tolotti, M., 2009. Other *Phytoflagellates* and groups of lesser importance. Ref. Module Earth Syst. Environ. Sci. 2009, 174–183.

Sarris, D., Matsakas, L., Aggelis, G., Koutinas, A.A., Papanikolaou, S., May 2014. Aerated vs non-aerated conversions of molasses and olive mill wastewaters blends into bioethanol by *Saccharomyces cerevisiae* under non-aseptic conditions. Ind. Crop. Prod. 56, 83–93.

Sati, S.C., Singh, L., 2014. Bioactivity of root endophytic freshwater hyphomycetes *Anguillospora longissima* (Sacc. &Syd.) Ingold. Sci. World J..

Shanan, S., Abd, H., Bayoumi, M., Saeed, A., Sandström, G., 2015. Prevalence of protozoa species in drinking and environmental water sources in Sudan. BioMed Res. Int..

Shankar, S., Rhim, J.-W., 2014. Effect of copper salts and reducing agents on characteristics and antimicrobial activity of copper nanoparticles. Mater. Lett. 132, 307–311.

Shayan, S.I., Agblevor, F.A., Bertin, L., Sims, R.C., 2016. Hydraulic retention time effects on wastewater nutrient removal and bioproduct production via rotating algal biofilm reactor. Biores. Technol. 211, 527–533.

Sheng, C., Nnanna, A.G.A., Liu, Y., Vargo, J.D., 2016. Removal of trace pharmaceuticals from water using coagulation and powdered activated carbon as pretreatment to ultrafiltration membrane system. Sci. Tot. Environ. 550, 1075–1083.

Si, Y., Zou, Y., Liu, X., Si, X., Mao, J., 2015. Mercury methylation coupled to iron reduction by dissimilatory iron-reducing bacteria. Chemosphere 122, 206–212.

Soares Gomes, F., VianaPontual, E., BreitenbachBarroso Coelho, L.C., GeudesPaiva, P.M., 2014. Saprophytic, symbiotic and parasitic bacteria: importance to environment. Biotechnol. Appl. Biocontrol. Adv. Res. 2 (5), 250–265.

Song, Z., Fathizadeh, M., Huang, Y., Chu, K.H., Yoon, Y., Wang, L., Xu, W.L., Yu, M., 2016. TiO_2 nanofiltration membranes prepared by molecular layer deposition for water purification. J. Membr. Sci. 510, 72–78.

Sonkusre, P., Cameotra, S.S., 2015. Biogenic selenium nanoparticles inhibit *Staphylococcus aureus* adherence on different surfaces. Colloids Surf. B 136, 1051–1057.

Sorensen, J.P.R., Lapworth, D.J., Marchant, B.P., Nkhuwa, D.C.W., Pedley, S., Stuart, M.E., Bell, R.A., Chirwa, M., Kabika, J., Liemisa, M., Chibesa, M., 2015. In-situ tryptophan-like fluorescence: a real-time indicator of faecal contamination in drinking water supplies. Water Res. 81, 38–46.

Spieck, E., Keuter, S., Wenzel, T., Bock, E., Ludwig, W., 2014. Characterization of a new marine nitrite oxidizing bacterium, *Nitrospina watsonii* sp. nov., a member of the newly proposed phylum "Nitrospinae". System. Appl. Microbiol. 37 (3), 170–176.

Sri-indrasutdhi, V., Boonyuen, N., Suetrong, S., Chuaseeharonnachai, C., Sivichai, S., Gareth Jones, E.B., 2010. Wood-inhabiting freshwater fungi from Thailand: *Ascothailandia grenadoidia* gen. et sp. nov., *Canalisporium grenadoidia* sp. nov. with a key to *Canalisporium* species (Sordariomycetes, Ascomycota). Mycoscience 51 (6), 411–420.

Stabili, L., Gravili, C., Tredici, S.M., Boero, F., Alifano, P., 2011. Association of a luminous *Vibrio* sp., taxonomically related to *Vibrio harveyi*, with Clytialinearis (Thornely, 1900) (Hydrozoa, Cnidaria). J. Experiment. Mar. Biol. Ecol. 396 (2), 77–82.

Sun, R., Gao, Y.X., Shen, K.Z., Xu, Y.B., Wang, C., Liu, H.Y., Dong, J.Y., 2011. Antimicrobial metabolites from the aquatic fungus *Delitschiacorticola*. Phytochem. Lett. 4(2), 101–105.

Sun, X.-F., Qin, J., Xia, P.-F., Guo, B.-B., Yang, C.-M., Song, C., Wang, S.-G., 2015. Graphene oxide–silver nanoparticle membrane for biofouling control and water purification. Chem. Eng. J. 281, 53–59.

Suzuki, T., Okamura, Y., Arakaki, A., Takeyama, H., Matsunaga, T., 2007. Cytoplasmic ATPase involved in ferrous ion uptake from Magnetotactic bacterium *Magnetospirillum magneticum* AMB-1. FEBS Lett. 581 (18), 3443–3448.

Talukder, M.R.R., Rutherford, S., Phung, D., Islamb, M.Z., Chu, C., 2016. The effect of drinking water salinity on blood pressure in young adults of coastal Bangladesh. Environ. Pollut. 214, 248–254.

Thuptimdang, P., Limpiyakorn, T., McEvoy, J., Prüß, B.M., Khan, E., 2015. Effect of silver nanoparticles on *Pseudomonas putida* biofilms at different stages of maturity. J. Hazard. Mater. 290, 127–133.

Tran, P.A., Webster, T.J., 2011. Selenium nanoparticles inhibit *Staphylococcus aureus* growth. Int. J. Nanomed. 6, 1553–1558.

Truong, H.Y.T., Chen, Y.W., Belzile, N., 2013. Effect of sulfide, selenite and mercuric mercury on the growth and methylation capacity of the sulfate reducing bacterium *Desulfovibrio desulfuricans*. Sci. Tot. Environ. 449, 373–384.

Tryggvesson, A., Ståhlberg, F.M., Töpel, M., Tanabe, N., Mogk, A., Clarke, A.K., 2015. Characterization of $ClpS_2$, an essential adaptor protein for the cyanobacterium *Synechococcus elongatus*. FEBS Lett. 589 (24, Part B), 4039–4046.

Vekeman, B., Hoefman, S., De Vos, P., Spieck, E., Heylen, K., 2013. A generally applicable cryopreservation method for nitrite-oxidizing bacteria. Syst. Appl. Microbiol. 36 (8), 579–584.

Völker, S., Schreiber, C., Kistemann, T., 2016. Modelling characteristics to predict *Legionella* contamination risk—surveillance of drinking water plumbing systems and identification of risk areas. Int. J. Hyg. Environ. Health 219 (1), 101–109.

von Sperling, M., 2007. Basic Principles of Wastewater Treatment, in Biological Wastewater Treatment Series. IWA Publishing, India, pp. 1–22.

Wahman, D.G., Speitel, Jr., G.E., 2015. Hydroxylamine addition impact to *Nitrosomonas europaea* activity in the presence of monochloramine. Water Res. 68, 719–730.

Wang, J., Zhou, W., Yang, H., Wang, F., Ruan, R., 2015. Trophic mode conversion and nitrogen deprivation of microalgae for high ammonium removal from synthetic wastewater. Biores. Technol. 196, 668–676.

Wang, L., He, H., Zhang, C., Sun, L., Liu, S., Wang, S., 2016. Antimicrobial activity of silver loaded MnO_2 nanomaterials with different crystal phases against *Escherichia coli*. J. Environ. Sci. 41, 112–120.

Wang, W., Yu, J., An, W., Yang, M., 2016a. Occurrence and profiling of multiple nitrosamines in source water and drinking water of China. Sci. Tot. Environ., 489–495, 551–552.

Wang, Y., Liu, X.H., Si, Y.B., Wang, R.F., 2016b. Release and transformation of arsenic from As-bearing iron minerals by Fe-reducing bacteria. Chem. Eng. J. 295, 29–38.

Weinrich, L., LeChevallier, M., Haas, C.N., 2016. Contribution of assimilable organic carbon to biological fouling in seawater reverse osmosis membrane treatment. Water Res. 101, 203–213.

West, D.M., Wu, Q., Donovan, A., Shi, H., Ma, Y., Jiang, H., Wang, J., 2016. N-nitrosamine formation by monochloramine, free chlorine, and peracetic acid disinfection with presence of amine precursors in drinking water system. Chemosphere 153, 521–527.

Wilson, M.A., Hodgson, D.A., Keely, B.J., 2004. Structural variations in derivatives of the Bacteriochlorophylls of Chlorobiaceae: impact of stratigraphic resolution on depth profiles as revealed by methanolysis. Org. Geochem. 35 (11-12), 1299–1307.

Wu, H., Niu, X., Yang, J., Wang, C., Lu, M., 2016a. Retentions of bisphenol A and norfloxacin by three different ultrafiltration membranes in regard to drinking water treatment. Chem. Eng. J. 294, 410–416.

Wu, Q., Ye, Y., Li, F., Zhang, J., Guo, W., 2016b. Prevalence and genetic characterization of *Pseudomonas aeruginosa* in drinking water in Guangdong Province of China. LWT Food Sci. Technol. 69, 24–31.

Xin, C., Addy, M.M., Zhao, J., Cheng, Y., Cheng, S., Mu, D., Liu, Y., Ding, R., Chen, P., Ruan, R., 2016. Comprehensive techno-economic analysis of wastewater-based algal biofuel production: a case study. Biores. Technol. 211, 584–593.

Xu, G., Xu, H., 2016. Can annual cyclicity of protozoan communities reflect water quality status in coastal ecosystems? Ecol. Indic. 67, 734–740.

Young, J.D., Shastri, A.A., Stephanopoulos, G., Morgan, J.A., 2011. Mapping photoautotrophic metabolism with isotopically nonstationary 13C flux analysis. Metabol. Eng. 13 (6), 656–665.

Yu, L., Hao, G., Gu, J., Zhou, S., Zhang, N., Jiang, W., 2015. Fe_3O_4/PS magnetic nanoparticles: synthesis, characterization and their application as sorbents of oil from waste water. J. Magnetism Magnetic Mater. 394, 14–21.

Yurekli, Y., 2016. Removal of heavy metals in wastewater by using zeolite nano-particles impregnated polysulfone membranes. J. Hazard. Mater. 309, 53–64.

Zhang, M., Zhang, K., De Gusseme, B., Verstraete, W., 2012. Biogenic silver nanoparticles (bio-Ag_0) decrease biofouling of bio-Ag_0/PES nanocomposite membranes. Water Res. 46 (7), 2077–2087.

Zhang, L., Qin, J., Zhang, Z., Li, Q., Huang, J., Peng, X., Qing, L., Liang, G., Liang, L., Huang, Y., Yang, X., Zou, Y., 2016a. Concentrations and potential health risks of methyl tertiary-butyl ether (MTBE) in air and drinking water from Nanning, South China. Sci. Tot. Environ. 541, 1348–1354.

Zhang, M., Sun, Y., Chen, L., Cai, C., Qiao, F., Du, Z., Li, E., 2016b. Symbiotic bacteria in gills and guts of Chinese mitten crab (*Eriocheir sinensis*) differ from the free-living bacteria in water. PLoS One 11 (1).

Zhao, F.Y., An, Q.F., Ji, Y.L., Gao, C.J., 2015. A novel type of polyelectrolyte complex/MWCNT hybrid nanofiltration membranes for water softening. J. Membr. Sci. 492, 412–421.

Zhen-Xing, T., Bin-Feng, L., 2014. MgO nanoparticles as antibacterial agent: preparation and activity. Braz. J. Chem. Eng. 31 (3), 591–601.

Zhou, Y., Zhang, Y., Jeppesen, E., Murphy, K.R., Shi, K., Liu, M., Liu, X., Zhu, G., 2016. Inflow rate-driven changes in the composition and dynamics of chromophoric dissolved organic matter in a large drinking water lake. Water Res. 100, 211–221.

Zhu, L.-J., Zhu, L.-P., Zhang, P.-B., Zhu, B.-K., Xu, Y.-Y., 2016. Surface zwitterionicalization of poly(vinylidene fluoride) membranes from the entrapped reactive core–shell silica nanoparticles. J. Colloid Interface Sci. 468, 110–119.

8

POLYMERIC MATERIALS FOR METAL SORPTION FROM HYDRIC RESOURCES

Arsénio de Sá, Ana S. Abreu, Isabel Moura, Ana Vera Machado

University of Minho, Department of Polymer Engineering (DEP), Institute of Polymers and Composites (IPC), Institute of Nanostructures, Nanomodelling and Nanofabrication (i3N), Campus of Azurém, Guimarães, Portugal

1 Introduction

Pure water is crucial for global population. However, there are many sources of pollutants that contribute to the upscaling contamination of hydric resources. Among the pollutants, heavy metals are the highest concern since they are responsible for several diseases and disorders in living beings. A wide range of industrial activities, such as mining, metal processing, petroleum refining, electroplating, battery manufacturing, textiles, tanneries, pigment manufacture, paint manufacture, printing, photography industries, and pesticides, are responsible sources for the releasing of metals into the environment (Ahmaruzzaman, 2011). Thus, it is of extreme importance to remove these pollutants from water for its consumption. Several physical and chemical methods, such as chemical precipitation, filtration, coagulation, membrane processes, ion exchange, and adsorption have been used to remove metal ions from water (Kołodyńska, 2013). Nevertheless, these methods have limitations. For instance, when the metal's concentration is in the range of 1–100 mg/L, precipitation and coagulation lose effectiveness and became more expensive. The high costs, associated to the complexity and low removal efficiency, limit the use of membrane processes for heavy metal removal (Mark et al., 2006). Therefore, among the techniques, the adsorption process can be highlighted.

Adsorption is a highly efficient method that provides flexibility in the design and it is easy to operate (Zhong et al., 2014). Activated carbon is one of the most widely used materials for this purpose, but it still suffers from costly regeneration and high attrition

Water Purification. http://dx.doi.org/10.1016/B978-0-12-804300-4.00008-3

rate (Pan et al., 2009). Due to their vast surface area, perfect mechanical rigidity, adjustable surface chemistry and pore size distribution, and feasible regeneration under mild conditions, polymeric adsorbents emerged as potential alternatives to activated carbon (Pan et al., 2009). Polymers with chelating properties are of great interest since they are suitable for metal-ion removal by ultrafiltration (Barakat and Schmidt, 2010) and can also be incorporated in a solid support, such as membranes (Bessbousse et al., 2012). Several polymers based on polysaccharides, specially chitosan, are reported as metal sorbents (Pontoni and Fabbricino, 2012), but more can be found with different chelating groups (Repo et al., 2013; Biçak et al., 2000; Hussein et al., 2014; Rahman et al., 2014; Reddy and Reddy, 2003).

Fig. 8.1 depicts the increased interest in the removal of several metal ions from hydric resources. It is the result of searching in Scopus database using the following combination of words, present in the publication title: "metal or symbol (i.e., lead or Pb)," "water," and "sorption or removal." Although it may only be

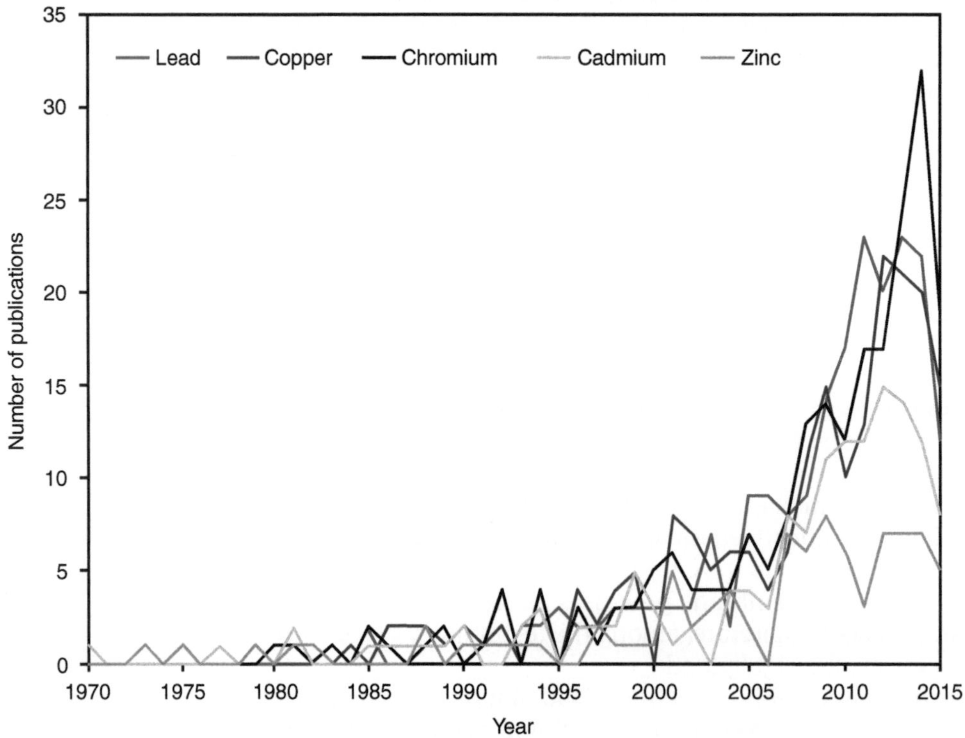

Figure 8.1. Evolution of the publication number along the years in the field of water remediation, specifically in lead, copper, chromium, and cadmium removal. Data obtained on July 27, 2015.

representative of this research field, since it is dependent on the searching keywords that were used, it is possible to observe an increasing number of publications from the 1990s to 2014. Comparing the five metal ions represented in the image, it is possible to notice that zinc, probably due to its well-known biochemistry, is the one that attracts less attention from researchers. The remaining metal ions, lead, copper, and chromium, are those that most concern the scientific society, being in 2014, chromium the one with the highest number of publications.

This chapter will focus on the adsorption process, which is a highly efficient method that provides flexibility in the design and is easy to operate (Zhong et al., 2014). It will be given a description of several phases related with this purification method. Moreover, it will also be described several parameters that can influence the adsorption process and several isotherm models used in the assessment of the sorption efficiency of adsorbent materials.

Polymeric adsorbents arise as potential alternatives to activated carbon, the most used adsorbent. For that reason, this chapter will give special emphasis to polymeric materials with chelating properties, most of them biopolymers obtained from natural wastes, that have been reported as adsorbents of metal ions from hydric resources, specifically for Cu(II), Cd(II), Zn(II), Cr(VI), and Pb(II), and their efficiency as metal adsorbents. Fig. 8.2 depicts a periodic table where lead (Pb) and the transition metals copper (Cu),

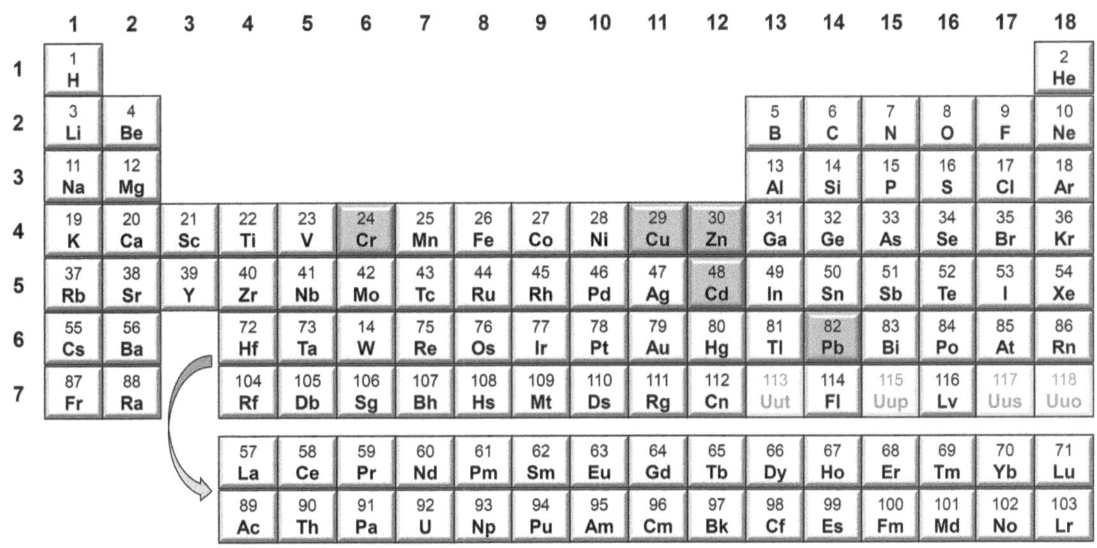

Figure 8.2. Periodic table. Elements discussed in the text are highlighted. Faded elements are under revision by IUPAC. Adapted from IUPAC, 2015. International Union of Pure and Applied Chemistry (IUPAC). Available from: http://www.iupac.org/

cadmium (Cd), zinc (Zn), and chromium (Cr) are highlighted. Since the sorption of heavy metals is related to the formation of coordination compounds, it is important to give same insights about coordination chemistry to better understand the sorption process.

2 The Adsorption Process

The adsorption process is one of the several strategies that can be used in wastewater treatment. The concept of this method is based on removing pollutants from the water by promoting their adsorption at the adsorbent surface. It can be applied for organic or inorganic pollutants and, in both cases, the sorption process and its efficiency can be evaluated using several isotherm models (Section 4).

Wastewater treatment by adsorption involves several steps that can be divided into four stages, according to Fig. 8.3.

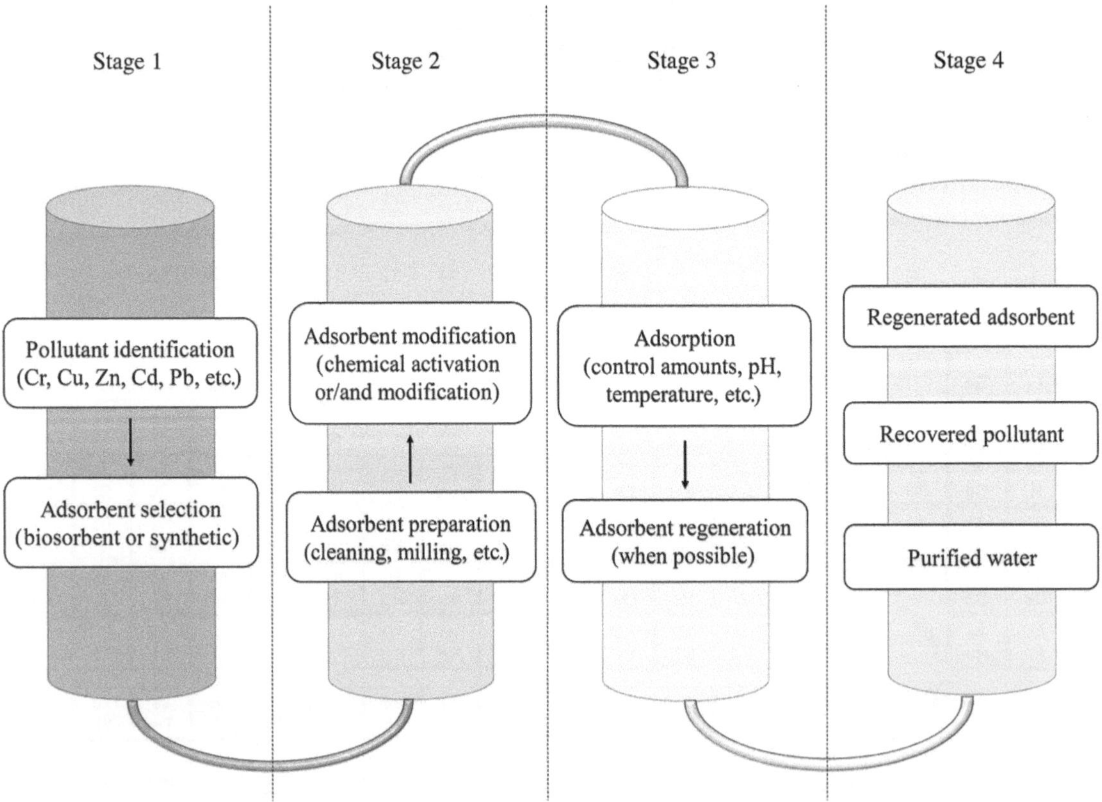

Figure 8.3. Adsorption process divided into stages, from identification of the pollutant to obtaining the toxin-free water.

Stage 1 corresponds to the identification of the problem (pollutant) and its solution (adsorbent). The pollution may be due to one or several toxins, which in the context of this chapter is a heavy metal, that upon identification will allow to select an adsorbent that can be obtained from natural resources, such as plants or crustacean shells wastes, or synthesized.

Stage 2 is associated with the preparation of the adsorbent that should accomplish several requirements. Adsorbents obtained from natural resources need to be cleaned before usage, in order to eliminate contaminants, such as dirt or undesired residues. In this stage, the adsorbents are also submitted to a milling process to obtain materials with controlled size. Additionally, the materials can be submitted to a chemical treatment that will activate functional groups on their surface or introduce new and/or more efficient functional groups. It is crucial to perform a full characterization of the adsorbent material in order to obtain information about its properties, such as chemical composition, surface morphology, and thermal stability, among others, thus its adsorption capacity can be easily predicted and understood.

Stage 3 corresponds to the sorption process itself. In this stage, the adsorption process is studied, testing different conditions, such as amounts of adsorbent and adsorbate, pH, and temperature, among others, in order to optimize the process. As part of this study, isotherm models are used to evaluate the adsorption efficiency. Also in this stage, the possibility of reuse the adsorbent material is evaluated. The coordination compounds formed during the adsorption are hydrolyzed allowing the recovery of the initial adsorbent and to obtain a concentrated solution of the pollutants, which can be properly eliminated or transformed in industries. If the material allows the regeneration, it is necessary to determine the number of cycles sorption/regeneration that can be performed without losing efficiency. It is also important to characterize the regenerated material in order to determine if its properties remain unchanged.

In the final stage, the purification is completed and heavy metal-free water is obtained, as well as the pollutant and the regenerated adsorbent, when regeneration is possible.

3 Insights in Coordination Chemistry

Specifically for inorganic pollutants, the adsorption purification consists of the coordination of the metal ions by the adsorbent and, for that reason, in this section basic insights into coordination chemistry will be given.

Transition metals, and other metals, such as lead or aluminum, are known for their Lewis acids metal ions that have the capacity to form coordination compounds. These coordination compounds correspond to Lewis acid–base pairs in which the Lewis base is a molecule or ion that can donate a pair of electrons, known as ligands.

Both Lewis acids and bases are classified as hard or soft accordingly to Pearson concept. Hard metals are those with high charge/ionic radii ratio and with high polarizing capacity (early to middle d block and f block) and are preferably coordinated by hard donors, which have high electronegativity and are not easily polarized (F, O, N, Cl, HO^-, and π-donors). Soft metals (zero-oxidation state or late d- and p-block metals), have low charge/ionic radii ratio and low polarizing capacity, are preferably coordinated by soft donors (I, S, P, H^-, CO, and alkenes), due to medium electronegativity are easily polarized and are π-acceptors.

Although of being cations, due to their electronegativity, metal ions do not form ionic bonds, since there are pair of electrons being shared with the donor atom. However, it is not shared equally by the two atoms, likewise it happens in a covalent bond. Thus, metal ions form coordination bonds, which are chemicals bonds with intermediate character between ionic and covalent bonds.

Due to the empty orbitals in their valence shell, metal ions can accept several pairs of electrons. Nevertheless, the number of ligands in a coordination compound, which is known by coordination number (CN), is limited by the metal's ionic radii, its electronic configuration, and the ligands' size. Thus, it is possible to found coordination compounds with CN ranging from 2 up to 12, being 6 the most common CN.

The ligands found in coordination compounds can be classified as mono- or polydentates ligands or chelators, accordingly to their denticity (number of donor atoms). Fig. 8.4 depicts the coordination of a metal ion by mono- and bidentate ligands. As it can be seen in Fig. 8.4B, when a polydentate ligand coordinates a metal ion they form a ring resulting in coordination compounds known as chelates. Nevertheless, polydentate ligands can themselves

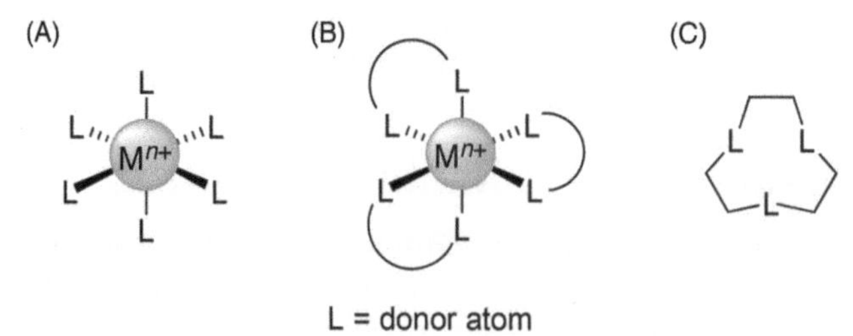

Figure 8.4. Schematic representation of monodentate ligands (A) and bidentate ligands (B) coordinating a metal ion with coordination number 6 and the basic structure of a macrocyclic chelator (C).

be cyclic compounds prior the coordination. They are known as macrocyclic chelators if they fulfill the requirements of having a minimum of nine atoms of which, at least, three of them are donor atoms (Fig. 8.4C). Thus, these type of ligands can be subdivided into open chain or macrocyclic chelators.

Another issue that is important to take into account in what concerns coordination compounds is their thermodynamic stability, usually associated to its formation constant ($\log\beta$). Taking into consideration the variation of the Gibbs free energy (ΔG) Eq. (8.1), where ΔH is the variation of enthalpy, T the temperature, and ΔS the entropy of the system, which gives insights about the spontaneity of the formation reaction, it possible to find out that only three parameters are affecting it. Thus, taking as an example the coordination of a given metal ion (M^{n+}) with CN = 6 by ammonia (NH_3) or ethylenediamine (en) [Eq. (8.2 and 8.3)] in aqueous solution, and performing the reaction at the same temperature, it is easy to understand that enthalpy and entropy assume higher importance. Nevertheless, ammonia and ethylenediamine are very similar and ΔH will not diverge considerably, but they have very different contribution to ΔS, which is in fact the most important parameter involved in the formation of coordination compounds. This effect is even more pronounce for chelators with higher denticity. When the chelator is a macrocyclic ligand, in addition to the entropic contribution, the fact that the chelator has less degrees of conformation, the formation constant is even higher. This is known as the macrocyclic effect.

$$\Delta G = \Delta H - T\Delta S \tag{8.1}$$

$$[M^{n+}(H_2O)_6]^{n+} + 6NH_3 \rightarrow [M^{n+}(NH_3)_6]^{n+} + 6H_2O \tag{8.2}$$

$$[M^{n+}(H_2O)_6]^{n+} + 3en \rightarrow [M^{n+}(en)_3]^{n+} + 6H_2O \tag{8.3}$$

4 Adsorption Isotherms

The phenomenon of retention/release or mobility of a certain substance from aqueous medium to a solid phase at a constant temperature and pH can be described by an invaluable curve, known as adsorption isotherm (Memon et al., 2007). In particular to metal ions, their distribution between the sorbent matrix and the solution, at equilibrium, is crucial for the determination of matrix's maximum adsorption capacity (Karnitz et al., 2007).

Several equilibrium isotherm models have been proposed for the analysis of the adsorption process, including Langmuir, Freundlich, Brunauer–Emmett–Teller, Redlich–Peterson,

Dubinin–Radushkevich, Temkin, Toth, Koble–Corrigan, Sips, Khan, Hill, Flory–Huggins, and Radke–Prausnitz isotherm. These models can be divided according to their fundamental approach: kinetics, thermodynamics, or potential theory (Kumar and Bandyopadhyay, 2006).

Since this chapter is dedicated to polymeric materials used to metal adsorption and these models are reviewed elsewhere (Kumar and Bandyopadhyay, 2006), only the Langmuir and Freundlich isotherm models will be briefly described.

4.1 The Freundlich Isotherm

The Freundlich isotherm, presented by Freundlich in 1906 (Gaballah et al., 1997), was the first isotherm model proposed for sorption processes. It can be applied for nonideal sorption on heterogeneous surfaces, as well as, multilayer sorption.

Eq. (8.4) represents the linear form of the Freundlich equation that can be fitted according to Eq. (8.5), where q is the metal's uptake per unit weight of sorbent (mg/g), C_e corresponds to the residual concentration at equilibrium of the metal ion in solution (mg/L), K_f is the sorption capacity (mg/g), and $1/n$ the sorption intensity (Karnitz et al., 2007). The slope, $1/n$, allows to understand the adsorption process. A variation in the slope between 0 and 1 is associated with a chemisorption process, which is more heterogeneous as the value gets closer to 0. When a slope above 1 is observed, it is consistent with a cooperative adsorption (Kumar and Bandyopadhyay, 2006).

$$q = K_f C_e^{1/n} \tag{8.4}$$

$$\log q = \log K_f + \frac{1}{n} \log C_e \tag{8.5}$$

Due to lack in fundamental thermodynamic basis, since there is no approach to Henry's law at vanishing concentrations, this represents a limitation of this isotherm model (Huang et al., 2013).

4.2 The Langmuir Isotherm

The Langmuir isotherm was first described by Irving Langmuir (Zein et al., 2010) when studying the gas–solid adsorption onto activated carbon. This model proposes a monolayer adsorption occurring at finite and definite localized sites. These adsorption sites are identical and equivalent, and there is no lateral interaction and steric hindrance between adsorbed molecules (Huang et al., 2012). Graphically, Langmuir isotherm displays a typical plateau that corresponds to the equilibrium saturation point at which a molecule occupies one site and there is no further adsorption (Kumar and Bandyopadhyay, 2006).

The nonlinear form of the Langmuir adsorption isotherm is expressed according to Eq. (8.6), in which q_e is the amount adsorbed at equilibrium (mg/g), Q_0 the maximum adsorption capacity of the adsorbent (mg/g), b the Langmuir isotherm constant (dm^3/mg), and C_e the concentration at the equilibrium (mg/L). Langmuir isotherm has four linear forms that are represented by Eqs. (8.7–8.10).

$$q_e = \frac{Q_0 b C_e}{1 + b C_e} \quad (8.6)$$

$$\frac{C_e}{q_e} = \frac{1}{bQ_0} + \frac{C_e}{Q_0} \quad (8.7)$$

$$\frac{1}{q_e} = \frac{1}{Q_0} + \frac{1}{bQ_0 C_e} \quad (8.8)$$

$$q_e = Q_0 - \frac{q_e}{bC_e} \quad (8.9)$$

$$\frac{q_e}{C_e} = bQ_0 - bq_e \quad (8.10)$$

5 Limitations to the Adsorption Process

The adsorption process is affected by several parameters, including the chemical nature of the adsorbent, its surface properties, and operational parameters (Pathak et al., 2015).

As it can be seen in Section 2, the coordination phenomenon of a metal ion is dependent on the type of ligand (mono or polydentate ligand) and on the functional groups present on those ligands. Thus, the chemical nature of the adsorbent is an aspect that should be taken into account to understand which functional groups will provide better results for a given metal ion. More suitable chelating groups will for sure increase the efficiency in removing the heavy metal. Nevertheless, it is also an important aspect of the adsorbents that they can be reused for the larger number of cycles sorption/regeneration as possible, and a highly efficient sorbent material can have its regeneration limited due to the high affinity for the metal ion. Thus, when new adsorbents are developed, it is important to reach an equilibrium between sorption efficiency and regeneration capacity.

In what concerns the surface of the adsorbent, there are several conditions affecting the adsorption process. The surface area of the adsorbent limits the contact area between adsorbent and adsorbate. Nevertheless, a large surface area is not important if the functional groups on the surface are not suitable for coordinating a metal ion. The pH of the surface is also important, since it can compromise the functional groups affinity for the metal ion. Other parameters, such as porosity, particle and bulk densities, and thermal properties can limit the adsorption process of a given material.

The adsorption process is also highly limited by the operation parameters used, such as amount of adsorbent, initial concentration of the adsorbate, particle size, pH of the medium, contact time, temperature, and stirring speed (Pathak et al., 2015). Increasing the amount of adsorbent in the adsorption results in higher number of binding sites for metal ions, which results in the enhancement of the removal percentage. Nevertheless, upon reaching an equilibrium, it is not possible to further increase the removal percentage by adding more adsorbent. The initial concentration of adsorbate influences the mass transfer between the bulk phase and the adsorbent. Thus, using higher initial concentration of adsorbate will result in higher loading capacities. The particle size of the adsorbent controls the surface area, and, as it was already referred, this parameter has high influence on the sorption process. The pH of the medium is of high importance in these type of processes, since it can influence the functional groups affinity for the metal ions and the chemical speciation of the cations. A controlled increasing of the pH can provide better sorption results. However, high pH may also promote the precipitation of other inorganic compounds rather than promoting adsorption of the metal ions. Each metal ion or chelator has its specific properties that influence the chemical processes occurring during adsorption, thus, not all systems respond in a similar way. These dissimilarities may result in systems that require longer time to reach an equilibrium. Therefore, it is of high importance that absorbent and adsorbate can be in contact for a time period suitable that makes possible the occurrence of the adsorption process. It is known that temperature influences the thermodynamic parameters of chemical reactions. At higher temperatures, the medium displays lower viscosity and the adsorbate has higher mobility in solution. This fact enhances the adsorption rates, which will influence the sorption process. Some authors observed in their works that the stirring speed used in the sorption assays influenced the adsorption process. This parameter can overcome the boundary-layer mass-transfer resistance that the adsorbate experiences, resulting in higher adsorption rates (Pathak et al., 2015).

As it is possible to notice, there are several parameters that can influence the adsorption capacity of a given adsorbent. Thus, different results can be achieved just by changing one parameter. Due to this, there are in the literature several works studying the same systems but reporting different efficiencies. This can be attributed to the problems in establishing a standard procedure for the determination of sorption efficiencies. However, it is of high concern that this standard can be established, so more accurate results' comparisons can be performed.

6 Polymeric Materials for Metal Sorption

As known, activated carbon is the most-used material as adsorption matrix for water remediation, however, due to its high cost other materials are attracting the researchers' attention. Among these materials, cellulose, chitin, and starch (Fig. 8.5), the most-abundant biopolymers on earth (Wan Ngah and

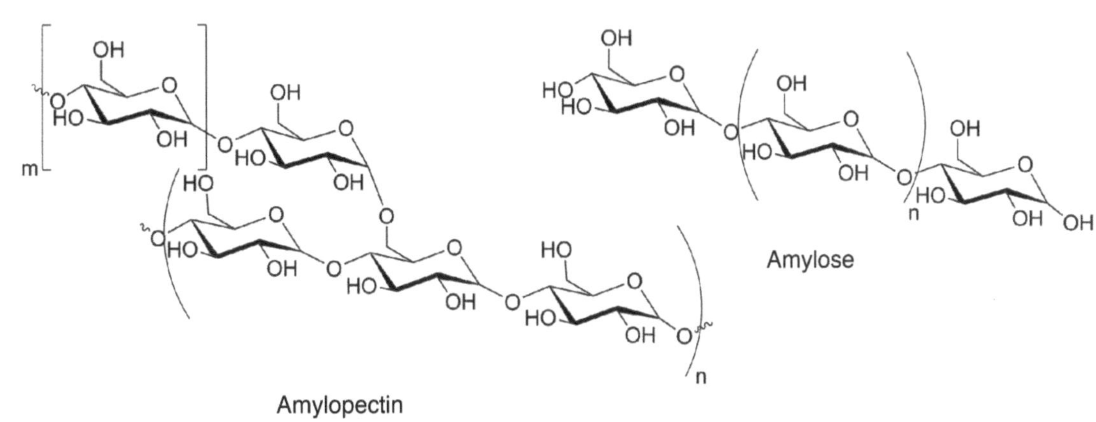

Figure 8.5. Chemical structure of biopolymers present in plants and crustaceous shells wastes used in adsorption studies.

Hanafiah, 2008; Wang and Huang, 2012), are ecofriendly adsorbents that have been tested for metal sorption. Cellulose is one of the main constituents of plants while chitin is found, for instance, in exoskeletons of crustaceous and starch, which is a mixture of amylopectin and amylose, is the energy storage material of living plants.

The excellent adsorption behavior of these polysaccharides is related to the high number of hydroxyl groups that confer high hydrophobicity, the presence of large number of functional groups and their high reactivity, and the flexibility of the polymer's chain (Wang and Huang, 2012).

The advantage of using such materials is related to their availability, associated low costs, being easily regenerated, ecofriendly nature, and the minimization of toxic sludge (Singha and Guleria, 2014).

Cellulosic materials for removal of toxic metal ions include agricultural wastes, such as tea and coffee waste, saw dusts, pinus bark, coconut husk, modified cellulosic materials, corncobs, sugarcane bagasse, agave bagasse, and different agricultural byproducts (Singha and Guleria, 2014).

Although plant wastes can be directly used in the removal of heavy metals from wastewaters, it has several disadvantages. Since these wastes contain soluble organic compounds that are released during the sorption process, they can influence the oxygen content of the water causing problems to aquatic life (Wan Ngah and Hanafiah, 2008). Thus, it is of high concern to avoid this environmental threat, which can be achieved by chemical modification of these wastes. Additionally, this modification can also be of high importance since, depending on the method applied, can result in materials with enhanced sorption capacity (Gaballah et al., 1997). Among others, treatment with alkaline solutions, acid solution, and oxidizing agents are commonly reported in the literature. Usually, treatment of plant wastes with alkaline solutions result in materials with higher chelation capacity than those treated with acids. This is due to the fact that bases promote deprotonation of functional groups, which consequently results in better metal-ion coordinating groups.

6.1 Chromium

Chromium, with the electronic configuration [Ar] $3d^5$ $4s^1$, belongs to group 6, fourth period, of the periodic table. It can be found in several oxidation states, and among them, Cr(III) and Cr(VI) are the most important. This metal is suggested as micronutrient for humans, but it is found in such low quantities in human body that its deficiency determination is based on observation of beneficial changes with chromium supplementation (Vincent, 2000). Cr(VI), due to it solubility, permeability through biological membranes, and interactions with biological molecules, such as proteins and nucleic acids, constitutes higher risk to human health than Cr(III) (El-Sherif et al., 2013). Cr(VI) is associated to several diseases, including liver, kidney, lung, and gastrointestinal cancers (Saygi et al., 2008). Among others, tannery industry widely uses Cr(III) and is responsible for its release in nature. However, it has been shown that Cr(III) can be oxidized to Cr(VI) in the tannery sludge (Apte et al., 2005) that increases health risks. Thus, it is necessary to remove this contaminant from hydric resources, and several studies with polymeric materials can be found in literature. Table 8.1 summarizes the sorption capacity of several materials that have been reported for Cr(III)/(VI) removal.

Gupta and Babu (2009) tested sawdust as biosorbent for removal of Cr(VI), finding a sorption capacity of 41.52 mg/g for this material. Baral et al. (2006) studied sawdust from *Shorea robusta* pretreated with formaldehyde and sulfuric acid. The authors observed that this material has a Cr(VI) sorption capacity of 9.55 mg/g. Sawdust from oak tree was also tested, by Argun et al. (2007), for Cr(VI) sorption. In this case, the authors found a sorption capacity of only 1.74 mg/g. The sorption of Cr(III) was studied by Li et al. (2007) using two plant wastes, sawdust from poplar trees, and peanut husk. In the case of sawdust, the authors did not preform any type of chemical modification, only washed the material with deionized water, and found that it shows a Cr(III) sorption capacity of 5.52 mg/g. In the case of the peanut husk, formalin and sulfuric acid were used to pretreat the wastes. In this case, a sorption capacity of 7.67 mg/g was found. Nasernejad et al. (2005), also for Cr(III), tested carrot residues pretreated with hydrochloric acid as metal-ion biosorbents, and found that this material has a maximum sorption capacity of 45.09 mg/g. Rao et al. (2014) used *Artimisia absinthium*'s seeds as biosorbents for Cr(VI), among others, and found that this material has a sorption capacity of 46.99 mg/g. Deseeded sunflower head was tested by Jain et al. (2013) for Cr(VI) removal, who had shown that, upon treatment with formaldehyde, this waste has a sorption capacity of 7.9 mg/g. A highest sorption capacity of 31.45 mg/g, at 319 K, was evidenced by Doke et al. (2012) when tested wood apple shell for the Cr(VI) biosorption. Chen et al. (2009) pretreated tobacco leaf with several solvents in order to extract its polyphenols. The material obtained was tested for Cr(VI) showing a sorption capacity of 113.2 mg/g. A phosphoric acid treatment, followed by

Table 8.1 Polymeric Materials Tested for Cr(III)/(VI) Removal Discussed in the Text

Material		Q_0 (mg/g)	References
Sawdust			
Sawdust		41.52	Gupta and Babu (2009)
S. robusta treated with formaldehyde in sulfuric acid		9.55	Baral et al. (2006)
Oak treated with hydrochloric acid		1.74	Argun et al. (2007)
Poplar trees		5.52	Li et al. (2007)
Peanut husk treated with formalin in sulfuric acid		7.67	Li et al. (2007)
Carrot treated with hydrochloric acid		45.09	Nasernejad et al. (2005)
A. absinthium seeds		46.99	Rao et al. (2014)
Sunflower head treated with formaldehyde		7.9	Jain et al. (2013)
Wood apple shell		31.45	Doke et al. (2012)
Tobacco-leaf treated with distilled water, ethanol, and acetone		113.2	Chen et al. (2009)
T. nudiflora's fruit peel treated with phosphoric acid		294.12	Bhattacharya et al. (2013)
Olive pomace		4.0	El-Sheikh et al. (2011)
L. japonica		59.35	Wang et al. (2008)
P. yezoensis Ueda		56.32	
Rice bran		58.89	
Wheat bran		40.48	
Pine needles treated with ammonia		477.2	Chauhan et al. (2005)
Cellulose-*g*-poly(GMA)		393.6	
Cellulose-*g*-poly(GMA-*co*-AAm)		983.6	
Cellulose-*g*-poly(GMA-*co*-AAc)		655.6	
Cellulose-*g*-poly(GMA-*co*-AN)		852.4	
Starch-*g*-poly(DMA-*co*-AAc)		6.70	Kolya et al. (2015)
CAPE		48.5	Saleh et al. (2015)
GCB		202.4	Wang and Ge (2014)
CGP		165.6	Karthik and Meenakshi (2014b)
CCGP		179.2	
α-Cyclodextrin	Crosslinked with epichlorohydrin	24.2	Sikder et al. (2013)
β-Cyclodextrin		21.0	
γ-Cyclodextrin		18.6	
PVA/PANI composite		111.2	Karthik and Meenakshi (2014a)

heat exposure, was the method adopted by Bhattacharya et al. (2013) in order to activate the surface of *Trewia nudiflora*'s fruit peel, which was tested for Cr(VI) sorption. The authors were able to prepare a material with a higher sorption capacity than those already mentioned, 294.12 mg/g. Olive pomace was described by El-Sheikh et al. (2011) to have a Cr(VI) sorption capacity of 4.0 mg/g that can be enhanced by pyrolysis of the material. Wang et al. (2008) studied Cr(VI) sorption by four biomaterials: *Laminaria japonica, Porphyra yezoensis Ueda*, rice bran, and wheat bran. While the first-three materials revealed, at 30°C, similar sorption capacity (59.35, 56.32, and 58.89 mg/g, respectively), wheat bran can only reach 40.48 mg/g. Chauhan et al. (2005) investigated the possibility of using cellulose extracted form pine needles and four derived grafted copolymers, cell-*g*-poly(GMA), cell-*g*-poly(GMA-*co*-AAm), cell-*g*-poly(GMA-*co*-AAc), and cell-*g*-poly(GMA-*co*-AN), for the sorption of Cr(VI) and Cu(II) (discussed further in the chapter), among other metal ions. The authors found that the cellulose itself has a sorption capacity of 477.2 mg/g for Cr(VI), which is compromised and decrease to 393.6 mg/g when glycidyl methacrylate (GMA) is grafted onto cellulose surface. Nevertheless, the copolymerization of acrylamide (AAm), acrylic acid (AAc), and acrylonitrile (AN) increased considerably the sorption capacity, achieving 983.6 mg/g in the case of cell-*g*-poly(GMA-*co*-AAm). Kolya et al. (2015) reported the grafting of poly(*N*,*N*-dimethylacrylamide-*co*-acrylic acid) [poly(DMA-*co*-AAc)] onto maize starch. This material revealed a sorption capacity of 6.70 mg/g. Sikder et al. (2013) preformed the crosslinking of the three most-common cyclodextrins (α, β, and γ-cyclodextrin) with epichlorohydrin in order to produce Cr(VI) sorption matrices. The authors showed that as the size of cyclodextrin increases (α to γ) their sorption capacity decreases, obtaining a maximum of 24.2 mg/g for the α-cyclodextrin derivative. Polyvinyl alcohol (PVA) was used by Karthik and Meenakshi (2014a) in the preparation of a composite with polyaniline (PANI) that showed a sorption capacity of 111.2 mg/g for Cr(VI). This group also reported the synthesis of two chitosan derivatives, chitosan–grafted-polyaniline (CGP) and glutaraldehyde crosslinked-chitosan–grafted-polyaniline (CCGP) (Karthik and Meenakshi, 2014b). These two polymers revealed higher sorption capacity than PVA/PANI composite, being 165.6 and 179.2 mg/g the sorption capacities determined for CGP and CCGP, respectively. More recently, Saleh et al. (2015) published a crosslinked tetrapolymeric anionic polyelectrolyte that was tested for Cr(III) removal. With this polymer, it was possible to obtain a sorption capacity of 48.5 mg/g. Wang and Ge (2014) produced the glutaraldehyde crosslinked chitosan biguanidine (GCB) polymer, for which a sorption capacity of 202.4 mg/g was determined.

6.2 Copper

Copper belongs to group 11, fourth period, of the periodic table. It has the electronic configuration [Ar] $3d^{10}$ $4s^1$ and can be found in two oxidation states, Cu(I) and Cu(II), being the latter the most common. Copper is an essential element that acts as electron donor/acceptor in metalloenzymes in mammalians (Stern et al., 2007). Normally, copper intake occurs via food or water ingestion. Several disorders are related to copper deficiency but high exposure is also a problem of high concern. High concentrations of Cu are associated, among others problems, to Alzheimer's and Wilson diseases (Llanos and Mercer, 2002). Table 8.2 presents some sorbents, either of natural sources or synthetic, proposed for the removal of Cu(II) from wastewaters and their sorption capacity.

There is a high number of studies involving natural sorbents. Wong et al. (2003), in their search for a Cu(II) biosorbent, reported the use of rice husk pretreated with tartaric acid. This material revealed maximum sorption capacity of 31.85 mg/g for Cu(II). Acar and Eren (2006) found that sawdust from the poplar tree Samsun clone can be used to remove Cu(II) from water, reaching a sorption capacity of 5.43 mg/g. However, the authors observed an enhancement to 13.5 mg/g upon pretreat the biosorbent with sulfuric acid. Šćiban et al. (2006) also proposed sawdust as Cu(II) biosorbent. In this case, the material was obtained from poplar and fir tree and tested with and without pretreatment with sodium hydroxide. The nontreated materials revealed similar sorption capacities, 2.54 and 2.35 mg/g, respectively for poplar and fir tree, while the pretreatment increased these values, with higher impact for the fir tree-base material, reaching 12.71 mg/g. Sawdust from poplar tress was also used by Li et al. (2007), without pretreatment, for Cu(II), revealing a sorption capacity of 6.585 mg/g. The same authors also tested peanut husk, which was treated with formalin and sulfuric acid, resulting in a Cu(II) sorbent with sorption capacity of 10.15 mg/g. Horsfall et al. (2006) used cassava (*Manihot sculenta* Cranz) tuber bark waste in the removal of Cu(II) from water. The authors tested the influence of two treatments in the sorption capacity of the material. Treating the cassava wastes with nitric acid allowed the material to reach a sorption capacity of 33.3 mg/g. However, when the cassava wastes were further treated with thioglycollic acid, its sorption capacity was enhanced to 90.9 mg/g. Özer et al. (2004) worked with wheat bran treated with sulfuric acid, observing an enhancement in its sorption capacity from 42.0 to 51.5 mg/g, by increasing the temperature from 20 to 60°C, at pH 5. Chemical

Table 8.2 Polymeric Materials Tested for Cu(II) Removal Discussed in the Text

Material		Q_0 (mg/g)	References
Sawdust			
Poplar tree		5.43	Acar and Eren (2006)
		2.54	Šćiban et al. (2006)
		6.58	Li et al. (2007)
Poplar tree treated with	Sulfuric acid	13.5	Acar and Eren (2006)
	Sodium hydroxide	6.93	Šćiban et al. (2006)
Fir tree		2.35	
Fir tree treated with sodium hydroxide		12.71	
Rice husk treated with tartaric acid		31.85	Wong et al. (2003)
Peanut husk treated with formalin in sulfuric acid		10.15	Li et al. (2007)
Cassava treated with	Nitric acid	33.3	Horsfall et al. (2006)
	Nitric acid and thioglycollic acid	90.9	
Wheat bran treated with sulfuric acid		51.5	Özer et al. (2004)
Sugarcane bagasse treated with	Sodium bicarbonate	114	Karnitz et al. (2007)
	Ethylenediamine	139	
	Trlethylenetetraamlne	133	
Pine needles treated with ammonia		179.6	Chauhan et al. (2005)
Cellulose-*g*-poly(NIPAM)		74.98	Kasgoz et al. (2013)
β-Cyclodextrin crosslinked with epichlorohydrin		111.11	Sikder et al. (2014)
Cellulose-*g*-poly(GMA)		134.8	Chauhan et al. (2005)
Cellulose-*g*-poly(GMA-*co*-AAm)		495.2	
Cellulose-*g*-poly(GMA-*co*-AAc)		912.8	
Cellulose-*g*-poly(GMA-*co*-AN)		213.6	
Cell-*g*-poly(GMA-PEI)		102	Tang et al. (2013)
Chitosan–tripolyphosphate beads		208.3	Wu et al. (2013)
Polyethyleneiminemethylene phosphonic acid		85.69	Ferrah et al. (2011)

modified biomass from sugarcane bagasse was studied by Karnitz et al. (2007) as biosorbent of several metal ions, including Cu(II). With these materials, it was possible to obtained sorption capacities ranging from 114 to 139 mg/g. Chauhan et al. (2005) studied the possibility of using cellulose extracted form pine needles for the sorption of several metal ions including Cu(II) and discovered that this material can reach a sorption capacity of 179.6 mg/g.

Synthetic materials also assume high importance in Cu(II) sorption. Kasgoz et al. (2013) reported a study with a cellulose-based copolymer, the cell-*g*-poly(*N*-isopropyl acrylamide) (NIPAM), in which the copolymer was used in combination with sodium dodecyl benzene sulfonate (SDBS) to remove Cu(II) from aqueous solutions. SDBS was responsible for the coordination of the metal ion and upon the metal complex formation adsorption takes place onto the copolymer surface. With this, the authors were able produce a system with a maximum sorption capacity of 74.98 mg/g. Chauhan et al. (2005) also studied four grafted copolymers derived from the cellulose extracted from pine needles, cell-*g*-poly(GMA), cell-*g*-poly(GMA-*co*-AAm), cell-*g*-poly(GMA-*co*-AAc), and cell-*g*-poly(GMA-*co*-AN). By grafting GMA onto cellulose surface the authors compromised the sorption capacity of their cellulose, reaching 134.8 mg/g. However, further copolymerization of acrylamide (AAm), acrylic acid (AAc), and acrylonitrile (AN) enhanced considerably the sorption capacity of the initial cellulose to 495.2, 912.8, and 213.6 mg/g, respectively. Cell-*g*-poly(GMA) was used by Tang et al. (2013) as an anchor for the immobilization of poly(ethylene imine) (PEI). In this case, an adsorption capacity of 102 mg/g was achieved. Sikder et al. (2014) prepared a water insoluble β-cyclodextrin crosslinked with epichlorohydrin for Cu(II) sorption. This material was found to have a sorption capacity of 111.11 mg/g. Wu et al. (2013) were able to produce a sorbent for Cu(II) that has 208.3 mg/g sorption capacity by forming porous chitosan–tripolyphosphate beads. Ferrah et al. (2011) found that the polymer polyethyleneiminemethylene phosphonic acid can be used to remove Cu(II) from water, showing a sorption capacity of 85.69 mg/g.

6.3 Zinc

Zinc (Zn), belongs to the fourth period, group 12 of the periodic table, and has an electronic configuration [Ar] $3d^{10}$ $4s^2$. Its most common oxidation state is Zn(II), which is considered an essential element for all organisms. In humans, Zn(II) is part of more than 300 metalloenzymes, thus there is a requirement for a regular uptake to maintain an ideal concentration for a proper biochemistry. Nonetheless, injection of high doses of Zn can result in multiple illnesses, such as gastrointestinal disorders, copper deficiency, and immunologic effects (Plum et al., 2010).

There are several anthropogenic activities responsible for releasing Zn(II) into the environment, such as paper production, electroplating, pipes galvanization, and metal alloys production (Sulaiman et al., 2011). Therefore, it is imperative to remove this pollutant from the environment.

In Table 8.3 are summarized several studies involving biopolymers and synthetic that are under investigation for Zn(II) removal from water. Sulaiman et al. (2011) treated Semantan bamboo (*Gigantochloa scortechinii*) with hydrochloric acid and tested its sawdust in Zn(II) removal from water. They found out that this bamboo can reach a maximum sorption capacity, at 30°C, of 107.52 mg/g. Pehlivan et al. (2006) observed that sugar beet pulp can be used in the Zn(II) removal from water and were able to determine a maximum sorption capacity of 35.6 mg/g. Wang et al. (2006) observed a sorption capacity of 14.17 mg/g when tested the efficiency of rice bran in removing Zn(II) from water. Mitić-Stojanović et al. (2012), upon performing acid (nitric acid) and alkaline (sodium hydroxide) washing procedures to *Lagenaria vulgaris*, tested the biomass in Zn(II) sorption. The authors noticed a sorption capacity of 6.41 mg/g. Conrad and Bruun Hansen (2007) investigated the possibility of use coir, fibers from *Coco nucifera*, in metal sorption. Their study revealed that this biomass has a maximum sorption capacity of 8.6 mg/g. Horsfall et al. (2006) tested cassava tuber bark in sorption experiments. The biomass was treated with nitric acid showing a maximum sorption capacity for Zn(II) of 22.2 mg/g. Nevertheless, upon further modification with thioglycollic acid, the resulting material had an enhancement on its efficiency to 88.3 mg/g. Singha and Guleria (2014) extracted cellulose fibers form okra plant and treated it with sodium hydroxide in order to prepare a biosorbent for Zn(II). The authors showed that this cellulose was able to reach a maximum sorption capacity of 16.85 mg/g. Šćiban et al. (2006), in the same study already referred in Section 6.2, proposed poplar and fir tree sawdust for Zn(II) sorption. The biomass were used either with or without sodium hydroxide treatment. The nontreated materials had the same sorption capacity, 0.98 mg/g. However, pretreating the materials with sodium hydroxide increased their efficiency to 15.82 and 13.40 mg/g for poplar and fir tree–derived materials, respectively. Paulino et al. (2013) modified cotton fiber with citric acid and tested the resulting material for removal of several metal ions, including Zn(II), from water. The authors observed that their material has a sorption capacity for Zn(II) of 4.53 mg/g. Nasernejad et al. (2005) tested carrot residues, which were pretreated with hydrochloric acid, as Zn(II) biosorbent. The maximum sorption capacity of this material was found to be 29.61 mg/g. Water hyacinth weed (*Eichhornia crassipes*) biomass, treated with nitric acid, was studied by Mahamadi and Nharingo (2010) as Zn(II)

Table 8.3 Polymeric Materials Tested for Zn(II) Removal Discussed in the Text

Material		Q_0 (mg/g)	References
Sawdust			
Semantan bamboo treated with hydrochloric acid		107.52	Sulaiman et al. (2011)
Poplar tree		0.98	Šćiban et al. (2006)
Poplar tree treated with sodium hydroxide		15.82	
Fir tree		0.98	
Fir tree treated with sodium hydroxide		13.40	
Sugar beet pulp		35.6	Pehlivan et al. (2006)
Rice bran		14.17	Wang et al. (2006)
L. vulgaris treated with nitric acid and sodium hydroxide		6.41	Mitić-Stojanović et al. (2012)
Coir		8.6	Conrad and Bruun Hansen (2007)
Cassava treated with	Nitric acid	22.2	Horsfall et al. (2006)
	Nitric acid and thioglycollic acid	88.3	
Cellulose extracted from okra plant		16.85	Singha and Guleria (2014)
Cotton fiber modified with citric acid		4.53	Paulino et al. (2013)
Carrot treated with hydrochloric acid		29.61	Nasernejad et al. (2005)
Water hyacinth weed treated with nitric acid		12.55	Mahamadi and Nharingo (2010)
Hazelnut shell		9.15	Demirbaş and Karadağ (2013)
Rapeseed		13.859	Paduraru et al. (2015)
A. fatua		23.54	Areco et al. (2013)
6-(2′-aminomethylpyridine)-6-deoxycellulose		4.64	Silva Filho et al. (2013)
Ch/IA/MAA hydrogel		105.5	Milosavljević et al. (2011)
PVA/EDTA resin		38.7	Zhang et al. (2010)
Chlorosulfonated styrene–DVB resin modified with TETA		173.26	Biçak et al. (2000)
Corn-cob cellulose		8.63	Rahman et al. (2014)
Corn-cob cellulose-*g*-PAN		2.67	
Polyamidoxime		190.99	
Amidoximated polymerized banana stem		85.89	Anirudhan et al. (2010)

biosorbent. These authors detected that water hyacinth has a sorption capacity of 12.55 mg/g for this metal ion. Demirbaş and Karadağ (2013) tested the waste of one of the most important agricultural products in Turkey, hazelnut (*Corylus avellana* L.). The authors performed Zn(II) sorption tests using hazelnut shell in order to evaluate its efficiency for this purpose and were able to determine a sorption capacity of 9.15 mg/g. Paduraru et al. (2015) evaluated the efficiency of rapeseed in the removal of Zn(II) from water, finding a maximum sorption capacity of 13.859 mg/g. Dead biomass of *Avena fatua* was used by Areco et al. (2013) in sorption experiments with several metal ions. Specifically for Zn(II), the authors found a maximum sorption capacity of 23.54 mg/g. Silva Filho et al. (2013) performed chemical modification on cellulose surface, inserting 2-aminomethylpyridine units, and obtained the 6-(2′-aminomethylpyridine)-6-deoxycellulose polymer. Upon obtaining this sorbent, the authors have determined its maximum sorption capacity for several metal ions, including zinc, founding a value of 4.64 mg/g for the latter. Milosavljević et al. (2011) prepared pH-sensitive hydrogels based on chitosan, itaconic acid, and methacrylic acid (Ch/IA/MAA) and tested them for Zn(II) sorption. The author found a maximum sorption capacity of 105.5 mg/g, at 25°C, for the hydrogel with Ch/IA/MAA ration 1:1.56:7.5. Zhang et al. (2010) prepared a resin based on PVA and chloromethylated polystyrene grafted by EDTA that was tested as Zn(II) sorbent. The authors concluded that the PVA/EDTA resin has a maximum sorption capacity of 38.7 mg/g. Biçak et al. (2000) modified chlorosulfonated styrene-divinyl benzene (DVB) resin with triethylene tetramine (TETA) in order to prepare a polymer with EDTA-like moieties that could be used to remove metal ions from water. The resulting polymer was tested as sorbent for several toxins and it was determined a maximum sorption capacity of 173.26 mg/g, specifically for Zn(II). Rahman et al. (2014) extracted cellulose from corn cob, performed several chemical modifications to its structure, and tested all the materials as metal sorbents. In the case of Zn(II), the starting cellulose achieved a sorption capacity of 8.63 mg/g. However, when acrylonitrile (AN) was grafted onto the cellulose surface (cellulose-g-PAN) the efficiency of the material dropped to 2.67 mg/g. The last modification performed by the authors, converting cellulose-g-PAN in the polyamidoxime ligand lead to a remarkable enhancement of the maximum sorption capacity to 190.99 mg/g. Anirudhan et al. (2010) performed chemical reactions with banana stem obtaining an amidoximated polymerized banana stem. This material was then used in Zn(II) sorption assays with which was possible to determine a maximum sorption capacity of 85.89 mg/g.

6.4 Cadmium

Cadmium is found in the fifth period, group 12, of the periodic table with electronic configuration [Kr] $4d^{10} 5s^2$. It has two oxidation states, Cd(I) and Cd(II). Cd is a nonessential element for humans, thus human body is not prepared for its proper metabolization. When exposure occurs, cadmium tends to accumulate in organs, resulting in kidneys' damage and provoking harmful effects on red blood cells and nervous system (Brown et al., 2000). Table 8.4 summarizes several polymeric sorbents that have been tested as Cd(II) sorption agents. Kumar and Bandyopadhyay (2006) proposed rice husk for sorption of Cd(II). These authors compared the efficiency of the raw material against it pretreated with different modifying agents. It was observed that by treating the rice husk with 0.5 M NaOH during 4 h the sorption efficiency was increased from 8.58 to 20.54 mg/g. Also with rice husk, Ye et al. (2010) obtained a sorption capacity of 73.96 mg/g for the raw material, while a pretreatment with 1 M NaOH during 3 days increased the value for 125.94 mg/g. The difference in the sorption capacity of both rice husk raw materials may be related to the origin and chemical composition of both wastes. Sawdust is one type of plant waste that attracts the attention from researchers. Memon et al. (2007) reported the use of *Cedrus deodara* wood sawdust. Like Kumar and Bandyopadhyay (2006) observed that the material pretreated with sodium hydroxide is more efficient than the raw material in the Cd(II) sorption. The authors were also able to produce a material with a maximum sorption capacity of 73.62 mg/g. Taty-Costodes et al. (2003) treated sawdust of *Pinus sylvestris* with formaldehyde, in acidic medium, in order to avoid the leaching of phenolic compounds that would stain the water, and obtained a matrix that showed sorption capacity of 9.29 mg/g. Horsfall et al. (2006) performed sorption experiments with cassava tuber bark. The authors treated the wastes with nitric acid and thioglycollic acid. When only nitric acid was applied, the material showed a sorption capacity of 5.88 mg/g. However, further treatment of the wastes with thioglycollic acid enhanced their sorption capacity almost five times, to 26.3 mg/g. Wheat bran, pretreated with sulfuric acid, was proposed by Özer and Pirinççi (2006). The sorption tests with this material showed the highest sorption capacity, 101 mg/g, being achieved at 25°C. Min et al. (2004) reported assays of Cd(II) sorption using juniper (*Juniperous monosperma*) fibers. The authors compared the efficiency of the raw fibers against those treated with sodium hydroxide during 24 h. The results demonstrated that pretreating the fibers improves the sorption capacity from 9.2 to 29.5 mg/g. Corncob was used by Leyva-Ramos et al. (2005) who studied the effect of oxidizing

Table 8.4 Polymeric Materials Tested for Cd(II) Removal Discussed in the Text

Material		Q_0/g)	References
Rice husk		8.58	Kumar and Bandyopadhyay (2006)
		73.96	Ye et al. (2010)
Rice husk treated with	Sodium hydroxide	20.24	Kumar and Bandyopadhyay (2006)
		125.94	Ye et al. (2010)
Rice husk treated with	Sodium bicarbonate	16.18	Kumar and Bandyopadhyay (2006)
	Epichlorohydrin	11.12	
Sawdust			
Cedrus deodar wood treated with sodium hydroxide		73.62	Memon et al. (2007)
P. sylvestris treated with formaldehyde in sulfuric acid		9.29	Taty-Costodes et al. (2003)
Cassava treated with	Nitric acid	5.88	Horsfall et al. (2006)
	Nitric acid and thioglycollic acid	26.3	
Wheat bran treated with sulfuric acid		101	Özer and Pirinççi (2006)
J. monosperma		9.2	Min et al. (2004)
J. monosperma treated with sodium hydroxide		29.5	
Corncob		5.12	Leyva-Ramos et al. (2005)
Corncob treated with	Nitric acid	19.3	
	Citric acid	55.2	
Sugarcane bagasse treated with	Sodium bicarbonate	196	Karnitz et al. (2007)
	Ethylenediamine	164	
	Triethylenetetraamine	313	
Mangosteen (dye removed)		3.15	Zein et al. (2010)
Mangosteen treated with	Calcium hydroxide	39.3	Huang et al. (2013)
	Thiosemicarbazide	39.37	Huang et al. (2012)
	Diethylenetriamine	20.49	Wang and Huang (2012)
Platanus leaf		3.69	Siswoyo et al. (2014)
Platanus leaf treated with	Citric acid	15.31	
	Agar encapsulated	6.89	
H. verticillata treated with hydrochloric acid		37.46	Li et al. (2013)
Mustard husk		42.85	Meena et al. (2008)
Chitosan from crab shell		138.3	Khan et al. (2015)
Glutaraldehyde crosslinked chitosan pyruvic acid		98.04	Boamah et al. (2015)
Chitosan modified with glycidylmethacrylate followed by 1,2-ethanedithiol		229.3	Khan et al. (2015)
Maize straw grafted succinic anhydride		196.1	Guo et al. (2015)
Polystyrene-diethylenetriamine		89.1	Zhang et al. (2013)
Poly(AAm-*co*-SMA)		24.05	Milosavljević et al. (2014)
Starch-*g*-poly(AA)		588	Abdel-Halim and Al-Deyab (2014)

the sorption matrix with nitric acid and citric acid on the sorption efficiency. It was detected that by using citric acid, the content of carboxylic sites on the corncob increased more than what was observed with nitric acid. The modifications increased the sorption capacity of the corncob from 5.12 to 19.3 mg/g, when pretreated with nitric acid, while citric acid pretreatment gave 55.2 mg/g. The introduction of carboxylate and amines groups into sugarcane bagasse was performed by Karnitz et al. (2007), by reacting the material with succinic anhydride followed by sodium bicarbonate, ethylenediamine, or triethylenetetraamine, to produce more-efficient sorption matrices. With carboxylate groups available to coordinate Cd(II) ions, the material revealed a sorption capacity of 196 mg/g, while the introduction of ethylenediamine decreased the efficiency to 164 mg/g. However, increasing the number of amine groups with triethylenetetraamine, the sorption capacity was enhanced to 313 mg/g. Mangosteen shell is also being tested by several research groups as metal-ion sorbents. Zein et al. (2010) found that, upon removal of the dyes, mangosteen shell has a Cd(II) sorption capacity of 3.15 mg/g. In an another study, Wang and Huang (2012) treated the mangosteen shell with diethylenetriamine and found that this increases, to 20.49 mg/g, the sorption capacity. In the same year, Huang et al. (2012) used mangosteen shell treated with thiosemicarbazide and obtained a sorption capacity of 39.37 mg/g. This result is very similar to the one found by Huang et al. (2013) with mangosteen shell treated with calcium hydroxide, 39.3 mg/g. Platanuś leaf was studied by Siswoyo et al. (2014) as it is, treated with citric acid and encapsulated in agar. The authors found that this waste has a sorption capacity of 3.69 mg/g without any treatment. However, treating the leaf powder with citric acid enhanced the sorption of Cd(II) to 15.31 mg/g, being this value reduced to 6.89 mg/g upon encapsulation of the treated material onto agar. Li et al. (2013) proposed the use of *Hydrilla verticillata* as Cd(II) sorbent and have determined a sorption capacity of 37.46 mg/g, upon washing the biomass with 3% HCl. Meena et al. (2008) tested untreated mustard husk for Cd(II) removal from water and found that it has a sorption capacity of 42.85 mg/g. Boamah et al. (2015) used a low molecular weight chitosan modified with pyruvic acid and crosslinked with glutaraldehyde as sorbent for Cd(II). This material was able to reach a sorption capacity of 98.04 mg/g. Khan et al. (2015) performed chemical modifications in chitosan (crab shell) by reacting it with glycidylmethacrylate followed by 1,2-ethanedithiol. This chitosan-based material was tested for Cd(II) removal and its efficiency was compared with the starting material. The unmodified chitosan showed a sorption capacity of 138.3 mg/g while the chemical modifications enhanced this value to 229.3 mg/g.

Guo et al. (2015) reported the grafting of succinic anhydride onto maize straw and subsequent activation with sodium carbonate. The polymer was tested for Cd(II) sorption, revealing a high sorption capacity of 196.1 mg/g. Zhang et al. (2013) modified polystyrene (PS) resin with diethylenetriamine (DETA) producing the new polymer PS-DETA. This new resin was tested for Cd(II) removal and showed maximum sorption capacity of 89.1 mg/g. A hydrogel of poly(acrylamide-*co*-sodium methacrylate) [poly(AAm-*co*-SMA)] was reported by Milosavljević et al. (2014) as Cd(II) sorbent with a sorption capacity of 24.05 mg/g. Abdel-Halim and Al-Deyab (2014) also prepared a hydrogel for Cd(II) removal. In this case, the authors preformed the grafting of polyacrylic acid [poly(AA)] onto starch obtaining the starch-*g*-poly(AA), which was further treated with epichlorohydrin to form the hydrogel. This material was found to have a sorption capacity of 588 mg/g.

6.5 Lead

Lead is a metal that belongs to group 14, sixth period, of the periodic table. It has the electronic configuration [Xe] $4f^{14}$ $5d^{10}$ $6s^2$ $6p^2$. Pb(II) is the most-common oxidation state of this metal. Lead is one of the most-toxic heavy metals and several health problems are associated to lead exposure, including, for instance, dysfunctions in several organs (Kim et al., 2014). A lot of work has been carried out, involving both polymers present in natural wastes and synthetic ones, in Pb(II)'s removal form water. Table 8.5 resumes several examples than can be found in literature. Wong et al. (2003), in the same work already referred to copper, tested rice husk treated with tartaric acid in Pb(II) sorption, and found out a maximum sorption capacity of 120.48 mg/g. Taty-Costodes et al. (2003), in their research with sawdust from *P. sylvestris*, were able to determine a sorption capacity of 9.29 mg/g of Pb(II) upon treating the material with formaldehyde in sulfuric acid. Li et al. (2007) found that sawdust, from poplar tree, can reach a Pb(II) sorption capacity of 21.05 mg/g while peanut husk, treated with formalin in sulfuric acid, can reach 29.14 mg/g. Noeline et al. (2005) treated pseudo stem of *Musa paradisiaca* L. (banana) with formaldehyde and sulfuric acid to produce a biosorbent material for Pb(II). This material, when tested for Pb(II) sorption, revealed a maximum sorption capacity of 91.74 mg/g. In the same study already referred for Cu(II) and Cd(II), Karnitz et al. (2007) also verified the efficiency of sugarcane bagasse with three different treatments for Pb(II) sorption. The authors did not find differences in the sorption capacity of the material treated with sodium bicarbonate and ethylenediamine, determining a value of 189 mg/g. However, increasing the

Table 8.5 Polymeric Materials Tested for Pb(II) Removal Discussed in the Text

Material		Q_0 (mg/g)	References
Rice husk treated with tartaric acid		120.48	Wong et al. (2003)
Sawdust			
P. sylvestris treated with formaldehyde in sulfuric acid		9.29	Taty-Costodes et al. (2003)
Poplar tree		21.05	Li et al. (2007)
Peanut husk treated with formalin in sulfuric acid		29.14	Li et al. (2007)
Pseudo stem of banana treated with formaldehyde and sulfuric acid		91.74	Noeline et al. (2005)
Sugarcane bagasse treated with	Sodium bicarbonate	189	Karnitz et al. (2007)
	Ethylenediamine	189	
	Triethylenetetraamine	313	
Wheat straw modified with urea		31.85	Farooq et al. (2010)
L. vulgaris treated with nitric acid and sodium hydroxide		26.94	Mitić-Stojanović et al. (2012)
Cashew nut shell		28.653	Coelho et al. (2014)
Apple pomace		16.39	Chand and Pakade (2013)
Custard apple		91.74	Isaac and Sivakumar (2013)
PVC membranes with 50% D2EHPA inclusion		60.78	Gherasim et al. (2011)
Poly(AAm-*co*-SMA) crosslinked with PEGDA		68.92	Kalagasidis Krušić et al. (2012)
Cotton-fiber modified with citric acid		21.62	Paulino et al. (2013)
Pigeon-peas hulls treated with sodium hydroxide and sulfuric acid		23.64	Ramana et al. (2012)
Water-hyacinth weed treated with nitric acid		26.32	Mahamadi and Nharingo (2010)
Cellulose acetate		9.4	Abdelwahab et al. (2015)
CA-*g*-(AA-*co*-AAm)		66.67	
EDTA/chitosan/PMMS		211	Tan et al. (2015)
Marula (*Sclerocarya birrea*) seed husk		20	Moyo et al. (2015)

number of amine groups with triethylenetetraamine resulted in higher sorption efficiency, reaching 313 mg/g. Farooq et al. (2010) used the straw of *Triticum aestivum* (wheat) modified with urea to remove Pb(II) from wastewater, finding a maximum sorption capacity of 31.85 mg/g. Mitić-Stojanović et al. (2012) performed acid (nitric acid) and alkaline (sodium hydroxide) washing to *L. vulgaris* in order to obtain a metal-free biomass that could be used to remove Pb(II) from water. This biomass showed a sorption capacity of 26.94 mg/g. Coelho et al. (2014) reported the use of cashew nut (*Anarcadium occidentale* L.) shell as biosorbent for Pb(II). The authors separated the shell form the nut and extracted the oils from the shell using hexane. The biomass obtained was then tested in the Pb(II) sorption, a sorption capacity of 28.653 mg/g was achieved with this material. Apple pomace was used by Chand and Pakade (2013), who tested its efficiency in Pb(II) removal from water. The authors were able to determine a sorption capacity of 16.39 mg/g. Isaac and Sivakumar (2013) performed Pb(II) sorption tests using custard apple (*Annona squamosa*) shell. The fruits waste was found to be able to remove Pb(II) from water, having a sorption capacity of 91.74 mg/g. Gherasim et al. (2011) prepared inclusions of bis-(2-ethylhexyl) phosphate (D2EHPA) onto polyvinyl chloride (PVC) membranes to be used in Pb(II) sorption. With membranes containing 50% D2EHPA, the authors achieved a maximum sorption capacity of 60.78 mg/g. Kalagasidis Krušić et al. (2012) used a hydrogel based on poly(AAm-*co*-SMA) crosslinked with poly(ethylene glycol) diacrylate (PEGDA) in Pb(II) sorption tests, finding a sorption capacity of 68.92 mg/g for this material. Paulino et al. (2013) produce a Pb(II) sorbent derived from natural sources by modification of the surface of natural cotton fibers with citric acid. The resulting material was tested in Pb(II) removal assays and a sorption capacity of 21.62 mg/g was noticed. Pigeon peas hulls were tested by Ramana et al. (2012), who treated the biomass with sodium hydroxide, to remove the lignin materials, followed by sulfuric acid. The sorption efficiency of the resulting biosorbent for Pb(II) was experienced, being possible to determine a sorption capacity of 23.64 mg/g. Mahamadi and Nharingo (2010) performed metal sorption tests with water-hyacinth weed (*E. crassipes*) biomass, treated with nitric acid. These authors detected that, for Pb(II), water hyacinth has a sorption capacity of 26.32 mg/g. Abdelwahab et al. (2015) performed sorption studies of Pb(II) using cellulose acetate (CA) and CA grafted with acrylic acid and acrylamide, CA-*g*-(AA-*co*-AAm). The authors found that CA can only reach a maximum sorption capacity of 9.4 mg/g while in the CA derivative case, the sorption capacity increased sevenfold, reaching a value of 66.67 mg/g. Tan et al. (2015) reported the preparation of a magnetic adsorbent (EDTA/chitosan/PMMS) for Pb(II) that showed maximum sorption capacity of 211 mg/g. Chitosan functionalized with EDTA moieties, responsible for the coordination of the metal ion, were obtained by reaction of EDTA-anhydride with chitosan in the presence of PEI-coated magnetic microspheres (PMMS). Moyo et al. (2015) tested marula seed husk as Pb(II) biosorbent. The biomass, only washed to remove dirt and without any chemical treatment, revealed a maximum sorption capacity of 20 mg/g.

7 Conclusions

The presence of heavy metals in drinking water is a problem of high concern to society, since it is a source of exposure to very harmful toxins that are related to several diseases and disorders.

Fortunately, the society, with special emphasis to scientific society, recognized that problem and started to create methodologies to improve the quality of our hydric resources. Several methods have been developed to remove heavy metal from water, being the adsorption method the one that has received higher attentions, due to its features.

Polymers, due to their properties, are assuming high importance as an alternative to activated carbon for water remediation. There are a high number of studies reporting the use of natural polymers, mainly cellulose-based materials, for the sorption of metal ions. Actually, this has several benefits including the fact that these polymers exist in high abundance in nature and are biodegradable, which contributes to sustainability. Moreover, the use of these materials opens a new window to a high-society concern, add value to the high amount of residues that are produced worldwide. Nevertheless, the sorption efficiency of these polymers can be enhanced with some chemical modifications that will provide better affinity to metal ions. Furthermore, new polymers can be developed with improved structure and appropriated features for this application, which has already being performed by several research groups.

However, it is important to remember that the sorption studies reported in literature do not follow a standard procedure, which limits the comparison of results reported by different authors. There are several parameters that influence the adsorption of metal ions, as it was referred earlier. Despite the fact that some of these parameters, such as pH or contact time, are dependent on the chemical nature of the adsorbent and adsorbate, there are some that can be standardized, including amounts of adsorbent used, initial concentration of adsorbate, particle size, temperature, or even stirring speed.

References

Abdel-Halim, E.S., Al-Deyab, S.S., 2014. Preparation of poly(acrylic acid)/starch hydrogel and its application for cadmium ion removal from aqueous solutions. React. Funct. Polym. 75, 1–8.

Abdelwahab, N.A., Ammar, N.S., Ibrahim, H.S., 2015. Graft copolymerization of cellulose acetate for removal and recovery of lead ions from wastewater. Int. J. Biol. Macromol. 79, 913–922.

Acar, F.N., Eren, Z., 2006. Removal of Cu(II) ions by activated poplar sawdust (Samsun Clone) from aqueous solutions. J. Hazard. Mater. 137, 909–914.
Ahmaruzzaman, M., 2011. Industrial wastes as low-cost potential adsorbents for the treatment of wastewater laden with heavy metals. Adv. Colloid Interface Sci. 166, 36–59.
Anirudhan, T.S., Divya, L., Bringle, C.D., Suchithra, P.S., 2010. Removal of copper(II) and zinc(II) from aqueous solutions using a lignocellulosic-based polymeric adsorbent containing amidoxime chelating functional groups. Sep. Sci. Technol. 45, 2383–2393.
Apte, A.D., Verma, S., Tare, V., Bose, P., 2005. Oxidation of Cr(III) in tannery sludge to Cr(VI): field observations and theoretical assessment. J. Hazard. Mater. 121, 215–222.
Areco, M.M., Saleh-Medina, L., Trinelli, M.A., Marco-Brown, J.L., Dos Santos Afonso, M., 2013. Adsorption of Cu(II), Zn(II), Cd(II) and Pb(II) by dead *Avena fatua* biomass and the effect of these metals on their growth. Colloid. Surf. B 110, 305–312.
Argun, M.E., Dursun, S., Ozdemir, C., Karatas, M., 2007. Heavy metal adsorption by modified oak sawdust: thermodynamics and kinetics. J. Hazard. Mater. 141, 77–85.
Barakat, M.A., Schmidt, E., 2010. Polymer-enhanced ultrafiltration process for heavy metals removal from industrial wastewater. Desalination 256, 90–93.
Baral, S.S., Das, S.N., Rath, P., 2006. Hexavalent chromium removal from aqueous solution by adsorption on treated sawdust. Biochem. Eng. J. 31, 216–222.
Bessbousse, H., Verchère, J.-F., Lebrun, L., 2012. Characterisation of metal-complexing membranes prepared by the semi-interpenetrating polymer networks technique. Application to the removal of heavy metal ions from aqueous solutions. Chem. Eng. J. 187, 16–28.
Bhattacharya, P., Banerjee, P., Mallick, K., Ghosh, S., Majumdar, S., Mukhopadhyay, A., Bandyopadhyay, S., 2013. Potential of biosorbent developed from fruit peel of *Trewia nudiflora* for removal of hexavalent chromium from synthetic and industrial effluent: analyzing phytotoxicity in germinating Vigna seeds. J. Environ. Sci. Health A 48, 706–719.
Biçak, N., Senkal, B.F., Melekaslan, D., 2000. Poly (styrene sulfonamides) with EDTA-like chelating groups for removal of transition metal ions. J. Appl. Polym. Sci. 77, 2749–2755.
Boamah, P.O., Huang, Y., Hua, M., Zhang, Q., Liu, Y., Onumah, J., Wang, W., Song, Y., 2015. Removal of cadmium from aqueous solution using low molecular weight chitosan derivative. Carbohydr. Polym. 122, 255–264.
Brown, P.A., Gill, S.A., Allen, S.J., 2000. Metal removal from wastewater using peat. Water Res. 34, 3907–3916.
Chand, P., Pakade, Y.B., 2013. Removal of Pb from water by adsorption on apple pomace: equilibrium, kinetics, and thermodynamics studies. J. Chem..
Chauhan, G., Guleria, L., Sharma, R., 2005. Synthesis, characterization and metal ion sorption studies of graft copolymers of cellulose with glycidyl methacrylate and some comonomers. Cellulose 12, 97–110.
Chen, Y., Tang, G., Yu, Q.J., Zhang, T., Chen, Y., Gu, T., 2009. Biosorption properties of hexavalent chromium on to biomass of tobacco-leaf residues. Environ. Technol. 30, 1003–1010.
Coelho, G.F., Gonçalves, J.R.A.C., Tarley, C.R.T., Casarin, J., Nacke, H., Francziskowski, M.A., 2014. Removal of metal ions Cd (II), Pb (II), and Cr (III) from water by the cashew nut shell *Anacardium occidentale* L. Ecol. Eng. 73, 514–525.
Conrad, K., Bruun Hansen, H.C., 2007. Sorption of zinc and lead on coir. Bioresour. Technol. 98, 89–97.
Demirbaş, Ö., Karadağ, A., 2013. Biosorption of zinc ions onto *Corylus avellana* L. Desalin. Water Treat. 53, 2692–2700.

Doke, K.M., Yusufi, M., Joseph, R.D., Khan, E.M., 2012. Biosorption of hexavalent chromium onto wood apple shell: equilibrium, kinetic and thermodynamic studies. Desalin. Water Treat. 50, 170–179.
El-Sheikh, A.H., Abu Hilal, M.M., Sweileh, J.A., 2011. Bio-separation, speciation and determination of chromium in water using partially pyrolyzed olive pomace sorbent. Bioresour. Technol. 102, 5749–5756.
El-Sherif, I.Y., Tolani, S., Ofosu, K., Mohamed, O.A., Wanekaya, A.K., 2013. Polymeric nanofibers for the removal of Cr(III) from tannery waste water. J. Environ. Manage. 129, 410–413.
Farooq, U., Khan, M.A., Athar, M., Sakina, M., Ahmad, M., 2010. Environmentally benign urea-modifed *Triticum aestivum* biomass for lead(II) elimination from aqueous solutions. CLEAN—Soil, Air, Water 38, 49–56.
Ferrah, N., Abderrahim, O., Didi, M.A., Villemin, D., 2011. Removal of copper ions from aqueous solutions by a new sorbent: polyethyleneiminemethylene phosphonic acid. Desalination 269, 17–24.
Gaballah, I., Goy, D., Allain, E., Kilbertus, G., Thauront, J., 1997. Recovery of copper through decontamination of synthetic solutions using modified barks. Metall. Mater. Trans. B 28, 13–23.
Gherasim, C.-V., Bourceanu, G., Timpu, D., 2011. Experimental and modeling studies of lead(II) sorption onto a polyvinyl-chloride inclusion membrane. Chem. Eng. J. 172, 817–827.
Guo, H., Zhang, S., Kou, Z., Zhai, S., Ma, W., Yang, Y., 2015. Removal of cadmium(II) from aqueous solutions by chemically modified maize straw. Carbohydr. Polym. 115, 177–185.
Gupta, S., Babu, B.V., 2009. Removal of toxic metal Cr(VI) from aqueous solutions using sawdust as adsorbent: equilibrium, kinetics and regeneration studies. Chem. Eng. J. 150, 352–365.
Horsfall, J.R.M., Abia, A.A., Spiff, A.I., 2006. Kinetic studies on the adsorption of Cd^{2+}, Cu^{2+} and Zn^{2+} ions from aqueous solutions by cassava (*Manihot esculenta* Cranz) tuber bark waste. Bioresour. Technol. 97, 283–291.
Huang, X.D., Liu, M.H., Lin, C.X., 2012. Adsorption of Cd^{2+} and Pb^{2+} on mangosteen shell modified by thiosemicarbazide. Environ. Sci. Technol. Chin. 35, 24–27.
Huang, K., Xiu, Y., Zhu, H., 2013. Removal of heavy metal ions from aqueous solution by chemically modified mangosteen pericarp. Desalin. Water Treat. 52, 7108–7116.
Hussein, M.A., Marwani, H.M., Alamry, K.A., Asiri, A.M., EL-Daly, S.A., 2014. Surface selectivity competition of newly synthesized polyarylidene(keto amine) polymers toward different metal ions. J. Appl. Polym. Sci., 131.
Isaac, C.P.J., Sivakumar, A., 2013. Removal of lead and cadmium ions from water using *Annona squamosa* shell: kinetic and equilibrium studies. Desalin. Water Treat. 51, 7700–7709.
Jain, M., Garg, V.K., Kadirvelu, K., 2013. Chromium removal from aqueous system and industrial wastewater by agricultural wastes. Biorem. J. 17, 30–39.
Kalagasidis Krušić, M., Milosavljević, N., Debeljković, A., Üzüm, Ö.B., Karadağ, E., 2012. Removal of Pb^{2+} Ions from water by poly(acrylamide-*co*-sodium methacrylate) hydrogels. Water Air Soil Pollut. 223, 4355–4368.
Karnitz, J.R.O., Gurgel, L.V.A., De Melo, J.C.P., Botaro, V.R., Melo, T.M.S., De Freitas Gil, R.P., Gil, L.F., 2007. Adsorption of heavy metal ion from aqueous single metal solution by chemically modified sugarcane bagasse. Bioresour. Technol. 98, 1291–1297.
Karthik, R., Meenakshi, S., 2014a. Adsorption study on removal of Cr(VI) ions by polyaniline composite. Desalin. Water Treat. 54, 3083–3093.
Karthik, R., Meenakshi, S., 2014b. Facile synthesis of cross linked-chitosan–grafted-polyaniline composite and its Cr(VI) uptake studies. Int. J. Biol. Macromol. 67, 210–219.

Kasgoz, H., Ozbas, Z., Esen, E., Sahin, C.P., Gurdag, G., 2013. Removal of copper(II) ions with a thermoresponsive cellulose-*g*-poly(*N*-isopropyl acrylamide) copolymer. J. Appl. Polym. Sci. 130, 4440–4448.

Khan, A., Badshah, S., Airoldi, C., 2015. Environmentally benign modified biodegradable chitosan for cation removal. Polym. Bull. 72, 353–370.

Kim, J., Lee, Y., Yang, M., 2014. Environmental exposure to lead (Pb) and variations in its susceptibility. J. Environ. Sci. Health C 32, 159–185.

Kołodyńska, D., 2013. Application of a new generation of complexing agents in removal of heavy metal ions from different wastes. Environ. Sci. Pollut. Res. 20, 5939–5949.

Kolya, H., Roy, A., Tripathy, T., 2015. Starch-*g*-Poly-(*N*,*N*-dimethyl acrylamide-*co*-acrylic acid): an efficient Cr(VI) ion binder. Int. J. Biol. Macromol. 72, 560–568.

Kumar, U., Bandyopadhyay, M., 2006. Sorption of cadmium from aqueous solution using pretreated rice husk. Bioresour. Technol. 97, 104–109.

Leyva-Ramos, R., Bernal-Jacome, L.A., Acosta-Rodriguez, I., 2005. Adsorption of cadmium(II) from aqueous solution on natural and oxidized corncob. Sep. Sci. Technol. 45, 41–49.

Li, Q., Zhai, J., Zhang, W., Wang, M., Zhou, J., 2007. Kinetic studies of adsorption of Pb(II), Cr(III) and Cu(II) from aqueous solution by sawdust and modified peanut husk. J. Hazard. Mater. 141, 163–167.

Li, G.-X., Yan, C.-Z., Zhang, D.-D., Zhao, C., Chen, G.-Y., 2013. Cadmium(II) biosorption from aqueous solutions using *Hydrilla verticillata*. Can. J. Chem. Eng. 91, 1022–1030.

Llanos, R.M., Mercer, J.F.B., 2002. The molecular basis of copper homeostasis copper-related disorders. DNA Cell Biol. 21, 259–270.

Mahamadi, C., Nharingo, T., 2010. Utilization of water hyacinth weed (*Eichhornia crassipes*) for the removal of Pb(II), Cd(II) and Zn(II) from aquatic environments: an adsorption isotherm study. Environ. Technol. 31, 1221–1228.

Mark, S.S., Crusberg, T.C., Dacunha, C.M., Iorio, A.A.D., 2006. A heavy metal biotrap for wastewater remediation using poly-γ-glutamic acid. Biotechnol. Progr. 22, 523–531.

Meena, A.K., Kadirvelu, K., Mishraa, G.K., Rajagopal, C., Nagar, P.N., 2008. Adsorption of Pb(II) and Cd(II) metal ions from aqueous solutions by mustard husk. J. Hazard. Mater. 150, 619–625.

Memon, S.Q., Memon, N., Shah, S.W., Khuhawar, M.Y., Bhanger, M.I., 2007. Sawdust—a green and economical sorbent for the removal of cadmium(II) ions. J. Hazard. Mater. 139, 116–121.

Milosavljević, N.B., Ristić, M.Đ., Perić-Grujić, A.A., Filipović, J.M., Štrbac, S.B., Rakočević, Z.L., Kalagasidis Krušić, M.T., 2011. Sorption of zinc by novel pH-sensitive hydrogels based on chitosan, itaconic acid and methacrylic acid. J. Hazard. Mater. 192, 846–854.

Milosavljević, N., Debeljković, A., Krušić, M.K., Milašinović, N., Üzüm, Ö.B., Karadağ, E., 2014. Application of poly(acrlymide-*co*-sodium methacrylate) hydrogels in copper and cadmium removal from aqueous solution. Environ. Prog. Sustain. Energy 33, 824–834.

Min, S.H., Han, J.S., Shin, E.W., Park, J.K., 2004. Improvement of cadmium ion removal by base treatment of juniper fiber. Water Res. 38, 1289–1295.

Mitić-Stojanović, D.-L., Bojić, D., Mitrović, J., Andjelković, T., Radović, M., Bojić, A.L., 2012. Equilibrium and kinetic studies of Pb(II), Cd(II) and Zn(II) sorption by *Lagenaria vulgaris* shell. Chem. Ind. Chem. Eng. Q. 18, 563–576.

Moyo, M., Guyo, U., Mawenyiyo, G., Zinyama, N.P., Nyamunda, B.C., 2015. Marula seed husk (*Sclerocarya birrea*) biomass as a low cost biosorbent for removal of Pb(II) and Cu(II) from aqueous solution. J. Ind. Eng. Chem. 27, 126–132.

Nasernejad, B., Zadeh, T.E., Pour, B.B., Bygi, M.E., Zamani, A., 2005. Camparison for biosorption modeling of heavy metals (Cr (III), Cu (II), Zn (II)) adsorption from wastewater by carrot residues. Process Biochem. 40, 1319–1322.

Noeline, B.F., Manohar, D.M., Anirudhan, T.S., 2005. Kinetic and equilibrium modelling of lead(II) sorption from water and wastewater by polymerized banana stem in a batch reactor. Sep. Sci. Technol. 45, 131–140.

Özer, A., Pirinççi, H.B., 2006. The adsorption of Cd(II) ions on sulphuric acid-treated wheat bran. J. Hazard. Mater. 137, 849–855.

Özer, A., Özer, D., Özer, A., 2004. The adsorption of copper(II) ions on to dehydrated wheat bran (DWB): determination of the equilibrium and thermodynamic parameters. Process Biochem. 39, 2183–2191.

Paduraru, C., Tofan, L., Teodosiu, C., Bunia, I., Tudorachi, N., Toma, O., 2015. Biosorption of zinc(II) on rapeseed waste: equilibrium studies and thermogravimetric investigations. Process Saf. Environ. Protect. 94, 18–28.

Pan, B., Pan, B., Zhang, W., Lv, L., Zhang, Q., Zheng, S., 2009. Development of polymeric and polymer-based hybrid adsorbents for pollutants removal from waters. Chem. Eng. J. 151, 19–29.

Pathak, P.D., Mandavgane, S.A., Kulkarni, B.D., 2015. Fruit peel waste as a novel low-cost bio adsorbent. Rev. Chem. Eng. 31, 361–381.

Paulino, Á.G., Da Cunha, A.J., Da Silva Alfaya, R.V., Da Silva Alfaya, A.A., 2013. Chemically modified natural cotton fiber: a low-cost biosorbent for the removal of the Cu(II), Zn(II), Cd(II), and Pb(II) from natural water. Desalin. Water Treat. 52, 4223–4233.

Pehlivan, E., Cetin, S., Yanık, B.H., 2006. Equilibrium studies for the sorption of zinc and copper from aqueous solutions using sugar beet pulp and fly ash. J. Hazard. Mater. 135, 193–199.

Plum, L.M., Rink, L., Haase, H., 2010. The essential toxin: impact of zinc on human health. Int. J. Environ. Res. Public Health 7, 1342–1365.

Pontoni, L., Fabbricino, M., 2012. Use of chitosan and chitosan-derivatives to remove arsenic from aqueous solutions—a mini review. Carbohydr. Res. 356, 86–92.

Rahman, M.L., Rohani, N., Mustapa, N., Yusoff, M.M., 2014. Synthesis of polyamidoxime chelating ligand from polymer-grafted corn-cob cellulose for metal extraction. J. Appl. Polym. Sci., 131.

Ramana, D.K.V., Reddy, D.H.K., Yu, J.S., Seshaiah, K., 2012. Pigeon peas hulls waste as potential adsorbent for removal of Pb(II) and Ni(II) from water. Chem. Eng. J. 197, 24–33.

Rao, R.A.K., Ikram, S., Uddin, M.K., 2014. Removal of Cr(VI) from aqueous solution on seeds of *Artimisia absinthium* (novel plant material). Desalin. Water Treat. 54, 3358–3371.

Reddy, K.H., Reddy, A.R., 2003. Removal of heavy metal ions using the chelating polymers derived by the condensation of poly(3-hydroxy-4-acetylphenyl methacrylate) with different diamines. J. Appl. Polym. Sci. 88, 414–421.

Repo, E., Warchoł, J.K., Bhatnagar, A., Mudhoo, A., Sillanpää, M., 2013. Aminopolycarboxylic acid functionalized adsorbents for heavy metals removal from water. Water Res. 47, 4812–4832.

Saleh, T.A., Haladu, S.A., Ali, S.A., 2015. A novel cross-linked pH-responsive tetrapolymer: synthesis, characterization and sorption evaluation towards Cr(III). Chem. Eng. J. 269, 9–19.

Saygi, K.O., Tuzen, M., Soylak, M., Elci, L., 2008. Chromium speciation by solid phase extraction on Dowex M 4195 chelating resin and determination by atomic absorption spectrometry. J. Hazard. Mater. 153, 1009–1014.

Šćiban, M., Klašnja, M., Škrbić, B., 2006. Modified softwood sawdust as adsorbent of heavy metal ions from water. J. Hazard. Mater. 136, 266–271.

Sikder, M.T., Kikuchi, T., Suzuki, J., Hosokawa, T., Saito, T., Kurasaki, M., 2013. Removal of cadmium and chromium ions using modified α, β, and γ-cyclodextrin polymers. Sep. Sci. Technol. 48, 587–597.
Sikder, M.T., Islam, M.S., Kikuchi, T., Suzuki, J., Saito, T., Kurasaki, M., 2014. Removal of copper ions from water using epichlorohydrin cross-linked β-cyclodextrin polymer: characterization, isotherms and kinetics. Water Environ. Res. 86, 296–304.
Silva Filho, E.C., Santos Junior, L.S., Fernandes Silva, M.M., Fonseca, M.G., Abreu Santana, S.A., Airoldi, C., 2013. Surface cellulose modification with 2-aminomethylpyridine for copper, cobalt, nickel and zinc removal from aqueous solution. Mater. Res. Ibero Am. J. Mater. 16, 79–87.
Singha, A.S., Guleria, A., 2014. Use of low cost cellulosic biopolymer based adsorbent for the removal of toxic metal ions from the aqueous solution. Sep. Sci. Technol. 49, 2557–2567.
Siswoyo, E., Endo, N., Mihara, Y., Tanaka, S., 2014. Agar-encapsulated adsorbent based on leaf of *platanus* sp to adsorb cadmium ion in water. Water Sci. Technol. 70, 89–94.
Stern, B.R., Solioz, M., Krewski, D., Aggett, P., Aw, T.-C., Baker, S., Crump, K., Dourson, M., Haber, L., Hertzberg, R., Keen, C., Meek, B., Rudenko, L., Schoeny, R., Slob, W., Starr, T., 2007. Copper and human health: biochemistry, genetics, and strategies for modeling dose-response relationships. J. Toxicol. Environ. Health B 10, 157–222.
Sulaiman, O., Ghani, N.S., Rafatullah, M., Hashim, R., Ahmad, A., 2011. Sorption equilibrium and thermodynamic studies of zinc (II) ions from aqueous solutions by bamboo sawdust. J. Dispers. Sci. Technol. 32, 583–590.
Tan, Z., Peng, H., Liu, H., Wang, L., Chen, J., Lu, X., 2015. Facile preparation of EDTA-functionalized chitosan magnetic adsorbent for removal of Pb(II). J. Appl. Polym. Sci. 132.
Tang, Y., Ma, Q., Luo, Y., Zhai FL., Che, Y., Meng, F., 2013. Improved synthesis of a branched poly(ethylene imine)-modified cellulose-based adsorbent for removal and recovery of Cu(II) from aqueous solution. J. Appl. Polym. Sci. 129, 1799–1805.
Taty-Costodes, V.C., Fauduet, H., Porte, C., Delacroix, A., 2003. Removal of Cd(II) and Pb(II) ions, from aqueous solutions, by adsorption onto sawdust of *Pinus sylvestris*. J. Hazard. Mater. 105, 121–142.
Vincent, J.B., 2000. The biochemistry of chromium. J. Nutr. 130, 715–718.
Wan Ngah, W.S., Hanafiah, M.A.K.M., 2008. Removal of heavy metal ions from wastewater by chemically modified plant wastes as adsorbents: a review. Bioresour. Technol. 99, 3935–3948.
Wang, Z., Ge, H., 2014. Adsorption of chromium(VI) from aqueous solution using a novel chitosan biguanidine. J. Dispers. Sci. Technol. 36, 1106–1114.
Wang, X.P., Huang, X.D., 2012. Adsorption of Cd^{2+} and Pb^{2+} from aqueous solutions by diethylenetriamine modified mangosteen shell. J. Sci. Teach. Coll. Univ. 32, 65–68.
Wang, X.S., Qin, Y., Li, Z.F., 2006. Biosorption of zinc from aqueous solutions by rice bran: kinetics and equilibrium studies. Sep. Sci. Technol. 41, 747–756.
Wang, X.S., Li, Z.Z., Sun, C., 2008. Removal of Cr(VI) from aqueous solutions by low-cost biosorbents: marine macroalgae and agricultural by-products. J. Hazard. Mater. 153, 1176–1184.
Wong, K.K., Lee, C.K., Low, K.S., Haron, M.J., 2003. Removal of Cu and Pb from electroplating wastewater using tartaric acid modified rice husk. Process Biochem. 39, 437–445.
Wu, S.-J., Liou, T.-H., Yeh, C.-H., Mi, F.-L., Lin, T.-K., 2013. Preparation and characterization of porous chitosan–tripolyphosphate beads for copper(II) ion adsorption. J. Appl. Polym. Sci. 127, 4573–4580.

Ye, H., Zhu, Q., Du, D., 2010. Adsorptive removal of Cd(II) from aqueous solution using natural and modified rice husk. Bioresour. Technol. 101, 5175–5179.

Zein, R., Suhaili, R., Earnestly, F., Indrawati, Munaf, E., 2010. Removal of Pb(II), Cd(II) and Co(II) from aqueous solution using *Garcinia mangostana* L. fruit shell. J. Hazard. Mater. 181, 52–56.

Zhang, Y., Li, Y., Yang, L.-Q., Ma, X.-J., Wang, L.-Y., Ye, Z.-F., 2010. Characterization and adsorption mechanism of Zn^{2+} removal by PVA/EDTA resin in polluted water. J. Hazard. Mater. 178, 1046–1054.

Zhang, Y., Ye, Z., Wang, L., Yang, L., Li, Y., 2013. Removal of Cd (II) by polystyrene-base chelating resins: adsorption properties and experiences of industrial wastewater treatment. Desalin. Water Treat. 52, 6481–6491.

Zhong, Q.-Q., Yue, Q.-Y., Li, Q., Gao, B.-Y., Xu, X., 2014. Removal of Cu(II) and Cr(VI) from wastewater by an amphoteric sorbent based on cellulose-rich biomass. Carbohydr. Polym. 111, 788–796.

9

PHOTOCATALYTIC DEGRADATION AND ADSORPTION TECHNIQUES INVOLVING NANOMATERIALS FOR BIOTOXINS REMOVAL FROM DRINKING WATER

Elvis Fosso-Kankeu*, Ajay Kumar Mishra**

**North-West University, School of Chemical and Minerals Engineering, Faculty of Engineering, Potchefstroom, South Africa; **College of Engineering, Science and Technology, University of South Africa, Nanotechnology and Water Sustainability Research Unit, Johannesburg, South Africa*

1 Occurrence of Biotoxins in Drinking Water

1.1 Occurrence of Cyanotoxins

Cyanobacteria are often referred to as "blue–green algae" because they appear similar to the green algae in morphology, habitat and photosynthetic ability (Taylor, 2003). They are however, classified as prokaryotes (true bacteria) because they do not have intracellular membrane-bound organelles, such as a true nucleus or mitochondrion, and they possess a cell wall composed of peptidoglycan and lipopolysaccharide layers instead of the cellulose of green algae. Morphological diversity ranges from single cells to small colonies of cells to simple branched filamentous (Whitton and Potts, 2000). Cyanobacteria can also survive extremely high and low temperatures and can grow in various environments, such as volcanic ash, desert sand, and rocks (Dor and Danin, 1996; Jaag, 1945), hot springs (Chorus and Bartram, 1999; Castenholz, 1973), mountain streams (Kann, 1988), Arctic and Antarctic lakes (Skulberg, 1996), and snow and ice (Laamanen, 1996;

Water Purification. http://dx.doi.org/10.1016/B978-0-12-804300-4.00009-5

Kol, 1968). Many species are capable of living in the soil and terrestrial habitats (Whitton, 1992) but the prominent habitats of cyanobacteria are limnic and marine environments. Cyanobacteria can be a nuisance in water. They release compounds, such as geosmin (*trans*-1, 10-dimethyl-*trans*-9-decalol); MIB (2-methylisoborneol); β-cyclocitral, IPMP (2-isopropylmethoxypyrazine), and IBMP (2-isobutylmethoxypyrazine). If released in large enough amounts, these substances cause taste and odor problems in water (Rae et al., 1999).

Most importantly, in the context of this study, they produce toxins that may harm the health of people (Chorus and Bartram, 1999). They produce cyanotoxins that may be harmful to humans and animals (WHO, 1999; Codd et al., 1997; Sivonen, 1996; Carmichael, 1992). They occur in eutrophic water, such as nutrients rich surface water (e.g., rivers), that substantial numbers of people in some rural and remote areas often use for drinking and other domestic purposes (Fosso-Kankeu et al., 2008). The production of secondary metabolite cyanotoxins can be influenced by environmental conditions, such as pH, intensity of light, temperature, nitrate, phosphate, dissolve oxygen, and CO_2 availability (Grobbelaar et al., 2004; Rapala and Sivonen, 1998). Within many species of cyanobacteria, there exist nontoxic strains and strains which can produce a particular toxin or many toxins (Grobbelaar et al., 2004; Chorus and Bartram, 1999). There are divergent opinions whether cyanotoxins are produced under stress conditions or not. According to study conducted by Sivonen (1990), high concentrations of toxins are produced under conditions which are favorable for cyanobacteria growth, while previous study by van der Westhuizen and Eloff showed that optimum conditions for growth did not coincide with those for toxin production by the *Microcystis aeruginosa* culture they studied. The primary metabolite cyanotoxin is lipopolysaccharide endtoxin, which is always produced by all cyanobacteria as they constitute part of their cell walls (Metcalf and Codd, 2004). Depending of the type of cyanotoxins (endotoxins, neurotoxins, and hepatotoxins), and the dose, the impact on human health may range from gastroenteritis to dermal irritation, liver cancer, paralysis (Prescott et al., 2002; Yu, 1995; Teixeria et al., 1993).

1.2 Occurrence of Mycotoxins

Fungi are divided in different groups namely filamentous fungi or moulds, mushrooms and yeasts. Various fungi species have been reported to produce mycotoxins as secondary metabolites, which are small and very stable. Most of the studies have evidenced the occurrence of mycotoxins in raw commodities and foodstuffs. The recurrent detection of fungi in water destined for various purposes including medical and drinking, has increased the concern of the impact of these toxins on human health. Fungi have been reported to grow and develop predominantly over surfaces and through solid materials; they are likely to grow in wide range of environment as they have the ability to extract nutrients from their environment and multiply when conditions, such as temperature and relative humidity are favorable. Observations of fungi are likely indications of the presence of mycotoxins as the occurrence of the later is governed entirely by the existence of conditions that favor the growth of the fungus concerned. When conditions favor the growth of toxigenic species, it becomes obvious that one or more mycotoxins for which the genus has the genetic potential will be produced. However, the occurrence of a given type of mycotoxin is dependent on the solubility of oxygen and the pH value. Various species of fungi identified in human foods have been associated with the presence of toxic mycotoxins at different concentrations often exceeding the level recommended by the existing international guidelines (Table 9.1).

Water distribution systems from various locations can easily be contaminated by fungi present in the air and soil; however, some fungi are naturally present in water as they are primarily adapted to aquatic environments, these fungi are from the phyla chytridiomycota and are called zoosporic; on the other hand

Table 9.1 FDA Action Level for Selected Mycotoxins in Human Food

Toxins	Human food	Level
Aflatoxin	Milk	0.5 ppb (Aflatoxin M1)
	Foods, peanut, and peanut products, Brazil and pistachio nuts	20 ppb
Vomitoxin	Finished wheat products	1 ppm
Total fumonisins (FB1, FB2, and FB3)	Degermed dry milled corn products	2 ppm
	Cleaned corn intended for popcorn	3 ppm
	Whole or partially degermed dry milled corn products	4 ppm
	Dry milled corn bran	4 ppm
	Cleaned corn intended for masa production	4 ppm

fungal groups, such as oomycetes and hyphomycetes are reported to naturally predominate in water (Paterson and Lima, 2005). Parts of pipe work with low circulation and areas of water systems with dead spaces are suitable for the growth of fungi; it is most likely for biofilms formed in the inner side of water storage tank to host fungi that can be easily released into water (Niemi et al., 1982). A relative increase of studies assessing the occurrence of fungi in drinking water distribution system has been recorded in the last decade, but mostly from developed countries; these studies have revealed that various species of fungi were present in tap water, bottled mineral water, drinking water sources (groundwater, spring water, and surface water), and water distribution systems (Hinzelin and Block, 1985; Rosenzweig et al., 1986; Cabral and Pinto, 2002; Hageskal et al., 2006; Yamaguchi et al., 2007; Pereira et al., 2009). The fungi are predominantly occurring in surface water than in groundwater and have been mostly identified in cold water and shower water than in hot tap water.

The fact that fungi can grow in water in quantity enough to affect human health especially severely immuno compromised patients in hospital has prompted frequent analysis of hospital water systems for the detection of fungi. A study conducted by Pires-Goncalves et al. aiming to identify and determine the diversity, occurrence, and distribution of fungi in water used at hemodialysis center had revealed the presence of fungi in 89% of water samples collected inside the hemodialysis unit with prevalence of moulds in tap water samples and of yeasts in dialysate samples. Filamentous fungi species including *Aspergillus* spp. and *Penicillium* spp., as well as yeast (*Candida* spp.) were detected in municipal water supplies of hemodialysis centers, treated water and dialysate from 85 hemodialysis unit in Greece; in that study they found a correlation between the count of yeasts, fecal coliforms, total heterotrophic bacteria, *Pseudomonas* spp., and total coliforms; while the counts of molds correlated with total heterotrophic bacteria and enterococci (Arvanitidou et al., 2000). The source of fungi in hospital water may be as a result of secondary contamination during handling or manipulation of equipment, but in most instances, the quality of water from the distribution system may be critical as residual contaminants that have escaped the treatment processes resurface and multiply.

A number of investigations carried out on tap water and on the water distribution systems often attest of the presence of various species of fungi in large quantity. Conducting the surveillance of mesophilic fungi in tap water, Goncalves et al. (2006) found highest counts of filamentous fungi in winter with a predominance of *Penicillium* (40.6%) and *Acremonium* (38.8%). Examining the presence of fungi in the water distribution system, Grabinska-Loniewska et al. (2007) found that the species *Cladosporium sphaerospermum*, *Aspergillus versicolor*, *Aspergillus parasiticus*, and *Penicillium frequentans* occurred mostly in section of the system made of cast iron and steel. The structure of the pipe is likely to promote the attachment of suspended particles and microorganisms, resulting in the formation of biofilm; assessing the occurrence of fungi within surface matrices of biofilms of a municipal water distribution system, Doggett (2000) found filamentous fungi and yeasts in various densities. Bottled water can be considered as stored and is therefore likely to support the growth or the regrowth of fungi; a host of fungi species including *Penicillium citrinium*, *P. glabrum*, *Cladosporium*, *Rhizopus*, *Aspergillus*, and *Phoma* were isolated from 126 samples of still bottled mineral water of eight different commercial brands in Argentina by Cabral et al. In a separate study, Criado et al. (2005) showed that the growth of fungi in bottled mineral water was mostly supported by the presence of organic matter released from polyethylene terephtalate (PET). Toxigenic strains of *A. alternata* and *P. citrinum* were identified to produce mycotoxins in vitro, while *P. citrinum* produced citrinin in mineral water.

2 Risks Associated to Biotoxins in Water

2.1 Cyanotoxins

They are divided according to their effects on human body. Chemical techniques have already enable to identify three neurotoxins with different modes of blocking neuronal signal transmission (anatoxin-a, anatoxin-a(s), and saxitoxins), one general cytotoxin which inhibits protein synthesis (cylindrospermopsin), and the much popular group of toxins termed hepatotoxins (microcystins and nodularins), which inhibit protein phosphatases. The last groups of cyanotoxins are dermatotoxins with a particular effect on skin, in this group are included aplysiatoxin and LPS endotoxin which is poorly studied and less known; endotoxin is also found in Gram-negative bacteria and is a constituent of the cell wall of almost all the cyanobacteria (Rapala et al., 2002; Metcalf and Codd, 2004).

The effects of cyanotoxins on human health are classified according to region of the body affected and generally target the liver (hepatotoxins), nerve (neurotoxins), cell (cytotoxins), or the skin (dermatotoxins).

Microcystins are potent hepatotoxins of animals and humans (Puiseux-Dao et al., 2005; Huynh-Delerme and Puiseux-Dao, 1998; Ohtani et al., 1992). After ingestion of contaminated water, microcystins are absorbed in the intestines (Kodak et al., 1993; Falconer, 1991) and transported to liver via a carrier-mediated transport system. In the liver, serine/threonine protein phosphatase 1 and 2A (Romanowska-Duda et al., 2002; Honkanen et al., 1996; Eriksson et al., 1990a,b) are inhibited. These enzymes in mammalian cells (Dondjin et al., 2003) are responsible for the dephosphorylation of amino acids serine and threonine. They control the function of hepatic cells. Microcystins are specific to hepatocytes and have difficulty to penetrate into the epithelial cells (Matsushima et al., 1990).

Studies on the mechanism of cell toxicity showed that microcystin interferes with cell structure and mitosis (Kaja, 1995; Falconer and Yeung, 1992). This action may explain the effect of microcystins as cancer promoters (Nishiwaki-Matsushima et al., 1992; Falconer, 1991), especially liver cancer in humans exposed to long-term low doses of microcystins through drinking water (Bouaicha et al., 1998; Mez et al., 1997; Ito et al., 1997; Carmichael, 1994).

In the same brain of thought, Fitzgeorge et al. (1994) demonstrated that the effect of microcystins on the liver could be cumulative. The consequence of an acute poisoning of these compounds is a rapid disorganization of the hepatic architecture (Eriksson et al., 1990a; Falconer et al., 1981), leading to intrahepatic haemorrhage, haemodynamic shock, heart failure, and death (Bhattacharya et al., 1997; Carmichael, 1992). Other organs affected are the kidneys and lungs (Hooser et al., 1990) and the intestines (Falconer and Humpage, 1996; Falconer, 1994).

To postulate the health impact of microcystin-LR on humans, studies have been done on mice, determining a lethal dose level of microcystins. The LD_{50} by intraperitoneal (i.p.) route ranges from 25 to 150 μg/kg body weight (bw) in mice. The oral LD_{50} (administered by gavage, i.e., dosing directly into the stomach through the mouth) is 5000 μg/kg bw in one strain of mice (Yoshida et al., 1997) and higher in rats (Fawell et al., 1994). To confirm the side of action of microcystins, sublethal doses of variously radiolabelled microcystins were administered through intravenous (i.v.) or i.p. injection in mice, immediately after which 70% of the toxins were localized in liver (Lin and Chu, 1994; Robinson et al., 1991; Meriluoto et al., 1990; Brooks and Codd, 1987).

On the basis of human life time exposure, the WHO (2004, 1999) recommended a maximum acceptable level (1 μg/L) for the most toxic variant of microcystins (microcystin-LR) in drinking water.

Other cyanotoxins also occurring in water include nodularin, which can cause gastro-enteritis, fever, pains in muscles and joints, nausea, vomiting, diarrhea, swollen liver, death by liver failure; cylindrospermopsin which can cause abdominal pains, vomiting, swollen liver, liver failure, pathological damage to the kidneys, spleen, thymus, and heart; anatoxin-a(S) which can cause muscle weakness, respiratory distress, exaggerated abdominal breathing, hyperactivity, hypersalivation, numbness about the lips, paralysis, and saxitoxins which is responsible for numbness around the lips, complete paralysis, death from respiratory failure.

2.2 Mycotoxins

The health risk associated with the occurrence of fungi and mycotoxins has always been neglected for certain obvious reasons among which, the capability of food processing operations to reduce the level of such contaminants to a certain level and the fact that the fungi growth is not always associated by toxin production, given that conditions which allow the toxin production are more restricted than those which promote cell growth (Northolt et al., 1996). The control of fungi occurrence and subsequently mycotoxins is however, limited as far as water distribution systems and storage are concerned; fungi with toxigenic potentiality have been often isolated from drinking water and water from haemodyalisis units. Mycotoxins are very toxics and carcinogenic capable to adversely affect the health of humans and animals at very low concentrations through acute and chronic effects. To monitor the incidence of these toxins and prevent outbreak, national guidelines in some countries have been made available for strict application in water supply industries, food industries, and other sectors administering water to humans or animals.

Although almost all the countries are aware of the potential impact of mycotoxins on the food safety, health, and agriculture, mycotoxin regulations are not available in all the countries, especially in Africa where only 15 countries were known to have specific mycotoxin regulations by 2003; therefore covering about 59% of the inhabitants of the continent (WHO, 2003). Most of the regulations are based on the ingestion of agricultural products and predominantly considered the occurrence of aflatoxins. According to the United States' Food and Drug Administration (FDA) Mycotoxin Regulatory Guidance (FDA, 2011), action levels for selected mycotoxins in human foods are below 4 ppm (Table 9.1), with the lowest level recommended for aflatoxin M1 in milk; this clearly implies that these toxins could constitute a risk to human health at very low concentrations. Similar levels have been recommended

Table 9.2 Human Diseases Associated with the Occurrence of Mycotoxins in Food

Diseases	Food	Etiological Agents	Toxins
Alimentary toxic aleukia	Cereal grains	*Fusarium* spp.	Trichothecenes
Balkan nephropathy and chronic interstitial nephropathy	Cereal grains	*Penicillium*	Ochratoxin A
Cardiac beriberi	Rice	*Aspergillus* and *Penicillium* spp.	Citreoviridin
Ergotism	Rye, cereal grains	*Claviceps purpurea*	Ergot alkaloids
Oesophagial tumors	Corn	*Fusarium verticillioides*	Fumonisin B1
Hepatocarcinoma (acute aflatoxicosis)	Cereal grains, peanuts	*Aspergillus flavus, A. parasiticus*	Aflatoxin B1
Human neural tube defects	Corn	*Fusarium verticillioides*	Fumonisin B1
Kashin Beck disease or "Urov disease"	Cereal grains	*Fusarium* spp.	Trichothecenes
Kwashiorkor	Cereal grains	*Aspergillus flavus* and A. parasiticus	Aflatoxin B1
Onyalai	Millet	*Phoma sorghina*	Undefined
Reye's syndrome	Cereal grains	*Aspergillus*	Aflatoxin B1
Testicular cancer	Various	*Penicillium*	Ochratoxin A

in animal feeds (Table 9.2) by the FDA, with little variation among the groups of animals depending of the type of commodity (Carlson, 2003). Lower levels of mycotoxins are recommended for younger or immature animals, which have weaker systems, while relative higher level of vomitoxin, fumonising, and zearalenone is tolerated compared to aflatoxin. Discomforts related to fungal contamination and mycotoxins occurrence have been extensively reviewed in literature and may range from esthetic concerns linked to unusual taste and odor generated by compounds produced by fungi, allergic effects of different degrees, chronic disease, such as cancer and acute, as well as chronic toxicity (Table 9.2).

Human disease resulting from exposure to mycotoxin is called mycotoxicosis and huge negative impact on the social and economic welfare of the country. The effects caused by mycotoxins are related to their chemical structures; zearalenone has a structure similar to naturally occurring estrogens and therefore interacts with human estrogen receptors once in the body (Miksicek, 1994; Kuiper et al., 1998; Leffers et al., 2001; Shier et al., 2001). Ingestion of zearalenone may cause hyperestrogenism in women resulting in numerous systemic disorders in the form of decreased libido, ovarian, and uterine dysfunctions, anovulation, infertility, and neoplasmic lesions in the entire reproductive system (Shier et al., 2001; Tiemann et al., 2003).

Water-borne diseases related to the presence of fungi in water is on the rise affecting generally immunocompromised patients who are undergoing chemotherapy, and organ and bone marrow transplants; the main culprit in most cases belong to the group of *Aspergillus* namely *A. fumigatus*. Fumonisins and aflatoxin B1 have been reported to be carcinogenic (Peraica et al., 1999; Miller et al., 2001; Marasas et al., 2001), and the later was identified as the causative agent of human liver cancer among individuals affected with hepatitis B or C virus in developing countries (Henry et al., 1999). As from the start of mycotoxicology in 1960, the host of mycotoxins related ailments reported are mostly from food ingestion; outbreak of food-borne disease associated with ingestion of high concentration of fumonisin was reported (Bhat et al., 1997) in India where the victims suffered from gastrointestinal pains and diarrhea. Incidence of acute liver cancer was also reported in India, Kenya, and Malaysia following aflatoxin intake (Bhat, 1991; Lye et al., 1995; Shephard, 2004). Oesophagial cancer was reported among a group of people in South Africa who were consuming beer made from moldy corn containing up to 118 mg/kg fumonisins (Rheeder et al., 1992; IARC, 1993). In the last decade few reports have exposed the so far hidden implications of mycotoxins in waterborne diseases, which could have been inaccurately attributed to other biological toxins in the past. Cases of allergies have been reported after showers and baths, waters contaminated with fungi *Phialophora richardsiae* were used by people in Sweden and Finland (Aslund, 1984). The occurrence of *Trichoderma viride* in Norwegian drinking water has been reported to cause asthma among children living in homes receiving inadequate supply of water (Hageskal et al., 2006; Vesper et al., 2006).

3 Application of Nanotechnology for Biotoxins Removal from Water

Biotoxins are relatively soluble and stable in water; they are not easily destroyed by the traditional chlorination method and ozonation is often required to effectively degrade these toxins. However, ozonation is a very corrosive and complex technology requiring sophisticated equipment therefore resulting in very high capital and operating costs not affordable by most of the countries in the developing world. Cheaper and sustainable techniques have been successfully used in water treatment plants or are under investigations for improvement of the processes aimed at the removal of biological toxins in water distribution systems or at point of use. Nanotechnology based adsorption techniques, as well as photocatalytic degradation is among the leading approaches so far considered.

3.1 Adsorption of Biotoxins

The adsorption process requires contact between the adsorbent and the adsorbate, and during that interaction, the binding may result from physical or chemical mechanisms. The physical mechanism generally involves weak forces such Van der Waal forces and the binding is therefore reversible; while the chemical mechanism requires higher activation energies and could be manipulated to be selective to a given adsorbate; because of the nature of the forces in presence, such as covalent and electrostatic forces, the binding is very strong and very difficult to reverse. The properties of the adsorbent and adsorbate, which could be the size, the shape, the polarity, and the charge distribution and dissociation constants, play an important role in adsorption (Huwig et al., 2001).

Biotoxins occur as extracellular or intracellular compounds therefore determining the effectiveness of the technique to be considered for their removal from the solution. Techniques, such as sand filtration, membrane filtration, coagulation, and sedimentation are suitable for the removal of particulate microbial cells, but not effective for the dissolved toxins. Rositano and Nicholson (1994) were not successful in attempting to remove biotoxins with coagulants including polyaluminium chloride, ferric chloride, and alum. Adsorption methods using silica-based inorganic compounds or carbon-based organic adsorbing agents are found to effectively remove biotoxins from water. Granular activated carbon (GAC) and powdered activated carbon (PAC) have shown good adsorption potential during application in the water treatment plants (Falconer et al., 1989; Jones et al., 1993; Donati et al., 1994; Newcombe et al., 2003; Campinas and Rosa, 2010).

Activated carbon is very porous with large surface area and good adsorptive capacity. The use of PAC for adsorption of cyanobacteria metabolites, such as variants of microcystins showed differences in the binding affinity of PAC among the variants; the size and the hydrophobicity of the microcystin molecules, the pore volume characteristic of the carbon, as well as the electrostatic interactions were reported to be the combination of factors responsible for the differences in the binding affinity of PAC (Donati et al., 1994; Pendleton et al., 2001; Cook and Newcombe, 2002).

Wood-based PAC has been shown to be the most effective among a range of activated carbon types in removing microcystins due to a higher fraction of mesopores (Donati et al., 1994; Fawell et al., 1993; Mohamed et al., 1999). The adsorption capacity can be affected by the presence of natural organic material (NOM) in water; previous studies have shown that adsorption sites on the surface of activated carbon can be occupied by NOM, which will therefore compete with toxins for the binding sites, reducing the adsorption capacity of the PAC. To mitigate such limitation in the water treatment, a multiple barriers treatment that uses a combination of coagulation, PAC and ultrafiltration (UF) is often applied and could remove both intracellular, as well as extracellular biotoxins. Efficient removal of MCLR, was achieved by Lee and Walker (2006) as they used PAC-UF system including polyethersulfone (PES) and cellulose acetate (CA) membranes.

A multiple barrier approach incorporating coagulation, PAC and UF was investigated by Dixon et al. (2011) for the removal of intracellular and extracellular cyanotoxins from two naturally occurring blooms in South Australia; although they achieved a total removal of cyanobacterial cells, the cyanotoxins were only partially removed from the water. The GAC is also effective for the removal of cyanotoxins from water, but its operational lifetime is reduced by excess loading of the NOM, resulting in lower removal of microcystins (Lambert et al., 1996).

A host of mycotoxin-adsorbing agents are available in literature but two main groups have been frequently used for the removal of mycotoxins from feed and water, namely aluminosilicates and activated carbons (AC). Most of the studies, which have focused on the attenuation of mycotoxins load to prevent mycotoxicosis, have predominantly used aluminosilicates as adsorbing agents (EFSA, 2009). The silicates minerals, such as bentonites, montmorillonites belonging to the subclass phyllosilicates while the zeolites belonging to the subclass tectosillicates of aluminosilicates, are the common adsorbents of aflatoxins (Ramos et al., 1996). Modified forms of aluminosilicates, such as hydrated sodium calcium aluminosilicates (HSCA) are reported to be the most studied mycotoxin-sequestering agents among the mineral clays (Diaz and Smith, 2005).

Early research study of the adsorption of micotoxin from solution was conducted by Masimango et al. (1978) who concluded, that mycelium and clay could adsorb aflatoxin B1 dissolved in liquid media.

The structure of mycotoxins influenced their adsorption from solution; adsorbing aflatoxins B1 and G1 from solution, Flores et al. (1999) found that the mineral aluminosilicates had higher affinity for G1, what they ascribed to the presence of an additional cyclic ester in G1, which provides higher electronic density to G1 molecules, forming more stable hydrogen bridges with respect to the cyclopentanone ring present in B1.

Dakovic et al. (2008) investigated the adsorption of aflatoxin B1 from aqueous solutions by natural montmorillonite and montmorillonite modified with copper ions and found that the copper modified montmorillonite was much effective in adsorbing aflatoxin B1.

Aluminosilicates adsorption mechanism consists of the attraction of the polar functional groups of the mycotoxin, which is trapped on their surface. Testing the in vitro adsorption of zearalenone by selected materials, such as crospovidone, montmorillonite, bentonite, sepiolite, and magnesium trisilicate, Ramos et al. (1996) found that the adsorption of zearalenone on the aluminosilicates followed the Freundlich isotherm model, which allowed them to suggest the involvement of different adsorption mechanisms. According to Phillips et al. (1991), specific silicate minerals can bind on aflatoxin by chelating the β-dicarbonyl moiety in aflatoxin with uncoordinated metal ions in the clay materials. It has also been reported (Grant and Phillips, 1998) that the adsorption of AfB1 onto HSCAS could occur as a result of heterogenous binding to different sites that might include the interlamellar region, edges, and the basal surfaces of HSCAS. Although such mechanism contributes to the high adsorption capacity of aluminosilicates, it leads to higher chemisorption selectivity for aflatoxins and show little or no beneficial effect against zearalenone, fumonisin B1, ochratoxin A and trichothecenes, including deoxynivelenol, T-2 toxin, or diacetoxyscirpenol (Diaz and Smith, 2005). The groups of AC derived from organic compounds and are available in the form of powder (PAC) and granule (GAC). AC has the ability to inactivate mycotoxins and have the potential to effectively remove almost all the mycotoxins in aqueous solution (Huwig et al., 2001). AC is a nonspecific and is therefore likely be ineffective in a complex environment as other elements will compete with mycotoxins for the binding sites.

Already in the early 1980s, Decker and Corby conducted an in vitro adsorption of aflatoxin B1 at neutral pH by using an efficient activated charcoal. The binding affinity of 19 AC, HSCAS and sepiolite was tested by Galvano et al. for the removal of ochratoxin A and deoxynivalenol from aqueous solution. They found that HSCAS and sepiolite were not effective in the removal of the mycotoxins from solution, while AC showed high in vitro affinity for both mycotoxins of different chemical structure and could therefore be considered as potential multimycotoxin-sequestering agents.

Jaynes et al. (2007) conducted an investigation comparing the adsorption capacity of reference clays and activated carbon to that of a commercial clay additive (Novasil) during the removal of aflatoxin B1 from water; they found that the AC, Novasil, and the reference clays deriving from montmorillonite had the same adsorption capacity as they all adsorbed ~200 g/kg AfB1 from water; while the other reference clay identified as sepiolite adsorbed only ~60 g/kg.

Emerging approaches considering nanotechnology for the removal of biotoxins offer promises in the control of the level of these toxins in water intended for human and animal use. The surface adsorption of the nanoparticles is the primary force behind the interactions between biomolecules and nonomaterials. Small size nanomaterials have unique surface adsorption energy with unsaturated surface chemical bonds stabilized by the binding of biomolecules (Patil et al., 2007; Nel et al., 2009). Interactions between nanoparticles and biomolecules may result into adsorption orientation effect, adsorption concentration effect, the configuration changing effect and the quantum size effect (Feng et al., 2008). Four classes of nanoscale materials are currently been evaluated as functional materials for water purification; these include zeolites, dendrimers, carbonaceous nanomaterials, and metal oxide nanoparticles. Dendrimers are polymeric molecules with uniform size, regular and highly branched three-dimensional architecture. Their properties make them suitable for direct application but also as substrates for metal nanoparticle synthesis; there are two commonly used dendrimers namely polyamidoamine dendrimers and poly(propylene imine) dendrimers (Fig. 9.1)

Polyamidoamine dendrimers have been used for the removal of natural organic matters from aqueous solutions through entrapment into their internal nanocavities; Li et al. (2012) synthesized

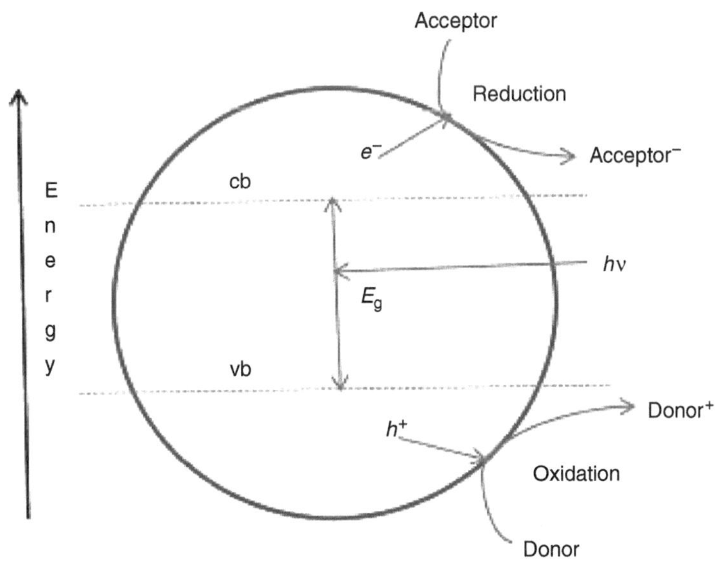

Figure 9.1. Mechanism of action of titanium dioxide during heterogenous photocatalysis.

an insoluble copolymer containing β-cyclodextrine and polyamidoamine units that were used as adsorbents for the removal of 4-chlorephenol, 2,4-dichlorophenol, 2,4,6-trichlorophenol from aqueous solutions. Zeolites are available as natural or synthesized nanostructured materials; natural zeolites consist of aluminosilicate minerals and characterized by their high ion exchange capacity and surface-to-volume ratio (Clariant and Company, 2013). Zeolites are suitable for the adsorption of mycotoxins that contain polar functional groups, such as aflatoxins however, a modification of their surface properties and control of hydrophobicity with high molecular weight quaternary ammonium is required for the adsorption of less polar mycotoxins, such as zearalenone and ochratoxin A (Phillips et al., 1995; Santin et al., 2002). Tomasevic-Canovic et al. (2003) modified the natural clinoptilolite–heulandite rich tuff with different amounts of octadecyldimethyl benzyl ammonium chloride and dioctadecyldimethyl ammonium chloride to achieve a successful adsorption of aflatoxin B1, zearalenone, ochratoxin A, and ergopeptine alkaloids.

Organic matters can be selectively removed from aqueous solutions by carbonaceous materials, which have high adsorption capacity. Fullerenes in ball shapes or carbon nanotubes (CNTs) are the major carbon-based nanomaterials. CNTs are carbon molecules composed of carbon lattices that can take the form of tubes with great flexibility and tensile strength (Yu et al., 2000). Investigating the adsorption capacity of wood-based activated carbon, sepiolite, kaolinite, talc, and CNTs for the removal of microcystin-RR and LR from aqueous solutions, Yan et al. (2006) found that the adsorption capacity of CNTs was four-fold higher than that of the other adsorbents. The removal of organic waste from wastewater by transition metal oxides can be through adsorption or catalytic degradation. Noble metals have high conductivity, extreme hardness, large density, high electron affinity, and a smaller atom with high ionization energies. Nanoscale metals or minerals have, such as nanoferric oxide, granular ferric hydroxide, nanoscale hematite, and nanoscale magnetite have been used to remove organic matters from aqueous solutions (Ganesh et al., 2005; Liu, 2006; Hagare et al. 2001; Hu et al., 2005; Yavuz et al., 2006). Metal oxides with strong adsorption ability, such as zirconium oxide, titanium oxide, tin oxide, cerium oxide, and zinc oxide have been utilized for the immobilization of biological molecules. Lee and Walker (2011) tested the potential of iron oxide nanoparticles for the removal of microcystin-LR from water; they found that the adsorption of microcystin-LR was mostly influenced by the pH of the solution and the mechanisms of adsorption included electrostatic and hydrophobic interactions (Table 9.3).

Table 9.3 Adsorption of Mycotoxins and Cyanotoxins on Various Adsorbents

Group of Toxins	Adsorbent	Parameters Investigated	Toxin	Efficiency	Modeling	References
Cyanotoxins	PAC	Multiple barrier approach for water treatment, removal of intra and extracellular toxins	Saxitoxin, microcystin-LR, microcystin-RR and microcystin-LA	>90%	NC	Dixon et al. (2011)
	PAC	Effects of time and water quality on the adsorption of cyanotoxins	Cylindrospermopsin, microcystin-LR, microcystin-RR, microcystin-LA, microcystin-YR	100% at 50 mg/L dosage of adsorbent	NC	
	GAC	Effect of temperature and bacterial concentration	Microcystin	100%	Homogeneous surface diffusion model (HSDM)	
	GAC and PAC	Influence of natural organic matter on the adsorption of microcystin-LR, effect of the combination of filtration techniques	Microcystin-LR	80%	Freundlich isotherm	Lambert et al. (1996)
	GAC	Effects of carbon properties and water characteristics	Microcystin-LR	45 μg/mg	Langmuir isotherm	
	Graphene oxide	Matrix effects of contaminated water on the ability of adsorbents	Microcystin-LR and microcystin-RR	1.7 mg/g MC-LR and 1.878 mg/g MC-RR	Langmuir model	
	PAC	Effects of pH and natural organic matter concentration	Saxitoxin	100% at 10 mg/L PAC dose	NC	
	Microsphere magnetite core and microsphere copper silicate nanotube shell	Adsorption capacity	Microcystin-LR	97.60%	NC	

(*Continued*)

Table 9.3 Adsorption of Mycotoxins and Cyanotoxins on Various Adsorbents (*cont.*)

Group of Toxins	Adsorbent	Parameters Investigated	Toxin	Efficiency	Modeling	References
Mycotoxins	Organozeolites	Effects of pH and nature of surfactant	Aflatoxin B1	100%	NC	Tomasevic-Canovic et al. (2003)
	Organozeolites	Effect of pH	Zearalenone, ochratoxin A, aflatoxin B1	100%	NC	
	Organophilic montmorillonite clay	Effect of surfactant impregnation on the adsorption capacity of clay	Zearalenone	0.39 mol/kg	Freundlich	
	Natural montmorillonite and copper modified montmorillonite	Effects of adsorbent pretreatment and pH of the solution	Aflatoxin B1	40.982 mg/g (Mont) and 66.225 (Cu-Mont)	Langmuir	Dakovic et al. (2008)
	Activated carbon, montmorillonite and novasil	Comparison of the adsorption capacity of adsorbents	Aflatoxin B1	200 g/kg	Isotherm model	Jaynes et al. (2007)
	Activated carbon (AC), hydrated sodium alluminosilicate, sepiolite	Relationships between adsorption ability and physicochemical parameters of AC	Ochratoxin A, deoxynivalenol,	Ac adsorbed 96.57% (OA) and 98.93%(DON)	NC	
	Modified montmorillonite nanocomposites	Effects pH, temperature, adsorbate and adsorbent concentrations	Zearalenone	8.83 mg/g	Langmuir and Freundlich isotherms	Feng et al. (2008)
	Octadecyldimethyl benzyl ammonium modified Clinoptilolite-Helandite rich stuff	Effects of pH and coverage of zeolitic surface	Ochratoxin A	100%	S-shaped isotherm	

NC, Not considered.

3.2 Photocatalytic Degradation of Biotoxins

Advance Oxidation Process is a promising alternative to conventional method and includes thermal processes with H_2O_2, and light-induced reactions, such as direct homogeneous photolysis (Vacuum UV included), UV/H_2O_2 photolysis, UV/ozone photolysis, heterogenous photolysis, radiolysis, and indirect electrolysis (Legrini et al., 1993; Rajeshwar, 1995). Water decontamination based on the heterogenous solar photocatalytic approach makes use of the near-ultraviolet (UV) band of the solar spectrum (wavelength shorter than 400 nm), to photo-excite a semiconductor catalyst in contact with water and in the presence of oxygen (Malato et al., 2009). Photocatalysis is an ecofriendly process that employs nontoxin photocatalysts capable of decomposing various kinds of organic compounds and biological pollutants in liquid due to their excellent functionality and long-term stability. According to the basic principles of heterogenous photocatalysis, semiconductors, such as CdS, ZnS, Fe_2O_3, ZnO, and TiO_2, can act as sensitizers for light induced processes due to their electronic structure which is characterized by a filled valence band (vb) and an empty conduction band (cb). According to Litter (1999), when a photon of energy higher or equal to the bandgap energy is adsorbed by a semiconductor particle, an electron from the vb is promoted to the cb with simultaneous generation of a hole (h^+) in the vb. The e^-_{cb} and the h^+_{vb} can recombine on the surface or in the bulk of the particle in a few nanoseconds (and the energy dissipated as heat) or can be trapped in surface states where they can react with donor (D) or acceptor (A) species adsorbed or close to the surface of the particle. A chemical equilibrium ensures photocatalytic reduction of adsorbed couples by cb electrons if they have redox potentials more positive than the V_{fb} of the cb, and be oxidized by vb holes if they have redox potentials more negative than the V_{fb} of the vb. This process is believed to follow the following mechanism (Rauf and Ashraf, 2009).

$$\text{Catalyst} + h_V = e^-_{cb} + h^+_{vb} \tag{9.1}$$

The e^-_{cb} is the electrons in the cb, while the h^+_{vb} is the electron vacancy in the vb. This electron and electron gap can then go to the surface of the nanoparticle where the proton can react to the H_2O on the surface, creating OH^- radicals, while the electron can bind with oxygen and cause a super oxide radicle (Rauf and Ashraf, 2009).

$$H_2O + h^+ \rightarrow OH^- + H^+ \tag{9.2}$$

$$O_2 + e^-_{cv} \rightarrow O_2^- \tag{9.3}$$

$$O_2^{-}+H_2O \rightarrow H_2O_2 \tag{9.4}$$

$$H_2O_2 \rightarrow 2OH^{-} \tag{9.5}$$

The superoxide and hydroxide radicals are thus used to oxidize the organic matter (Rauf and Ashraf, 2009).

The process is relatively cheap and sustainable using mostly solar energy and can therefore be easily applied for disinfection of water sources in developing countries were high levels of solar radiation are available throughout the year. During photocatalysis it is considered that oxidizing species, such as the hydroxyl radicals are mainly responsible for the attack on organic molecules, that they oxidize and mineralize to yield CO_2, H_2O, and inorganic ions. This process is attractive for several reasons (Malato et al., 2009)—the process takes place at ambient temperature and without overpressure—oxidation of the substances into CO_2 and other inorganic species is complete—the oxygen necessary for the reaction can be directly obtained from atmosphere—the catalyst is cheap, innocuous, and can be reused—the catalyst can be attached to different types of inert matrices; the energy for photo-exciting the catalyst can be obtained from the Sun. Solar photocatalysis has been extensively applied in the industries, healthcare environments, and households for the inactivation of a wide range of microorganisms, such as bacteria, viruses, protozoa, and fungi in drinking water (Matsunaga et al., 1988; Watts et al., 1995; Sichel et al., 2007; Rincon and Pulgarin, 2007; Benabbou et al., 2007; Guillard et al., 2008; Bui et al., 2008; Pigeot-Remy et al., 2011).

Solar phocatalysis disinfection of water combines the effect of Sun and the catalyst to kill the microorganisms in the water. The inhibitory effect of the Sun may result from the irradiation of sunlight and solar heating to kill the pathogens in contaminated drinking water (Helali et al., 2014). According to Berney et al. (2006), inactivation of microbial species by solar disinfection is caused by disrupting a sequence of normal cellular functions. The impact on these functions have been eventually identified as (McGuigan et al., 2012) ATP synthesis and efflux pump activity both cease shortly after the start of exposure. These are followed by gradual loss of membrane potential and a reduction in glucose uptake. Finally, the cytoplasmic membrane of the bacterial cells becomes permeable. The irradiation of UV-A and UV-B wavelength has been shown to mostly target the *Escherichia coli* DNA (Eisenstark, 1987). Although a number of factors including dissolved oxygen concentration (Reed, 1997), salt concentration (Moss and Smith, 1981), and postirradiation growth conditions (Khaengraeng and Reed, 2005) contribute in the inaction of *E. coli*, irradiation wavelengths dominate the inactivation rates of *E. coli* (Webb and Brown, 1979; Wegelin et al., 1994). It has therefore been suggested (McGuigan et al., 2012) that the DNA damage dominates *E. coli* inactivation under some conditions, while membrane damage may be the dominant route to cell death under others. Different strategies have been adopted to enhance the solar treatment of water.

Physical and chemical approaches have been contemplated and include solar mirrors, consideration of flow reactors, chemical additives, thermal enhancement, and heterogenous photocatalysis. The principles of these enhancement methods vary to a certain extend: (1) solar mirrors allow to concentrate solar radiation in the water stored container that can be a bottle or a reactor; solar collectors or reflectors such aluminium foil (Kehoe et al., 2001), compound parabolic collectors, parabolic, and V-groove (McLoughlin et al., 2004; Navntoft et al., 2008) have been shown to enhance the natural solar radiation and improve the disinfection efficiencies that were mostly attributed to the increase in water temperature of the system. (2) Variation of the design of flow reactors for improvement of solar disinfection has been investigated with a particular focus on the increase of the optical inactivation component of sunlight inactivation using solar collectors and reflectors (Fujishima et al., 2000; Vidal and Diaz, 2000), incorporation of photocatalyst (Fernandez-Ibanez et al., 2009; Dunlop et al., 2002), increase of thermal inactivation using flat plates painted black to adsorb solar radiation solar heating support (Jorgensen et al., 1998). (3) A host of chemical additives have been considered to improve SODIS inactivation; these include among others the lime juice/pulp (Harding and Schwab, 2012), a combination of sodium percarbonate, citric acid, and ascorbate (Fisher et al., 2012).

The photocatalytic disinfection of water can be accelerated in the presence of the catalysts in the process now known as heterogenous photocatalysis that has been described previously. The process has been so far effective for the inactivation of microorganisms however, some organic molecules also identified as "emerging contaminants" which have been found to be of serious concern with regard to their potential health effects and their continuous introduction to the environment. This new generation of contaminants include pharmaceuticals, antibiotics, analgesics, hormones, and biotoxins which have been the focus of much research, investigating the potential of solar photocatalysis for their degradation in the environment (Petrovic et al., 2003; Klavarioti et al., 2009).

Water authorities have shown concerns about the occurrence of biotoxins in the water systems due to the fact that they often occur in small amount as the byproducts of cell microbial destruction, their toxicity and the relative high cost of techniques required for their detection and quantification.

The photocatalytic degradation of biotoxins has been extensively investigated using several types of semiconductors, but TiO_2 has been must effective in achieving a rapid and effective decomposition of the cyanotoxins (Robertson et al., 1998; Robertson et al., 1999a; Robertson et al., 1999b; Feitz et al., 1999; Feng et al., 2006). In the series of attacks directed at the cyanotoxins during photocalytic decomposition, it is suggested that hydroxyl radical attack, UV transformation direct oxidation of the substrate are the major pathways. In a photocatalytic degradation and mineralization study carried out by Triantis et al. (2012), the nitrogen doped TiO_2 photocatalyst, was used under UV-A, solar, and visible light for the degradation of microcystin-LR. A comparative investigation of the potential of kronos, commercial Degussa P25 TiO_2 and reference TiO_2 nanopowders to degrade microcystin-LR showed that although all the kronos were effective in toxin elimination under UV-A irradiation, Degussa P25 TiO_2 achieved the higher toxin degradation (99%).

Zhang et al. (2014) have investigated the photocatalytic degradation of cylindrospermopsin using nanoparticles composed of anatase-brookite-rutile polymorphic titanium dioxide synthesized through the modified sol–gel method. They found the photocatalytic process to be effective for the degradation of cylindrospermopsin in clear synthetic solution and were positively influenced by the physicochemical properties of the synthesized polymorphic TiO_2, such as high surface area, small particle size, formation of heteronanostructure, and bandgap. Environmental conditions have been found to have a considerable influence on the photocatalytic degradation of cyanotoxins factors, such as pH of the solution have been predominantly reported. Using thin transparent TiO_2 photocatalytic films, Antoniou et al. (2009) reported that the acidic pH was more efficient for the degradation of microcystin-LR due to toxin interaction with the catalyst surface increased adsorption into the porous films. The effects of basic pH and the presence of high carbonate concentration on the titanium dioxide photocatalytic degradation of cylindrospermopsin was investigated by Zhang et al. (2015) and they did not find a major difference of the degradation yield under high or neutral pH, although the high pH appeared to inhibit the sulfate elimination. The high concentration of carbonate was found to inhibit the hydroxyl radical formation shifting the degradation mechanism toward a photocatalytic reaction controlled by carbonate radical.

Although the photocalytic degradation of mycotoxins has not been extensively studied as the cyanotoxins, existing studies have shown that a couple of mycotoxins are susceptible to degradation by the photocatalytic process. Using bentonite as photocatalyst, Combos and Serban studied its effect on the photocatalytic degradation of B1, B2, G2 aflatoxins, and T-2 mycotoxins under the irradiation of ultraviolet light. They found that the photocatalytic degradation effectiveness varied according to the micotoxin in presence and followed the pseudo-first-order rate law; the bentonite was found to activate the photocatalytic degradation of mycotoxin. Hilda-Elizabeth et al. (2005) investigated the titanium dioxide photocatalytic degradation of fumonisins B1 and B2 under ultraviolet radiation and found that 97% of the fumonisins in the samples were degraded after 120 min of exposure to the photocatalytic treatment.

As part of the photocatalytic reaction a numerous UV-detectable byproducts can be generated as a result of the nonselective action of hydroxyl radicals, these byproducts are often very stable and will persist (Table 9.4) even after a prolonged photocatalytic treatment (Lawton et al., 1999).

Five main types of the photocatalytic degradation byproducts have been suggested by Malato et al. (2009). Hydroxylated and dehydrogenated products, the latter of these derived from the loss of dehydrogenated substitute in the original contaminant if it had one, products from the oxidation of the alkali chain if it had one products derived from the opening of the aromatic ring in aromatic contaminants products of decarboxylation—products of isomerization and cyclation. The toxicity of six byproducts have been assessed using brine shrimp and protein phosphatase bioassays and no obvious toxic effect was detected for most of them (Liu et al., 2002) however, the accumulation of these byproducts can cause number of challenges including increase of organic matter, as well as inhibition of further treatment through clouding of the photocatalyst.

4 Conclusions

Biotoxins, such as cyanotoxins and mycotoxins have shown to be particularly dangerous even at very low dose for humans, and they occur more often in water used in hospital, as well as in household. They are very stable and subsist to conventional treatment, such as chlorination and flocculation. Over the years, there have been several attempts to improve the conventional methods for the effective removal of the so-called imminent pollutants. The

Table 9.4 Susceptible Bioproducts Formed During the Photocatalytic Degradation of Cyanotoxins

Type of Toxin	Semiconductor	Parameters investigated	Efficiency	By-products	References
Microcystin-LR	Nitrogen-doped TiO_2	Comparison of performance of semiconductors	2.49 $\mu Mmin^{-1}$	TOC, NO_3^-, NH_4^+, NO_2^-	Triantis et al. (2012)
	Degussa P25 TiO_2		3.23 $\mu Mmin^{-1}$		
	Kronos TiO_2		0.603 $\mu Mmin^{-1}$		
	TiO_2 nanopowder		2.61 $\mu Mmin^{-1}$		
Cylindrospermopsin	Polymorphic TiO_2(Anatase, Brukite, and Rutile)	Physicochemical properties of synthesized PM-TiO_2 nanoparticles, photocatalyst loading, pH, presence of natural organic matter	4 μg/g	NA	Zhang et al. (2014)
Cylindrospermopsin	Degussa P25 TiO_2	Temperature, pH, and presence of NOM	90%	NA	
	Hombikat UV-100 TiO_2		95%		
Microcystin-LR	Anatase TiO_2 immobilized on stainless steel	pH, adsorption, and initial toxin concentration	100% after 60 min	CO, CO_2, NH_3, H_2O, C9H10N, *m/z* 1011.5, *m/z* 1029.5	
Cylindrospermopsin	Degussa P25 TiO_2	Reaction byproducts and pathways of degradation	100% after 18 min	*m/z* 432.12, *m/z* 448.11, *m/z* 392.12, *m/z* 375.1	
Microcystin-LR	Nonmetal-(C, S, and N) codoped mesoporous anatase brookite heterojunction TiO_2	Effect of irradiation	100%	NI	
Cylindrospermopsin	Aeroxide TiO_2 P_{25}	Effect of basic pH and carbonate ion	100%	*m/z* 432.12, *m/z* 448.11, *m/z* 392.12, *m/z* 375.1, *m/z* 420.12	
Microcystin-LR	TiO_2 thin film	Effects of UV-A radiation, solution pH, initial toxin concentration, coated surface area of the TiO_2 films and their structural properties	100% after 180 min	*m/z* 1011.5, *m/z* 1027.5, *m/z* 1009.5, *m/z* 965.5, *m/z* 1029.5, and *m/z* 1063.5	Antoniou et al. (2009)

discovery of nanomaterials has raised some hopes as they have proven to be effective as adsorbents and catalysts due to their particular physicochemical properties, such as the large surface area they provide. The photocatalytic process is energy efficient using relatively cheap and mild catalyst that could operate under mild conditions; such a process could therefore be very beneficial for the sustainability of the environment. Photocatalytic degradation exhibits effective inhibition of the toxicity of biotoxins, however, the process may suffer a number of challenges such inhibition by high concentrations of normal organic matters and incomplete degradation of the biotoxins which result in the formation of byproducts to be dealt with in complementary processes. Some improvement of the photocatalytic degradation process is therefore required for effective implementation at large scale in water treatment plants.

References

Antoniou, M.G., Nicolaou, P.A., Shoemaker, J.A., de la Cruz, A.A., Dionysiou, D.D., 2009. Impact of the morphological properties of thin TiO_2 photocatalytic films on the detoxification of water contaminated with the cyanotoxin, microcystin-LR. Appl. Catal. B Environ. 91, 165–173.

Arvanitidou, M., Spaia, S., Velegraki, A., Pazarloglou, M., Kanetidis, D., Pangidis, P., Askepidis, N., Katsinas Ch, Vayonas, G., Katsouyannopoulos, V., 2000. High level of recovery of fungi from water and dialysate in haemodialysis units. J. Hosp. Infect. 45, 225–230.

Aslund, P., 1984. Skin irritations cause by moulds. Var Foda. 30, 327–330, (in Swedish).

Benabbou, A.K., Derriche, Z., Felix, C., Lejeune, P., Guillard, C., 2007. Photocatalytic inactivation of *Escherichia coli*: effect of concentration of TiO_2 and microorganism, nature and intensity of UV irradiation. Appl. Catal. B Environ. 76, 257–263.

Berney, M., Weilenmann, H.U., Egli, T., 2006. Flow-cytometric study of vital cellular functions in *Escherichia coli* during solar disinfection (SODIS). Microbiol. Mol. Biol. Rev. 152, 1719–1729.

Bhat, R.V., 1991. Aflatoxins: successes and failures of three decades of research. ACIAR Proc. 36, 77–82.

Bhat, R.V., Sherry, P.H., Amruth, R.P., Sudershan, R.V., 1997. A foodborne disease outbreak due to the consumption of mouldy sorghum and maize containing fumonisis mycotoxins. J. Toxicol. Clin. 35, 249–255.

Bhattacharya, R., Sugendram, K., Dangi, R.S., Rao, P.V., 1997. Toxicity evaluation of freshwater cyanobacterium *Microcystis aeruginosa* PCC 7806, II, nephrotoxicity in rats. Biomed. Environ. Sci. 10, 93–101.

Bouaicha, N., Via-Ordorika, L., Vandevelde, T., Fauchon, N., Puiseux-Dao, S., 1998. Toxic cyanoprokaryotes in resource waters: monitoring of their occurrence and toxin detection, workshop molecular methods for safe drinking water. Interlaken, Paris, France, pp. 1–9.

Brooks, W.P., Codd, G.A., 1987. Distribution of *Microcystis aeruginosa* peptide toxin and interactions with hepatic microsomes in mice. Pharmacol. Toxicol. 60 (3), 187–191.

Bui, T.H., Felix, C., Pigeot-remy, S., Herrmann, J.M., Lejeune, P., Guillard, C., 2008. Photo-catalytic inactivation of wild and hyper-adherent *E. coli* strains in presence of suspended or supported TiO_2. Influence of the isoelectric point, of the particle size and of the adsorptive properties of titania. J. Adv. Oxidat. Technol. 11, 510.

Cabral, D., Pinto, V.E.F., 2002. Fungal spoilage of bottled mineral water. Int. J. Food Microbiol. 72, 73–76.

Campinas, M., Rosa, M.J., 2010. Removal of microcystins by PAC/UF. Sep. Purif. Technol. 71, 114–120.

Carlson, M.P., 2003. Use of Feed Contaminated with Fungal (Mold) Toxins (Mycotoxins). University of Nebraska, Lincoln Extension, Institute of Agriculture and Natural Resources. Availbale from: http://ianrpubs.unl.edu/live/g1514/build/g1514.pdf

Carmichael, W.W., 1992. A status report on planktonic cyanobacteria (blue–green algae) and their toxins. Office of Research and Development United States Environmental protection agency/600/R-92/079, pp. 32–33.

Carmichael, W.W., 1994. The toxins of cyanobacteria. Sci. Am. 270, 1–86.

Castenholz, R.W., 1973. Ecology of blue–green algae in hot springs. In: Carr, N.G., Whitton, B.A. (Eds.), The Biology of Blue–Green Algae. Blackwell Scientific Publications, Oxford, pp. 379–414.

Chorus, I., Bartram, J., 1999. Toxic Cyanobacteria in Water: A Guide to Their Public Health Consequences, Monitoring and Management. Published by E & FN Spon on behalf of the World Health Organization, London.

Clariant Company, 2013. Advanced Materials. Available from: www.advancedmaterials.clariant.com

Codd, G.A., Ward, C.J., Bell, S.G., 1997. Cyanobacterial toxins occurrence, modes of action, health effects and exposure routes. In: Applied toxicology: approaches through basic science. Seiler, J.P., Vilanova, E. (Eds.), Applied Toxicology: Approaches Through Basic Science. Archives of ToxicologySpringer, Berlin, pp. 399–410, Suppl. 19.

Cook, D., Newcombe, G., 2002. Removal of microcystin variants with powdered activated carbon. Water Sci. Technol. Water Supply 2, 201–207.

Criado, M.V., Pinto, V.E.F., Badessari, A., Cabral, D., 2005. Conditions that regulate the growth of moulds inoculated into bottled mineral water. Int. J. Food Microbiol. 99, 343–349.

Dakovic, A., Matijasevic, S., Rottinghaus, G.E., Ledoux, D.R., Butkeraitis, P., Sekulic, Z., 2008. Aflatoxin B1 adsorption by natural and copper modified montmorillonite. Colloids Surf. B 66, 20–25.

Diaz, D.E., Smith, T.K., 2005. Mycotoxin sequestering agents: practical tools for the neutralisation of mycotoxins. In: Diaz, D.E. (Ed.), The Mycotoxin Blue Book. Nottingham University Press, Nottingham, UK, pp. 323–338.

Dixon, M.B., Richards, Y., Ho, L., Chow, C.W.K., O'Neill, B.K., Newcombe, G., 2011. A coagulation-powdered activated carbon-ultrafiltration—multiple barrier approach for removing toxins from two Australian cyanobacterial blooms. J. Hazard. Mater. 186, 1553–1559.

Doggett, M.S., 2000. Characterization of fungal biofilms within a municipal distribution system. Appl. Environ. Microbiol. 66 (3), 1249–1251.

Donati, C., Drikas, M., Hayes, R., Newcombe, G., 1994. Microcystin-LR adsorption by powdered activated carbon. Water Res. 28 (8), 1735–1742.

Dondjin, P., Eniyule, C., Geun, S.C., Jungae, L., Soyoung, J., Min, S.K., 2003. Production and characterization of monoclonal antibodies against microcystin-LR. Korean Chem. Soc. 24 (1), 1–3.

Dor, I., Danin, A., 1996. Cyanobacteria desert crusts in the Dead Sea Valley, Israel. (Suppl. 117), Algological Studies. Arch. Hydrobiol. 83, 197–206.

Dunlop, P.S.M., Byrne, J.A., Manga, N., Eggins, B.R., 2002. The photocatalytic removal of bacterial pollutants from drinking water. J. Photochem. Photobiol. A 148, 355–363.
EFSA, 2009. Review of mycotoxin-detoxifying agents used as feed additives: mode of action, efficacy and feed/food safety. CFP/EFSA/FEEDAP/2009/01.
Eisenstark, A., 1987. Mutagenic and lethal effects of near-ultraviolet radiation (290–400 nm) on bacteria and phage. Environ. Mol. Mutagen. 10, 317–337.
Eriksson, J.E., Toivola, D., Meriluoto, J., Karaki, H., Han, Y., Hartshorne, O., 1990a. Hepatocyte deformation induced by cyanobacterial toxins reflects inhibition of protein phosphatases. Biochem. Biophys. Res. Commun. 173, 1347–1353.
Eriksson, J.E., Gronberg, L., Nygard, S., Slotte, J.P., Meriluoto, J., 1990b. Hepatocellular uptake of ^{3}H-dihydromicrocystin-LR, a cyclic peptide toxin. Biochem. Biophys. Acta 1025, 60–66.
Falconer, I.R., 1991. Tumor promotion and liver injury caused by oral consumption of cyanobacteria. Environ. Toxicol. Water Qual. 6 (2), 177–184.
Falconer, I.R., 1994. Health problems from exposure to cyanobacteria and proposed safety guidelines for drinking and recreational water. In: Codd, G.A., Jefferies, T.M., Keevil, C.W., Potter, P. (Eds.), Detection methods for cyanobacterial toxins. Royal Society of Chemistry, Cambridge, pp. 3–10.
Falconer, I.R., Humpage, A.R., 1996. Tumour promotion by cyanobacterial toxins. Phycologia 35 (6), 74–79.
Falconer, I.R., Yeung, D.S.K., 1992. Cytoskeletal changes in hepatocytes induced by Microcystis toxins and their relation to hyperphosphorylation of cell proteins. Chem. Biol. Interac. 81, 181–196.
Falconer, I.R., Jackson, A.R.B., Langley, J., Runnegar, M.T.C., 1981. Liver pathology in mice in poisoning by the blue–green algae *Microcystis aeruginosa*. Aust. J. Biol. Sci. 34, 179–187.
Falconer, I.R., Runnegar, M.T.C., Buckley, T., Huyn, Y.L., Bradshaw, P., 1989. Using activated carbon to remove toxicity from drinking water containing cyanobacterial blooms. J. Am. Water Works Ass. 81, 102–105.
Fawell, J., Hart, J., James, H., Parr, W., 1993. Water Suppl. 11 (3–4), 109–121.
Fawell, J.K., James, C.P., James, H.A., 1994. Toxins from blue–green algae: toxicological assessment of microcystin-LR and a method for it's determination in water. Water Res. Centre, Medmenham, UK, 1–46.
FDA, 2011. FDA Regulatory Guidance for Mycotoxins. A Guide for Grain Elevators, Feed Manufacturers, Grain Processors and Exporters. National Grain and Feed Association.
Feitz, A.J., Waite, T.D., Jones, G.J., Boyden, B.H., Orr, P.T., 1999. Photocatalytic degradation of the blue green algal toxin microcystin-LR in a natural organic-aqueous matrix. Environ. Sci. Technol. 33, 243–249.
Feng, X., Rong, F., Fu, D., Yuan, C., Hu, Y., 2006. Photocatalytic degradation of trace-level of microcystin-LR by nano-film of titanium dioxide. Chin. Sci. Bull. 51, 1191–1198.
Feng, J., Shan, M., Du, H., Han, X., Xu, Z., 2008. In vitro adsorption of zearalenone by cetyltrimethyl ammonium bromide-modified montmorillonite nanocomposites. Microporous Mesoporous Mater. 113, 99–105.
Fernandez-Ibanez, P., Sichel, C., Polo-Lopez, I., de Cara-Garcia, M., Tello, J.C., 2009. Photocatalytic disinfection of natural well water contaminated with *Fusarium solani* using TiO_2 slurry in solar CPC photo-reactors. Catal. Today. 144, 62–68.
Fisher, M.B., Iriarte, M., Nelson, K.L., 2012. Solar water disinfection (SODIS) of *Escherichia coli*, *Enterococcus* spp., and MS2 coliphage: effects of additives and alternative container materials. Water Res. 46, 1745–1754.

Fitzgeorge, R.B., Clark, S.A., Keevil, C.W., 1994. Detection methods for cyanobacterial toxins. In: Codd, G.A., Jeffries, T.M., Keevil, C.W., Potter, E. (Eds.). Royal Society of Chemistry, Cambridge, p. 69.

Flores, C.M., Dominguez, J.M., Diaz-De-Leon, J., 1999. Modeling and experimental comparison of the differential adsorption of B1 and G1 aflatoxins on mineral aluminosilicate surfaces. J. Environ. Pathol. Toxicol. Oncol. 18, 213–220.

Fosso-Kankeu, E., Jagals, P., Du Preez, H., 2008. Exposure of rural households to toxic cyanobacteria in container-stored water. Water SA 34 (5), 631–636.

Fujishima, A., Rao, T.N., Tryk, D.A., 2000. Titanium dioxide photocatalysis. J. Photochem. Photobiol. C 1, 1–21.

Ganesh, R., Leong, L.Y.C., Skandan, G., Jain, M., Liu, B., Larson, L.E., 2005. Use of nanoscale materials for water treatment to achieve multiple water quality goals. In: Proceedings of Water Quality Technology Conference and Exposition. Quebec City, Quebec, Canada.

Goncalves, A.B., Russell, R., Paterson, M., Lima, N., 2006. Survey and significance of filamentous fungi from tap water. Int. J. Hyg. Environ. Health 209, 257–264.

Grabinska-Loniewska, A., Konillowicz-Kowalska, T., Wardzynska, G., Boryn, K., 2007. Occurrence of fungi in water distribution system. Polish J. Environ. Stud. 16 (4), 539–547.

Grant, P.G., Phillips, T.D., 1998. Isothermal adsorption of aflatoxin B1 on HSCAS clay. J. Agric. Food Chem. 46, 599–605.

Grobbelaar, J.U., Elsabe, B., Van den Heever, J.A., Oberholster, A.-M., Oberholster, P.J., 2004. Toxin production by cyanobacteria: scope and dynamics of toxins produced by cyanophytes in the freshwater of South Africa and the implementations of human and others users, WRC Report No: 1029/1/04, pp. 1–9.

Guillard, C., Bui, T.H., Felix, C., Moules, V., Lina, B., Lejeune, P., 2008. Microbiological disinfection of water and air by photocatalysis. CR Chim. 11, 107–113.

Hagare, P., Thiruvenkatachari, R., Ngo, H.H., 2001. A feasibility study of using hematite to remove dissolved organic carbon in water treatment. Sep. Sci. Technol. 36, 2547–2559.

Hageskal, G., Knutsen, A.K., Gaustad, P., de Hoog, G.S., Skaar, I., 2006. Diversity and significance of mold species in Norwegian drinking water. Appl. Environ. Microbiol. 72, 7586–7593.

Harding, A.S., Schwab, K.J., 2012. Using limes and synthetic psoralens to enhance solar disinfection of water (SODIS): a laboratory evaluation with norovirus, *Escherichia coli* and MS2. Am. J. Trop. Med. Hyg. 86, 566–572.

Helali, S., Polo-Lopez, M.I., Fernandez-Ibanez, P., Ohtani, B., Amano, F., Malato, S., Guillard, C., 2014. Solar photocatalysis: a green technology for *E. coli* contaminated water disinfection. Effect of concentration and different types of suspended catalysts. J. Photochem. Photobiol. A: Chem. 276, 31–40.

Henry, S.H., Bosch, F.X., Troxell, T.C., Bolger, P.M., 1999. Reducing liver cancer—global control of aflatoxin. Science 286, 2453–2454.

Hilda-Elizabeth, C.-V., Pandiyan, T., Magda, C.-M., Guillermina, B., Silvia-Denise, P.-B., 2005. Photo-degradation of fuminisins B-1 and B-2, toxins of the fungus *Fusarium verticillioides* (Sacc.) Nirenberg from corn (*Zea mays* L), by ultraviolet radiation with titanium dioxide. Revista Mexicana de Fitopatologia. 23 (3), 246–252.

Hinzelin, F., Block, J.C., 1985. Yeasts and filamentous fungi in drinking water. Environ. Technol. Lett. 6, 101–106.

Honkanen, R.E., Chaplan, F.R., Patterson, G.M.L., Abercombie, J., 1996. Development of a protein phosphatase-based assay for the detection of phosphatase inhibitors in crude whole cell/animal extracts. Toxicon 34, 307–308.
Hooser, S.B., Beasley, V.R., Basgall, E.J., Carmichael, W.W., Haschek, W.M., 1990. microcystin-LR induced ultrastructural changes in rats. Vet. Pathol. 27 (1), 9–15.
Hu, J., Chen, G., Lo, I.M.C., 2005. Removal and recovery of Cr(VI) from wastewater by maghemite nanoparticles. Water Res. 39, 4528–4536.
Huwig, A., Freimund, S., Kappeli, O., Dutler, H., 2001. Mycotoxin detoxication of animal feed by different adsorbents. Toxicol. Lett. 122, 179–188.
Huynh-Delerme, C.S., Puiseux-Dao, S., 1998. Toxines d'origine algale inhibitrices de serine-threonine phosphatases. CR. Soc. Biol. 192 (3), 1–22.
IARC, 1993. Some naturally occurring substances: food items and constituents, heterocyclic aromatic amines and mycotoxins. IARC Monogr. Eval. Carcinog. Risk Hum. 56, 445–466.
Ito, E., Kondo, F., Teraro, K., Harada, K.-I., 1997. Neo-plastic nodular formation in mouse liver induced by repeated intraperitoneal injections of microcystin-LR. Toxicon 35, 1453–1457.
Jaag, O., 1945. Untersuchungen uber die vegetation und biologie der Algen des nackten gesteins in den Alpen, im Jura und im schweizerischen Mitteland. Kryptogamen-flora der Schweiz, Band ix, heft 3. Kommissionsverlag buchdruckerei buchler and Co., Bern.
Jaynes, W.F., Zartman, R.E., Hudnall, W.H., 2007. Aflatoxin B1 adsorption by clays from water and corn meal. Appl. Clay Sc. 36, 197–205.
Jones, G., Minatol, W., Craig, K., Naylor, R., 1993. Removal of low level cyanobacterial peptide toxins from drinking water using powdered and granular activated carbon and chlorination—results of laboratory and pilot plant studies. In: Proceedings of the 15th Federal AWWA Convention, April 18–23, Gold Coast, QLD, Australia.
Jorgensen, F.A.J., Nohr, K., Sorensen, H., Boisen, F., 1998. Decontamination of drinking water by direct heating in solar panels. J. Appl. Microbiol. 85, 441–447.
Kaja, K., 1995. Toxicology of microcystins. In: Watanabe, M., Harada, K., Carmichael, W., Fujiki, H. (Eds.), Toxic Microcystis. CRC Press, Boca Raton, FL, pp. 175–202.
Kann, E., 1988. Zur autokologie benthischer cyanophyten in reinen europaischen seen und Fliessgewassern. Arch. Hydrobiol. (Suppl. 80). Algol. Stud. 50-53, 473–495.
Kehoe, S.C., Joyce, T.M., Ibrahim, P., Gillespie, J.B., Shahar, R.A., McGuigan, K.G., 2001. Effect of agitation, turbidity, aluminium foil reflectors and container volume on the inactivation efficiency of batch-process solar disinfectors. Water Res. 35, 1061–1065.
Khaengraeng, R., Reed, R.H., 2005. Oxygen and photoinactivation of *Escherichia coli* in UVA and sunlight. J. Appl. Microbiol. 99, 39–50.
Klavarioti, M., Mantzavinos, D., Kassinos, D., 2009. Removal of residual pharmaceuticals from aqueous systems by advanced oxidation processes. Environ. Int. 35, 402–417.
Kodak, B.G., Kenefick, S.L., Fritz, D.L., Rousseaux, C.G., Prepas, E.E., Hrudey, S.E., 1993. Occurrence and toxicological evaluation of cyanobacterial toxins in Alberta lakes and farm dugouts. Water Res. 27 (3), 495–506.
Kol, E., 1968. Kryobiologie. I. Kryovegetation. In: Elster, H.J., Ohle, W., (Eds.), Die Binnengewasser, band ××i∨. E. Schweizerbart'sche verlagsbuchhandlung, Stuttgart, p. 216.

Kuiper, G.J., Lemmen, J.G., Carlsson, B., Corton, C., Safe, S.H., van der Saag, P.T., 1998. Interaction of estrogenic chemicals and phytoestrogens with estrogen receptor β. Endocrinology 139, 4252–4263.
Laamanen, M., 1996. Cyanoprokaryotes in the Baltic Sea ice and winter plankton. Arch. Hydrobiol. (Suppl. 117), Algol. Stud. 83, 423–433.
Lambert, T.W., Holmes, C.F.B., Hrudey, S.W., 1996. Adsorption of microcystin-LR by activated carbon and removal in full scale treatment. Water Res. 30, 1411–1422.
Lawton, L.A., Robertson, P.K.J., Cornish, B.J.P.A., Jaspars, M., 1999. Detoxification of microcystins (cyanobacterial hepatotoxins) using TiO_2 photocatalytic oxidation. Environ. Sci. Technol. 33, 771–775.
Lee, J., Walker, H.W., 2006. Effect of process variables and natural organic matter on removal of microcystin-LR by PAC/UF. Environ. Sci. Technol. 40 (23), 7336–7342.
Lee, J., Walker, H.W., 2011. Adsorption of microcystin-LR onto iron oxide nanoparticles. Colloids Surf. A Physicochem. Eng. Aspects 373, 94–100.
Leffers, H., Haesby, M., Vendelbo, B., Skakkebaek, N.E., Jorgensen, M., 2001. Oestrogenic potencies of zeranol, oestradiol, diethylstilbestrol, bisphenol-A and genistein: implications for exposure assessment of potential endocrine disrupters. Hum. Reprod. 16, 1662–1670.
Legrini, O., Oliveros, E., Braun, A.M., 1993. Chem. Rev. 93, 671.
Li, N., Mei, Z., Wei, X., 2012. Study on sorption of chlorophenols from aqueous solutions by an insoluble copolymer containing β-cyclodextrin and polyamidoamine units. Chem. Eng. J. 192, 138–145.
Lin, J.R., Chu, F.S., 1994. Kinetics of distribution of microcystin-LR analysis. J. Agric. Food Chem. 42 (4), 1035–1040.
Litter, M.I., 1999. Heterogenous photocatalysis transition metal ions in photocatalytic systems. Appl. Catal. B Environ. 23, 89–114.
Liu, B.H., 2006. Contaminants removal from water using nanoparticles. MS thesis. Department of Environmental Science, California State University, Fullerton.
Liu, I., Lawton, L.A., Cornish, B.J.P.A., Robertson, P.K.J., 2002. Mechanistic and toxicity studies of the photocatalytic oxidation of microcystin-LR. J. Photochem. Photobiol. A Chem. 148, 349–354.
Lye, M.S., Ghazali, A.A., Mohan, J., Alwin, N., Mair, R.C., 1995. An outbreak of acute hepatic encephalopathy due to severe aflatoxicosis in Malaysia. Am. J. Trop. Med. Hyg. 53, 68–72.
Malato, S., Fernandez-Ibanez, P., Maldonado, M.I., Blanco, J., Gernjak, W., 2009. Decontamination and disinfection of water by solar photocatalysis: recent overview and trends. Catal. Today 147, 1–59.
Marasas, W.F.O., Miller, J.D., Riley, R.T., Viscontini, A., 2001. Fumonisins—occurrence, toxicology, metabolism and risk assessment. In: Summerell, B.A., Leslie, J.F., Backhouse, D., Bryden, W.L., Burgess, L.W., (Eds.), *Fusarium*: Paul E, Nelson Memorial Symposium APS Press, St Paul, Minnesota, pp. 332–359.
Masimango, N., Remacle, J., Ramaut, J.L., 1978. The role of adsorption in the elimination of aflatoxin B1 from contaminated media. Eur. J. Appl. Microbiol. 6, 101–105.
Matsunaga, T., Tomoda, R., Nakajima, T., Nakamura, N., Komine, T., 1988. Continuous-sterilization system that uses photosemiconductor powders. Appl. Environ. Microbiol. 54, 1330.
Matsushima, R., Yoshizawa, S., Watanabe, M.F., Harada, K., Furusawa, M., Carmichael, W.W., Fujiki, H., 1990. Biochem. Biophys. Res. Commun, 171–867.
McGuigan, K.G., Conroy, R.M., Mosler, H.-J., du Preez, M., Ubomba-Jaswa, E., Fernandez-Ibanez, P., 2012. Solar water disinfection (SODIS): a review from bench-top to roof-top. J. Hazard. Mater. 235–236, 29–46.

McLoughlin, O.A., Kehoe, S.C., McGuigan, K.G., Duffy, E.F., Al Touati, F., Gernjak, W., Oller, I., Malato, S., Gill, L.W., 2004. Solar disinfection of contaminated water: a comparison of three small-scale reactors. Sol. Energy 77, 657–664.

Meriluoto, J.A., Nygard, S.E., Dahlem, A.M., Eriksson, J.E., 1990. Synthesis, organotropism and hepatocellular uptake of two tritium-labelled epimers of dihydromicrocystin-LR, a cyanobacterial peptide toxin analog. Toxicon 28 (12), 1439–1446.

Metcalf, J.S., Codd, G.A., 2004. Cyanobacterial Toxins in the Water Environment. Foundation for Water Research, Marlow, UK.

Mez, K., Beattie, K.A., Codd, G.A., Hanselmann, K., Hauser, B., Naezeli, H., Preisig, H.R., 1997. Identification of a microcystin in a benthic cyanobacteria linked to cattle deaths on alpine pastures in Switzerland. Eur. J. Phycol. 32, 111–117.

Miksicek, R.J., 1994. Interaction of naturally occurring non-esteroidal estrogens with expressed recombinant human estrogen receptor. J. Esteroid. Biochem. Mol. Biol. 49, 153–160.

Miller, J.D., Simon, J.W., Blackwell, B.A., Greenhalgh, R., Taylor, A., 2001. Deoxynivalenol: a 25 year perspective on a trichothecene of agricultural importance. In: Summerell, B.A., Leslie, J.F., Backhouse, D., Bryden, W.L., Burgess, L.W. (Eds.), Fusarium: Paul E, Nelson Memorial Symposium APS Press, St. Paul, Minnesota, pp. 310–320.

Mohamed, Z.A., Chrmichael, W.W., An, J., El-Sharouny, H.M., 1999. Env. Toxicol. 15, 197–201.

Moss, S.H., Smith, K.C., 1981. Membrane damage can be a significant factor in the inactivation of *Escherichia coli* by near-ultraviolet radiation. Photochem. Photobiol. 33, 203–210.

Navntoft, C., Ubomba-Jaswa, E., McGuigan, K.G., Fernandez-Ibanez, P., 2008. Enhancement of batch solar disinfection (SODIS) using non-imaging reflectors. J. Photochem. Photobiol. B 93, 155–161.

Nel, A.E., Madler, L., Velegol, D., Xia, T., Hoek, E.M., Somasundaran, P., Klaessig, F., Castranova, V., Thompson, M., 2009. Understanding biophysicochemical interactions at the nano-bio interface. Nat. Mater. 8, 543–557.

Newcombe, G., Cook, D., Brooke, S., Ho, L., Slyman, N., 2003. Treatment options for microcystin toxins: similarities and differences between variants. Environ. Technol. 24, 299–308.

Niemi, R.M., Kunth, S., Lundstrom, K., 1982. Actinomycetes and fungi in surface waters and potable water. Appl. Environ. Microbiol. 43, 378–388.

Nishiwaki-Matsushima, R., Ohta, T., Nishiwaki, S., Suganuma, M., Kohyama, K., Ishiwaki, T., Carmichael, W.W., Fujiki, H.J., 1992. Liver tumor promotion by the cyanobacterial cyclic peptide toxin microcystin-LR. J. Cancer Res. Clin. Oncol. 118, 420–424.

Northolt, M.D., Frisvad, J.C., Samson, R.A., 1996. Occurrence of food-borne fungi and factors for growth. In: Samson, R.A., Hoekstra, E.S., Frisvad, J.C., Filtenborg, O. (Eds.). Introduction of food-borne fungi. Centraalbureau voor schimmelcultures, Baarn, The Netherlands, pp 243–251.

Ohtani, I., Moore, R.E., Runnegar, M.T.C., 1992. Cylindrospermopsin: a potent hepatotoxin from the blue–green algea *Cylindrospermopsis raciborskii*. J. Am. Chem. Soc. 114, 7941–7942.

Paterson, R.R.M., Lima, N., 2005. Fungal contamination of drinking water. Lehr, J., Keeley, J., Lehr, J., Kingery III, T.B. (Eds.), Water Encyclopedia. John Wiley & Sons, Inc, New York.

Patil, S., Sandberg, A., Heckert, E., Self, W., Seal, S., 2007. Protein adsorption and cellular uptake of cesium oxide nanoparticles as a function of zeta potential. Biomaterials 48, 4600–4607.

Pendleton, P., Schumann, R., Wong, S.H., 2001. Microcystin adsorption by activated carbon. J. Colloid Interface Sci. 240, 1–8.
Peraica, M., Radic, B., Lucic, A., Pavlovic, M., 1999. Toxic effects of mycotoxins in human health. Bull. WHO 77, 754–766.
Pereira, V.J., Basilio, M.C., Fernandes, D., Domingues, M., Paiva, J.M., Benoliel, M.J., Crespo, M.T., San Romao, M.V., 2009. Occurrence of filamentous fungi and yeasts in three different drinking water sources. Water Res. 43, 3813–3819.
Petrovic, M., Gonzalez, S., Barcelo, D., 2003. Analysis and removal of emerging contaminants in wastewater and drinking water. Trends Anal. Chem. 22, 685.
Phillips, T.D., Sarr, B.A., Clement, B.A., Kubena, L.F., Harvey, R.B., 1991. Prevention of aflatoxicosis in farm animals via selective chemisorption of aflatoxin. In: Bray, G.A., Ryan, D.H. (Eds.), Mycotoxins, Cancer and Health. Louisiana State University Press, Baton Rouge, pp. 223–237.
Phillips, T.D., Sarr, B.A., Grant, P.G., 1995. Nat. Toxins 3, 204.
Pigeot-Remy, S., Simonet, F., Errazuriz-Cerda, E., Lazzaroni, J.C., Atlan, D., Guillard, C., 2011. Photocatalysis and disinfection of water: identification of potential bacterial targets. Appl. Catal. B Environ. 104, 390–398.
Prescott, L.M., Harley, J.P., Klein, D.A., 2002. Microbiology, fifth ed. McGraw-Hill Companies, Inc, New York, pp. 799–801.
Puiseux-Dao, S., Bouaicha, N., Diogene, G., 2005. Maito-toxin, okadaic acid and microcystins: toxins disturbing signal transduction and phosphorylations. In: Rochat, H., Martin-eauclaire, M.F. (Eds.), Animal toxins tools in cell biology. Chapman & Hill, Weinheim, Germany.
Rae, B., Moolan, R.W., Clark, R.C., 1999. Algal toxins in drinking water supplies. Water Research Commission Report No. 549/1/99, Pretoria, South Africa.
Rajeshwar, K., 1995. J. Appl. Electrochem. 25, 1067.
Ramos, A.J., Fink-Gremmels, J., Hernandez, E., 1996. Prevention of toxic effects of mycotoxins by means of non-nutritive adsorbent compounds. J. Food Prot. 59, 631–641.
Rapala, J., Sivonen, K., 1998. Assessment of environmental conditions that favor hepatotoxic and neurotoxic *Anabaena* spp. strains cultured under light limitation at different temperatures, microbial ecology, 36, Springer-Verlag, New York, Inc, pp. 181–192.
Rapala, J., Lahti, K., Rasasen, L.A., Esala, A.-L., Niemela, S.I., Sivonen, K., 2002. Endotoxins associated with cyanobacteria and their removal during drinking water treatment. Water Res. 36, 2627–2635.
Rauf, M.A., Ashraf, S.S., 2009. Fundamental principles and application of heterogeneous photocatalytic degradation of dyes in solution. Chem. Eng. J. 151, 10–18.
Reed, R.H., 1997. Solar inactivation of faecal bacteria in water: the critical role of oxygen. Lett. Appl. Microbiol. 24, 276–280.
Rheeder, J.P., Marasas, W.F.O., Thiel, P.G., Sydenham, E.W., Shephard, G.S., van Schalwyk, D.J., 1992. *Fusarium moniliforme* and fumonisins in corn in relation to human esophageal cancer in Transkei. Phytopathology 82, 353–357.
Rincon, A.G., Pulgarin, C., 2007. Fe^{3+} and TiO_2 solar-light-assisted inactivation of *E. coli* at field scale: implications in solar disinfection at low temperature of large quantities of water. Catal. Today 122, 128–136.
Robertson, P.K.J., Lawton, L.A., Cornish, B.J.P.A., Jarpars, M., 1998. Processes influencing the destruction of microcystin-LR by TiO_2 photocatalysis. J. Photochem. Photobiol. A Chem. 116, 215–219.
Robertson, P.K.J., Lawton, L.A., Cornish, B.J.P.A., 1999a. The involvement of phycocyanin pigment in the photodecomposition of the cyanobacterial toxin, microcystin-LR. J. Porphyrins Phthalocyanines 3, 544–551.

Robertson, P.K.J., Lawton, L.A., Munch, B., Cornish, B.J.P.A., 1999b. The destruction of cyanobacterial toxins by titanium dioxide photocatalysis. J. Adv. Oxid. Technol. 4, 20–26.

Robinson, N.A., Pace, J.G., Matson, C.F., Miura, G.A., Lawrence, W.B., 1991. Tissue distribution, excretion and hepatic biotransformation of microcystin-LR in mice. J. Pharmacol. Exp. Ther. 256 (1), 176–182.

Romanowska-Duda, Z., Mankiewicz, J., Tarczynska, M., Walter, Z., Zalewski, M., 2002. The effect of toxic cyanobacteria (blue–green algae) on water plants and animal cells, polish. J. Environ. Stud. 11, 561–566.

Rosenzweig, W.D., Minnigh, H.A., Pipes, W.O., 1986. Fungi in potable water distribution systems. J. AWWA 78, 53–55.

Rositano, J., Nicholson, B.C., 1994. Water treatment techniques for removal of cyanobacterial toxins from water. Australian Centre for Water Quality Research. Salisbury, South Australia.

Santin, E., Paulillo, A.C., Maiorka, P.C., Alessi, A.C., Krabbe, E.L., Maiorka, A., 2002. Avian Pathol. 31, 73.

Shephard, G.S., 2004. Mycotoxins worldwide: current issues in Africa. In: Barug, D., van Egmond, H., Lopez-Garcia, R., van Ossenbruggen, T., Visconti, A. (Eds.), Meeting the Mycotoxin Menace, Wageningen Academic Publishers, Wageningen, the Netherlands. pp. 81–88.

Shier, W.T., Shier, A.C., Xie, W., Mirocha, C.J., 2001. Structure-activity relationships for human estrogenic activity in zearalenone mycotoxins. Toxicon 39, 1435–1438.

Sichel, C., Tello, J.C., de Cara-Garcia, M., Fernandez-Ibanez, P., 2007. Effect of UV solar intensity and dose on the photocatalytic disinfection of bacteria and fungi. Catal. Today 129, 152–160.

Sivonen, K., 1990. Effects of light, temperature, nitrate, orthophosphate, and bacteria on growth of and hepatotoxin production by *Oscillatoria agardhii* strains. Appl. Environ. Microbiol. Am. Soc. Microbiol. 56 (9), 2658–2666.

Sivonen, K., 1996. Cyanobacterial toxins and toxin production. Phycologia 35 (Suppl.6), 12–14.

Skulberg, O.M., 1990. Terrestrial and limnic algae and cyanobacteria. In: Elvebakk, A., Prestrud, P. (Eds.), A catalogue of Svalbard plants, fungi, algae and cyanobacteria. Part 9, Norsk polarinstitutt Skrifter. 198, pp. 383–395.

Taylor, H., 2003. Surface waters. In: Mara, D., Horan, N. (Eds.), Water and wastewater microbiology. Academic Press, London.

Teixeria, M.G.L.C., Costa, M.C.N., Carvalho, V.L.P., Pereira, M.S., Hage, E., 1993. Gastroenteritis epidemic in the area of the Itaparica Dam, Bahia, Brazil Bull. Pan Am. Health Org. 27, 244–253.

Tiemann, U., Tomek, W., Schneide, F., Vanselow, J., 2003. Effects of the mycotoxins α- and β-zearalenol on regulation of progesterone synthesis in cultured granulosa cells from porcine ovaries. Reprod. Toxicol. 17, 673–681.

Tomasevic-Canovic, M., Dakovic, A., Rottinghau, G., Matijasevic, S., Duricic, M., 2003. Surfactant modified zeolites—new efficient adsorbents for mycotoxins. Micropor. Mesopor. Mater. 61, 173–180.

Triantis, T.M., Fotiou, T., Kaloudis, T., Kontos, A.G., Falaras, P., Dionysiou, D.D., Pelaez, M., Hiski, A., 2012. Photocatalytic degradation and mineralization of microcystin-LR under UV-A, solar and visible light using nanostructured nitrogen doped TiO_2. J. Hazard. Mater. 211–212, 196–202.

Vesper, S.J., Mckinstry, C., Yang, C., Haugland, R.A., Kercsmar, C.M., Yike, I., Schluchter, M.D., Kirchner, H.L., Sabolewski, J., Allan, T.M., Dearborne, D.G., 2006. Specific molds associated with asthma in water-damaged homes. J. Occup. Environ. Med. 48, 852–858.

Vidal, A., Diaz, A.I., 2000. High-performance, low-cost solar collectors for disinfection of contaminated water. Water Environ. Res. 72, 271–276.

Watts, R.J., Kong, S., Orr, M.P., Miller, G.C., Henry, B.E., 1995. Photocatalytic inactivation of coliform bacteria and viruses in secondary wastewater effluent. Water Res. 29, 95–100.

Webb, R.B., Brown, M.S., 1979. Action spectra for oxygen-dependent and independent inactivation of *Escherichia coli*-Wp2s from 254-nm to 460-nm. Photochem. Photobiol. 29, 407–409.

Wegelin, M., Canonica, S., Mechsner, K., Fleischmann, T., Pesaro, F., Metzler, A., 1994. Solar water disinfection: scope of the process and analysis of radiation experiments. J. Water SRT—Aqua. 43, 154–169.

Whitton, B.A., 1992. Diversity and taxonomy of the cyanobacteria. In: Mann, N.H., Carr, N.G. (Eds.), Photosynthetic Prokaryotes. Plenum press, New York, pp. 1–51.

Whitton, B.A., Potts, M., 2000. The Ecology of Cyanobacteria. Kluwer Academic Publishers, printed in the Netherlands, pp. 1–11.

WHP, 1999. Toxic cyanobacteria in water: a guide to their public health consequences, monitoring and management, laboratory analysis of cyanotoxins. In: Chorus, I., Bartram, J.E., (Eds.), Spon, London and New-York.

WHO, 2003. Worldwide regulations for mycotoxins in food and feed 2003. Available from: ftp://ftp.fao.org/docrep/fao/007/y5499e/y5499e00.pdf

WHO, 2004. Guidelines for drinking water quality, third ed. WHO Library Cataloguing in Publication data; Geneva.

Yamaguchi, M.U., Rampazzo, R.C.P., Yamada-Ogatta, S.F., Nakamura, C.V., Ueda-Nakamura, T., Filho, B.P.D., 2007. Yeasts and filamentous fungi in bottled mineral water and tap water from municipal supplies. Braz. Arch. Biol. Technol. 50 (1), 1–9.

Yan, H., Gong, A., He, H., Zhou, J., Wei, Y., Lv, L., 2006. Adsorption of microcystins by carbon nanotubes. Chemosphere 62, 142–148.

Yavuz, C.T., Mayo, J.T., Yu, W.W., Prakash, A., Falkner, J.C., Yean, S., Cong, L.L., Shipley, H.J., Kan, A., Tomson, M., Natelson, D., Colvin, V.L., 2006. Low-field magnetic separation of monodisperse FeO_4 nanocrystals. Science 314, 964–967.

Yoshida, T., Makita, Y., Tsutsumi, T., Yoshida, F., Sekijima, M., Tamura, S.-I., Ueno, Y., 1997. Acute oral toxicity of microcystin-LR, a cyanobacterial hepatotoxin, in mice. Nat. Toxins 5, 91–95.

Yu, S.Z., 1995. Primary prevention of hepatocellular carcinoma. J. Gastroent. Hepatol. 10, 674–682.

Yu, M.F., Lourie, O., Dyer, M.J., Moloni, K., Kelly, T.F., Ruoff, R.S., 2000. Strength and breaking mechanism of multiwalled carbon nanotubes under tensile load. Science 287, 5453–5637.

Zhang, G., Nadagouda, N.M., O'shea, K., El-Sheikh, M.S., Ismail, A.A., Likodimos, V., Falaras, P., Dionysiou, D.D., 2014. Degradation of cylindrospermopsin by using polymorphic titanium dioxide under UV-Vis irradiation. Catal. Today 224, 49–55.

Zhang, G., He, X., Nadagouda, N.M., O'shea, K.E., Dionysiou, D.D., 2015. The effect of basic pH and carbonate ion on the mechanism of photocatalytic destruction of cylindrospermopsin. Water Res. 73, 353–361.

10

NANOTECHNOLOGY-BASED MEMBRANE-SEPARATION PROCESS FOR DRINKING WATER PURIFICATION

Santanu Sarkar*, Ankur Sarkar, Chiranjib Bhattacharjee****

**Environment Research Group, R&D, Tata Steel Ltd, Jamshedpur, Jharkhand, India; **Jadavpur University, Chemical Engineering Department, Kolkata, West Bengal, India*

1 Introduction

The scarcity of drinking water has become an alarming threat globally due to rapid increase of the global population and consequently in last century, it has been estimated the rate of water consumption has been increased up to seven times to its initial rate (Human Development Report, 2006). According to a survey of the World Water Council, about 4 million people will suffer water scarcity by 2030 due to rapid increase of water use in domestic purpose, agriculture, industry, energy sector, and so on (Urban Urgency, 2007). In 2010, another study by WHO has shown that almost 1.1 billion people are facing insufficiency of drinking water whereas more than 2.5 billion people are suffering from improper sanitization (Progress on Sanitation and Drinking-Water, 2010). In early 2007 the United Nations reported that in every year almost 2.2 million people die due to waterborne infections, and among those are children under the age of 5 (Coping with Water Scarcity, 2007). Therefore, scarcity of fresh or drinking water may cause the end of human race. Thus, it is the high time to protect the available fresh-water resources and create such types of water resources to mitigate rising demand of drinking water worldwide. It is well known that the sources of fresh water are very limited and it is very tough to generate new resource for drinking water,

Water Purification. http://dx.doi.org/10.1016/B978-0-12-804300-4.00010-1

therefore the development and adaptation of water purification technologies is very much essential.

In current practice, there are different technologies available for water purification purposes but some of them are able to treat the water up to the quality for the drinking purposes. Most of the water-treatment processes are highly energy and/or chemical consuming in nature. During water purification with the help of such methods, toxic by-products are generated and hence, the disposal of those materials is very challenging indeed. Among those treatment processes, membrane-separation processes are often used for the treatment of drinking water, as these techniques are less complicated to other methods and at the same time it neither consumes any chemical nor produces any toxic by-products. During the membrane-separation process, pressure is applied as a driving force and after certain time interval; the membrane surface is cleaned by means of some physical or nonhazardous chemical processes to enhance the performance of the membrane. Nowadays, pressure-driven membrane processes are the most widely used in water purification applications in both domestic and commercial purposes (Ulbricht, 2006). The membrane-separation technology is commercially available and classified according to their purpose of use, such as for separation of suspended solids, protozoa, and bacteria microfiltration (MF) is commonly used, for removal of virus and colloid ultrafiltration (UF) is used, and for the reduction of hardness, heavy metals, and dissolved organic matter nanofiltration (NF) is frequently used. Moreover, for the desalination and production of ultrapure water, reverse osmosis (RO) (Ulbricht, 2006) is commonly used in both industrial and domestic purposes. Though membrane technology has several advantages over other common water purification process, it has some limitations. The most common disadvantage is the fouling tendency of the membrane during the operation. The fouling of a membrane is caused due to the pore blockage by foreign particles present in water stream. To make the membrane-separation process more acceptable for purification and protection of drinking water, the production membrane with low cost, higher selectivity, fouling resistance, stable performance, and so on are required.

In the last decade, adaptation of nanotechnology has become a promising pathway to overcome the limitations of normal membrane separation. Moreover, different commercial applications of nanotechnology have been implemented in the field of water purification and eventually such types of applications are rapidly increasing. The presence of nanoparticles improve the antifouling properties, permeability, catalytic activity, flux recovery, stability both thermal and chemical, and so on. This chapter would portray a general overview of different type of nanomaterial-based membrane and their various applications in the domain of water purification. Moreover, a detailed discussion of present and future perspectives for the commercialization of such modified membrane has been elaborated here.

2 The Brief History of Membranes

The membrane can be mainly classified depending on the composition fabrication material and according to that, membrane can be manufactured from both organic and inorganic materials. A brief description of membrane fabrication and their possible applications in different water-treatment processes has been illustrated later in the chapter.

2.1 Inorganic Membranes

Inorganic membranes are fabricated from particularly inorganic material and such type of membrane is very robust in nature. The inorganic membranes are broadly classified into ceramic membranes and hence, a detailed description along with its utilization has been described in this chapter.

2.1.1 Ceramic Membranes

The ceramic membrane has been prepared long way before and further development in the same domain has been made in the year of 1980 for industrial applications (Bhave, 1991). Such types of membranes are prepared via sol–gel process and those are structurally asymmetric which consist of dense layer on the porous support made of alumina, silica, zirconia, mullite, oxide mixtures, and sintered metals (Kazemimoghadam, 2010; Hoek and Jawor, 2002). Interestingly, such porous structure is formed by deposition of particle with gradual decrement in size after that the complete structured is sintered at high temperature to achieve desired porosity (Bhave, 1991; Hoek and Jawor, 2002). Moreover, the properties of active surface layer eventually depend on the selection of size and type of the particle (Bhave, 1991) and depending on the purposes, those properties can be changed with introduction of different materials. Such as to improve the mechanical strength of ceramic membrane, mullite ($3Al_2O_3$, $2SiO_2$) can be used. This structure is formed during the high-temperature calcination of kaoline clay and desired porosity is achieved by controlled leaching of silica (Benito et al., 2007). Simultaneously, membrane surface can be coated with metals, metal oxides, and zeolites to achieve desired thickness, pore structure, and surface characteristics of ceramic membrane (Hassan et al., 1995). Those aforementioned characteristics make ceramic membrane robust in nature therefore these type of membranes are very suitable in industrial, oily, and hazardous wastewater treatment (Benito et al., 2007). The major advantage is that the fouled ceramic membranes can be washed chemically or thermally to recover flux, thus such category membrane has a long lifetime (Faibish and Cohen, 2001a). Due to the mentioned advantages, there are huge applications of ceramic membranes for MF of drinking water in domestic application. However, for large-scale implementation of such type of membranes, the cost is very high, therefore the applications of ceramic types are very limited in commercial scale, that is, purification of municipal drinking water and wastewater treatment (Hoek and Jawor, 2002; Faibish and Cohen, 2001a,b), and so on. Henceforth, development of cost-effective ceramic membrane is very urgent for commercialization.

2.2 Organic Membranes

2.2.1 Integrally Skinned Asymmetric Polymeric Membranes

The development of porous polymeric membranes (Ulbricht, 2006) have fulfilled the rapid demand for treatment of water and wastewater. The whole membrane is consists of same material though it has two asymmetric layers, a thin-porous support layer and above the support layer, less-porous skin layer is present. The skin layer is responsible for the separation process and less-resistive support layer provides mechanical strength to the top layer. Hence, the top layer is considered as the active surface of the membrane (Ulbricht, 2006), which is responsible for the filtration process. Two types of membranes, one is flat sheet membrane and another is hollow fiber membrane in both MF and UF range can be formed via phase inversion method. The phase inversion method is a technique of controlled interaction between polymeric solvent and nonsolvent solution. After casting of the membrane with polymeric solution, it is immersed into nonsolvent for phase inversion to form solid membrane and to provide extra mechanical strength to the membrane surface; the casting can be done on the nonwoven fabric (Ulbricht, 2006; Young and Chen, 1995; Guillen et al., 2010; Mulder, 1996). The properties of the membrane greatly depend on the type of polymer and its concentration, solvent, nonsolvent, casting time, phase inversion time, bath conditions, and so on (Mulder, 1996). Moreover, during the phase inversion humidity should be maintained as much low as possible and ambient temperature is favorable.

In the field of polymeric membrane cellulose acetate (CA) and its associate derivatives, that is, cellulose diacetate, triacetate, and regenerated cellulose have been introduced for membrane fabrication. The use of such type of materials has some advantages, such as the easier fabrication, inexpensive, and low fouling tendency (Sivakumar et al., 2005) though, there are some limitations, such as CA membrane can be implemented below 30°C and within pH range of 3–5 only (Mulder, 1996). Moreover, same types of the membranes are sensitive to free chlorine and due to biodegradable nature of CA biofilm formation takes place on the membrane surface (Cheryan, 1998). To overcome mentioned limitations in the use of CA some other polymers, such as polysulfone (PS), polyethersulfone (PES), sulfonated PSf or PES, polyacrylonitrile (PAN), polypropylene (PP), polytetrafluoroethylene (PTFE, a.k.a., Teflon), and polyvinylidine fluoride (PVDF) are commonly used to manufacture MF, UF, NF, RO membranes (Hoek and Jawor, 2002) and such type of the membrane show excellent permeability, selectivity, and stability during water purification. Similar type but having different molecular weight cut off membrane can be fabricated with minute alternation in casting solution thus polymeric membrane can be used to separate wide range of pollutants from drinking water as well wastewater. Moreover, polymeric membrane is more cost effective than the ceramic membrane.

2.2.2 Thin-Film Composite Membranes

In the field of the membrane science, based on the knowledge of the polymeric membrane, the development of thin-film composite (TFC) membrane is one of the major innovations so far. The TFC membrane has dense selective layer works as separation barrier over the porous polymeric support (Mulder, 1996; Cadotte et al., 1980) and such type of membrane is called “composite.” The remarkable thing is that TFC is registered trademark of Koch Membrane Systems, Inc. in the United States and other countries. Both the layers of TFC membrane can be selected independently according to the requirement, thus performance of the TFC membrane is better than the normal polymeric membrane (Rao et al., 1997). Moreover, in this type of membrane, expensive monomers can be used in very little amount to form selective layer rather than to make the whole membrane. Two layers of the TFC membrane are prepared from different polymers, and commonly, the porous support layer is formed via phase inversion method which is very similar to polymeric membrane, whereas dense layer is introduced via interfacial polymerization, for example, in situ polycondensation or coating (dip, spray, spin) followed by crosslinking (Mulder, 1996; Cadotte et al., 1980). For further improvement of crosslinking, thermal, UV, and chemical curing is performed to reach desired stability, permeability, and selectivity of TFC (Mulder, 1996). Aforementioned methods are applied to prepare TFC RO/NF membranes. The common polymers for the development of TFC membrane are polyurea, polyamide (PA), polyurea-amide, polyether-amide, and so on (Cadotte et al., 1980; Petersen, 1993) and such type of membrane having high salt selectivity, water permeability. Hence, this type of membrane can be utilized in RO application. In 1960, DuPont and Monsanto successfully developed asymmetric hollow fiber TFC RO membrane TRC using polyamide for seawater desalination (Schafer et al., 2005).

Further development in TFC membrane has established that microstructural supports can be made of PSf, PES, sulfonated PSf and PES, polyether ketones, PVDF, sulfonated PVDF, PAN, and so on (Cheryan, 1998). Among those polymers polysulfone is extensively used polymer for preparation of the support of TFC RO membranes (Mulder, 1996; Choi et al., 2006a). The performance of TFC membrane based on permeability is greatly influenced by the porosity of the support layer and hence, to increase the porosity some additives, for example, poly(ethylene glycol) and polyvinylpyrrolidone (PVP) are added to PSf or any polymeric casting solutions to fabricate membrane support. It is assumed that thin film is formed over PSf membrane via interfacial polymerization to prepare TFC membrane having molecular weight cut off of about 60 kDa (Cheryan, 1998). Several types of TFC membranes are commercially available but the exact chemistry of commercial TFC membrane is very complicated and still it is unknown.

There are several steps present in interfacial polymerization during the fabrication of TFC membranes (Cadotte et al., 1980). Initially the support membrane is prepared according to simple membrane fabrication method, then it is immersed into an aqueous solution of the first reactant, for example, diamine. After allowing sufficient time, the substrate is placed in contact with an organic solution of triacyl halide and during this period, the interphase polymerization takes place only on the upper surface of the membrane. Due to polymerization, a very thin but dense selective layer is formed having high water permeability and after formation of this layer, the polymerization immediately stopped. The selection of the organic solution should be done in such a way that it is immiscible with the aqueous solution, therefore the reaction takes place only at the interface of the two solutions. Several research studies have confirmed that during the preparation of TFC membrane, the best polymerization reaction could be possible between the amine monomer 1,3-diaminobenzene or m-phenylenediamine (MPD) and 1,3,5-tricarbonyl chloride or trimesoyl chloride (TMC), other di/tri-acid chlorides and sometimes chemical combinations may change according to the required surface specification. The TFC NF membrane can be fabricated by polymerization of piperazine or polypiperazine derivatives with TMC, other di/tri-acid chlorides, or other combinations. It is assumed that properties of all commercial TFC NF/RO membranes depends on the different support membranes, interfacial polymerization additives, physical/chemical posttreatments, and so on (Cadotte et al., 1980). The fabrication of TFC membrane is a controlled chemical reaction process, therefore the lifetime and performance of the membrane should be enhanced which can be done by making the membrane surface antifouling in nature. The common practices for enhancing the antifouling property of TFC membrane are surface modifications via graft polymerization with the help of plasma exposure, UV-photo initiation, or redox initiation (Mulder, 1996). In a recent study, Kim et al. (2010) fabricated nanostructured RO membranes and to enhance its antifouling property they have used plasma-induced graft polymerization. During the treatment, the membrane surface was exposed to plasma at atmospheric pressure in presence of hydrophilic, water-soluble monomer, poly(methacrylic acid) (PMAA). After such modification, the PMAA-modified membrane surface became rougher than unmodified RO membrane, moreover the modified membrane had better antifouling property and water permeability with negligible change in salt rejection. Therefore, surface modification of TFC membrane can enhance the performance, as well as the lifetime of the membrane. Hence, such modified TFC membrane can be utilized in water purification process in commercial scale.

3 Nanoparticle-Based Membrane

The particles of any material, which have a size of 1–100 nm, are called nanoparticles. Due to their uniform characteristics, the nanoparticles have a wide range of potential applications in the field of biomedical, optical electronic, and so on. The most-lucrative property of the nanoparticles is that they provide high surface area, antibacterial activity, good adsorption capability, and so on and thus they have been utilized for surface modification for different purposes for last few years. Similarly, they have been incorporated in the field of membrane science to enhance the performances of membrane. The details of development of different types of membrane has been described earlier however, to minimize the limitations of the membrane separation, the nanoparticles of different materials have been successfully introduced in the last decades. Here, a detail description of nanomaterial-based membrane for water purification has been elucidated.

3.1 Nanoparticle-Based Ceramic Membranes

3.1.1 Zeolite-Based Ceramic Membranes

In the earlier section, it has been already mentioned that ceramic membrane is advantageous due its better antifouling nature, higher mechanical stability under high pressures, and chemical stability than the polymeric membrane. In earlier days ceramic membrane was mainly used in gas-separation process but the molecular dynamic study has proven that zeolite membrane can be implemented in aqueous medium also (Lin and Murad, 2001). However, based on dynamic study such type of membrane has been used as an RO for desalination of brackish water, as well as for the treatment of different water streams (Li et al., 2004, 2007a; Duke et al., 2009; Lu et al., 2010; Li and Lee, 2009; Liu et al., 2008). Moreover, due to high chemical stability of ceramic membrane, different disinfectants can be used during water-treatment process. Zeolites are nanometer-scale crystalline structure of naturally occurring aluminosilicate minerals. Amorphous silicate, aluminosilicate, or aluminophosphate are some of the zeolites which are commonly used to prepare ceramic membrane. The zeolite membranes are formed in different ways, that is, via hydrothermal synthesis (Kazemimoghadam, 2010; Kumakiri et al., 2000), in situ layer-by-layer crystallization, and dry-gel conversion (Noack et al., 2005). The membrane made of aluminosilicate is chemically inert; extremely thermally and chemically stable (Liu et al., 2008). Zeolite crystals are structurally tetrahedral which control the movement of containment of ions and water molecules (Lobo, 2003). In the tetrahedral structure, either Al or Si occupies the vertex and is connected with four oxygen atoms. Hence, this type of structure generates cavities inside the framework, which controls the movement of water molecules and that movement is also controlled by Al to Si ratio in the matrix. Different types of zeolites can be produced synthetically and some common zeolites, which are used in membranes fabrication, are MFI-type, sodalite (SOD), and Linde Type A (LTA) (Dyer, 1988).

In the water purification system using ceramic membrane two major parameters are pore size and framework density. The pore size depends on the ion selectivity of the membrane, whereas framework density depends on required water permeability. To change the structural properties of zeolites different atoms except Si and Al can be replaced with help of ion exchange to incorporate alternate charge into the tetrahedral structures of zeolites.

As a result the structure of channels present in the zeolite-based membrane changes and the separation process greatly depends on (Li et al., 2007b) the relative density of the framework structure (Dyer, 1988). The framework density is defined as the number of Si or Al atoms present per 1000 Å and such density for MFI, SOD, and LTA are 18.4, 16.7, and 14.2, respectively (Baerlocher and McCusker, 2015). The factors for the water treatment by the zeolite-based membrane are chemical stability, hydrophilic properties, and occurrence of intercrystalline defects and those parameters are controlled by the ratio of Si:Al in zeolite cage. In MFI-type zeolites water permeability and selectivity for water increase with decrease of Si:Al ratio though simultaneously the defects increase, therefore this ratio should be maintained in such a way that both parameters become optimized (Noack et al., 2005). The working principle of zeolite-based membrane is molecular sieving, competitive adsorption, or ion exchange (Kazemimoghadam, 2010). Smaller ions pass by diffusion quickly through porous structures. If the membrane surface is negatively charged, the cationic adsorption takes place on the surface of zeolite membranes, which may enhance performance of the membrane. However, recently nanoscale zeolite–coated RO membranes have been developed having ultra-thin, dense layer. Li et al. (2004) has used MFI-type zeolite-based RO membranes having dense layer thickness of 3 μm for desalination with 0.1 M NaCl feed solution at 2.07 MPa. They got the water flux of 0.112 kg.m^2/h and 76.7% rejection of Na^+ ions and the membrane performance was tested under more-concentrated NaCl solution having Na^+ ejection of 58.1%. The reduction of rejection was attributed due to the formation of double layer inside of the zeolite material in presence of high ionic strength of working solution. Other two operating parameters are transmembrane pressure (TMP) and the temperature of the working solution. At higher TMP, the water permeation increases and ion permeation decreases, subsequently separation performance of the membrane enhances. Nonetheless, at higher temperature, both water and salt permeation increases as at higher temperature the viscosity of the feed solution decreases and increases the diffusivity of water molecules and salt ions at the same time (Li et al., 2004). Similarly, in another case, Duke et al. (2009) prepared MFI-type membranes with the help of dip-coating method, during the process nanoparticle films were formed over alumina support under hydrothermal condition. Such type of membrane was utilized for seawater desalination. Due to the presence of alumina the hydrophobicity and charge of the surface changed but the surface charge did not vary with Si:Al ratio (Duke et al., 2009).

Liu et al. adopted in situ crystallization for the fabrication an α-alumina supported MFI-type zeolite-based membrane and the crystallization occurred on the inner surface of tubular ceramic membranes. Such type of membrane was very useful for the removal of organics from water stream (Liu et al., 2008) and in RO application with 0.1 M NaCl solution at 2.76 MPa, the membrane had shown Na^+ rejection of 99.4%. Na^+ was separated via size exclusion of hydrated ions, as well as Donnan exclusion at pore entries. In the water-treatment process the modified membrane exhibited a water flux of 0.33 kg m^2/h with an organic rejection of 96.5% which was highly appreciable. For practical implementation, certain modification is required for such type of membrane to improve water flux as well.

Kumakiri et al. fabricated A-type zeolite membranes via hydrothermal synthesis over porous α-alumina substrate and multiple coating could be possible as per the separation performance requirement (Kumakiri et al., 2000). Such type of membrane is very much effective for the separation of ethanol from water and the performance of the membranes tested with 10 wt.% ethanol feed solution at 1.47 MPa and 30°C. The membrane selectivity for such solution was 44% and the flux varied linearly with TMP though no such effect of TMP on selectivity was observed. The maximum TMP for the stable operation was 4.90 MPa. Higher mechanical strength is required for high TMP operation. According to that principle, Kazemimoghadam (2010) fabricated composite polycrystalline hydroxyl SOD membranes using tubular mullite supports with the help of hydrothermal growth. Most of the crystallization took place in the temperature range of 80–100°C. The performance of the same membrane was tested with variation of feed flow rate, temperature, and TMP. Flux increased with TMP, temperature, and feed rates as with increase of temperature viscosity of the working solution decreased and higher feed rate enhanced turbulence. Therefore, such type of membrane is applicable in high temperature, pressure, and fouling applications.

The controlling of the thickness of the zeolite coating over the porous support is one of the challenging tasks as the chances of generation of defects is higher at lower thickness of zeolite. However, lesser thickness generates better water permeability and such type of membrane can be used as an alternative of polymeric RO membrane. The most difficult task is the fabrication of defect-free ultra-thin zeolite films, therefore a major modification of traditional hydrothermal synthesis methods is required (Kumakiri et al., 2000). As per the requirement, Öztürk and Akata adopted e-beam lithography to form mono and double layers of zeolite-A nanocrystals (~250 nm) atop silicon wafers for nanoelectronics applications (Ozturk and Akata, 2009). However, such type of technology may be, very expensive to cover membrane surface by zeolite.

3.1.2 Nanoparticle-Based Reactive Ceramic Membranes

The semiconductor-based nanoparticles are photocatalytically active in nature, for example, titania, zinc oxide, ferric oxide, and so on. Those nanoparticles are activated in presence of some external energy sources, such as UV illumination. When those particles are present on the membrane surface, then such type of membrane becomes reactive in nature in presence of UV irradiation. These types of membranes are very effective for the degradation of organic compounds present in the water stream (Bahnemann, 2004; Pramauro et al., 1998; Prevot, 1999; Hoffmann et al., 1995). Carey et al. (1976) first introduced the concept of photocatalysis to degrade polychlorobiphenyls (PCBs). Nanoparticle having semiconductor electronic properties has valence band and an empty conduction band. One electron is transferred from valence band to conduction band when the absorbed energy of the nanoparticle is greater than band-gap energy, as well as generation of one hole in the valence band (Bahnemann, 2004; Krishna et al., 2005). These generated electrons and holes react with components of solution, as a result; those pollutants are degraded with help of photocatalytic action. Sometimes, hydroxyl radicals are generated from the water molecule with the help of electron and then that hydroxyl radical oxidize pollutants present in the water stream into other by-products (Bahnemann, 2004). The photocatalytic degradation of contaminants using TiO_2 nanoparticle suspension have better performance as in such cases the availability of active surface area which is exposed to UV irradiation is maximum (Prevot et al., 2001). However, the major limitation of the suspension mode of operation is the recovery and regeneration of nanoparticles after photocatalytic degradation.

Titania nanoparticles are applied for the fabrication of ceramic membrane using silica (Ilisz et al., 2003; Dong et al., 2009), alumina (Choi et al., 2006b), zeolites (Fukahori et al., 2003), and activated carbon (Matos et al., 2001), which are very stable under UV light and in the presence of reactive oxygen species. The sol–gel method can be adopted to produce high-efficient films and composites having crystalline size of 8–10 nm of anatase crystal of titania. This size range favors recombination of electron-hole pairs during catalytic reaction under UV illumination by allowing the maximum number of active sites per unit mass of catalyst (Zhang et al., 1998). The dip-coating method can also be adopted to form thin layer of the nanoparticles on the surface of the ceramic membrane. Multiple layers provide larger active surface area for the photocatalytic reaction. Moreover, the permeability of the coated membrane with TiO_2 nanoparticles is better than the bare membrane (Choi et al., 2006b). In the presence of ozone, the catalytic activity of the nanoparticles increases significantly and such process is very useful for the removal of natural organic matter and organic compound in water and wastewater treatment. In the presence of O_3, atomic oxygen is generated during the reaction, which expedites the rate of photo oxidation. This combined treatment effectively degraded phenols, aromatic hydrocarbons, and humic substances for water streams (Beltran et al., 2003; Legube and Leitner, 1999; Liotta et al., 2009; Gracia et al., 1996). Karnik et al. (2005) have shown the potential application catalytic membranes in presence of ozone for disinfection of drinking water. The ceramic membrane composed of alumina, zirconia, and titania can be coated with iron-oxide nanoparticles via the layer-by-layer technique and the permeability of the coated membrane is similar to the original membrane. Such membrane shows enhanced catalytic activity in presence of ozone and degrades trihalomethanes and halogenic acetic acids, up to 90 and 85%, respectively. In this case, the formation of hydroxyl radical is speeded up in presence of ozone and that hydroxyl radical is responsible for the degradation (Ma and Graham, 1999, 2000).

Sometimes photocatalysis fails to degrade organics, organisms with complex and dense cell-wall structures, such as bacterial endospores because of the faster rate of recombination of the produced electron and hole pairs during the reaction (Krishna et al., 2005; Liu et al., 2004). Moreover, another limitation of photocatalysis is the reduction of active surface area of photocatalyst due to immobilization of nanoparticles on the membrane surface. Several research works have been carried out to increase the photocatalytic activity of nanoparticles by enhancing active surface area with the help of doping of foreign particles (Liu et al., 2004; Qiu and Burda, 2007; Bessekhouad et al., 2004; Brunet et al., 2009). Krishna et al. successfully delayed the recombination of electron and hole with help of multiwalled carbon nanotubes (CNTs) as CNTs provide large surface area and substantial photon-generated electron trapping capacity. Titania-coated CNTs can successfully degrade *Bacillus cereus* spores whereas commercially available titania alone fails to do so (Krishna et al., 2005).

The earlier discussion reveals that zeolite and photocatalyst-coated ceramic membranes face some challenges for the commercialization of water-treatment process, such as high manufacturing cost and lower packing density than polymeric membranes.

In addition, photocatalytic water treatment requires energy for UV illumination to activate catalyst surface and to minimize energy emission solar-induced photocatalytic water treatment has been investigated (Augugliaro et al., 2002) and such technique has immense potential for application in water-treatment process in future.

3.2 Inorganic–Organic Membranes

The organic membranes are polymeric membranes (PS, PES, PVDF, and so on), however, in presence of inorganic particles (zeolite, TiO_2, Ag, ZnO, CNT, and so on) it becomes inorganic–organic membrane. To nullify limitations of the organic membrane and to fulfill some specific requirements, the incorporation of inorganic materials has been made in last two decades. The current chapter has tried to highlight the fabrication techniques and applications of such type of inorganic–organic membranes in the next subsequent sections.

3.2.1 Mixed-Matrix Membranes

The ease of fabrication and economic make mixed-matrix membranes (MMM) is very lucrative in the application of water purification. Moreover, such membranes have good mechanical strength and functional properties due to the presence of inorganic materials. In the late 1990s, Zimmerman et al. (1997) first proposed MMM to overcome the limitations of polymeric membranes for gas separations. MMM have better prospects in different types of water treatment as these type of membranes show higher selectivity, targeted functionalities, and improved thermal, chemical, and mechanical stability. However, due to the incorporation of inorganic particles the defect formation also increases in the polymeric matrix during the fabrication of the membrane. Therefore, the composition of the membrane should be maintained in such a way that the defect formation must be minimized.

The selectivity of the membrane for the water treatment can be enhanced with addition of inorganic particles having diameter in micrometer (Aerts et al., 2001, 2006; Wara et al., 1995; Genne et al., 1996; Bottino et al., 2002) and additionally inorganic materials help to increase the interconnectivity between pores and enhance the mechanical strength of the membrane (Aerts et al., 2006). Therefore, the presence of inorganic materials enhances the robustness of any polymeric membrane. Such morphological and mechanical changes of MMM are desirable during high-pressure water-treatment process. One of the best example is the Zirfon UF MMM, composed of polymer PSf and 40 wt.%

zirconia (ZrO_2) particles, which has shown better permeability than normal PSf membrane but having similar particle retention (Aerts et al., 2006; Genne et al., 1996) and with increase of zirconia particle (~0.9 μm) content mechanical stability of membrane increases (Aerts et al., 2001). Wara et al. (1995) reduced macrovoids by introducing ceramic alumina particles (~0.34 μm) in CA membranes during fabrication via phase inversion, therefore the selectivity of the membrane increased.

In recent years, nanocomposite MMM is the emerging field in membrane science. The main feature of the nanoparticles is that such particles provide high surface area per unit volume. Sol–gel is the most-common technique to prepare nanoparticles (Sharma et al., 2009; Fernandes et al., 2009) though other methods are inert gas condensation, pulsed laser ablation, spark discharge generation, ion sputtering, spray pyrolysis, laser pyrolysis, photothermal synthesis, thermal plasma synthesis, flame synthesis, low-temperature reactive synthesis, flame spray pyrolysis, mechanical alloying/milling, mechanochemical synthesis, and electrodeposition (Sharma et al., 2009).

The nanoparticles have some unique properties, such as disinfectant, good absorbent, and degradation of pollutants from different water systems (Ulbricht, 2006; Sharma et al., 2009; Stoimenov et al., 2002; Savage and Diallo, 2005; Yue and Economy, 2005). Stoimenov et al. (2002) have found that magnesium oxide (MgO) particles have good antibacterial property, it can easily inactivate Gram-positive bacteria, Gram-negative bacteria, and spore cells. Similarly, Sharma et al. (2008) identified alumina nanoparticles were a good adsorbent to adsorb nickel [Ni(II)] for aqueous medium. The nanoparticles of iron oxide, aluminum oxide, and titanium oxide are very much effective to adsorb heavy metals (Sharma et al., 2009) and zero-valent iron nanoparticles can effectively remove halogenated hydrocarbons, radionuclides, and organic compounds (Darab et al., 2007). As nanoparticles have very high surface area for adsorption, it can be used as an alternative of activated carbon (Sharma et al., 2009). Nanocomposite membranes not only remove different contaminants from different water bodies by means of adsorption and degradation or any other means but also such type of membranes have enhanced flux recovery, better selectivity and antifouling property, and an increased thermal and mechanical stability (Taurozzi et al., 2008; Yan et al., 2005; Maximous et al., 2009; Fan et al., 2008a,b; Morones et al., 2005; Bae and Tak, 2005; Cao et al., 2006; Li et al., 2009; Hosseini et al., 2007). The fabrication of nanocomposite membrane is very similar to its fully polymeric counterpart. The nanoparticle can be impregnated inside the membrane surface or surface coating of membrane can be done using nanoparticles.

The activity of metallic nanoparticles can be enhanced in presence of another metallic nanoparticle and such combinations are called bimetallic nanoparticles (e.g., Fe/Pd, Fe/Ni, Mg/Pd). During the pollutant degradation by bimetallic nanoparticles, one metal donates an electron, which takes part in the degradation of the target compound while the second metal works as a promoter in the reaction through hydrogenation (Smuleac et al., 2009; Shih et al., 2009; Han et al., 2008; Meyer and Bhattacharyya, 2007; Zhu and Lim, 2007; Hadnagy et al., 2007).

The presence of nanoparticles can enhance membrane performance in the presence of alumina nanoparticles (~10 nm; 19 wt.%) in PVDF mixed matrix UF membranes, such as enhancement of pore density without altering size, hydrophilicity, water permeability, fouling resistance, flux recovery, and mechanical stability (Yan et al., 2005). In connection with that, Maximous et al. (2009) fabricated PES UF MMM with alumina nanoparticles (~0.48 nm; 0.01–0.05 wt.%) having better antifouling and flux recovery properties. In another research work Fan et al. (2008a) added polyaniline (PANI) nanofibers to commercial UF membranes (1–15 wt.%) to increase the water permeability, selectivity, surface wettability, antifouling, and flux recovery of the membrane (Maximous et al., 2009).

During the water-purification process, the membrane surface is very prone to the growth of bacterial film, which causes biofouling of the membrane and such fouling tendency, can be reduced with help of nanoparticles having antimicrobial properties (Baker and Dudley, 1998; Lee et al., 2005). According to the studies, silver nanoparticles have excellent antibacterial properties (Morones et al., 2005; Marambio-Jones and Hoek, 2010) thus; silver nanoparticle coatings are now widely applied to introduce antibacterial property to the coated surfaces (Marambio-Jones and Hoek, 2010). Taurozzi et al. (2008) identified that the formation of biofilm could be successfully reduced in presence nanosilver coating over the membrane surface during the membrane-separation process. The control release of silver ions from the membrane surface can be achieved only when silver nanoparticles are incorporated in membranes in the zerovalent state (Tarabara, 2009). The silver nanoparticles can be introduced in PSf membrane during membrane fabrication via ex situ reduction of the nanoparticles prior to addition to the casting solution and with in situ reduction during casting. The fabricated composite MMM has better water permeability, with negligible reduction in solute rejection

(Taurozzi et al., 2008). Therefore, the presence of silver nanoparticles not only increase antibacterial property of the membrane, but also enhance the membrane performance in water-treatment process. However, the lifetime of such MMM is needed to identify the commercial drinking-water purification.

In another way, to enhance the performance of the membrane CNTs have been successfully used (Choi et al., 2006a; Brunet et al., 2008; Lin et al., 2007). Not only silver nanoparticles but also CNTs exhibit good antimicrobial activity; thus, CNTs are an alternative for the fabrication of improved disinfection or antifouling membranes. The main disadvantage for the production of CNTs-blended MMM is the solubilization problem during the preparation of casting solution with CNTs. CNTs are insoluble in water and organic solvents which affects large-scale fabrication of MMM (Ajayan, 1999). Lin et al. (2007) recommend that CNTs should be functionalized with polymer groups so that CNTs become structurally similar to the bulk polymer matrix and then CNTs disperse in polymer to ensure homogenous membrane properties. Hydrophilicity of the membrane surface increases with the presence of acid-treated CNTs due to the carboxylic acid groups and pore size increases with increase of CNTs concentration up to 1.5 wt.% and after that pore size decreases and becomes smaller than pure PSf membrane at CNTs concentration of 4 wt.%. However, water permeability and rejection are still increasing with nanotube additions beyond 1.5 wt.% because of the improved hydrophilicity in presence of CNTs. Also, the resulting membrane has better antifouling property (Brunet et al., 2008). However, CNTs-blended membranes do not display the desired antimicrobial activity because the contact time between organisms and the CNTs is not sufficient as CNTs are entrapped inside the polymer matrix. Further development is required to expose CNTs as much as possible to the working solution for antimicrobial applications.

Inorganic nanoparticles enhance the thermal and mechanical stability of MMM by reducing the impacts of heating and membrane compaction. During the compaction, irrecoverable flux decline occurs during the initial stages of membrane-separation process. Ebert et al. (2004) fabricated poly(vinylidene fluoride) (PVDF) membranes with titania nanoparticles as inorganic fillers via phase inversion to demonstrate increased stability of MMM and such MMM exhibited higher thermal stability and less compaction. They have found that 83% decrease in pore volume in case of pure PVDF membranes, whereas only 17% decrease in PVDF/titania MMM during the compaction of the membrane. The similar type of earlier observation has been reported in case of silica and zeolite nanocomposite-PSf-supported RO membranes

(Pendergast et al., 2010). In general, the nanocomposite PSf-based MMMs have higher initial water permeation with less flux declination during compaction. Scanning electron microscopy images confirms that the nanocomposite membrane resists the impacts of compaction by maintaining open surface pores better than the pure PSf polymeric membrane.

Overall, the fabrication nanoparticle containing MMM is very easy and MMM potentially provide novel functionalities, enhanced performance, and higher mechanical, as well as thermal stability. Although such type of membranes are very effective in water treatment, due to unviability of commercial nanoparticles, the industrial-scale fabrication of such membrane has not started yet. When industrial-scale nanoparticle production grows, costs of such type of inorganic fillers will come down and many of the research-level innovations may make their own way into the marketplace.

3.2.2 Thin-Film Nanocomposite Membranes

In Section 2.2.2, detailed development of TFC membrane has been elaborated and this section is devoted for the description of thin-film nanocomposite (TFN) membranes. To avail the benefits of the presence of nanoparticles on the membrane surface, for example, enhanced separation performance, reduced fouling, antimicrobial activity, and other novel functionality, TFN membrane can be prepared with the help of interfacial polymerization processes or surface attachment via self assembly (Jeong et al., 2007; Lind et al., 2009a,b; Lee et al., 2008; Bae et al., 2006; Luo et al., 2006).

Zeolite nanoparticles enhance the molecular sieving properties of TFN RO membranes (Jeong et al., 2007; Lind et al., 2009b). Jeong et al. (2007) fabricated zeolite-polyamide thin films over the PSf support membrane via interfacial polymerization by dispersing zeolite nanoparticles in the TMC solution. Such membrane had 80% better water permeability than identically casted TFC membranes at zeolite loading of 0.4 wt.%. Due to the porous structure of zeolite, the pure-water permeability increases even for pore-filled zeolites and the permeability increases more for pore-opened zeolites.

Lind et al. (2009a) fabricated TFN membranes containing LTA nanoparticles of 0.2 wt.% with help of interfacial polymerization and the presence of zeolite nanoparticles enhanced permeability, negative surface charge, and thicker membranes irrespective of particle size used. They observed that larger nanoparticles enhanced surface properties, whereas smaller nanoparticles increased permeability by increasing average pore size. Moreover, in all TFN membranes better crosslinked between porous structure has been observed than pure polyamide TFC membrane and that observation is responsible for the enhancement of TFN membrane permeability. For further enhancement of performance of TFN membranes, Lind et al. (2009b) casted composite membrane with sodium and silver exchanged LTA nanoparticles. The permeability and NaCl rejection significantly increased in case of both type of TFN membranes and in presence of silver-based zeolite the membrane surface became more hydrophilic and antibiofouling in nature. Lee et al. (2008) used TiO_2 nanoparticle to prepare PA-based TFN NF membranes through interfacial polymerization and they found that the pure-water flux increased with titania concentration up to 5 wt.%, salt rejection decreased. Moreover, the presence of titania nanoparticle makes the membrane surface reactive in nature under UV illumination and the antifouling property is also improved. However, the presence of nanoparticles cased certain defects in coating film and that should be taken care during membrane fabrication.

Similar to the MMM, CNTs can be used for the surface modification of TFC membrane to produce TFN membrane. As CNTs have antibacterial activity, Kang et al. (2007) and Brady-Estevez et al. (2008) used single-walled CNTs-blended PVDF TFN membrane (5-μm pore size) for successful removal of viral and bacterial pathogens from water at low-pressure membrane-separation process. According to them such type of membrane could easily destroy *Escherichia coli* cells (~2 μm) and fluorescence-based viability test proved that almost 80% of the bacteria are killed after 20 min contact time between working solution and membrane surface whereas similar normal TFC membrane failed to do so. The present study also confirmed that with increase in the thickness of CTNs layer, the contact time for the inactivation of the viral pathogen got reduce and consequently, full virus removal (5–7 log removal) was observed with a 6 μm skin layer; 3.2-log removal was seen with a thin 2-μm layer. Such type of membrane offers a possible opportunity to disinfect drinking water.

In the field of membrane science the enhancement of hydrophilicity of the membrane surface signifies, reduction of fouling tendency. Luo et al. (2005, 2006) fabricated both TiO_2 nanoparticle coated and impregnated PES UF membrane. They found that contact angle reduced from 39.6 to 19.2 degrees for coated membrane and the contact angle reduced from 79.6 to 41.2 degrees for 10 wt.% of nanoparticle impregnated membrane. Bae et al. (2006) also confirmed that in the presence of titania the polymeric membranes showed better antifouling property thus reduction in initial sharp flux declination had been observed. However, pore size and water permeability slightly decreased in the presence of TiO_2 nanoparticles.

It is certain that the nanoparticles having higher water permeability than the polymer matrix can eventually increase the permeability of the nanocomposite membrane made of that polymer and nanoparticles, as the nanoparticles provide less-resistive flow paths through the cross-section of membrane. On the other hand, impermeable nanoparticles certainly reduce the water permeability of a membrane. However, impermeable fillers can increase membrane permeability and decrease rejection as during the fabrication of the membrane it creates defects on the membrane surface. Sometimes such defects can be compromised to make the membrane surface superhydrophilic or antimicrobial in nature to reduce membrane fouling significantly. As discussed earlier, there are several advantages of TFNs but antimicrobial and zeolite nanoparticles are expensive. Therefore, a rigorous research work in such field is highly encouraged to minimize the production cost of nanoparticles, as well as TFNs so that it should commercialize in very near future.

3.3 Biologically Inspired Membranes

The cell is the smallest unit of any living inhabitancy on the globe. The outer surface of any cell is made up of semipermeable membrane and that membrane selectively controls the transportation of any material to or from the cell. Such membrane has high selectivity than any type of membrane as discussed earlier. Current research fraternity has identified the mechanism of cellular membrane and tried to develop artificially for utilization in real practice. This type of modified membrane is called biologically inspired membrane and the rapid development in this field has been observed in the last decade. A brief description of the development, mechanisms, and application of such modified membrane has been portrayed in this chapter.

3.3.1 Aquaporin Membranes

In a biological film, aquaporins are the protein channels that control water flux and such type of system is called biological membrane. In 1993, Agre et al. first discovered such type of protein channels and they named those channels as Aquaporin-1 (AQP1). Due to their innovative work, they were awarded with the Nobel Prize (Agre et al., 1993). Aquaporins are found widely in human tissues and passive water is transported rapidly across cell membranes via these channels. A single transmembrane protein is ~120 kDa in size, and composed of four channels (Kaufman et al., 2010). The physiological plumbing of water in different organs of our body takes place via these channels by selective, rapid diffusion caused by osmotic gradients (Agre, 2003; Meinild et al., 1998). Based on the earlier findings Zhu et al. (2004) successfully simulated water permeation in AQP1. Two factors, such as osmotic permeability (pf) and diffusion permeability (pd) are involved in water transport in AQP1. The ratio of pf/pd defines the number of effective steps; a water molecule must move in order to permeate through a channel.

The selectivity of aquaporin channels is very high and that concept can be introduced in water treatment by membrane separation. According to the study of the Kaufman et al. (2010), the rate of water molecule transport by single aquaporin is 2–8 $\times$ 10^9 molecules per second. Therefore, in the presence of biological lipid bilayers containing aquaporins on RO membrane, the rate of transport of water and selectivity is very much higher than any commercial RO membrane and according to that, a RO membrane with 75% coverage of aquaporins could have a hydraulic permeability in the range of 2.5 $\times$ 10^{-11} m Pa/s (Kaufman et al., 2010). In a biological membrane, the transport is occurred due to osmotic pressure. Not only RO membrane but Kaufman et al. also formed lipid bilayers over dense NF membrane via vesicle fusion and such modified NF membrane can be operated under similar transmembrane pressure of RO membrane. The NF membrane was chosen due to its high permeability and low surface roughness that allowed for minimal distortion of the lipid bilayer.

Another similar type of surface modification of the membrane has been done by Kumar et al. (2007) and they included Aquaporin-Z from *E. coli* bacterial cells over polymeric membrane. As earlier described that the aquaporin has high water permeation and high selectivity and additionally it can be purified easily and multiply using recombinant *E. coli* strain. The modified membrane has shown an order of magnitude increase in water permeability over a purely polymeric membrane. Kumar et al. have obtained the full rejection of glucose, glycerol, salt, and urea using that membrane. These observations encourage the use of modified membrane by aquaporins for water-treatment applications.

Very few literatures are found on water treatment by biologically inspired membrane, hence further research work is required to optimize the formation of lipid layers over the membrane surface and their resulting permeability and selectivity. The rigorous study can only help for commercialization of such unique aquaporin-based membranes. There are several limitations in the large-scale production of aquaporin, thus synthetic approach for the production and purification of aquaporin is very much essential indeed.

3.3.2 Vertically Aligned Nanotube Membranes

The water transport through CNTs is two to five times higher than theoretical predictions by the "Hagen–Poiseuille" equation (Ahadian and Kawazoe, 2009). From the molecular dynamic simulations, it has been observed that due to atomic smoothness and molecular ordering of CNTs water molecules are passed through CNTs in a one-dimensional single-file procession (Hummer et al., 2001; Kalra et al., 2003). Therefore, aligned CNT membranes require low driving pressure, as well as energy than the normal membranes for the transportation of water molecule across the membrane surface. However, such process is not for the desalination of water as the process is driven by osmotic pressure via the *thermodynamic restriction* (Song et al., 2003). As CNTs have high mechanical stability, CNTs-blended membranes have longer lifetimes than conventional membrane (Salvetat et al., 1999a,b). The composite membrane with CNTs as a selective layer can achieve high flux by molecular sieving action at the membrane surface. Such type of CNTs-blended, high-selective, high-flux membranes are essential to operate cost-effective and energy-saving water-treatment process.

Chemical vapor deposition (CVD) is the most common method till date to produce uniform, vertically aligned nanotube arrays (Gao et al., 1999; Choi et al., 2000; Yoshikawa et al., 2008; Mauron et al., 2002). Vertically aligned nanotubes array can be synthesized over the substrate of titanium, nickel, silicon, aluminum, and so on (Gao et al., 1999; Choi et al., 2000; Yoshikawa et al., 2008; Mau ron et al., 2002) with the help of CVD processes. Sometimes CNTs are functionalized with negatively charged group to control the movement of ions. It has been observed during the production of CNTs via CVD process, that generally multiwalled nanotubes are perpendicularly aligned whereas single-walled nanotubes are curved. Holt et al. successfully produced gap-free sub-2-nm diameter-aligned double-walled CNT using catalytic CVD process with the help of microelectromechanical fabrication system, and produced CNTs which had pore densities of 0.25×10^{12} pores/cm^2 area. That value pore density is the highest value to date. Water flux through these CNT membranes was three times higher than theoretical measurement by Hagan–Poiseuille predictions (Holt et al., 2006). These nanoporous membranes have extreme selectivity, without altering water permeation. As the production of aligned CNT membranes with some promising characteristics is a controlled and expensive job, such type of membrane fabrication is not yet commercialized.

In another way, Casavant et al. (2003) produced single-walled CNTs via magnetic-alignment approach for macroscopic-film formation. Srivastava et al. (2004) established that aligned CNT membranes could separate heavy hydrocarbons from petroleum during crude oil post distillation and successfully remove microbial contaminants, namely, *E. coli* (2–5 μm), *Staphylococcus aureus* (~1 μm), and the poliovirus (~25 nm) from drinking water. Moreover, antibiofouling property of the membrane is improved in the presence of CNT, hence the lifetime of the CNT-aligned membrane is higher than the normal membrane. Due to the strong mechanical and thermal stability, CNT filters can be cleaned and reused, whereas conventional water-filtration membranes are typically disposed off at the end of single use as it is permanently damaged.

CNTs show promising alternative of biological aquaporin channels as CNTs can be produced on a large scale. However, the difficulty of fabrication of large membrane with CNT as selective layer stands in the way of commercial application. Higher cost of production and lack of stability for both aquaporin- and CNT-based membranes prevent such modified membranes in large-scale application, however, as the time will progress the scale-up issues will get resolved gradually.

3.3.3 Isoporous Block-Copolymer Membranes

Block copolymers are macromolecules composed of multiple block–polymeric species having the ability to self-assemble into highly ordered structures when placed in a selective solvent (Rodriguez-Abreu and Lazzari, 2008; Dove, 2008; Peinemann et al., 2007; Smart et al., 2008). The nature of self-assembly of block copolymer provides the narrow pore size distributions and high porosities, as well as sharp molecular weight cut-off of the membrane. The membrane fabrication with block copolymers can be easily done via phase-inversion method. Accordingly, Peinemann et al. (2007) fabricated asymmetric porous membranes composed solely of block copolymer materials via phase-inversion method and such complex fabrication method is greatly influenced by the both thermodynamic and kinetic factors. It is very difficult to maintain perpendicular arrangement of block copolymers throughout the cross-section membranes and this arrangement is done by several factors, such as solvent composition, selectivity, and concentration (Peinemann et al., 2007). In their one-step process, Peinemann et al. (2007) fabricated membrane having dense layer thickness of aligned nanocylinders of 200–300 nm, with a pore density of 240×10^{12}, and an effective pore diameter of 8 nm, that membrane had water flux of 201 $m^{-2} \cdot h^{-1}$ at 0.5 bar TMP with 82% rejection of albumin. Peinemann et al. (2009) has shown that one-step fabrication holds promise for large-scale production and block- (di- and tri-) copolymer membranes, which are very suitable for separation applications, including UF and NF. Phillip et al. (2009) fabricated 100-μm thick "doubly reactive" block-polymer membrane with tuneable selectivity, narrow pore-size distribution (~14 nm), and a 40% void fraction and sometimes such membranes were combined with selective etching of a single block (Chen et al., 2007). The water permeability was lower than the predicted value; however, it increased with applied pressure.

The block-copolymeric membrane can be fabricated similar to the surface modification of the polymeric membrane with the help of block copolymers. Application of coating of block copolymer reduces the cost of production of such membrane, which leads to a large-scale production of block-copolymer membrane (Fasolka and Mayes, 2001). Yang et al. (2006) fabricated NF membrane by coating conventional support with 80 nm thick block–copolymer film having cylindrical pores of 15-nm diameter and such membrane is capable to remove viruses from the water. Moreover, this fabrication process is advantageous as the thickness of copolymer layer can be altered as per the requirement. However, there are some issues during the scale up of this process as pores are damaged during the films transformation. In this approach, the advantage is that the pore-size distribution can be varied from 10 to 40 nm.

To avoid such limitation, direct formation of self-assembly of block copolymers over functional supports has been attempted by Fukunaga et al. (2003) and the surface composition and roughness influences the orientation of the self-assembly. In another case Li et al. (2010) formed nanoporous films via spin coating followed by solvent evaporation on ceramic, as well as polymeric support and they introduced polyacrylic acid to guide self-assembly of their diblock copolymer system and to maintain desired pore structure. Multiblock or star copolymers enable more predictable alignment atop commercial substrates than block polymers. These types of copolymers reduce thermodynamic loss due to conformational entropy and provide an opportunity for the formation of more-specific structures (Stratford et al., 2005). Chen et al. prepared tunable porous structures with self-assembling ABA triblock amphiphilic copolymers into a highly ordered honeycomb film (Yao et al., 2009). These types of triblock copolymers have both hydrophobic and hydrophilic properties, which influence the pore structure. Accordingly, the regularity of pores decreases with increasing hydrophilic block content and pore diameter increased with increasing hydrophobic block length (Yao et al., 2009). The generalized study about the block copolymers has not been found yet, therefore further understanding of the mechanisms of formation of copolymer membrane and their working principle is essential in the field of membrane science. However, membranes of block copolymers have a potential for the application in water-treatment process with some fine tunings.

4 Pathway for Commercialization of Nanomaterial-Based Membrane

In the last few decades, several developments have been made to commercialize membrane-based water-purification process. Rigorous lab-scale studies have been performed to overcome several limitations of membrane-separation process and to enhance the performance of the membrane. As a result, nanomaterial-based membrane has been developed to make the water of drinkable quality. Based on the earlier description nanotechnology-based membrane-separation process can be divided into two major categories; first, based on the performance enhancement and second, state of commercial readiness of such type of membrane. Again, those mentioned categories are dependent on different parameters. The measurable parameters for the performance enhancements of membrane are permeability, selectivity, and robustness over the current state-of-the-art. Permeability is the measure of productivity of membrane-separation process whereas selectivity is the inherent property of any individual membrane and the robustness is the measure of chemical, mechanical, and thermal stability, as well as fouling resistance. Commercial readiness is the measure of the extent of commercial application of any membrane and the commercialization of any membrane depends on the anticipated material cost, manufacturing scalability, and apparent time for commercialization. The acceptance criteria of nanotechnology-based membrane in real application greatly depends on those six parameters. According to several literatures, nanostructured ceramic membrane has good selectivity and robustness but other parameters are not up to the mark, thus commercialization of such type of membrane is still lagging behind. Among all types of inorganic–organic membrane, nanoparticles and zeolite, incorporated TFNs are in the doorstep of commercial application due to their high productivity, robustness, and scalability. However, the main concern is the cost of production for such type of membranes. In the last decades, some major development has been observed in the field of biologically inspired membranes and those types of membranes have high selectivity and productivity but unfortunately, other parameters are not up to the mark. Therefore, a major evolution is required in such category of membrane for commercialization in future.

Ceramic membrane is always very robust in nature, thus reactive or catalytic and zeolite-coated ceramic membranes can be fabricated with small changes in regular practice of inorganic membrane-fabrication methods and modified membrane gives improved performance. However, unfortunately, still these innovations are in lab scale only because ceramic-based membrane has higher capital cost and lower membrane area density with comparison to any polymeric membrane. As ceramic membrane surface can withstand high temperature and corrosive environment, the surface can be modified to be reactive in nature without hampering its original property. Such type of modification has been done up to lab-scale level, therefore further development can help to produce commercially viable systems. The negligence in improvement of the productivity for such membranes has been observed and the selectivity can be increased by making the surface susceptible for catalytic degradation of target compounds. Moreover, reactive surfaces have shown promising characteristics for biofouling resistant. Though, the materials and production cost of nanomaterial-based ceramic membrane is very similar to normal ceramics used for laboratory and pilot-scale testing, no such type of membrane for commercialization is available as of now.

Both zeolitic and ceramic material has immense stability therefore zeolite-coated membrane has shown some promising enhancement in the molecular selectivity of ceramic membranes. Due to the high stability of inorganic materials, such types of membranes have a great future in desalination and purification of water, which can be used in both drinking and industrial purposes. However, further development is required for the synthesis of thinner zeolite films to improve water permeability without hampering the selectivity of the membrane. Such development is required because commercially available membrane has higher flux than zeolite-coated ceramic membrane. Nonetheless, such membrane posses a more chemically and thermally stable alternative to the polymeric membranes, which are typically applied for high transmembrane pressure and complex water-treatment process. Though, there are several advantages, the commercial viability of zeolite-coated ceramic membrane is still lagging behind as the cost of materials to produce fully zeolitic coatings is more than typical polymer membrane materials.

As we discussed earlier, the production technique of MMM and TFNs is almost similar to the polymeric membrane. The modification of polymeric membrane results in significant performance enhancements and it has been observed that there is an appreciable amount of increase of productivity in compression with current "state of the art" membranes. The membrane selectivity does not change in case of MMM and zeolite TFNs, however, there is some decrement in selectivity for nanoparticles as some defects are generated on the membrane surface during the composite membrane fabrication. Due to the presence of inorganic filler, hydrophilicity and fouling resistance are improved thus robustness of polymeric membrane is enhanced. Moreover, the major concern for such type of inorganic–organic membrane is the cost of the filler materials; hence, the production cost of such modified polymeric membrane is much higher than normal one. However, the production cost of zeolite-based TFNs is less expensive, as well as their scalability is higher than MMMs. Nowadays both MMM and TFN are commercially available for water-treatment process though more commercialization is observed in case of TFN than MMM considering the factor of cost effectiveness.

In recent years, the development of biologically inspired membranes is one of the greatest achievement by scientists in the field of membrane science. Such type of membrane shows better performance enhancements than any other type of membrane. However, these category membranes are in completely development stage at lab scale and it will take appreciable amount of time for commercialization. Several studies have confirmed that aquaporin-based membranes have higher flux than the normal polymeric membranes thus, the productivity of such membrane has been enhanced in presence of aquaporin. Similarly, aligned nanotubes and isoporous block–copolymer membranes have shown enhanced productivity. Not only the productivity is increased for the biologically inspired membrane but also those materials have enhanced the selectivity with exceptionally narrow-pore distributions, which is clearly compressible with any type of normal membrane. In maximum cases, there is no change in robustness for both aquaporin and nanotube-based membranes. The pure block-copolymer membranes are mechanically unstable as block copolymer structure is unstable by itself; therefore initial step toward the commercialization of biologically inspired membranes is the increment of robustness of the membrane. The production cost of aquaporins and nanotubes are expensive thus fabrication of membrane with such type of material is very expensive in real practice which is another limitation for large-scale production. It is very difficult to cover large membrane surface with uniform coating of aquaporins and aligned CNT films thus, currently both types of membranes are produced at lab scale only. Block copolymer materials are costly and overall production cost of copolymerized membrane may be reduced if the selective layer of the membrane is prepared with block copolymer. Therefore, some minor changes are required in current polymeric membrane fabrication system to produce block-copolymer membranes and such type of membrane may be commercialized in very near future. Another type biologically inspired membrane is aligned block-copolymer membrane, which is at the initial stage of development. Block-copolymer membranes having some performance enhancements are more cost effective than the aquaporin and CNT membranes. Biologically inspired membranes have the greatest separation performance than any other membrane, however, their cost and robustness are not established yet, therefore commercialization of such membranes is very challenging indeed. Similar problems were faced at the early-stage development of conventional polymeric membrane but it is now at the doorstep of commercialization.

It can be easily summarized that biologically inspired membranes show the greatest potential performance enhancements but those are farthest from commercial reality. On the other hand, zeolite-based TFN membranes offer moderate performance enhancement and appear nearest to commercial viability.

5 Conclusions

In this chapter we have already described several aspects of next generation nanomaterial-based membrane-treatment process for water purification in the earlier section. The three major categories of membranes, highlighted over here are nanoparticle-based ceramic membranes, inorganic–organic membranes, biologically inspired membranes. Each development in the field of nanomembrane has shown the promising pathway of improved energy-efficient water-purification process. In the presence of nanoparticles, a significant improvement has been observed in seawater desalination by membrane technology. In addition, due to the presence of nanoparticles the fouling tendency both physical and biological has been minimized, hence the lifetime of the membrane has increased. Several pollutants and pathogens are destroyed under the photocatalytic action of nanoparticles. MMM offer enhanced separation performance, fouling resistance, and mechanical stability for filtration applications. Both aquaporin and CNT-blended membrane produce dramatically higher water flux than the conventional polymeric membrane. Block-copolymer membrane has also shown some promising stage. Nowadays, several applications of RO membrane in drinking-water treatment for both commercial and household purpose have been observed and in some cases nanoparticle-based membranes have already been applied. Each of the aforementioned innovative nanomaterials promise unique performance enhancements in water purification via membrane-separation process and each has unique hurdles to overcome before its commercialization.

References

Aerts, P., Greenberg, A.R., Leysen, R., Krantz, W.B., Reinsch, V.E., Jacobs, P.A., 2001. The influence of filler concentration on the compaction and filtration properties of Zirfon®-composite ultrafiltration membranes. Sep. Purif. Technol. 22-23, 663–669.

Aerts, P., Kuypers, S., Genne, I., Leysen, R., Mewis, J., Vankelecom, I.F.J., Jacobs, P.A., 2006. Polysulfone-ZrO_2 surface interactions. The influence on formation, morphology and properties of zirfon-membranes. J. Phys. Chem. B 110, 7425–7430.

Agre, 2003. Agre, Aquaporin Water Channels, Nobel Lecture, Stockholm, SE.

Agre, P., Sasaki, S., Chrispeels, M.J., 1993. Aquaporins: a family of water channel proteins. Am. J. Physiol. Renal 265, F461.

Ahadian, S., Kawazoe, Y., 2009. An artificial intelligence approach for modeling and prediction of water diffusion inside a carbon nanotube. Nanoscale Res. Lett. 4, 1054–1058.

Ajayan, P., 1999. Nanotubes from carbon. Chem. Rev. 99, 1787–1800.

Augugliaro, V., Baiocchi, C., Bianco Prevot, A., García-López, E., Loddo, V., Malato, S., Marcí, G., Palmisano, L., Pazzi, M., Pramauro, E., 2002. Azo-dyes photocatalytic degradation in aqueous suspension of TiO_2 under solar irradiation. Chemosphere 49, 1223–1230.

Bae, T.H., Tak, T.M., 2005. Effect of TiO_2 nanoparticles on fouling mitigation of ultrafiltration membranes for activated sludge filtration. J. Membr. Sci. 249, 1–8.

Bae, T.H., Kim, I.C., Tak, T.M., 2006. Preparation and characterization of fouling-resistant TiO_2 self-assembled nanocomposite membranes. J. Membr. Sci. 275, 1–5.

Baerlocher, C., McCusker L.B., 2015. Database of Zeolite Structures. Available from: http://www.iza-structure.org/databases/

Bahnemann, D., 2004. Photocatalytic water treatment: solar energy applications. Sol. Energ. 77, 445–459.

Baker, J.S., Dudley, L.Y., 1998. Biofouling in membrane systems–a review. Desalination 118, 81–89.

Beltran, F., Rivas, F., Montero-de-Espinosa, R., 2003. Ozone-enhanced oxidation of oxalic acid in water with cobalt catalysts. 1. Homogeneous catalytic ozonation. Ind. Eng. Chem. Res. 42, 3210–3217.

Benito, J., Sanchez, M., Pena, P., Rodríguez, M., 2007. Development of a new high porosity ceramic membrane for the treatment of bilge water. Desalination 214, 91–101.

Bessekhouad, Y., Robert, D., Weber, J., Chaoui, N., 2004. Effect of alkaline-doped TiO_2 on photocatalytic efficiency. J. Photochem. Photobiol. A 167, 49–57.

Bhave, R.R., 1991. Inorganic Membranes: Synthesis, Characteristics, and Applications. Van Nostrand Reinhold, New York, NY.

Bottino, A., Capannelli, G., Comite, A., 2002. Preparation and characterization of novel porous PVDF-ZrO_2 composite membranes. Desalination 146, 35–40.

Brady-Estevez, S., Kang, S., Elimelech, M., 2008. A single-walled-carbon-nanotube filter for removal of viral and bacterial pathogens. Small 4, 481–484.

Brunet, L., Lyon, D.Y., Zodrow, K., Rouch, J.C., Caussat, B., Serp, P., Remigy, J.C., Wiesner, M.R., Alvarez, P.J.J., 2008. Properties of membranes containing semi-dispersed carbon nanotubes. Environ. Eng. Sci. 25, 565–575.

Brunet, L., Lyon, D.Y., Hotze, E.M., Alvarez, P.J.J., Wiesner, M.R., 2009. Comparative photoactivity and antibacterial properties of C-60 fullerenes and titanium dioxide nanoparticles. Environ. Sci. Technol. 43, 4355–4360.

Cadotte, J.E., Petersen, R.J., Larson, R.E., Erickson, E.E., 1980. New thin-film composite seawater reverse-osmosis membrane. Desalination 32, 25–31.

Cao, X.C., Ma, J., Shi, X.H., Ren, Z.J., 2006. Effect of TiO_2 nanoparticle size on the performance of PVDF membrane. Appl. Surf. Sci. 253, 2003–2010.

Carey, J., Lawrence, J., Tosine, H., 1976. Photodechlorination of PCB's in the presence of titanium dioxide in aqueous suspensions. Bull. Environ. Contam. Toxicol. 16, 697–701.

Casavant, M.J., Walters, D.A., Schmidt, J.J., Smalley, R.E., 2003. Neat macroscopic membranes of aligned carbon nanotubes. J. Appl. Phys. 93, 2153–2156.

Chen, L., Phillip, W.A., Cussler, E.L., Hillmyer, M.A., 2007. Robust nanoporous membranes templated by a doubly reactive block copolymer. J. Am. Chem. Soc. 129, 13786–13787.

Cheryan, M., 1998. Ultrafiltration and Microfiltration Handbook. Technomic Publishing Co., Inc., Lancaster, UK.

Choi, Y.C., Shin, Y.M., Lee, Y.H., Lee, B.S., Park, G.S., Choi, W.B., Lee, N.S., Kim, J.M., 2000. Controlling the diameter, growth rate, and density of vertically aligned carbon nanotubes synthesized by microwave plasma-enhanced chemical vapor deposition. Appl. Phys. Lett. 76, 2367–2369.

Choi, J.H., Jegal, J., Kim, W.N., 2006a. Fabrication and characterization of multi-walled carbon nanotubes/polymer blend membranes. J. Membr. Sci. 284, 406–415.

Choi, H., Stathatos, E., Dionysiou, D., 2006b. Sol-gel preparation of mesoporous photocatalytic TiO_2 films and TiO_2/Al_2O_3 composite membranes for environmental applications. Appl. Catal. B Environ. 63, 60–67.

Coping with Water Scarcity, 2007. Challenge of the twenty-first century, United Nations (UN) Water Food and Agricultural Association (FAO).

Darab, J.G., Amonette, A.B., Burke, D.S.D., Orr, R.D., Ponder, S.M., Schrick, B., Mallouk, T.E., Lukens, W.W., Caulder, D.L., Shuh, D.K., 2007. Removal of pertechnetate from simulated nuclear waste streams using supported zerovalent iron. Chem. Mater. 19, 5703–5713.

Dong, W., Lee, C., Lu, X., Sun, Y., Hua, W., Zhuang, G., Zhang, S., Chen, J., Hou, H., Zhao, D., 2009. Synchronous role of coupled adsorption and photocatalytic oxidation on ordered mesoporous anatase TiO_2-SiO_2 nanocomposites generating excellent degradation activity of RhB dye. Appl. Catal. B Environ. 95, 197–207.

Dove, A., 2008. Controlled ring-opening polymerisation of cyclic esters: polymer blocks in self-assembled nanostructures. Chem. Commun. 2008, 6446–6470.

Duke, M., O'Brien-Abraham, J., Milne, N., Zhu, B., Lin, J., Diniz da Costa, J., 2009. Seawater desalination performance of MFI type membranes made by secondary growth. Sep. Purif. Technol. 68, 343–350.

Dyer, A., 1988. An Introduction to Zeolite Molecular Sieves. John Wiley & Sons Ltd, Chichester, UK.

Ebert, K., Fritsch, D., Koll, J., Tjahjawiguna, C., 2004. Influence of inorganice fillers on the compaction beaviour of porous polymer based membranes. J. Membr. Sci. 233, 71–78.

Faibish, R., Cohen, Y., 2001a. Fouling-resistant ceramic-supported polymer membranes for ultrafiltration of oil-in-water microemulsions. J. Membr. Sci. 185, 129–143.

Faibish, R., Cohen, Y., 2001b. Fouling and rejection behaviour of ceramic and polymer- modified ceramic membranes for ultrafiltration of oil-in-water emulsions and microemulsions. Colloids Surf. A 191, 27–40.

Fan, Z.F., Wang, Z., Duan, M.R., Wang, J.X., Wang, S.C., 2008a. Preparation and characterization of polyaniline/polysulfone nanocomposite ultrafiltration membrane. J. Membr. Sci. 310, 402–408.

Fan, Z.F., Wang, Z., Sun, N., Wang, J.X., Wang, S.C., 2008b. Performance improvement of polysulfone ultrafiltration membrane by blending with polyaniline nanofibers. J. Membr. Sci. 320, 363–371.

Fasolka, M.J., Mayes, A.M., 2001. Block copolymer thin films: physics and applications. Annu. Revi. Mater. Res. 31, 323–355.

Fernandes, D.M., Silva, R., Hechenleitner, A.A.W., Radovanovic, E., Melo, M.A.C., Pineda, E.A.G., 2009. Synthesis and characterization of ZnO, CuO and a mixed Zn and Cu oxide. Mater. Chem. Phys. 115, 110–115.

Fukahori, S., Ichiura, H., Kitaoka, T., Tanaka, H., 2003. Capturing of bisphenol A photodecomposition intermediates by composite TiO_2-zeolite sheets. Appl. Catal. B Environ. 46, 453–462.

Fukunaga, K., Hashimoto, T., Elbs, H., Krausch, G., 2003. Self-assembly of a lamellar ABC triblock terpolymer thin film. Effect of substrates. Macromolecules 36, 2852–2861.

Gao, Y., Liu, J., Shi, M., Elder, S.H., Virden, J.W., 1999. Dense arrays of well-aligned carbon nanotubes completely filled with single crystalline titanium carbide wires on titanium substrates. Appl. Phys. Lett. 74, 3642–3644.

Genne, I., Kuypers, S., Leysen, R., 1996. Effect of the addition of ZrO_2 to polysulfone based UF membranes. J. Membr. Sci. 113, 343–350.

Gracia, R., Aragües, J., Ovelleiro, J., 1996. Study of the catalytic ozonation of humic substances in water and their ozonation byproducts. Ozone Sci. Eng. 18, 195–208.

Guillen, G., Pan, Y., Li, M., Hoek, E.M.V., 2010. A review of nonsolvent induced phase separation membrane formation and characterization. Desalination.

Hadnagy, E., Rauch, L., Gardner, K., 2007. Dechlorination of polychlorinated biphenyls, naphthalenes and dibenzo-p-dioxins by magnesium/palladium bimetallic particles. J. Environ. Sci. Health A Tox. Hazard. Subst. Environ. Eng. 42, 685–695.

Han, Y., Li, W., Zhang, M., Tao, K., 2008. Catalytic dechlorination of monochlorobenzene with a new type of nanoscale Ni (b)/Fe (b) bimetallic catalytic reductant. Chemosphere 72, 53–58.

Hassan, M.H., Way, J.D., Thoen, P.M., Dillon, A.C., 1995. Single-component and mixed-gas transport in a silica hollow-fiber membrane. J. Membr. Sci. 104, 27–42.

Hoek, E., Jawor, A., 2002. Nano-filtration separations. CRC Press, New York, NY.

Hoffmann, M., Martin, S., Choi, W., Bahnemann, D., 1995. Environmental applications of semiconductor photocatalysis. Chem. Rev. 95, 69–96.

Holt, J.K., Park, H.G., Wang, Y.M., Stadermann, M., Artyukhin, A.B., Grigoropoulos, C.P., Noy, A., Bakajin, O., 2006. Fast mass transport through sub-2-nanometer carbon nanotubes. Science 312, 1034–1037.

Hosseini, S.S., Li, Y., Chung, T.S., Liu, Y., 2007. Enhanced gas separation performance of nanocomposite membranes using MgO nanoparticles. J. Membr. Sci. 302, 207–217.

Human Development Report, 2006. United Nations Development Programme (UNDP) (Chapter 4).

Hummer, G., Rasaiah, J.C., Noworyta, J.P., 2001. Water conduction through the hydrophobic channel of a carbon nanotube. Nature 414, 188–190.

Ilisz, I., Dombi, A., Mogyorósi, K., Dékány, I., 2003. Photocatalytic water treatment with different TiO_2 nanoparticles and hydrophilic/hydrophobic layer silicate adsorbents. Colloids Surf. A 230, 89–97.
Jeong, B.H., Hoek, E.M.V., Yan, Y.S., Subramani, A., Huang, X.F., Hurwitz, G., Ghosh, A.K., Jawor, A., 2007. Interfacial polymerization of thin film nanocomposites: a new concept for reverse osmosis membranes. J. Membr. Sci. 294, 1–7.
Kalra, A., Garde, S., Hummer, G., 2003. Osmotic water transport through carbon nanotube membranes. Proc. Natl. Acad. Sci. USA 100, 10175–10180.
Kang, S., Pinault, M., Pfefferle, L.D., Elimelech, M., 2007. Single-walled carbon nanotubes exhibit strong antimicrobial activity. Langmuir 23, 8670–8673.
Karnik, B., Davies, S., Baumann, M., Masten, S., 2005. Fabrication of catalytic membranes for the treatment of drinking water using combined ozonation and ultrafiltration. Environ. Sci. Technol. 39, 7656–7661.
Kaufman, Y., Berman, A., Freger, V., 2010. Supported lipid bilayer membranes for water purification by reverse osmosis. Langmuir 26, 7388–7395.
Kazemimoghadam, M., 2010. New nanopore zeolite membranes for water treatment. Desalination 251, 176–180.
Kim, M., Lin, N., Lewis, G., Cohen, Y., 2010. Surface nano-structuring of reverse osmosis membranes via atmospheric pressure plasma-induced graft polymerization for reduction of mineral scaling propensity. J. Membr. Sci. 354, 142–149.
Krishna, V., Pumprueg, S., Lee, S.H., Zhao, J., Sigmund, W., Koopman, B., Moudgil, B.M., 2005. Photocatalytic disinfection with titanium dioxide coated multi-wall carbon nanotubes. Process Saf. Environ. 83, 393–397.
Kumakiri, I., Yamaguchi, T., Nakao, S., 2000. Application of a zeolite A membrane to reverse osmosis process. J. Chem. Eng. Jpn. 33, 333–336.
Kumar, M., Grzelakowski, M., Zilles, J., Clark, M., Meier, W., 2007. Highly permeable polymeric membranes based on the incorporation of the functional water channel protein aquaporin Z. Proc. Natl. Acad. Sci. USA 104, 20719–20724.
Lee, D., Cohen, R.E., Rubner, M.F., 2005. Antibacterial properties of Ag nanoparticle loaded multilayers and formation of magnetically directed antibacterial microparticles. Langmuir 21, 9651–9659.
Lee, H.S., Im, S.J., Kim, J.H., Kim, H.J., Kim, J.P., Min, B.R., 2008. Polyamide thin-film nanofiltration membranes containing TiO_2 nanoparticles. Desalination 219, 48–56.
Legube, B., Leitner, K., 1999. Catalytic ozonation: a promising advanced oxidation technology for water treatment. Catal. Today 53, 61–72.
Li, L., Lee, R., 2009. Purification of produced water by ceramic membranes: material screening, process design and economics. Sep. Purif. Technol. 44, 3455–3484.
Li, L., Dong, J., Nenoff, T., Lee, R., 2004. Desalination by reverse osmosis using MFI zeolite membranes. J. Membr. Sci. 243, 401–404.
Li, L., Liu, N., McPherson, B., Lee, R., 2007a. Enhanced water permeation of reverse osmosis through MFI-type zeolite membranes with high aluminum contents. Ind. Eng. Chem. Res. 46, 1584–1589.
Li, Y., Chung, T., Kulprathipanja, S., 2007b. Novel Ag^+-zeolite/polymer mixed matrix membranes with a high CO_2/CH_4 selectivity. AIChE J. 53, 610–616.
Li, J.F., Xu, Z.L., Yang, H., Yu, L.Y., Liu, M., 2009. Effect of TiO_2 nanoparticles on the surface morphology and performance of microporous PES membrane. Appl. Surf. Sci. 255, 4725–4732.
Li, X., Fustin, C., Lefèvre, N., Gohy, J., De Feyter, S., De Baerdemaeker, J., Egger, W., Vankelecom, I., 2010. Ordered nanoporous membranes based on diblock

copolymers with high chemical stability and tunable separation properties. J. Mater. Chem. 20, 4333–4339.

Lin, J., Murad, S., 2001. A computer simulation study of the separation of aqueous solutions using thin zeolite membranes. Mol. Phys. 99, 1175–1181.

Lin, Y., Meziani, M., Sun, Y., 2007. Functionalized carbon nanotubes for polymeric nanocomposites. J. Mater. Chem. 17, 1143–1148.

Lind, M., Ghosh, A., Jawor, A., Huang, X., Hou, W., Yang, Y., Hoek, E., 2009a. Influence of zeolite crystal size on zeolite-polyamide thin film nanocomposite membranes. Langmuir 25, 10139–10145.

Lind, M.L., Jeong, B.H., Subramani, A., Huang, X.F., Hoek, E.M.V., 2009b. Effect of mobile cation on zeolite-polyamide thin film nanocomposite membranes. J. Mater. Res. 24, 1624–1631.

Liotta, L., Gruttadauria, M., Di Carlo, G., Perrini, G., Librando, V., 2009. Heterogeneous catalytic degradation of phenolic substrates: catalysts activity. J. Hazard. Mater. 162, 588–606.

Liu, G., Zhang, X., Xu, Y., Niu, X., Zheng, L., Ding, X., 2004. Effect of $ZnFe_2O_4$ doping on the photocatalytic activity of TiO_2. Chemosphere 55, 1287–1291.

Liu, N., Li, L., McPherson, B., Lee, R., 2008. Removal of organics from produced water by reverse osmosis using MFI-type zeolite membranes. J. Membr. Sci. 325, 357–361.

Lobo, R.F., 2003. Handbook of Zeolite Science and Technology. Marcel Dekker, Inc., New York, NY.

Lu, J., Liu, N., Li, L., Lee, R., 2010. Organic fouling and regeneration of zeolite membrane in wastewater treatment. Sep. Purif. Technol. 72, 203–207.

Luo, M.L., Zhao, J.Q., Tang, W., Pu, C.S., 2005. Hydrophilic modification of poly(ether sulfone) ultrafiltration membrane surface by self-assembly of TiO_2 nanoparticles. Appl. Surf. Sci. 249, 76–84.

Luo, M.L., Tang, W., Zhao, J.Q., Pu, C.S., 2006. Hydrophilic modification of poly(ether sulfone) used TiO_2 nanoparticles by a sol–gel process. J. Mater. Process. Tech. 172, 431–436.

Ma, J., Graham, N., 1999. Degradation of atrazine by manganese-catalysed ozonation: influence of humic substances. Water Res. 33, 785–793.

Ma, J., Graham, N., 2000. Degradation of atrazine by manganese-catalysed ozonation: influence of radical scavengers. Water Res. 34, 3822–3828.

Marambio-Jones, C., Hoek, E.M.V., 2010. A review of the antibacterial effects of silver nanomaterials and potential implications for human health and the environment. J. Nanoparticle Res. 12, 1–21.

Matos, J., Laine, J., Herrmann, J., 2001. Effect of the type of activated carbons on the photocatalytic degradation of aqueous organic pollutants by UV-irradiated titania. J. Catal. 200, 10–20.

Mauron, P., Emmenegger, C., Zuttel, A., Nutzenadel, C., Sudan, P., Schlapbach, L., 2002. Synthesis of oriented nanotube films by chemical vapor deposition. Carbon 40, 1339–1344.

Maximous, N., Nakhla, G., Wan, W., Wong, K., 2009. Preparation, characterization and performance of Al_2O_3/PES membrane for wastewater filtration. J. Membr. Sci. 341, 67–75.

Meinild, A., Klaerke, D., Zeuthen, T., 1998. Bidirectional water fluxes and specificity for small hydrophilic molecules in aquaporins 0–5. J. Biol. Chem. 273, 32446.

Meyer, D., Bhattacharyya, D., 2007. Impact of membrane immobilization on particle formation and trichloroethylene dechlorination for bimetallic Fe/Ni nanoparticles in cellulose acetate membranes. J. Phys. Chem. B 111, 7142–7154.

Morones, J.R., Elechiguerra, J.L., Camacho, A., Holt, K., Kouri, J.B., Ramirez, J.T., Yacaman, M.J., 2005. The bactericidal effect of silver nanoparticles. Nanotechnology 16, 2346–2353.

Mulder, M., 1996. Basic Principles of Membrane Technology. Kluwer Academic Publishers, London.

Noack, M., Kölsch, P., Seefeld, V., Toussaint, P., Georgi, G., Caro, J., 2005. Influence of the Si/Al-ratio on the permeation properties of MFI-membranes. Micropor. Mesopor. Mat. 79, 329–337.

Ozturk, S., Akata, B., 2009. Oriented assembly and nanofabrication of zeolite A monolayers. Micropor. Mesopor. Mat. 126, 228–233.

Peinemann, K.V., Abetz, V., Simon, P.F.W., 2007. Asymmetric superstructure formed in a block copolymer via phase separation. Nat. Mater. 6, 992–996.

Peinemann, K.V., Abetz, P.F., Simon, W., Johannsen, G., 2009. Isoporous membrane and method of production thereof, US Patent, US0173694.

Pendergast, M.M., Nygaard, J.M., Ghosh, A.K., Hoek, E.M.V., 2010. Using nanocomposite membrane materials to understand and control reverse osmosis membrane compaction. Desalination 261, 255–263.

Petersen, R.J., 1993. Composite reverse-osmosis and nanofiltration membranes. J. Membr. Sci. 83, 81–150.

Phillip, W.A., Amendt, M., O'Neill, B., Chen, L., Hillmyer, M.A., Cussler, E.L., 2009. Diffusion and flow across nanoporous polydicyclopentadiene-based membranes. ACS Appl. Mater. Interfaces 1, 472–480.

Pramauro, E., Prevot, A., Vincenti, M., Gamberini, R., 1998. Photocatalytic degradation of naphthalene in aqueous TiO_2 dispersions: effect of nonionic surfactants. Chemosphere 36, 1523–1542.

Prevot, B., 1999. Analytical monitoring of photocatalytic treatments. Degradation of 2, 3, 6- trichlorobenzoic acid in aqueous TiO_2 dispersions. Talanta 48, 847–857.

Prevot, A., Baiocchi, C., Brussino, M., Pramauro, E., Savarino, P., Augugliaro, V., Marci, G., Palmisanos, L., 2001. Photocatalytic degradation of acid blue 80 in aqueous solutions containing TiO_2 suspensions. Environ. Sci. Technol. 35, 971–976.

Progress on Sanitation and Drinking-Water, 2010. World Health Organization (WHO)/United Nations International Children's Fund (UNICEF) Joint Monitoring Programme for Water Supply and Sanitation.

Qiu, X., Burda, C., 2007. Chemically synthesized nitrogen-doped metal oxide nanoparticles. Chem. Phys. 339, 1–10.

Rao, A.P., Desai, N.V., Rangarajan, R., 1997. Interfacially synthesized thin film composite RO membranes for seawater desalination. J. Membr. Sci. 124, 263–272.

Rodriguez-Abreu, C., Lazzari, M., 2008. Emulsions with structured continuous phases. Curr. Opin. Colloid Interface Sci. 13, 198–205.

Salvetat, J.P., Briggs, G.A.D., Bonard, J.M., Bacsa, R.R., Kulik, A.J., Stockli, T., Burnham, N.A., Forro, L., 1999a. Elastic and shear moduli of single-walled carbon nanotube ropes. Phys. Rev. Lett. 82, 944–947.

Salvetat, J.P., Kulik, A.J., Bonard, J.M., Briggs, G.A.D., Stockli, T., Metenier, K., Bonnamy, S., Beguin, F., Burnham, N.A., Forro, L., 1999b. Elastic modulus of ordered and disordered multiwalled carbon nanotubes. Adv. Mater. 11, 161–165.

Savage, N., Diallo, M.S., 2005. Nanomaterials and water purification: opportunities and challenges. J. Nanopart. Res. 7, 331–342.

Schafer, A.I., Fane, A.G., Waite, T.D., 2005. Nanofiltration Principles and Applications. Elsevier Advanced Technology, New York, NY.

Sharma, Y.C., Srivastava, V., Upadhyay, S.N., Weng, C.H., 2008. Alumina nanoparticles for the removal of Ni(II) from aqueous solutions. Ind. Eng. Chem. Res. 47, 8095–8100.

Sharma, Y.C., Srivastava, V., Singh, V.K., Kaul, S.N., Weng, C.H., 2009. Nano-adsorbents for the removal of metallic pollutants from water and wastewater. Environ. Technol. 30, 583–609.

Shih, Y., Chen, Y., Chen, M., Tai, Y., Tso, C., 2009. Dechlorination of hexachlorobenzene by using nanoscale Fe and nanoscale Pd/Fe bimetallic particles. Colloid Surf. A 332, 84–89.

Sivakumar, M., Mohan, D.R., Rangarajan, R., 2005. Studies on cellulose acetate-polysulfone ultrafiltration membranes II. Effect of additive concentration. J. Membr. Sci. 268, 208–219.

Smart, T., Lomas, H., Massignani, M., Flores-Merino, M.V., Perez, L.R., Battaglia, G., 2008. Block copolymer nanostructures. Nano Today 3, 38–46.

Smuleac, V., Bachas, L., Bhattacharyya, D., 2009. Aqueous-phase synthesis of PAA in PVDF membrane pores for nanoparticle synthesis and dichlorobiphenyl degradation. J. Membr. Sci. 346, 310–317.

Song, L., Hu, J.Y., Ong, S.L., Ng, W.J., Elimelech, M., Wilf, M., 2003. Emergence of thermodynamic restriction and its implications for full-scale reverse osmosis processes. Desalination 155, 213–228.

Srivastava, A., Srivastava, O.N., Talapatra, S., Vajtai, R., Ajayan, P.M., 2004. Carbon nanotube filters. Nat. Mater. 3, 610–614.

Stoimenov, P.K., Klinger, R.L., Marchin, G.L., Klabunde, K.J., 2002. Metal oxide nanoparticles as bactericidal agents. Langmuir 18, 6679–6686.

Stratford, K., Adhikari, R., Pagonabarraga, I., Desplat, J.C., Cates, M.E., 2005. Colloidal jamming at interfaces: a route to fluid-bicontinuous gels. Science 309, 2198–2201.

Tarabara, V.V., 2009. Nanotechnology applications for clean water. In: Savage, N., Diallo, M., Duncan, J., Street, A., Sustich, R. (Eds.), William Andrew Inc., Norwich, NY, pp. 59–75.

Taurozzi, J.S., Arul, H., Bosak, V.Z., Burban, A.F., Voice, T.C., Bruening, M.L., Tarabara, V.V., 2008. Effect of filler incorporation route on the properties of polysulfone–silver nanocomposite membranes of different porosities. J. Membr. Sci. 325, 58–68.

Ulbricht, M., 2006. Advanced functional polymer membranes. Polymer 47, 2217–2262.

Urban Urgency, 2007. Water Caucus Summary. World Water Council (WWC), Marseille, France.

Wara, N.M., Francis, L.F., Velamakanni, B.V., 1995. Addition of alumina to cellulose acetate membranes. J. Membr. Sci. 104, 43–49.

Yan, L., Li, Y.S., Xiang, C.B., 2005. Preparation of poly(vinylidene fluoride)(PVDF) ultrafiltration membrane modified by nano-sized alumina (Al_2O_3) and its antifouling research. Polymer 46, 7701–7706.

Yang, S.Y., Ryu, I., Kim, H.Y., Kim, J.K., Jang, S.K., Russell, T.P., 2006. Nanoporous membranes with ultrahigh selectivity and flux for the filtration of viruses. Adv. Mater. 18, 709–712.

Yao, X., Yao, H.W., Li, Y.T., Chen, G., 2009. Preparation of honeycomb scaffold with hierarchical porous structures by core-crosslinked core-corona nanoparticles. J. Colloid Interface Sci. 332, 165–172.

Yoshikawa, N., Asari, T., Kishi, N., Hayashi, S., Sugai, T., Shinohara, H., 2008. An efficient fabrication of vertically aligned carbon nanotubes on flexible aluminum foils by catalyst- supported chemical vapor deposition. Nanotechnology 19, 245607.

Young, T.-H., Chen, L.-W., 1995. Pore formation mechanism of membranes from phase inversion process. Desalination 103, 233–247.

Yue, Z.R., Economy, J., 2005. Nanoparticle and nanoporous carbon adsorbents for removal of trace organic contaminants from water. J. Nanopart. Res. 7, 477–487.

Zhang, Z., Wang, C., Zakaria, R., Ying, J., 1998. Role of particle size in nanocrystalline TiO_2-based photocatalysts. J. Phys. Chem. B 102, 10871–10878.

Zhu, B., Lim, T., 2007. Catalytic reduction of chlorobenzenes with Pd/Fe nanoparticles: reactive sites, catalyst stability, particle aging, and regeneration. Environ. Sci. Technol. 41, 7523–7529.

Zhu, F.Q., Tajkhorshid, E., Schulten, K., 2004. Theory and simulation of water permeation in aquaporin-1. Biophys. J. 86, 50–57.

Zimmerman, C., Singh, A., Koros, W., 1997. Tailoring mixed matrix composite membranes for gas separations. J. Membr. Sci. 137, 145–154.

11

FATE OF ANTIBIOTICS DURING WATER TREATMENT: IMPACT ON ANTIMICROBIAL RESISTANCE IN ENVIRONMENTAL AND CLINICAL STRAINS

Florica Marinescu*, Irina Gheorghe, Ani Ioana Cotar**, Denisa Alexandra Florea†, Liliana Neagu**,†, Doina Maria Cirstea‡, Veronica Lazăr**, Mariana Carmen Chifiriuc****

**National Institute for Research and Development in Environmental Protection, Institute of Biology, Bucharest, Romania; **University of Bucharest, Research Institute of the University of Bucharest, Bucharest, Romania; †Politehnica University of Bucharest, Department of Biomaterials and Medical Devices, Faculty of Medical Engineering, Bucharest, Romania; ‡National Institute of Public Health, Institute of Biology, Bucharest, Romania*

Antimicrobial resistance has particularly important implications for both human health and ecology. In the recent decades, the widespread and uncontrolled use of antibiotics in human and veterinary medicine, as well as in agriculture, had a major impact on bacterial communities, leading to the selection of antibiotic resistant bacteria, for which an effective treatment currently may be difficult to find (Levy, 1992; Moura et al., 2010).

Antibiotics are considered an important category of water pollutants because of potential adverse effects on both aquatic ecosystems and human health due to the disturbance of the ecological balance of waters microbiota, which can be followed by increasing of the incidence of antibiotic resistant bacteria (Aminov, 2009; Martinez, 2009a).

Antibiotic resistant bacteria do not always constitute a direct risk factor for public health but through their resistance genes which are added to those from the environment and by the

Water Purification. http://dx.doi.org/10.1016/B978-0-12-804300-4.00011-3

possibility to transfer their resistance genes to other species and even genera, including those with pathogenic potential for humans may lead to the emergence of new mechanisms of clinical resistance (Wright, 2010). The investigation of antimicrobial resistance reservoirs should therefore include microorganisms from the environment (Abriouel et al., 2008).

Horizontal gene transfer by conjugation may result in the occurrence of multidrug resistant (MDR) bacteria, which have multiple resistance genes to different classes of antibiotics acquired without barriers of species or genus. The antimicrobial resistance is exacerbated by the production of bacterial biofilms on medical biomaterials or living tissues (Lazăr, 2007, 2011).

Currently, the classic definition of a pathogen has expanded, comprising besides the classic attributes of virulence, pathogenicity, and epidemic potential, the resistance to antibiotics (Baquero et al., 2008). Annually, significant quantities of antibiotics are used globally to treat human and animal infectious diseases. Also, antibiotics are used in animal husbandry as inhibitors and in aquaculture to prevent occurrence of infections, as well as in animal feeding, as growth factors. It is estimated that in the last 50 years were produced for use in human and veterinary medicine, and in agriculture over one million tons of antibiotics. Although the amount of antibiotics used for veterinary use is known, the World Health Organization estimates that about half of the total amount of antibiotics produced globally is used in the field (http://www.who.int/mediacentre/factsheets/antibiotic-resistance/en/). All of these characteristics contribute to the release of significant amounts of antibiotics in the environment, with major impact on bacterial communities, leading to the selection of antibiotic-resistant bacteria in natural ecosystems (Martinez, 2009b).

In the recent years, the international authorities have made considerable efforts to improve monitoring the circulation of antimicrobial resistant bacterial strains by recognizing the importance of investigation of natural reservoirs of resistance genes to achieve an effective epidemiological surveillance.

1 The Origin, Transfer, and Accumulation of Antibiotic Resistance Genes in the Aquatic Environment

1.1 The Origin of Antimicrobial Resistance Genes

The emergence of antimicrobial resistance is an evolutionary phenomenon caused by the selective pressure of the environmental factors (Greenwood, 1998). From a human perspective, microbial drug resistance is the process by which pathogens have developed defense mechanisms against the pharmacological compounds used to treat infections. Bacteria have shown a remarkable capacity to develop resistance to each new antibiotic introduced for medical use by developing new functional resistance mechanisms. Antimicrobial resistant bacteria can survive and even multiply in the presence of therapeutic levels of the antibiotic (Mihăescu et al., 2007a). Under stress conditions, both commensal and pathogenic bacteria increase the mutation rate, making them permutable. In such conditions, the bacteria express and replicate the genetic information, including antibiotic resistance AR genes located on plasmids, transposons orintegrons (Cambray et al., 2010). On the other hand, bacteria have a greater ability to preserve the genetic material that confers an evolutionary selective advantage, and to preserve beneficial mutations even in the presence of DNA repair mechanisms that tend to correct them.

The AR can be natural (intrinsic/innate) or acquired via mutations in endogenous genes and by acquisition of exogenous resistance, genes using horizontal gene transfer mechanisms.

Intrinsic resistance is the innate ability of a bacterial species to resist activity of a particular antimicrobial agent through its inherent structural or functional characteristics, which allow tolerance of a particular drug or antimicrobial class. This can also be called "insensitivity" since it occurs in organisms that have never been susceptible to that particular drug. Such natural insensitivity can be duc to: (1) lack of affinity of the drug for the bacterial target; (2) inaccessibility of the drug into the bacterial cell; (3) extrusion of the drug by chromosomally encoded active exporters; (4) innate production of enzymes that inactivate the drug.

The intrinsic resistance is found in all strains of a particular species, and its genetic support is represented by the bacterial chromosome and is vertically transmitted to offsprings through genetic transfer. For example, mycoplasma, bacteria lacking cell walls, are naturally resistant to the antibiotics which act at this level: β-lactam antibiotics, whereas anaerobic bacteria are intrinsically resistant to aminoglycosides (Chifiriuc et al., 2011). Usually, the natural resistance of these bacteria is due to the low permeability of the cell membrane to antibiotics. The bacterial cell envelope is forming a physical permeability barrier, which severely limits free diffusion of molecules (Piddock, 2006). Another cause of natural resistance is represented by the lack of cellular targets for the action of antibiotic. The bacteria of *Chlamydia* genus are not susceptible to inhibitors of cell wall synthesis (penicillins), because of the lack of peptidoglycan, although they possess the penicillin-binding proteins (PBPs) required for the synthesis of peptidoglycan.

In Gram-positive bacteria, the thickness and relative rigidity of peptidoglycan cell wall confer protection against antibiotics. The mureinic wall has a porosity of 1.1 nm, thus the free diffusion of the antibiotic molecules is possible for those having less than 1200 Da. The peptidoglycan component of the bacterial cell wall provides an excellent selective target for the antibiotics, and it is essential for the growth and survival of most bacteria. Consequently, enzymes involved in synthesis and assembly of the peptidoglycan component of the bacterial cell wall provide excellent targets for selective inhibition. The presence of mutations in the penicillin-binding domain of PBPs results in decreased affinity to beta-lactam antibiotics. Alterations among PBPs result in ampicillin resistance among *Enterococcus faecium*, and penicillin resistance among *Streptococcus pneumoniae*.

Gram-negative bacteria possess an outer membrane (OM) consisting of an inner layer containing phospholipids and an outer layer containing the lipid A moiety of lipopolysaccharides (LPS). This composition of the OM slows down drug penetration, and transport across the OM is achieved by porin proteins that form water-filled channels. Drug molecules can penetrate the OM employing one of the following modes: by diffusion through porins, by diffusion through the bilayer or by self-promoted uptake. The mode of entry employed by a drug molecule largely depends on its chemical composition. For example, hydrophilic compounds either enter the periplasm through porins (e.g., beta-lactams) or self-promoted uptake (aminoglycosides). Antibiotics, such as beta-lactams, chloramphenicol, and fluoroquinolones enter the Gram-negative OM via porins. As such, changes in porin copy number, size or selectivity will alter the rate of diffusion of these antibiotics.

Target inaccessibility due to the impermeability of structures as the OM of Gram-negative bacteria, represents a passive resistance. However, target inaccessibility can be also an active process (e.g., the activation of efflux pumps, energy dependent, which eliminate the antibiotic against a concentration gradient) (Nikaido and Pages, 2011).

Knowledge of the intrinsic resistance of a pathogen of concern is important in practice to avoid inappropriate and ineffective therapies. For bacterial pathogens, which are naturally insensitive to a large number of classes of antimicrobials, such as *Mycobacterium tuberculosis* and *Pseudomonas aeruginosa*, this consideration can pose a limitation in the range of options for treatment and thus consequently further increase the risk for emergence of acquired resistance.

Acquired resistance occurs when a particular microorganism gets the ability to resist the activity of a particular antimicrobial agent to which it was previously susceptible.

Acquired resistance to antimicrobials can be determined by: (1) chromosomal genetic determinants; (2) exogenous DNA segments (transposons) integrated in the genome by recombination; (3) the presence of R (resistance) plasmids, which may harbour one or more antimicrobial resistance genes conferring multidrug resistance.

Chromosomally mediated resistance (primary resistance) is due to a mutation in the gene encoding the target of the antibiotic or a membrane transport system that controls the penetration of the antibiotic in the bacterial cells (Lazăr, 2007). Errors occurred in the replicative DNA synthesis escaping from DNA repair systems lead to spontaneous mutation frequency of a pair of bases/10^7–10^{10} cells. Nevertheless the rate of spontaneous mutations generating resistance is small because is necessary the occurrence of multiple mutations to achieve a primary resistance. Therefore, it was considered that the emergence of antibiotic resistant strains through mutational processes during therapy is unlikely. Although the mutation is a rare event, the very rapid rate of growth of bacteria, is favoring relatively quickly expression of resistance in a cell population. Once the resistance gene mutation has been stabilized, it can be directly transferred to next generation of cells (the vertical genes flow).

Micro evolutionary genetic changes (point mutations) can alter the susceptibility of organisms to antibiotics by changing the structure of the target (e.g., the β-lactamase enzymes of some microorganisms are modified by point mutation and thus the enzyme activity spectrum extends. Resistance to sulfonamides is the result of a single amino acid change in the sequence of pteridine synthetase enzyme. Similarly, a point mutation changes aribosomal protein, which has as result the resistance to streptomycin (Mihăescu et al., 2007b).

Macroevolutionary genetic changes are involving the rearrangement of large segments of DNA by transposition. Whether the changes are involving moving, and the resistance genes are under control of a new promoter, the organism can become resistant. Cross-resistance means the phenomenon whereby acquiring a single mechanism of resistance, confers resistance to a whole class of antibiotics. For example, in *Staphylococcus aureus* resistance to methicillin confers resistance to all β-lactam antibiotics. In the case of efflux, the mechanism of resistance is nonspecific because may confer cross-resistance to a wide variety of antibiotics from different classes. The definition of multidrug resistance

(MDR) was proposed by the European Centre for Disease Prevention and Control (ECDC) and the Centers for Disease Control and Prevention (CDC) to create a standardized international terminology, and formulated by a group of international experts (Magiorakos et al., 2012) as being the acquired nonsusceptibility to at least one agent in three or more antimicrobial categories. The extensively drug-resistant (XDR) was defined as nonsusceptibility to at least one agent in all but two or fewer antimicrobial categories (i.e., bacterial isolates remain susceptible to only one or two categories) and PDR was defined as nonsusceptibility to all agents in all antimicrobial categories (Magiorakos et al., 2012).

1.2 Antibiotic Resistance Genes Transfer Mechanisms in the Aquatic Environment

1.2.1 The Potential Routes for the Introduction of Antibiotics and Genes Encoding Antibiotic Resistance in the Environment

The environmental habitats especially water bodies, such as rivers and streams are ideal vectors for the AR dissemination (Lupo et al., 2012). Here, the propagation of the bacteria harboring AR genes can occur spatially along the river. Furthermore, the dispersion of these bacteria in the environment favors the interaction with the autochtonous microbiota, creating new scenarios for the evolution of AR (Lupo et al., 2012). Antibiotic-resistant bacteria can get access into the aquatic ecosystems through untreated waste water, effluent of WWTPs and due to leakage from agricultural areas where manure was stored/applied (Hayes et al., 2004; Mispagel and Gray, 2005; Li et al., 2009; Haggard and Bartsch, 2009; Munir and Xagoraraki, 2011). Also, these AR genes may contaminate water resources used for production of drinking water, and could be transferred to bacteria in drinking water or human food chains (Chee-Sanford et al., 2001; Shar et al., 2009). AR can be disseminated through vertical and/or horizontal gene transfer (Rysz, 2007). The vertical gene flow means transmission of genetic information, including resistance genes to subsequent generations of cells. In contrast with horizontal transfer of resistance genes, the vertical transfer of genes is a slowly evolutionary mechanism for spreading of AR genes. Horizontal gene flow means the transfer of genetic information between bacterial cells without barriers of gender and species through other ways than by cell division. Horizontal genetic exchange is the primary mechanism of evolution and occurrence of AR; is achieved by conjugation, transduction, or transformation. This continuous flow of genes seems to be a defining characteristic of the bacterial world and have greatly contributed to extreme metabolic versatility of bacteria in almost all natural environments (Jurcoane et al., 2006).

1.2.2 Horizontal Gene Transfer of Antibiotic Resistance in the Aquatic Environment

The horizontal genetic exchange, the major route for global dissemination of AR (Sorensen et al., 2005), is responsible for the transfer of 60–90% of resistance (R) plasmids of Gram-negative bacteria (Levy and Miller, 1998). Many of the AR genes are carried on plasmids, transposons, or integrons that can act as vectors that transfer these genes to other members of the same bacterial species, as well as to bacteria in another genus or species.

Numerous studies have showed that horizontal transfer of resistance genes between bacteria occur in a variety of aquatic habitats: rivers, lakes, wastewater, sediment, and sludge, wastewater treatment plants (Cernat et al., 2002b,c; Guardabassi and Dalsgaard, 2004; Shakibaie et al., 2009). These habitats may provide sufficient amounts of nutrients and surface colonization factors for bacterial biofilms and bacterial populations of high density, thereby creating a microenvironment favorable for the horizontal transfer of genes by conjugation between bacterial cells. Also Wiedenbeck and Cohan (2011) showed that the adaptation of bacteria to antibiotics accelerates the selective pressure of AR genes acquired by horizontal transfer from the donor to the recipient bacterial cells.

Horizontal gene transfer of AR among bacteria in the aquatic environments may occur via three main mechanisms: transformation, transduction, or conjugation: conjugation involves transfer of DNA via sexual pilus and requires cell–cell contact. DNA fragments (present on plasmids, transposons, or integrons integrated in the structure of plasmids or transposons) that contain resistance genes from resistant donors can then make previously susceptible bacteria express resistance as coded by these newly acquired resistance genes; transformation involves uptake of short fragments of naked DNA by naturally transformable bacteria. In the case of bacteria that have not naturally ability to uptake DNA from the external environment, this feature may be induced (e.g., the presence of calcium) (Zhang et al., 2009a); transduction involves transfer of DNA from one bacterium into another via bacteriophages.

After the new DNA is introduced whether via conjugation, transduction, or transformation, it is incorporated into the cell and results in the emergence of a new, resistant genotype.

Studies of environmental bacterial communities have demonstrated that contamination of freshwater with heavy metals correlated positively with a higher abundance of class I integrons

(Wright et al., 2008; Cernat et al., 2002a). Also, pollution and nitrogen sediment (quaternary ammonium salts), detergents and/or antibiotics favor the selection of bacteria carrying class 1 integrons and *qac* gene cassette, which is responsible for bacterial resistance to quaternary ammoniums by efflux pumps (Gaze et al., 2005, 2011). These integrons are derived from Tn_{402} transposon, and may insert into larger Tn_{21} transposon, carrying genes associated with resistance to sulfonamide (*sul I*) and trimethoprim (*dfr1*).

Class 2 integrons are exclusively associated with Tn_7 transposon, easily integrate into chromosome of *E. coli* and other *Proteobacteria*, therefore contribution to the spread of AR genes in large bacterial communities.

Class 3 integrons are also thought to be located in a Tn and are less prevalent than Class 2.

Classes4 and 5 integrons are associated with trimethoprim resistance genes in *Vibrio* species.

Resistance (R) plasmids are circular extrachromosomal DNA molecules that carry genes conferring adaptive characters, beneficial to bacterial hosts, allowing their survival under certain environmental conditions, such as AR and tolerance to heavy metals, important features for pathogenic bacteria. A bacterial cell may carry several resistance genes responsible for multidrug-resistance or resistance to all currently used antibiotics. R plasmid genes may be part of a transposon, being bounded by the insertion sequences, and have a very strong mobility being able to move from one site to another structure or between the plasmid and chromosome (Bennett, 2008). The number of the resistance genes in a plasmid is variable. R plasmids can be transferred from one cell to another by conjugation, transformation, or transduction.

Genetic determinants of AR and transfer can exist together on the same plasmid or separately on different plasmids.

Bacteriophages are a major constituents of aquatic ecosystems (Weinbauer, 2004) and play an important role in the horizontal gene transfer in the aquatic environment (Lupo et al., 2012). Recent studies (Parsley et al., 2010; Colomer-Lluch et al., 2011) have demonstrated the presence of genes coding for the synthesis of β-lactamases (bla_{TEM}, $bla_{CTX\text{-}M}$) and of genes encoding methicillin resistance (*mecA*) in the viral metagenome of the activated sludge and wastewater from wastewater treatment plants and wastewater into the river receptor, confirming that transduction can be responsible for the dissemination of resistance genes in these aquatic habitats. In addition, the presence of the *mecA* gene in the DNA of phages in aquatic environments is a major problem for the public health because of the threat posed by infections with *S. aureus* resistant to methicillin

(MRSA), both in hospitals settings and in the communities (Campanile et al., 2011).

Resistance genes located in the chromosome or in different mobile genetic elements are encoding a variety of AR mechanisms: (1) the enzymatic inactivation of the antibiotics; active efflux of the antimicrobial agent; reduced uptake of the antimicrobial agent due to decreased permeability of OM; a mutation in the antimicrobial agent's target, which reduces the binding of the antimicrobial agent; posttranscriptional or posttranslational modification of the antimicrobial agent's target, which reduces binding of the antimicrobial agent; the presence of an alternative enzyme to that is inhibited by the antimicrobial agent.

1.2.3 Factors Influencing the Dissemination of Antibiotic Resistance Genes in Aquatic Environments

The major factors that may influence the dissemination of AR genes in aquatic environments are: the spatial distribution and the density of the donor and recipient strains; the metabolic activity of bacterial cells depends on the availability of nutrients (Naik et al., 1994); the mineral content of sediment/soil and microbial adsorption capacity; the temperature and the pH; the selective pressure, determined by the presence of antibiotics and/or heavy metals.

In order to transfer gene of AR to be achieved through conjugation is a necessary direct contact between two cells (donor and receptor). In the aquatic environments, bacteria tend to increase the contact surface and adhere to the substrate or surface, forming monospecific or multispecific bacterial biofilms (Cernat and Lazăr, 2011) and high density bacterial populations. The number of adherent bacteria in biofilms of aquatic microbiota from rivers is 500–500,000 times higher than planktonic bacteria as biofilms growth ensures the survival and adaptation to changes in environmental conditions, such as higher level of resistance of cells from biofilm and a better protection of cells from stress factor. Among the cells forming the biofilm the possible genetic transfer (including R plasmids) are facilitated by the proximity of cells, a process that leads to the occurrence of resistance in a population of cells initially sensitive to certain antibiotics, a multiclonal or multispecific community and so the resistance is amplified. Additionally, the cells contained in biofilms show a high rate of mutations (hipermutators cells), influenced by the stress factors with adaptive role (Lazăr and Chifiriuc, 2010). Also, the extracellular matrix of the biofilm from the aquatic habitats creates favorable conditions for plasmids transfer and genetic transformation (Molin and

Tolker-Nielsen, 2003). It has a crosslinked structure and contains water (95–99% of the biofilm) and various substances: poliosids, some of them with adhesive part, the capsular polysaccharides, DNA fragments, proteins, such as fimbriae involved in bacterial conjugation (Cernat and Lazăr, 2011).

Quorum sensing (QS) is an important feature of microbial communities, including biofilm and represents the regulation of gene expression in response to fluctuations in cell-population density. QS signaling is mediated by extracellular signaling molecules called autoinducers, that increase in concentration as a function of cell density. The detection of a minimal threshold stimulatory concentration of an autoinducer leads to an alteration in gene expression. In planktonic populations, the chemical signals produced by cells are not sufficiently high to determine changes in the gene expression. In the biofilm matrix, where bacterial cells are in close proximity, the signaling molecules accumulate in sufficient concentrations in direct correlation with cell density up to a critical concentration. Once a critical concentration has been reached, autoinducers can activate or repress a number of target genes. The final result is the simultaneous modification of gene expression and activation of certain cellular phenotypes in the whole bacterial population (Lazăr, 2011). QS systems are coordinating several important physiological processes (e.g., competence induction for genetic transformation, the conjugative plasmids transfer, the antibiotics and pigments biosynthesis, the sporulation, the motility, biofilm formation and development) (Davey and O'Toole, 2000). The pathogenic or opportunistic bacteria are also using the QS systems for coordinated expression of certain virulence factors involved in infectious process evolution in host organisms. These signaling systems increase survival of their producing bacteria since the coordinated expression of virulence factors (e.g., the capsule production by *P. aeruginosa* responsible for mucoid phenotype and adhesion to the susceptible host cells, the synthesis of elastase and protease which play specific roles in infectious process evolution) occur only when the bacterial population reaches a critical density adequate to make them to survive to defense mechanisms of the host (Kovacikova and Skorupski, 2002).

Recent studies have suggested that bacterial biofilms constituted in the water at the air–liquid interface (Costerton et al., 1995) are favorable environments for the conjugative transfer of genes (Cunliffe and Murrell, 2009), because this interface facilitates attaining of high density of donor and recipient bacterial cells (Król et al., 2011), probably due to increased concentration of oxygen in the area. Physical–chemical properties of the air–liquid interface may also directly promotes the adhesion of bacterial cells to each other and to the inert substratum, thus increased level of cells in that area due to aerobic conditions may explain the increased proportion of the adherent cells and hence increased cell to cell contact.

In a study, performed in 2011 by Krol and coworkers, the influence of specific factors on the transfer of plasmid PB10 (broad host range plasmid that contains multiple AR genes) from the *E. coli* K-12 cells in biofilms has been investigated. The results have shown that the transfer occurred with plasmid PB10 is at higher rate in biofilms formed at the air–liquid interface compared with submerged biofilms, mainly due to differences in oxygen concentration and increased density of the donor and recipient cells. Similarly, Nguyen et al. (2010) showed that chromosomal markers of transfer between *Mycobacterium smegmatis* strains have frequently occurred at the air–liquid interface.

The main factor, which, contributes to the horizontal transfer of AR genes is represented by the selective pressure exerted by the intensive use of antibiotics for human and veterinary treatment (Jehl et al., 2004). High selective pressure facilitates acquisition of AR genes, which is leading to increased incidence of resistant bacteria and their global dissemination (Luo et al., 2005).

Bacteria in the environments polluted by heavy metals show an increased prevalence of AR phenotypes compared to the control areas (Wright et al., 2008). The *tetA* gene, which encodes tetracycline resistance, was identified in a *Serratia marcescens* strain isolated from a stream polluted with heavy metals (Thompson et al., 2007), providing an indirect evidence of coselection. The resistance plasmids may be responsible not only for AR determination but also for simultaneously resistance to heavy metals, which is favoring the maintenance and spread of AR due to the selective pressure exerted by heavy metals (Derore et al., 1994; Alonso et al., 2001; Baker-Austin et al., 2006).

2 Antibiotic Resistance in the Fresh and Sewage Waters

Any human activity represents a potential source of direct or indirect (through atmospheric and soil pollution) pollution of waters. Currently there is a continuation in the reduction of wet zones, increasing usage of fertilizers and pesticides in agriculture, as well as intensification of mining and industry, and increasing of severe industrial accidents. Moreover, annually, significant quantities of antibiotics are used to treat human and animal infectious diseases. Also, antibiotics are used in animal husbandry at subinhibitory concentrations and in aquaculture to prevent disease and in animal feed as growth promoters (Kümmerer, 2003; Davies and Davies, 2010).

Aquaculture constitutes a major source for antibiotics in environment and the risk of transfer of resistance to human pathogens is high. A way to limit the occurrence of infections is to administrate antibiotics and other chemical substances to fish from aquaculture. The antibiotics enter in the water and destroy local ecosystems. From economical reasons, quinolones, sulfonamides, and tetracyclines are the most popular antibiotics used in aquaculture, although others, such as macrolides or beta-lactams are occasionally used (http://www.who.int/mediacentre/factsheets/antibiotic-resistance/en/). The majority of fish species grown in aquaculture is poikylothermic and is adapted to survive to low temperatures (Heuer et al., 2009). Nevertheless, some zoonotic bacteria, such as *Aeromonas*, *Salmonella*, and *Mycobacterium* may also infect humans and transport AR genes (Weir et al., 2012). *E. coli* is the most frequently human pathogen and exposure to antibiotics in aquaculture environment may stimulate high resistance. In case of cultivated mollusca, pathogenic bacteria, such as *Vibrio* and *Salmonella* may acquire resistance through horizontal transfer. For example, the fish pathogenic agents *Vibrio* and *Lactococcus* may transfer resistance genes to tetracycline. AR genes in association with antibiotic resistant bacteria have been identified as contaminants, spread in treated drinking waters and residual waters. In conclusion, high concerns occurred due to the transport of antibiotic resistant bacteria or their resistance genes between aquatic compartments which may increase spread of AR in nonresistant bacterial communities through horizontal gene transfer mechanisms. The conventional processes of disinfection of wastewater and drinking water in general is offering efficient ways of reducing antibiotic resistant bacteria transport minimizing the risk of gene transfer through conjugation. Nevertheless, even if antibiotic resistant bacteria present in a treated water are completely inactivated during a disinfection process, the possibility of existence of intact DNA in the resulted residue could confer new resistance genotypes to bacterial population from downstream through natural transformation and/or transduction, which do not necessitate live donor cells. Different antibiotics have been used during the last 20 years and continue to be registered for use in aquaculture in United Kingdom, Norway, and Canada. These are including β-lactams (amoxicillin), macrolides (eritromycin), quinolones (nalidixic acid), fluoroquinolones (floxacin), sulfonamides, and tetracyclines (oxitetracycline). The vaccines have replaced the antibiotics as control way against bacterial pathogens from aquaculture. Accelerated increase of aquaculture has conducted to large use without restrictions of prophylactic antibiotics, especially in countries under development determining situation detrimental for environment and human health. This process has led to the appearance of antibiotic resistant bacteria in the environment from aquaculture, increasing of AR of pathogens characteristic to fish, as well as to modification of bacterial microbiota in sediments and water. Usage of large quantities of antibiotics, which have mixed with fish food is creating problems of health and increases opportunities for presence of residual antibiotics in fish meat or in fish products. Thus, there are necessary efforts at global level to promote a more judicious use of prophylactic antibiotics in aquaculture because the studies are indicating that their use without restriction is nonfavorable to fish, terrestrial animals, human health, and environment (Cabello, 2006). Use of antimicrobial drugs in aquaculture has some differences comparatively with their use in terrestrial animals. In aquaculture, antimicrobial substances are regularly added in the food, which is then introduced directly in water. In some countries, the regulations concerning antibiotic use are strict and only a few antibiotics are licensed for aquaculture use. Nevertheless, an important part of aquaculture at global level occurs in countries, which do have permissive regulations. Moreover, many governments have established maximum mandatory limits of residues for products in aquaculture. The public health risk associated with residues from antimicrobial substances is depending of exposure to these substances. Intensive fish culture has promoted spread of bacterial diseases, which have also conducted to intensive use of antimicrobial substances. The concerns regarding the consequences of antibiotic consumption on human health has encouraged development of strict regulations, which control antibiotic use. High proportions of antibiotic resistant bacteria, which persist in the sediments may represent a threat for fish farms, because these can serve as sources of AR genes for pathogenic agents near the farms.

Other important pollution sources with antibiotics are represented by agriculture and urban wastewater from hospitals and pharmaceutical industry (Fig. 11.1).

Waste deposits are still representing risks because leaching may determine pollution of ground and surface waters, and are usually redirected toward a municipal purification station (Renou et al., 2008). The incineration is a zero risk solution regarding the reduction of AR bacteria, although there are compromises concerning the air quality. Whether is adequately used, incineration can provide an alternative source of energy. The waste derived from hospitals and drug producers should be pretreated to partial eliminates the antibiotics and other drugs, as well as AR bacteria

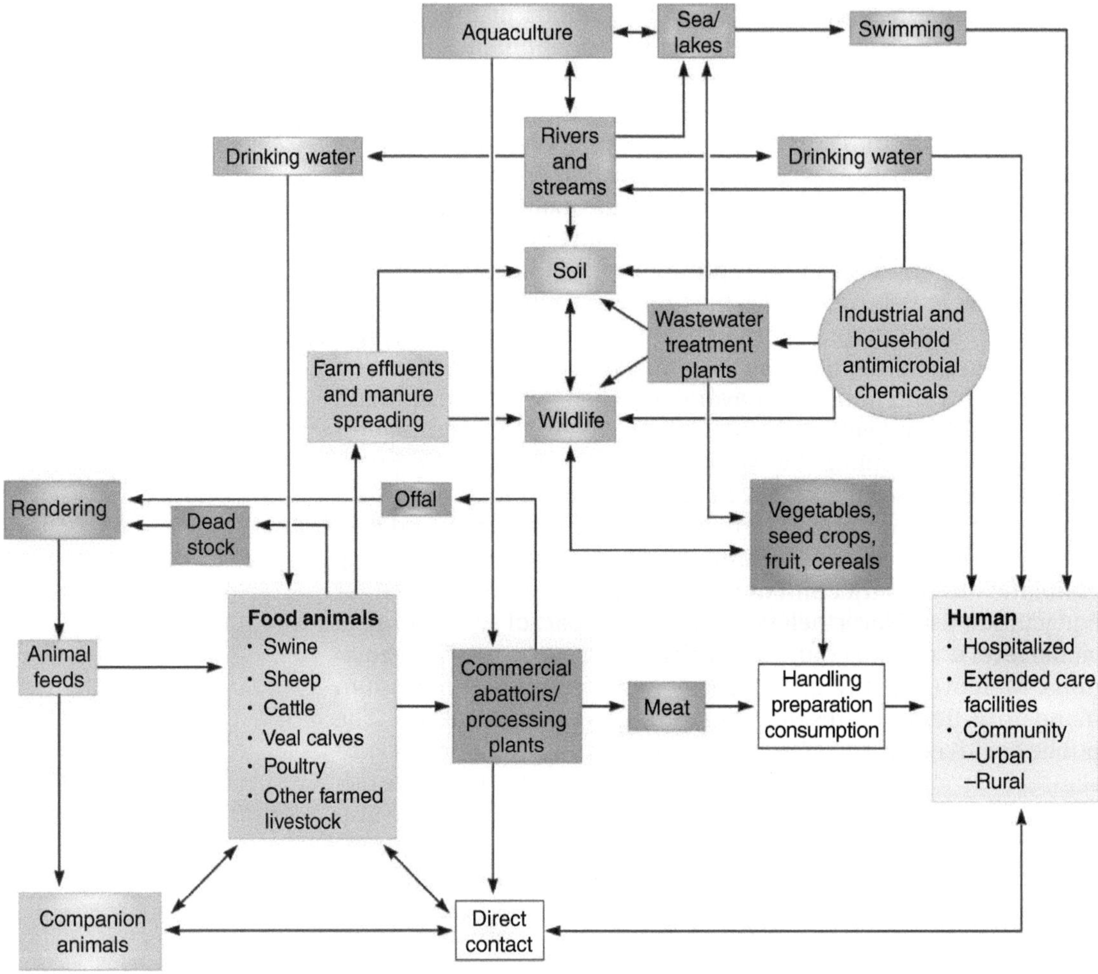

Figure 11.1. Dissemination of antibiotics and antibiotic resistance within agriculture, community, hospital, wastewater treatment, and associated environments (Davies and Davies, 2010).

before discharge in public sewerage systems. Production sites were identified as potential hot points for development of AR, with a level reaching concentrations of milligram per liter in some cases. Larsson et al. (2007) have discovered an extremely high level of fluoroquinolones in the effluent treated by a purification station, which have served approximately 90 producers of drugs in India. In the same area, the contamination with antibiotics was severe as the presence of antibiotics was found in the soil and local drinking waters (Fick et al., 2009), and mobile genetic elements associated with AR were found downstream in high significant concentrations (Kristiansson et al., 2011). Some industries treat their own waste until discharging to purification station, whereas others are discharging the waste in their own station with or without pretreatment (e.g., pH-adjustment, chelation, precipitation). The cycles of production from manufacture stations/sites of pharmaceuticals are very variable, and many drugs are produced in a discontinued way. Thus, the composition of residual water can vary very much in time. For this variation in the composition it is necessary a distinct treatment, thus the purification station which receives the waste from drug producers will have as benefit pretreatment or is necessary to establish the limits of discharging of antibiotic. The waste derived from industrial production will need several technologies of treatment. A major challenge is represented by the fact that high levels of antibiotics concentration in purification stations will exert a strong selective pressure for antibiotic resistant bacteria. From this reason, the active mud is not recommended for waste contaminated with antibiotics due to increased density of microbial populations.

Generally, uncontrolled use of antibiotics and metals is increasing in agriculture and livestock and has as result enrichment of manure with bacteria resistant to antibiotics (Zhu et al., 2013), which has effect on the environment especially on the soils (Wu et al., 2010).

Discharges in the natural ecosystems of large quantities of antibiotics and resistance genes in evolutionary terms are recent events (Martinez, 2009a). However, both types of pollution may have impact on the structure and activity of microbial populations in the environment and can lead to the selection of antibiotic-resistant bacteria and dissemination of AR determinants among pathogenic bacteria populations.

Antibiotic resistant bacteria and a high number of antibiotic resistance genes were detected in the aquatic environment (Bell et al., 1983; Huys et al., 2000; Balotescu et al., 2003; Ram et al., 2007; Tao et al., 2010) as a result of uncontrolled use of urban wastewater and wastewater in agriculture (Goñi-Urriza et al., 2000a; Kim and Aga, 2007; Auerbach et al., 2007; García-Delgado et al., 2007; Gulkowska et al., 2008; Dolliver and Gupta, 2008). AR among bacterial strains isolated from the aquatic environments is achieved through different molecular mechanisms: modifying the target of the antibiotic through production of an alternative target (dfrA1, A5, A7, A12, A15, A17, A18, spears, II, III, and A) and synthesis of a target protein unable to bind the antibiotic; presence of the efflux pump (cmlA1 and A5, flo_R, otrB, tetA, A, (41) B, C, D, E, G, J, Y, Z, 33, 39) which reduce penetration of antibiotics through membrane cells by structural or functional alteration in the membrane cell; the enzymatic inactivation of the antibiotics (aacC1, C2, C3, and C4;

aadA1, A2, A5, A13, and B; ampC; aphA1, D; mphA, sat1, and 2, strA and B); modification of the antibiotic target [ermA, B, C, E, F, T, V, X, mecA, penA, otrA, tetB (P), M, O, Q, S, T, W, vanA, and B].

2.1 Antibiotic Resistance in Bacterial Strains Isolated from Natural Aquatic Environments

Surface waters and wastewaters are suitable ecological systems for maintaining and development of natural reservoir of virulence and AR genes, constituting the so-called resistome (Wright, 2007). In the database of AR genes (ARDB—AR Database, http://ardb.cbcb.umd.edu/) recently developed, were recorded over 13,000 AR genes (args) identified in more than 600 bacterial strains resistant to antibiotics (Liu and Pop, 2009).

Genetic determinants of resistance of potentially pathogenic bacteria represent only a small part of the resistance genes identified in bacterial world, which means that most of AR genes are present in nonpathogenic bacteria in the environment (Davies and Davies, 2010). A link between the so-called resistome and the emergence of AR in the medical clinics seems plausible given possible contacts between aquatic bacteria and opportunistic pathogens, such as *P. aeruginosa, Acinetobacter baumannii, Stenotrophomas maltophilia* (Baquero et al., 2008; Martinez, 2009a).

The potential for transmission of resistance in the aquatic environments is linked to the presence of mobile genetic elements in certain bacterial species involved in efficient transfer of resistance and virulence genes. In this context, the integrons play an important role in spreading of AR genes, because contain gene cassettes responsible for providing MDR to many antibiotics of different classes, as well as resistance to other compounds, such as pollutants (heavy metals, biocides).

Integrons (Fig. 11.2) contain a coding sequence for a site specific, recombinase called integrase (Int), a site of recombination for genes sequence recognition (att), and a promoter (P), which activate the expression of genes downstream from the recombination site (att). Resistance genes can be acquired by using the site-specific recombination, with the help of integrase, which is designed to integrate the gene cassette. The integration of gene cassette essentially occurs at a specific locus immediately adjacent to the *intI* referred to as the primary recombination site attI. A large majority of cassettes are promotorless. The expression of genes they carry is then ensured by a dedicated Pc promoter, which is embedded either in the *intI* gene or in the attI site and orientated toward the integration site. Successive integration at the attI site results in the streamlined assembly of different gene

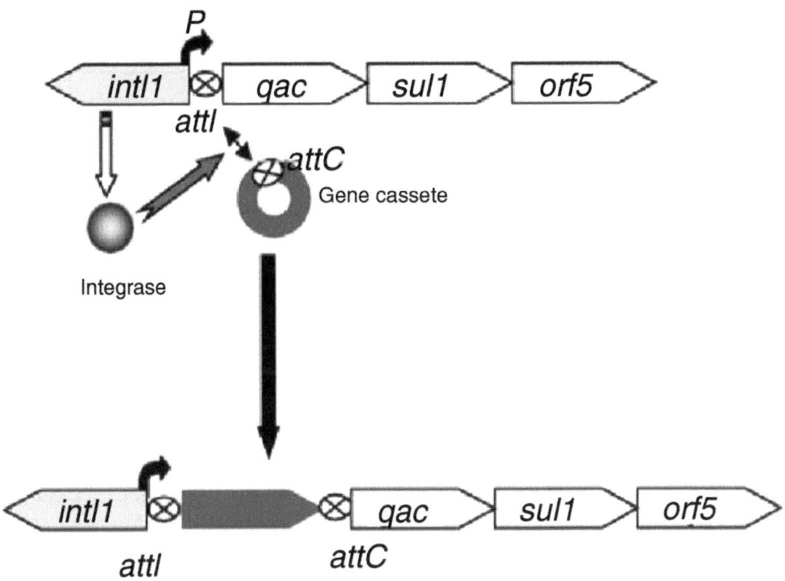

Figure 11.2. The structure of Class 1 integrons.

cassettes arrays. Genes cassette are generally constituted by a single operon reading frames (ORF) immediately followed by a recombination site termed attC (formerly called 59-base element), which is specifically recognized by IntI. After integration, recombination sequences are restored, thus enabling further integration of other gene cassettes. Through gene cassettes, the resistance genes complement can be rearranged and moved permanently from one replicon to another.

2.2 Antibiotic Resistance in Bacterial Strains Isolated from Wastewater and Sewage Treatment Plants

From the ecological point of view, the wastewater is one of the most complex environment in which there are many interactions among physiological groups of bacteria. Commonly, the wastewater may contain various human and animal pathogens (Zarnea, 1994). Usage of water for specific human population needs through development of modern sanitary-hygienic equipment and its use in agriculture (crop, livestock) and industry has conducted to changes in its physicochemical and microbiological properties becoming wastewater due to contamination or pollution. Most of the water used in various human activities is discharged into the aquatic environment, which is the main collector. Human may supply with water from every types in which this can be found in nature, but mostly from ground or surface water. Surface water is the most accessible source of water, which allows taking on large amount of water, despite of the seasonal variation of water flow. Ground waters are an important source because unlike surface waters, ground waters are usually little or unpolluted and can be made potable using minimum measures, sometimes only with disinfection or without any preparation.

Water supply is performed using two different types of installations: central and local. Central installations are consisting in water distribution from the source to the consumer through a network of pipelines. For local installation, each consumer is obtaining necessary water directly from the water source. A central system of water supply/supplying is composed of 4 sectors for capture, treating, storing, and distribution. Treating sector is mandatory for surface waters and is consisting in a number of installations through which the water passes successively taking off its impurities. Many of these plants are automatically functioning, different installations switch off immediately when treated water through that installation does not meet the established requirement.

Stations of water treatment have different structures depending on sizes, complexity, and technology used. Nevertheless, the steps and principles of treatment are the same most of the time. The water is captured from accumulation lakes, rarely from rivers, from zone of sanitary protection. The raw water pass through the following procedures until became drinking water: (1) sifting (passing through a strainer) is the first phase of water preparation. During this phase in the station of screening the water successively is passing through a strainer with large and small eyes, and afterwards through microstrainers. During this phase the floating bodies, fish, plankton, and bulky suspensions are eliminated; (2) sedimentation—is produced in a decantation apparatus, which can be linear or circular. In these decanters, the water is staying a period of time, while suspensions are gravitational deposited on the bottom of decanter, from where there are periodically removed. The suspensions are reduced around 40–60 %, thus the process is amplified through flocculation and coagulation. For this purpose in the water are introduced compounds, such as aluminum sulfate, iron chloride or sulfate, chalk, and so on. Thus, the electrical charged particles are connected and larger aggregates are formed, natural charged, which are precipitating and determines increasing of sedimentation efficiency until 60–80%. The remaining suspensions are eliminated through filtration; (3) the filtration represents the procedure of advanced cleaning of water, consisting in passing of water through a mass of porous material called filterable stratum. Filtration is used as the last phase of cleaning in the process of obtaining drinking water. The most widespread are slowly filtered and quickly filtered. These are sand basins on which the water is gravitational passing, and is coming out pure. The filters are periodically washed to remove the retained mass of impurities. At "quickly filter" the filtration process is mechanical, but at the "slowly filter" the process is mechanical–biological because at the filters' surface is forming a stratum colonized with algae, bacteria, and protozoa, which actively contributes to retention of impurities through chemical, enzymatic mechanisms; (4) disinfection is the next step in the process of water treatment and is mandatory in quickly filtration. Disinfection is the operation of destroying of pathogenic germs in water and of reduction of commensally ones.

Water disinfection procedures maybe classified in: (1) physical procedures, which are using some energy types with high penetration power, such as UV, X, gamma waves, ultrasound field, infrared radiations, microwaves; (2) chemical procedures, which use chemical reagents heavy oxidative, such as chlorine, chlorine dioxide, ozone, oxygen water, or ions of heavy metals (silver, copper). Disinfection process is performed in two phases: (1) disinfectant penetration through cellular wall and (2) cytoplasmic enzymes denaturation.

Bacterial inactivation is performed by membrane cell degradation and its permeability increases, followed by oxidation of enzymes involved in respiration. For viruses, the disinfectant action is manifested at nucleic acids, whereas at protozoa the phenomenon that takes place is not very well known.

Biological methods of wastewater treatment plants are classified into two categories, depending on environmental conditions (presence or absence of oxygen) and the nature of microorganisms involved in the treatment: (1) aerobic treatment technologies which include the use of biological filters, the activated sludge or biooxidation ponds; (2) anaerobic treatment technologies, which use anaerobic degradation of methane-tanks or septic ponds.

The complexity of the wastewater treatment process is determined by the physico–chemical and microbiological water quality requirements and of discharging into receiving rivers.

The most effective method for removing organic substances from wastewater is biological treatment. This method is based on metabolic reactions of a mixed population of bacteria, fungi, protozoa, and even some lower metazoans forming sludge. The efficiency of the cleaning process and composition of biomass depend on several factors: the type and concentration of wastewater pollutants, temperature, dissolved oxygen concentration, poisonings acute/chronic bodies, due to the discharge of toxic pollutants, and sometimes retention, operation of the plant and the technological variant of applied methods.

Environmental conditions in the wastewater treatment stimulate the multiplication of bacteria that degrade organic matter and the presence of antibiotics and/or other stress factors (heavy metals, pharmaceuticals etc.) may lead to the selection of multiresistant bacteria that can be released into waters of rivers and lakes (García-Delgado et al., 2007).

Antibiotics residues from different sources (housekeeping, pharmaceutical industry, hospitals) enter in urban wastewater together with other coselection factors (metals and surfactants). At least 56 antibiotics from 6 different classes have been detected in sewerage systems from Eastern Asia, North America, Europe, Australia (Zhang and Li 2011). The elimination ways of these antibiotic residues are including adsorbtion, biodegradation, disinfection, and separation through membrane (Zhang and Li, 2011). Other ways, such as hydrolysis, photolysis, and volatilization are also contributing to elimination depending of antibiotic properties (Zhang and Li, 2011). The tetracyclines are eliminated through adsorbtion, whereas the beta-lactams are mainly degraded through hydrolysis reactions produced by bacteria or physical–chemical processes, and the erythromycin and ciprofloxacin are hard to eliminate through biodegradation process in the active mud (Li et al., 2010). The most direct way for AR bacteria elimination and their AR genes is through solid separation, such as sedimentation. Nevertheless, next steps of biological treatment may conduct to selective growth of antibiotic resistant bacteria (Zhang et al., 2009b).

The advanced treatment technologies (separation through membrane) could be applied to retain bacterial cells, including their genetic material (Riquelme Breazeal et al., 2013). Moreover, the ozone was proposed to be used in disinfection process, being considered efficient in destroying AR genes (Dodd, 2012).

Reutilization of wastewaters became a worldwide strategy. Nevertheless, it is very important performing of an attentively evaluation on application of recovered waters and to establish adequate protection measures to avoid unwanted consequences. Wastewaters are usually disinfected through ultraviolet radiations or chlorination, which may kill resistant bacteria, but their resistance genes are more difficult to be destroyed (Auerbach et al., 2007; Kim et al., 2010; McKinney and Pruden, 2012; Munir et al., 2011). Li and Zhang (2012) have reported that chlorination may reduce many types of antibiotics, including ampicillin, chlortetracycline, sulfametoxazole, sulfadiazene, ofloxacin, and trimethoprim, but ozone disinfection is efficiently reducing a widespread types of antibiotics inclusive their active metabolites (Dodd et al., 2010).

The studies recently performed have shown that chlorination is efficient for AR bacteria inactivation, most of the bacteria being completely inactivated at the lowest chlorine concentration. This, chlorination is considered efficient in antimicrobial resistance combating, but the major concern must be orientated to the potential risk of appearance of AR genes in wastewaters after chlorination.

The accumulation of microorganisms on surfaces, as a result of the formation of biofilms, increases the potential of organic pollutants biotransformation (Lazăr, 2003). Flocs sludge constituted in the process of biological treatment of wastewater, is an example of bacterial biofilm with a high adsorption capacity and mineralization of organic matter, as the combination of many bacteria in the biofilm are responsible for performing more diverse and efficient metabolic activities from the economic point of view than the cells in suspension (Fig. 11.3). In the sediments from wastewater and sewage plants, high cell density creates optimal conditions for biofilm formation, as well as for horizontal gene transfer between adjacent resistant bacteria (Levy, 1992). Many authors revealed that wastewater treatment stations represent a significant reservoir of enteric multiresistant bacteria (Iversen et al., 2002)

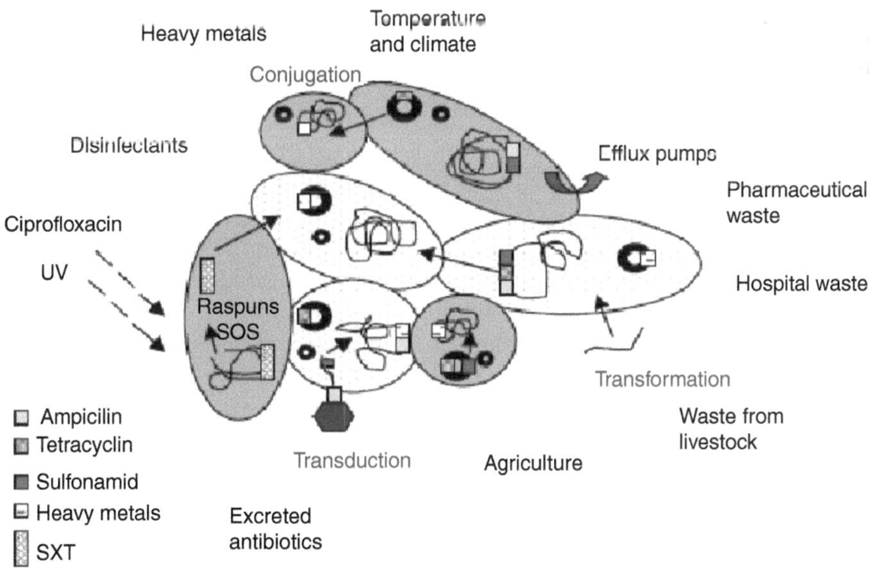

Figure 11.3. The horizontal genetic transfer mechanisms and external factors favorable to the emergence of resistant bacteria in a sewage treatment plant (Jury et al., 2010).

and a favorable environment for horizontal gene transfer, due to the high concentration of nutrients and increased density of microorganisms (Goñi-Urriza et al., 2000b; Rahube and Yost, 2010).

The characteristics of bacterial strains resistant to antibiotics in wastewater treatment plants containing high concentrations of antibiotics and the effects of wastewater discharges on the environment are still unknown. Currently it is speculated that the high level of multiple AR strains in wastewater and surface water may be due to the efflux systems showing no specificity to the antibiotics with a specific chemical structure (Li et al., 2009).

Recent studies (Mispagel and Gray, 2005) have shown that through the oxidation of the wastewater basin are discharged resistant coliform bacteria containing 18 integrons and plasmids, which is demonstrating the existence of horizontal gene transfer. The authors also looked for the presence of three MDR bacterial species: *S. maltophilia* and *P. aeruginosa* exhibiting resistance to ceftriaxone and ciprofloxacin-resistant *E. coli*, through capturing the pathogens that survive to the treatment process. They have studied the presence of *Salmonella* serogroup D species in wastewater after the treatment process. Resistance genes to type C and D tetracyclines have been normally detected in enteric bacteria in pools whose primary and secondary oxidation products escape directly into the environment, so it can be considered that these genes have human origin. It turned out that both antibiotics and their AR genes are not removed/eliminated during the treatment process by oxidation.

In the last 3 decades, literature has emphasized the implications of *Aeromonas* spp. in the human pathology. These species have been described as being involved in the intestinal infections in children and adults (acute gastroenteritis, cholera/dysentery, diarrhea, chronic ulcerative colitis etc.), but also in extra intestinal infections in immunocompromised hosts. Intrinsic and acquired resistance analysis in *A. hydrophila* strains isolated from the aquatic environment in the Danube Delta showed high resistance to penicillin, mediated by the presence of β-lactamase and also to tetracyclines, clotrimoxazol, norfloxacin, and chloramphenicol (Balotescu et al., 2003). Koksal et al. (2007) have shown that a high prevalence of *Aeromonas* sp. strains resistant to antibiotics in drinking water in Istanbul associated with a major risk to public health. A study in Greece showed a rate of over 91.3% of multidrug-resistant Gram-negative bacteria with faecal and nonfaecal origin isolated from drinking water (Papandreou et al., 2000).

In the wastewater treatment plant in Constanta Sud the studies have been conducted on the influence of various physical–chemical parameters on the microbiota of the activated sludge and have demonstrated that strains of *E. coli* from manure present multiple AR (Aonofriesei and Petroşanu, 2007). Also, *E. coli* strains isolated from rivers showed a great diversity of resistance plasmids (with size ranging from 3.8 to 50 kbp), which can be transferred by conjugation at high frequency (10^{-4}) (Cernat et al., 2002a).

Borjesson (2009) was the first author, who described the presence of *S. aureus* and multidrug-resistant strains of *S. aureus* (MRSA) in municipal wastewater, indicating that wastewater can be a potential reservoir for MRSA.

An ecotaxonomic study conducted by Huys et al. (2000) had the purpose to compare the tolerance to antibiotics in cultivated heterotrophic oxytetracycline-resistant strains isolated from two areas: hospital wastewater and wastewater from a hatchery, and showed the presence of some representative strains of *Acinetobacter* spp. and *S. maltophilia*, such as oxytetracycline-resistant *Acinetobacter veronii* biovar HG8 Sobria. The authors also indicated that the AR profile in a specific environment in a country does not reflect the resistance profiles in the same type of environment in another country since the profiles are largely influenced by the microbial complexity of each geographical site.

Tennstedt et al. (2003) showed that plasmids carrying the *bla* gene, present in the bacterial strains isolated from sewage treatment plants are commonly associated with transposons and integrons, often bearing other determinants of resistance, such as: (1) *aad* or *aac* gene encoding acetyltransferase and aminoglycoside-nucleotidiltransferase, respectively, (2) *cml* gene encoding chloramphenicol efflux protein; (3) *cat* gene which encodes chloramphenicol acetyltransferase.

3 Conclusions

The knowledge of the contribution of waste waters to the constitution and maintainance of the environmental reservoir of resistance is of major importance for the development of efficient strategies for the management of emergence and spread of resistance genes. The highest potential risk for public health is the development of a reservoir of AR genes, transferable to aquatic bacteria, which may be transmitted through horizontal gene transfer to other bacteria and in the end to human pathogens. Nevertheless, the appropriate evaluation of AR risk in the aquatic environment is still difficult to be performed due to the lack of data, as well as to the complex ways of gene flows among different aquatic species and environmental compartments. One strategy to reduce the consumption of antibiotics is to develop alternatives to antibiotics, such as probiotics, phage therapy, essential oils, and other plant-derived compounds, not only for the clinical sector, but also for aquaculture, agriculture, and farming.

Acknowledgements

The results presented in this study were supported by the Human Resources grant (PN-II-RU-TE- 2014-4-2037).

References

Abriouel, H., Omar, N.B., Molinos, A.C., López, R.L., Grande, M.J., Martínez-Viedma, P., Ortega, E., Cañamero, M.M., Galvez, A., 2008. Comparative analysis of genetic diversity and incidence of virulence factors and antibiotic resistance among enterococcal populations from raw fruit and vegetable foods, water and soil, and clinical samples. Int. J. Food Microbiol. 123, 38–49.

Alonso, A., Sanchez, P., Martinez, J.L., 2001. Environmental selection of antibiotic resistance genes. Environ. Microbiol. 3, 1–9.

Aminov, R.I., 2009. The role of antibiotics and antibiotic resistance in nature. Environ. Microbiol. 11, 2970–2988.

Aonofriesei, F., Petroşanu, M., 2007. Dynamics of microfauna in activated sludge and its relationship with physico–chemical conditions at WWTP, Constanța Sud (Ro). Proc. Rom. Acad. Series B. 3, 189–195.

Auerbach, E.A., Seyfried, E.E., McMahon, K.D., 2007. Tetracycline resistance genes in activated sludge wastewater treatment plants. Water Res. 41, 1143–1151.

Baker-Austin, C., Wright, M.S., Stepanauskas, R., McArthur, J.V., 2006. Co-selection of antibiotic and metal resistance. Trends Microbiol. 14, 176–182.

Balotescu, C., Israil, A., Radu, R., Alexandru, I., Dobre, G., 2003. Aspects of constitutive and acquired antibioresistance in *Aeromonas hydrophila* strains isolated from water sources. Roum. Arch. Microbiol. Immunol. 62 (3–4), 179–189.

Baquero, F., Martinez, J.L., Canton, R., 2008. Antibiotics and antibiotic resistance in water environments. Curr. Opin. Biotechnol. 19, 260–265.

Bell, J.B., Elliott, G.E., Smith, D.W., 1983. Influence of sewage treatment and urbanization on selection of multiple resistance in fecal coliform populations. Appl. Environ. Microbiol. 46 (1), 227–232.

Bennett, P.M., 2008. Plasmid encoded antibiotic resistance: acquisition and transfer of antibiotic resistance genes in bacteria. Brit. J. Pharmacol. 153, 347–357.

Borjesson, S., 2009. Antibiotic resistance in wastewater: methicillin-resistant *Staphylococcus aureus* (MRSA) and antibiotic resistance genes. Linkoping University Medical Dissertations.

Cabello, F.C., 2006. Heavy use of prophylactic antibiotics in aquaculture: a growing problem for human and animal health and for the environment. Environ Microbiol 8, 1137–1144.

Cambray, G., Guerout, A.M., Mazel, D., 2010. Integrons. Annual Rev. Genetics 44, 141–166.

Campanile, F., Bongiorno, D., Falcone, M., Vailati, F., Pasticci, M.B., Perez, M., Raglio, A., Rumpianesi, F., Scuderi, C., Suter, F., Venditti, M., Venturelli, C., Ravasio, V., Codeluppi, M., Stefani, S., 2011. Changing Italian nosocomial-community trends and heteroresistance in *Staphylococcus aureus* from bacteremia and endocarditis. Eur. J. Clin. Microbiol. Infect. Dis 31 (5), 739–745.

Cernat R., Lazăr V., 2011. Taxonomia şi Biodiversitatea Microorganismelor. Editura Universităţii Bucureşti, Bucureşti, p. 147–150, 153.

Cernat, R., Herlea, V., Coipan, E., Lazăr, V., Cotar, A., 2002a. Plasmid profile and fingerprinting of some heavy-metal and multiple antibiotic resistant *E. coli* strains isolated from river water. Bul. Societ. Nat. de Biologie Celulară 30, 285.
Cernat, R., Lazăr, V., Balotescu, C., Cotar, A., Coipan, E., Cojocaru, C., 2002b. Plasmid-mediated transfer of antibiotic resistance into some antibioresistant enterobacterial strains isolated from various aquatic sources. Bacteriol Virusol Parazitol Epidemiol. 47 (3–4), 179–184.
Cernat, R., Herlea, V., Coipan, E., Lazăr, V., Cotar, A., 2002c. Plasmid-mediated transfer of antibiotic resistance into some antibioresistant enterobacterial strains isolated from various aquatic sources. Bul. Societ. Nat. de Biologie Celulară 30, 283–284.
Chee-Sanford, J.C., Aminov, R.I., Krapac, I.J., Garrigues-Jeanjean, N., Mackie, R.I., 2001. Occurrence and diversity of tetracycline resistance genes in lagoons and groundwater underlying two swine production facilities. Appl. Environ. Microbiol. 67 (4), 1494–1502.
Chifiriuc, C., Lazăr V., Mihăescu G., 2011. Microbiologie şi virologie medicală. Ed. Universităţii Bucureşti, p. 395–431.
Colomer-Lluch, M., Imamovic, L., Jofre, J., Muniesa, M., 2011. Bacteriophages carrying antibiotic resistance genes in fecal wastes 1 from cattle, pigs and poultry. Antimicrob. Agents. Chemother. 55 (10), 4908–4911.
Costerton, J.W., Lewandowski, Z., Caldwell, D.E., Korber, D.R., Lappin-Scott, H.M., 1995. Microbial biofilms. Annu. Rev. Microbiol. 49, 711–745.
Cunliffe, M., Murrell, J.C., 2009. The sea-surface microlayer is a gelatinous biofilm. ISME J. 3, 1001–1003.
Davey, M.E., O'Toole, G.A., 2000. Microbial biofilms: from ecology to molecular genetics. Microbiol. Mol. Biol. Rev. 64 (4), 847–867.
Davies, J., Davies, D., 2010. Origins and Evolution of Antibiotic Resistance. Microbiol. Mol. Biol. Rev. 74 (3), 417–433.
Derore, H., Top, E., Houwen, F., Mergeay, M., Verstraete, W., 1994. Evolution of heavy-metal resistant transconjugants in a soil environment with a concomitant selective pressure. FEMS Microbiol. Ecol. 14 (3), 263–273.
Dodd, M.C., 2012. Potential impacts of disinfection processes on elimination and deactivation of antibiotic resistance genes during water and wastewater treatment. J. Environ. Monit. 14, 1754–1771.
Dodd, M.C., Rentsch, D., Singer, H.P., Kohler, H.P.E., von Gunten, U., 2010. Transformation of beta lactam antibacterial agents during aqueous ozonation: reaction pathways and quantitative bioassay of biologically-active oxidation products. Environ. Sci. Technol. 44, 5940–5948.
Dolliver, H., Gupta, S., 2008. Antibiotic losses in leaching and surface runoff from manure-amended agricultural land. J. Environ. Qual. 37, 1227–1237.
Fick, J., Söderström, H., Lindberg, R.H., Phan, C., Tysklind, M., Larsson, D.G.J., 2009. Contamination of surface, ground, and drinking water from pharmaceutical production. Environ. Toxicol. Chem. 28, 2522–2527.
García-Delgado, M., Rodríguez-Cruz, M.S., Lorenzo, L.F., Arienzo, M., Sánchez-Martín, M.J., 2007. Seasonal and time variability of heavy metal content and of its chemical forms in sewage sludges from different wastewater treatment plants. Sci. Total. Environ. 382 (1), 82–92.
Gaze, W.H., Abdouslam, N., Hawkey, P.M., Wellington, E.M., 2005. Incidence of class1 integrons in a quaternary ammonium compound-polluted environment. Antimicrob. Agents Chemother. 49, 1802–1807.
Gaze, W.H., Zhang, L., Abdouslam, N.A., Hawkey, P.M., Calvo-Bado, L., Royle, J., Brown, H., Davis, S., Kay, P., Boxall, A.B., Wellington, E.M., 2011. Impacts of anthropogenic activity on the ecology of class1 integrons and integron-associated genes in the environment. ISME J. 5, 1253–1261.

Goñi-Urriza, M., Capdepuy, M., Arpin, C., Raymond, N., Caumette, P., Quentin, C., 2000a. Impact of an urban effluent on antibiotic resistance of riverine *Enterobacteriaceae* and *Aeromonas* spp. Appl. Environ. Microbiol. 66 (1), 125–132.

Goñi-Urriza, M., Pineau, L., Capdepuy, M., Roques, C., Caumette, P., Quentin, C., 2000b. Antimicrobial resistance of mesophilic Aeromonas spp. isolated from two European rivers. J. Antimicrob. Chemother. 46 (2), 297–301.

Greenwood, D., 1998. Resistance to antimicrobial agents: a personal view. J. Med. Microbiol. 47, 751–755.

Guardabassi, L., Dalsgaard, A., 2004. Occurrence, structure, and mobility of Tn1546-lik elements in environmental isolates of vancomycin-resistant enterococci. Appl. Environ. Microbiol. 70, 984–990.

Gulkowska, A., Leung, H.W., So, M.K., Taniyasu, N., Yamashita, N., Yeung, L., Richardson, B., Lei, A.P., Giesy, J.P., Lam, P., 2008. Removal of antibiotics from waste water by sewage treatment facilities in Hong Kong and Shenzhen, China. Water Res. 42 (1–2), 395–403.

Haggard, B., Bartsch, L., 2009. Net changes in antibiotic concentrations downstream from an effluent discharge. J. Environ. Qual. 38, 343–352.

Hayes, J., English, L., Carr, L., Wagner, D., Joseph, S., 2004. Multiple-antibiotic resistance of *Enterococcus* spp. isolated from commercial poultry production environments. Appl. Environ. Microbiol. 70 (10), 6005–6011.

Heuer, O.E., Kruse, H., Grave, K., Collignon, P., Karunasagar, I., Angulo, F.J., 2009. Human health consequences of use of antimicrobial agents in aquaculture. Clin. Infect. Dis. 49, 1248–1253.

Huys, G., Rhodes, G., McGann, P., Denys, R., Pickup, R., Hiney, M., Smith, P., Swings, J., 2000. Characterization of oxytetracycline-resistant heterotrophic bacteria originating from hospital and freshwater fishfarm environments in England and Ireland. Syst. Appl. Microbiol. 23 (4), 599–606.

Iversen, A., Kuhn, I., Franklin, A., Mollby, R., 2002. High prevalence of vancomycin-resistant enterococci in Swedish sewage. Appl. Environ. Microbiol. 68, 2838–2842.

Jehl, F., Chemarat, M., Weber, M., Berard, A., 2004. From the Antibiogram to Prescription. second ed. bioMerieux. S.A., France, pp. 80–110.

Jurcoane, Ş., Cornea, P., Stoica, I., Vassu, T., 2006. Tratat de biotehnologie, vol. II, editura Tehnică, Bucureşti.

Jury, K.L., Vancov, T., Stuetz, R.M., Khan, S.J., 2010. Antibiotic resistance dissemination and sewage treatment plants. Curr. Res. Technol. Educ. Topics Appl. Microbiol. Microb. Biotechnol. 509–510.

Kim, S., Aga, D.S., 2007. Potential ecological and human health impacts of antibiotics and antibiotic-resistant bacteria from wastewater treatment plants. J. Toxicol. Environ. Health 10, 559–573.

Kim, S., Park, H., Chandran, K., 2010. Propensity of activated sludge to amplify or attenuate tetracycline resistance genes and tetracycline resistant bacteria: a mathematical modeling approach. Chemosphere 78, 1071–1077.

Koksal, F., Oguzkurt, N., Samasti, M., Altas, K., 2007. Prevalence and antimicrobial resistance patterns of Aeromonas strains isolated from drinking water samples in Istanbul, Turkey. Chemotherapy 53 (1), 30–35.

Kovacikova, G., Skorupski, K., 2002. Regulation of virulence gene expression in Vibrio cholerae by quorum sensing: HapR functions at the aphA promoter. Mol. Microbiol. 46 (4), 1135–1147.

Kristiansson, E., Fick, J., Janzon, A., Grabic, R., Rutgersson, C., Weijdegård, B., 2011. Pyrosequencing of antibiotic-contaminated river sediments reveals high levels of resistance and gene transfer elements. PLoS One 6, 2.

Król, J.E., Nguyen, H.D., Rogers, L.M., Beyenal, H., Krone, S.M., Top, E.M., 2011. Increased transfer of a multidrug resistance plasmid in *Escherichia coli* biofilms at the air-liquid interface. Appl. Environ. Microbiol. 77 (15), 5079–5088.

Kümmerer, K., 2003. Significance of antibiotics in the environment. J. Antimicrob. Chemother. 52, 5–7.

Larsson, D.G.J., de Pedro, C., Paxeus, N., 2007. Effluent from drug manufactures contains extremely high levels of pharmaceuticals. J. Hazard. Mater. 148, 751–755.

Lazăr, V., 2003. Aderenţa microbiană. Edit. Acad. Rom., Bucureşti, pp. 88–146, 170–178.

Lazăr, V., 2007. Microbiologie medicală. Note de curs şi principii de diagnostic microbiologic. Editura Universităţii Bucureşti, Bucureşti, pp. 205–215, 247–249.

Lazăr, V., 2011. Quorum sensing in biofilms-hou to destroy the bacterial citadels or their cohesion/power? Anaerobe 17, 1095–8274.

Lazăr, V., Chifiriuc, C., 2010. Architecture and physiology of microbial biofilms. Rom. Arch. Microb. Immunol. 69 (2), 95–107.

Levy, S.B., 1992. The Antibiotic Paradox: How Miracle Drugs are Destroying the Miracle. Plenum Press, New York, pp. 137–181.

Levy, S., Miller, R., 1998. Gene transfer in the environment. McGraw-Hill Book, New York.

Li, D., Yang, M., Hu, J., Zhang, J., Liu, R., Gu, X., Zhang, Y., Wang, Z., 2009. Antibioti resistance profile in environmental bacteria isolated from penicillin production wastewater treatment plant and receiving river. Environ. Microbiol. 11 (6), 1506–1517.

Li, D., Yu, T., Zhang, Y., Yang, M., Li, Z., Liu, M., Qi, R., 2010. Antibiotic resistance characteristics of environmental bacteria from an oxytetracycline production wastewater treatment plant and the receiving river. Appl. Environ. Microbiol. 76 (11), 3444–3451.

Li, B., Zhang, T., 2012. pH significantly affects removal of trace antibiotics in chlorination of municipal wastewater. Water Research. 46 (11), 3703–3713.

Liu, B., Pop, M., 2009. Antibiotic resistence genes database. Nucleic Acids Res. 37, 443–447.

Luo, N., Pereira, S., Sahin, O., Lin, J., Huang, S., Michel, L., Zhang, Q., 2005. Enhanced in vivo fitness of fluoroquinolone-resistant *Campylobacter jejuni* in the absence of antibiotic selection pressure. Proc. Natl. Acad. Sci. USA 102, 541–546.

Lupo, A., Coyne, S., Berendonk, T.U., 2012. Origin and evolution of antibiotic resistance: the common mechanisms of emergence and spread in water bodies. Front. Microbiol. 3, 18.

Magiorakos, A.-P., Srinivasan, A., Carey, R.B., Carmeli, Y., Falagas, M.E., Giske, C.G., Harbarth, S., Hindler, J.F., Kahlmeter, G., Olsson-Liljequist, B., Paterson, D.L., Rice, L.B., Stelling, J., Struelens, M.J., Vatopoulos, A., Weber, J.T., Monnet, D.L., 2012. Multidrug resistant, extensively drug-resistant and pandrug-resistant bacteria: an international expert proposal for interim standard definitions for acquired resistance. Clin. Microbiol. Infect. 18, 268–281.

Martinez, J.L., 2009a. Environmental pollution by antibiotics and by antibiotic resistance determinants. Environ. Pollut. 157, 2893–2902.

Martinez, J.L., 2009b. The role of natural environments in the evolution of resistance traits in pathogenic bacteria. Proc. Royal Soc. B 276, 2521–2530.

McKinney, C.W., Pruden, A., 2012. Ultraviolet disinfection of antibiotic resistant bacteria and their antibiotic resistance genes in water and wastewater. Environ. Sci. Technol. 46, 13393–13400.

Mihăescu. G., Chifiriuc, C., Dițu M., 2007a. Microbiologie generală. Editura Universității București, București. pp. 178–196.
Mihăescu, G., Chifiriuc, M.C., Dițu, L.M., 2007b, Antibiotice și substanțe chimioterapeutice antimicrobiene. Editura Academiei Române, București.
Mispagel, H., Gray, J.T., 2005. Antibiotic resistance from wastewater oxidation ponds. Water Environ. Federation 77 (7), 2996–3002.
Molin, S., Tolker-Nielsen, T., 2003. Gene transfer occurs with enhanced efficiency in biofilms and induces enhanced stabilisation of the biofilm structure. Curr. Opin. Biotechnol. 14, 255–261.
Moura, A., Henriques, I., Smalla, K., Correia, A., 2010. Wastewater bacterial communities bring together broad-host range plasmids, integrons and a wide diversity of uncharacterized gene cassettes. Res. Microbiol. 161, 58–66.
Munir, M., Xagoraraki, I., 2011. Levels of antibiotic resistance genes in manure, biosolids, and fertilized soil. J Environ. Qual. 40, 248–255.
Munir, M., Wong, K., Xagoraraki, I., 2011. Release of antibiotic resistant bacteria and genes in the effluent and biosolids of five wastewater utilities in Michigan. Water Res. 45, 681–693.
Naik, G.A., Bhat, L.N., Chopade, B.A., Lynch, J.M., 1994. Transfer of broag-host rang antibiotic resistance plasmids in soil microorganism. Curr. Microbiol. 28, 209–215.
Nguyen, K.T., Piastro, K., Gray, T.A., Derbyshire, K.M., 2010. Mycobacterial biofilms facilitate horizontal DNA transfer between strains of *Mycobacterium smegmatis*. J. Bacteriol. 192, 5134–5142.
Nikaido, H., Pages, J.M., 2011. Broad-specificity efflux pumps and their role in multidrug resistance of Gram-negative bacteria. FEMS Microbiol. Rev.
Papandreou, S., Pagonopoulou, O., Vantarakis, A., Papapetropoulou, M., 2000. Multiantibiotic resistance of Gram-negative bacteria isolated from drinking water samples in southwest Greece. J. Chemother. 12 (4), 267–273.
Parsley, L.C., Consuegra, E.J., Kakirde, K.S., Land, A.M., Harper, Jr., W.F., Liles, M.R., 2010. Identification of diverse antimicrobial resistance determinants carried on bacterial, plasmid, or viral metagenomes from an activated sludge microbial assemblage. Appl. Environ. Microbiol. 76, 3753–3757.
Piddock, L.J., 2006. Clinically relevant chromosomally encoded multidrug resistance efflux pumps in bacteria. Clin. Microbiol. Rev. 19, 382–402.
Rahube, T.O., Yost, C.K., 2010. Antibiotic resistance plasmids in wastewater treatment plants and their possible dissemination into the environment. Afr. J. Biotechnol. 9 (54), 9183–9190.
Ram, S., Vajpayee, P., Shanker, R., 2007. Prevalence of multi antimicrobial agent resistant, shiga toxin and enterotoxins producing *Escherichia coli* in surface water of the river Ganga. Environ. Sci. Technol. 41, 7383–7388.
Renou, S., Givaudan, J.G., Poulain, S., Dirassouyan, F., Moulin, P., 2008. Landfill leachate treatment: review and opportunity. J. Hazard. Mater. 150, 468–493.
Riquelme Breazeal, M.V., Vikesland, P.J., Novak, J.T., Pruden, A., 2013. Effect of wastewater colloids on membrane removal of microconstituent antibiotic resistance genes. Water Res. 47, 130–140.
Rysz, M., 2007. Antibiotic Resistance Vector Transport, Reservoir Amplification and Attenuation. Rice University, Houston, TX.
Shakibaie, M.R., Jalilzadeh, K.A., Yamakanamardi, S.M., 2009. Horizontal transfer of antibiotic resistance genes among gram negative bacteria in sewage and lake water and influence of some physico-chemical parameters of water on conjugation process. J. Environ. Biol. 30 (1), 45–49.
Shar, A.H., Kazi, Y.F., Soomro, I.H., 2009. Antibiotic susceptibility of termo-tolerant *Escherichia coli* 2 isolated from drinking water of Khairpur City, Sindh, Pakistan. Pak. J. Biol. Sci. 12 (8), 648–652.

Sorensen, S.J., Bailey, M., Hansen, L.H., Kroer, N., Wuertz, S., 2005. Studying plasmid horizontal transfer in situ: a critical review. Natur. Rev. Microbiol. 3, 700–710.
Tao, R., Ying, G.G., Su, H.G., Zhou, H.W., Sidhu, J.P.S., 2010. Detection of antibiotic resistance and tetracycline resistance genes in Enterobacteriaceae isolated from the Pearl rivers in South China. Environmental. Pollution. 158, 2101–2109.
Tennstedt, T., Szczepanowski, R., Braun, S., Pühle, A., Schlüter, A., 2003. Occurrence of integron-associated resistance gene cassettes located on antibiotic resistance plasmids isolated from a wastewater treatment plant. FEMS Microbiol. Ecol. 45, 239–252.
Thompson, S.A., Maani, E.V., Lindell, A.H., King, C.J., McArthur, J.V., 2007. Novel tetracycline resistance determinant isolated from an environmental strain of *Serratia marcescens*. Appl. Environ. Microbiol. 73, 2199–2206.
Weinbauer, M.G., 2004. Ecology of prokaryotic viruses. FEMS Microbiol. Rev. 28, 127–181.
Weir, M., Raji'c, A., Dutil, L., Cernicchiaro, N., Uhland, F., Mercier, B., et al., 2012. Zoonotic bacteria, antimicrobial use and antimicrobial resistance in ornamental fish: a systematic review of the existing research and survey of aquaculture-allied professionals. Epidemiol. Infect. 140, 192–206.
Wiedenbeck, J., Cohan, F.M., 2011. Origins of bacterial diversity through horizontal genetic transfer and adaptation to new ecological niches. FEMS Microbiol. Rev. 35, 957–997.
Wright, G.D., 2007. The antibiotic resistome: the nexus of chemical and genetic diversity. Nat. Rev. Microbiol. 5, 175–186.
Wright, G.D., 2010. Antibiotic resistance in the environment: a link to the clinic? Curr. Opin. Microbiol. 13, 589–594.
Wright, M.S., Baker-Austin, C., Lindell, A.H., Stepanauskas, R., Stikes, H.W., McArthur, J.V., 2008. Influence of industrial contamination on mobile genetic elements: class 1 integron abundance and gene cassette structure in aquatic bacterial communities. ISME J. 2, 417–428.
Wu, N., Qiao, M., Zhang, B., Cheng, W.D., Zhu, Y.G., 2010. Abundance and diversity of tetracycline resistance genes in soils adjacent to representative swine feedlots in China. Environ. Sci. Technol. 44, 6933–6939.
Zarnea, G., 1994. Tratat de Microbiologie generală, vol. V: Ecologia microorganismelor. Ed. Acad. Rom., Bucureşti, pp. 24–38, 1009–1050.
Zhang, T., Li, B., 2011. Occurrence, transformation, and fate of antibiotics in municipal wastewater treatment plants. Crit. Rev. Environ. Sci. Technol. 41, 951–998.
Zhang, X.X., Zhang, T., Fang, H.H.P., 2009a. Antibiotic resistance genes in water environment. Appl. Microbiol. Biotechnol. 82, 397–414.
Zhang, Y., Marrs, C.F., Simon, C., Xi, C., 2009b. Wastewater treatment contributes to selective increase of antibiotic resistance among *Acinetobacter* spp. Sci. Tot. Environ. 407, 3702–3706.
Zhu, Y.G., Johnson, T.A., Su, J.Q., Qiao, M., Guo, G.X., Stedtfeld, R.D., 2013. Diverse and abundant antibiotic resistance genes in Chinese swine farms. Proc. Natl. Acad. Sci. USA 110, 3435–3440.

NANOSTRUCTURATED MEMBRANES FOR THE MICROBIOLOGICAL PURIFICATION OF DRINKING WATER

Roxana Cristina Popescu*,, Mariana Oana Mihaela Fufă*,†, Alexandru Mihai Grumezescu*, Alina Maria Holban*,‡**

**Department of Science and Engineering of Oxide Materials and Nanomaterials, Faculty of Applied Chemistry and Materials Science, University Politehnica of Bucharest, Bucharest, Romania; **"Horia Hulubei", National Institute of Physics and Nuclear Engineering, Department of Life and Environmental Physics, Magurele, Bucharest, Romania; †National Institute for Laser, Plasma and Radiation Physics, Lasers Department, Măgurele, Bucharest, Romania; ‡University of Bucharest, Department of Botanic-Microbiology, Faculty of Biology, Bucharest, Romania*

1 Introduction

Water purification for human consumption purposes consists in the removal of different contaminants as chemicals (i.e., pollutants, toxic metals), biological contaminants (algae, bacteria, fungi, parasites, viruses), suspended solids, and gases.

There are several methods used in the water purification process, which include: (1) physical processes, such as filtration, sedimentation, or distillation; (2) biological processes, such as sand filters, active carbon; (3) chemical processes, such as flocculation, chlorination, the use of ultraviolet light.

Regarding the physical approaches used in water purification, these differ in the phenomenon on which the processes are based. For instance, filtration is based on the separation of solids from fluids, by interposing a porous medium (filter) which retains the solid particles and allows the fluid to pass on the other side. On the other hand, the sedimentation approach uses the gravitational force which determines the solids to form a deposit at the bottom of the tube containing contaminated water, while distillation involves the transformation of the liquid (water) into a vapor

Water Purification. http://dx.doi.org/10.1016/B978-0-12-804300-4.00012-5

phase, process which is based on the difference in the volatility of the compounds.

Slow sand filters represent one example of biological approaches used in water purification, which implies using 1–2 m deep tubes filled with sand, which retain the impurities present in the filtered water. Activated carbon (charcoal), another utilized approach, is a microporous carbon with high surface area and enhanced adsorption properties.

In scientific terms, flocculation is a process in which colloids in suspension become destabilized after the addition of a clarifying agent; regarding the water purification process, the flocculation phenomenon can refer to the destabilization and coagulation of contaminants present in the water. Water chlorination is commonly encountered in the treatment of water supply and consists in the addition of chlorine or hypochlorite to kill microbes and to prevent the spreading of waterborne diseases.

The use of electromagnetic light, especially with short wavelength (in ultraviolet range) is commonly used in disinfection, due to the fact that it produces gaps in the structure of nucleic acids of microorganisms disabling their cellular functions. This method is often used in water decontamination.

This chapter focuses mainly on the most investigated physical methods which involve the filtration process. Compared to the other approaches used in water treatment, membrane water filtration presents some advantages, such as: (1) the continuity of the operation; (2) it does not require the use of any chemicals; (3) it does not imply high energy consumption; (4) the possibility of scaling up, be integrated/integrate other processes and possibility of automation (Street et al., 2014).

There are several parameters influencing the properties and efficiency of the membranes used in water filtration: (1) the size of membrane pores/the size of the contaminant molecules/particles; (2) the positive/negative charge of the membrane surface, respectively the polarity of the contaminant molecules; and (3) the adsorbing capacity of the membrane surface (Street et al., 2014).

Depending on the pore size of the used filters (membranes), contaminants with different sizes can be removed and a better purification can be achieved while decreasing the pore size; thus, the available methods are: (1) microfiltration (pore size about 0.1 μm), which removes bacteria and suspended solids in the water; (2) ultrafiltration (pore size about 0.01 μm), which in addition to microfiltration removes viruses; (3) nanofiltration (pore size about 0.001 μm), which removes most of the organic molecules and some multivalent ions (divalent ions from hard water); and (4) reverse osmosis (pore size about 0.0001 μm), which removes all organic molecules and minerals in water, resulting ultrapure water (Fig. 12.1).

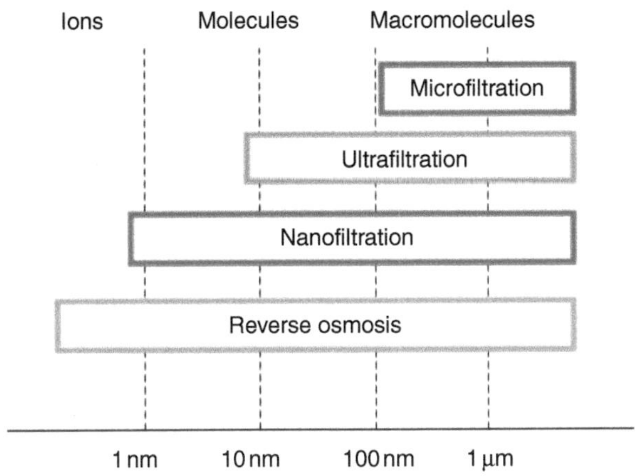

Figure 12.1. Different types of membranes and their filtering principles depending on the pore.

When analyzing the properties of these types of membranes, it is clear that the most advantageous are the ones with smaller pore size. However, for cost-efficiency purposes, all of these filtering techniques can be used to obtain drinkable water, when used in combination with other treatments. For example, membranes with micropores can remove sand, silt, clays, *Giardia lamblia* and *Crypotosporidium* cysts, algae, and some bacterial species, when used in combination with disinfecting agents. In the case of membranes designed for ultrafiltration, even if they can remove the majority of microorganisms and viruses in water, it is recommended to use additional disinfection. However, between the advantages of such membranes we can recall: (A1) filtration by means of size-exclusion mechanisms; (A2) the quality of the removal process in constant; (A3) possibility of automation; (A4) the installation integrating this type of filter and the filtering process are compact. The main disadvantage may refer to fouling phenomenon (see Section 12.3). Nanofiltration is the first step to a perfect filtration of water, because it removes all microorganisms, viruses, and organic matter. Other important advantage is given by their ability to remove alkalinity and hardness from water. The most advanced filtration is made by reverse osmosis membranes, which additionally can remove most of inorganic contaminants, resulting ultrapure water. Between the advantages of this type of filtration, we mention: (A1) complete elimination of contaminants; (A2) the filtration sensibility is not severely influenced by the flow; and (A3) possibility of automation. Besides these advantages, the technique presents some limitations, which refer to the high costs, the process management, and the need of a pretreatment and fouling predisposition.

There are several examples of membranes whose principle of filtration is based on the polarity of the contaminant molecules and their ability to retain these species by means of electrostatic interactions. However, dipole molecules (with both positive and negative charged groups), such as water molecules, can easily orient and thus permeate the membrane (Street et al., 2014).

Regarding the membranes with adsorbing properties, the retaining of contaminant molecules is also based on electrostatic interactions: due to the fact that the membrane surface is negatively charged, positively ions in water are adsorbed, forming an electrical double layer at the surface of the membrane (Street et al., 2014). There are several factors influencing the adsorption ability of the contaminants, like their nature, concentration and mass distribution. Also, Ca^{2+} concentration influences the filtering properties because the interaction with the negatively charged surface of the membrane is preferentially favored, compared to the contaminant molecules, which are usually bigger in size and have different spatial conformations. Also, the physical and chemical properties of the membrane are important for the process (Street et al., 2014).

Depending on their structure and chemical composition, membranes can also be grouped as: (1) isotropic, having uniform physical nature and chemical composition in cross-section; and (2) anisotropic, which are nonuniform in cross-section. Examples of anisotropic membranes for water filtration are given by layer-by-layer approaches (Wang et al., 2015, 2016a; Qin et al., 2016; Xu et al. 2015; Zhang et al., 2015a; Diep et al., 2015; Gu et al., 2015; Kaner et al., 2015). The main advantage of this type of membranes is given by the antifouling property determined by different approaches in surface modification (see Section 12.3 for details).

The progress in water purification has been mainly conditioned by the scientific progress made in the identification of pathogens and other contaminants in water, the introduction of new regulations regarding water quality, but also the development of materials science, smart materials, and nanomaterials. Today, there are many examples of commercial membranes for water purification; however, there is still a permanent need for improvement regarding their properties, such as: (1) antifouling properties; (2) chemical stability; (3) mechanical stability; and (4) thermal stability.

2 Microbiological Contaminants of Waters

When considering the hazards related to drinking water contamination is usually investigated the contaminant type, its physicochemical versatility and biological impact. As previously mentioned, there are various agents responsible for water pollution, including chemical, biological or suspended agents. Regarding the chemical contamination of drinking water, recent studies have reported an alarming data with respect to the presence of arsenic (Butts et al., 2015; Chang, 2015; Mahanta et al., 2016), bromate (Dongmei et al., 2015), cadmium (Bhowmik et al., 2015; Handayani et al., 2016), cesium (Chen et al., 2015a), chromium (Bahadir et al., 2015; Catalani et al., 2015), fluoride (Bretzler and Johnson, 2015; Chuah et al., 2016), lead (Trueman and Gagnon, 2016; Vasile et al., 2016), manganese (Gerke et al., 2016), mercury (Huber and Leopold, 2015), methyl tertiary-butyl ether (Zhang et al., 2016a), organophosphate ester residues (Ding et al., 2015), perfluorooctanoic acid (Happonen et al., 2016), multiple nitrosamines (Wang et al., 2016b), and sodium hydroxide (Lendowski et al., 2015) within the drinkable water sources.

Given the current concern in water purification process and the progress reported in novel technologies, the scientific community gathers their resources toward the development and implementation of modern strategies for the removal of water contaminants. In this respect, there is a plethora of research studies that aims the successful use of nanotechnology-related materials and investigative techniques especially for chemical pollutants removal.

Comprehensive studies are instead required when the concerned contaminant is represented by living organisms, considering thus both the structural and functional complexity of microorganisms and the intricate related disorders. Within the following paragraphs, we will particularly discuss the aspects related to biological contamination of drinking water, including herein the current worldwide concern regarding the associated healthcare conditions.

Since the structural features and adaptive functional behaviors related to bacterial strains are under an impressive constant changing, they represent the most widespread microorganisms entailed in drinking water contamination. Given the complex pathologies related to bacterial contamination (due to both microorganism-related phenotypic complexity and human-related genetic diversity), bacterial contaminants are an actual concern of worldwide scientific community, being identified either as free floating or as biofilm-forming organisms.

Enterobacteriaceae group contains the most predominant bacterial strains that have been reported as microbial pathogenic contaminants for drinking water sources, since their particular phenotype and genotype encouraged an increased prevalence and alarming healthcare conditions. According to UK governmental Environment Agency (2002), several *Enterobacteriaceae* species responsible for water pollution are reported as *coliform* bacteria, described as Gram-negative, nonspore-forming and β-galactosidase-containing strains with rod shape, with aerobic and facultative anaerobic growth. Potential contamination of water sources with *coliform* bacteria occurs due to improper or insufficient treatment processes and to intrinsic distribution system damage (including industrial water distribution networks and domestic plumbing systems).

Escherichia coli is the representative strain of Gram-negative bacilli, that physiologically colonize the gastrointestinal tract but it can also be pathologically associated with several clinical conditions (in case of auspicious anatomical environment and immunosuppressed background) (Kaper et al., 2004; Sergeeva et al., 2015). When considering the physiological occurrence of *E. coli*, it is not a novelty for the worldwide scientific community that this bacterium is associated with fecal contamination of water. A plethora of research studies confirmed its potential role as a water pollutant in countries all over the world, but when considering the alarming phenomena of multidrug resistance and microorganism mutagenicity there is an immediate demand to develop novel detection tools and antibacterial fighting strategies. Thus, various recent studies turned their attention toward the intimate investigation of *E. coli*-related genetic elements, such as integrons (Canal et al., 2016; Wu et al., 2015), resistance genes (Ferro et al., 2016) and plasmids (Soto-Alonso et al., 2015)—that provide long-term and inheritable resistance to common antibiotics and disinfection strategies used for clean waters (Abhirosh et al., 2011; Coleman et al., 2013; Farkas et al., 2013; Patel et al., 2011; Veldman et al., 2014).

The performed studies regarding global prevalence of *E. coli* water contamination have reported alarming data, if considering the reported number of resistant strains in many countries from Africa (Abia et al., 2015; Ateba and Mbewe, 2011; Sorensen et al., 2015), North America (Chandrasekaran et al., 2015; Coleman et al., 2013; Wang and Fiessel, 2009), South America (Avigliano and Schenone, 2015; Rodriguez-Alvarez et al., 2015), Asia (Dada et al., 2013; Patel et al., 2011; Sukkua et al., 2015), Australia (Daly et al., 2013), and Europe (Khan et al., 2016; Popa et al., 2015; Vantarakis et al., 2013). Most-common healthcare conditions associated with consumption of *E. coli*-polluted water are gastroenteritis and intestinal diseases, urinary tract infection and bloodstream infection (Fhogartaigh and Dance, 2013; Kratochwill et al., 2015; Leistner et al., 2014; Saxena et al., 2015).

Others bacterial strains that belong to Enterobacteriaceae family that have been reported as water microbial pollutants include microbial members of *Citrobacter* spp. (Machado and Bordalo, 2014; Zhang et al., 2012), *Enterobacter* spp. (Bai et al., 2015; Bergeron et al., 2015), *Hafnia* spp. (Callon et al., 2016), *Klebsiella* spp. (Christensen et al., 2012; Struve and Krogfelt, 2004; Oliveira et al., 2014), *Salmonella* spp. (Coulliette et al., 2013; Levantesi et al., 2012), *Serratia* spp. (Ebrahim et al., 2015), and *Yersinia* spp. (Hakalehto et al., 2014).

Enterococcaceae family is a typical Gram-positive representative for fecal contamination of drinking water sources, as the bacterial members are nonsporulating cocci with aerobic and facultative anaerobic development that occur in human and warm-blooded animals feces. *Enterococcus faecalis* and *E. faecium* are human-derived commensal strains of the gut microbiota that particularly tolerate hard environmental conditions, including extreme temperatures (10°C up to 45°C), hypotonic or hypertonic and acidic or alkaline growth media (Bhardwaj et al., 2013; Huycke et al., 1998). Given the ineffectiveness of common water treatments against intestinal enterococci and the innate resistance to most commercial antibiotics (Ahmad et al., 2014; Santos et al., 2015; Yang et al., 2015), itemized research studies were required in order to reveal phenotypic and genotypic features of various strains of *Enterococcus* spp. (Chen et al., 2015b; Lerma et al., 2014; Nam et al., 2013; Rehaiem et al., 2014). Alarming data regarding global prevalence of *E. faecalis* and *E. faecium* as water contaminants were lately reported, but the recently performed research studies also revealed other pollutant enterococcal strains, namely *E. mundtii* (Schelegueda et al., 2015), *E. hirae* (Larsson et al., 2014), *E. casseliflavus* (Tang et al., 2016), *E. avium* and *E. raffinosus* (Cardoso de Oliveira and Pinhata, 2008). Especially, when considering the current concern regarding the complex opportunistic character associated with these bacteria, novel efficient detection tools and strategies to combat both planktonic and biofilm-embedded bacteria are urgent requirements.

Regarding the resistant and persistent bacterial strains associated with biofilm development, the encountered issues are even more intricate, since the colony-forming bacteria (which normally serve as a nourishment source for free-living amoebae)can resist to the phagocytosis process of this polyphyletic protozoa. Thus, resistant and more virulent bacterial strains with an increased potential for opportunistic human conditions (in case of water ingestion) can develop within amoebae cultures. *Acanthamoeba, Hartmannella, Naegleria,* and *Vahlkampfia* genera are most-prevalent amoeba within global water sources and the associated pathologies in humans mainly include encephalitis and keratitis (Baig, 2015; Bonilla-Lemus et al., 2014; Le Calvez et al., 2012; Loret and Greub, 2010). Clinical relevant and concerning data regarding opportunistic bacterial species associated with free-living amoebae genera have been reported for *Aeromonas salmonicida* (Zeybek and Binay, 2014), *Legionella oakridgensis* (Brzuszkiewicz et al., 2013), *Legionella pneumophila* (Cervero-Aragó et al., 2014; Ji et al., 2014), *Pseudomonas aeruginosa* (Cateau et al., 2014a; Tosetti et al., 2014), *Toxoplasma gondii* (Winiecka-Krusnell et al., 2009), *Chlamydia* spp. (Nunes and Gomes, 2014; Vouga et al., 2015), and *Mycobacterium* spp. (Calvo et al., 2013; Drancourt, 2014). Furthermore, several recent studies have reported similar behavior for fungi and viruses, whose development and water contamination can be associated with free-living amoebae: *Fusarium oxysporum* (Cateau et al., 2014b) and *Acaulopage* spp. (Michel et al., 2014), respectively, giant viruses (La Scola, 2014; Yutin et al., 2014).

3 Membranes in Water Purification

Depending on the nature of the materials used for the membrane fabrication, we can utilize both organic (polymeric) (Yüksel et al., 2016; Wagh et al., 2015; Goetz et al., 2016; Diep et al., 2015) and inorganic (Athanasekou et al., 2015; Da et al., 2016; Bet-moushoul et al., 2016; Song et al., 2016; Zhang et al., 2015b) membranes, but they can be also fabricated using composite materials (Liang et al., 2016; Haider et al., 2016; Al-Hobaib et al., 2016; Zhang et al., 2016b; He et al., 2015; Wanjale et al., 2016; Suhartono and Tizaoui, 2015; Yin et al., 2016; Kaminska et al., 2015; Chen et al., 2015c; Olukman and Şanli, 2015; Liu et al., 2015, 2016a; Zhu et al., 2016; Zha et al., 2016; Lu et al., 2016; Yurekli, 2016).

The most important properties of membranes utilized in water filtration are those which directly determine their performances and refer to: (1) the water flux; (2) the rejection of contaminant molecules; and (3) the antifouling properties of the membrane.

The water flux is directly influenced by the pore size of the membrane that is used in the water purification system, but also by the applied pressures. Thus, in order to assure high water fluxes in nanoporous membranes, very high pressures of the liquid must be assured. Other alternative is represented by the use of fibrous membranes, which assure highly specific surface areas. Recently, Zhou et al. (2016), obtained electrospun membranes from cellulose acetate with different fiber diameters. For 2-μm fibers, the resulted membranes showing a rejection rate of 99.8% for polystyrene particles dispersed in water. The best results were obtained for 800-nm fiber membranes which were evaluated using natural river water: the resulted turbidity reached 0.135 nephelometric turbidity unit for an initial flow of 20,455 L/m^2 h, at 14 psi pressure. Goetz et al. (2016), obtained superhydrophilic electrospun cellulose acetate membranes by coating their surface with chitin nanocrystals and, thus, obtaining very high water fluxes of 14,217 L/m^2 h at 0.5 bar.

The most important property of membranes for water filtering purposes is their capacity to reject contaminant molecules. As we have previously stated, the membrane rejects more contaminants (type of contaminants) as the dimension of its pores decreases (the best performances are obtained for reverse osmosis membranes and nanofiltration membranes). Nanometric size pore membranes can assure very high rejection of natural organic matter and medium or no rejection of salts: for example, 1-nm pore size TiO_2 nanofiltration membranes with 48 L/m^2 h barwater permeability showed a rejection of 43% for Na_2SO_4, 35% for $MgSO_4$, 96% for methylene blue, respectively, 99% for natural organic matter (Song et al., 2016). While reverse osmosis membranes assure high rejection efficiency, the water fluxes through this type of membranes are very low. Sanyal et al. (2015) modified commercial NF 270 and NF 90 nanofiltration membranes (Filmtec, Dow Chemical Company, Midland, MI) using a layer-by-layer assembly of poly(allylamine hydrochloride) and polyacrylic acid to achieve high perchlorate rejection (93%, at 5 bar pressure), obtaining similar properties with commercial reverse osmosis membranes (BW 30 and SW 30 membranes, Filmtec, Dow Chemical Company, Midland, MI).

The antifouling property is the ability of a certain surface to remove or to prevent the attachment of any molecules or microorganisms on wetted surfaces (Yebra et al., 2004). In order to improve the antifouling properties of membranes used in water filtration, different approaches for surface modification can be made, such as: capping amphiphilic polymers, developing coatings with star-shaped copolymers having high surface energy, grafting polymer brushes, and so on. The antifouling properties of the polymer/copolymer are determined by the functional groups available for interaction (Chen et al., 2013): quaternary ammonium groups determine the antibacterial properties of the polymers, highly hydrophilic polyethylene glycol (PEG) derivates reduce the protein adhesion, zwitterionic polymers prevent the nonspecific adhesion of microorganisms, while Fluor functional groups determine a non-sticking surface and low surface energy. For example, Gao et al. (2015) modified polyether sulfone (PES) membranes with an amphiphilic copolymer composed of PEG and poly dimethyl siloxane (PDMS). Its low attachment properties are influenced by the mixture of hydrophilic and low surface energy chains. A layer-by-layer technique (Chen et al., 2013) was employed to obtain polymeric reverse osmosis membranes with antifouling properties, which are determined by steric hindrance phenomenon. Kim et al. (2012), used polyethylene glycol methyl ether methacrylate (PEGMA) and methyl methacrylate (MMA) to form star-shaped polymers around polyhedral oligomeric silsesquioxane (POSS) cores. The high antifouling properties come from the shape of the copolymers, which possess higher properties, compared to linear structured polymers.

4 Nanostructurated Membranes and Nanofiltration

The most utilized methods in the design of membranes are: (1) phase inversion (Marbelia et al., 2016; Li et al., 2016; Hendrix et al., 2013); (2) interfacial polymerization (Cheng et al., 2015; Mahdavi et al., 2016; Miao et al., 2015; Arribas et al., 2014; Zhou et al., 2014; Wang et al., 2014a); (3) stretching (Kim et al., 2008; Wang et al., 2014b); (4) track-etching (Kravets et al., 2005; Choi et al., 2009; Tripathi et al., 2013); (5) electrospinning (Zhou et al., 2016; Goetz et al., 2016; Peng et al., 2015).

Wei et al. (2014) developed a composite hollow-fiber membrane based on polyethyleneimine (PEI) for nanofiltration. The obtaining process involved a modified interfacial polymerization process, the resulted membranes showing a nanoporous morphology, as proved by scanning electron microscopy (SEM). The inner surface of the membrane had uniform, dense and smooth structure. The zeta potential measurements showed that, at neutral pH, the membrane has a positive charge, mainly due to the protonation of the amine groups. Also, the as-obtained membranes proved good mechanical properties (mechanical strength). Different parameters in the obtaining process (like precursors concentration, aqueous phase immersion time, reaction time or heat-treatment temperature) were varied and the salt separation properties were tested. The separation properties of the membrane were evaluated in terms of pure water permeability, molecular weight cut-off, salt rejection, and permeate flux, respectively dye removal. For the first test, the pure water flux varied between 25.5 and 45.3 L/m^2 h, while pressures ranged between 0.2 and 0.6 MPa. Molecular weight cut-off was determined for polyethylene glycol (PEG) with different molecular weights (200–4000 Da), showing a value of about 500 Da. The salt rejection experiment was done for different solutions ($MgSO_4$, Na_2SO_4, $MgCl_2$, NaCl) showing that the rejection efficiency varies with the polarity, valence and steric hindrance. The rejection of $MgCl_2$ from a solution of 1 g/L, at 0.4 MPa, was done with an efficiency of 97%, the permeate flux being approximately 35.3 L/m^3 h. Dye molecules are frequently used in applications involving liquids purification, because their discharge can be easily monitored by macroscopic observing. In this case, the dyes differed in polarity and molecular weight (steric hindrance, by default). Thus, the rejection efficiency increased with the molecular weight of the dye, while for dyes with similar molecular weights, the rejection of cationic molecules is favored.

The interfacial polymerization method was also used by Akbari et al. (2016), who obtained thin-film membranes based on 2,5-diaminobenzene sulfonic acid (2,5-DABSA), a novel sulfonated aromatic diamine monomer, designed in order to obtain membranes with higher performances (higher water flux—61.2 L/m^2 h). This property was determined by the high hydrophobicity of the material and antifouling properties.

Carbon-based nanomaterials have been recently investigated with increased interest for water purification due to their exceptional properties, such as: (1) chemical, mechanical and thermal stability; (2) conductive properties; (3) antibacterial properties; and (4) high surface area (Manawi et al., 2016). Table 12.1 presents some examples of filtering membranes developed for water purification incorporating carbon-based nanomaterials.

Moreover, if considering the tunable features of novel nanomaterials and their size-dependent intrinsic antimicrobial behaviors, as in case of silver nanoparticles (AgNPs) and titanium dioxide nanostructures (particles and tubes), new composite and hybrid systems were lately successfully developed and evaluated as potential membrane for water purification process. A brief summary regarding latest potential candidates based on AgNPs and TiO_2 for water purification applications is included in Table 12.2. Besides, it is worthy to mention that the attractive properties of novel technologies and the current concern regarding drinking water contamination turned the attention of worldwide scientific community toward various nanotechnology-derived materials with this particular potential application, including especially inorganic oxides, such as ZnO (Colmenares et al., 2015; Pawar et al., 2016), Fe_3O_4 (Pinakidou et al., 2016; Habibi-Yangjeh and Akhundi, 2016; Zhao et al., 2016), and SiO_2 (Liu et al., 2016b; Puthai et al., 2016; Rakhshan and Pakizeh, 2015).

Table 12.1 Carbon Nanomaterials–Based Filtering Membranes for Water Purification

Type of Carbon Nanomaterials	Membrane System Description	Pore Size/Type of Filtration	System Evaluation	References
Carbon nanotubes	Vertically aligned carbon nanotubes (CNTs) membrane	4.8 nm/ nanofiltration	3 times higher water flux than ultrafiltration (UF) membrane; 70,000 times higher water transport than UF membrane; 15% less permeate flux reduction and 2 log less bacterial attachment than the UF membrane	Baek et al. (2014)
	Oxidized pristine multiwalled CNTs (MWCNTs)–cellulose acetate/polyethylene glycol		Morphology and roughness characterization using SEM and AFM; increase of thermal stability using TGA; MWCNTs improved salt rejection performance (99.8%)	Sabir et al. (2015)
	MWCNTs/polyaniline (PANI)/ polyethersulfone (PES)		High water permeability of 1400 L/m^2 h (LMH)/bar (30 times higher than PES membrane); 80% efficiency of removal for natural organic matter; 100% water flux recovery and 65% total fouling ratio	Lee et al. (2016)
	Polyhydroxybutyrate–calcium alginate/carboxyl MWCNTs electrospun nanofibers	Nanofiltration	Morphology characterizing using SEM; improved tensile mechanical property and hydrophilic property; enhanced dye Brilliant blue adsorption capacity (32.95 L/m^2 h flux and 98.20% rejection rate); (compared to PHB–CaAlg membrane)	Guo et al. (2016)
Graphene	Magnesium silicate–modified reduced graphene oxide (MgSi@RGO) nanosheets—polyacrylonitrile (PAN) membrane	Ultrafiltration	High rejection for PEG molecules with different molecular weights (PEG200, PEG400, PEG600, and PEG1000)	Liang et al. (2016)
	m-Phenylenediamine (MPD)-organic trimesoyl chloride (TMC)–GO		SEM, TEM, AFM for morphology evaluation; permeate flux under 300 psi increases from 39 to 59.4 L/m^2 h, decreased rejection of NaCl from 95.7 to 93.8%, decreased rejection of Na_2SO_4 from 98.1 to 97.3% for membranes with increasing concentration of GO from 0 to 0.015 wt.%	Yin et al. (2016)
	GO-AgNPs-cellulose acetate membrane		Improved relative flux drop under continuous filtration, after 24 h; antibacterial properties against *E. coli*	Sun et al. (2015a,b)
Carbon nanofibers	Mesoporous carbon nanofibers (CNFs)		Long fibrous morphology, high surface area and mesoporous structure proved using SEM, TEM, XRD, FTIR; high absorption of methylene blue (MB) and methyl orange (MO) from aqueous solution	Li et al. (2016)
	Cellulose triacetate (CTA)-carboxylated CNFs		Increased hydrophilicity with the carboxylated CNFs; water flux 15.6 L/m^2 h, forward osmosis use for sea water desalinization; higher tensile strength compared to simple cellulose triacetate membranes	Dabaghian and Rahimpour (2015)

Table 12.2 AgNPs and TiO_2-Based Membranes for Water Purification

Type of Inorganic Nanomaterial	Membrane System Description	Type of Filtration	System Evaluation	References
AgNPs	AgNPs immobilized onto aminated polyethersulfone (PES) membrane	Microfiltration	FT-IR and XRD analysis confirmed the composite nature; FE-SEM data revealed an asymmetric porous structure; TGA evaluation showed a good thermal stability; long-term release of Ag^+ was evaluated by ICP-OES; antibacterial activity was in vitro assessed against *E. coli* strain	Haider et al. (2016)
	AgNPs loaded into polyacrylonitrile (PAN) membrane	Microfiltration	Increased hydrophilicity was confirmed by water contact angle measurements; negative zeta potential was revealed; enhanced water permeability was assessed in PRO test mode; antibacterial activity was in vitro assessed against *E. coli* and *B. subtilis* strains; antifouling activity was in vitro assessed against *C. testosteroni* strain	Liu et al. (2016a,b)
	AgNPs-GO (graphene oxide) composite onto cellulose acetate membrane	Microfiltration	FT-IR and Raman spectroscopy revealed the composite nature; TEM data showed uniform distribution of spherical AgNPs within the GO matrix; SEM images revealed a dense and smooth composite with intimate well-packed layered structure; enhanced hydrophilicity was confirmed by water contact angle measurements; antibacterial and antifouling activity was in vitro assessed against *E. coli* strain	Sun et al. (2015a,b)
	AgNPs-MWCNTs composite coated onto a glass-fiber membrane filter	Microfiltration, ultrafiltration	FE-TEM analysis showed uniform distribution of sphere-like AgNPs onto the MWCNTs; increased porosity was evaluated by porosimetry analysis; enhanced hydrophilicity was confirmed by contact angle measurements; filtration performance was assessed against cox A9 virus, *E. coli* and *S. aureus* strains; antibacterial activity was in vitro assessed against *E. coli* and *S. aureus* strains	Kim et al. (2015)
	AgNPs embedded into a hybrid membrane based on polyvinyl alcohol (PVA) and mercaptopropyltriethoxysilane (MPTES)	Microfiltration	XRD, FT-IR, and UV-Vis spectrophotometry analysis confirmed the hybrid nature; antibacterial activity was in vitro assessed against *E. coli* and *S. aureus* strains	Zhao et al. (2015)

(*Continued*)

Table 12.2 AgNPs and TiO_2-Based Membranes for Water Purification (*cont.*)

Type of Inorganic Nanomaterial	Membrane System Description	Type of Filtration	System Evaluation	References
TiO_2	Nd^{3+}-doped TiO_2 NPs loaded into heteropolyphosphotungstic acid (HPA)	Ultrafltration	XRD and FT-IR spectroscopy showed crystalline composite nature; BET porosimetry revealed an high surface area depending on increased Nd^{3+} and decreased HPA compositions; SEM and HR-TEM analysis showed uniform distribution of polyhedral TiO_2 NPS within HPA matrix; cleaning performance was assessed against 4-chlorophenol after sunlight photoirradiation by mass spectral analysis	Thomas et al. (2016)
	Composite membrane based on TiO_2 NPs embedded into polystyrene (PS) nanofiber	Nanofiltration	SEM images showed increased roughness and TiO_2-dependent diameter and morphology; XRD spectra revealed crystalline composite; improved hydrophilicity was evaluated by contact angle measurements; adsorption and desorption studies revealed a TiO_2-dependent kinetics of Cu^{2+}	Wanjale et al. (2016)
	Monoliths based on TiO_2 NTs and carbon	Ultrafiltration	XRD analysis revealed crystalline composite nature of monoliths; SEM data showed uniform distribution of TiO_2 NTs and porous microstructure (micropores and mesopores); purification performances were assessed on wastewater by capturing the methylene blue during adsorption-lifting-irradiation experiment	Zhang et al. (2016c)
	TiO_2-MWCNTs-polysulfone (PS) composite membrane	Ultrafiltration	XRD analysis revealed crystalline composite nature; SEM images showed uniform porous morphology; enhanced hydrophilicity was evaluated by contact angle measurements; antifouling behavior was assessed by zeta potential measurements; membrane performances were evaluated in terms of increased water flux and humic acid filtration	Esfahani et al. (2015)

5 Conclusions and Perspectives

Microbial contamination of drinking water is an alarming increasing phenomenon that attracted worldwide attention and recently special funding, due to complex opportunistic related pathologies. Bacterial pollution is the most common biological contamination of drinking water sources and distribution systems, but significant concerns were also reported regarding the impact role of viruses and fungi as important water pollutants. When considering the current strategies applied for the treatment of drinking water, novel nanostructured materials, devices, and processes are under extensive research. Thus, experimental platforms based on nanostructured organic or inorganic, composite or hybrid materials have been successfully produced and assessed as potential unconventional and high-performance systems for water purification process. Specific membranes containing canonized materials are under development and their impact in water purification is estimated to be unlimited, these approaches being investigated as an alternative to less efficient methods for current needs in terms of water supply.

References

Abhirosh, C., Sherin, V., Thomas, A.P., Hatha, A.A.M., Mazumder, A., 2011. Potential public health significance of faecal contamination and multidrug-resistant *Escherichia coli* and *Salmonella* serotypes in a lake in India. Public Health 125 (6), 377–379.

Abia, A.L.K., Ubomba-Jaswa, E., Momba, M.N.B., 2015. Impact of seasonal variation on *Escherichia coli* concentrations in the riverbed sediments in the Apies River, South Africa. Sci. Total Environ. 537, 462–469.

Ahmad, A., Dada, A.C., Usup, G., Heng, L.Y., 2014. Occurrence of *Enterococcus* species with virulence markers in an urban flow-influenced tropical recreational beach. Marine Poll. Bull. 82 (1–2), 26–38.

Akbari, A., Aliyarizadeh, E., Mojallali Rostami, S.M., Homayoonfal, M., 2016. Novel sulfonated polyamide thin-film composite nanofiltration membranes with improved water flux and anti-fouling properties. Desalination 377, 11–22.

Al-Hobaib, A.S., Al-Sheetan, K.M., El Mir, L., 2016. Effect of iron oxide nanoparticles on the performance of polyamide membrane for ground water purification. Mater. Sci. Semicond. Process. 42 (Part 1), 107–110.

Arribas, P., Khayet, M., García-Payo, M.C., Gil, L., 2014. Self-sustained electro-spun polysulfonenano-fibrous membranes and their surface modification by interfacial polymerization for micro- and ultra-filtration. Sep. Purifi. Technol. 138, 118–129.

Ateba, C.N., Mbewe, M., 2011. Detection of *Escherichia coli* O157:H7 virulence genes in isolates from beef, pork, water, human and animal species in the northwest province, South Africa: public health implications. Res. Microbiol. 162 (3), 240–248.

Athanasekou, C.P., Moustakas, N.G., Morales-Torres, S., Pastrana-Martínez, L.M., Figueiredo, J.L., Faria, J.L., Silva, A.M.T., Dona-Rodriguez, J.M., Romanos, G.E.,

Falaras, P., 2015. Ceramic photocatalytic membranes for water filtration under UV and visible light. Appl. Catal. B 178, 12–19.
Avigliano, E., Schenone, N.F., 2015. Human health risk assessment and environmental distribution of trace elements, glyphosate, fecal coliform and total coliform in Atlantic Rainforest mountain rivers (South America). Microchem. J. 122, 149–158.
Baek, Y., Kim, C., Seo, D.K., Kim, T., Lee, J.S., Kim, Y.H., Ahn, K.H., Bae, S.S., Lee, S.C., Lim, J., Lee, K., Yoon, J., 2014. High performance and antifouling vertically aligned carbon nanotube membrane for water purification. J. Membr. Sci. 460, 171–177.
Bahadir, Z., Bulut, V.N., Hidalgo, M., Soylak, M., Marguí, E., 2015. Determination of trace amounts of hexavalent chromium in drinking waters by dispersive microsolid-phase extraction using modified multiwalled carbon nanotubes combined with total reflection X-ray fluorescence spectrometry. Spectrochim. Acta B 107, 170–177.
Bai, X., Ma, X., Xu, F., Li, J., Zhang, H., Xiao, X., 2015. The drinking water treatment process as a potential source of affecting the bacterial antibiotic resistance. Sci. Total Environ. 533, 24–31.
Baig, A.M., 2015. Pathogenesis of amoebic encephalitis: are the amoebae being credited to an 'inside job' done by the host immune response? Acta Trop. 148, 72–76.
Bergeron, S., Boopathy, R., Nathaniel, R., Corbin, A., LaFleur, G., 2015. Presence of antibiotic resistant bacteria and antibiotic resistance genes in raw source water and treated drinking water. Int. Biodeter. Biodegr. 102, 370–374.
Bet-moushoul, E., Mansourpanah, Y., Farhadi, K., Tabatabaei, M., 2016. TiO_2 nanocomposite based polymeric membranes: a review on performance improvement for various applications in chemical engineering processes. Chem. Eng. J. 283, 29–46.
Bhardwaj, S., Bhamre, K., Dhawale, J., Patil, M., Divase, S., 2013. *Enterococcus faecium* and *Enterococcus faecalis*, the nosocomial pathogens with special reference to multi-drug resistance and phenotypic characterization. Int. J. Pharmaceut. Sci. Pract. 2 (1), 1–10.
Bhowmik, A.K., Alamdar, A., Katsoyiannis, I., Shen, H., Ali, N., Ali, S.M., Bokhari, H., Schäfer, R.B., Eqani, S.A.M.A.S., 2015. Mapping human health risks from exposure to trace metal contamination of drinking water sources in Pakistan. Sci. Total Environ. 538, 306–316.
Bonilla-Lemus, P., Caballero Villegas, A.S., Carmona Jiménez, J., Vázquez, A.L., 2014. Occurrence of free-living amoebae in streams of the Mexico Basin. Exp. Parasitol. 145 (Suppl.), 28–33.
Bretzler, A., Johnson, C.A., 2015. The Geogenic Contamination Handbook: addressing arsenic and fluoride in drinking water. Appl. Geochem. 63, 642–646.
Brzuszkiewicz, E., Schulz, T., Rydzewski, K., Daniel, R., Gillmaier, N., Dittmann, C., Holland, G., Schunder, E., Lautner, M., Eisenreich, W., Lück, C., Heuner, K., 2013. *Legionella oakridgensis* ATCC 33761 genome sequence and phenotypic characterization reveals its replication capacity in amoebae. Int. J. Med. Microbiol. 303 (8), 514–528.
Butts, C.D., Bloom, M.S., Neamtiu, I.A., Surdu, S., Pop, C., Anastasiu, D., Fitzgerald, E.F., Gurzau, E.S., 2015. A pilot study of low-moderate drinking water arsenic contamination and chronic diseases among reproductive age women in Timiş County, Romania. Environ. Toxicol. Pharmacol. 40 (3), 1001–1004.
Callon, C., Ndob, A.M., Savignard, A., Rifa, E., Montel, M.C., 2016. *Escherichia coli* O26:H11 growth in co-culture with an antagonistic strain of *Hafnia alvei* according to pH and temperature. Food Control 62, 117–124.

Calvo, L., Gregorio, I., García, A., Fernández, M.T., Goñi, P., Clavel, A., Peleato, M.L., Fillat, M.F., 2013. A new pentaplex-nested PCR to detect five pathogenic bacteria in free living amoebae. Water Res. 47 (2), 493–502.

Canal, N., Meneghetti, K.L., Ponzi de Almeida, C., da Rosa Bastos, M., MunerOtton, L., Corção, G., 2016. Characterization of the variable region in the class 1 integron of antimicrobial-resistant *Escherichia coli* isolated from surface water. Braz. J. Microbiol..

Cardoso De Oliveira, A.J.F., Pinhata, J.M.W., 2008. Antimicrobial resistance and species composition of *Enterococcus* spp. isolated from waters and sands of marine recreational beaches in Southeastern Brazil. Water Res. 42 (8–9), 2242–2250.

Catalani, S., Fostinelli, J., Gilberti, M.E., Apostoli, P., 2015. Application of a metal free high performance liquid chromatography with inductively coupled plasma mass spectrometry (HPLC–ICP-MS) for the determination of chromium species in drinking and tap water. Int. J. Mass Spectrom. 387, 31–37.

Cateau, E., Delafont, V., Hechard, Y., Rodier, M.H., 2014a. Free-living amoebae: what part do they play in healthcare-associated infections? J. Hosp. Infect. 87 (3), 131–140.

Cateau, E., Hechard, Y., Fernandez, B., Rodier, M.H., 2014b. Free living amoebae could enhance *Fusarium oxysporum* growth. Fungal Ecol. 8, 12–17.

Cervero-Aragó, S., Sommer, R., Araujo, R.M., 2014. Effect of UV irradiation (253.7 nm) on free *Legionella* and *Legionella* associated with its amoebae hosts. Water Res. 67, 299–309.

Chandrasekaran, R., Hamilton, M.J., Wang, P., Staley, C., Matteson, S., Birr, A., Sadowsky, M.J., 2015. Geographic isolation of *Escherichia coli* genotypes in sediments and water of the Seven Mile Creek—a constructed riverine watershed. Sci. Total Environ. 538, 78–85.

Chang, J.S., 2015. Biotransformation of arsenite and bacterial aox activity in drinking water produced from surface water of floating houses: arsenic contamination in Cambodia. Environ. Pollut. 206, 315–323.

Chen, L., Therien-Aubin, H., Wong, M.C.Y., Hoek, E.M.V., Ober, C.K., 2013. Improved antifouling properties of polymer membranes using a "layer-by-layer" mediated method. J. Mater. Chem. B 1, 5651–5658.

Chen, G.R., Chang, Y.R., Liu, X., Kawamoto, T., Tanaka, H., Kitajima, A., Parajuli, D., Takasaki, M., Yoshino, K., Chen, M.L., Lo, Y.K., Lei, Z., Lee, D.J., 2015a. Prussian blue (PB) granules for cesium (Cs) removal from drinking water. Sep. Purif. Technol. 143, 146–151.

Chen, H., Kong, L., Wang, Y., 2015b. Enhancing the hydrophilicity and water permeability of polypropylene membranes by nitric acid activation and metal oxide deposition. J. Membr. Sci. 487, 109–116.

Chen, X., Song, Y.Q., Xu, H.Y., Menghe, B.L.G., Zhang, H.P., Sun, Z.H., 2015c. Genetic relationships among *Enterococcus faecalis* isolates from different sources as revealed by multilocus sequence typing. J. Dairy Sci. 98 (8), 5183–5193.

Cheng, X.Q., Shao, L., Lau, C.H., 2015. High flux polyethylene glycol based nanofiltration membranes for water environmental remediation. J. Membr. Sci. 476, 95–104.

Choi, J.-H., Park, S.-K., Ng, H.-Y., 2009. Membrane fouling in a submerged membrane bioreactor using track-etched and phase-inversed porous membranes. Sep. Purif. Technol. 65, 184–192.

Christensen, S.C.B., Nissen, E., Arvin, E., Albrechtsen, H.J., 2012. Influence of *Asellus aquaticus* on *Escherichia coli, Klebsiella pneumoniae, Campylobacter jejuni* and naturally occurring heterotrophic bacteria in drinking water. Water Res. 46 (16), 5279–5286.

Chuah, C.J., Lye, H.R., Ziegler, A.D., Wood, S.H., Kongpun, C., Rajchagool, S., 2016. Fluoride: a naturally-occurring health hazard in drinking-water resources of Northern Thailand. Sci. Total Environ. 545–546, 277–279.

Coleman, B.L., Louie, M., Salvadori, M.I., McEwen, S.A., Neumann, N., Sibley, K., Irwin, R.J., Jamieson, F.B., Daignault, D., Majury, A., Braithwaite, S., Crago, B., McGeer, A.J., 2013. Contamination of Canadian private drinking water sources with antimicrobial resistant *Escherichia coli*. Water Res. 47 (9), 3026–3036.

Colmenares, J.C., Kuna, E., Jakubiak, S., Michalski, J., Kurzydłowsk, K., 2015. Polypropylene nonwoven filter with nanosized ZnO rods: promising hybrid photocatalyst for water purification. Appl. Catal. B 170–171, 273–282.

Coulliette, A.D., Enger, K.S., Weir, M.H., Rose, J.B., 2013. Risk reduction assessment of waterborne *Salmonella* and *Vibrio* by a chlorine contact disinfectant point-of-use device. Int. J. Hyg. Environ. Health 216 (3), 355–361.

Da, X., Chen, X., Sun, B., Wen, J., Qiu, M., Fan, Y., 2016. Preparation of zirconia nanofiltration membranes through an aqueous sol–gel process modified by glycerol for the treatment of wastewater with high salinity. J. Membr. Sci. 504, 29–39.

Dabaghian, Z., Rahimpour, A., 2015. Carboxylated carbon nanofibers as hydrophilic porous material to modification of cellulosic membranes for forward osmosis desalination. Chem. Eng. Res. Des. 104, 647–657.

Dada, N., Vannavong, N., Seidu, R., Lenhart, A., Stenström, T.A., Chareonviriyaphap, T., Overgaard, H.J., 2013. Relationship between *Aedes aegypti* production and occurrence of *Escherichia coli* in domestic water storage containers in rural and sub-urban villages in Thailand and Laos. Acta Trop. 126 (3), 177–185.

Daly, E., Kolotelo, P., Schang, C., Osborne, C.A., Coleman, R., Deletic, A., McCarthy, D.T., 2013. *Escherichia coli* concentrations and loads in an urbanised catchment: The Yarra River, Australia. J. Hydrol. 497, 51–61.

Diep, J., Tek, A., Thompson, L., Frommer, J., Wang, R., Piunova, V., Sly, J., La, Y.-H., 2015. Layer-by-layer assembled core–shell star block copolymers for fouling resistant water purification membranes. Polymer.

Ding, J., Shen, X., Liu, W., Covaci, A., Yang, F., 2015. Occurrence and risk assessment of organophosphate esters in drinking water from Eastern China. Sci. Total Environ. 538, 595–965.

Dongmei, L., Zhiwei, W., Qi, Z., Fuyi, C., Yujuan, S., Xiaodong, L., 2015. Drinking water toxicity study of the environmental contaminant—bromate. Regul. Toxicol. Pharmacol. 733, 802–810.

Drancourt, M., 2014. Looking in amoebae as a source of mycobacteria. Microb. Pathog. 77, 119–124.

Ebrahim, S., Reda, M., Hussien, A., Zayed, D., 2015. CdTe quantum dots as a novel biosensor for *Serratia marcescens* and lipopolysaccharide. Spectrochim. Acta A 150, 212–219.

Environment Agency, 2002. The Microbiology of Drinking Water—Part 1—Water Quality and Public Health. Available from: www.gov.uk

Esfahani, M.R., Tyler, J.L., Stretz, H.A., Wells, M.J.M., 2015. Effects of a dual nanofiller, nano-TiO_2 and MWCNT, for polysulfone-based nanocomposite membranes for water purification. Desalination 372, 47–56.

Farkas, A., Butiuc-Keul, A., Ciatarâş, D., Neamţu, C., Crăciunaş, C., Podar, D., Drăgan-Bularda, M., 2013. Microbiological contamination and resistance genes in biofilms occurring during the drinking water treatment process. Sci. Total Environ. 443, 932–938.

Ferro, G., Guarino, F., Cicatelli, A., Rizzo, L., 2016. β-Lactams resistance gene quantification in an antibiotic resistant *Escherichia coli* water suspension treated by advanced oxidation with UV/H_2O_2. J. Hazard. Mater..

Fhogartaigh, C.N., Dance, D.A.B., 2013. Bacterial gastroenteritis. Medicine 41 (2), 693–699.

Gao, F., Zhang, G., Zhang, Q., Zhan, X., Chen, F., 2015. Improved antifouling properties of poly(ether sulfone) membrane by incorporating the amphiphilic comb copolymer with mixed poly(ethylene glycol) and poly(dimethylsiloxane) brushes. Ind. Eng. Chem. Res. 54, 8789–8800.

Gerke, T.L., Little, B.J., Maynard, J.B., 2016. Manganese deposition in drinking water distribution systems. Sci. Total Environ. 541, 184–193.

Goetz, L.A., Jalvo, B., Rosal, R., Mathew, A.P., 2016. Superhydrophilic anti-fouling electrospun cellulose acetate membranes coated with chitin nanocrystals for water filtration. J. Membr. Sci. 510, 238–248.

Gu, J.-E., Lee, J.S., Park, S.-H., Kim, I.T., Chan, E.P., Kwon, Y.-N., Lee, J.-H., 2015. Tailoring interlayer structure of molecular layer-by-layer assembled polyamide membranes for high separation performance. Appl. Surf. Sci. 356, 659–667.

Guo, J., Zhang, Q., Cai, Z., Zhao, K., 2016. Preparation and dye filtration property of electrospun polyhydroxybutyrate–calcium alginate/carbon nanotubes composite nanofibrous filtration membrane. Sep. Purif. Technol. 161, 69–79.

Habibi-Yangjeh, A., Akhundi, A., 2016. Novel ternary g-$C_3N4/Fe_3O_4/Ag_2CrO_4$ nanocomposites: magnetically separable and visible-light-driven photocatalysts for degradation of water pollutants. J. Mol. Catal. A 415, 122–130.

Haider, M.S., Shao, G.N., Imran, S.M., Park, S.S., Abbas, N., Tahir, M.S., Hussain, M., Bae, W., Kim, H.T., 2016. Aminated polyethersulfone-silver nanoparticles (AgNPs-APES) composite membranes with controlled silver ion release for antibacterial and water treatment applications. Mater. Sci. Eng. C 62, 732–745.

Hakalehto, E., Nyholm, O., Bonkoungou, I.J.O., Kagambega, A., Rissanen, K., Heitto, A., Barro, N., Haukka, K., 2014. Development of microbiological field methodology for water and food-chain hygiene analysis of *Campylobacter* spp. and *Yersinia* spp. in Burkina Faso, West Africa. Pathophysiology 21 (3), 219–229.

Handayani, E.M., Komalasari, I., Kurin, R., 2016. Determination of assigned values for cadmium, cobalt and manganese in drinking water proficiency testing. Procedia Chem. 16, 72–77.

Happonen, M., Koivusalo, H., Malve, O., Perkola, N., Juntunen, J., Huttula, T., 2016. Contamination risk of raw drinking water caused by PFOA sources along a river reach in south-western Finland. Sci. Total Environ. 541, 74–82.

He, L., Dumée, L.F., Feng, C., Velleman, L., Reis, R., She, F., Gao, W., Kong, L., 2015. Promoted water transport across graphene oxide–poly(amide) thin film composite membranes and their antibacterial activity. Desalination 365, 126–135.

Hendrix, K., Vaneynde, M., Koeckelberghs, G., Vankelecom, I.F.J., 2013. Synthesis of modified poly(ether ether ketone) polymer for the preparation of ultrafiltration and nanofiltration membranes via phase inversion. J. Membr. Sci. 447, 96–106.

Huber, J., Leopold, K., 2015. Nanomaterials-based strategies for enhanced mercury trace analysis in environmental and drinking waters. Trends Anal. Chem..

Huycke, M.M., Sahm, D.F., Gilmore, M.S., 1998. Multiple-drug resistant Enterococci: the nature of the problem and an agenda for the future. Emerg. Infect. Diseases 4 (2), 239–249.

Ji, W.T., Hsu, B.M., Chang, T.Y., Hsu, T.K., Kao, P.M., Huang, K.H., Tsai, S.F., Huang, Y.L., Fan, C.W., 2014. Surveillance and evaluation of the infection risk of free-living amoebae and *Legionella* in different aquatic environments. Sci. Total Environ. 499, 212–219.

Kaminska, G., Bohdziewicz, J., Calvo, J.I., Prádanos, P., Palacio, L., Hernández, A., 2015. Fabrication and characterization of polyethersulfone nanocomposite membranes for the removal of endocrine disrupting micropollutants from wastewater. Mechanisms and performance. J. Membr. Sci. 493, 66–79.

Kaner, P., Johnson, D.J., Seker, E., Hilal, N., Altinkaya, S.A., 2015. Layer-by-layer surface modification of polyethersulfone membranes using polyelectrolytes and AgCl/TiO_2 xerogels. J. Membr. Sci. 493, 807–819.

Kaper, J.P., Nataro, J.P., Mobley, H.L.T., 2004. Pathogenic *Escherichia coli*. Nat. Rev. Microbiol. 2, 123–140.

Khan, S., Beattie, T.K., Knapp, K.W., 2016. Relationship between antibiotic- and disinfectant-resistance profiles in bacteria harvested from tap water. Chemosphere 152, 132–141.

Kim, J., Kim, S.S., Park, M., Jang, M., 2008. Effects of precursor properties on the preparation of polyethylene hollow fiber membranes by stretching. J. Membr. Sci. 318, 201–209.

Kim, D.-G., Kang, H., Han, S., Lee, J.-C., 2012. The increase of antifouling properties of ultrafiltration membrane coated by star-shaped polymers. J. Mater. Chem. 22, 8654–8661.

Kim, J.P., Kim, J.H., Kim, J., Lee, S.N., Park, H.O., 2015. A nanofilter composed of carbon nanotube-silver composites for virus removal and antibacterial activity improvement. J. Environ. Sci..

Kratochwill, L., Powers, M., McGraw, M.A., King, L., O'Neill, J.M., Venkat, A., 2015. Factors associated with ciprofloxacin-resistant *Escherichia coli* urinary tract infections in discharged ED patients. Am. J. Emerg. Med. 33 (10), 1473–1476.

Kravets, L., Dmitriev, S., Gilman, A., Drachev, A., Dinescu, G., 2005. Water permeability of poly(ethylene terephthalate) track membranes modified by DC discharge plasma polymerization of dimethylaniline. J. Membr. Sci. 263, 127–136.

La Scola, B., 2014. Looking at protists as a source of pathogenic viruses. Microb. Pathog. 77, 131–135.

Larsson, J., Lindberg, R., Aspán, A., Grandon, R., Westergren, E., Jacobson, M., 2014. Neonatal piglet diarrhoea associated with enteroadherent *Enterococcus hirae*. J. Comp. Pathol. 151 (2–3), 137–147.

Le Calvez, T., Trouilhé, M.C., Humeau, P., Moletta-Denat, M., Frère, J., Héchard, Y., 2012. Detection of free-living amoebae by using multiplex quantitative PCR. Mol. Cell. Probe. 26 (3), 116–120.

Lee, J., Ye, Y., Ward, A.J., Zhou, C., Chen, V., Minett, A.I., Lee, S., Liu, Z., Chae, S.-R., Shi, J., 2016. High flux and high selectivity carbon nanotube composite membranes for natural organic matter removal. Sep. Purif. Technol. 163, 109–119.

Leistner, R., Bloch, A., Sakellariou, C., Gastmeier, P., Schwab, F., 2014. Costs and length of stay associated with extended-spectrum β-lactamase production in cases of *Escherichia coli* bloodstream infection. J. Glob. Antimicrob. Resist. 2 (2), 107–109.

Lendowski, L., Färber, H., Holy, A., Darius, A., Ehrich, B., Wippermann, C., Küfner, B., Exner, M., 2015. Accidental contamination of a German town's drinking water with sodium hydroxide. Int. J. Hygiene Environ. Health 218 (3), 366–369.

Lerma, L.L., Benomar, N., Sánchez Valenzuela, A., del Carmen Casado Muñoz, M., Gálvez, A., Abriouel, H., 2014. Role of EfrAB efflux pump in biocide tolerance and antibiotic resistance of *Enterococcus faecalis* and *Enterococcus faecium* isolated from traditional fermented foods and the effect of EDTA as EfrAB inhibitor. Food Microbiol. 44, 249–257.

Levantesi, C., Bonadonna, L., Briancesco, R., Grohmann, E., Toze, S., Tandoi, V., 2012. *Salmonella* in surface and drinking water: occurrence and water-mediated transmission. Food Res. Int. 45 (2), 587–602.

Li, L., Chen, M., Dong, Y., Dong, X., Cerneaux, S., Hampshire, S., Cao, J., Zhu, L., Zhu, Z., Liu, J., 2016. A low-cost alumina–mullite composite hollow fiber ceramic membrane fabricated via phase-inversion and sintering method. J. Eur. Ceram. Soc. 36, 2057–2066.

Liang, B., Zhang, P., Wang, J., Qu, J., Wang, L., Wang, X., Guan, C., Pan, K., 2016. Membranes with selective laminar nanochannels of modified reduced graphene oxide for water purification. Carbon 103, 94–100.

Liu, T.-Y., Tong, Y., Liu, Z.-H., Lin, H.-H., Lin, Y.-K., Van der bruggen, B., Wang, X.-L., 2015. Extracellular polymeric substances removal of dual-layer (PES/PVDF) hollow fiber UF membrane comprising multi-walled carbon nanotubes for preventing RO biofouling. Sep. Purif. Technol. 148, 57–67.

Liu, X., Foo, L.X., Li, Y., Lee, J.Y., Cao, B., Tang, C.Y., 2016a. Fabrication and characterization of nanocomposite pressure retarded osmosis (PRO) membranes with excellent anti-biofouling property and enhanced water permeability. Desalination 389, 137–148.

Liu, Y., Fu, R., Sun, Y., Zhou, X., Baig, S.A., Xu, X., 2016b. Multifunctional nanocomposites Fe_3O_4@SiO_2-EDTA for Pb(II) and Cu(II) removal from aqueous solutions. Appl. Surf. Sci. 369, 267–276.

Loret, J.F., Greub, G., 2010. Free-living amoebae: biological by-passes in water treatment. Int. J. Hyg. Environ. Health 213 (3), 167–175.

Lu, P., Liang, S., Qiu, L., Gao, Y., Wang, Q., 2016. Thin film nanocomposite forward osmosis membranes based on layered double hydroxide nanoparticles blended substrates. J. Membr. Sci. 504, 196–205.

Machado, A., Bordalo, A.A., 2014. Prevalence of antibiotic resistance in bacteria isolated from drinking well water available in Guinea-Bissau (West Africa). Ecotoxicol. Environ. Saf. 106, 188–194.

Mahanta, R., Chowdhury, J., Nath, H.K., 2016. Health costs of arsenic contamination of drinking water in Assam, India. Econ. Anal. Policy 49, 30–42.

Mahdavi, H., Razmi, F., Shahalizade, T., 2016. Polyurethane TFC nanofiltration membranes based on interfacial polymerization of poly(bis-MPA) and MDI on the polyethersulfone support. Sep. Purif. Technol. 162, 37–44.

Manawi, Y., Kochkodan, V., Hussein, M.A., Khaleel, M.A., Hilal, N., 2016. Can carbon-based nanomaterials revolutionize membrane fabrication for water treatment and desalination? Desalination 391, 69–88.

Marbelia, L., Bilad, M.R., Piassecka, A., Jishna, P.S., Naik, P.V., Vankelecom, I.F.J., 2016. Study of PVDF asymmetric membranes in a high-throughput membrane bioreactor (HT-MBR): influence of phase inversion parameters and filtration performance. Sep. Purif. Technol. 162, 6–13.

Miao, J., Zhang, R., Bai, R., 2015. Poly (vinyl alcohol)/carboxymethyl cellulose sodium blend composite nanofiltration membranes developed via interfacial polymerization. J. Membr. Sci. 493, 654–663.

Michel, R., Walochnik, J., Scheid, P., 2014. Article for the "Free-living amoebae special issue": isolation and characterisation of various amoebophagous fungi and evaluation of their prey spectrum. Exp. Parasitol. 145 (Suppl. S131–136), 131–136.

Nam, S., Kim, M.J., Park, C., Park, J.G., Maeng, P.J., Lee, G.C., 2013. Detection and genotyping of vancomycin-resistant *Enterococcus* spp. by multiplex polymerase chain reaction in Korean aquatic environmental samples. Int. J. Hyg. Environ. Health 216 (4), 421–427.

Nunes, A., Gomes, J.P., 2014. Evolution, phylogeny, and molecular epidemiology of *Chlamydia*. Infect. Genet. Evol. 23, 49–64.

Oliveira, S., Moura, R.A., Silva, K.C., Pavez, M., McCulloch, J.A., Dropa, M., Matté, M.H., Mamizuka, E.M., Sato, M.I.Z., Pestana de Castro, A.F., Lincopan, N.,

2014. Isolation of KPC-2-producing *Klebsiella pneumoniae* strains belonging to the high-risk multiresistant clonal complex 11 (ST437 and ST340) in urban rivers. J. Antimicrob. Chemother. 69 (3), 849–852.

Olukman, M., Şanli, O., 2015. A novel in situ synthesized magnetite containing acrylonitrile and 2-hydroxyethyl methacrylate grafted poly(vinyl alcohol) nanocomposite membranes for pervaporation separation of acetone/water mixtures. Chem. Eng. Process. 98, 60–70.

Patel, C.B., Vajpayee, P., Singh, G., Upadhyay, R.S., Shanker, R., 2011. Contamination of potable water by enterotoxigenic *Escherichia coli*: qPCR based culture-free detection and quantification. Ecotoxicol. Environ. Saf. 74 (8), 2292–2298.

Pawar, R.C., Son, Y., Kim, J., Ahn, S.H., Lee, C.S., 2016. Integration of ZnO with g-C3N4 structures in core–shell approach via sintering process for rapid detoxification of water under visible irradiation. Curr. Appl. Phys. 16, 101–108.

Peng, C., Zhang, J., Xiong, Z., Zhao, B., Liu, P., 2015. Fabrication of porous hollow γ-Al_2O_3 nanofibers by facile electrospinning and its application for water remediation. Micropor. Mesopor. Mater. 215, 133–142.

Pinakidou, F., Katsikini, M., Simeonidis, K., Kaprara, E., Paloura, E.C., Mitrakas, M., 2016. On the passivation mechanism of Fe_3O_4 nanoparticles during Cr(VI) removal from water: a XAFS study. Appl. Surf. Sci. 360, 1080–1086.

Popa, M., Dumitrel, G.A., Mirel, G., Popa, D.V., 2015. Anthropogenic contamination of water from Galda River—Alba County, Romania. Agric. Agric. Sci. Procedia 6, 446–452.

Puthai, W., Kanezashi, M., Nagasawa, H., Wakamura, K., Ohnishi, H., Tsuru, T., 2016. Effect of firing temperature on the water permeability of SiO_2–ZrO_2 membranes for nanofiltration. J. Membr. Sci. 497, 348–356.

Qin, Z., Ren, X., Shan, L., Guo, H., Geng, C., Zhang, G., Ji, S., Liang, Y., 2016. Nacrelike-structured multilayered polyelectrolyte/calcium carbonate nanocomposite membrane via Ca-incorporated layer-by-layer-assembly and CO_2-induced biomineralization. J. Membr. Sci. 498, 180–191.

Rakhshan, N., Pakizeh, M., 2015. Removal of triazines from water using a novel OA modified SiO_2/PA/PSf nanocomposite membrane. Sep. Purif. Technol. 147, 245–256.

Rehaiem, A., Belgacem, Z.B., Edalatian, M.R., Martínez, B., Rodríguez, A., Manai, M., Pérez Guerra, N., 2014. Assessment of potential probiotic properties and multiple bacteriocin encoding-genes of the technological performing strain *Enterococcus faecium* MMRA. Food Control 37, 343–350.

Rodriguez-Alvarez, M.S., Weir, M.H., Pope, J.M., Seghezzo, L., Rajal, V.B., Salusso, M.M., Morañ, L.B., 2015. Development of a relative risk model for drinking water regulation and design recommendations for a peri urban region of Argentina. Int. J. Hyg. Environ. Health 218 (7), 627–638.

Sabir, A., Shafiq, M., Islam, A., Sarwar, A., Dilshad, M.R., Shafeeq, A., Zahid Butt, M.T., Jamil, T., 2015. Fabrication of tethered carbon nanotubes in cellulose acetate/polyethylene glycol-400 composite membranes for reverse osmosis. Carbohydr. Polym. 132, 589–597.

Santos, T., Capelo, J.L., Santos, H.M., Oliveira, I., Marinho, C., Gonçalves, A., Araújo, J.E., Poeta, P., Igrejas, G., 2015. Use of MALDI-TOF mass spectrometry fingerprinting to characterize *Enterococcus* spp. and *Escherichia coli* isolates. J. Proteom. 127 (Part B), 321–331.

Sanyal, O., Sommerfeld, A.N., Lee, I., 2015. Design of ultrathin nanostructured polyelectrolyte-based membranes with high perchlorate rejection and high permeability. Sep. Purif. Technol. 145, 113–119.

Saxena, T., Kaushik, P., Mohan, M.K., 2015. Prevalence of *E. coli* O157:H7 in water sources: an overview on associated diseases, outbreaks and detection methods. Diagn. Microbiol. Infect. Dis. 82 (3), 249–264.

Schelegueda, L.I., Vallejo, M., Gliemmo, M.F., Marguet, E.R., Campos, C.A., 2015. Synergistic antimicrobial action and potential application for fish preservation of a bacteriocin produced by *Enterococcus mundtii* isolated from *Odontesthes platensis*. LWT Food Sci. Technol. 64 (2), 794–801.

Sergeeva, O.V., Bogdanova, A.A., Sergiev, P.V., 2015. What do we know about ribosomal RNA methylation in *Escherichia coli*? Biochimie 117 (Special Issue: Regulatory RNAs), 110–118.

Song, Z., Fathizadeh, M., Huang, Y., Chu, K.H., Yoon, Y., Wang, L., Xu, W.L., Yu, M., 2016. TiO_2 nanofiltration membranes prepared by molecular layer deposition for water purification. J. Membr. Sci. 510, 72–78.

Sorensen, J.P.R., Lapworth, D.J., Marchant, B.P., Nkhuwa, D.C.W., Pedley, S., Stuart, M.E., Bell, R.A., Chirwa, M., Kabika, J., Liemisa, M., Chibesa, M., 2015. In-situ tryptophan-like fluorescence: a real-time indicator of faecal contamination in drinking water supplies. Water Res. 81, 38–46.

Soto-Alonso, G., Cruz-Medina, J.A., Caballero-Pérez, J., Arvizu-Hernández, I., Ávalos-Esparza, L.M., Cruz-Hernández, A., Romero-Gómez, S., Rodríguez, A.L., Pastrana-Martínez, X., Fernández, F., Loske, A.M., Campos-Guillén, J., 2015. Isolation of a conjugative F-like plasmid from a multidrug-resistant *Escherichia coli* strain CM6 using tandem shock wave-mediated transformation. J. Microbiol. Methods 114, 1–8.

Street, A., Sustich, R., Savage, N., Duncan, J., 2014. Nanotechnology Applications for Clean Water: Solutions for Improving Water Quality. Elsevier Science, USA.

Struve, C., Krogfelt, K.A., 2004. Pathogenic potential of environmental *Klebsiella pneumoniae* isolates. Environ. Microbiol. 6, 584–590.

Suhartono, J., Tizaoui, C., 2015. Polyvinylidene fluoride membranes impregnated at optimised content of pristine and functionalised multi-walled carbon nanotubes for improved water permeation, solute rejection and mechanical properties. Sep. Purif. Technol. 154, 290–300.

Sukkua, K., Rattanachuay, P., Sukhumungoon, P., 2015. Potential routes of transmission of an emerging hospital strain: enterotoxigenic *Escherichia coli* O169:HUT from Southern Thailand. Asian Pac. J. Trop. Dis. 5 (11), 881–883.

Sun, X.-F., Qin, J., Xia, P.-F., Guo, B.-B., Yang, C.-M., Song, C., Wang, S.-G., 2015a. Graphene oxide–silver nanoparticle membrane for biofouling control and water purification. Chem. Eng. J. 281, 53–59.

Sun, X.F., Qin, J., Xia, P.F., Guo, B.B., Yang, C.M., Song, C., Wang, S.G., 2015b. Graphene oxide–silver nanoparticle membrane for biofouling control and water purification. Chem. Eng. J. 281, 53–59.

Tang, S., Yin, H., Chen, S., Peng, H., Chang, J., Liu, Z., Dang, Z., 2016. Aerobic degradation of BDE-209 by *Enterococcus casseliflavus*: isolation, identification and cell changes during degradation process. J. Hazard. Mater. 308, 335–342.

Thomas, J., Radhika, S., Yoon, M., 2016. Nd^{3+}-doped TiO_2 nanoparticles incorporated with heteropolyphospho tungstic acid: a novel solar photocatalyst for degradation of 4-chlorophenol in water. J. Mol. Catal. A 411, 146–156.

Tosetti, N., Croxatto, A., Greub, G., 2014. Amoebae as a tool to isolate new bacterial species, to discover new virulence factors and to study the host–pathogen interactions. Microb. Pathog. 77, 125–130.

Tripathi, B.P., Dubey, N.C., Stamm, M., 2013. Functional polyelectrolyte multilayer membranes for water purification applications. J. Hazard. Mater. 252-253, 401–412.
Trueman, B.F., Gagnon, G.A., 2016. A new analytical approach to understanding nanoscale lead-iron interactions in drinking water distribution systems. J. Hazard. Mater., available online 5 March 2016.
Vantarakis, A., Smaili, M., Detorakis, I., Vantarakis, G., Papapetropoulou, M., 2013. Diachronic long-term surveillance of bacteriological quality of bottled water in Greece (1995–2010). Food Control 33 (1), 63–67.
Vasile, M., Loots, H., Jacobs, K., Verheyen, L., Sneyers, L., Verrezen, F., Bruggeman, M., 2016. Determination of 210Pb, 210Po, 226Ra, 228Ra and uranium isotopes in drinking water in order to comply with the requirements of the EU 'Drinking Water Directive'. Appl. Radiat. Isot. 109, 465–469.
Veldman, K., Kant, A., Dierikx, C., van Essen-Zandbergen, A., Wit, B., Mevius, D., 2014. Enterobacteriaceae resistant to third-generation cephalosporins and quinolones in fresh culinary herbs imported from Southeast Asia. Int. J. Food Microbiol. 177, 72–77.
Vouga, M., Diabi, H., Boulos, A., Baud, D., Raoult, D., Greub, G., 2015. Antibiotic susceptibility of *Neochlamydia hartmanellae* and *Parachlamydia acanthamoebae* in amoebae. Microb. Infect. 17 (11–12), 761–765.
Wagh, P., Parungao, G., Viola, R.E., Escobar, I.C., 2015. A new technique to fabricate high-performance biologically inspired membranes for water treatment. Sep. Purif. Technol. 156 (Part 2), 754–765.
Wang, D., Fiessel, W., 2009. Evaluation of media for simultaneous enumeration of total coliform and *Escherichia coli* in drinking water supplies by membrane filtration techniques. J. Environ. Sci. 20 (3), 273–277.
Wang, H., Ding, S., Zhu, H., Wang, F., Guo, Y., Zhang, H., Chen, J., 2014a. Effect of stretching ratio and heating temperature on structure and performance of PTFE hollow fiber membrane in VMD for RO brine. Sep. Purif. Technol. 126, 82–94.
Wang, Q., Zhang, G.-S., Li, Z.-S., Deng, S., Chen, H., Wang, P., 2014b. Preparation and properties of polyamide/titania composite nanofiltration membrane by interfacial polymerization. Desalination 352, 38–44.
Wang, M., Wang, Z., Wang, X., Wang, S., Ding, W., Gao, C., 2015. Layer-by-layer assembly of aquaporin Z-incorporated biomimetic membranes for water purification. Environ. Sci. Technol. 49, 3761–3768.
Wang, L., Wang, N., Li, J., Li, J., Bian, W., Ji, S., 2016a. Layer-by-layer self-assembly of polycation/GO nanofiltration membrane with enhanced stability and fouling resistance. Sep. Purif. Technol. 160, 123–131.
Wang, W., Yu, J., An, W., Yang, M., 2016b. Occurrence and profiling of multiple nitrosamines in source water and drinking water of China. Sci. Total Environ. 551–552, 489–495.
Wanjale, S., Birajdar, M., Jog, J., Neppalli, R., Causin, V., Karger-Kocsis, J., Lee, J., Panzade, P., 2016. Surface tailored PS/TiO_2 composite nanofiber membrane for copper removal from water. Colloid. Interface Sci. 469, 31–37.
Wei, X., Wang, S., Shi, Y., Xiang, H., Chen, J., Zhu, B., 2014. Characterization of a positively charged composite nanofiltration hollow fiber membrane prepared by a simplified process. Desalination 350, 44–52.
Winiecka-Krusnell, J., Dellacasa-Lindberg, I., Dubey, J.B., Barraga, A., 2009. *Toxoplasma gondii*: uptake and survival of oocysts in free-living amoebae. Exp. Parasitol. 121 (2), 124–131.
Wu, H., Xia, S., Bu, F., Qi, J., Liu, Y., Xu, H., 2015. Identification of integrons and phylogenetic groups of drug-resistant *Escherichia coli* from broiler carcasses in China. Int. J. Food Microbiol. 211, 51–56.

Xu, G.-R., Wang, S.-H., Zhao, H.-L., Wu, S.-B., Xu, J.-M., Li, L., Liu, X.-Y., 2015. Layer-by-layer (LBL) assembly technology as promising strategy for tailoring pressure-driven desalination membranes. J. Membr. Sci. 493, 428–443.

Yang, J.X., Li, T., Ning, Y.Z., Shao, D.H., Liu, J., Wang, S.Q., Liang, G.W., 2015. Molecular characterization of resistance, virulence and clonality in vancomycin-resistant *Enterococcus faecium* and *Enterococcus faecalis*: a hospital-based study in Beijing, China. Infect. Genet. Evol. 33, 253–260.

Yebra, D.M., Kiil, S., Dam-Johansen, K., 2004. Antifouling technology—past, present and future steps towards efficient and environmentally friendly antifouling coatings. Prog. Org. Coat. 50, 75–104.

Yin, J., Zhu, G., Deng, B., 2016. Graphene oxide (GO) enhanced polyamide (PA) thin-film nanocomposite (TFN) membrane for water purification. Desalination 379, 93–101.

Yüksel, E., Karakeçili, A., Demirtaş, T.T., Gümüşderelioğlu, M., 2016. Preparation of bioactive and antimicrobial PLGA membranes by magainin II/EGF functionalization. Int. J. Biolo. Macromol. 86, 162–168.

Yurekli, Y., 2016. Removal of heavy metals in wastewater by using zeolite nano-particles impregnated polysulfone membranes. J. Hazard. Mater. 309, 53–64.

Yutin, N., Wolf, Y.I., Koonin, E.V., 2014. Origin of giant viruses from smaller DNA viruses not from a fourth domain of cellular life. Virology 466–467, 38–52.

Zeybek, Z., Binay, A.R., 2014. Growth ability of Gram negative bacteria in free-living amoebae. Exp. Parasitol. 145 (Suppl.), 21–26.

Zha, S., Zhang, G., Dawson, N., Yu, J., Liu, N., Lee, R., 2016. Study of PVDF/Si-R hybrid hollow fiber membranes for removal of dissolved organics from produced water by membrane adsorption. Sep. Purif. Technol. 163, 290–299.

Zhang, X., Lü, X., Zong, Z., 2012. *Enterobacteriaceae* producing the KPC-2 carbapenemase from hospital sewage. Diagn. Microbiol. Infect. Dis. 73 (2), 204–206.

Zhang, X., HU, J., Chang, Q., Wang, Y., Zhou, J.-E., Zhao, T., Jiang, Y., Liu, X., 2015a. Influences of internal coagulant composition on microstructure and properties of porous YSZ hollow fibre membranes for water treatment. Sep. Purif. Technol. 147, 337–345.

Zhang, Z.-B., Wu, J.-J., Su, Y., Zhou, J., Gao, Y., Yu, H.-Y., Gu, J.-S., 2015b. Layer-by-layer assembly of graphene oxide on polypropylene macroporous membranes via click chemistry to improve antibacterial and antifouling performance. Appl. Surf. Sci. 332, 300–307.

Zhang, L., Qin, J., Zhang, Z., Li, Q., Huang, J., Peng, X., Qing, L., Liang, G., Liang, L., Huang, Y., Yang, X., Zou, Y., 2016a. Concentrations and potential health risks of methyl tertiary-butyl ether (MTBE) in air and drinking water from Nanning, South China. Sci. Total Environ. 541, 1348–1354.

Zhang, Q., Lu, Z., Jin, S., Zheng, Y., Ye, T., Yang, D., Li, Y., Zhu, L., Zhu, L., 2016b. TiO_2 nanotube-carbon macroscopic monoliths with multimodal porosity as efficient recyclable photocatalytic adsorbents for water purification. Mater. Chem. Phys. 173, 452–459.

Zhang, X., Lang, W.-Z., Yan, X., Lou, Z.-Y., Chen, X.-F., 2016c. Influences of the structure parameters of multi-walled carbon nanotubes (MWNTs) on PVDF/PFSA/O-MWNTs hollow fiber ultrafiltration membranes. J. Membr. Sci. 499, 179–190.

Zhao, G., Chen, J., Huang, Y., Bryaskova, R., 2015. Novel antibacterial hybrid materials based on polyvinyl alcohol and mercaptopropyltriethoxysilane with embedded silver nanoparticles (PVA/AgNps/MPTES). C.R. Chim. 18, 586–592.

Zhao, H., Huang, M., Wu, J., Wang, L., He, H., 2016. Preparation of Fe_3O_4@PPy magnetic nanoparticles as solid-phase extraction sorbents for

preconcentration and separation of phthalic acid esters in water by gas chromatography–mass spectrometry. J. Chromatogr. B 1011, 33–44.
Zhou, C., Shi, Y., Sun, C., Yu, S., Liu, M., Gao, C., 2014. Thin-film composite membranes formed by interfacial polymerization with natural material sericin and trimesoyl chloride for nanofiltration. J. Membr. Sci. 471, 381–391.
Zhou, Z., Lin, W., Wu, X.-F., 2016. Electrospinning ultrathin continuous cellulose acetate fibers for high-flux water filtration. Colloids Surf. A 494, 21–29.
Zhu, L.-J., Zhu, L.-P., Zhang, P.-B., Zhu, B.-K., Xu, Y.-Y., 2016. Surface zwitterionicalization of poly(vinylidene fluoride) membranes from the entrapped reactive core–shell silica nanoparticles. J. Colloid Interface Sci. 468, 110–119.

13

GOLD NANOPARTICLES: ADVANCES IN WATER PURIFICATION APPROACHES

Kostantinos Pantapasis*, Alexandru Mihai Grumezescu**

**University Politehnica of Bucharest, Department of Biomaterials and Medical Devices, Faculty of Medical Engineering, Bucharest, Romania; **Department of Science and Engineering of Oxide Materials and Nanomaterials, Faculty of Applied Chemistry and Materials Science, University Politehnica of Bucharest, Bucharest, Romania*

1 Introduction

The thematic of this chapter is water, the most valuable resource on which life is dependent. This source is endangered by the pollution of the environment with different chemical toxins, inorganic, and organic pollutants, heavy and light metals. Most of the pollutants are being produced through artificial processes by humans, and we ought to monitor their levels in nature, in order to defend ourselves, as well as the ecosystem. The last decades, through the technological advancement and the use of nanotechnology combinatorial to biology and chemistry, monitoring the pollutants has become easier and more efficient. The use of nanoparticles (NPs) in environmental monitoring is being utilized and as a result, we can see how the maximum accepted levels of arsenic or lead have been reduced drastically the last 50 years from 200 and 100 ppb, respectively, to 10 ppb both. NPs have a size between 1 and 100 nm, due to their small size they present unique physical properties, mainly due to the strong confinement of electrons. Furthermore, due to their small size they present a very high surface to volume ratio, leading to a very high sensitivity and fast response making them the best option of material for sensor devices (Pradeep and Anshup, 2009).

The misuse of industrial waste and the general hazardous human life style lead to a rising demand for clean water, thus many researchers looked toward developing many different and efficient nanomaterials that can be efficiently used in that purpose.

Water Purification. http://dx.doi.org/10.1016/B978-0-12-804300-4.00013-7

A large number of NPs, such as quantum dots (QDs), magnetic nanoparticles (MNPs), dendrimers, carbon nanotubes (SWNTs and MWNTs), TiO_2 naocomposites, oxide NPs, Ag, and zeolites are used by now in water purification processes which involve filtration, disinfection, or desalination (Amin et al., 2014). Among these nanonaterials, AuNPs meet the most of the properties required for sensitive and specific determination and removal of many different categories of pollutants (Tiwari, 2015).

The continuously rising interest for AuNPs as nanoprobes used for the detection of many chemically differentiated pollutants, such as heavy metals, organic compounds, pesticides, toxicants (Koedrith et al., 2015), and pathogens (Tamer et al., 2013) is attributed to their high selective capacity to bind to specific natural or synthetic, organic or metallic compounds. Thus, they create new stable and dispersible gold nanostructures with selective or SERS properties, which after allow highly sensitive and selective target detection at great detection limits. The AuNPs exhibit unique optical properties as total of all collective oscillations of their surface electrons named surface plasmon resonance (SPR). The coherent electron oscillations excited by electromagnetic radiation, which appear at the interface between metal and dielectric are known as surface plasmons (SP). The properties of these interactions of light with the metal are dictated by the metal properties, frequency of electromagnetic radiation and the dielectric environment. When the excitation radiation frequency resonates with the SP frequency, at the metal-dielectric interface, the phenomenon of SPR will occur (Yu, 2013). The shape, size, and the composition of the AuNPs are factors, which determine the SPR frequency. For example, the resonance frequency of spherical AuNPs lies in the visible region of the EM spectrum. Due to the high surface area of AuNPs related to the volume ratio the SPR frequency of these NPs is highly depending on the dielectric nature of its interface with the surrounding medium (Maier and Atwater, 2005). Thus, changes to the environment of AuNPs, such as medium refractive index (dielectric nature), aggregation, and the absorption or desorption surface of chemical reagents, are shifting the SPR frequency resulting in colorimetric changes used further as detection signals (Wang et al., 2010). The SPR of AuNPs may cause an enhancement of the electromagnetic field of their local medium known as surface-enhanced Raman scattering (SERS). SERS AuNPs are known as extraordinary sensitive NPs used in detection and determination assays (Kneipp et al., 2006).

AuNPs are also good catalyzers used in many chemical transformations, such as hydrocarbon oxidation, hydrogenation-dehydrogenation, C–C coupling and redox. The oxidation of CO at low temperature assisted by supported AuNPs on various metal oxides is one of the model reaction used for the investigation of catalytic properties of AuNPs (Pradeep and Anshup, 2009). Another model reaction is the 4-nitrophenol reduction to 4-aminophenol in the presence of AuNPs and borohydride. This reaction takes place at room temperature, in aqueous solution and the AuNPs can be free or supported (immobilized) (MaríaJesús Lerma-García et al., 2014). Core–shell and alloyed heterometallic NPs containing AuNPs are an important type of bimetallic catalysts. For example, noble metal alloy NPs, such as Au–Pd NPs stabilized with PVP, have been employed as catalysts in the oxidation process of 2-buten-1-ol, benzyl alcohol, 1,4-butanediol, 1-butanal, and 2-butanol in aqueous solution and under mild conditions, showing high catalytic activities at the ratio 1:3 Au/Pd. Another example is the synergistic effect of Au–Ag bimetallic alloy as nanocatalyst in the oxidation of CO (Jiang and Xu, 2011).

From another point of view today the AuNPs can be produced by cheap and green synthesis methods, which represent an attractive choice to use them at a large scale. Their solutions are stable for long time and without special storage requirements. Furthermore, most of the AuNPs can be obtained and used straight in sequent analysis after they were synthesized. There are many reports on one-step citrate (Cit) reduction of gold salts resulting in AuNPs synthesis based on the classical methods of Turkevich and Frens (Zhong et al., 2015). Usually in this method, Cit is used as reducing and stabilizer factor in the synthesis protocol of AuNPs. The AuNPs obtained by this method are colloidal, their diameter size vary with many factors as pH, time and Cit concentration, are stable and exhibit high catalytic properties (MaríaJesús Lerma-García et al., 2014). To avoid the formation of toxic by-products in the synthesis process of AuNPs many ecofriendly synthesis have been reported. For example, Gopinath et al. (2016) reported the one-pot biosynthesis of multifaceted AuNPs using filtered extract from dried fruit of *Tribulus terrestris* in presence of chloroauric acid and Park et al. (2011) obtained AuNPs with various shapes (spheroidal, decahedral, and icosahedral) from gold chloride trihydrate ($HAuCl_4 \cdot 3H_2O$) and heparin (Park et al., 2011). G. Ghodake et al. achieved a green synthesis of AuNPs using amino acid asparagine (ASP) and after used them for the selective detection of arsenic in water while Boopathi et al. (2012) used as reducing agent tetraphenylborate and Xun et al. (2015) used polyacrylamide (PAM). Even more, AuNPs have been synthesized using sunlight irradiation directly from an aqueous solution containing hydrogen tetrachloroaurate trihydrate and *N*-cholyl-L-valine (NaValC) as a self-reducing and stabilizing agent (Annadhasan et al., 2015).

Beside their easy green synthesis, the AuNPs can be also functionalized with many biological and nontoxic compounds, such as amino thiol surfactants (Azzam et al., 2014), xanthine (Pu et al., 2014), and olelyamine (Bae et al., 2014). The functionalization of AuNPs is very important as the functional groups bonded on AuNPs surfaces offer stability and new properties to these NPs, which can be further employed for water pollutants determination.

The possibility to remove the NPs used for the detection of heavy metals or other water pollutants after they have been used is another urging factor for the water quality protection. In this regard, there are many reports on the possibility to determine AuNPs in environmental samples by using fluorescence or absorption-based methods (Zuber et al., 2016).

Trace metals are essential for human, animals, and plants as most of them are involved in physiological processes and the insufficient level of these metals can cause serious affections, such as iron deficiency anemia or in case of manganese deficiency bone malformations. However, chronic exposure to them or ingestion of excessive quantities can also lead to many affections or even death. Therefore, the nanotechnology can play an important role in the trace metals determination and removal providing new technologies and sensing tools. There are many conventional methods established concerning trace metal ions determination and their removal from polluted waters including reverse osmosis, chemical and electrochemical precipitation, adsorption, and ion exchange. Other methods, such as anodic stripping voltammetry, coupled plasma optical emission spectroscopy (ICP-OES), atomic absorption spectroscopy (AAS), fluorescence spectrometry, and inductively coupled plasma mass spectrometry (ICP-MS) are highly selective and sensitive toward trace metal ions but these methods are costly and time consuming. The need of rapid and selective on-spot techniques urges the development of quick and simple optical sensing methods. In this regard, colorimetric detection of metal ions has attracted large attention in many applications. This method is easy, quick, and the results can be observed at once with naked eye. Many colorimetric methods are highly sensitive and selective with good reproducibility and repeatability. AuNPs used as sensors can be coated with other metals, such as Ag, Pt, and Fe, functionalized with different ligands (thiolated) (Azzam et al., 2014) or other chemical compounds like heparine for example, forming to bimetallic NPs or to AuNPs/complexes, respectively. Most of the colorimetric assays are based in the induced and controlled aggregation or antiagregation (Hormozi-Nezhad and Abbasi-Moayed, 2014) of AuNPs in aqueous solution in the presence of target metallic ions. AuNPs aggregation may be labeled or label-free and depends on the AuNPs surface functionalization. The aggregation usually occurs due to the interactions between the detectable metal ions and the functionalization moieties or due to the penetration of cations in the interspatial gaps formed between the AuNPs complexes in aqueous solution.

The organic compounds represent another category of very toxic and persistent water pollutants. The most used methods for the removal of organic compounds from water samples based on the AuNPs technology are electrochemical, adsorption, and air-stripping methods.

2 Trace Elements

2.1 Iron

Iron (Fe) is one of the most abundant metal in the Earth's crust where it can be found mostly in its combined forms as the iron ions Fe^{2+} and Fe^{3+} combine very quickly with other compounds and form hydroxides, oxides, sulfides, and carbonates. Iron metal is known and has been used since ancient times. Iron compounds today have many industrial applications, such as steel production. Fe^{2+} can be detected by smell and taste in distilled water at 40 μg/L while in well-water concentrations lower than 0.3–3 mg/L are acceptable. Iron (II) salts are unstable in drinking water supplies and precipitate as a rust colored soil of iron (III) hydroxide.

Iron in the form of ferric cations Fe (III) represents an important trace element for many organisms and for humans. The iron minimum daily requirement range is between 10 and 50 mg/day depending on age and sex. With the oxygen, it forms the oxygen proteins transporters myoglobin and hemoglobin. It is also a component for the active site of redox enzymes. Iron deficiency leads to serious affections, such as iron deficiency anemia. Contrarily, chronic iron over uptake determines hematochromatosis (genetic disorder) and when ingested at doses of 40 mg/kg body weight iron is lethal due to hemorragetic necrosis (WHO, 2008).

The toxicity of iron is mostly discussed concerning its compounds. For example, CDC is giving a list with the exposure limits for different compounds, such as soluble iron salts such as, $FeSO_4$ (ferrous sulfate, Iron (II) sulfate), $FeCl_2$ (ferrous chloride, iron (II) chloride), $Fe(NO_3)_3$ (ferric nitrate, iron (III) nitrate) $Fe(SO_4)_3$ (ferric sulfate, iIron (III) sulfate), and $FeCl_3$ (ferric chloride, iron (III) chloride) are to the limit of TWA 1 mg/m^3. Another example represent the iron ion present in iron pentacarbonyl, chemical formula $Fe(CO)^5$ for which the exposure limit NIOSH REL exprimated in TWA is 0.1 ppm (0.23 mg/m^3) and ST 0.2 ppm (0.45 mg/m^3) (CDC, 2011a).

Conventional methods to detect and determinate iron are mostly spectrophotometric, such as ICPMS, atomic absorption spectrometry (AAS) or inductively coupled plasma atomic emission spectrometry (ICP-AES). Voltammetry has been also used for this purpose.

As the necessity to determine the iron presence in wastewater increased, the use of high sensitive chemosensors combining the binding of a metal with a nanomaterial became an attractive solution.

The fabrication of nanogold-Schiff base chemosensor with high selectivity for Fe^{3+} in a mixture of water/ethanol has been reported by Jimoh et al. (2015). AuNPs were coated with L-thiolated Schiff base ligand (L) and the AuNP/L obtained were used for iron determination in solutions.

Initially Cit-coated AuNPs were fabricated by Cit reduction of $HAuCl_4 \cdot 3H_2O$ in deionised water. AuNPs/Cit were further functionalized with 2-[(4-mercaptophenyl)imino methyl] phenol Schiff base (L) by one-step addition of this compound to the AuNPs/Cit, Fig. 13.1.

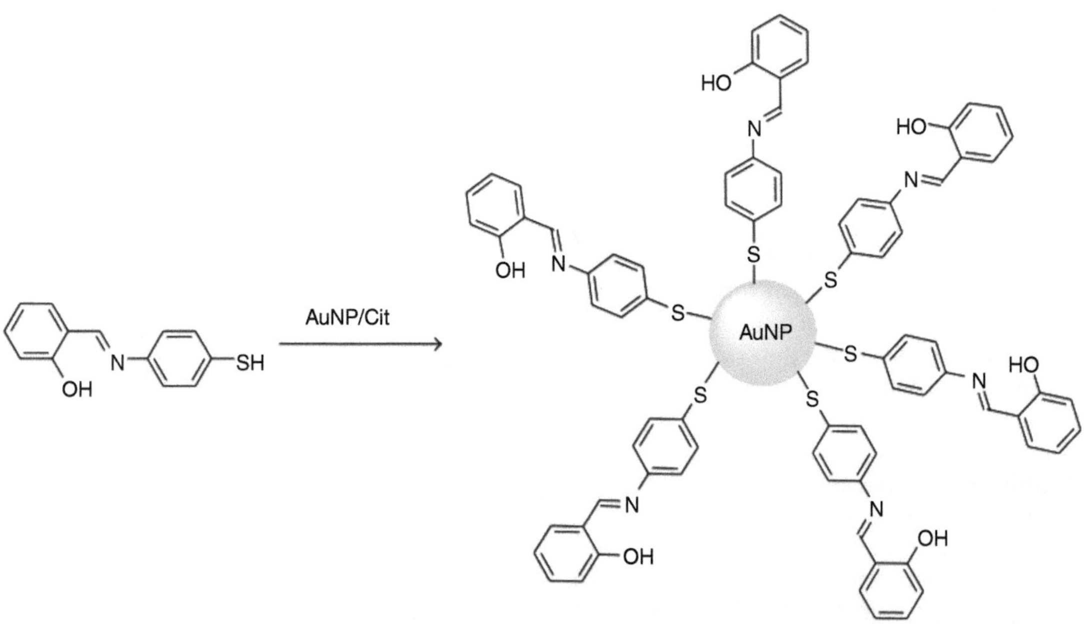

Figure 13.1. One-step functionalization of AuNPS/Cit with thiolatedbidentate Schiff base ligand (Jimoh et al., 2015).
Reprinted from an open access source.

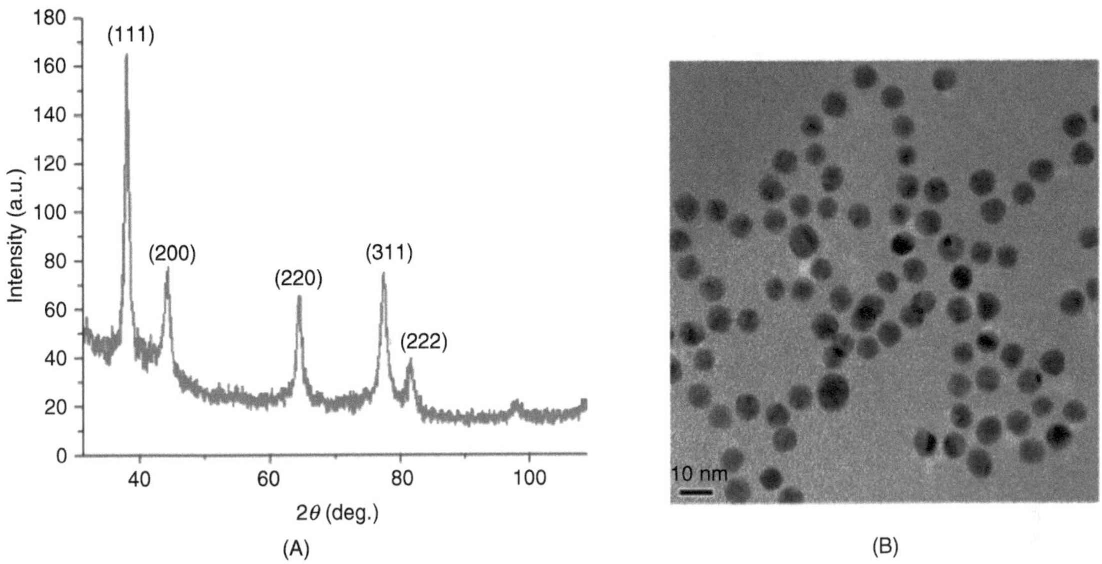

Figure 13.2. (A) XRD pattern and (B) TEM image of the synthesized AuNP/L (Jimoh et al., 2015). Reprinted from an open access source.

The AuNPs/L synthesized well dispersed, spherical with 8–10 nm diameters and uniform distributed as shown by TEM images and XRD pattern, Fig. 13.2.

As shown by fluorescent emission and UV-absorption studies, the AuNPs/L obtained exhibit high selectivity for ferric ions of 10 μM dimension, Fig. 13.3. In Fig. 13.3, the peak of SPR for

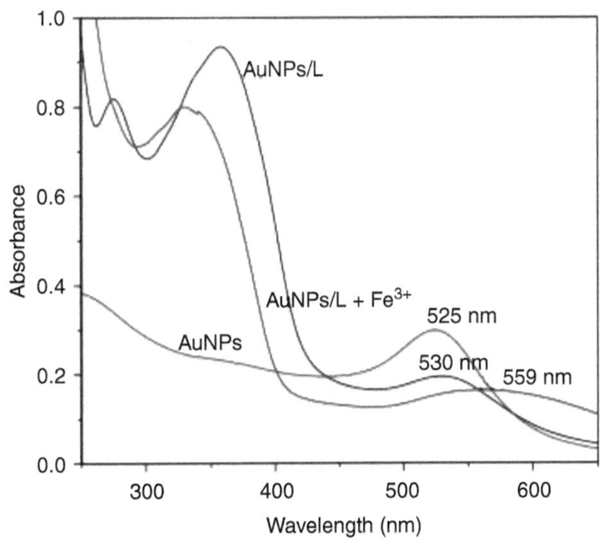

Figure 13.3. UV-Vis absorption spectra of AuNP/Cit, AuNP/L, and AuNP/L + Fe^{3+} (Jimoh et al., 2015). Reprinted from an open access source.

bare AuNPs is at 525 nm, which after the ligand addition moves to 530 nm. Furthermore, by adding the Fe^{3+} ions, the plasmonic absorption peak moves to 559 nm.

Solutions of Fe^{3+} with gradual concentration increase were added and the UV-spectra measured showed a quick change of the peak from 530 to 559 for higher concentrations indicating the formation of AuNPs aggregates induced by the iron ions, Fig. 13.4.

Photoluminescence studies showed that the Fe^{3+} added extinguished the fluorescent emission because of chelation-enhanced quenching, Fig. 13.5.

Studies upon selectivity of AuNPs for iron ions were effectuated in the presence of other cations. The results showed that the presence of Cu^{2+}, Zn^{2+}, Hg^{2+}, and Al^{3+} lead to a partial quenching of the fluorescent emission while the presence of F^{3+} a maximum of fluorescence extinguish was observed, Fig. 13.6.

This method has a detection limit of 1.2 μM and interferences from other metal ions were not observed. Absorption and emission titration showed a binding mode 1:1 with a binding constant 8.5×10^5 and 2.9×10^5 M^{-1}, respectively.

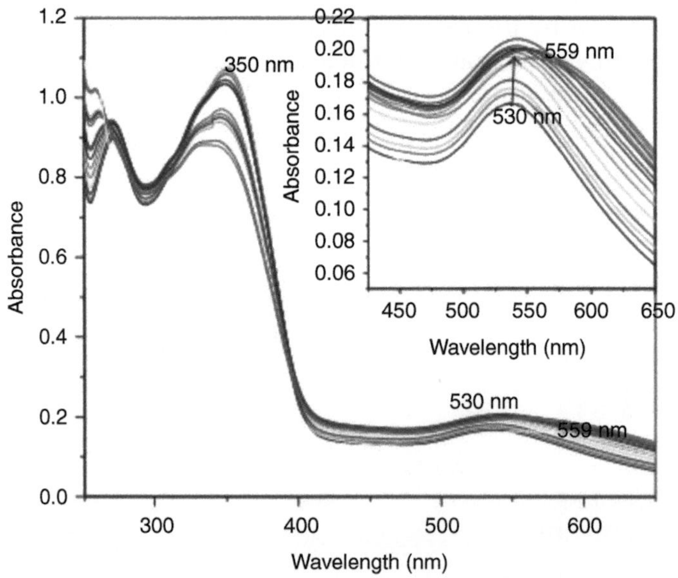

Figure 13.4. Evolution of the UV-Vis spectra of AuNP/L (10 μM) upon the addition of $Fe(NO_3)^3$ in a (9:1) $EtOH:H_2O$ mixture. Inset: exploded view of the shift of the plasmonic absorption peak from 530 to 559 nm with increasing ferric ion concentration (Jimoh et al., 2015). Reprinted from an open access source.

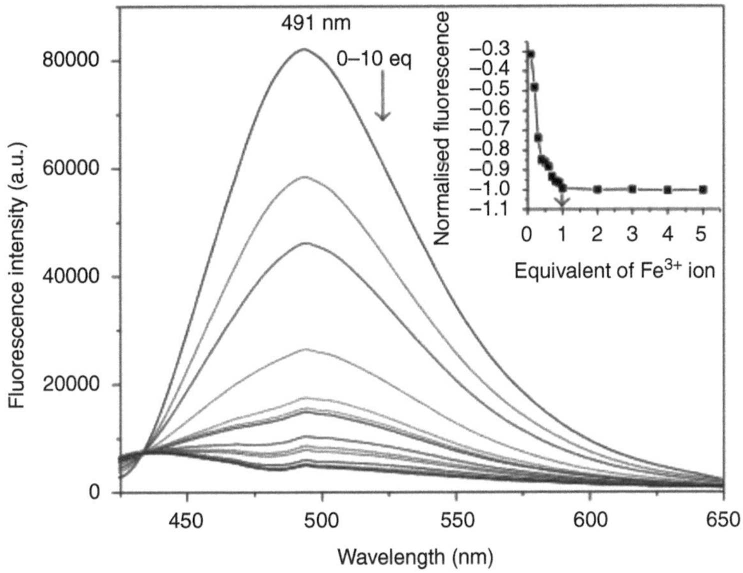

Figure 13.5. Fluorescence titration of AuNP/L (10 μM) H_2O:EtOH (1:9) (=390 nm). Inset: mole ratio plot of the emission at 491 nm (Jimoh et al., 2015). Reprinted from an open access source.

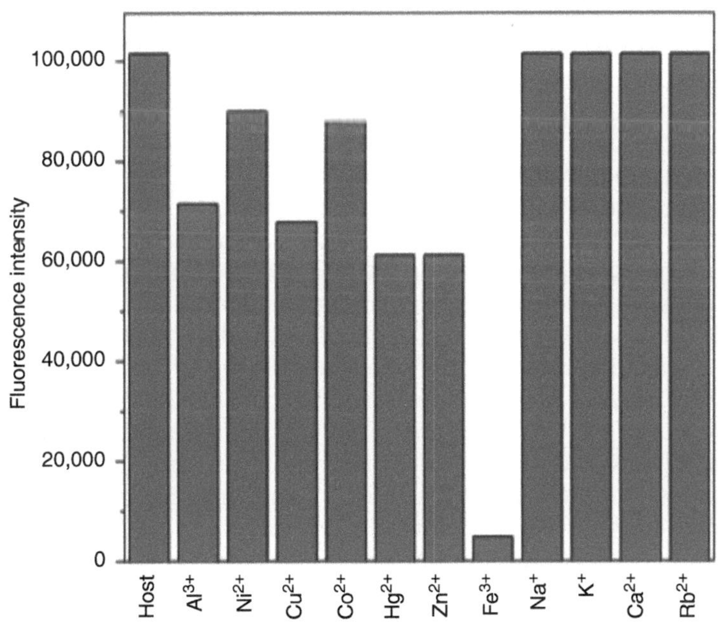

Figure 13.6. Metal-ion selectivity of AuNP/L; bars indicate the fluorescence intensity (excitation at 390 nm and emission at 491 nm). Nitrate salts of various metal ions (10.0 equivalents) were added to AuNP/L (10 μM) in H_2O:EtOH (1:9) (Jimoh et al., 2015). Reprinted from an open access source.

2.2 Copper

Copper is an essential trace metal as structural component or cofactor of many proteins and enzymes of the human body. However, like any other trace elements its excessive accumulation manifests high toxicity causing kidneys and liver damage, neurodegenerative diseases, and oxidative stress. In fresh water copper ions may naturally occur in concentrations between 0.20 and 30 ng/mL. In 2007, EPA revised the criteria concerning the concentration limits of copper in fresh water at 13 ng/mL for acute conditions and 9 ng/mL for chronic conditions (EPA, 2007).

In this regard the technologies used for the identification and removal of copper ions become very important especially the colorimetric methods which are simple, fast, selective, easily to be read out, portable, and low-cost.

There are many reports on colorimetric methods using AuNPs for Cu^{2+} detection in aqueous solutions based on the aggregation followed by antiaggregation of AuNPs with various functionalizations in the presence of Cu^{2+} ions. Yang et al. (2007) used AuNPs functionalized by L-cysteine. In this method, AuNPs of 16 nm average diameter coated with sodium Cit were obtained in aqueous solution. The method of Cit reaction is used to achieve AuNPs with intense SPR band, which can be better used in colorimetric assays. Further, these particles were functionalized with L-cysteine, which limits the aggregation or flocculation creating an electrostatic barrier between the particles. By adding Cu^{2+} to the L-cysteine functionalized AuNPs, the color of the solution changed immediately from red to purple which suggests the bonding of L-cysteine to the copper ions and the aggregation of the AuNP located in the close area. The AuNPs aggregation enhances with the increase of Cu^{2+} concentration shifting the band of surface plasmon absorption to a longer wavelength. The quantitative determination of Cu^{2+} by this method is lower than 10^{-5} M concentration range.

Possible interferences of other metal ions in the selectivity for Cu^{2+} of the AuNPs L-cysteine functionalized were investigated with Ag^{+}, Co^{3+}, Cr^{3+}, Cd^{2+}, Fe^{3+}, Zn^{2+}, and Ni^{2+}. From all these ions, only the Ni^{2+} affected the determination of Cu^{2+} but the effect was pronounced only at a larger than 50 mM concentration of Ni^{2+} ions.

Lu et al. (2012), also investigated the colorimetric detection of Cu^{2+} by Cit stabilized AuNPs in the presence of L-cysteine. Like other thiol compounds, L-cysteine can bind through Au—S bonds to the AuNPs surface causing their aggregation due to electrostatic interactions. In addition, in the presence of O_2, Cu^{2+} catalyzes the oxidation reaction of thiol compounds leading to disulfide compounds. The detection limit of Cu^{2+} in water was 10 nM. The selectivity of the method was investigated using other metal ions which presence is relevant in environmental water, such as Ag^+, Al^{3+}, Ba^{2+}, Ca^{2+}, Cd^{2+}, Co^{2+}, Fe^{3+}, Hg^{2+}, K^+, Mg^{2+}, Mn^{2+}, Na^+, Ni^{2+}, Pb^{2+}, and Zn^{2+}. The method showed high selectivity for Cu^{2+} against all these metal ions, except Co^{2+}, which showed slight interference but still, lower than the aggregation degree of AuNPs produced by Cu^{2+}. Thus, this method is highly specific for Cu^{2+}.

Besides the colorimetric methods used to detect Cu^{2+} based in the aggregation of AuNPs in aqueous solution there are also colorimetric strategies based on the size dependence of the NPs. Lou et al. (2011) detected trace of Cu^{2+} in an aqueous solution using label-free probes of AuNPs coated with silver Ag/AuNPs. This colorimetric method is based in the property of Cu^{2+} ions to accelerate the leaching rate of probes by the ions of thiosulfate ($S_2O_3^{2+}$) leading to a sudden decrease in the SPR adsorption due to the decreasing size of the Ag/AuNPs. A partial dissolving of the NPs was observed when the probes Ag/AuNPs reacted with ($S_2O_3^{2-}$) due to easy oxidation of Ag by the O_2 dissolved in solution leading to Ag $(S_2O_3)_2^{3-}$ complexes. In the presence of $S_2O_3^{2+}$, Cu^{2+} also oxidize Ag and Au. As a result, the size of the probe NPs decreased, the SPR absorption also dramatically decreased making possible the quantification of copper ions from aqueous solution. This colorimetric method is rapid, highly selective, and sensitive. The sensitivity toward copper ions is 1.0 nM with a linear detection range between 5 and 800 nM and has been successfully used in real application for Cu^{2+} recovery from tap water (82–115% with RSD 0.66–1.47%) and pond water (81–119% with RSD 1.01–1.68%).

2.3 Cobalt

Cobalt is a naturally occurring element and it can be found only in small amounts in water, soil, animals, and plants. It is used in the preparation of magnetic alloys, batteries, inks, pigments, prosthetics, and electronics or as oxidation catalysts in chemical reactions.

Cobalt is one of the essential trace elements for healthy humans and animals as it is the active center of the coenzymes named cobalamine (vitamin B12).

In excessive accumulation of Co^{2+} in organism, health effects may occur, such as pneumonia, asthma and wheezing, allergies to Co^{2+}, skin rashes, liver damage, and renal failure. The OSHA permissible exposure limit (PEL) in the workplace (TWA) for Co^{2+} is 0.1 mg/m^3 and NIOSH recommended exposure limit (REL) TWA is 0.05 mg/m^3. The immediately dangerous to life and health (IDLH) value is 20 mg/m^3 (CDC, 2011b).

Traditional methods used to determine the presence of Co^{2+} are atomic adsorption spectrometry (AAS), voltammetry and ICP-MS with limited possibility for on-site monitoring. Thus, once again colorimetric assays are taken in consideration as facile, rapid, and sensitive methods for the determination of Co^{2+}.

AuNPs stabilized with further capping or functionalization have been used in many investigations to determinate Co^{2+} by colorimetric assays. Usually in these methods, labeled AuNPs are used to form ion-ligand complexes inducing their aggregation. However, these methods necessitate complex modifying processes (labeling process) and specific equipments. As shown for the detection of Cu^{2+} the label-free colorimetric method took advantage over the labeled one as the metal ions themselves improve the leaching of AuNPs or induce their aggregation.

Zhang et al. (2012) developed a label-free AuNPs method for determination of Co^{2+} at room temperature in aqueous solution. They used AuNPs stabilized with ($S_2O_3^{2-}$) and ethylenediamine (en). Briefly the method consists in the reaction of Co^{2+} with (en) with the formation of $Co(en)_3^{2-}$ which further was oxidized by the O_2 dissolved in the aqueous solution giving $Co(en)^{33+}$. Subsequently the ($S_2O_3^{2-}$) absorbed on the surface of AuNPs reacted with $Co(en)_3^{3+}$ leading to positively charged ion $(en)_2CoS_2O^{3+}$. The final result was the aggregation of AuNPs expressed by change of color of the aqueous solution from red to dark blue.

The calibration graph was linear within the range of 0.1–0.7 mM for Co^{2+} and the detection limit calculated was 0.04 mM (2.36 ppb) (3*s* rule, where *s* is the standard deviation of a blank solution and $n = 5$).

The selectivity of the method was tested in the same way by comparing the sensitivity of the method for Co^{2+} with that of the following ions: Ag^{+}, Al^{3+}, Ca^{2+}, Cd^{2+}, Cr^{3+}, Fe^{3+}, Hg^{2+}, K^{+}, Mg^{2+}, Mn^{2+}, Ni^{2+}, Pb^{2+}, and Zn^{2+}. The method was selective toward Co^{2+} except for Cu^{2+}.

An even more rapid, sensitive end economical colorimetric method for the determination of Co^{2+} was reported by Raghav and Srivastava (2015). They used Cit stabilized AuNPs without any further capping or functionalization, in aqueous solution, by adding increasing concentrations of cobalt chloride solution. In this way metal ion, Co^{2+} was added to the colloidal solution of AuNPs in ratio 1:1 v/v. After 15 min incubation, the probes were analyzed and the Co^{2+} was qualitatively and quantitatively evaluated. With the addition of Co^{2+} in equal volume to the solution of monodispersed AuNPs a progressive change of the initial red color from red to pinkish purple, purple and blue was observed. The changing in the color is due the formation of aggregates with chain like structures induced by the interactions between the negatively charged AuNPs and the divalent cations. These structures were analyzed by spectrophotometry and the length scale of the complexation was directly correlated to the Co^{2+} concentration with linear range from 0.1 to 15 μM, leading to a semiquantitative visual scale as: red (<1 μM)—pinkish purple (0.1–2 μM)—purple (5–15 μM)—blue colored beyond 0.25 mM Co^{2+} concentration. The method showed high sensitivity over other metal ions, such as K^+, Na^+, Ca^{2+}, Mn^{2+}, Mg^{2+}, Fe^{2+}, Cu^{2+}, and Ni^{2+}.

Recently Annadhasan et al. (2015) reported the green synthesis of AuNPs under sunlight irradiation and their use for Co^{2+} and Ni^{2+} colorimetric detection in environmental samples. The AuNPs were obtained from $HAuCl_4$ solution in the presence of sodium salt of *N*-cholyl-L-valine (NaValC). Quantitative analyses were separately proceeded by adding increasing concentrations of the Co^{2+} or Ni^{2+}. A gradually decrease in the color of the solutions as a function of metal ions concentrations was observed for both Co^{2+} and Ni^{2+}. After the studies carried out by different methods, such as UV-visible, TEM, DLS studies, and zeta potential measurements the authors concluded in the possible aggregation of AuNPs in aqueous solution through metal–ligand interactions. Consequently the changes in the color of the solution for Co^{2+} from faded pink to pale pink and for Ni^{2+} from pink to purple and after to violet could be directly correlated with the increasing concentrations of the two metal ions in the solutions. For both Co^{2+} and Ni^{2+} the linearity was achieved in the interval from 5 to 40 nM and the LoD was 10 nM. The sensitivity of the method for real samples was tested. The LoD with Co^{2+} for drinking water samples was 20 and 10 nm for tap water samples while for Ni^{2+} was 15 and 10 nM, respectively.

Comparative with cobalt environmentally found in traces, the nickel is an abundant natural element. It is used in the fabrication of stainless steel and, in combination with other metals, to form alloys. Nickel alloys are used to make coins, heat exchangers, and valves. Other nickel compounds are used for batteries, to color ceramics, or as catalysts (ATSDR, 2011). In water, nickel may occur as Ni^{2+} or $NiCO_3$ and it can be dissolved or bounded in different complexes.

Today nickel is also recognized as essential trace element for bacteria, archaea, fungi, algae, and plants playing an important role as constituent of some enzymes, such as urease, NiFe hydrogenase, glyoxalase I, and superoxide dismutase (Astrid Sigel and Roland, 2008). The human body needs nickel as trace element included in multiple vitamins for osteoporosis treatment and to increase iron absorption. Possible over exposure from air, water, and food can lead to very serious affections of blood vessels, heart, skin, lungs, and immune system (Di Gioacchino et al., 2014). High concentrations of Ni^{2+} compounds are carcinogenetic. The standard exposure limits NIOSH REL are TWA 0.015 mg/m^3 and OSHA PEL TWA 1 mg/m^3 (not applied for nickel carbonyl) (CDC, 2011c).

Beside, the model of aggregation method of Annadhasan et al. (2015) described prevoiusly, a platform based on dual combination of AuNPs complexes designed for multimodal colorimetric detection and SERS sensing of Ni^{2+}, was achieved by Krpetić et al. (2012). AuNPs functionalized with L-carnosine were mixed together with AuNPs functionalized with nitrilotriacetic acid (NTA). The particle core sizes were different thus an asymmetric and symmetric assembly was achieved. As in aqueous solution between the two different types of functional groups, there are no unspecific interactions, in the absence of Ni^{2+}, both of functionalized AuNPs were stable and aggregation-free. With the addition of Ni^{2+} into the AuNPs functionalized mixture, Ni^{2+} is coordinated toward specific sites present on the nanoparticle surfaces and a fast aggregation occurs. The aggregation induces both an increase in the SERS intensity from Raman reporter and a change in the extinction profile. Comparative, the LoD for each modality were: 0.5 ppm for the measurements with SERS and of ~5 ppm extinction measurements.

The surface-sandwich method using NPs to enhance the sensitivity of SPR has been used for the detection of proteins (Kwon et al., 2012) and nucleic acids (D'Agata et al., 2011; Lee et al., 2011; Wang et al., 2011). This model has been applied by Kim et al. (2012) for Ni^{2+} detection in water probes. Enhanced selectivity was achieved using two selective toward Ni^{2+} different ligands, the polyhistidine (polyHis), and the *N*-[5-(3′-maleimidopropylamido)-1 carboxypentyl]iminodiacetic acid (maleinimido-C3-NTA) which were designed in a surface sandwich assay. The AuNPs employed were quasispherical and their functionalization was made with polyhistidine (polyHis). The polyHis/AuNPs were specifically adsorbed onto SPR chip surface functionalized with NTA and in the presence of Ni^{2+}.

This new sensing platform designed is selective and sensitive and offer the in situ detection of Ni^{2+} at the very low concentration of 50 parts per trillion (211 pM). The selectivity for Ni^{2+} was tested performing various measurements in the presence of other ions such Pn^{2+}, Cu^{2+}, and Zn^{2+}. The results showed a good sensitivity toward Ni^{2+} except Zn^{2+}.

2.4 Manganese

Manganese is one of the most wide spread and abundant metals in the environment and is mostly used in iron and steel production. Combined it is used in products with catalytic or disinfectant properties. Manganese oxide is used in ceramics and fertilizer production. It is also an essential trace element for plants, animals, and the human body where it is found in kidney, liver, pancreas, and bones. It plays an important role in the calcium absorption, carbohydrate metabolism, and blood sugar regulation. Manganese is also essential for normal nerve and brain function (Xun et al., 2015; Pawlas and Sas-Nowosielska, 2015). Low levels of this metal are associated with bone malformations, infertility in adults, weakness. However, it also may create health problems at too high concentration in the body. As demonstrated, it may cause sleeping disorders, fatigue, headaches, and neurasthenia (Hu et al., 2014). Higher level than 14 mg/L in drinking water can cause manganism, a Parkinson syndrome resembling disease (Jiang et al., 2006). As a result, there are strict health and environmental regulations, which control the possible water and human body contamination. For example, the concentration of manganese in bottled drinking water established by FDA should not exceed 0.05 mg/L (ATSDR, 2012). In the United States the concentration of Mn^{2+} in drinking water allowed by the Environmental Protection Agency should not exceed 0.05 mg/L (0.9 mmol/L).

As mentioned earlier, colorimetric methods for the detection of metallic ions gained the attention of many researchers over the last years due to their relative facility and high selectivity. Colorimetric assays for the detection of Mn^{2+} have been also successfully applied. Recent reports refer the use of AuNPs obtained by green strategies, which further can be involved in colorimetric detection by cross-linking aggregation in the presence of Mn^{2+}. Narayanan and Park (2014) reported the use of AuNPs using as stabilizer L-dopa, (L-3,4-dihydroxyphenylalanine) which present multifunctional groups, such as carboxylic, amino, and hydroxyl and can reduce Au^{3+} to metal NPs. This AuNPs/dopa sensor has a high selectivity toward Mn^{2+}. The results showed that at 700–500 nm, the extinction ratio of absorbance is linear against the concentration of Mn^{2+} in the range of (5–100 μM) and the calibration curve is given by

$$A_{700} / A_{550} == 0.09561 + 0.00751 \times C_{\mathrm{Mn}} (R = 0.99)$$

Based on the same principle, Xun et al. (2015) reported the synthesis of AuNPs functionalized with PAM and demonstrated that the presence of Mn^{2+} determinates the cross-linking of the NPs. The linearity of the system is obtained in the range 8.0–500 mmol/L and the LoD 5.0 mmol/L.

The increased require in clean water emerged the researchers toward more selective and sensible technologies using novel materials and technologies coupled with AuNPs nanotechnology. An example is the use of fiber gratings (FB) as fiber Bragg gratings

(FBGs) coated with a thin layer of AuNPs. The FBGs are FB with a periodic refractive index modulation less than 100 μM. They act as filters and reflect a spectrally narrowband portion of the incoming broadband light. FBGs have been employed for chemical sensing as solution concentration or refractive index sensors. In the last years, NPs of noble metals have been used to enhance the sensitivity of fiber-optic sensors. Akki et al. (2011) achieved the fabrication of a FBG chemical sensor for the detection and determination of Mn^{+} concentration in water. Further, the reflected spectrum sensitivity of the FBG was enhanced by a thin layer of AuNPs to form a thin layer above the etched grating area.

The head of the FBG sensor was coated by immersing for 4 h the etched portion in the aqueous solution of AuNPs of 4–12 nm diameter. To determine the concentration of Mn^{2+} standard aqueous solutions of Mn^{2+} with different concentrations (0.04–0.09 ppm) were prepared and the reflected spectra of FBG for each concentration were recorded.

The graph of the Bragg wavelength (λ_B) can be represented as function of the concentration of Mn^{2+} solutions and the linearity of the shift in (λ_B) can be observed.

2.5 Vanadium

Vanadium (V) is an abundant naturally occurring element and its average concentration in the Earth's crust is estimated at about 100 mg/kg. It can be found in fossil fuel or minerals deposits and it is used to produce steels for high-speed tools, superconductive magnets (as catalyst in vanadium pentoxide form) or in ceramics. Large amounts of vanadium ions released in water, soil, and air, are from industrial sources (ATSDR, 2015a).

Vanadium is a trace element essential in rats' growth, in roots and stems of plants (Akoumainaki Ioannidou et al., 2015). Vanadium is used as dietary supplement for humans in vanadyl sulfate form to improve the control of glucose in type 2 diabetes but its role is yet not clear. Contrarily, there are many adverse effects observed in the use of vanadium compounds in mammals, such as DNA breakdown, congenital aberrations, nausea, stomach cramps in people taking medical compounds including vanadium for treatment of diabetes. Laboratory studies with mice showed a decrease in the red blood cells number, neurological effects, and lung cancer.

People may be exposed to vanadium through ingestion, inhalation, and eye or skin contact. The surface water concentration of vanadium geographically differs and can be from 0.04 to 220 μg/L. The average concentrations of vanadium in tap water are about

0.001 mg/L and daily vanadium intakes from food can range from 0.01 to 0.02 mg (ATSDR, 2015a).

The NIOSH REL vanadium dust exposure limit for 15 min, to all vanadium compounds but not for vanadium carbide and vanadium metal, is 0.05 mg V/m^3 and OSHA PEL 0.5 mg (for V_2O_5 m^{-3}) (CDC, 2011d). ATSDR inhalation minimal risk level (MRL) are 0.0008 mg/m^3 for acute exposure and 0.0001 mg/m^3 for chronic exposure.

Many methods have been used to investigate vanadium traces, such as chromatographic, spectrometric, and electrochemical assays. Also, biosensors based on screen printed electrodes (SPCEs) enzymes modified have been reported. Alvarado-Gámez et al. (2013) investigated the determination of vanadium ions in water in a chronoamperometric assay. Chronoamperometric biosensors SPCEs were coated through direct electrochemical deposition of AuNPs (AuNPs-SPCEs) on their surfaces. Further, the alkaline phosphatase (ALP) enzyme (metalloenzyme sensitive to vanadium) was entrapped with polyvinyl alcohol (PVA) on the surface of AuNPs-SPCEs (ALP-biosensor). The substrate used was riboflavin 5-monophosphate sodium salts as vanadium ions affect the enzymatic activity over this compound. In the determination process, the enzyme first hydrolyzed riboflavin 5-monophosphate to riboflavin giving rise to an oxidation current. The addition of vanadium as inhibitor of ALP activity results in a decrease of the current proportional to the amount of vanadium present in the solution which permits the quantitatively estimation of vanadium. The detection limit of the method is of 3.9 ± 0.4 μM. The reproducibility was about 11.6% (n = 4) and the repeatability about 7.6% (n = 4). Possible interferences occur in the presence of Mo (VI).

2.6 Tungsten

Tungsten, known as Wolfram (W) is also a naturally occurring trace element that can be found almost exclusively as chemical compounds in sediments and soil in concentrations between 0.2 and 2.4 mg/kg (El-Shahawi and Al Khateeb, 2012). Its alloys are used in many industrial applications, such as the fabrication of X-ray tubes, filaments for the light bulbs, gas tungsten arc welding, and superalloys. Tungsten provenience in water is mainly from its dissolution after soil and rocks erosion. The soluble tungsten may enter groundwater at pH greater than 7 (ATSDR, 2015b).

Tungsten is also known as the heaviest element with biological functionality. Tungstate anion may occur in soil due tungsten metal oxidation from where it can be took over by some prokaryote organisms, such as *archea* bacteria, to sustain certain enzymes activities (sometimes as substitute of molybdate). An enzyme known to be dependent on tungsten is formate dehidrogenase found in *Clostridium thermoaceticum* bacteria (Kletzin and Adams, 1996).

Although the toxicity of tungsten has been less investigated, there are limitations to tungsten exposure. NIOSH REL exposure limit TWA is 5 mg/m^3 and short-term exposure limit (ST) is 10 mg/kg for insoluble tungsten while OSHA stats the time weighted average (TWA) of 1 mg/m^3 and the short-term exposure limit (STEL) of 3 mg/m^3 for soluble tungsten (CDC, 2011e). Also, in 2012 it has been reported that as trace element tungsten may be toxic for animals and humans since 5 μg/kg were enough to cause animal embryos death (El-Shahawi and Al Khateeb, 2012).

However, the effects of tungsten upon the environment and human health have been poor investigated therefore activities in these regard along with the regulation of waste disposal are strongly required in the next future.

Most of the techniques used to analyze the presence of tungsten are spectrophotometric methods like inductively coupled plasma atomic emission spectroscopy (ICP–AES) and atomic absorption but polarographic and voltammetric methods have been also reported (Johannesson et al., 2013). After their report on the use of chronoamperometric biosensors SPCEs for vanadium determination in tap water samples in 2013, the group of Alvarado-Gámez, investigated the same method for tungsten analysis with AuNPs coated biosensors (Alvarado-Gámez et al., 2015). The method differs in the substrate used to investigate ALP activity and its inhibition in the presence of tungsten thus both the activity of ALP supported by 2-phospho-L-ascorbic acid trisodium salt and the possibility to determinate tungsten ions in the solution were investigated.

The method had a limit of detection of 0.29 ± 0.01 μM, with the quantitation limit of 0.58 ± 0.02 μM. Unfortunately at low concentrations of 1.0 μM interferences of metals like As(V), Se(IV), Al(III), Sn(II), and Fe(III) have been recorded.

3 Organic Compounds

The presence of organic compounds in water represents a great matter because of their persistence in the natural medium, toxicity, and the possibility to invade the food chain of animals and human leading to various and serious health hazards. The conventional methods used to remove organic compounds dissolved in water or organic liquids include electrochemical treatment, air-stripping, adsorption, and ultrafiltration (Qian et al., 2013).

Between the many organic compounds found in polluted waters the phenols represent one of the most toxic chemicals for the environment and humans. In this regard many of these compounds have been registered and included in legislations regarding the environment safety and health hazardous. For example, the phenol (C_6H_5OH) with the structural formula is known also with the synonyms and trade names as phenyl alcohol carbolic acid or hydroxybenzene is present in CDC NIOSH Pocket Guide to Chemical Hazards catalog with the CAS No. 108-95-2 and the RELs allowed are (TWA) 5 ppm (19 mg/m^3) C 15.6 ppm (60 mg/m^3) (15 min, skin). Some of the health effects in case of exposure to phenol are kidney and liver damage, dermatitis, skin burns, and nervous system disturbance (CDC, 2011f)

Many of the phenols and substituted phenols are accumulative, toxic, and environmentally persistent chemical compounds and so their elimination or recovery is the outmost importance.

The nitrophenols, substituted phenols, are organic compounds extensively employed in many agricultural, industrial, and defense applications. They are used in a large variety of industrial processes, such as explosive manufacture, pesticides, pharmaceuticals, preservatives, rubber, dyes, and pigments. They also may occur after microbial hydrolysis of organophosphorous pesticides (Thabile Ndlovu et al., 2010). For example, *p*-nitrophenol (one of the most toxic water pollutant) can be produced in the enzymatic hydrolysis process of parathion organophosphorous insecticide (Guo et al., 2015).

Nitrophenols can be detected through chromatographic methods which can be also associated with spectrophotometry and mass spectrometry but these techniques are expensive regarding their analytical cost, need time for the sample preparation and for the analysis.

Here two of the isomers of nitrophenol ($C_6H_5NO_3$) are discussed regarding their toxic activity and the methods used for their detection in water samples. Their structural formula is given in Fig. 13.7.The o-nitrophenol (or 2-nitrophenol) is a nitrophenol mainly used in the production of rubber chemicals, fungicides, dyes, and paint colorings. Exposure to this compound may cause

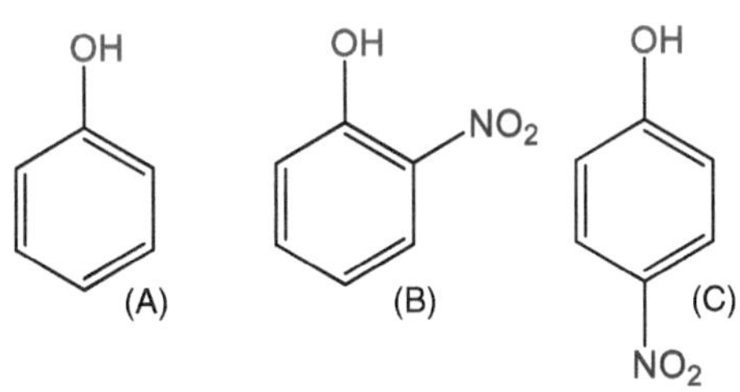

Figure 13.7. Structural formula of (A) phenol, (B) o-nitrophenol, and (C) *p*-nitrophenol.

headache, nausea, and cyanosis. It can be also absorbed through intact skin. The *p*-nitrophenol (or 4-nitrophenol) ($C_6H_5NO_3$) is severely toxic. The allowed limit EPA provided for *p*-nitrophenol in drinking water is 60 ppb (about 0.43 μM).

The catalytic activity of AuNPs free or supported has been the subject of many reports concerning the use of Au NPs as catalysts for the reduction of *p*-nitrophenol to 4-aminophenol by excess of borohydride since this method represent one of the model reaction frequently used to investigate the catalytic properties of AuNPs. For example, AuNPs synthesized using different phenolic compounds, as described by MaríaJesús Lerma-García et al. (2014), were employed to reduce *p*-nitrophenol to 4-aminophenol in the presence of $NaBH_4$ in excess.

In an experiment carried out by Lin et al. (2013), the catalytic properties of Au/Al_2O_3 were investigated for *p*-nitrophenol reduction by $NaBH_4$. The catalyst Au/Al_2O_3 was prepared at room temperature in a simple method by adsorbing the colloidal AuNPs onto Al_2O_3 (Lin et al., 2013). The complete adsorption of AuNPs is dependent on the solution pH, which has to be lower than the isoelectric point of Al_2O_3. The UV-VIS spectra shown in Fig. 13.8,

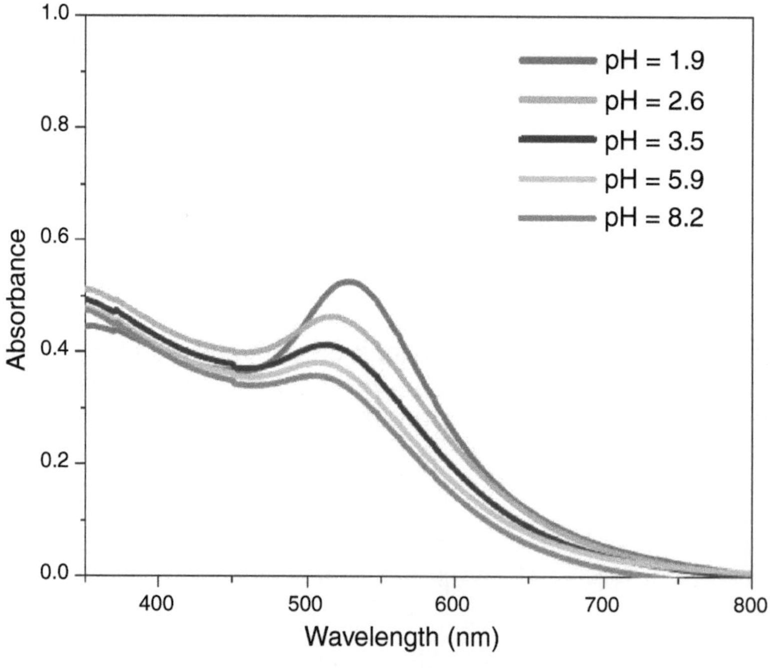

Figure 13.8. UV-Vis spectra recorded after adjusting the pH value of a 5.7 nm Au colloid solution by different amounts of aqueous HCl (Lin et al., 2013). Reprinted from an open access source.

reflect the dependence of the NPs growth on the pH adjustment. For example, by decreasing the values of the solution pH from 8.2 to 1.9 the absorption band of the 5.7 nm AuNP used became red-shifted and sharper, indicating their growth in very acidic solutions (Lin et al., 2013).

The growth of NPs was confirmed also by TEM, as shown in Fig. 13.9. The initial average size of the AuNPs was 5.7 nm remaining the same with the pH adjustment up to 5.9 and the AuNPs could be adsorbed into the catalyst in contrast with the condition of pH adjustment to 1.9 when the average size of the NPs was 8.7 nm.

Further investigation on the influence of the AuNPs size on *p*-nitrophenol reduction showed that the optimal size of the AuNPs used to synthesize the Au/Al_2O_3 catalyzer was at 3.4 nm. As authors concluded this catalyzer showed high activity at pH 5.9 and

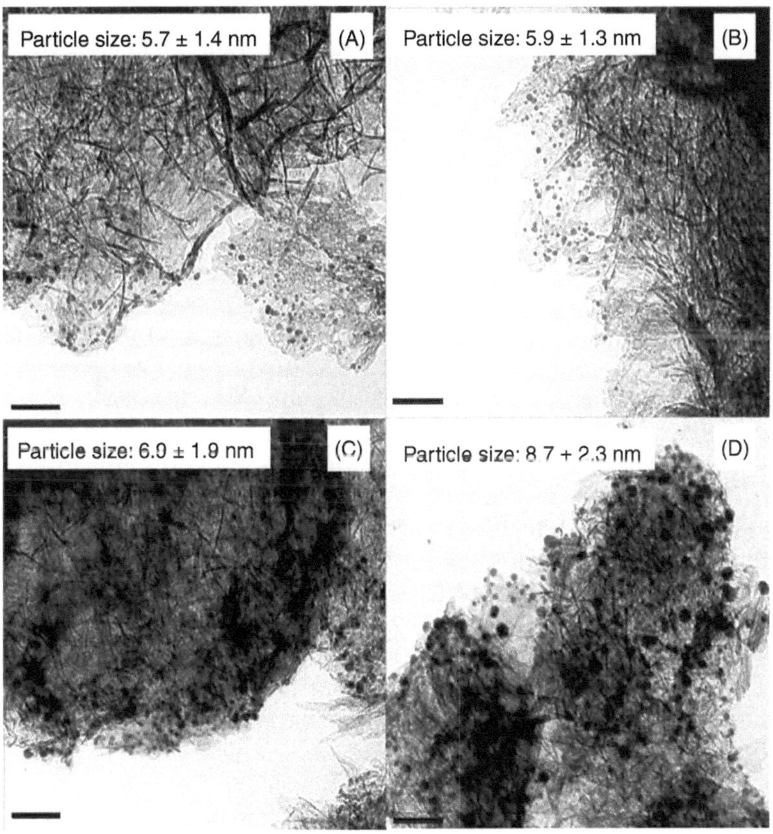

Figure 13.9. TEM images of Au/Al_2O_3 catalysts with Au colloid solutions adjusted to different pH values. (A) pH = 5.9; (B) pH = 3.5; (C) pH = 2.6; (D) pH = 1.9. The scale bars represent 50 nm. The size of the original Au colloid before pH adjustment is 5.7 nm (Lin et al., 2013). Reprinted from an open access source.

with 3.4 nm AuNPs with a turnover frequency (TOF) of 2.06×10^{-2} (moles/g s) (Lin et al., 2013).

Pozun et al. (2013), demonstrated that not only monometallic NPs but also bimetallic NPs can be used as catalyzers in the reduction reaction of *p*-nitrophenol to *p*-aminophenol by $NaBH_4$. For this purpose using poly(amido)amine dendrimers (DENs), they carried out a precise, uniform, and simple method to synthesis different bimetallic NPs of Au/Cu, Pd/Au, and Pt/Au. In this way, DENs are monodispersed and the changes in the elemental metallic combination do not interfere in their size distribution. Further, the authors demonstrated that the catalytic properties of DENs particles are correlated with the binding energy, which makes possible the approximation of the reaction rates of *p*-nitrophenol reduction (Pozun et al., 2013).

Electrochemical methods for the detection of nitrophenols in waste water using AuNPs or materials containing AuNPs are used on a large scale because they are low cost, require cheap instrumentation, are simple and easy to operate offering real time detection. Many research concentrate in the use of AuNPs as they are known for their increased electrode area and high potential to enhance conductivity. Thabile Ndlovu et al. (2010), carried out a reproductible and simple method for the electrochemical determination of o-nitrophenol. Using cyclic voltammetry (CV) they managed to codeposit AuNPs from a solution of $HAuCl_4$ and a dendrimer of generation 2 (G2) poly(propyleneimine) (PPI) on a glassy carbon electrode (GCE) (Thabile Ndlovu et al., 2010). The modified electrode GCE/PPI-AuNP was further electrochemically characterized by CV and square wave voltammetry (SWV) comparative with GCE and GCE/PPI electrodes. After they analyzed the response of o-nitrophenol at different concentrations in PBS (pH 5).

The modified electrode was stable and with a good reproducibility, demonstrated by the calibration plot. A good linearity was obtained in the range from 6.1×10^{-7} to 6.25×10^{-4} mol/L of o-nitrophenol. The platform showed a detection limit of 4.5×10^{-7} mol/L.

Tang and Chen (2011), carried out an experiment in which two different electrodes were employed to detect the presence of *p*-nitrophenol in acidic media. The electrodes used were a glass electrode coated with indium tin oxide (ITO) and a screen-printing carbon (SPC) electrode. Both electrodes were coated with AuNPs through potential-sweeping electrodeposition method to improve their electrode performance. Experimental studies demonstrated that the electrode performance of *p*-nitrophenol reduction depends on the size distribution and particle density of electrodeposited AuNPs. Therefore, the AuNP-ITO and AuNP-SPC modified electrodes were fabricated with AuNPs with narrow size distribution, moderate particle size, and high particle density. Each electrode was tested to detect for 30 times, 50 μM of *p*-nitrophenol through CV in acidic solution due the necessity of protons in the *p*-nitrophenol reduction. The ratio of ip (*p*-nitrophenol cathodicpeak currents) and ip_{max} (*p*-nitrophenol maximum cathodic peak current) from the 30 detections. The relative standard deviation (RSD) of AuNP-ITO electrode was 0.65% while for AuNPs-SPC electrode was 2.7% indicating that the AuNP-ITO electrode presented a better stability than the AuNP-SPC electrode (Tang and Chen, 2011).

The experiments made to test the possible interferences of other organic compounds or inorganic and metal ions shown a good selectivity of the tow modified electrodes concerning the inorganic ions, such as $\left(CO_3^{-2}\right)(Cl^-)$, (NO^{3-}), $\left(PO_4^{-3}\right)$, $\left(SO_4^{-2}\right)$, the metal ions Co^{2+}, Fe^{2+}, Ni^{2+}, and Zn^{2+} and the two organic compounds tested aniline and phenol. Unfortunately, the metal ions of Hg^{2+} and Cu^{2+} seriously interfere on *p*-nitrophenol detection.

An important mention is underlined in this report (experimental data not shown) concerning the high interference of *m*-nitrophenol, o-nitrophenol, and nitrobenzene on the *p*-nitrophenol detection with electrode modified with AuNP these compound presenting over-lapped reductive waves with that of *p*-nitrophenol at the electrode (Tang and Chen, 2011).

The determination of *p*-nitrophenol was also possible by combining the layer-by-layer assembly method and the molecular imaging to produce molecularly imprint polymers (MIPs). The method was used by Xu et al. (2013). First, silica microspheres were functionalized using AuNPs and thiol groups and assembled on the surface of a gold electrode through layer by layer method. A solution of *p*-nitrophenol and pyrrole was prepared in which the electrode was immersed for electropolymerization. The process led to the creation of a polymer-modified surface. The electropolymerization determined the formation of a polymer-modified surface, which after the removal of the template and the silica spheres was characterized using differential pulse voltammetry (DPV) method and electrochemical impedance spectroscopy. The results showed a very high recognition capacity of the platform for *p*-nitrophenol compared with other similar species and a linear dependence of DPV peak current to the *p*-nitrophenol concentration in the range from 0.1 μM to 1.4 mM and a limit detection 0.1 μM (S/N = 3).

The high demand of better and more sustainable methods used to detect and remove the toxic waste from water driven many scientists to study and borrow concepts from the natural world. A great example in this regard is the elucidation of inside structure of *Morpho* butterfly, the lamellar ridge architecture, which lead to many applications including chemical (Tan et al., 2011) or gas (Han et al., 2014) sensors with high selectivity/sensitivity and infrared imaging devices of high speed (Zhang et al., 2015).

There are many reports on attempts to mimic the blue vivid color or the architecture of this butterfly wings like those of Chung et al. (2012) and Lakhtakia and Martín-Palma (2012) who used the multilayer deposition on irregular substructures, or to fabricate a replica of the butterfly scales, like Keiichiro et al. (2005) who used focused ion-beam chemical vapor-deposition (FIB–CVD). Other researchers, Chen et al. (2011) used directly single wing scales (SWS) from the original butterfly *Morphodidius* replicating them with ZrO_2.

To better understand the importance of this structure in the technology of water and air purification we have to consider briefly the data regarding this natural phenomenon.

The coloring effect of *Morpho* (Rhetenor) butterfly wings is attributed to structural coloration and back scattering more than a natural pigmentation. This phenomenon can be explained by the presence of subwavelength microstructures, on the wing scales, which reflect the incident light at specific wavelengths in multiple instances. The microstructures present in the scales are cross-ribs, microribs, ridges, and ridge-lamellae and are arranged similar with a multilayer interference (Okada et al., 2012).

In Fig. 13.10 the first two images (A) and (B) are from an optical microscope, the image (C) is from an electron transmission electron micrograph. The width of a lamella is about 300 nm and its thickness 100 nm. In Fig. 13.10 the lamellae are shown with an inclination at about 10 degree to the x–z plane.

This Christmas tree like structure of the wing scales is proved to be very important in the electrochemical detection of *p*-nitrophenol. In this regard, Guo et al. (2015) used gold nanograins with lamellar-ridge architecture lamellar-ridge-Au to modify a GCE which further was used for the detection of *p*-nitrophenol . This architecture is very important because it offers a large surface area and an efficient mass transport (Guo et al., 2015). The lamellar-ridge-Au were synthesized using as architectural pattern natural wing-scales of *Morph* butterfly which after functionalization were coated with the AuNPs through electroless deposition method. The sites with specific molecular recognition for *p*-nitrophenol molecules were achieved by surface molecular imprinting method preparing a solution of *p*-toluenethiol, L-dodecanethiol, and molecular template *p*-nitrophenol , which was incubated with the gold

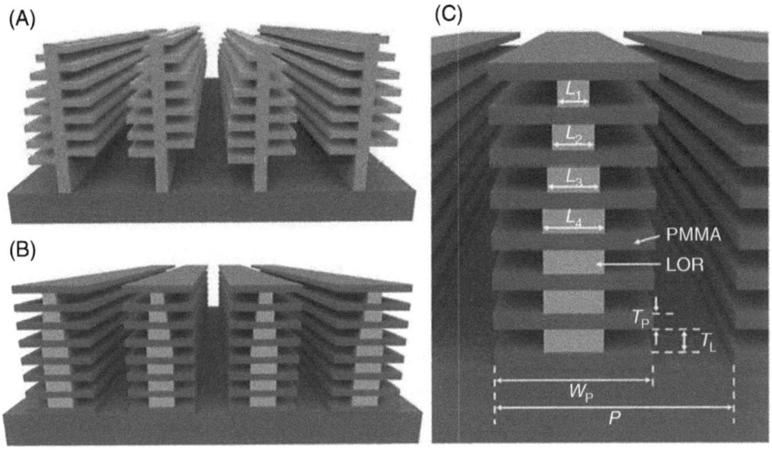

Figure 13.10. Schematic diagram for the *Morpho* butterfly wing scales. (A) The original configuration similar to real wing scales with Christmas-tree shape and off-set lamellae layers. (B) The designed scales to be fabricated with aligned lamellae structures of PMMA/LOR alternate layers. (C) Definitions of dimension symbols used in the text (Zhang and Chen, 2015). Reprinted from an open access source.

samples. By removing the molecules used as templates through CV numerous sites for *p*-nitrophenol recognition were left on the surface. This final architecture was selectively used to determine *p*-nitrophenol through DPV. The selectivity was investigated regarding the possible interference of the isomers *m*-nitrophenol and o-nitrophenol. The imprinted sensor showed prominent selectivity for *p*-nitrophenol with a linear range of 2.5×10^{-8} M and a detection limit of 2.0×10^{-8} M (S/N = 3), 240 times lower compared with the detection limit of bare (nonimprited) lamellar ridge-Au which was 4.8×10^{-6} M.

3.1 Ethylene Glycol

Ethylene glycol (EG) is involved mostly as precursor or raw material in the production of many chemical products, such as polyethylene terephthalate used in fabrication of plastic bottles. High volumes of EG are used as coolant, antifreeze agents, fuel, and as component of detergents and cosmetics. While a moderately toxic for humans and environment the overexposure to EG or to its toxic derivatives may lead to metabolic acidosis, damage to liver, central nervous system, kidneys, and heart, thus EG was the subject of many researches. Taking in consideration their previous experience in the use of bare gold electrodes as templates and the catalytic effect of AuNPs, Łuczak and Bułat (2015), proposed the fabrication of a gold electrode, which was modified with AuNPs and used for the electrooxidation of EG. The AuNPs surface was functionalized with organic compounds containing S through a self-assembling method and the self-assembled layers (SAMs) were used to modify the gold electrode. The electrode obtained was tested upon EG electrooxidation in a solution with pH7 and proved to be highly catalytic. The relation ship between the EG concentration (C_{EG}) in the solution and the electrical response of EG at the potential of peak maximum (j_p) was linear in the range between 0.1 and 0.7 M. The detection sensitivity counted as j_p/C_{EG} at $v = 0.1$ V/s was about 5 A/cm^2 mol dm^3 and the detection limit 0.046 μM (Łuczak and Bułat, 2015).

3.2 Pesticides

Industrial development and increased consumption of agricultural products, such as grains, corn, fruits, and vegetables, often out of their natural period of development, have led to environmental pollution, air, and water, with chemical products known as chemical fertilizers and pesticides. Pesticides are a class of biocides, meant for crop protection products by damaging insects, fungi, weeds, and other plant pathogens. Their potential toxicity to humans and other species, particularly for birds and aquatic species, it is of outmost importance. Pesticides class includes many biological or chemical products and can be classified in many ways: by target organism (algaecides, fungicides, bactericides, acaricides/miticides, insecticides, herbicides, etc.) or by chemical structure. There are some prominent chemical families of insecticide including carbamates (e.g., carbaryl), organochlorines (OCPs, e.g., endosulfan), and organophosphorous (OP, e.g., organothiophosphates, such as chlorpyrifos, malathion, methidathion, fenthion).

Organophosphate (OP) insecticides are highly toxic for humans as they act like inhibitors of acetyl-cholinesterase affecting the nervous system. Thus, some of this products like parathion and chlorpyrifos are no longer registered for any use by the United States Environmental Protection Agency (EPA). Kim et al. (2015) used AuNPs as probes for signal-enhancement in the detection of OP compounds. The optical properties of AuNPs biosensors are size-dependent and with the addition of imidazole to a solution of OP and AuNPs, the AuNPs agglomerate and solution color changed from red to purple with an increase of the peak absorbance at 660–670 nm. The method can be used toward diazinon, edifenphos, and iprobenfos at ppb concentrations.

4 Conclusions

As mentioned before through the use of nanotechnology there has been an advancement on the methods used for pollutant monitoring and elimination, allowing the development of ultrasensitive detection—imaging sensors and devices. There has already been used a large number of NPs, such as silver and AuNPs, QDs, carbon nanotubes, and MNPs in different applications. Each kind of nanoparticle presents different magnetic, fluorescence and optical properties and can be easily coated or functionalized with various materials or biologic molecules, offering highly specific binding properties. However, only AuNPs present all the required properties mentioned before. They can be easily obtained with green synthesis methods, fast, at low costs and they present the best SPR, offering the highest sensitivity and consequently the highest potential.

Even though there has been a vast advance, the methods used for pollutant elimination must reach molecular limits, high efficiency molecular capturing agents that are able to catch the contaminants during single encounter must be produced at low costs and with environmental friendly procedures. Water, as vital necessity of all living beings on Earth, is a very precious and non-negotiable natural public supply. On the other hand, the pollutants present in water today are mainly of human activity provenience, therefore further cumulative efforts for developments of green technologies applied in all our activities are required. In the field of nanotechnology used for water purification, an innovative design of AuNPs with enhanced optical and catalytic properties synthesized based on accumulated knowledge and data, only by green methods will present a necessity.

Abbreviations

ATSDR Agency for toxic substances and disease registry
CDC Centers for disease control and prevention
EPA Environmental protection agency
FDA Food and drug administration
IDLH Immediately dangerous to life and health
MRL Minimal risk level
NIOSH National institute for occupational safety and health of CDC
OSHA Occupational safety and health administration
PEL Permissible exposure limits during any 8-h workshift of a 40-h workweek
REL Recommended exposure limits
ST NIOSH short-term exposure limit
STEL OSHA short-term exposure limit
TWA Time-weighted average concentration for up to a 10-h workday during a 40-h workweek

References

Akki, J.F., et al., 2011. Detection and determination of manganese concentration in water using a fiber Bragg grating coupled with nanotechnology. Appl. Optics 50 (32), 6033–6038.

Akoumianaki Ioannidou, A., et al., 2015. Effect of vanadium on dry matter and nutrient concentration in pennyroyal (*Mentha pulegium* L). Bulletin UASVM Horticulture 72 (2), 295–298.

Alvarado-Gámez, A.L., et al., 2013. Vanadium determination in water using alkaline phosphatase based screen-printed carbon electrodes modified with gold nanoparticles. J. Electroanal. Chem. 693, 51–55.

Alvarado-Gámez, A., et al., 2015. A chronoamperometric screen printed carbon biosensor based on alkaline phosphatase inhibition for W(VI) determination in water, using 2-phospho-L-ascorbic acid trisodium salt as a substrate. Sensors 15 (2), 2232.

Amin, M.T., Alazba, A.A., Manzoor, U., 2014. A review of removal of pollutants from water/wastewater using different types of nanomaterials. Adv. Mater. Sci. Eng. 2014, 24.

Annadhasan, M., Kasthuri, J., Rajendiran, N., 2015. Green synthesis of gold nanoparticles under sunlight irradiation and their colorimetric detection of Ni^{2+} and Co^{2+} ions. RSC Adv. 5 (15), 11458–11468.

Astrid Sigel, H.S., Roland, K., 2008. Nickel and its surprising impact in nature. Metal Ions Life Sci. 2, 824–826.

ATSDR, 2011. Nickel. Toxic Substances Portal. http://www.atsdr.cdc.gov/substances/toxsubstance.asp?toxid=44

ATSDR, 2012. Toxic substances portal—manganese agency for toxic substances & disease registry. https://www.atsdr.cdc.gov/toxprofiles/TP.asp?id=102&tid=23

ATSDR, 2015a. Toxicological Profile for Vanadium. Toxic Substances Portal. http://www.atsdr.cdc.gov/ToxProfiles/tp.asp?id=276&tid=50

ATSDR, 2015b. Public health statement for tungsten. Toxic Substances Portal—Tungsten. https://www.atsdr.cdc.gov/phs/phs.asp?id=804&tid=15

Azzam, E.M.S., et al., 2014. Detection of heavy metal ions using synthesized amino thiol surfactants assembled on gold nanoparticles. J. Dispersion Sci. Technol. 35 (2), 175–184.

Bae, D.R., et al., 2014. Low-ppm-level colorimetric acid detection using gold nanoparticles with electro-steric stabilization. J. Nanosci. Nanotechnol. 14 (12), 9477–9484.

Boopathi, S., Senthilkumar, S., Phani, K.L., 2012. Facile and one pot synthesis of gold nanoparticles using tetraphenylborate and polyvinylpyrrolidone for selective colorimetric detection of mercury ions in aqueous medium. J. Anal. Methods Chem. 2012.

CDC, 2011a. Iron pentacarbonyl. NIOSH Pocket Guide to Chemical Hazards. https://www.cdc.gov/niosh/npg/npgd0345.html

CDC, 2011b. Cobalt metal dust and fume. https://www.cdc.gov/niosh/npg/npgd0146.html

CDC, 2011c. Nickel metal and other compounds. NIOSH Pocket Guide to Chemical Hazards. https://www.cdc.gov/niosh/npg/npgd0445.html

CDC, 2011d. Vanadium dust. NIOSH Pocket Guide to Chemical Hazards. https://www.cdc.gov/niosh/npg/npgd0653.html

CDC, 2011e. Tungsten NIOSH HomeNIOSH Pocket Guide to Chemical Hazards. https://www.cdc.gov/niosh/npg/npgd0647.html

CDC, 2011f. Phenol. NIOSH Pocket Guide to Chemical Hazards. https://www.cdc.gov/niosh/npg/npgd0493.html

Chen, Y., et al., 2011. Tunable three-dimensional ZrO_2 photonic crystals replicated from single butterfly wing scales. J. Mater. Chem. 21 (39), 15237–15243.
Chung, K., et al., 2012. Flexible, angle-independent, structural color reflectors inspired by *Morpho* butterfly wings. Adv. Mater. 24 (18), 2375–2379.
D'Agata, R., et al., 2011. Direct detection of point mutations in nonamplified human genomic DNA. Anal. Chem. 83 (22), 8711–8717.
Di Gioacchino, M., et al., 2014. Nickel oral hyposensitization in patients with systemic nickel allergy syndrome. Ann. Med. 46 (1), 31–37.
El-Shahawi, M.S., Al Khateeb, L.A., 2012. Spectrofluorometric determination and chemical speciation of trace concentrations of tungsten species in water using the ion pairing reagent procaine hydrochloride. Talanta 88, 587–592.
EPA, 2007. Aquatic life ambient freshwater quality criteria—copper. National Service Center for Environmental Publications, CAS Registry No: 7440-50-8.
Gopinath, V., et al., 2016. Anti-Helicobacter pylori, cytotoxicity and catalytic activity of biosynthesized gold nanoparticles: multifaceted application. Arab. J. Chem, (in press).
Guo, X., et al., 2015. Electrochemical detection of *p*-nitrophenol on surface imprinted gold with lamellar-ridge architecture. Sensor Actuat.B Chem. 220, 33–39.
Han, Z., et al., 2014. Unparalleled sensitivity of photonic structures in butterfly wings. RSC Adv. 4 (85), 45214–45219.
Hormozi-Nezhad, M.R., Abbasi-Moayed, S., 2014. A sensitive and selective colorimetric method for detection of copper ions based on anti-aggregation of unmodified gold nanoparticles. Talanta 129, 227–232.
Hu, R., Zhang, L., Li, H., 2014. A highly sensitive and selective colorimetric sensor for the detection of Mn^{2+} based on supramolecular silver nanoparticle clusters. New J. Chem. 38 (6), 2237–2240.
Jiang, H.-L., Xu, Q., 2011. Recent progress in synergistic catalysis over heterometallic nanoparticles. J. Mater. Chem. 21 (36), 13705–13725.
Jiang, Y.-M., et al., 2006. Effective treatment of manganese-induced occupational parkinsonism with *p*-aminosalicylic acid: a case of 17-year follow-up study. J. Occup. Environ. Med. 48 (6), 644–649.
Jimoh, A.A., et al., 2015. Schiff base ligand coated gold nanoparticles for the chemical sensing of Fe(III) ions. J. Nanomater. 2015, 7.
Johannesson, K.H., et al., 2013. Controls on tungsten concentrations in groundwater flow systems: the role of adsorption, aquifer sediment Fe(III) oxide/ oxyhydroxide content, and thiotungstate formation. Chem. Geol. 351, 76–94.
Keiichiro, W., et al., 2005. Brilliant blue observation from a *Morpho*-butterfly-scale quasi-structure. Japn. J. Appl. Phys. 44 (1L), L48.
Kim, E.J., Chung, B.H., Lee, H.J., 2012. Parts per trillion detection of Ni(II) ions by nanoparticle-enhanced surface plasmon resonance. Anal. Chem. 84 (22), 10091–10096.
Kim, M.S., Kim, G.W., Park, T.J., 2015. A facile and sensitive detection of organophosphorus chemicals by rapid aggregation of gold nanoparticles using organic compounds. Biosens. Bioelectron. 67, 408–412.
Kletzin, A., Adams, M.W.W., 1996. Tungsten in biological systems. FEMS Microbiol. Rev. 18 (1), 5–63.
Kneipp, K., Kneipp, H., Bohr, H.G., 2006. Single-molecule SERS spectroscopy. In: Kneipp, K., Moskovits, M., Kneipp, H. (Eds.), Surface-Enhanced Raman Scattering: Physics and Applications. Springer Berlin Heidelberg, Berlin, Heidelberg, pp. 261–277.
Koedrith, P., et al., 2015. Recent trends in rapid environmental monitoring of pathogens and toxicants: potential of nanoparticle-based biosensor and applications. Sci. World J. 2015, 12.

Krpetić, Ž., et al., 2012. Importance of nanoparticle size in colorimetric and sers-based multimodal trace detection of Ni(II) ions with functional gold nanoparticles. Small 8 (5), 707–714.

Kwon, M.J., et al., 2012. Nanoparticle-enhanced surface plasmon resonance detection of proteins at attomolar concentrations: comparing different nanoparticle shapes and sizes. Anal. Chem. 84 (3), 1702–1707.

Lakhtakia, A., Martín-Palma, R.J., 2012. Bioinspiration, biomimetics, and bioreplication for harvesting solar energy. Frontiers in Optics 2012/Laser Science XXVIII. Optical Society of America, Rochester, New York.

Lee, K., Drachev, V.P., Irudayaraj, J., 2011. DNA—gold nanoparticle reversible networks grown on cell surface marker sites: application in diagnostics. ACS Nano 5 (3), 2109–2117.

Lin, C., et al., 2013. Size effect of gold nanoparticles in catalytic reduction of *p*-nitrophenol with $NaBH_4$. Molecules 18 (10), 12609.

Lou, T., et al., 2011. Colorimetric detection of trace copper ions based on catalytic leaching of silver-coated gold nanoparticles. ACS Appl. Mater. Interfaces 3 (11), 4215–4220.

Lu, C.-H., et al., 2012. Ultrasensitive detection of Cu^{2+} with the naked eye and application in immunoassays. NPG Asia Mater 4, e10.

Łuczak, T., Bułat, K., 2015. Gold surface functionalization with S-containing organic compounds and gold nanoparticles for ethylene glycol electrooxidation. Electroanalysis, n/a–n/a10.

Maier, S.A., Atwater, H.A., 2005. Plasmonics: localization and guiding of electromagnetic energy in metal/dielectric structures. J. Appl. Phy. 98 (1), 011101.

MaríaJesús Lerma-García, M.Á., Simó-Alfonso, E.F., Ríos, Á., Zougagh, M., 2014. Synthesis of gold nanoparticles using phenolic acids and its application in catalysis. J. Mater. Environ. Sci. (6), 1919–1926.

Narayanan, K.B., Park, H.H., 2014. Colorimetric detection of manganese(II) ions using gold/dopa nanoparticles. Spectrochim. Acta Mol. Biomol. Spectrosc. 131, 132–137.

Okada, N., et al., 2012. Rendering *Morpho* butterflies based on high accuracy nano-optical simulation. J. Optics 42 (1), 25–36.

Park, Y., et al., 2011. Detection of malathion, fenthion and methidathion by using heparin-reduced gold nanoparticles. J. Nanosci. Nanotechnol. 11 (9), 7570–7578.

Pawlas, N., Sas-Nowosielska, H., 2015. Heavy metals in the cell nucleus—role in pathogenesis. Acta Biochim. Pol. 62 (1), 7–13.

Pozun, Z.D., et al., 2013. A systematic investigation of *p*-nitrophenol reduction by bimetallic dendrimer encapsulated nanoparticles. J. Phys. Chem. C 117 (15), 7598–7604.

Pradeep, T., Anshup, 2009. Noble metal nanoparticles for water purification: a critical review. Thin Solid Films 517(24), 6441–6478.

Pu, W.D., et al., 2014. Visual detection of Hg^{2+} Based on Hg^{2+}-xanthine complex preventing aggregation of gold nanoparticles. J. Nanosci. Nanotechnol. 14 (6), 4297–4302.

Qian, H., et al., 2013. Gold nanoparticles for cleaning contaminated water. J. Chem. Technol. Biotechnol. 88 (5), 735–741.

Raghav, R., Srivastava, S., 2015. Gold nanoparticles based colorimetric detection of cobalt (II) ions. Sensor Lett. 13 (3), 254–258.

Tamer, U., et al., 2013. Gold-coated iron composite nanospheres targeted the detection of *Escherichia coli*. Int. J. Mol. Sci. 14, 6223–6240.

Tan, Y., et al., 2011. Versatile fabrication of intact three-dimensional metallic butterfly wing scales with hierarchical sub-micrometer structures. Angew. Chem. Int. Ed. 50 (36), 8307–8311.

Tang, Y.-Y., Chen, P.-Y., 2011. Gold nanoparticle-electrodeposited electrodes used for *p*-nitrophenol detection in acidic media: effect of electrodeposition parameters on particle density, size distribution, and electrode performance. J. Chin. Chem. Soc. 58 (6), 723–731.

Thabile Ndlovu, O.A.A., Rui, W., Krause, Mamba, B.B., 2010. Electrochemical detection of o-nitrophenol on a poly(propyleneimine)-gold nanocomposite modified glassy carbon electrode. Electrochem. Sci. 5(8), 1179–1186.

Tiwari, B.P.a.A., 2015. A brief review on the application of gold nanoparticles as sensors in multi dimensional aspects. J. Environ. Sci.,Toxicol. Food Technol. (1), 1–7.

Wang, L., et al., 2010. Nanoparticle-based environmental sensors. Mater. Sci. Eng. 70 (3–6), 265–274.

Wang, Y., Dostalek, J., Knoll, W., 2011. Magnetic nanoparticle-enhanced biosensor based on grating-coupled surface plasmon resonance. Anal. Chem. 83 (16), 6202–6207.

WHO, 2008. Iron in Drinking-water. WHO Guidelines for Drinking-water Quality.

Xu, G., et al., 2013. Selective recognition and electrochemical detection of *p*-nitrophenol based on a macroporous imprinted polymer containing gold nanoparticles. Microchim. Acta 180 (15), 1461–1469.

Xun, L., et al., 2015. Polyacrylamide-protected gold nanoparticles for the determination of manganese ions. Anal. Methods 7 (23), 9906–9911.

Yang, W., et al., 2007. Fast colorimetric detection of copper ions using L-cysteine functionalized gold nanoparticles. J. Nanosci. Nanotechnol. 7 (2), 712–716.

Yu, C.W.C., 2013. Detection of chemical pollutants in water using gold nanoparticles as sensors: a review. Anal. Chem. 32 (1), 1–14.

Zhang, S., Chen, Y., 2015. Nanofabrication and coloration study of artificial *Morpho* butterfly wings with aligned lamellae layers. Sci. Rep. 5, 16637.

Zhang, Z., et al., 2012. Label-free colorimetric sensing of cobalt(II) based on inducing aggregation of thiosulfate stabilized gold nanoparticles in the presence of ethylenediamine. Analyst 137 (2), 400–405.

Zhang, F., et al., 2015. Infrared detection based on localized modification of *Morpho* butterfly wings. Adv. Mater. 27 (6), 1077–1082.

Zhong, G., Liu, J., Liu, X., 2015. A fast colourimetric assay for lead detection using label-free gold nanoparticles (AuNPs). Micromachines 6 (4), 462.

Zuber, A., et al., 2016. Detection of gold nanoparticles with different sizes using absorption and fluorescence based method. Sensors Actuat. B: Chem. 227, 117–127.

14

NANOMATERIALS AND FOOD-PROCESSING WASTEWATER

Rajani Bharati*, Suresh Sundaramurthy*, Chandrakant Thakur**

**Green Catalysis and Process Technology Research Lab, Department of Chemical Engineering, MANIT, Bhopal, Madhya Pradesh, India; **Department of Chemical Engineering, NIT, Raipur, Chhattisgarh, India*

1 Introduction

In recent times, application of nanotechnology has increased in various fields including the food industry. Nanotechnology has been used in food industry for various purposes, such as for green packing, pesticide reducing, tracking, tracing, brand protection, texture, flavor, antimicrobial packaging, improved food storage, enhanced nutrient delivery, contamination sensor, bacteria identification and elimination, wastewater treatment and reuse. Many future applications of nanotechnology in food industry are at research stage, one of the research areas is the application of nanotechnology in the food industry, is wastewater treatment. Water is used for various purposes in food industry, due to the scarcity of water, recent research on treatment of water and its reuse has been increased. Application of nanotechnology is one of the promising approaches which are used for wastewater treatment and reuse in food industry. Potential of application of nanotechnology in wastewater treatment is very wide, full exploitation of which has been not done yet, many researcher are working on it. A very small application of nanotechnology in food industry has been done in various developing countries, most of the research in this field has been done in the United States, Japan, and China by 2013.

The main applications of nanotechnology which have been done are encapsulation and emulsion formation, sensor development, utilization, food contact materials, enhance the stability and texture of food, and so on (Bereka, 2015). Research in this field has attracted government and institution investments. Application of nanotechnology to enhance nutrients in food industry is the recent research area.

Water Purification. http://dx.doi.org/10.1016/B978-0-12-804300-4.00014-9

Water, raw materials, and energy are the main resources which are used in food industry. Water is required in huge quantity in food-processing industry and other purposes, such as washing, cleaning equipments, an efficient transportation conveyor of raw materials, main agent used in sanitizing plant machinery and areas, and so on, prevention of water pollution, reduction of wastewater volume, treatment and reuse of wastewater has immerged as a serious issue in the food industry on which research has to done. Food-processing industry has been differentiated into four sectors which are fruit and vegetables; meat, poultry, and seafood; beverage and bottling; and dairy operations. Water is needed in huge quantity in all of these sectors. The main problem of these sectors is the treatment of wastewater and its safe disposal in water bodies. For food industry sweet tap water, groundwater is required for food processing, to get sweet water food industry needs to expand huge money to purify water. For food processing in food industry high quality of water is needed, the parameters for high quality are sensory, chemical, and microbiological nature; it should be potable water:

1. Food industry follows many strategies to minimize fresh-water consumption and wastewater volume. Adopting such kind of technology and process which requires less quantity of water. Use of steam for cleaning and washing cans and canneries (Carroad and Rose, 1977).
2. Adapting many strategies which reduce uncontrolled water use by modifying cleaning and layout design.
3. Proper water management to recycle and reuse water.

Many techniques are used to increase nutrients, such as N, P, K, protein in food product in food industry; application of nanotechnology to increase nutrients has become one of the effective technologies which are ecofriendly and cost effective also. In recent years researchers are working on the potential of nanotechnology to enhance nutrient delivery, nanoencapsulating is used to improve solubility of vitamins, antioxidants, healthy omega oils, and other nutraceuticals.

In the past, for wastewater-treatment practice, wastewater was collected from different processes into one treatment plant. But in the recent trend, it was found that treatment of wastewater from each process stream separately is a better practice than the previous one, because it results in less energy consumption, more efficiency, and more economical strategy.

Nutrient-rich effluent from food industry has become a very big challenge, pond system are not much effective to reduce nutrient in wastewater, it has become a very serious environmental issue, discharge of nutrient-rich water in water bodies or surface water, recent environmental issue is developing effective technology to

Table 14.1 Rates for Water Used in Various Food Industries (Pollution from Food Processing Factories and Environmental Protection)

Industry	Range of gal/ton Product
Fruits and Vegetables	
Green beans	12,000–17,000
Peaches and pears	3,600–4,800
Other fruits and vegetables	960–8,400
Food and Beverages	
Beer	2,400–3,840
Bread	480–960
Meat packing	3,600–4,800
Milk products	2,400–4,800
Whiskey	14,400–19,200

Source: http://www.unido.org/fileadmin/import/32129_25PollutionfromFoodProcessing.7.pdf

reduce nutrients in treated water of spread wastewater on land. The traditional techniques which were used to reduce nutrients are advanced pond-treatment systems, stripping techniques to reduce nutrient concentration, land-application strategies involving nutrient budgeting models to minimize environmental degradation and enhance fodder quality, and deferred irrigation. There are many products and food-processing processes in the food industry the volume of wastewater discharged and their characteristics are different for different product and production processes. In conventional wastewater-treatment techniques biological-treatment process is in practice for almost all food-processing effluents. High BOD, SS, and oil concentrations and smells from acidification, high nutrient concentration are contained in effluents from food industry. Flow rates for water used for various food industries have been given in Table 14.1.

There are many food-processing industries, some are breweries and beer, sake, oils and fats, milk and dairy products, fruit and vegetable food-processing sector, meat, poultry, and seafood sector, dairy sector, can-cooker products, and so on. Different food industries follow different steps in their food processing; general steps followed in food-processing industry have been shown in Fig. 14.1.

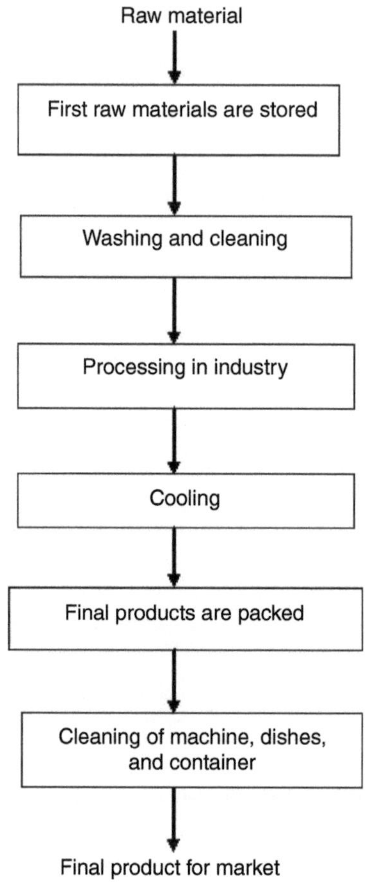

Figure 14.1. General steps in food-processing industry.

1. Fruit and vegetable food-processing sector
   The main steps in processing fruits and vegetables include:
   a. First fruits and vegetables are washed to make them clean.
   b. Unnecessary parts of fruits and vegetables, such as leaves and seeds are removed.
   c. They are bleached.
   d. They are cooled.
   e. Final products are packed.
   f. Cleaning of machine, dishes, and container is done.
2. Meat, poultry, and seafood sector
   The main steps in processing of meat, poultry, and seafood sector include:
   a. First rendering, bleeding is done then skin is removed.
   b. Viscera is removed.

c. Washing and cooling is done.
d. Final product is packed.
e. Cleaning of machines and containers is done.

3. Beverage and fermentation sector
The main steps in processing beverages are:
a. First raw materials are stored then used for processing in industry.
b. Mixing is done.
c. Fermentation is followed by cooking.
d. Cooling is done.
e. Beverage is filled in bottles, and bottles are packaged.
f. Cleaning of machines and containers is done.

Various operations from which wastewater are generated from milk-based food industry have been shown in Table 14.2.

Table 14.2 Various Operations from which Wastewater are Generated from Milk-Based Food Industry (Environmental Guidelines for the Dairy Processing Industry)

Operations	Processes	Sources of Waste
Preparation stages	Storage of milk	Drainage from tankers, Spills and leaks from pipes and tanks, spills from foaming, cleaning operations
	Pasteurization	Liquid leaks, water from recovery processes of downgraded product, cleaning operations, foaming, water from removing operation of deposits from the surfaces of pasteurization and heating equipment
	Homogenization	Water from liquid leaks, cleaning operations,
	Separations	water from foaming, cleaning operations, such as centrifuge and reverse osmosis
Product-processing stages	Market milk	Water from washing cleaning operations of product, water from removing sludge from clarifier or separators, water from cleaning of machine
	Production of cheese	Water from various operations, such as adding salt in cheese making, partial separation of whey from curd, overfilling vats, cleaning
	Production of butter	Waster from various operation, such as cleaning washing and adding salt and vacreation
	Manufacturing of power	Water from various operations, such as cleaning of evaporators, bagging losses, start-up and shut-down losses, plant malfunction, stack losses

Source: http://www.epa.vic.gov.au/~/media/Publications/570.pdf

Wastewater volume and qualities and raw materials used in different food industries have been discussed in this section.

Cornstarch, malts, and rice are used as raw materials to make beer in beer-brewing process. The characteristics and volume of wastewater changes for different brewing processes. A huge volume of effluents containing high BOD and SS is generated from brewing processes. After fermentation water is used for washing, and reused. Some portion of water is reused for food, medicine, and other uses. The main source of wastewater from these industries is consuming water in processes, water for washing and rinsing, washing filter equipments. Ten times the volume than beer production wastewater volume is discharged from this industry. Rapeseed, corn, and soybean are used as raw materials in vegetable oil production. Wastewater from oils and fats industry contains dust and odors, lecithin, carbohydrates, proteins, fats, gummy substances, fatty acids, pigments, and so on. Water is used in this industry for various operations, such as cooling, production process, bottle washing, boiler, sanitary, and so on. Large part of wastewater is generated in this industry by deacidification and deodorization operations in the refining process. The main source of wastewater from milk and dairy industries is; water is used for washing equipment, machines, floor, leakage from pipe lines, water from foaming, cleaning operations, such as centrifuge and reverse osmosis. Cooling, air conditioning, boilers, sanitation, equipments used are cleaned by chemicals and, before and after the cleaning, washed and rinsed with water, and so on. Fruits and vegetables food-processing industry generate wastewater mainly by washing vegetables and fruits to remove dirt and cleaning of machines and container, and so on. Meat, poultry, and seafood sector produces very complicated effluent to treat. The killing and rendering processes generates blood as by-products and effluents, which contain high BOD. Effluents produced from these operations contain pathogenic organisms which can spread. Wastewaters from these industries contain protein and nitrogen. Beverage and fermentation industries produce wastewater in huge volume in comparison to other food-processing sectors. It contains high BOD. Dairy sector produces wastewater by many operations but major portion of wastewater is produced by start-up and shutdown operations. The wastewater from milk industry contains milk as impurities with water. Dairy sector generates wastewater from equipment and tank cleaning. Wastewater from dairy sector contains 90% milk. Wastewater from food industry contains high pathogenic levels which must be disinfected before discharge into water body. Wastewater from dairy sector contains high COD, total nitrogen, orthophosphates, total suspensions, fats, and pH. Effluents from dairy sector contain organic nitrogen in high concentration, some ions are as follows: NH^{4+}, NO^{2-}, and NO^{3-}, dairy wastewater contains Na, K, Ca, Mg, Fe, Co, Ni, and Mn elements, organic and inorganic phosphorus.

Food nanotechnology is a new area of application in food industry. In recent years, applications of nanotechnology have increased for many purposes in food industry, these are: improved taste, flavor, color, texture and consistency of foodstuffs. Application of nanotechnology has increased nutraceuticals bioavailability, it has also improved health supplements. Application of nanotechnology has provided better material for food packaging and application of nanotechnology has increased antimicrobial properties of food products. Nanotechnology provides better nanosensors for monitoring the condition of food when food has to be transported and stored. In the recent years, researchers are working in new area of food industry, these are, to increase nutrients in food products, to improve separation of proteins by biotechnology, application of nanotechnology to trace biological and chemical contaminants, and nanoencapsulation of nutraceuticals. The application of nanotechnology has increased to improve the quality of food products. Researchers are working on tracing toxic effects and risks on human health and environment which can be due to the application of nanotechnology in food products and food industry. Application of nanotechnology in wastewater treatment to increase efficiency of nutrient removal and possible impact of nanomaterial on environment after application of nanomaterial in wastewater treatment is the recent issue on which scientists and researchers are working.

Nanotechnology can be applied in agricultural and food industries for the improvement of various fields in this sector. This has been investigated by the United States Department of Agriculture (USDA) published in September 2003. Many types of nanotechniques and materials are invented to improve food products. Nanotechnology is also used to enhance many qualities of food products, such as it improves the flavor of food, texture, decreases time of processing, improves heat resistivity, enhances life of food, reduces cost needed for production of food, increases bioavailability of nutrients for food (Chaudhry et al., 2008; Weiss et al., 2006). The scope for recycling and reuse of water in the food industry is in practice in many countries, such as the Netherlands. It seemed possible to reduce the use, that by recycling and reuse, use of water can reduce by 20–50%.

Reuse of water in the food industry cannot be done before proper treatment because water used in the food industry needs high quality otherwise it could create adverse effect on human health and hygiene of food, so government has made strict regulations for water-quality parameters. Water can be used after treatment for other purposes, such as washing and cleaning but government has not made different guidelines and regulations for reuse of water for other purposes, and government has not decided water-quality parameters for other purposes which are different from drinking water. The adoption of different technologies for treatment of wastewater and its reuse in industry are judged on the basis of three necessary factors, these are economic, environmental, and technological, while before adopting nanotechnology, many points have to be considered, such as it should be safe for environment and human health, it should meet water quality decided by various legislations.

Wastewater is generated from the food-processing industry for various processes and unit operations, these are washing and cleaning of dishes, container and machines, bleaching, heating operations, pasteurizing, cooling, generation of steam, sanitation and disinfection purposes. In daily dish processing factories, 400 products are made. Some of them are cakes, breads, noodles, and so on. The water consumption per unit of raw material changes with different types of products. The raw materials for this sector are different for different products. Some are of 9,000–10,000 tons of flour; 3,000–4,000 tons soybean; and 130–140 tons alimentary yam paste powder to produce 6,000–7,000 tons of noodles; 5,000–6,000 tons loaf bread; 4,000–5,000 tons sweet bun; 2,000–3,000 tons alimentary yam paste; 8,000–9,000 tons soybean curd; and 1,000–1,500 tons fried bean curd.

The food industry has been forced to adopt various technologies which reduce less impact on environment. The food-processing industry has been given many guidelines to make product which should be healthy and safe for consumer. Main resources which are needed in the food-processing industry are water, raw materials, and energy. The food-processing industry has been a huge water user. Water is used as an ingredient in food production, it is used for cleaning in food production processes, also used as conveyor of raw materials for transportation of raw materials, and also used for sanitizing purposes of plant machines and containers. Water consumption is a part of food-processing industry, pollution prevention, and source reduction practices have emerged as biggest challenge in front of processing industry. The main challenges related to wastewater generated from food industry are following.

Wastewater, from food industry contains biochemical oxygen demand (BOD), COD, total suspended solids (TSS), excessive nutrients, such as nitrogen and phosphorus compounds; pathogenic organisms, residual chlorine, and pesticide. This is a very big challenge to treat wastewater and decrease these contaminants from wastewater by adopting technology which should be ecofriendly and cheap. Food industries need to adopt wastewater treatment technologies which are not conventional. It is the requirement of the food industry to apply radiation to kill pathogenic microorganisms. Food industry needs to redesign their effluent treatment plant to make them zero-emission systems.

Water quality is judged by following characteristics which are given here:

1. TSS
2. relative proportion of sodium to other cations
3. residual carbonates
4. organic loading
5. nutrient
6. microbiological aspects
7. BOD
8. chemical oxygen demand (COD)
9. dissolved oxygen (DO)

In Section 2, we will discuss characteristics of wastewater generated by various food industries. Wastewater characteristics of different food industries are different for different industries.

2 Food-Processing Waste Stream

Food-processing industry discharges 45% of the total organic pollution as wastewater and solid wastes. These organic pollutants create disposal and environmental issues. Many costly and energy-consuming advanced technologies are required to treat and dispose the total organic pollutants. These organic pollutants create high organic loading in organic carbon and nutrient sources in the environment. These wastes are naturally rich in nutrients. Most of the conventional treatment technologies for the food-processing wastes are outdated and cause large losses of valuable nutrient. The main challenges which the food industries are facing are BOD; TSS; excessive

Table 14.3 Comparison of Nutrient Concentrations (%) in Various Wastewater with Farm Dairy Wastewater (Longhurst and Luo, 2007)

Source	DM	N	P	K
Farm dairy effluent	0.8	0.45	0.006	0.035
Feed pad				
Slurry	4.0	0.150	0.030	0.100
Effluent postseparation	0.3	0.025	0003	0.030
Separated solids	20	0.45	0.08	0.200
Stand-off pad				
Solids	25	0.20	0.15	0.200
Self-feed pad				
Scraped manure	15	0.20	0.03	0.075
Herd home				
Bunker manure	18	0.50	0.20	0.750

nutrient loading, namely, nitrogen and phosphorus compounds; pathogenic organisms, residual chlorine, pesticide, and so on. Antimicrobials are used in water which is used for washing in food industries to reduce pathogens in washing and wastewater. Nutrient concentration for various farm dairy effluents has been given in Table 14.3. Quality indicators of wastewaters from the oil-producing facilities have been given in Table 14.4. Characterization of the overall effluents generated in dairy factories has been given in Table 14.5. Wastewater characteristics of wheat-starch processing have been given in Table 14.6. Wastewater characteristics for different food industries have been given in Tables 14.7 and 14.8. Coca-Cola company requirements for treated wastewater quality have been given in Table 14.9.

Nutrients (nitrogen, phosphorus, potassium)—nutrients are found in wastewater in the form of nitrogen (N) which is found in organic and mineral forms, ammonia (NH_3), ammonium (NH^{4+}), nitrate (NO^{3-}), nitrite (NO^{2-}), and urea (NH_2CONH_2). Wastewater

Table 14.4 Various Effluent Parameters from Oil-Producing Process

S. No.	Effluent Parameters	Measuring Unit	Measuring Unit Values (De Facto)	Values (Normal)
1	Iron (general)	mg/dm^3	1.62	0.3
2	Ammonia	mg/dm^3	10.0	2.0
3	Phosphates	mg/dm^3	102.79	3.5
4	Suspended substances	mg/dm^3	104.4	15
5	Dry residue	mg/dm^3	1571.0	200
6	Synthetic surface active substances (anionic)	mg/dm^3	0.88	0.5
7	Fats	mg/dm^3	13.0	1.0
8	Biological oxygen demand-5 (BOD-5)	mg/dm^3	36.5	15
9	pH value	mg/dm^3	2.1	6.5–8.5
Heavy Metals				
Iron		12.63		
Copper		0.468		
Manganese		0.444		
Nickel		0.126		
Zinc		0.620		
Chromium		0.026		
Lead		0.468		
Cadmium		Nil		

All values are expressed in mg/L except pH and color.
Source: https://www.kth.se/polopoly_fs/1.650938!/JPSU16p123.pdf

Table 14.5 Characterization of the Overall Effluents Generated in Dairy Factories

Origin	COD	BOD	Fates	N	P	pH	TS	VS	TSS	VSS
DF	4000	2600	400	55	35	8–11	—	—	675	635
DF	4000	2160	—	200	60	5–9	5100	4300	—	500
DF	2926	1580	294	36	21	6.7	2750	1880	—	—
DF	633	260	—	106	—	8.9	710	447	240	—
DF	2209	1112	60	—	—	7.2	—	—	278	—
DF	4500	2300	—	56	33	7.2	2540	1093	816	—
DF	—	285	—	296	—	8.1	—	—	943	—
DF	2125	1250	—	70	100	9.8	1500	—	280	250
DF	—	241	—	191	50.9	8.5	—	—	—	804
DF	4500	2300	350	60	50	—	—	—	800	—
DF	4000	2000	—	60	—	—	—	—	800	—
DF	1750	—	—	75	9.1	—	—	—	400	355

Composition is given in mg/L, except for pH. DF, Dairy factory.
Source: http://citeseerx.ist.psu.edu/viewdoc/download?doi=10.1.1.616.5454&rep=rep1&type=pdf

Table 14.6 Wastewater Characteristics of Wheat-Starch Processing

Items	Values
pH	4.4–5.4
TS (mg/L)	12,800–16400
DS (mg/L)	10,140–12,000
SS (mg/L)	2,600–4,400
COD (mg/L)	18750–20,600
BOD (mg/L)	9,400–13,200
TOC (mg/L)	4,600–5,800
T–N (mg/L)	500–600
T–P (mg/L)	170–190

Source: https://www.env.go.jp/earth/coop/coop/document/male2_e/007.pdf

Table 14.7 Wastewater Characteristics for Different Food Industries (Rajagopal et al., 2013)

Industry	TS (mL/L)	TP (mg/L)	TN (mg/L)	BOD (mg/L)	COD (mg/L)
Food processing	—	3 ± 1	50 ± 5	500–4,000	1,100–8,100
Palm oil mill	40 ± 5	—	750 ± 10	25 ± 5	50 ± 5
Sugar-beet processing	6,000–6,100	2.7 ± 0.5	10 ± 1	—	6,000–6,600
Dairy	100–1,700	—	—	800–1,000	1,400–2,500
Corn milling	600–700	125 ± 5	174 ± 5	3,000–3,100	4,000–5,000
Potato chips	5,000–5,100	100 ± 5	250 ± 10	5,000–5,100	6,000–6,100
Baker's yeast	500–600	3 ± 1	275 ± 10	—	6,000–6,100
Winery	100–200	40–60	300–400	—	20,000–21,000
Dairy	200–2,750	—	10–100	600–6,300	400–16,000
Cheese dairy	1,600–3,900	60—100	400–800	—	22,000–4,0000
Olive mill	75,500 ± 500	—	460 ± 10	—	130,100 ± 500
Cassava starch	830 ± 50	90 ± 10	525 ± 25	6,300 ± 500	10,500 ± 500

Contains flour, soybean, tomato, pepper, and salt. BOD, Biological oxygen demand; COD, chemical oxygen demand; TN, total nitrogen; TP, total phosphorus; TS, total solids.

Table 14.8 Effluent Characteristics for Food-Processing Industry (Varnekar et al., 2013)

Parameters	Raw Wastewater
pH	4.0–4.30
Color (visual)	Creamish white
Total acidity/alkalinity as $CaCO_3$	980–1000
Total suspended solids	2200–2220
Total dissolved solids	1600–1650
Total solids	3800–4000
Chemical oxygen demand (COD)	1100–1200
Biological oxygen demand (BOD, 5 day −20°C)	6800–7000
Sulfide as (S^{-2})	260–300
Sulfates as (SO_4^{-2})	200–300
Total phosphates as (PO_4^{-2})	3.2 ± 0.5
Total nitrogen as N	16.4 ± 0.5
Oil and grease	100–120
Sodium as Na^+	82 ± 5
Potassium as K^+	27 ± 3
Chloride	182 ± 5
Heavy Metals	
Parameters	**Raw Wastewater**
Iron	12.63 ±1
Copper	0.468 ± 0.01
Manganese	0.4444 ± 0.01
Nickel	0.126 ± 0.01
Zinc	0.620 ± 0.01
Chromium	0.026 ± 0.001
Lead	0.468 ± 0.001
Cadmium	Nil

Table 14.9 Colddrink Company Requirements for Treated Wastewater Quality (Liters/Liter of Product Produced)

Parameters	Values
5-day BOD	<49.5
pH level	6.5–8.1
Total suspended solids	<49.5
Total dissolved solids	<4.9
Total nitrogen	<4.8
Total phosphorus	<1.8

All values have been expressed in mg/L except pH and color.

from food industries also contains pathogens which need to be treated (Thawale et al., 2006). Concentrations of nutrients in different dairy effluents have been given in Table 14.10.

3 Nanoscale Materials: Definition and Properties

Materials which have one or more dimensions of the order of 100 nm or less than 100 nm are called nanomaterials. Physicochemical and biological properties of nanomaterials, such as surface to volume ratio, quantum effects are different than large-scale materials. One billionth of a meter is equal to one nanometer. Nanotechnology is the study and application of matters which are in nanosize; control of material for nanoscale size is also a very-important phenomenon in nanotechnology. Any type of material which has one or more dimensions in nanosize is a nanomaterial. In other words, nanomaterial word represents a natural, incidental, or synthesized material which contains particles whether in unbound state or in an aggregate state or in an agglomerate state where, 50% or more particles in number are in dimension is in the range 1–100 nm. Materials which have two dimensions in nanoscale are nanowire and nanotube. Materials which are one dimensional in nanoscale are thin film or coating. Nanomaterials show different properties than their bulk counterparts nanomaterials due to their relative surface area is high and quantum effects.

Table 14.10 Concentration of Nutrients in a Range of Different Dairy Effluents

		Concentration (mg/L)					
References	Country	Total N	Ammonium (NH_4–N)	Nitrate (NO_3–N)	Organic N	Total P	Total K
Bolan et al. (2004)	NZ	135 ± 5				22 ± 5	231 ± 5
Cameron et al. (1997)	NZ	190 ± 5				30	220
Di et al. (1998)	NZ	363 ± 5	95 ± 5	0.5 ± 0.05			
Di and Cameron (2002)	NZ	246 ± 5	58 ± 5			55 ± 5	
Hawke and Summers (2003)	NZ	80 ± 5	36 ± 5	0.2 ± 0.02		31 ± 5	53 ± 5
Jacobs et al. (2008)	Australia	146 ± 5				34 ± 5	439 ± 5
Longhurst et al. (2000)	NZ	14,269 ± 500	48 ± 5	2 ± 0.5	219 ± 5	69 ± 5	370 ± 5

NZ, New Zealand.
Adapted from Wang et al. (2004).

Materials which are synthesized commercially in nanoscale dimensions with their unique properties and their specific composition are called engineered nanomaterials. In solution stability of nanoparticles is the key factors, for application of nanoparticles in the form of liquid fluid, stability of nanoparticles in the solution should be large, stability of nanoparticle in solution depends on functional group on nanoparticles, and different functionalized nanoparticles have different stability and solubility in solution for different solvent. Functionalized metal nanoparticles have more stability than nonfunctionalized metal nanoparticles.

In solution, metal nanoparticles are affected by van der Waals force because they are close to each other in the solution. Due to van der Waals force, nanoparticles have the tendency to aggregate in the absence of other forces; aggregation of nanoparticles reduces its surface area which reduces its catalytic activities. The functionalized nanoparticles which are organometallic, ligands are used to change the properties and size of the nanoparticle catalysts. Stability of nanoparticles can also be increased by coating which prevents nanoparticles to interact with each other.

The term nano was introduced by Greek word for dwarf. The term nanotechnology was first introduced in 1974 by Norio Taniguchi and it was defined by Richard Feynman in 1959. Nanotechnology is the understanding and control of matter at dimensions of range from 1 to 100 nm. Due to their unique phenomena and properties it has many applications in various fields, such as encompassing nanoscale science, engineering, and technology. Nanotechnology involves imaging, measuring, modeling. Particles with dimensions in the range of 1–100 nm which show unique properties are called nanomaterials. It has been introduced by the Food and Drug Administration (FDA) in 2006. Nanotechnology in the food industry was first introduced by the United States Department of Agriculture (USDA) in September 2003. In the recent years many researches are going on the invention of new techniques for more control on food-character traits, to increase processing functionalities, such as improved flavor, texture, speed of processing, heat tolerance, shelf life, traceability, safety, the bioavailability of nutrients, and cost-effective food (Chaudhry et al., 2008; Weiss et al., 2006). The particles which are in the range between 1 and 100 nm in dimensions are called nanoparticles. In nanotechnology, nanoparticles play role as a whole unit in transportation and in showing their unique properties. Nanoparticles are classified in respect to their diameter size.

There are two methods to prepare nanomaterials; one is top-down and second is bottom-up method. In top-down methods, physical machining is applied on materials to make it nanometer size range by applying processes, such as grinding, milling, etching, and lithography. Bottom-up technology for food industry includes; improving high water binding capacity by dry-milling technology. It also applies in self-assembly and self-organization. In bottom-up methods, large structures are made by atom by atom of molecule by molecule approach. This approach follows various methods, such as chemical synthesis, self-assembly, and positional. Nanoparticles are more chemically active than larger particles of the same materials because nanoparticles have more surface area than larger particles of same materials. This property of nanoparticles has increased the efficiency and application of nanotechnology in various processes of food industry in processing, packing, and wastewater treatment in food industries.

Water plays a very-important role in the food industry. It is used in the food industry for many purposes in processes in cleaning, washing, and so on. Food industries need a huge amount of water that is why treatment and reuse of water is very important in the food industry. Water plays a very key role to maintain hygiene in the food industry. So it is very necessary to use pure water or treated water to purify it. Maintaining standard parameter of water is very necessary for food safety. In accordance with Regulation (EC) No. 178/2002 on General Food Law, introduced the term food as follows:

> *Food or foodstuff represent the substance or product, which could be processed, partially processed or unprocessed, and which are used to consume by humans. According to this definition drinking water has been included under the definition of food.*

Nanotechnology is used in the food industry for increasing production efficiency, to improve processing, improve packing material, and to improve effluent treatment performance plants. For all these applications of nanotechnology we need to adopt new nanotechnology techniques, tools, addition, and synthesis of nanomaterials.

Nanotechnology has many applications in the food industry. These are as follows, such as nanotechnology is used for making nanosensors which are used for food safety and traceability of contamination in food stuff (Chaudhry et al., 2008), nanotechnology and devices made by these technology is used for water treatment and for nanofiltration, it is used for improving food-packing materials for food safety, it is used to improve quality and flavor of food by adding ingredients, it is used to be added as food additives.

In this chapter we will discuss application of nanotechnology in effluent treatment from food industry where nanotechnology has been used for making devices and materials to increase efficiency of treatment processes and sensitivity of devices has improved. Nanofiltration materials have also improved for water treatment. With the application of nanotechnology in food industry it is also necessary to estimate toxicity impact of nanomaterials on environment and on human health. Applications of nanotechnology in the food industry create environmental contamination. Sometimes during the application of nanotechnology some accident could happen due to which nanomaterial could enter in animal feed and can affect human heath directly or indirectly.

Nanotechnology has become a multidisciplinary research area with increasing demand in industrial application. So it is necessary to define term nanotechnology and nanomaterial which should have common meaning in every research area and industry. The International Standards Organization (ISO) subjected that there must be a standard terminology and a general language for scientific, technical, commercial, and regulatory purposes. The ISO/TC229 Technical Committee on Nanotechnologies established a working group to advice on this specific issue. According to this, nanotechnology is:

> *Understanding and control of matter and processes at the nanoscale, typically, but not exclusively, below 100 nanometers in one or more dimensions where the onset of size-dependent phenomena usually enables novel applications, where one nanometer is one thousand millionth of a meter.*

This definition is general and can be used in every field of nanotechnology from academic to industry. Nanoobject has been introduced by the EU Commission recommendation of 07/02/2008 on a code of conduct for nanosciences and nanotechnologies research which contains all nanomaterials, nanostructured materials, nanoparticles, and their aggregation at the nanoscale, nanosystems, and nanoproducts.

3.1 Types of Nanomaterials Used in Water Treatment

Nanomaterial for wastewater treatment has been divided in three classes these have been discussed further.

3.1.1 Nanosorbents

Nanosorbents are nanomaterials which have dimensions in the range of 1–100 nm which are used to absorb some inorganic and organic molecules. In effluent treatment nanosorbents are used for the separation of contaminants. They have wide application in wastewater treatment due to large surface area to volume ratio. There are many nanomaterials which are used in application for wastewater treatment. These are carbon nanotubes, metal nanoparticles, and zeolites.

3.1.2 Carbon Nanotubes

Carbon nanotubes have two dimensions in nanosize with hollow graphitic layers in cylinder shape. Two types of carbon nanotubes are more in practice; these are multiwalled carbon nanotubes and single-walled carbon nanotubes. A carbon nanotube shows different important mechanical, electrical, and structural properties. They have been used for electronics, biosensing, and chemical sensing. They have large surface area that is why they are used as absorbents. Carbon nanotubes are used to remove heavy metals and organics from wastewater. Silver, magnesium oxide, and copper nanoparticles are used for effluent treatment as disinfectant agent. Zeolites, carbon nanotubes, and activated carbon are very effective for wastewater treatment. These nanoparticles can be used as a disinfectant in wastewater.

3.1.3 Zeolites

Zeolites are porous materials with high surface area. They contain silicone, aluminum, and oxygen. Zeolite is used in effluent treatment for ion-exchange methods, separation methods, as catalysis. Zeolites are used to treat wastewater by removing heavy metals, such as silicone, aluminum, and oxygen from it. Research is going on the surface modification of zeolite to improve its efficiency to remove radioactive materials, such as 137Cs and 90Sr from nuclear wastewater.

3.1.4 Nanosponges

Nanomaterials with polymeric properties are used as nanosponges to reduce organic contaminant. They are found in powder form with large surface area. By adding metal nanoparticles, such as silver and copper, antibacterial properties of these polymeric nanosponges can improve.

3.1.5 Nanocatalysts

Many metals and metal oxides with nanosize and have zerovalent metallic properties show catalysis properties. By adding two metals, bimetallic nanoparticles are made to improve their catalytic properties. Nanocatalysts are applied for various treatment methods in wastewater treatment to improve efficiency of processes. There are many nanocatalysts on which research work is going on and how to apply them to improve efficiency of treatment processes at low cost and low environmental impact. Some catalysts on which lot of research works have been done are TiO_2, Pt, Fe, and Fe/Pd, Fe/Ni, Fe/Co. TiO_2, ZnO, and many other metal compound nanosize catalysts are used for photocatalytic reaction for disinfection and for pollutant removal. Sometimes doping of metal and metal ions is done on nanocatalysts to improve their catalytic properties. Nanocatalysts have high surface area so they more active and effective than other conventional catalyst. Immobilization is also done to improve catalyst's properties.

3.1.6 Nanostructured Filtration Membranes

Polymeric membranes with nanostructured and nanosized pores are used for filtration. Nanotechnology is applied in membrane filtration to improve the properties and age of the membranes. Nanostructured membranes are used for desalination in wastewater treatment. Pore size of nanostructured membrane is more than reverse osmosis membranes. Use of nanostructured membrane is more economical than RO membrane, because they need less energy than RO.

Nanomaterials are used to absorb many types of pollutants. Research is going on to their efficiency and making them cheap at low risk for environment. Researchers are working on issue, such as the effect of nanomaterial on environment after use when this material will be disposed. So safety assessment of nanomaterial is the new area in which researchers are working in wastewater treatment. In the past more research has been done on advantages of nanotechnology, but much less research has been done on nanotoxicology. Silver nanoparticles have been found toxic on health and its toxicity depends and changed with both size and shape. Titania nanoparticles show less toxicity depending on the type of TiO_2 used. The cytotoxicity of titanium dioxide depends on size and crystalline structure (Chellappa et al., 2015). Toxicity of nanoparticles is still the area about which much information is not available.

Catalyst synthesized by nanomaterials are heterogeneous catalysts. Nanocatalysts increase efficiency and speed of reaction because they provide more surface area so that more reaction can occur on catalyst. It is easy to separate nanomaterial catalysts and recycle them than bulk size material because they have high retention catalytic properties. Nanomaterials are used for two purposes in catalytic reactions; one is active site of catalyst and other is a supporter for catalyst to increase their surface area. Incorporation of enzymes into nanostructure materials is called nanobiocatalyst. Nanobiocatalysts are very useful in the food industry as they enable generation of bioenergy through treatment of wastewater of the food industry.

3.1.7 Nanometrology

We need appropriate methods for assessment of risk, effect, and exposure of nanomaterials on human or animal health. There are some physiochemical properties of nanoparticles which need to be measured in the food sector, these properties are as follows:

- particle size
- particle distribution
- surface area
- surface charge and topography
- composition and purity
- hydrophobicity and solubility
- chemical reactivity and bioactivity
- dispersion/aggregation state

Particle size and distribution, surface area, and aggregation state are the most important properties among all the above properties which are being studied for nanomaterials. With decreasing the particle size, number of molecules increases which increase the surface to volume ratio which increases chemical interaction. Nanoparticles have been divided into different types by their quality to carry different ingredients and their reaction for required environmental conditions. Mainly nanoparticles have been divided into two types, inorganic and organic, based on the chemical characteristics of the nanoparticle.

3.1.8 Inorganic Nanoparticles

Inorganic nanoparticles are used in the food industries for changing additives to improved food quality, such as food colorant, UV protection barrier during food packaging, prevention of prepared food, as antimicrobial. TiO_2, silver nanoparticles are examples of inorganic nanoparticles.

3.1.9 Organic Nanoparticles

Nutrient value of food products can be increased by using organic nanoparticles through improvement or alteration of food functionality. Organic nanoparticles are used to deliver vitamins or other nutrients in food without affecting the taste or appearance. To encapsulate the nutrients and to improve nutrients bioavailability, organic nanoparticles are used. The characterization of nanoparticles is a very important step before its application and it does require special knowledge and precautions because there are many uncertainties in these properties, such as particle aggregation, size, purity, batch variations, and so on. There are many tests which are used for characterization of nanomaterials. These are SEM, XRD, TEM, EDX, FTIR, atomic force microscopy (AFM), BET, and so on.

4 Developing Environmental Regulations Pertinent to Nanotechnology

Nanotechnology is applied in the food-processing industry to increase processing, health, and packaging functionalities—flavor, texture, shelf life, transportability, reduced costs, and improving nutritional quality and quantity of processed foods. Therefore, it is required to study and research food safety and food-quality aspects. The potential application of nanotechnology needs to be analyzed in respect to safety, toxicology, environmental impact, economics, and consumer acceptance of nanotechnology.

4.1 Toxicology and Safety Aspect of Nanoparticles

There are many risks and effects of nanoparticles on human bodies. Some of them are given here.

1. Nanoparticles are more chemically reactive than larger particles;
2. Have more access to human bodies than larger particles;
3. Enhanced toxicity due to greater bioavailability;
4. Can compromise human immune system response; and
5. Long-time application and contact can create pathological effects.

Nanomaterials have different desired properties than bulk counterpart (Buzea et al., 2007). Sometimes same properties create some adverse effect on health. Risk assessment of nanoparticles results that some nanomaterials affect badly on human health and on environment. There are less data which show how much exposure and dosage will have a negative impact effect on human body and on environment.

Four steps are followed during risk assessment of nanoparticles. These are given in Fig. 14.2.

1. Identification of hazard materials and elements;
2. Characterization and identification of properties of hazard materials;
3. Assessment of exposure time of those hazard materials with humans; and
4. Assessment of type and effect of risk on environment and humans.

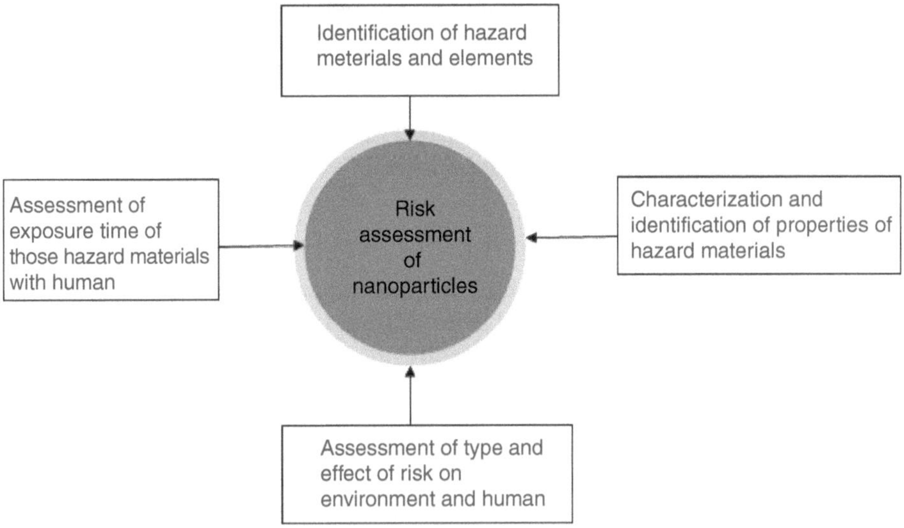

Figure 14.2. Risk assessments of nanoparticles.

4.2 Scientists and Industrial Perspective

Scientists have an opinion that application of nanotechnology has more potential benefits and less risks than the public. Scholar and coworkers studied and recommended that most experts have data for benefit of nanotechnology for its application in various field including curing diseases and cleaning environment, and there are unavailability of data which shows nanotechnology has negative effect on human health and on environment, we assist the risk of nanotechnology and their negative impact on human body and environment hypothetically. Negative effects of nanotechnology on human health are more this is the public opinion. Industry survey in Switzerland and in Germany has concluded that there is a doubt that companies applying nanotechnology do not explain this issue properly.

A number of expert reports have been made to measure risk of application nanotechnology on human health and on environment. The FSAI's Scientific Committee has also subjected the guideline for application of nanotechnology in the agrifood sector. Application of nanotechnology has a potential to create some challenges in terms of safety, regulatory, and ethical considerations; it also has many potential benefits to manufacturers and consumers (Chau et al., 2007). The European Union has introduced a regulatory framework for covering application of nanotechnology in food industry by principles of the General Food Law. Major gaps had been found in the previous law which was improved by a review undertaken by the UK Food Standards Agency, but there was some uncertainty in that law. So in some areas, specific supporting instruments (guidelines, test protocols, standards) were needed, existing legislation in practice.

4.3 Provisions of the General Food Law

According to General Food Law, unsafe food (as defined in Article 14, Reg. (EC) 178 of 2002) cannot be placed in the market. The European Communities (Food and Feed Hygiene) Regulations, 2005, S.I. No. 910 of 2005. According to this regulation, there is a general obligation on food and feed business to carry out their activities in accordance with HACCP principles. These contain pointing out the points in those operations where food hazards may occur, and pointing out and implementing required measures to control the hazards at these critical points. These obligations will apply to the use of nanoparticles in food production and processing. It is subjected that it is required to risk managers, including food business operators, to make work plan according the food law for application of nanotechnology in the food industries. According to the Risk Assessment (BfR, 2008), 66% of respondents subjected that the overall benefits of nanotechnology measured overweighed than the adverse effect of it on environment and health.

5 Nanoparticle Use in Pollution Control

Nanotechnology has much application to reduce pollution from industries. It is used for reduction, treatment, and recycle of solid waste and wastewater. Nanotechnology is also applied to reduce pollution from air. Nanoparticles are used for various purposes, such as absorption of contaminants from water, leaching, as nanocatalysts and as photocatalysts. Nanocarbon tubes zeolites are used for water treatment. TiO_2, ZnO are used as photocatalysts. Nanomaterials are also used as sensor. Application of nanotechnology has a very wide range of applications; it is used to increase the efficiency and effectiveness of processes at low cost and low environmental impact. In this chapter we will discuss application of nanotechnology to cover wastewater treatment in detail.

5.1 Food-Processing Waste Stream

5.1.1 Nanotechnology and Water Purification

Nanotechnologies are applied for water purification. Many researches are going on for application of nanotechnology in effluent treatment in food-processing industries. A US company has made 2-nm diameter aluminum-oxide nanofibers for water filtration; lanthanum nanoparticles have been used as an absorbent to absorb phosphates from aqueous environments. Research at the Centre for Biological and Environmental Nanotechnology (CBEN) had subjected that nanoscale iron-oxide particles are very effective in removing arsenic from groundwater.

6 Nanoparticles for the Food-Processing Wastewater Purification

The applications of nanotechnology for the food sector have been divided into four sectors; these are given here:

1. application of nanotechnology for nanosized, nanoencapsulated, or engineered nanoparticle additives;
2. application of nanotechnology to improve food ingredients by nanostructured materials;
3. application of nanotechnology to improve, active or intelligent materials for food packaging or in food-contact materials or surfaces;
4. application of nanotechnology to make nanotechnology-based devices and materials for using in nanofiltration and water treatment; and
5. application of nanotechnology to make nanosensors for food safety and traceability and contaminant detection (Chaudhry et al., 2008).

Some important nutrients used in food industries and raw materials, that is, proteins, starch, and fats, undergo structural changes at the nanometer and micrometer scales during normal food processing.

The purpose of use of nanotechnology for wastewater treatment is to make treatment more cost effective, more efficient, eco-friendly and should meet water-quality parameters specified for pure water. The following benefits must be gained by the application of nanotechnology for effluent treatment in the food industry:

1. reduced operating costs
2. meets wastewater effluent targets for reuse and recycle
3. ease of operations
4. compact footprint
5. environmentally friendly

These are the food and beverage processor responsibilities to the society and customer with regard to food safety, nutrition, food labeling, plant security, and quality-management protocol. Additionally, municipal, regional, and federal governments must be satisfied by industries, food safety standards, and practices must fulfill the strict regulatory guidelines to save environment and society from harmful practices. Wastewater management of industries must meet all municipal, regional, and federal guidelines for the safe treatment and disposal of wastewater from food industries. Nanoparticles have unique physicochemical properties, such as high surface area, quantum effect, which can increase resistivity of separation material against corrosion that makes them particularly attractive as separation and reactive media for water purification. Characterizations of the interactions of the nanoparticles with the bacterial contaminant are done by AFM.

6.1 Nanomaterials in Wastewater Treatment

There are many types of nanomaterials which are used in wastewater treatment. Some of these are dendrimers, zeolites, carbonaceous nanomaterials, membrane of nanomaterials, metal-containing nanoparticles, and so on.

For water purification, nanotechnology has lots of potentials for the efficient removal of pollutants and germs. To remove pollutant from effluent we used nanoparticles, nanomembrane, and nanopowder to increase efficiency of biological- and chemical-treatment processes. Contaminates which are removed by nanaotechnology are heavy metals, such as cadmium, copper, lead, mercury, nickel, zinc; nutrients, such as phosphate, ammonia, nitrate and nitrite, cyanide, organics; algae, such as cyanobacterial toxins and viruses, bacteria, parasites, antibiotics, and so on. Nanomaterials results are better than other techniques used in effluent treatment because of its high surface area (surface/volume ratio). It is expected that this would be used in future at large scale for water purification. It has been shown in literature that the application of ultrasonic irradiation for short time period for coliform bacteria treatment before Ag nanoparticle treatment at low concentration enhanced antibacterial effect. So both can be combined to increase the effectiveness of treatment processes. Application of nanotechnology for wastewater treatment has the potential to address and solve many environmental and cost problems in developing countries including controlling of epidemic diseases. Application of nanotechnology has potential to solve many issues, such as health, energy, and water.

Nanomaterials are used to make separation and reactive media because it increases quality of media by improving its reactivity and performance (Bellona and Drewes, 2007). Nanoparticles are used as bioremediate and disinfect wastewater. TiO_2

are used as a nanocatalyst for antimicrobial activity; magnetic nanoparticles are used as an adsorbent for heavy-metal removal and for removing organic compounds; contaminants from wastewater are removed by nanocatalyst (Hillie et al., 2006). Magnetite nanoparticles are used for adsorption combined with magnetic separation in wastewater treatment, these approaches are cheap, easy to handle, ecofriendly. This method was used to remove heavy metals, such as chromium from wastewater and magnetite nanoparticles were collected by magnetic separation for reuse.

Membranes for filtration of wastewater are made by nanomaterials which are:

1. Carbon nanotubes, nanoparticles, and dendrimers nanomaterials are used to make membranes of nanostructured materials for wastewater treatment.
2. Metal nanoparticles and other nanomaterials are used to make nanoreactive membranes (Theron et al., 2008) for wastewater treatment.

Metal nanoparticles are used to increase removal rate of pollutants from wastewater. Photocatalysts work as a disinfectant in wastewater treatment in the presence of UV light using TiO_2 as a catalyst.

7 *Escherichia coli* and Total Coliforms from Food-Processing Wastewater

Nanomaterials are used to make contamination sensor: which is used to detect the presence of *E. coli* bacteria in water by flashing light in water. Application of nanotechnology has immerged as an effective tool in bacteria identification and elimination. Nanocarbohydrate particles are used to bind bacteria with them so they can be detected and eliminated. The approach of recycling and reusing water in the food industry has attracted the focus of researchers due to limited sources of water, in reusing of water in food industry, microbiological risks could be very serious factors which need to be considered. Pathogenic microorganisms are very harmful, for human health and they can be reason for food and waterborne diseases so it is the prime task to identify and destroy those pathogenic microorganisms which could be present in water before reuse.

The microbiological quality parameters or indicators for purified water for reuse in food industries are very high. *E. coli* must be 0 in a 100-mL water sample. Many researchers have shown that the existence of some pathogens is not well correlated with the presence of indicator microorganisms. Many times indicators cannot detect microorganisms present in water for more perseverance in water than *E. coli*, for example, *Yersinia enterocolitica*. Sometimes contaminations which are not traceable by sensor and indicator and which are caused by pathogens, *E. coli*, and *Enterococci* are very sensitive to chlorine. *Clostridium perfringens* has been used as a new indicator to find the presence of *Cryptosporidium* and *Giardia* spp. in water. Researchers have found that combination of amoxicillin and silver nanoparticles are more effective and efficient for destroying *E. coli* cells than when they were applied separately, so it has been concluded that this combination has more antibacterial properties than separate use of both of them.

7.1 Improving Effect of Ag-Nanoparticle Combined with Ultrasonic Radiation

Ultrasound is very effective to destroy the bacterial cell walls, The bacterial cells produce isoenzymes that biocatalyze hydrolytic reactions. Biocatalyzed hydrolytic reactions degrade the organic material into smaller and biodegradable fractions. So this technology is used for the degradation of organic materials by hydrolysis for sludge treatment. The main function of ultrasound treatment is destruction of both bacterial cell membranes and degradation of organics. Ultrasound targets the bacterial cell walls, the bacterial cell generates enzymes, than these enzymes are used to catalyze hydrolytic reactions and by this whole processes bacterial cells are destroyed. This mechanism degrades organic material into smaller biodegradable fractions.

By microbiological test antibacterial effect of combined Ag-nanoparticle and ultrasonic irradiation on *E. coli* cells can be found, for this purpose *E. coli* cells need to be isolated from wastewater. In this way by microbiological test it has been found by researchers that antibacterial effect increases when both Ag-nanoparticle and ultrasonic irradiations are combined.

8 Purification Methods

There are many methods which are used to treat wastewater from food industries these methods will be discussed further.

Various methods for treatment of wastewater from food industry have been summarized in Table 14.11. Various treatment methods for food-processing water and their advantages and limitations have been given in Table 14.12. Performance of two ponds–treatment system to treat wastewater has been given in

Table 14.11 Wastewater from Various Food Industries and their Treatment Methods

Wastewater	pH	Major Pollutants BOD	Major Pollutants COD	Major Pollutants SS	Major Pollutants Oil	Major Pollutants N	Major Pollutants P	Treatment Methods
Beverage		Present	Present					AS, AD
Vegetable oil		Present	Present		Present in major quantity			OS, AS, AD
Milk/dairy product		Present	Present					AS
Starch		Present	Present	Present in major quantity				AS
Daily dishes		Present	Present					AS
Confectionary		Present	Present					AS
Brewery		Present	Present	Polluted				AS, AD

Treatment methods: AS, aerobic biological treatment, AD, anaerobic biological treatment, OS, oil separation, in case advance treatment, filtration, activated carbon absorber, and membrane separation are provided in addition to above unit operations.
Source: https://www.env.go.jp/earth/coop/coop/document/male2_e/007.pdf

Table 14.12 Various Treatment Methods for Food-Processing Water and their Advantages and Limitations (Casani et al., 2005)

Water-Treatment Methods	Limitations	Advantages
Chemical		
Application of chlorine as disinfectant	Due to formation chlorine gas it does need more safety precautions otherwise it would be harmful for human health and for equipment	It is an economic method , chlorine can be found easily and it is easy to monitor free residuals present in water
Chlorine dioxide	There are some hazardous by-products which could be harmful for health	Easy to handle, need small amount of concentration[a], more corrosion resistance
Chloramines	Requires long contact times	Stable and more effective
Ozone as disinfectant	Poor on-line testing efficiency, toxic when inhaled	Need short contact times, more safe for treatment of drinking water and direct food applications[b]
Hydrogen peroxide	Need high dosage and long contact times, reacts with catalase from blood, which may result in undesirable appearance of foods[c]	Easy to handle and cheap
Physical		
Membrane processes	Costly technology	High-quality purifying of water and, by-products can be recovering by this process[d]
UV	Need pretreatment[e], costly, uncertain to determining the UV dose	No residual in treated water

[a] When compared to chlorine, for example, treating poultry-processing water.
[b] US Food and Drug Administration.
[c] For example, for poultry chillier water.
[d] Proteins from Surimi processing and proteins and fat from poultry retentive waters (Hart et al., 1988).
[e] Codex Alimentarius (1999).

Table 14.13. Biological treatment of wastewater from milk-based food industry has been summarized in Table 14.14.

8.1 Wastewater Treatment Methods for Milk-Based Food Industry

Methods which are used in milk-based food industries are: various methods are used to treat wastewater generated from milk-based food industries some of them are biofiltration, sequencing

Table 14.13 Performance of Two Ponds Treatment System to Treat Wastewater (Bolan et al., 2004)

Characteristic	Anaerobic Pond Inlet (g/m^3)	Oxidation Pond Outlet (g/m^3)	Percentage Removal (%)
TSS	320 ± 25	85 ± 5	74 ± 5
COD	950 ± 25	453 ± 25	52 ± 5
BOD	210 ± 5	91 ± 5	57 ± 5
NH_4–N	115 ± 5	95 ± 5	8.3 ± 1
NO_3–N	5 ± 1	15 ± 1	
Total P	28 ±	24 ± 5	14 ± 1
Dissolved P	21 ± 1	18 ± 1	14 ± 1
Total K	175 ± 5	168 ± 5	4.0 ± 1

Table 14.14 Biological Treatment of Wastewater From Milk-Based Food Industry

Type of Wastes	Biological Treatment	Detention Time	Treatment Efficiency					
			BOD5	COD	TKN	TOC	VOM	VFA
Dairy wastewater	ASP	19.8 h	99%	—	91%	—	—	—
Dairy wastewater	SBR	6 h	—	90%	—	—	—	—
Whey-processing plant	ASP	19.8 h	99%	—	91%	93%	—	—
	Completely mixed anaerobic reactor	2 days	87%	—	—	—	—	—
Dairy wastewater	ASP	24 h	70%	71%	—	—	64%	—
	TF	—						90%
	UASB	—	96%	97%	—	—	—	—

Source: http://citeseerx.ist.psu.edu/viewdoc/download?doi=10.1.1.616.5454&rep=rep1&type=pdf

batch reactor, land treatment and pond systems, activated sludge process, and so on.

Choosing of treatment method or combination of methods to meet the quality standard for treatment for reuse or recycling is a big challenge.

9 Use of Purified Water

9.1 Water Reuse Practices within the Food Industry

Reuse of food industry's wastewater for irrigation was in practice. But due to water scarcity and large amount with high quality, water requirement for food processing has increased researchers interest to develop technologies which treat effluent water from food industries to meet standard quality parameters for purified water to use this water within industries. Due to the variety of processes and raw materials used in food industries the constituents in effluent of food industry are complicated so it is necessary to consider this factor when purified water has to be reused. In the past, purified water from food processing was reused for only nonfood and cleaning purposes, such as cleaning and washing of machines, dishes, container, cooling and chilling purposes, and fire-extinguishing purposes.

There are many applications of treated water in food industries. Some are recycling and reuse of water in food processing, washing and cleaning, cooling and chilling, these uses are in practice in butter-manufacturing industry; reuse of water for evaporators and condenser in the dairy-based food industry; and reuse of recovered water from pulping and pressing processes for washing orange peel. Some other uses are washing and cleaning of vegetables and fruits, fluming of unprepared products, and scalding water of meat and poultry. It is necessary to choose those types of technologies which treat water for recycling and reuse should have at least tap-water quality. It has been concluded by Lazarova et al. that selection of treatment technologies must be cost effective, safe, should meet processed-water quality, existing standards, reuse applications, and wastewater-treatment capacity.

Selection of technology for microbial decontamination of process water in the food industry is studied by Casani et al. (2005). The methods for microbial contamination removal have been divided in two categories: chemical and physical. Factors affecting wastewater-treatment solutions to implement in the food industry have been summarized in Table 14.15. Wastewater treatment

Table 14.15 Factors Affecting Wastewater Treatment Solutions to Implement in the Food Industry (Casani et al., 2005)

Factors	Drivers	Problems	Solutions
Environmental	There is difference between demand and supply of water need reuse and recycle of water and this increased concern about the impact of industrial activity on the environment	Industrialization, population growth, increasing life standards	Need wastewater management
Economic	Need to spend more money for fresh water, increasing costs for wastewater treatment	Industries need short payback time's policies, not possible to measure the true cost of using water by accounting system	Estimation of cost should be based on true cost of water
Legislation	Strict wastewater quality discharge regulations, different water-quality parameters for different purposes	Standards versus guidelines, complex process, regulation are too strict and not flexible, lack of relevant training among regulations	Finding minimum required quality need to concentrate on relevant aspect, for example, risk assessment, and support R & D on water reuse need collaboration with scientist and food processions
Technological (treatment)	Availability of easy to handle purification processes and technology, use of by-products, and waste after treatment	Lack of easy access to evaluate different methods, lack of guidelines for process water treatment, demanding procedures for quality control of processes	Need to develop effective, efficient, cheap, and easy to handle methods, research on developing microbial resistance wastewater treatment methods
Water-quality assessment	Need for effective methods and efficient instrument	Variation water-quality parameters and methods, unclear relationship between indicators and health, current methods are only suited for verification purposes, no guarantee to maintain sufficient water-quality parameters	Developing new and sensitive indicators, research on making sensitive and cheap indicators and instruments
Social	Awareness regarding water quality and water management	Negative perception about reuse ability of water	Make communication between scientists, media, common people, industries

(*Continued*)

Table 14.15 Factors Affecting Wastewater Treatment Solutions to Implement in the Food Industry (Casani et al., 2005) (*cont.*)

Factors	Drivers	Problems	Solutions
Food industry	High water consumption, high wastewater generation, potential for water reuse within factories	Water reuse need to spend much money for treatment , need to adopt proper water management system and technologies, need trained and skilled people to handle technology	Implementation of proper wastewater management system, risk assessment promotes communication between scientists, media, and consumers
Academic			Create awareness among people by education, Motivate and educate people for use resources, skill people to handle technology and quality tools, skill people to use and reuse water properly, teach people about wastewater management, collaboration with regulatory agencies and research institutions, open mind for applied research in collaboration with industry

methods used in food industry for reused water have been summarized in Table 14.16.

10 Conclusions

From the present literature it has been concluded that in recent days, the available conventional technologies for wastewater treatment are not sufficient to meet water-quality parameters for food industries, to reuse and recycle treated wastewater in food industry, need of high-quality parameters of water which cannot be met by available conventional methods, so application of nanotechnology has immerged as an effective, efficient, cheap, and ecofriendly solution to treat water. Food industry is a very big consumer of water, it requires water in huge amount for various processes, such as cleaning and washing of equipments, machines,

Table 14.16 Wastewater Treatment Methods Used in Food Industry to Reuse Water (Casani et al., 2005)

Food Product/Processing	Type of Water	Water-Treatment Methods	References
Egg	Washing water	Alkaline detergent + sanitizer	Leclair et al. (1994)
Fresh vegetable and salad	Wastewater collection at the balance tank	Bioreactor + UF + RO + UV	Goodwin and Catley (2001)
Leafy vegetables	Washing water	Settling + filtration	
Potato processing	Washing water	sedimentation	
Pickling cucumbers	Hydrocooling water, pickling brines	Chlorine dioxide, addition of NaOH	Reina et al. (1995)
Sweet-cherry processing	Processing brines	Filtration + treatment with activated carbon + addition of SO_2 and lime	
Milk production	Bottle washing water, vapor condensate	sand filtration + gas chlorine, prefiltration + NF + UV	Wallis (1979)
Fish processing	Refrigeration brines in fishing stories vessels	UF	Welsh and Zall (1983)
Shrimps processing	Water recovered from steam cooker	UF	Senstad and Almas (1985)
Surimi production	First washing water	Centrifugation + UF	Pavia and Tyagi (1972)
Meat processing	Chiller shower water	Prefiltration + NF + RO + UV	
Swine processing	Carcass washing, dehairing and polishing water	Secondary treatment + chlorination	Miller et al. (1994)
Sausage production	Chiller spraying water	NF + UV + chlorination	Fahnrich et al. (1998)
Frankfurter production	Chiller brine	MF	
Poultry processing	Carcass washing water	Screening + diatomaceous earth filtration + ozonation	
Beer production	Press water from Brewer's spent grains	Centrifugation	
Ice manufacturing	Sterilization of water	Ozone	Anonymous (1995)
Bottle washing	Wash water	Prefiltration + NF + RO + UV	
Bottle cleaning machines	Washing water	RO + UV	Scharnagl et al. (2000)
Food production	Sterilization of spring water	Ozonation	Franz and Gagnaux (1971)

MF, Microfiltration; NF, nanofiltration; RO, reverse osmosis; UF, ultrafiltration; UV, ultrafiltration radiation.

dishes, containers, chilling, bottle washing, and so on. It is also used in huge amounts in food processing as a prime ingredient or element. The increasing demand of water in food industries and limited sources of water has forced food industries to reuse and recycle water by adopting proper treatment processes and technologies to meet high water-quality parameters. Due to limited source of water, use of fresh water for food industries has become very expensive practice, so reuse and recycle of water is also a cheap practice for food industries. Food industry is a wide industry, there is so much complexity and variation in characteristics of wastewater found in food industries effluents, so selection of technology to treat wastewater from food industries and meeting high quality parameters is a very complicated task. In the present condition nanotechnology has emerged as an efficient, cheap, effective, and ecofriendly approach to treat wastewater. There are many applications of nanotechnology for treatment of wastewater from food industries, such as making more sensitive quality indicators, nanostructured membranes, application of nanocatalyst for wastewater treatment, application of nanomaterials for absorbent to increase surface area and removal efficiency for contaminant, nanofiltration, and so on. There is no clear opinion of scientists about toxicity of application of nanotechnology on environment.

References

Anonymous, 1995. New method for improvement of the quality of crushed ice. Fleischerei 46 (5), 86–88, (in German).

Bereka, M.M., 2015. Nanotechnology in food Industry; Advances in food processing, Packaging and food safety. Int. J. Curr. Microbiol. App. Sci. 4 (5), 345–357.

Bellona, C., Drewes, J.E., 2007. Viability of a low-pressure nanofilter in treating recycled water for water reuse applications: a pilot-scale study. Water Res. 38, 2795–2809.

Bolan, N.S., Wong, L., Adriano, D.C., 2004. Nutrient removal from farm effluents. Bioresource Technol. 94, 251–260.

Buzea, C., Blandio, I.I.P., Robbie, K., 2007. Nanomaterials and nanoparticles: sources and toxicity. Biointerphase. 2 (4), 17–172.

Cameron, K.C., Di, H.J., McLaren, R.G., 1997. Is soil an appropriate dumping ground for our wastes? Aust. J. Soil. Res. 35 (5), 995–1035.

Carroad, P.A., Rose, W.W., 1977. Water recycle improves disc cleaning of tomatoes. Food Technol. 31 (3), 92–96.

Casani, S., Rouhanyb, M., Knøchel, S., 2005. A discussion paper on challenges and limitations to water reuse and hygiene in the food industry. Water Res. 39, 1134–1146.

Chau, C.F., Wu, S.H., Yen, G.C., 2007. The development of regulations for food nanotechnology. Trends Food Sci. Technol. 18, 269–280.

Chaudhry, Q., Scotter, M., Blackburn, J., Ross, B., Boxall, A., Castle, L., Aitken, R., Watkins, R., 2008. Applications and implications of nanotechnologies for the food sector. Food Additives Cont. 25 (3), 241–258.

Chellappa, M., Anjaneyulu, U., Manivasagam, G., Vijayalakshami, U., 2015. Preparation and evaluation of the cytotoxic nature of TiO2 nanoparticles by direct contact method. Int. J. Nanomedicine. 10 (1), 31–41.

Codex Alimentarius, 1999. Codex Alimentarius Commission: Codex Committee on Food Hygiene. Discussion Paper on Proposed Draft Guidelines for the Hygienic Reuse of Processing Water in Food Plants. Joint FAO/WHO Food Standards Programme, 32nd Session, Washington, DC, USA.

Di, H.J., Cameron, K.C., 2002. Nitrate leaching and pasture production from different nitrogen sources on a shallow stoney soil under flood-irrigated dairy pasture. Aust. J. Soil. Res. 40, 317–334.

Fahnrich, A., Mavrov, V., Chmiel, H., 1998. Membrane processes for water reuse in the food industry. Desalination 119 (1–3), 213–216.

Franz, J., Gagnaux, A., 1971. Sterilization with ozone under extremely high sterilization requirements. Wasser Luft Betrieb 15 (11), 393–396, (in German).

Goodwin, S., Catley, G., 2001. Membranes boost water re-use. Water Waste Treat. 13-14, 16.

Hart, M.S., Huxsoll, C.C., Tsai, L.-S., Ng, K.C., 1988. Preliminary studies of microfiltration for food processing water reuse. J. Food Prot. 51 (4), 269–276.

Hawke, R.M., Summers, S.A., 2003. Land application of farm dairy effluent: results from a case study, Wairarapa, New Zealand. New Zeal. J. Agr. Res. 46, 339–346.

Hillie, T., Munasinghe, M., Hlope, M, Deraniyagala, Y., 2006. Overview and comparison of conventional water treatment technologies and nano-based treatment technologies. Workshop on Nanotechnology, Water and Development, 10-12 Chennai, India. Meridian-Institute.

Leclair, K., Heggart, H., Oggel, M., Barlett, F.M., McKellar, R.C., 1994. Modelling the inactivation of Listeria monocytogenes and Salmonella typhimurium in simulated eggwash water. Food Microbiol. 11, 345–353.

Longhurst, R.D., Roberts, A.H.C., O'Connor, M.B., 2000. Farm dairy effluent: a review of published data on chemical and physical characteristics in New Zealand. New Zeal. J. Agr. Res. 43 (1), 7–14.

Longhurst, R.D., Luo, J., 2007. On-off farm winter management practices: potential environmental benefits and issues. In: Currie, L.D., Yates, L.J. (Eds.), Design Sustainable Farms, Critical Aspects of Soil and Water Management. Occasional Report No. 20. Fertilizer and Lime Research Centre, Massey University, Palmerston North, New Zealand, pp. 403–410.

Miller, A.J., Schultz, F.J., Oser, A., Hallman, J.L., Palumbo, S.A., 1994. Bacteriological safety of swine carcasses treated with reconditioned water. J. Food Sci. 59 (4), 739–741, 746.

Pavia, E.H., Tyagi, A.D., 1972. Safe and economical reuse of transport water in fishmeal and oil industry. Mech. Eng. 94 (6), 66–68.

Rajagopal, R., Saady, N.M.C., Torrijos, M., Thanikal, J.V., Hung, Y.T., 2013. Sustainable agro-food industrial wastewater treatment using high rateanaerobic process. Water 5, 292–311.

Reina, L.D., Flemming, H.P., Humphries, E.G., 1995. Microbiological control of cucumber hydrocooling water with chlorine dioxide. J. Food Prot. 58 (5), 541–546.

Scharnagl, N., Bunse, U., Peinemann, K.-V., 2000. Recycling of washing waters from bottle cleaning machines using membranes. Desalination 131, 55–63.

Senstad, C., Almas, K.A., 1985. Use of chitosan in the recovery of protein from shrimp processing waste. In: Proceedings of the third International Conference on Chitin/Chitosan, Muzzarelli ed., Ancona, Italy.
Theron, J., Walker, J.A., Cloete, T.E., 2008. Nanotechnology and water treatment: applications and emerging opportunities. Crit. Rev. Microbiol. 34, 43–69.
Varnekar, P., Satyanarayan, S., Satyanarayan, S., 2013. Treatment of foodprocessing industry wastewater by a coagulation/flocculation process. Int. J. Chem. Phys. Sci. 2, 63–72.
Wallis, A.P.L., 1979. Water recovery systems for processing dairies. J. Soc. Dairy Technol. 32 (3), 141–145.
Wang, H., Magesan, G., Bolan, N., 2004. An overview of the environmental effects of land application of farm effluents. N. Z. J. Agric. Res. 47, 389–403.
Welsh, F.W., Zall, R.R., 1983. Use of a model scale system for recycling spent fishery brines. J. Food Prot. 46 (12), 1026–1031.
Weiss, J., Takhistov, P., Mcclements, D.J., 2006. Functional materials in food nanotechnology. J. Food Sci. 71, R107–R116.

NANOCOMPOSITE FILTRATION MEMBRANES FOR DRINKING WATER PURIFICATION

Priscila Anadão
University of São Paulo, Cidade Universitária, Department of Metallurgical and Materials Engineering, Engineering School, São Paulo, SP, Brazil

1 Introduction

Nanotechnology has been recognized worldwide as one of the key subjects in science and technology in the 21st century because it is able to promote benefits, which will affect society radically by means of its applications in a wide range of sectors. This technology has an important role in the development of new materials, the nanomaterial ones, to substitute the existing products aiming to improve their performance. Nanomaterials have at least one of their dimensions larger than 1 nm and smaller than 100 nm, which makes them able to develop unique properties deriving from these dimensions (Rejeski and Lekas, 2008; Robinson, 2009; Sellers et al., 2008; Takemura, 2008).

The environmental pollution reduction area, especially in the water and wastewater treatment, will be benefited by using nanomaterials due to the expansion of the types of materials and creation of new applications and technologies (Anadão et al., 2011; Li et al., 2008; Wang et al., 2010).

Membrane technology has been largely employed in this sector by using a filtration technique, since it represents a substantial construction cost and operation reduction in treatment plants, besides being competitive in relation to the conventional treatment systems. Another important characteristic of this technology is its efficiency and specificity, which allows reaching the limits of pathogens and chemical contaminants in potable water and treated wastewater. The regulatory authorities have made the evaluation parameters of treated water quality increasingly stricter; conversely, a significant increase of the contamination level of the fountainhead is observed. Besides, scholars also express

Water Purification. http://dx.doi.org/10.1016/B978-0-12-804300-4.00015-0

concerns with regard to aquifers, which means that the adoption of more enhanced techniques for water treatment is necessary, in order to eliminate the problems caused by growing environmental degradation.

One of the factors that contributed to the filtration membrane popularization is the fact that the modules need a reduced space to implement a treatment plant, which allows its use in large metropolises that have few available areas for building treatment plants (Anadão, 2010).

In the water and wastewater membrane filtration, be it micro-, ultra-, nanofiltration, or reverse osmosis, polymer membranes are employed; the preparation generally used is the wet-phase inversion method. By using a ternary system composed of polymer/solvent/nonsolvent, this process occurs by dissolving polymer in an appropriate solvent or a solvent mixture, casting this solution in a thin film form and later immersing it in a nonsolvent bath. In this last step, solvent, and nonsolvent exchange occurs, causing various phase transformations and resulting in the final membrane.

Aiming to reduce filtration costs and make the process competitive, countries need to invest massively in research and development of the membrane, besides meeting regulatory norms. The shortage of water not only derives from climate problems, but also from population growth, associated to the growing water pollution levels and the absence of adequate treatment in large scale. This requires the development of an industrial technology that guarantees, on the one hand, an accentuated production rate and, on the other hand, a better quality at a lower cost, therefore requiring research into new materials to prepare the membrane and provide the required flow-rate.

One of the strategies to improve certain membrane properties is the addition of a fourth component to the phase inversion technique. Several properties are reported, such as improvement of solution rheology, mechanical resistance, lifetime, pore interconnectivity, and macropore suppression, resulting in superior permeability combined with better retention properties.

The addition of this material therefore originates a mixed matrix membrane in which there is a dispersed phase in a polymer matrix. By adding this component, improvements in several properties are reported, such as mechanical, thermal, chemical, among others. Moreover, by using a dispersed material, preferable paths can be created, giving rise to selective permeation due to the barrier imposed on undesirable permeation of certain solutes, thus enhancing the separation performance (Ghaemi et al., 2011).

Researches focused on membrane preparation with improved properties to be employed in water filtration treatment are convenient in this context. Therefore, the aim of this chapter is to present the membrane processes frequently employed for drinking water purification, as well as their different types of nanocomposite membranes. Theories related to each membrane process, such as micro, ultra, nanofiltration, and reverse osmosis, will be discussed and the concepts related to nanocomposite technology will be presented. After that, the most recent scientific and technological advances in the development of nanocomposite membranes for water filtration will be shown. Finally, the future research toward producing these high-performance membranes will be also argued.

2 Membrane Processes

Membrane is a selective barrier between two phases that, subjected to a driving force, allows the passage of one or more selected species of the feed phase. The species retained by the membrane constitute the concentrated phase; whereas the species that pass through it consists the permeate phase (Fig. 15.1).

The membrane processes generally used in water purification are microfiltration, ultrafiltration, nanofiltration, and reverse osmosis. Their principles will be discussed as follows.

2.1 Microfiltration

Microfiltration is a separation technique able to remove particles with a micron size, such as bacteria, yeast cells, colloids,

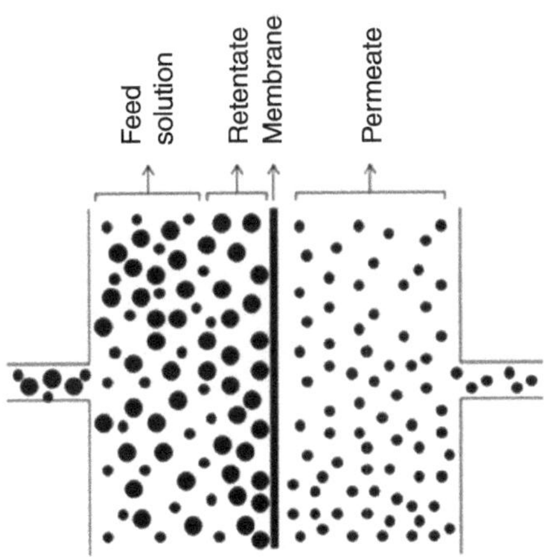

Figure 15.1. Schematic representation of permeate and retentate in a filtration operation.

particles from suspensions, or gases by passing through a membrane with has pore diameter between 0.1 and 10 μm. The fluid goes through tortuous paths whereas the particles are rejected on the filter surface. The driving force of this process is the pressure gradient formed between the two sides of the membrane.

This is the largest industrial market in the membrane separation area, corresponding to 40% of the total sales of separation processes in Europe and in the USA. In 1997, in the USA, microfiltration represented a 400-million dollar income, with an annual growth rate of 6.6%. In 2010, membranes for liquid filtration exceeded 1.2 billion dollars. Factors, such as federal regulations, industrial patterns, and need of equipment replacement, besides decrease of operational costs, lever the microfiltration membrane commerce.

The efficiency of microfiltration is guaranteed by the following membrane properties:

- uniformity of pore size, since membranes will not retain anything smaller than their pores;
- pore density;
- layer thickness.

The impact of these parameters in the flux can be examined by the Poiseuille law:

$$Q/A = (\pi \Delta p / 8 \mu \delta) \sum n_i d_i^{4}$$

in which Q/A is the volumetric flux rate by membrane area unit, Δp is the pressure drop, μ is the solution viscosity, δ is the thickness of the pore active layer, and d_i is the individual pore diameter of the area unit A.

There are two microfiltration types. The first one is perpendicular filtration (Fig. 15.2), in which the feed solution is forced through the membrane by an applied pressure. Both feed solution and permeate are perpendicular to the membrane surface; then, all fluid which enters the filter emerges as permeate, a 100% conversion occurring in the first passage. Suspended particles are continuously dragged through the membrane, being retained on the surface or inside the pores. The particle deposition leads to a continuous flux resistance, resulting in a continuous reduction of the permeate flux rate.

As filtration is perpendicular, solid concentration must be lower than 0.5%. Due to this restriction, there is a second type of filtration, the parallel one (Fig. 15.2). This filtration is carried out with a feed solution in a parallel direction to the membrane surface, producing two types of solutions, a clean filtrate and a concentrate that contains the majority of the solution particles. With

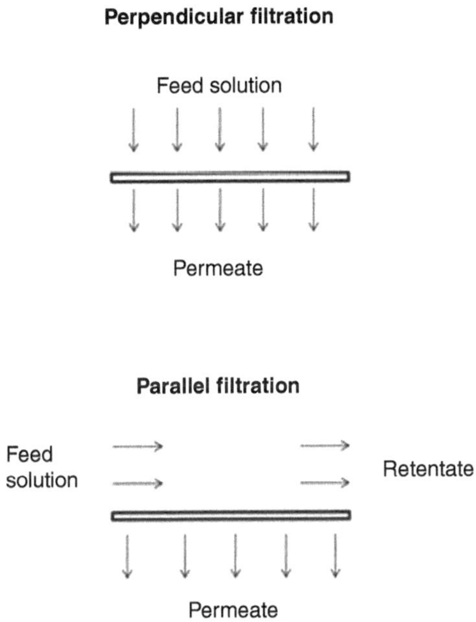

Figure 15.2. Schematic representation of perpendicular and parallel filtration.

a high feed solution flux rate (with a rate of 10–100 times higher than the permeate rate), the retained material is dragged off the membrane surface, decreasing fouling probability and extending the membrane lifetime indefinitely. Less than 20% of the feed solution goes through the filter.

This process is efficient in green algae separation, which cause smell and taste after contact. Besides, the removal of viruses is also effective.

2.2 Ultrafiltration

Ultrafiltration (UF) is the second most used process to purify water. This process is based on size exclusion, able to fractionate liquid mixtures, concentrate, or purify by sieving. UF membranes have pores between 10 and 1,000 Å, retaining molecules with molecular mass between 300 and 500,000 Da. This means solutes between 0.002 and 0.1 μm in size, except for sugar, biomolecules, polymers, and colloidal particles (1 Da is equivalent to 1/16 of the oxygen atom mass) and only allowing smaller molecules to pass through the pores. Its driving force is the difference of transmembrane pressure, operated at 2 and 10 bars. Generally, its operation is in the parallel mode, although there are some works in the literature describing operation in the perpendicular mode.

The UF volumetric flux (permeate volume/membrane area *x* time) is given by:

$$J = \frac{P(\Delta p - \Delta \pi)}{l}$$

in which *P* is the permeability coefficient, Δp is the difference of hydrostatic pressure, $\Delta \pi$ is the difference of osmotic pressure between the feed solution and the permeate, and *l* is the membrane thickness. In the SI system, the flux unity is micrometers per second; however, by multiplying by 3.6, a more conventional unity is used, liter per meter square hour.

In low-pressure situations, flux is linearly dependent on pressure variation. As the retained species become more concentrated on the membrane surface, the independence of pressure becomes a rule and velocity is controlled by mass transference and no longer by pressure. At average pressures, these factors are combined.

The solution rejection (R_s) is the measurement of the membrane capacity.

$$R_s = 1 - \frac{c_p}{c_r}$$

Where c_p is the solution concentration in the permeate and c_r is the concentrate concentration.

Besides separation by molecular size, the interaction between membrane and solute is important. For this reason, membrane selection is important since the membrane surface can influence concentration-polarization and fouling formations. UF membranes are frequently described by the rejection of the nominal molecular mass (MWCO) defined by the species with the lowest molecular mass the membrane is able to reject at 90%; that is, those with R_s value with at least 0.9.

The rejection curve is constructed by R_s value measurements of several similar compounds with chemically different molecular masses. Such profile can be characterized by an average rejection value and a dispersion coefficient (Δ), which is the measurement of the precision of the separation. Low Δ values represent the ability of fractionate solutes more efficiently but the conventional values are between 0.7 and 2.8 and the Δ isoporous membranes value is 0.25. The MWCO value is not so useful because the polarization-concentration layer and fouling act as a filter, saddling another retention to the process, besides there being pressure variations, and also varying polarized layers.

The molecules generally used to determine MWCO are: dextran, carbowax fractions, and polymeric surfactants. The proteins are not efficient to be used because their size depends on pH, ionic force, and interactions with solvent molecules. In the case of other molecules commonly employed, factors, such as molecular orientation, molecular configuration, pH, ionic force, operation conditions, and chemistry of the feed solution also affect the MWCO membrane determination. In some cases, the precision of nominal rejection is determined on the basis of standard deviation of the molecular mass, resulting in 30–80% of rejection.

As the diffusivity of the liquid phase is low, the concentration-polarization and fouling are topics that need attention in the UF study. The concentration-polarization is present in all pressure-driven processes, concerning the selected solute on the surface with c_w concentration. This c_w concentration is determined by the balance between the solute dragged to the surface by convective flux of the solvent and the solute that leaves the surface by a diffusion process. The concentration-polarization leads to an increase of small increments in the flux as pressure is increased and until a gel layer is formed and the flux no longer increases, which is called limiting flux.

Fouling regards the deposition of some components of the feed solution on the surface or inside the pores. If the c_w concentration reaches a point in which the solute precipitates or forms a thixotropic gel, this gel layer can provide an additional resistance to this membrane and the flux equation must be rewritten:

$$J = P\frac{(\Delta p - \Delta\pi)}{R_m + R_c}$$

in which $R_m = 1/P$ is the membrane resistance to the mass transport and R_c is the resistance corresponding to the cake or gel layer. R_c can be higher than R_m and the flux will not depend on the membrane permeability.

It should be emphasized that the use of higher pressure values can cause a reduction in the flux and not in its increase, as the equation would predict, since higher pressures favor retained material compaction onto the membrane surface, increasing the concentration-polarization layer and the possibility of fouling formation.

UF is seen as a highly efficient technique for potable water production, if conjugated with reverse osmosis systems, acting as a pretreatment. In China, such operation has been used in seawater treatment for potable water production in the Liaoning province, yielding 28,000 m^3 water/day by using hollow fiber modules.

2.3 Nanofiltration and Reverse Osmosis

Osmosis is a natural phenomenon in which water goes through a semipermeable membrane from the side with the lowest concentration to the side with the highest concentration, until the equilibrium of the chemical potential is reestablished. At equilibrium, the pressure difference between the two sides of the membrane is equal to the osmotic pressure difference.

To revert, water flux, a pressure difference higher than the osmotic pressure gradient is applied; as a result, the separation of water and solutes is possible. This phenomenon is called reverse osmosis or frequently hyperfiltration.

Reverse osmosis is a liquid–liquid separation process that uses a dense semipermeable membrane, highly permeable to water and extremely impermeable to microorganisms, colloids, dissolved salts and organics. By applying a pressure higher than the osmotic pressure, the solvent goes through the reverse osmosis membrane, which acts as a flux barrier to solutes, allowing solvent and solute separation.

The driving thermodynamic forces involved in the solute and solvent transport through the membrane: $-\Delta\mu_s$ = (solute chemical potential gradient) e – $\Delta\mu_w$ (water chemical potential gradient). Since these amounts are not measurable, the driven forces to the reverse osmosis are normally related with the differences in concentrations (chemical activity) and pressure through the membrane.

Solution separation and solvent permeation depend on several variables: membrane material that will define the affinity between membrane and solute, types of solute, feed solution rate, concentration of dissolved solutes in the feed solution, structure of the barrier layer of the membrane, transmembrane pressure (ΔP), temperature, pH, and suspended solid concentration if they exist in the solution.

Aiming at higher efficiency, the process should follow these requirements: high intrinsic permeability or low-salt intrinsic permeability, large membrane surface area, thin free-defect membrane, and low loss of pressure through the membrane.

Reverse osmosis can be classified into three types, according to the pressure used in the process:

- High-pressure reverse osmosis: 5.6–10.5 MPa, used in desalination to obtain inorganic salt rejection (95–99.9% NaCl rejection) and moderate organic compounds rejection;
- Low-pressure reverse osmosis: 1.4–4.2 MPa, also employed in saline water desalination, reaches high inorganic rejection (95–99.9% NaCl rejection);

- Nanofiltration: 0.3–1.4 MPa, employed in situations of partial demineralization, with high bi or trivalent ion rejection and low rejection of monovalent ions (between 30% and 85%).

Reverse osmosis membrane pores have dimension between 5 and 20 Å, whereas nanofiltration membranes have greater dimensions, separating intermediate particles between UF and reverse osmosis.

When a membrane has charged groups in its structure, high salt separation can be reached because of Donnan dialysis. For example, for a negatively charged membrane, rejection follows this sequence: $Na_3PO_4 > Na_2SO_4 > NaCl$. Nanofiltration membranes normally have negative charged groups and provide selective separations.

Reverse osmosis is established by three parameters:

water flux, J_w:

$$J_w = \text{water flux} = \frac{\text{volume, mass, or molar rate of permeation}}{\text{membrane area}}$$

$$J_w = K(\Delta P - \Delta\pi)$$

K (g/cm²·s·atm) is the hydraulic membrane constant, ΔP is the applied pressure (atm), and $\Delta\pi$ is the osmotic pressure (atm). The J_w unit is g/cm²·s.

The salt flux, J_s, is:

$$J_s = B(C_1 - C_2)$$

in which B is the salt permeability constant (cm/s) and $C_1 - C_2$ is the salt concentration difference through the membrane (g/cm³). The salt flux is independent on the applied pressure, whereas water flux is dependent, that is, the membrane becomes more selective with the increase of pressure.

To calculate the solute osmotic pressure in low concentration, van't Hoff equation can be used:

$$\pi = cRT$$

in which c is the molar concentration that can include solute dissociation effects, R is the gas constant in appropriate units, and T is the absolute temperature. For higher concentrations, more sophisticated models are used.

The solute rejection, R, is:

$$R = 1 - \left(\frac{c''}{c'}\right)$$

in which c'' is the solute concentration in the permeate and c' is the solute concentration in the feed solution.

Around 50% of the total sales concerning reverse osmosis are related to brackish and seawater. In 2007, sales reached 1.8 billion dollars, being this increase explained by the reduction of the operational costs and low energy consumption. In China, this market is very promising for the use of reverse osmosis, since there are few natural resources and these are not geographically well distributed.

Seawater desalination is possible nowadays due to the development of new composite membranes with high salt rejection, to the reduction of energy consumption, to the increase of the installed capacity by membrane arrangement that allows a better hydraulic optimization and to the increase of pumping energy efficiency. There are several plants in operation in different parts of the world, which employ spiral and hollow fiber membranes.

Seawater is generally pretreated by processes, such as sedimentation, pH adjustment, sand filtration, coloration, and microfiltration. This pretreatment is necessary to maintain the high performance of the reverse osmosis membranes. Hence, a large number of researches address several types of pretreatment, such as coagulation, direct filtration, UF, microfiltration, and activated carbon sorption, besides the connection with other membrane separation processes, such as nanofiltration and electrodialysis (Anadão, 2010).

3 Nanocomposites

3.1 A Brief History About Nanocomposites

On December 29, 1959, Richard Feynman, a physicist, presented a lecture whose title was "There is plenty of room at the bottom" at the American Physical Society. This lecture is considered as a milestone of nanotechnology, since Feynman suggested the use of nanotechnology in the information storage, as well as a number of new techniques to support this technology (Zyvez, 2012). Thereafter, due to the theoretical and experimental results, the technological and scientific field of nanometric scale is becoming stronger. In this backdrop, 1 trillion of dollars is the amount expected for the worldwide nanotechnology market in the next 5 years (National Science Foundation, 2015).

Until the early 1970s, since minerals are heavier and cheaper than polymers, these fillers were mixed with polymers as fillers commercially to reduce costs. By the 1970s, the 1973 and 1979 crisis provoked the increase of the petroleum price. In the 1970s, the petroleum price increased vertiginously and successively during and after the 1973 and 1979 crises. Hence, an expansion of the ceramic raw materials as fillers was a result of the petroleum price increased besides the polypropylene introduction in commercial scale. It was also observed the development of composites containing with mica, glass spheres and fibers, talc, calcium carbonate. At the same time, researches related to the interactions between fillers and polymers were initiated (Rabello, 2010).

A higher understanding of the mechanisms of interactions with different types of polymers was obtained in the next 2 decades, raising mineral from the simple filler condition to the functional filler condition.

Works, which guided to the development of thermoplastic nanocomposites began to appear in the literature in the end of the 1980s, although polymer nanocomposites were discovered before this decade (Carter et al., 1950). These thermoplastic nanocomposites were developed by Toyota by using polyamide 6 and organophilic clays specially prepared for their preparation (Kawasumi, 1989; Okada et al., 1988).

As a result of this scenario, these nanocomposites were applied in timing belt covers of Toyota vehicles, in collaboration with UBE Company, a Japanese polyamide 6 industry. A very small clay amount (5 wt.% special clay) was responsible for the improvement of the properties of this material as compared to that of pure polyamide 6. The final nanocomposite presented an increase of 40% in the rupture tension, 60% of the tensile modulus and 126% in the flexion modulus. Thermal distortion temperature was also augmented from 65 to 152°C in comparison with pure polymer.

Since then, a great number of companies have presented thermoplastic nanocomposites, the majority of them with polyamide and polypropylene for automotive applications. The need of obtaining sufficient volume of this material justifies the choice for the automotive sector. Moreover, General Motors, that is the industry that employs the most nanocomposites in the world, used only 245 tons of these materials in 2003 (Cox et al., 2004; Patterson, 2004).

Gas barrier is another application that has gained attention in the last years, by using polyamide and polyester as the matrix (Conway, 2004; Goldman and Copsey, 2004). Since, although their performance is better, the competition with other materials is harmed when the cost–benefit ratio is analyzed, researches should focus the development of a large-scale commercial application of these materials.

3.2 Nanocomposite Definition

Hence, it is opportune to define the concepts regarding nanomaterials. The nanometric definition considers that the material dimensions are between 1 and 100 nm. Hence, the nanomaterials are the materials that have one, two, or three external dimensions in the nanometric domain. Among the nanomaterials, the isodimensional nanoparticles contain their three dimensions in the nanometric domain; the nanosheets have one nanometric dimension while the nanofibers are the nanomaterials that present two nanometric dimensions (Gaffet, 2011).

Nanocomposite materials are two-phase systems that consist of a matrix and dispersed particles of nanometer scale. Generally, nanocomposite filtration membranes present a polymer matrix and inorganic dispersed particles, which can be particles (e.g., calcium carbonate), fibers (e.g., glass fibers), or plate-shaped particles (e.g., mica). These polymer matrices reinforced with the nanoparticles combine the functionalities of polymer matrices, which include low weight and easy formability, with the unique features of the inorganic nanoparticles. Therefore, the nanocomposites obtained by incorporation of these types of materials report improvements in several areas, such as optical, mechanical, electrical, magnetic, rheological, and fire retardancy properties. However, some drawbacks are often described in literature as weight increase, brittleness, and opacity (Choudalakis and Gotsis, 2009; Kangoa et al., 2013).

3.3 Polymers

Since nanocomposite membranes are generally made of polymer, it is important to present its concepts. Polymers are constituted by large molecules and are often called as macromolecules. The linkage between their atoms is covalent. Long and flexible chains compose most of polymers and rough sketch is generally made of carbon atoms (Fig. 15.3). These carbon atoms have two valence electrons not shared in the bonds, being part of the linkages between other atoms or radicals.

Mero is the small repetitive unit that composes these chains and this word is derived from the Greek word meros, which means part. Therefore, monomer represents one part and several meros is a polymer. These chains are composed of small repetitive units called mero. The polymer is considered a homopolymer when

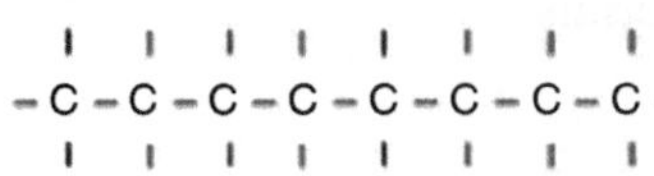

Figure 15.3. An example of an organic polymer chain.

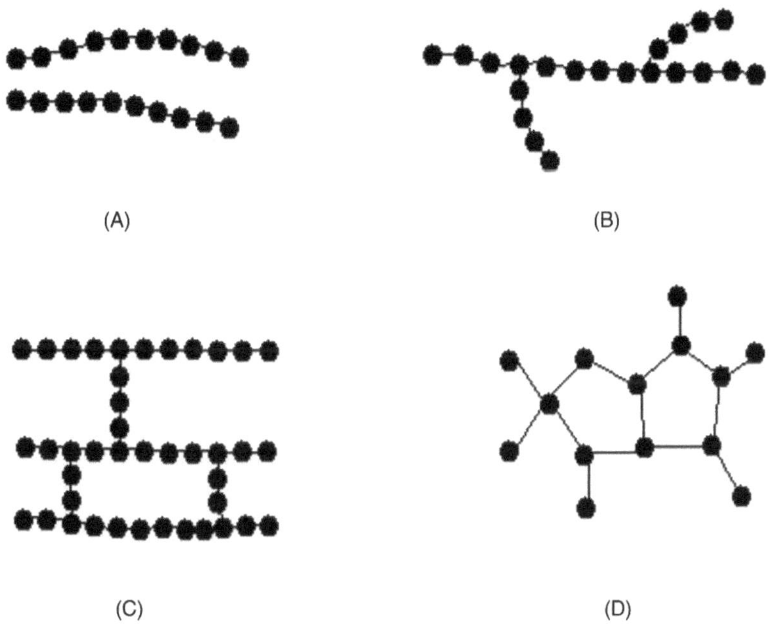

Figure 15.4. Types of polymers, according their molecular structure. (A) Linear, (B) ramified, (C) with crossed bonds, and (D) network.

all meros of the polymer are equal and a copolymer is a polymer composed of two or more different meros.

With respect to the polymer molecular structure, polymers can be: linear, which is composed of meros united in a single chain; ramified, which presents lateral ramification linked to the main chain; polymer with crossed bonds, with united linear chain by covalent bonds and network, which presents trifunctional meros that has three active covalent bonds, forming 3D networks (Fig. 15.4).

According to their structures, polymers can be amorphous or semicrystalline. Amorphous polymers have polymer chains arranged in a disorganized manner due to the great number of radicals bonded to the main chain, which cannot allow a closed stacking. On the other hand, semicrystalline polymers are polymers with linear chains and small groups, which are grouped in a more oriented form, forming crystals.

The polymer structure can originate two types of polymers, which are thermoplastics and thermofixes. The main feature of thermoplastic polymers is the possibility of conforming it mechanically several times with reheating due to the shear of the intermolecular bonds. As a general rule, thermoplastic polymers are linear and ramified and thermofix polymers are network.

Temperature does not soften thermofix polymers as these macromolecules present crossed bonds in the 3D structure. Hence, they are not recyclables (Callister, 1994).

3.4 Nanocomposite Production Methods

Two methods are generally employed in nanocomposite preparation. The first method developed is the in situ polymerization. In this method, the used medium for clay dispersion is the monomer and the polymerization between the clay mineral layers is performed when favorable conditions are imposed. Heat or radiation, diffusion of an adequate initiator or a fixed catalyzer inside the layers before the filling step by the monomer can initiate polymerization.

Polymerization is the driving force for complete exfoliation. Since the nanoparticles have high surface energy, attraction is exerted by these particles on the monomer units toward the inside of the galleries until the moment that equilibrium is reached. The polymerization reactions occur and displace equilibrium between the nanoparticles and there is the diffusion of new polymer chains between these particles.

Solution dispersion is another method used to produce nanocomposites. A solvent medium is used to disperse nanoparticles, where the polymer (or prepolymer in the case of insoluble polymers) is soluble. As weak forces maintain the nanoparticles united, an adequate solvent can disperse these materials due to the disorganization, which promotes the increase of entropy. After that, polymer is adsorbed in the nanoparticles and the nanoparticles are reunited and filled with polymer when the solvent is evaporated (Ke and Stroeve, 2005; Souza et al., 2006).

4 Nanocomposite Membranes for Water Filtration

Researches into nanocomposite filtration membranes have received great attention in this decade and then, a great number of papers and patents have been published in several journals. Hereinafter, examples of the investigative advances related to this area will be presented, according to their preparation method: in situ polymerization and solution dispersion. For a better understanding, nanocomposite membranes are presented according to their polymers.

4.1 In Situ Polymerization

Silica nanoparticles produced from organically functionalized silicon alkoxide precursors were incorporated into a polyamide (PA) film that was onto a polysulfone (PSf) support. These membranes presented improved features compared to the neat PA membrane. The used technique was successful since a good interaction was established between polymer and silica by means of absorption of diamine monomers by silica and consequent formation of silica-PA particles. It was observed that the higher the silica content, the higher the aggregate size (Jadav et al., 2010).

PA reverse osmosis membranes were also prepared with aluminosilicate single-walled nanotubes (SWNT). A membrane with desirable reverse osmosis characteristics was obtained, since a rugous structure with "leaf-like" outgrowths and the "ridge-and-valley" structure was obtained. Furthermore, the hydrophilicity and consequently, water flux and pure water permeance were enhanced while sustaining high rejection of monovalent and divalent ions, typical of PA reverse osmosis membrane (Baroña et al., 2013).

Amine functionalized multiwalled carbon nanotubes were used to form PA thin-film nanocomposite membranes. By increasing carbon nanotube concentration, surface hydrophilicity was improved as surface properties were changed. Besides that, membranes presented a high water permeability and acceptable salt rejection; the most permeable nanocomposite membrane had a water flux approximately 160% higher than a neat membrane (Amini et al., 2013).

Monodispersed spherical mesoporous nanosilica was synthesized by hydrothermal method and was used in the PA reverse osmosis membrane preparation. By increasing silica content, water flux increased from 19 to 53 L/m^2 h and all solute rejection rates exceeded 96%, showing their great potential to be applied in reverse osmosis desalination. The great hydrophilicity and spherical morphology of mesoporous silica were a good ally to improve water flux membrane (Bao et al., 2013).

Mesoporous silica was again used to prepare PA thin film nanocomposite membrane. By using an appropriate content, membrane showed a good performance. For example, a 0.03 wt.% content led to a maximum water flux, while maintaining the rejection to Na_2SO_4 at a relative high level besides presenting an enhanced-antifouling ability and good long-term stability due to the interaction between the mesoporous silica and the polymer (Wu et al., 2013b).

Two types of silica [LUDOX HS-40 (16 nm) and a 3 nm-size silica] were employed in the preparation of PA thin film nanocomposite membrane. It was observed that the nanoparticle changed the PA network structure, as well as pore structure and transport properties. With the increase of the silica content, the number of pores increased and these structures were more thermally stable in comparison with neat PA membrane. The membrane containing 1–2 wt.% silica exhibited the best performance in terms of separation efficiency and productivity flux (Jadav and Singh, 2009).

LUDOX HS-40 was used in another study that prepared PA nanocomposite films from trimesoyl chloride-*m*-phenylene diamine based PAs. For the membranes containing 4.5% silica, the surface of silica particle interacted well with the PA chains and the surface coated silica take the form of core–shell morphology. The excess of silica caused a silica–polymer phase separation (Singh and Aswal, 2008).

NaA zeolite was dispersed within PA thin films. Reverse osmosis membranes with improved permeability and interfacial properties were obtained. Important properties of conventional membrane polymers as flexibility, ease of manufacture, high-packing density modules were combined with the zeolite properties which are tunable hydrophilicity, charge density, pore structure, and antimicrobial capability along with better chemical, thermal, and mechanical stability. Besides that, water molecules flowed preferentially through super-hydrophilic nanoparticle pores, while solute rejection remained comparable to pristine PA membrane (Jeong et al., 2007).

NaY zeolite-PA thin film nanocomposites were prepared on a polysulfone porous substrate. The use of this nanoparticle contributed to modify the membrane separation properties, enhancing the water permeability of the membrane due to the porous nature of the zeolite. As an example, the use of 0.1 wt.% zeolite increased 80% compared to the neat membrane. Nevertheless, by increasing the zeolite content, water permeability was reduced (Ma et al., 2012a).

A different type of zeolite, NaX, was incorporated into PA thin film reverse osmosis by using two levels of monomer concentrations and nanozeolite loadings. The nanocomposite membranes presented higher thermal stability and more water permeability than the pure PA membranes. Moreover, it was observed that RMS roughness, contact angle, and solid–liquid interfacial free energy were improved, film thickness was decreased and pore size and water flux were increased. The nanocomposite membranes with high concentration of monomers showed a high water flux and low salt rejection (Fathizadeh et al., 2011).

Another study showed four different types of nanocomposite reverse osmosis membranes formed by PA or zeolite A-PA thin films over a neat polysulfone or zeolite A-PSf support membrane. The three nanocomposite membranes presented several enhanced properties, such as superior separation performance, more hydrophilic surfaces, higher water permeability and salt rejection, and improved resistance to physical compaction (Pendergast et al., 2013).

PA thin film nanocomposites containing poly(methyl methacrylate)-multiwalled carbon nanotubes were prepared successfully presenting a 99% Na_2SO_4 and a 62% higher water flux than that of neat membranes, which confirmed that nanocomposite formation improved both selectivity and permeability (Shen et al., 2013).

Another study showed multiwalled carbon nanotube PA solvent-resistant nanocomposite membranes prepared using a microporous polypropylene membrane appropriately hydrophilized which had higher solvent fluxes compared to pure PA membranes. Nanofiltration membranes were also prepared using porous polyethersulfone substrate that presented better solute rejections and higher fluxes compared with commercially available nanofiltration membranes. The better properties were obtained due to the formation of a transport corridor between an individual multiwalled carbon nanotube and the surrounding polymer chain that provided a low-resistance pathway for solvent and water transport (Roy et al., 2011).

Alumina nanoparticles, synthesized by an aqueous sol–gel method, were used to produce PA thin film nanocomposite. A good dispersion of the nanoparticles onto the surface was obtained and a decrease of the water contact angle was observed which means that adding alumina nanoparticles can improve hydrophilicity. The performance on water flux and salt rejection revealed that the performance of nanocomposite membranes was superior to pure membrane. Although the permeate flux was improved, the enhancement of the solute rejection was not so significant (Saleh and Gupta, 2012).

Aquaporin-Z, which is a water channel protein with excellent water permeability and solution rejection, was used to prepare PA thin film nanocomposite membranes via interfacial polymerization. Cross-flow reverse osmosis experiments were carried out and it was observed that the nanocomposite membranes achieved significantly higher water permeability (~4 L/m^2 h bar) with comparable NaCl rejection (~97%) at an applied pressure of 5 bar. This permeability was approximately 40% higher compared to a commercial brackish water reverse osmosis membrane (BW30) and an order of magnitude higher than a seawater reverse osmosis membrane (SW30HR), demonstrating their great potential for desalination applications (Zhao et al., 2012b).

MCM-41 silica nanoparticles are another filler used to prepare PA thin film nanocomposite membranes for water purification. Porous and nonporous nanoparticles were added in concentrations varying from 0 to 0.1 wt.%. A good dispersion of these particles in the PA matrix was observed and the increase of the MCM-41 particle concentration resulted in an increase of hydrophilicity, roughness, and zeta potential. The permeate water flux was also increased while high rejections of NaCl and Na_2SO_4 were maintained (97.9 and 98.5 L/m^2 h, respectively). The pores of the zeolite contributed to an increase of water permeability (Yin et al., 2012).

Both silver nanoparticles and multiwalled carbon nanotubes were added in a PA thin film membranes by means of interfacial polymerization. These particles were well dispersed in the PA matrix. Furthermore, membrane permeability was enhanced due to the diffusive tunnels of the carbon nanotubes. Surface hydrophilicity, antibacterial, and antifouling properties were also improved by using the silver nanoparticles (Kim et al., 2012).

Hydrophilized ordered mesoporous carbon (H-OMC), prepared from silica template and argon atmospheric plasma was applied to increase PA membrane surface hydrophilicity. The results revealed that 5 wt.% H-OMC increased membrane hydrophilicity, wettability, and pure water permeability and decreased bovine serum albumin adsorption. By increasing H-OMC loading up to 6%, rejection of monovalent ion (NaCl) was reduced from 68% to 50%, but the rejection of divalent ions (Na_2SO_4) slightly decreased from 94% to 90%. Besides that, it was found that the use of this nanoparticle did not modify solution permeability in NaCl solution and slightly increased the solution permeability with Na_2SO_4 (Kim and Deng, 2011).

A strategy presented the preparation of inorganic-PA nanocomposite membranes using three different metal alkoxides which were: titanium tetraisopropoxide, bis(triethyoxysilyl)ethane, and phenyltriethoxysilane. These membranes presented a superior performance compared to the pure PA membrane. Greater amounts of metal alkoxide increased both pore size and water flux. Finally, the best results were obtained with the membrane prepared with phenyltriethosylane, which exhibited water flux twice bigger than that of pure PA membrane (Kong et al., 2011).

Another used filler was in situ hydrophilic surface modifying macromolecules in PA thin film membrane. It has been shown that these particles contributed to an excellent performance for desalination of salty water. Moreover, the addition of silver nanoparticles also improved antibacterial properties, whose potential was revealed by halo tests with *Escherichia coli* (Rana et al., 2011).

A silica-fluoropolyamide nanofiltration membrane was also reported in the literature. A high performance (water flux with 15.2 L/m^2 h) and high salt rejection (85% Na_2SO_4) were observed. Moreover, the membrane rejection to inorganic electrolyte solutions decreased in the order: Na_2SO_4, $MgSO_4$, $MgCl_2$, and NaCl, which were typical characteristics of negatively charged membranes (Hu et al., 2013).

A poly(arylene ether sulfone)/modified silica nanocomposite reverse osmosis membrane for seawater desalination was prepared by interfacial polymerization. The salt rejection and water permeation were 97.7% and 34.5 L/m^2 h, respectively. After chlorine aging, these membranes showed a decrease of 15% in salt rejection and an increase in water permeation of 4 L/m^2 h because the functional amino groups of silica protected the membrane surface from chlorine attack and improved membrane hydrophilicity (Kim et al., 2013a). Zeolites were also employed with this polymer aiming to enhance chlorination resistance and membrane performance. Prior chlorination test, salt rejection, and water flux were 98.8% and 37.8 L/m^2 h and after that, salt rejection decreased by only 12.7% and water flux increased by 2.5 L/m^2 h. Such values were a result of the high degree of cross-linking in reverse osmosis. Moreover, this structure helped to protect the active layer structure from degradation and to improve chlorine membrane resistance (Kim et al., 2013b).

Multiwalled carbon nanotubes/polyester thin film nanocomposite were prepared via interfacial polymerization and it was found that the carbon nanotube loading was very important to determine the final membrane properties. By controlling this parameter and also the kind of surfactant and reaction time, membranes with both high permeation flux and excellent selectivity could be obtained. These membranes also presented favorable values of Na_2SO_4 rejection and good long-term stability (Wu et al., 2013a).

Poly(vinyl alcohol) and aluminosilicate SWNT were used to prepare a thin film nanofiltration nanocomposite membrane onto a polysulfone PSf support. No large defects and cracks were observed. Whereas, the thickness of the PVA layers increased, membrane roughness was decreased and therefore, membrane hydrophilicity was increased which resulted in an improvement of the water flux and rejection of divalent ions (97%) and monovalent ions (59%). Higher rejection of divalent ions compared to monovalent ones can be ascribed to Donnan potential and steric effects, being then useful for the removal of hardness, as well as partial desalination (Baroña et al., 2012).

Polyvinylidene fluoride (PVDF) was hydrophilizated by using polyacrylic acid (PAA) and titanium dioxide nanoparticles. Antifouling properties were improved due to the immobilization of TiO_2 nanoparticles. Furthermore, the presence of hydrophilic functional groups of PAA and photocatalytic characteristic of titanium dioxide nanoparticles reduced hydrophobic adsorption of whey protein and modified membrane surface (Madaeni et al., 2011).

4.2 Solution Dispersion

SiO_2–GO (silica–graphene oxide) nanohybrids were used to produce PSf nanocomposite membranes by dispersing these particles in *N*-methyl-2-pyrrolidone, casting the dispersion onto a glass plate and immersing the resulting film into deionized water. As a result, an increase of hydrophilicity was observed, as well as an enhancement of water flux rate (twice), egg albumin rejection (>98%) and antifouling ability in comparison with the pure PSf membranes due to the SiO_2–GO sandwich structure, which helped its dispersion in the PSf matrix (Wu et al., 2014).

PSf hollow fiber membranes prepared with titanium dioxide showed enhanced humic acid removal as antifouling properties were improved. With the increase of air gap, a higher permeation flux was obtained due to the increased pore size, although a lower rejection of humic acid was also achieved (Hamid et al., 2011).

PSf/titanium dioxide UF membranes were successfully prepared and it was observed that the rheological behavior of the solutions changed by adding these nanoparticles from Newtonian viscous behavior to non-Newtonian viscous behavior with an increase of the viscosity which was responsible for important differences in morphology, porosity, and other properties. The addition of 2 wt.% titanium dioxide was responsible for an excellent water permeability, hydrophilicity, mechanical strength, and good antifouling properties with almost unchanged retentions. These improvements can be ascribed to the interactions between titanium oxide nanoparticles and polymer matrix. However, higher nanoparticle contents provoked a serious particle aggregation that resulted in a decline of the nanocomposite membrane properties. Moreover, these nanoparticles helped the treatment of emulsified oil wastewater (Yang et al., 2007).

Titanium dioxide and 2-hydroxyethyl methacrylate (HEMA) grafted with titanium dioxide were used to prepare PSf nanocomposite UF membranes in order to produce membranes without agglomeration of nanoparticles and leakage of these nanoparticles in the water treatment process. The modified titanium dioxide membranes showed better hydrophilicity, compatibility, and dispersibility in organic solvent and matrix polymer. Moreover, these membranes revealed superior water flux and antipollution performance in separation compared to pure PSf and titanium dioxide membranes (Zhang et al., 2013b).

An interesting comparison was carried out by using different nanofillers (silver, copper, silica, zeolite, and silver-zeolite) in the preparation of PSf nanocomposite membranes. The membrane with zeolite was the only one that presented improvements in water permeability and solute selectivity. The other membranes were more permeable, but presented a less selectivity since they have defects associated with a low degree of interaction between polymer and filler. Other properties were studied as: protein fouling resistance, which were higher for the membranes with zeolite and silver and bacterial fouling resistance, which were higher for silver and silver-zeolite based membranes due to their antimicrobial properties. It was concluded that zeolite-PSf and silver-zeolite PSf offered the best combination of permeability, selectivity, and fouling resistance compared to a poly(acrylonitrile) (PAN) membrane (Hoek et al., 2011).

Polysulfone-montmorillonite membranes have also been demonstrating great potential to be applied in water filtration. Polysulfone Udel P-1700, gently supplied by Solvay Advanced Polymers, was used; the solvent was *N*-methyl-2-pirrolidone and the nonsolvent was distilled water at room temperature. Sodium montmorillonite (MMT) from Wyoming (Sigma commercial name), with cation exchange capacity equal to 95 meq/100 g, was used without any treatment. This research studied different clay contents, which were 0.0, 0.5, and 3.0 wt.% in the preparation of 25 wt.% PSf membranes were prepared under vigorous mechanical stirring. These dispersions were cast onto a glass support and were immersed in distilled water bath. Next, membranes were completely detached from the glass surface. It was possible to note from small-angle X-ray diffraction results that, since MMT presented a basal spacing of 14.8 Å and both diffractograms presented a basal spacing of 13 Å, the formation of intercalated nanocomposites was revealed. A complex morphology with partial intercalation/exfoliation structure, besides PSf intercalation, was found in the membrane prepared with 3.0 wt.% clay. Scanning electron micrographs (SEM) was used to evaluate the dispersion of the clay particles in the PSf matrix. Clay particles could not be observed in the SEM cross-section micrographs of any of the membranes, which imply that there was a good dispersion in the PSf matrix. Furthermore, the cell-like cross-section morphology was not altered. Thus, nanocomposite membranes could be used in the same filtration applications to which PSf membranes were designated. The profile of the thermogravimetric curves was not changed by adding clay, either, in comparison with PSf properties and an increase in the onset temperature of decomposition was denoted, being explained by the barrier properties conferred by the clay mineral layers, which hampered oxygen diffusion into the material, and by the labyrinth effect, related to the tortuous path the oxygen molecules have to go through. Hence, the two combined effects delayed volatilization. It was also observed an improvement of the mechanical properties by the nanocomposite formation. Elongation ate break was improved presumably due to a rearrangement of the clay mineral layers in the direction of deformation. At last, the increase of the clay content increased hydrophilicity and the free energy of hydration values of the nanocomposite membranes were near to the value of true hydrophilic materials (-113 mJ/m^{-2}) (Anadão et al., 2010).

The influence of the clay content and the polysulfone molar mass (PSf P-1700 and P-3500) on nanocomposite membrane properties was also investigated. The increase of the basal spacing with the increase of clay content could be caused by the intercalation of PSf chains. All nanocomposite membranes showed intercalated clay mineral layers and aggregated clay mineral layers while exfoliated portions could be seen in the membranes with lower clay contents. Surfaces depicted the decrease in the number of pores and pore diameter with the increase of clay content for both types of PSf and the cross-section showed that the cellular morphology of the pure PSf membranes was not altered by clay addition. Thermal resistance was improved by using a higher clay content and polysulfone molar mass, since mass loss was decreased and the onset temperature of decomposition was increased. Finally, by increasing clay content up to 4 wt.% and by using higher polysulfone molar mass, mechanical resistance was enhanced (Anadão et al., 2013). Moreover, the membranes with low clay contents had dense surfaces as well as elongated macrovoids in the membranes. With the increase in solution viscosity, in the case of higher clay contents and higher molecular weight PSf, bulbous macrovoids were formed and membrane porosity was higher. The increase in both features augmented water sorption and reduced linear expansion. The PSf P-3500 membranes presented improvements in the contributions related to the acid-base interactions whereas the PSf P-1700 hydrophilicity values may be due to the roughness irregularity. By increasing PSf molecular weight and the clay content, the membranes revealed reduced permeabilities and higher selectivity. The membranes with higher clay contents and higher molecular weight PSf presented higher rejection values even with the use of low molecular weight solutes, which characterizes them as nanofiltration. The membranes with lower clay contents and lower molecular weight PSf were closer to the selectivity range characteristic of UF membranes (Anadão et al., 2014).

Solution dispersion in NMP was again used to prepare PSf nanocomposite membranes by using an unmodified (Cloisite Na) and organically modified (Cloisite 30B and 93) clays. The nanocomposite formation was observed by only using Cloisite 30 B, with an exfoliated/intercalated morphology whose structure led to enhanced mechanical properties and wettability, also allowing its use in UF operations (Monticelli et al., 2007).

The preparation and characterization of PSf/clay nanocomposite membranes with LiCl as a pore forming additive was also presented in the literature. A good dispersion of MMT in the PSf matrix was presented. It was also observed an improvement of membrane hydrophilicity and an increase of the ratio of large pore in the skin layer, of the pure water flux and of the membrane porosity. By increasing the clay content from 0 to 6 wt.%, the pure water flux increased from 199 to 263 L/m^2 h. The tensile strength reached a maximum value when the clay content was 3 wt.%. Clay had high stability in the nanocomposite membranes (Ma et al., 2012b).

The effect of different clay content in the preparation of flat sheet PSf/clay nanocomposite membranes with PEG 400 as a pore-forming agent was studied. Asymmetric structures with a good level of clay dispersion in the form of exfoliated platelets were obtained. Clay addition increased the ratio of large pore in the skin layer, porosity, and pure water flux and had little effect on the hydrophilicity of membranes. However, besides these improvements, the mechanical properties of membranes were weakened (Ma et al., 2012c).

Also aiming to reduce fouling, mesoporous silica was added in polyethersulfone (PES)/*N,N*-dimethylformamide (DMF) solution to form nanocomposite membranes. The final dispersion was immersed in fresh distilled water to form nanocomposite membranes. The addition of the mesoporous silica enlarged the pore size in the sublayer and enhanced the interconnectivity of pores between the sublayer and the bottom layer. The introduction of the hydrophilic mesoporous silica resulted in an increase of hydrophilicity, porosity, water uptake, and thermal stability. Besides that, the use of 2% particle showed the best properties; water flux was 180.2 L/m^2 h and BSA rejection was 96.1%. The best antifouling properties were also found in the 2% membrane. However, the increase of filler content decreased the membrane flux as particles agglomerated in the PES matrix (Huang et al., 2012).

Nano-SiO_2 was used to produce PES nanocomposite UF membranes. Regarding cross-section membrane structure, the connectivity of the pores between the sublayer and the bottom layer was increased by adding these particles, therefore, pure water fluxes and raw water fluxes were increased. Moreover, hydrophilicity and antifouling ability were also improved. After a cleaning method, it was found that water flux recovery ratio was higher for SiO_2/PES membranes than for PES membrane (Shen et al., 2011).

PES UF membranes were prepared with titanium oxide particles by solution dispersion. Although pure water flux was lower compared with pristine PES membrane, antifouling property and long-term flux were significantly enhanced (Rahimpour et al., 2008).

Degussa P25 titanium oxide nanoparticles were used in the PES UF membranes in order to overcome the agglomeration of nanoparticles in the polymer matrix. A system composed of PES/polyvinylpyrrolidone/dimethylacetamide/isopropanol was employed. The incorporation of modified nanoparticles into PES UF membranes improved fouling behavior. These nanocomposite membranes exhibited a flux recovery with an increase of 60%. The degree of dispersion had significant effects on free surface energy, roughness, surface pore size, protein absorption resistance, and hydrophilicity, this last one was improved around 18%. Nevertheless, higher concentrations of titanium oxide did not prevent agglomerations and had a poorer performance (Razmjou et al., 2011).

PES-titania membranes were prepared from casting solution consisting of various compositions of polymer solvent (*N,N*-dimethylformamide and ethanol) and titania additive. It was observed that the addition of ethanol did not improve the rate of particle dispersion but the cluster size was increased because of a double effect: reduction of the dieletric constant of solution and the depletion of the suspension field determined by the action of the polymer chains. A structural change from a sponge-like to a finger-like structure was observed. Both ethanol addition and increase of titania concentration provoked an increase of macrovoid dimensions, resulting in a more porous membrane. A higher titania content and the addition of ethanol resulted in an increase of membrane hydrophilicity and consequently, the enhancement of membrane permeation. Moreover, the addition of ethanol improved fouling resistance and rejection potential (Sotto et al., 2011).

Titanium dioxide was again used in PES membranes and it was found that the membrane structure was not altered by nanoparticle addition. Besides that, hydrophilicity, thermal stability, mechanical strength, and antifouling ability were improved by nanocomposite formation. The 0.5 wt.% nanoparticle content was responsible for the formation of the best membrane while the addition of higher contents resulted in defective pore structure of the membranes and therefore, the decline of the properties, such as permeability and mechanical strength. Thermal and mechanical experiment results revealed that there was a good compatibility between the polymer matrix and the titanium oxide nanoparticles (Wu et al., 2008).

Hollow fiber PES membranes were prepared with mechanically and chemically modified titania nanoparticles. Poor dispersion and migration of titania nanoparticles toward outer layer were observed for mechanically modified titania nanoparticles. The flux recovery reduction under constant pressure filtration for mechanically and chemically modified PES hollow fiber membrane was higher than that of control membrane. Both types of titania increased the glass transition temperature, membrane porosity and pore size, stiffness, and hydrophilicity whereas, tensile strength and elongation at break decreased (Razmjou et al., 2012).

Three types of titania nanoparticles (P25, PC105, and PC500) with various sizes were applied for the preparation of PES nanofiltration membranes. The use of these nanoparticles increased membrane hydrophilicity and pure water flux, also affecting the mean pore radius and overall porosity of the membranes. The increase of titania content also increased the nanoparticle aggregation, being this behavior more pronounced in the PC105 and PC500 membranes which caused membrane pore clogging and reduced pure water flux. The addition of titania also promoted a reduction of the macrovoid size and a slight increase of width of the membrane pores. In addition, the length of macrovoids elongated and finger-like structure was not changed. Because of the increase of hydrophilicity, the biofouling resulted from whey filtration was decreased. A reduction of the nanoparticle size, because of the high surface area and the higher adsorption of water, promoted a more flux recovery percentage. The P25 titania nanoparticles showed a low tendency to the aggregation and did not block the membrane pores, leading to a higher pure water flux and water flux recovery ratio for P25 prepared membranes. Finally, the 4 wt.% P25/PES membrane presented the highest pure water flux and flux recovery ratio and revealed superior antifouling properties (Vatanpour et al., 2012a).

The size and distribution of titania, alumina, and zirconia on the properties of PES nanocomposite UF membranes were evaluated. An abrupt increase in the particle size was observed due to their dispersions into the polymeric solution, signifying that the particle showed a significant larger size, different from the primary nanosize. Moreover, the presence of these particles resulted in a higher affinity of modified membranes as a consequence of water affinity. Membrane fouling resistance was also reduced with the addition of these particles as membrane hydrophilicity increased due to their addition. Strong correlation between membrane fouling resistance and the spatial distribution of the particles inside the membrane structure was observed. Finally, the rejection potential of the nanocomposite membranes was hardly affected by the metal-oxide doping, being similar to the control membrane (Arsuaga et al., 2013).

PES nanofiltration membranes were improved by adding multiwalled carbon nanotubes coated by anastase titanium dioxide nanoparticles in their preparation. A finger-like and porous structure was obtained and the agglomeration of nanoparticles was very low. Moreover, this nanoparticle helped to improve membrane hydrophilicity and pure water flux. It was also found that 0.1 wt.% titanium dioxide coated multiwalled carbon nanotube membrane had the best antibiofouling properties (evaluated by whey solution filtration) due to its lowest surface roughness and synergistic photocatalytic activity induced by incorporated nanoparticles (Vatanpour et al., 2012b).

Poly vinylidene chloride (PVDC) was used with modified MCM-41 particles (a long chain surfactant molecules) to prepare filtration membranes. The addition of low contents of this particle ($\leq$3%) enhanced several properties as: mechanical properties, thermal stability, hydrophilicity, membrane flux, and hydrodynamic characteristics due to the interactions between PVDC and MCM-41, as MCM-41 presents a high surface area and PEO is highly hydrophilic (Jomekian et al., 2011).

A novel cellulose acetate (CA) membrane was reported in the literature. Titanium oxide was added to CA solutions in different concentrations. An increase of the membrane porosity was observed because of increasing mean pore size. Thermal stability was improved due to the interactions between these nanoparticles and CA. Its addition also promoted an increase of water permeability (Abedini et al., 2011).

Multiwalled carbon nanotubes were also added in the CA water filtration membranes at very small amounts that were well and randomly dispersed in the CA matrix. The addition of the carbon nanotubes at higher concentrations decreased the number of macrovoids and the increase of carbon nanotube loading decreased pore size. The lowest used carbon nanotube content contributed to an increase of 54% in water flux with a minimum decrease in salt rejection (6%) compared to the neat CA membrane. However, higher carbon nanotube contents decreased the permeate rates since porosity and surface area were both reduced (Badawi et al., 2014).

Lithium chloride monohydrate and titanium dioxide were added in the preparation of PVDF membranes. Lower titanium nanoparticle concentrations produced membranes with smaller mean pore size and more apertures inside the membrane with enhanced hydrophilicity. Lithium chloride monohydrate was responsible for reducing the thermodynamic miscibility, resulting in an increase of the rate of liquid–liquid demixing process (Yuliwati and Ismail, 2011).

PVDF/sulfonated polyethersulfone (PES) blend membrane was modified using titanium dioxide. Although the flux was reduced in the presence of the titanium nanoparticles, membrane hydrophilicity and long-term stability were improved. Moreover, a dramatic photo-bactericidal effect on *E. coli* was found in the nanocomposite membranes (Rahimpour et al., 2011).

PVDF UF membranes were modified by dispersing titania particles in a PVDF/*N*-methyl-2-pirrolidone solution and the used nonsolvent was water. This research investigated the effect of these nanoparticles and type of coagulants on membrane permeability. The PVDF membranes presented a lower fouling after titania addition. Moreover, the membranes immersed in pure water coagulation bath exhibited a finger-like structure whereas, the membranes immersed in a water/isopropanol mixture presented a sponge-like structure (Oh et al., 2009).

The effect of the titanium oxide nanoparticle size was investigated on the morphology and performance of PVDF membranes. The antifouling properties were improved in the membranes with the smaller nanoparticles, having a smaller pore size on their surfaces and more apertures inside them. Moreover, a stronger effect on the crystallization of PVDF was observed (Cao et al., 2006).

Flat sheet PVDF nanocomposite UF membranes were prepared with titanium dioxide in a *N,N*-dimethylacetamide medium. The main objective of this research was to obtain a membrane with high rejection and flux for micellar enhanced methylene blue UF. The interaction effects of several parameters as polymer concentration, membrane casting thickness and titanium dioxide concentration toward rejection of micellar enhanced methylene blue and permeation flux were evaluated. It was found that the optimum membrane had a polymer concentration of 18 wt.%, casting thickness of 400 μm and titanium dioxide of 1.62 wt.%. The obtained value of permeation flux and rejection was 53.28 L/m^2 h and 99.02%, respectively (Ngang et al., 2012).

The influence of titanium oxide on the compaction behavior of PVDF and polyamideimide (PAI) membranes was studied. PAI membranes with 40 wt.% titanium oxide could withstand a temperature treatment at 180°C with only minor decrease in fluxes and its pore size distribution remained practically unchanged. PVDF/titanium oxide membranes were less susceptible to compaction than pure PVDF membranes during pressure treatment at 30 bar. Better permeate quality and higher fluxes of the PVDF nanocomposite membranes under vapor permeation experiments for the dehydration of ethanol were observed (Ebert et al., 2004).

Poly(acrylic acid)-grafted PVDF (G-PVDF)/titania nanocomposite hollow fiber UF membranes were successfully prepared. The nanocomposite membranes presented extremely high water permeation, improved antifouling property, rejection efficiency, and stability compared to the pure polymer membrane. As an example, by using 1 wt.% titania content, the water flux of the nanocomposite membrane reached a four times higher value than that of PVDF membrane. Furthermore, the incorporation of titania nanoparticles did not weaken the mechanical strength of the membranes (Zhang et al., 2013a).

Titanium dioxide nanoparticles were added to the polyimide (PI) solution before spinning to form a hollow fiber membrane. The nanoparticle distribution across the membrane wall varied from convex to concave profiles depending on elongation draw ratio which means that high elongation draws resulted in particles moving radially inward to the bore fluid side and radially outward to the outer coagulant side. Therefore, properties as membrane morphology, particle distribution, and separation performance could be controlled by the manipulation of elongational draw ratio with the aid of shear stress induced by dope flow within the spinneret and die swell out of the spinneret (Xiao et al., 2006).

Poly(phthalazine ether sulfone ketone) (PPESK)/titanium dioxide nanocomposite UF membranes with enhanced performance for high temperature condensed water treatment were prepared by solution dispersion. The introduction of the nanoparticles delayed phase separation process, leading to a suppression of the macrovoids structure in the sublayer and a thicker skin layer. Besides that, the fillers, due to the nodular shapes having ridges and valleys, also affected surface morphology; final membranes presented a larger porosity and a little changed molecular weight cut-off. The hydrogen bond between titanium dioxide and polymer improved mechanical strength and thermal stability of the membranes. Membrane hydrophilicity was also enhanced by increasing titanium dioxide content, as there was hydroxyl groups on water-polymer interface in an aqueous environment. The enhanced hydrophilicity reduced the interaction between organic pollutants and membrane surface and improved permeability, antifouling properties, filtration resistance, and flux recover property during high temperature condensed water treatment (Zhao et al., 2012a).

5 Conclusions and Future Research Directions

It is possible to observe that researches focused on the development of nanocomposite membranes for water treatment were extensively published in this last decade, showing a constant progress in this area. The addition of nanoparticles presented several improvements not only in the membrane properties (hydrophilicity, porosity, permeability, rejection, and membrane structure) but also in the final material properties, which corresponded to thermal, mechanical, and stability enhancements. The possibility of control of the filtration type by using fixed nanoparticle contents represents an important issue in the membrane technology.

As we can observe, great number of papers described the use of titanium oxide and there are also an expressive number of studies that reported the use of other fillers as silica, multiwalled carbon nanotubes and clays.

However, the challenges, listed here, still need to be overcome:

- there is not an exact correlation between the final membrane structure and its performance. The contributions of each property (hydrophilicity, porosity, and morphology) to the membrane performance are still unclear and need to be elucidated;
- nanoparticle aggregation in the polymer matrix is often reported by several researchers. Therefore, closer attention needs to be paid to the modification of these particles in order to become them more compatible with the chosen polymer;
- few studies report the production of the nanocomposite membranes in large scale. Hence, researches should focus this topic.

Finally, if the mentioned problems are solved, membrane filtration area can take a huge leap in direction of higher water filtration productivity with a great quality.

References

Abedini, R., Mousavi, S.M., Aminzadeh, R., 2011. A novel cellulose acetate (CA) membrane using TiO_2 nanoparticles: preparation, characterization and permeation study. Desalination 277, 40–45.

Amini, M., Jahanshahi, M., Rahimpour, A., 2013. Synthesis of novel thin film nanocomposite (TFN) forward osmosis membranes using functionalized multi-walled carbon nanotubes. J. Membr. Sci. 435, 233–241.

Anadão, P., 2010. Ciência e tecnologia de membranas. Artliber Editora, São Paulo.
Anadão, P., et al., 2010. Montmorillonite as a component of polysulfone nanocomposite membranes. Appl. Clay Sci. 48, 127–132.
Anadão, P., et al., 2011. Montmorillonite/carbon nanocomposites prepared from sucrose for catalytic applications. Appl. Clay Sci. 53, 288–296.
Anadão, P., et al., 2013. Influence of the clay content and the polysulfone molar mass on nanocomposite membrane properties. Appl. Surf. Sci. 275, 110–120.
Anadão, P., et al., 2014. Polysulphone/montmorillonite nanocomposite membranes: effect of clay addition and polysulphone molecular weight on the membrane properties. J. Membr. Sci. 485, 187–199.
Arsuaga, J.M., et al., 2013. Influence of the type, size, and distribution of metal oxide particles on the properties of nanocomposite ultrafiltration membranes. J. Membr. Sci. 438, 131–141.
Badawi, N.E., et al., 2014. Novel carbon nanotube–cellulose acetate nanocomposite membranes for water filtration applications. Desalination 344, 79–85.
Bao, M., et al., 2013. Preparation of monodispersed spherical mesoporous nanosilica–polyamide thin film composite reverse osmosis membranes via interfacial polymerization. Desalination 309, 261–266.
Baroña, G.N.B., Choi, M., Jung, B., 2012. High permeate flux of PVA/PSf thin film composite nanofiltration membrane with aluminosilicate single-walled nanotubes. J. Colloid Interf. Sci. 386, 189–197.
Baroña, G.N.B., et al., 2013. Interfacial polymerization of polyamide-aluminosilicate SWNT nanocomposite membranes for reverse osmosis. Desalination 325, 138–147.
Callister, W.D., 1994. Materials Science and Engineering: An Introduction. John Wiley & Sons, New York.
Cao, X., et al., 2006. Effect of TiO_2 nanoparticle size on the performance of PVDF membrane. Appl. Surf. Sci. 253, 2003–2010.
Carter, L.W., Hendricks, J.G., Bolley, D.S., 1950. US Patent 2,531,396. Elastomer reinforced with a modified clay. National Lead Co.
Choudalakis, G., Gotsis, A.D., 2009. Permeability of polymer/clay nanocomposites: a review. Eur. Polym. J. 45, 967–984.
Conway, R., 2004. Using nanotechnology to provide active and passive barriers in beverage packaging application. Proceedings of 4th World Congress in Nanocomposites, September 2004. San Francisco.
Cox, H. et al., 2004. Nanocomposite systems for automotive applications. Proceedings of 4th World Congress in Nanocomposites, San Francisco.
Ebert, K., et al., 2004. Influence of inorganic fillers on the compaction behavior of porous polymer based membranes. J. Membr. Sci. 233, 71–78.
Fathizadeh, M., Aroujaliana, A., Raisi, A., 2011. Effect of added NaX nano-zeolite into polyamide as a top thin layer of membrane on water flux and salt rejection in a reverse osmosis process. J. Membr. Sci. 375, 88–95.
Gaffet, E., 2011. Nanomatériaux : Une revue des définitions, des applications et des effets sur la santé. Comment implémenter un développement sûr. CR Phys. 12, 648–658.
Ghaemi, M., et al., 2011. Preparation, characterization and performance of polyethersulfone/organically modified montmorillonite nanocomposite membranes in removal of pesticides. J. Membr. Sci. 382, 135–147.
Goldman, A.Y., Copsey, C.J., 2004. Multilayer barrier liner material with nanocomposites for packaging applications. 4th World Congress in Nanocomposites, San Francisco.
Hamid, N.A.A., et al., 2011. Morphological and separation performance study of polysulfone/titanium dioxide (PSF/TiO_2) ultrafiltration membranes for humic acid removal. Desalination 273, 85–92.

Hoek, E.M.V., et al., 2011. Physical–chemical properties, separation performance, and fouling resistance of mixed-matrix ultrafiltration membranes. Desalination 283, 89–99.

Hu, D., Xu, Z.-L., Wei, Y.-M., 2013. A high performance silica–fluoropolyamide nanofiltration membrane prepared by interfacial polymerization. Sep. Pur. Technol. 110, 31–38.

Huang, J., et al., 2012. Fabrication of polyethersulfone-mesoporous silica nanocomposite ultrafiltration membranes with antifouling properties. J. Membr. Sci. 423–424, 362–370.

Jadav, G.L., Singh, P.S., 2009. Synthesis of novel silica-polyamide nanocomposite membrane with enhanced properties. J. Membr. Sci. 328, 257–267.

Jadav, G.L., Aswal, V.K., Singh, P.S., 2010. SANS study to probe nanoparticle dispersion in nanocomposite membranes of aromatic polyamide and functionalized silica nanoparticles. J. Colloid Interf. Sci. 351, 304–314.

Jeong, B.-H., et al., 2007. Interfacial polymerization of thin film nanocomposites: a new concept for reverse osmosis membranes. J. Membr. Sci. 294, 1–7.

Jomekian, A., Mansoori, S.A.A., Monirimanesh, N., 2011. Synthesis and characterization of novel PEO–MCM-41/PVDC nanocomposite membrane. Desalination 276, 239–245.

Kangoa, S., et al., 2013. Surface modification of inorganic nanoparticles for development of organic–inorganic nanocomposites—a review. Progr. Polym. Sci. 38, 1232–1261.

Kawasumi, M., 1989. US Patent 4,810,734. Process for producing composite material. Kabushiki Kaisha Toyota Chuo Kenkyusho.

Ke, Y.C., Stroeve, P., 2005. Polymer-layered silicate and silica nanocomposites. Elsevier, Amsterdam.

Kim, E.-S., Deng, B., 2011. Fabrication of polyamide thin-film nano-composite (PA-TFN) membrane with hydrophilized ordered mesoporous carbon (H-OMC) for water purifications. J. Membr. Sci. 375, 46–54.

Kim, E.-S., et al., 2012. Development of nanosilver and multi-walled carbon nanotubes thin-film nanocomposite membrane for enhanced water treatment. J. Membr. Sci. 394-395, 37–48.

Kim, S.G., et al., 2013a. Preparation, characterization and performance of poly(aylene ether sulfone)/modified silica nanocomposite reverse osmosis membrane for seawater desalination. Desalination 325, 76–83.

Kim, S.G., et al., 2013b. Nanocomposite poly(arylene ether sulfone) reverse osmosis membrane containing functional zeolite nanoparticles for seawater desalination. J. Membr. Sci. 443, 10–18.

Kong, S., et al., 2011. Enhanced performance of inorganic-polyamide nanocomposite membranes prepared by metal-alkoxide-assisted interfacial polymerization. J. Membr. Sci. 366, 382–388.

Li, Q., et al., 2008. Antimicrobial nanomaterials for water disinfection and microbial control: Potential applications and implications. Water Res. 42, 4591–4602.

Ma, N., et al., 2012a. Zeolite-polyamide thin film nanocomposite membranes: towards enhanced performance for forward osmosis. J. Membr. Sci. 405–406, 149–157.

Ma, Y., et al., 2012b. Preparation and characterization of PSf/clay nanocomposite membranes with LiCl as a pore forming additive. Desalination 303, 39–47.

Ma, Y., et al., 2012c. Preparation and characterization of PSf/clay nanocomposite membranes with PEG 400 as a pore forming additive. Desalination 286, 131–137.

Madaeni, S.S., Zinadini, S., Vatanpour, V., 2011. A new approach to improve antifouling property of PVDF membrane using in situ polymerization of PAA functionalized TiO_2 nanoparticles. J. Membr. Sci. 380, 155–162.

Monticelli, O., et al., 2007. Preparation and properties of polysulfone-clay composite membranes. J. Appl. Polym. Sci. 103, 3637–3644.
National Science Foundation, 2015. Investments in nanotechnology. Available from: http://www.nsf.gov
Ngang, H.P., et al., 2012. Preparation of mixed-matrix membranes for micellar enhanced ultrafiltration based on response surface methodology. Desalination 293, 7–20.
Oh, S.J., Kim, N., Lee, Y.T., 2009. Preparation and characterization of PVDF/TiO_2 organic–inorganic composite membranes for fouling resistance improvement. J. Membr. Sci. 345, 13–20.
Okada, A. et al., 1988. US Patent 4,739,007. Composite material and process for manufacturing same. Kabushiki Kaisha Toyota Chou Kenkyusho.
Patterson, T., 2004. Nanocomposites—our revolutionary breakthrough. Proceedings of 4th World Congress in Nanocomposites, San Francisco.
Pendergast, M.M., Ghosh, A.K., Hoek, E.M.V., 2013. Separation performance and interfacial properties of nanocomposite reverse osmosis membranes. Desalination 308, 180–185.
Rabello, M.S., 2010. Aditivação de Polímeros. Artliber Editora, São Paulo.
Rahimpour, A., et al., 2008. Coupling TiO_2 nanoparticles with UV irradiation for modification of polyethersulfone ultrafiltration membranes. J. Membr. Sci. 313, 158–169.
Rahimpour, A., et al., 2011. TiO_2 entrapped nano-composite PVDF/SPES membranes: preparation, characterization, antifouling and antibacterial properties. Desalination 278, 343–353.
Rana, D., et al., 2011. Development of antifouling thin-film-composite membranes for seawater desalination. J. Membr. Sci. 367, 110–118.
Razmjou, A., Mansouri, J., Chen, V., 2011. The effects of mechanical and chemical modification of TiO_2 nanoparticles on the surface chemistry, structure and fouling performance of PES ultrafiltration membranes. J. Membr. Sci. 378, 73–84.
Razmjou, A., et al., 2012. The effect of modified TiO_2 nanoparticles on the polyethersulfone ultrafiltration hollow fiber membranes. Desalination 287, 271–280.
Rejeski, D., Lekas, D., 2008. Nanotechnology field observations: scouting the new industrial west. J. Clean. Prod. 16, 1014–1017.
Robinson, D.K.R., 2009. Co-evolutionary scenarios: an application to prospecting futures of the responsible development of nanotechnology. Technol. Forecast. Soc. 76, 1222–1239.
Roy, S., et al., 2011. Facile fabrication of superior nanofiltration membranes from interfacially polymerized CNT-polymer composites. J. Membr. Sci. 375, 81–87.
Saleh, T.A., Gupta, V.K., 2012. Synthesis and characterization of alumina nano-particles polyamide membrane with enhanced flux rejection performance. Sep. Pur. Technol. 89, 245–251.
Sellers, K., et al., 2008. Nanotechnology and the environment. CRC Press, Boca Raton.
Shen, J.N., et al., 2011. Preparation and characterization of PES–SiO_2 organic–inorganic composite ultrafiltration membrane for raw water pretreatment. Chem. Eng. J. 168, 1272–1278.
Shen, J.N., et al., 2013. Preparation and characterization of thin-film nanocomposite membranes embedded with poly(methyl methacrylate) hydrophobic modified multi-walled carbon nanotubes by interfacial polymerization. J. Membr. Sci. 442, 18–26.
Singh, P.S., Aswal, V.K., 2008. Characterization of physical structure of silica nanoparticles encapsulated in polymeric structure of polyamide films. J. Colloid Interf. Sci. 326, 176–185.

Sotto, A., et al., 2011. Effect of nanoparticle aggregation at low concentrations of TiO_2 on the hydrophilicity, morphology, and fouling resistance of PES–TiO_2 membranes. J. Colloid Interf. Sci. 363, 340–350.

Souza, M.A., Pessan, L.A., Rodolfo, Jr., A., 2006. Nanocompósitos de poli(cloreto de vinila) (PVC)/argilas organofílicas. Polímeros 16, 257–262.

Takemura, M., 2008. Japan's engagement in health, environmental and societal aspects of nanotechnology. J. Clean. Prod. 16, 1003–1005.

Vatanpour, V., et al., 2012a. TiO_2 embedded mixed matrix PES nanocomposite membranes: influence of different sizes and types of nanoparticles on antifouling and performance. Desalination 292, 19–29.

Vatanpour, V., et al., 2012b. Novel antibifouling nanofiltration polyethersulfone membrane fabricated from embedding TiO_2 coated multiwalled carbon nanotubes. Sep. Pur. Technol. 90, 69–82.

Wang, J., et al., 2010. Amino-functionalized Fe_3O_4@SiO_2 core–shell magnetic nanomaterial as a novel adsorbent for aqueous heavy metals removal. J. Colloid Interf. Sci. 349, 293–299.

Wu, G., et al., 2008. Preparation and characterization of PES/TiO_2 composite membranes. Appl. Surf. Sci. 254, 7080–7086.

Wu, H., Tang, B., Wu, P., 2013a. Optimization, characterization and nanofiltration properties test of MWNTs/polyester thin film nanocomposite membrane. J. Membr. Sci. 428, 425–433.

Wu, H., Tang, B., Wu, P., 2013b. Optimizing polyamide thin film composite membrane covalently bonded with modified mesoporous silica nanoparticles. J. Membr. Sci. 428, 341–348.

Wu, H., Tang, B., Wu, P., 2014. Development of novel SiO_2–GO nanohybrid/polysulfone membrane with enhanced performance. J. Membr. Sci. 451, 94–102.

Xiao, Y., et al., 2006. Evolution of nano-particle distribution during the fabrication of mixed matrix TiO_2-polyimide hollow fiber membranes. Chem. Eng. Sci. 61, 6228–6233.

Yang, Y., et al., 2007. The influence of nano-sized TiO_2 fillers on the morphologies and properties of PSF UF membrane. J. Membr. Sci. 288, 231–238.

Yin, J., et al., 2012. Fabrication of a novel thin-film nanocomposite (TFN) membrane containing MCM-41 silica nanoparticles (NPs) for water purification. J. Membr. Sci. 423–424, 238–246.

Yuliwati, E., Ismail, A.F., 2011. Effect of additives concentration on the surface properties and performance of PVDF ultrafiltration membranes for refinery produced wastewater treatment. Desalination 273, 226–234.

Zhang, F., et al., 2013a. Sol–gel preparation of PAA-*g*-PVDF/TiO_2 nanocomposite hollow fiber membranes with extremely high water flux and improved antifouling property. J. Membr. Sci. 432, 25–32.

Zhang, G., et al., 2013b. Novel polysulfone hybrid ultrafiltration membrane prepared with TiO_2-*g*-HEMA and its antifouling characteristics. J. Membr. Sci. 436, 163–173.

Zhao, S., et al., 2012a. Thermostable PPESK/TiO_2 nanocomposite ultrafiltration membrane for high temperature condensed water treatment. Desalination 299, 35–43.

Zhao, Y., et al., 2012b. Synthesis of robust and high-performance aquaporin-based biomimetic membranes by interfacial polymerization-membrane preparation and RO performance characterization. J. Membr. Sci. 423–424, 422–428.

Zyvez, 2012. Richard Feynman's lecture "There is plenty of room in the bottom". Available from: http://www.zyvex.com/nanotech/Feynman.html

16

NANOTECHNOLOGY DEPOLLUTION OF HEAVY METALS PRESENT IN POTABLE WATER

Paul Cătălin Balaure, Dragoş Gudovan, Iulia Alexandra Gudovan

University "Politehnica" of Bucharest, Faculty of Applied Chemistry and Materials Science, "Costin Neniţescu" Department of Organic Chemistry, Bucharest, Romania

1 Introduction: The Toxicological Concern and the Nanotechnology Solution

Heavy metals toxicity is well known, the intake of those toxic elements being associated with serious illnesses and even death in some cases of chronic exposure. The struggle for cleaner water sources has become a top priority for all countries, but up to a point clean, pure water, remains a facility of the developed countries and is in some cases a parameter that is associated with the quality of life in a social environment. Although groundwater is generally thought of as being cleaner than surface water this is typically true with respect to biological contamination but not to heavy metal contamination. The level of heavy metal pollution in a country depends on the level and structure of the economical activity in that country and on country's institutional and legal capacity to address it. There is a global trend of an increasing level of contamination with heavy metals of groundwater systems and this is a very concerning issue since once contaminated these freshwater resources are very difficult and costly to clean. According to a recent study (Hu et al., 2014) in China, the country with the fastest growing economy, only 40% of the groundwater is suitable for drinking or fishing and 28% of it is improper to be used even for industrial purposes.

Arsenic is highly toxic and carcinogenic, its availability increasing the chance of exposure (Jaishankar et al., 2014). Common sources of contamination include arsenical pesticides, mining or inappropriate disposal of arsenical chemicals (Jaishankar

Water Purification. http://dx.doi.org/10.1016/B978-0-12-804300-4.00016-2

et al., 2014). Lead is also a highly toxic metal. Common sources of contamination are gasoline and house paint, but the sources expanded with the industrial growth to batteries, toys, pipes, and so on (Jaishankar et al., 2014). Mercury's danger comes from the high rate of bioaccumulation. Major sources include but are not limited to mining, incineration, discharges of industrial wastewater (Jaishankar et al., 2014). Cadmium is highly toxic because once it is absorbed by the human body it accumulates throughout life (Jaishankar et al., 2014). Its use in rechargeable batteries became restricted, thus limiting the sources of exposure but not removing them completely (Jaishankar et al., 2014). Chromium's toxicity comes from its abundance in nature and thus the likeliness of human exposure (Jaishankar et al., 2014). Aluminum seems to be also highly toxic, its toxicity being pH dependant (lower pH, higher toxicity). Aluminum replaces the Mg^{2+} and Fe^{3+} ions, the lesions caused this way being similar to those observed in Alzheimer patients (Jaishankar et al., 2014).

While higher concentrations of heavy metal ions are fatal, lower concentrations of metal ions seem to have a serious impact over the quality of life in general. Some metals cause irreversible diseases or conditions in time, conditions that usually appear later on in life. Pamphlett and Jew (2013) performed a study on the patients suffering from motor neuron disease and their study showed that those persons had an unusually high accumulation of heavy metals in the neurons, which could be causing the disease. Chang et al. (2014) reached the conclusion that children seem to be more sensitive than adults to intake of heavy metals by consumption of leafy vegetables. Thus, although the daily intake of vegetables was lower for children than for adults, the bioaccumulation of heavy metals (Hg, Cd, Pb, Cr, As) was higher in the organism of children than in adult organisms. Therefore, care should be taken in regards to the exposure of children to heavy metals. Their study is also supported by Kim et al. (2015). They showed by testing pregnant women's blood that the placenta acts as a barrier in the case of cadmium, thus preventing the contamination of the fetus, but this is not true in the case of mercury and methylmercury, those two species managing to penetrate the placenta and get into the blood stream of the fetus. Their study was centered on fish as a contamination source, but the same results can easily be extrapolated to other potential sources, such as potable water. Chen et al. (2015) conducted a recent study about the effect of heavy metal contamination on the development of drug resistance in bacteria. The heavy metals contribute to the developing of drug resistance even in very low concentrations, concentrations which are not toxic for human consumption. Their results highlight the need for further purification and reduction of contamination with heavy metals of the potable water, the reduction below the threshold of human toxicity being no longer sufficient.

Nanomaterials and nanotechnology in general can come to the aid of the quality of life. The high surface area and specific surface to volume ratio impart unique properties to this class of materials as compared to their bulk counterparts and therefore nanomaterials quickly found a large variety of possible applications including water treatment. On the other hand, for this revolutionary technique to become sustainable, nanomaterials involved in removal of water contaminants should be reusable for several depollution cycles without losing their properties. Otherwise, along with solving the problem of water quality another one is created, the problem of the nanomaterial waste management. Even with multicycle usable nanomaterials, the problem of recycling waste nanomaterials still remains and better solutions to the safe use of nanotechnology need to be found in the future.

The nanomaterials used in the depollution of heavy metals from potable waters can be classified in five main categories, categories which will be expanded further in the following sections: carbon-based materials (the such called "carbonaceous" materials, which are nanomaterials containing carbon nanostructures and/or composites containing carbon nanostructures), membranes for nanofiltration, ligand-based materials (materials reversible forming coordinative bonds with heavy metal ions present in water), nanostructured metal oxides, natural or nature-based materials (hydroxyapatite, chitosan, peanut husks, and composites containing those elements).

2 Carbon-Based Materials

Graphene-oxide (GO) nanosheets appear to be a very interesting and easy to use method of depollution of heavy elements, as Gopalakrishnan et al. (2015) are claiming. The study they present comes with an easy to use synthetic method, and an efficient way to remove Pd(II), Ni(II), and Cr(VI) from pharmaceutical industry effluents.

Tofighy and Mohammadi (2015) synthesized an interesting new composite material, carbon nanotubes—mullite, and studied different types of this composite material with the aim to use it for water purification from copper ions. The material was synthesized by chemical vapour–deposition method, the method being easily transferable to large-scale applications in the industrial synthesis. The composite material as such was tested as adsorbent for copper ions. A second sample of the carbon nanotubes—mullite composite was modified with chitosan by simply dipping the sample in an acidic solution of chitosan. A third composite sample was oxidized on treatment with nitric acid. Another type of composite material was prepared by functionalization with chitosan of the above-oxidized sample. Their study showed that the most-promising material was the material functionalized with chitosan after being previously oxidized by nitric acid. The group postulates that the copper adsorption rates are proportional with the number of functional groups on the surface, thus explaining the better properties of the chitosan-modified composite after acid treatment.

Henriques et al. (2016) found a new way to enhance the performance and selectivity of GO for mercury removal by synthesizing a GO foam. The interesting fact about these foams is that a relatively small quantity of this material, 10 mg/L, can remove 96% of mercury after 24 h of treatment from natural matrices with higher content of mercury. The group tested the foams on river and seawater samples. The matrix effects can be sometimes difficult to predict and the importance of testing the final product on real is not to be trifled with. The fact that Henriques et al. (2016) tested their outcome on real samples is an important and noticeable result in their study.

Venkata Ramana et al. (2013) deposited silver nanoparticles inside multiwalled carbon nanotubes (MWCNTs) in order to obtain suitable sorbents for the removal of Cu(II) and Cd(II) from water. For the synthesis of MWCNTs, the catalytic chemical vapour–deposition method was used. The obtained MWCNTs were oxidized with sodium hypochlorite and silver nanoparticles were eventually deposited on the surface of the oxidized MWCNTs using a chemical reduction method (Alimohammadi et al., 2012). Oxidation mostly led to formation of acidic carboxylic groups but also to some basic hydroxyl groups on the external and internal surface of MWCNTs pores. These ionizable functional groups are mainly responsible for the increased adsorption capacity of these silver crosslinked MWCNTs. The optimum adsorption performance was achieved at pH 6.0 for Cu(II) and at pH 7.0 for Cd(II), respectively. The pH dependence of the adsorption capacity was explained in terms of pH_{PZC} (point of zero charge) of these particles. The material reaches an adsorption equilibrium after 100 min of exposure to the medium. The regeneration of the adsorption capacity can easily be achieved on treatment with 0.25 M aqueous HCl. The adsorption rate is also temperature dependent, due to the exothermic adsorption. The pH dependency of the adsorption rate is useful for the regeneration cycle of the adsorbent, but on the other hand due to this feature the water to be treated may require pH adjustment in some cases.

Deng et al. (2013) successfully synthesized a magnetic graphene oxide (MGO) nanocomposite by precipitating iron oxide nanoparticles on the surface of the GO. They synthesized this material with the aim of simultaneous removal from water samples of Cd(II) and ionic dyes, methylene blue (MB) and orange G (OG) being chosen as models for cationic and anionic dyes, respectively. The effect of pH on removal efficiencies was studied in the pH range 3–8, since at pH values higher than 8, $Cd(OH)_2$ precipitated in the system. The pH_{PZC} value of MGO was found to be 3.5, meaning that at pH values less than 3.5, the MGO surface is positively charged, while above pH 3.5 it is negatively charged. Furthermore, it was observed that adsorption of Cd(II) and of the cationic MB dye was favored with the increase in the pH, while the adsorption of the anionic OG dye decreased with increasing pH. These observations are consistent with an electrostatic mechanism for the fixation of the pollutants on the adsorbent surface. The reported capacity for Cd removal in the case of a monocomponent system was found to be 91.29 mg/g of adsorbent. In Cd(II)–MB and Cd(II)–OG binary systems, the sorption capacity of Cd(II) decreased with increasing MB concentration due to competition between the positively charged MB dye and cadmium ions for binding to the negatively charged adsorbent surface, while it was not affected by the concentration of the anionic dye OG. The equilibrium for Cd(II) sorption was reached in 165 min. The above sorption capacity of Cd(II) was determined using ultrapure water. However, in tap-water samples the sorption capacity was only 65.39% of the aforementioned value and this was probably due to the increased ionic strength in tap water. Therefore, this finding highlights that proper studies should involve real-water samples and that the results from distilled water are not directly traceable to real-water samples.

A novel nanocomposite between GO, silver, and cobalt ferrite was studied by Ma et al. (2015) with the intended purpose to develop a material for the efficient removal of Pb(II) from water. This group developed this new nanomaterial having in mind that wastewater containing lead may also contain dangerous bacteria, and hence a nanomaterial that can remove lead and exhibits bactericidal activity at the same time would be very useful. The antibacterial properties are based on the Ag moiety, the adsorption properties rely on GO, while the cobalt ferrite enables facile work up via magnetic separation. Multifunctional materials seem to be highly sought, due to the fact that if a nanomaterial cannot be obtained from renewable sources or produced through a green-chemistry procedure, at least it should justify its meaningfulness by having more than one intended purpose.

Gollavelli et al. (2013) had a similar approach, but in a more environmental friendly way, using a solvent-free synthesis and a one-pot approach. They reported microwave-assisted synthesis of smart magnetic graphene using GO and ferrocene as starting materials. The sorbent proposed by this group was studied on aqueous solutions containing Cr(VI), As(V), and Pb(II) ions, ions extremely harmful for the human health. The tested concentration was 5 ppm, and it was removed to 1 ppb with 99% adsorption efficiency. Although the adsorption capacity of this material is not even comparable to other materials [4.86 mg/g for Cr(VI), 3.26 mg/g for As(V) and 6.00 mg/g for Pb(II)], it is a viable environmentally friendly adsorbent, presenting also disinfection properties toward *Escherichia coli* and a low toxicity, making this material a viable choice for the purification of drinking water.

Carbon nanotubes (German et al., 2014; Qu et al., 2013; Liu et al., 2013) represent a novel approach in water purification and, due to the reduction in the costs of manufacture, a highly feasible one. The functionalization step of carbon nanotubes is important in some types of applications, and since the possibilities for functionalization are seemingly endless, the research in this field is always open. The important fact in all endeavors will be to emphasize the applicability in each case, the main goal of applicative research being to eventually deliver industrially transferable research. The next table summarizes the main achievements in heavy metal depollution of water by using carbon-based materials (Table 16.1).

3 Membranes for Nanofiltration

Membranes for nanofiltration constitute a different approach in heavy metal removal from potable waters. The main advantage membranes have over other types of purification systems is that the regeneration step is done automatically and continuously, up to a certain point where membrane fouling becomes significantly high and the membrane has to be replaced. However, membranes usually have a very long life, depending on the quality of the feed water. If the concentration of ions is too high, membrane fouling will increase exponentially.

Al-Rashdi et al. (2013) studied a commercial membrane, NF270, and tried to optimize the membrane parameters in order to obtain the best possible heavy metals (Cu, Cd, Mn, Pb, As) rejection. The group studied several parameters affecting the rejection of heavy metal ions, such as feed and permeate pH, effect of metal concentration and the effect of multicomponent solution versus

Table 16.1 Performances of Carbon-Based Materials Used for Decontamination of Heavy Metals from Drinking Water

Type of Adsorbent Material	Adsorption (mg Adsorbed Species/ Gram Adsorbent)	Other Advantages	References
Graphene-oxide nanosheets		Simultaneous removal of Ni(II), Cr(VI), and Pb(II)	Gopalakrishnan et al. (2015)
Carbon nanotubes—mullite composite material	Cu(II) 60.34 mg/g	Removal of Cu(II)	Tofighy and Mohammadi (2015)
Graphene-oxide foam	Hg(II) 35 mg/g	96% removal rate after 24 h contact time; drinking-water application; effects tested on real-water samples	Henriques et al. (2016)
Multiwalled carbon nanotubes functionalized with Ag	Cu(II) 58.02 mg/g Cd(II) 54.92 mg/g	Cu(II) and Cd(II) simultaneous removal; fast removal rate, 100 min; regeneration checked after four cycles	Venkata Ramana et al. (2013)
Magnetic graphene oxide nanocomposite	Cd(II) 91.29 mg/g	Simultaneous removal of Cd(II) and ionic dyes	Deng et al. (2013)
Graphene oxide/ silver cobalt ferrite nanocomposite		Environmental-friendly solution; dual use material, depollution and sanitation purpose	Ma et al. (2015)
Graphene oxide	Cr(VI) 4.86 mg/g As(V) 3.26 mg/g Pb(II) 6.00 mg/g	Dual purpose, depollution and sanitation; environmental-friendly synthesis	Gollavelli et al. (2013)

monocomponent solution. Rejection was higher at feed pH values below the isoelectric point of the membrane for all metals tested except for arsenic. At low metal concentrations, the membrane was able to reject almost completely the copper ions, but rejection capacity decreased with increasing metal concentration. At a concentration level of 1000 mg/L, an operating pressure of 4 bar and a feed pH = 1.5 ± 0.2, the membrane rejected 99% of cadmium, 89% of manganese, and 74% of lead, respectively.

He et al. (2014) developed a hollow fiber polysulfone membrane embedded with zirconium nanoparticles with the aim of utilizing it to the purification of arsenic from contaminated waters. The hollow fibers were fabricated by the phase-inversion technique using a spinning apparatus which has a major potential for industrial applications. An increase in amount of loaded zirconium nanoparticles resulted in an enhancement in the water flux due to increased membrane hydrophilicity and porosity. The authors found the membrane to be working successfully in a quite large pH range (from 2.0 to 9.0), and coexisting ions, such as carbonate, fluoride, silicate, and phosphate do not affect the arsenate uptake significantly. The membrane can be regenerated using sodium-hydroxide washing followed by washing with aqueous sulfuric acid. After a regeneration cycle, the adsorption rate is maintained. The main mechanism through which adsorption is achieved is by anion exchange between arsenate and sulfate ions. At the initial concentration of 92.65 ppb, the adsorption equilibrium was reached in 10 h, while increasing the initial contaminant concentration to 51.65 ppm it took 48 h for the absorption equilibrium to be attained. The membrane exhibited low toxicity when tested on MCF7 human breast-cancer cell lines.

Polyvinylidene fluoride (PVDF) membranes functionalized with polyacrylic acid and with iron/iron-oxide nanoparticles were chosen by Gui et al. (2015) for selenium removal from power plant scrubber water. Gui et al. (2013) developed an elegant synthetic procedure starting with preformed PVDF membranes. In situ polymerization of acrylic acid using ammonium persulfate as initiator and *N,N′*-methylenebisacrylamide (NNMA) as crosslinker led to functionalization of PVDF membrane pores by formation of crosslinked polyacrylic acid (PAA) chains inside the pores. Next, the PVDF–PAA membrane was soaked in NaCl solution when the protons of the carboxyl groups in the PAA chains were released their place being taken by sodium cations. A second cation exchange took place by immersing the membrane in $FeCl_2$ solution, when Na^+ ions were replaced by Fe^{2+} ions. Eventually, Fe^{2+} ions were reduced to metallic iron nanoparticles by immersing the membrane in $NaBH_4$ solution. The decontamination mechanism is based on reduction of selenate ions by iron when elemental selenium, iron oxides and iron selenide are formed. The maximum reduction capacity was 32.1 ± 2.0 mg Se/g Fe this value being obtained in convective experiments. However, formation of a shell of iron oxides surrounding the iron nanoparticles impedes on the reduction reaction since neither selenium oxyanions nor electrons can pass through this iron oxides layer. Therefore, the iron surface should be regenerated by acid treatment which dissolves the iron oxides. Moreover, selenium can also be recycled for use in electronic industry during a membrane posttreatment that involves digestion and purification. Competing ions seem to adversely affect the performance of the membrane, making this method highly selective but less effective in the case of high levels of coexisting ions. However, this decontamination technique appears as promising and provides many opportunities for further improvement and optimization.

Azamat et al. (2015) studied the permeation of Zn^{2+} through functionalized graphene nanosheet and boron nitride nanosheet membranes. Based on the results, both membranes are capable of removing zinc ions from aqueous solutions, but the boron nitride nanosheet can remove zinc more efficiently. By increasing the voltage applied on the membrane more ions could pass and the residence time of the ions decreases.

NF270 and NF90 commercial membranes were the choice of Sousa and Teixeira (2015) for the removal of silver nanoparticles and colloidal silver from water in order to prevent silver spread into the environment. The authors proved that it is possible to remove both silver nanoparticles and the dissolved silver released by nanoparticles at the same time using the same filtration membrane. While silver nanoparticles were completely removed, dissolved silver was removed by more than 92%. The membranes removed silver nanoparticles through size exclusion, while a mechanism involving complexation with salts and adsorption on natural organic matter was postulated for the removal of dissolved silver. The authors studied the effect of ionic strength and of the presence of natural organic matter on the membrane performances. Increasing of the ionic strength resulted in reduced rejection of dissolved silver as a consequence of the compression of the electric double layer. The effect of natural organic matter was studied concerning type (hydrophilic or hydrophobic) and concentration of the organic material. In the case of the hydrophilic NF270 membrane, the type of the natural organic material was showed to be more important than its concentration which was not the case with the hydrophobic NF90 membrane for which similar effects on the permeate flux were observed by varying either the nature or the concentration of the organic materials present in the tested water samples. For instance, the permeability of the NF270 decreased in the presence of the hydrophilic tannic acid most likely due to membrane fouling during nanofiltration, while the hydrophobic NF90 membrane was almost unaffected.

Zhong et al. (2013) developed a carbonaceous nanowire membrane for the removal of heavy metal ions from water samples. The carbonaceous nanowires were obtained through the hydrothermal dehydration and carbonization of glucose, using tellurium nanowire as templating agent. Eventual solvent evaporation coroborated with self-assembly converted the unidimensional nanowires into the final product, the bidimensional membrane. The group used Cd^{2+} as a model heavy metal ion and the adsorption capacity was 70.0 mg/g in the pH range 5–7, a value that is much greater than the adsorption capacities corresponding to many other adsorbents like charcoal, hematite, or chitin. The initial Cd^{2+} concentration used in the study was 10 mg/L. The membrane can be regenerated and reused after acid leaching of the ions trapped. The membrane provides efficient retention of heavy metals and high water permeation as well. Another strong point of this decontamination method is its sustainability provided by the green fabrication of functional nanomaterials using biomass as a renewable starting material.

Ghaemi et al. (2015) used polyethersulfone membranes and enhanced them by embedding various types of surface modified Fe_3O_4 nanoparticles in order to increase the benefits obtained through nanofiltration. Three different types of nanoparticles were used for enhancement, silica–coated nanoparticles, metformin modified silica–coated nanoparticles, and amine modified silica–coated nanoparticles. The best results for copper removal (about 92% removal) were obtained by the metformin modified nanoparticles embedded in the membrane. The membrane showed acceptable reusability being regenerated through elution with ethylenediaminetetraacetic acid (EDTA) which effective chelates Cu(II) ions. This study shows that the surface modification plays an important role in the adsorption of heavy metals from water.

Nanofiltration is an important technique in the removal of heavy metal ions, the potential for automation being greater than in other means of purification. Also the possibility of self-regeneration or some sort of regeneration cycle seems to be another asset in this type of systems. Nanofiltration membranes can be incorporated into modular portable water filtration devices suitable for household use which is exceptionally important for those area severely affected by pollution like Bangladesh and India, where a huge part of the global population resides (German et al., 2014). Global access to clean potable water should not be a struggle taken lightly, given the fact that the main purpose of industrialization and economic progress should be the improvement of the quality of life around the globe (Qu et al., 2013).

4 Ligand-Based Materials

Ligand-based materials are nanomaterials that are tailored to form a coordinative bond with the heavy metal ion that is to be captured. The interesting fact about this bond is that it can be reversed, the triggering factor being the pH of the medium. These type of materials absorb at one pH domain and then release the load in another pH range. This property is extremely useful when the nanomaterial needs regeneration.

Awual et al. (2015a) developed an interesting type of nanocomposite material designed to target Se(IV) selectively through a coordinative ligand-based bond. The group incorporated *N,N′*-di(3-carboxysalicylidene)-3,4-diamino-5-hydroxypyrazole (DSDH), the ligand pH-responsible organic moiety, into an inorganic framework silicon-based mesoporous material. The detection limit was 1.14 μg/mL, while the maximum sorption capacity was 111.12 mg/g. This type of ligand-based material was specifically designed to be highly selective for Se(IV) ions and this selectivity was put to test by the group and proven to be as claimed. The group tested the selectivity of the nanocomposite by using a synthetic solution containing Na^+, K^+, Ca^{2+}, Mg^{2+}, Mn^{2+}, Ni^{2+}, Co^{2+}, Zn^{2+}, NO_3^-, Cl^-, SO_4^{2-}, and PO_4^{3-} each at a concentration level of 20 mg/L. The presence of these ions did not adversely affect the properties of the sorbent. The sorbent can be desorbed and reutilized by pH change, using 0.10 M NaOH solution. The sorbent remained highly active, as the group claims, after eight cycles of sorption–desorption.

The same research group led by Awual et al. (2014) developed another ligand-based dual conjugate adsorbent for Co(II) and Cu(II) adsorption from water using the same silica-based monolithic structure in which they incorporated the active ligand moiety, 1E, 1′E, 1″E, 1‴E (tetrakis(3-carboxysalicylidene))naphthalene-1,2,5,5-tetramine (TSNT). Color change occurs as soon as the metal complex is formed so the system can also act as a colorimetric detector for the targeted species, Co(II) and Cu(II). The sorption rate is highly dependent on the pH of the medium. The study of sorption was conducted over the pH range 2.0–9.5. No sorption of Cu(II) was observed at pH 2.0, while the sorption of Co(II) was drastically reduced when the pH dropped below 5.0. The maximum sorption capacity of the nanomaterial was 205.33 mg/g for Co(II) and 199.20 mg/g for Cu(II), respectively. The selectivity of the method was tested in the presence of several interfering cations (Al^{3+}, Fe^{3+}, Zn^{2+}, Ni^{2+}, Mn^{2+}, Mg^{2+}, Ca^{2+}, Hg^{2+}, K^+, Na^+) and anions (sodium dodecyl sulfate, citrate, acetate, NO_3^-, CO_3^{2-}, PO_4^{3-}, SO_4^{2-}) and no significant modification in either the visible color or in the reflectance absorbance spectra could be evidenced. Elution with aqueous hydrochloric acid resulted in adsorbent regeneration. A drop in performance from 99.6 to 85.8% after 10 regeneration cycles was noticed. The drop in efficiency is not as radical as to consider that the material is to be discarded, and the data show that it may very well sustain another 10 or 20 cycles before complete consumption (Fig. 16.1).

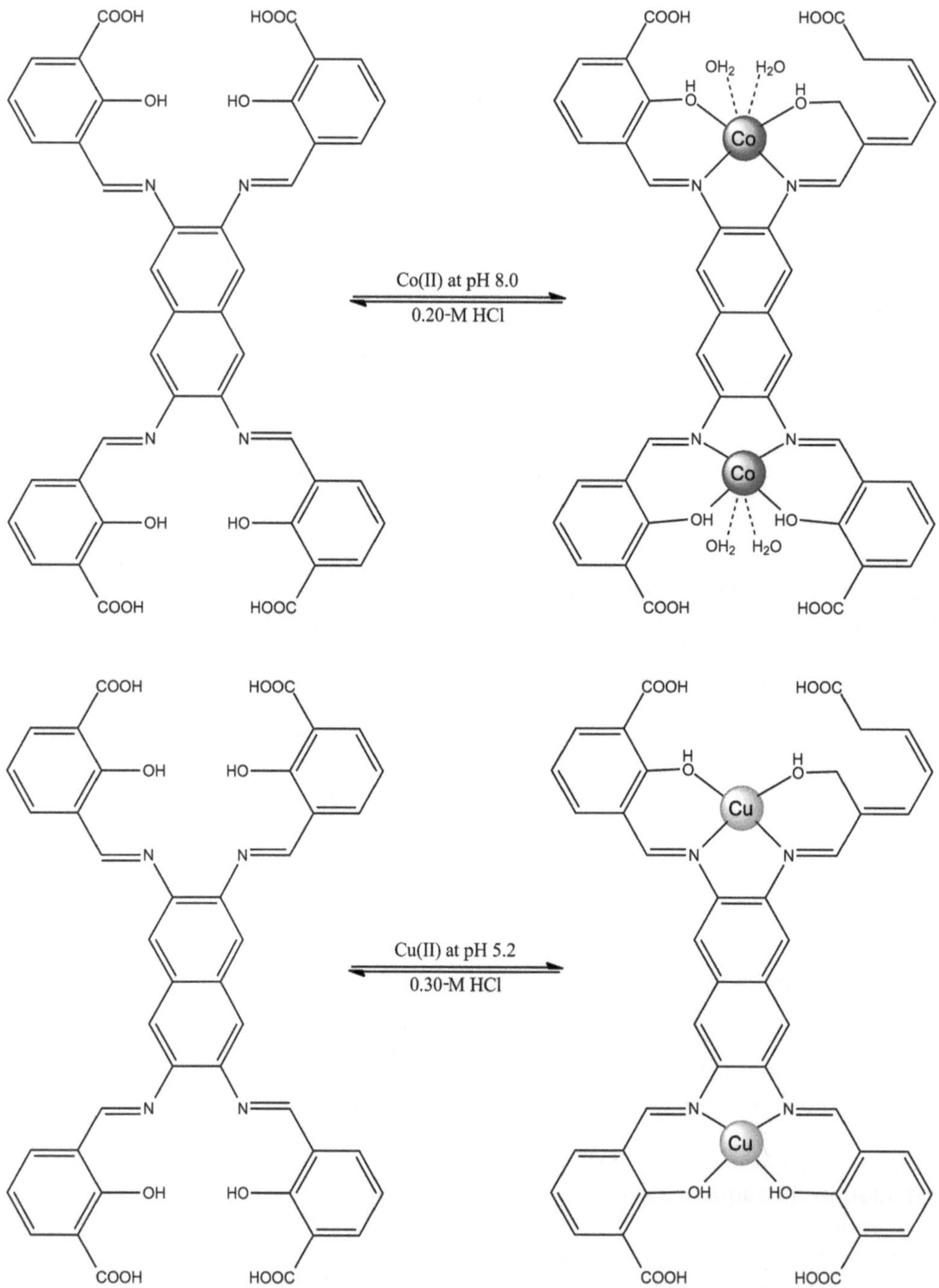

Figure 16.1. Ligand-based copper and cobalt removal from water (Awual et al., 2014).

The DSDH ligand was used also by Shahat et al. (2015) who anchored it as a facial adsorbent onto mesoporous silica. Mesoporous silica monoliths were synthesized using tetramethylorthosilicate (TMOS) as silica source and the lyotropic liquid crystalline phase of the triblock copolymer Pluronic F108 as soft template (Awual et al., 2013a). Immobilization of the DSDH and preparation of the mesoporous facial adsorbent was achieved using an indirect method named the building-block approach (Awual et al., 2013b). First, an ethanolic solution of dilauryl dimethyl ammonium bromide (DDAB) and the previously fabricated mesoporous silica monoliths were combined and immobilized (El-Safty et al., 2013). Then, the DDAB-mesoporous silica material was added into an ethanolic solution of the DSDH ligand; the mixture was stirred until becoming saturated with DSDH and eventually ethanol was removed under vacuum. The equilibrium isotherms for the sorption of Co(II) ions by this hybrid nanoadsorbent fit the Langmuir model giving a maximum sorption capacity of 157.73 mg/g which was in good agreement with the experimental results. Co(II) ions can be eluted from the adsorbent using a 0.2-M HCl solution. After as much as eight adsorption–regeneration cycles, the adsorbent lost less than 8% of its initial sorption capacity.

Awual et al. (2015b) developed a nanoconjugate adsorbent for selective removal of Co(II) ions, using the same strategy to incorporate a ligand-based adsorbent inside a mesoporous silica framework. The ligand anchored was *N,N′*-(octane-1,8-diylidene) di(2-hydroxy-3,5-dimethylaniline) (DHDM). The group studied the influences of potential interfering ions, contact time and initial Co(II) concentration on the Co(II) sorption. The sorption equilibrium was reached in 40 min, and at equilibrium 99% of Co(II) ions were adsorbed from a solution having the initial concentration of 0.5 mg/L. Co(II) sorption efficiencies higher than 97% were obtained even in the sorption experiments carried out in multimetal ion containing solutions (Na^+, K^+, Li^+, Hg^{2+}, Ba^{2+}, Ca^{2+}, Zn^{2+}, Mg^{2+}, Pb^{2+}, Cd^{2+}, Ni^{2+}, Sr^{2+}). The maximum adsorption capacity was found to be 165.83 mg/g and was calculated using the Langmuir model. The sorbent is reusable and a sorption capacity of 92% can be maintained even after nine cycles of adsorption–regeneration. The regeneration of this medium is extremely easy, a solution of 0.10 M HCl and a contact time of only 15 min leaching 99% of the adsorbed species (Fig. 16.2).

Awual and Hasan (2014) also developed a new ligand-based nanoadsorbent embedded in mesoporous silica for Pb(II) removal from water. The ligand proposed for the adsorption of Pb(II) was 4-dodecyl-6-((4-(hexyl-oxy)phenyl)diazenyl) benzene-1,3-diol (DPDB). The authors studied the effects of pH, contact time,

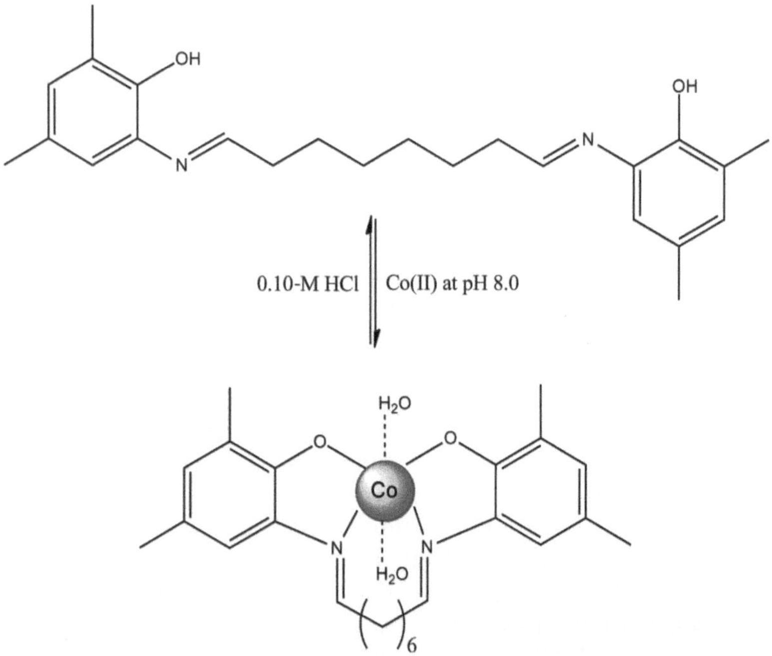

Figure 16.2. Image of ligand-based cobalt removal from water (Awual et al., 2015b).

and coexisting ions on Pb(II) removal. The best sorption behavior was in the pH range 5–7. Pb(II) was almost completely removed in 40 min (fast sorption kinetics) for an initial concentration of 0.25 mg/mL. The maximum sorption capacity was 195.31 mg/L. Lead complexation generates a colorimetric signal response that can be easily visualized with naked eye. The authors determined a limit of detection as low as 0.18 μg/L. The adsorption never dropped lower that 95% in the presence of competing ions (K^+, Na^+, Ag^+, Ni^{2+}, Zn^{2+}, Al^{3+}, Cd^{2+}, Mn^{2+}, Co^{2+}, Cu^{2+}, Hg^{2+}, Fe^{3+}, Ca^{2+}, and Mg^{2+}), which proves the high selectivity of the conjugate adsorbent.. The sorbent can be reused and the Pb(II) ions can be removed by using a 0.2 M HCl solution. The sorbent shows a minor decrease in efficiency after ten cycles of regeneration.

One of the major advantages of ligand-based nanomaterials is the fact that these type of conjugate materials can be tailored to suit the need for sensitive materials and ion selective sorption and desorption. All these materials have very good regenerative capabilities, and slight changes in efficiency can be observed after several cycles of regeneration. Besides the sorption behavior, the capability of these sorbents to change color when being in contact with the target ion makes them highly accurate sensors even at low concentrations of the analyte. Due to the high possibility of reuse combined with a very-small fouling rate over several regeneration cycles, nanomaterials in this class are suited not only for research purposes but also for industrial applications.

5 Nanostructured Metals/Metal Oxides

This particular type of adsorbent material for heavy metals covers the magnetic oxide materials and the nonmagnetic ones, also core–shell structured ones and other types of structure, hierarchical or not. It is a very large class of materials, but not all of them have gained the attention of researchers from the field of water purification from heavy metals, some of them not having the required sorption properties for heavy metal ions. Core–shell systems also have another interesting property, the fact they can be used as plasmonic sensors (Thatai et al., 2014) for heavy metal ions analysis.

Alimohammadi et al. (2013) conducted a very interesting research on the subject of the treatment of drilling fluid (mud) with the purpose of Cd removal. They used magnetic nanoparticles, core–shell systems, by immobilizing polyacrylic acid on the surface of the Fe_3O_4 nanoparticles, and afterward this system was functionalized with amino groups by using diethylenetriamine via carbodiimide chemistry. This group characterized their material thoroughly, the effects of pH, contact time, sorbent dosage, shaking rate, being emphasized. The optimal adsorption was found at pH 6, shaking rate from 240 to 480 rpm, the optimum contact time being short, after 2 min 90% of the concentration being absorbed on the nanoparticles. The equilibrium sorption data were fitted well into the Freundlich isotherm model, an adsorption capacity (K_f) of 2.683 mg/g being determined. The very short purification time renders these nanomaterials economically attractive since speed is an important parameter.

One-dimensional conducting nanofibers of polyaniline (PANI) have attracted much interest due to high surface area to volume ratio, high conductivity, good redox properties, and stability. These properties can be further enhanced by embedding metallic nanoparticles which can synergistically add other interesting electronic, catalytic, magnetic, or optical properties to create PANI/metallic nanoparticles composites. Among metallic nanoparticles Fe(0) nanoparticles have been proposed and effectively used for water treatment due to their large specific surface area corroborated with their excellent reducing capacity. However, zero-valent iron nanoparticles are highly sensitive to moisture and air oxygen in the surrounding medium, thereby being oxidized and undergoing reactivity loss. Therefore, embedding of Fe(0) nanoparticles within the conducting polymer matrix protects them from premature oxidation. Bhaumik et al. (2014) have developed a very simple and efficient method to produce PANI/Fe(0) nanofibers to remove highly toxic As(V) and Cr(VI) as well as an organic contaminant, that is, the azo dye Congo red (CR) from water. Their synthetic procedure consists in chemical polymerization of the aniline monomer using $FeCl_3$ as oxidizing agent. After the polymerization reaction was completed, the byproduct $FeCl_2$ and excess $FeCl_3$ remained in the reaction mixture. The authors exploited these compounds as iron source to produce Fe(0) nanoparticles by chemical reduction. To this purpose, the authors simply added a solution of $NaBH_4$ to the preformed PANI nanofibers. As(V) removal relies on the electrostatic interaction between the negatively charged $H_2AsO_4^-$ and $HAsO_4^{2-}$ species and the positively charged PANI (below the isoelectric pH of 8.4). Another possible removal mechanism postulated by the authors involves anion exchange of doped chloride ions of PANI with the aforesaid As(V) anions. Similarly, the mechanism of Cr(VI) removal is based on the ion exchange properties of the doped (Cl^-) PANI, but also on the reducing properties of both PANI and Fe(0) nanoparticles which reduce highly toxic Cr(VI) to the much less toxic Cr(III) ions which are subsequently adsorbed as Cr(III) oxides or hydroxides. The authors determined the maximum sorption capacities as being 42.37 mg/g for As(V) and 434.7 mg/g for Cr(VI), respectively. The specific surface area (SBET) of the PANI/Fe(0) composite nanofibers was 42.58 m^2/g.

Huang et al. (2013) also used zero-valent iron nanoparticles for Cd(II), Cr(IV), and Pb(II) removal from water based on chemical reduction and magnetic separation. They used an industrially available deposition technique, the electrospray technique, to deposit zero-valent iron particles, and studied both the material which was deposited through electrospray and the material deposited in a more classic alternative and debated on the results. In a short amount of time, less than 1 min, 90% of the metal ions are absorbed, and further increase of the reaction time does not change the sorption properties, except for Pb(II), where the efficiency increases with the increase of contact time between the solution and the sorbent. The removal of nanoparticles from the medium afterward takes some time, 5 min being an optimum time for the magnetic separation to occur. The use of the electrospray technique versus nonelectrospray technique did not seem to alter the results significantly.

Electrospun Fe_2O_3 nanofibers were investigated by Nalbandian et al. (2016) as adsorbent for the chromate anion. The electrospinning solution consisted of iron(III) 2-ethylhexano-isopropoxide as iron source, acetic acid and 4–5%wt. polyvinylpyrrolidone (PVP). After electrospinning, the Fe/PVP nanofibers were subjected to a thermal annealing process during which the PVP polymer was removed and pure Fe_2O_3 nanofibers crystallized. Electrospinning technique is a highly versatile technique, since it is not only a research technique, but it can also be very easily adapted for mass production and for industrial use. Moreover, this type of synthesis allows the control of nanofibers diameter which is tunable by varying specific parameters, such as PVP polymer content, applied voltage, and solution feed rate. The nanofiber diameter is very important, the sorption efficiency being inversely proportional with the diameter of the nanofiber. Fibers of 23 nm diameter were obtained by the authors working at a polymer content of 4%, an applied voltage of 20 kV, and a feed rate of 0.3 mL/h. With a CrO_4^{2-} sorption capacity of 90.9 mg/g these Fe_2O_3 nanofibers outperformed commercially available Fe_2O_3 nanoparticles (49.3 mg/g).

Many water-decontamination systems successfully tested on laboratory scale are not applied on a larger scale because of ineffective knowledge transfer (Michalak et al., 2013). German et al. (2014) built a large-scale application for remote locations and installed it in many locations across India, Nepal, and Cambodia. They proposed a franchise that would significantly reduce the cost of pure water for the arsenic-affected communities and also increase the development of the area by providing jobs for the persons responsible for the regeneration of the sorbent. They developed a hybrid anion exchanger (HAIX) impregnated with hydrated iron-oxide nanoparticles that are highly reliable, multiuse, robust, and can be utilized for remote small-scale communities in an economic fashion. The decontamination process is based on the known affinity of hydrated ferric oxides (HFOs) for both As(III) and As(V) at groundwater pH. Regeneration was achieved by changing the pH value. Hence, by raising pH, the iron-oxide functional groups acquire negative charges thereby repelling arsenite and arsenate anions. This type of endeavor is a small step toward providing clean potable water to the whole population of the world, but a highly significant one, due to the large number of devices already installed which serve a significant number of people from countries which are severely affected by clean potable water shortage.

Simeonidis et al. (2015) tested Fe_3O_4 nanoparticles in a dynamic way, for in flux water depollution of Cr(VI). The authors concluded that Cr(VI) uptake was due to the reduction of Cr(VI) to Cr(III) by the Fe^{2+} sites of the ferrite nanoparticles. The superparamagnetic response of ferrite nanoparticles was exploited to easily collect the chromium-loaded nanoparticles. Taking into account the very small dimensions, the magnetic separation yield as a function of the intensity of the applied external magnetic field and of the residence time into the magnetic separation filter is a crucial issue. For a residence time of 1 h which corresponds to a volume velocity of 33 mL/min total removal of nanoparticles from the treated water was achieved using a relative low magnetic field (600 Oe). Another important parameter of chromium removal is the pH of the treated water. Removal experiments were carried out in the pH range 5–8. Although, the removal capacity increased with decreasing pH, the highest removal efficiency being attained at pH 5, at this pH value a significant amount of Cr(VI) (41.1%) was not reduced by magnetite and remained adsorbed on the nanoparticles surface in this highly toxic form with the possibility of redissolution. On the other hand, at the normal water acidity (pH 7), chromium was found in the inofensive, insoluble form of $Cr(OH)_3$. To obtain a residual Cr(VI) concentration below the current regulation limit of 10 μg/L at pH 7, the contact time in the reaction tank should be 2 h. Under these working parameters, the removal capacity of magnetite nanoparticles was 2 μg Cr(VI)/mg. Another important advantage of this purification method is that it can operate at household scale with a very low cost of about 5 €/year for an initial concentration of Cr(VI) in the treated water of 50 μg/L.

Hierarchically structured nanomaterials seem to have an advantage by having a higher sorption rate, as it is claimed by Zhu et al. (2013) who developed a simple template-free solvothermal synthesis of α-Fe_2O_3 flower-like hollow spheres using a glycerol–ethanol system. The same type of sorbent, flower-like structured, was studied independently by Cao et al. (2012), using a solvothermal microwave–assisted synthesis and providing a low-cost, efficient, environmentally friendly method for the removal of As(V) and Cr(VI) from water. The maximum adsorption capacities were 51 and 30 mg/g for As(V) and Cr(VI), respectively, which is clearly far superior to the sorption capacity of commercially available bulk α-Fe_2O_3.

Maghemite nanoparticles were also studied by Ahmadi et al. (2014) for the removal of Pb(II), Zn(II), and Cd(II) ions from water. The effects of several parameters, such as pH, temperature, initial concentration of the contaminant ions, and adsorbent dosage on the removal efficacy were studied. In the case of Cd(II) the initial ion concentration and pH have been proven as being the most-important factors for the removal efficiency. The efficiency of Pb(II) removal was most affected by the adsorbent dosage and pH.

The authors predicted maximum adsorption capacities of 10.55, 4.49, 1.75 mg/g for Pb(II), Zn(II), and Cd(II), respectively. Under optimized working conditions, that is, pH of 7.5, temperature of 45°C, initial ion concentration of 50 mg/L, and adsorbent dosage of 4.74 g/L, an efficiency of 100% was obtained for Pb(II) removal.

Xu et al. (2014) proposed a system consisting of monodisperse Fe_3O_4/ZnO microrods and compared the results obtained with their composite microrods with results obtained for the pure materials. The authors put their results on the basis of the fact that the presence of Fe_3O_4 nanoparticles in the ZnO network provides additional surface active sites for the adsorption of toxic metal ions, a very reasonable conclusion backed up by the experimental results.

Many studies were centered toward the very good applications of maghemite nanoparticles for heavy metal ions removal, but newer tendencies try to improve the properties of the maghemite network by incorporating other materials (Xu et al., 2014; Lemine et al., 2014). Lemine et al. (2014) proposed a simple method to prepare Fe_2O_3 powder doped with nanocrystalline Ni. The XRD data showed, after Rietveld analysis, a mixture of phases, α-Fe_2O_3 and $NiFe_2O_4$. The authors studied the influence of Ni content on heavy metals adsorption and the best results were obtained for Ni-doped αFe_2O_3 with 15% wt. Ni. The maximum sorption capacities obtained in this study were, 65, 43, 42, and 4.6 mg/g for Cd(II), Co(II), Ni(II), and Cr(VI) accordingly.

Titanium dioxide is a metal oxide considered, among others, by Guo et al. (2013) for heavy metal ions removal from water. They prepared a 3D TiO_2 nanostructure consisting of nanoparticle core and needle-like surface. The diameter of the 3D system could be tuned in the self-assembly process. The sorption capacity was tested by the authors for the removal of As(V) and Cr(VI). Maximum adsorption capacities of 59.7 and 21.92 mg/g for As(V) and Cr(VI), respectively, were determined. A short removing time (less than 1 h) was reported. The conclusion of their study is that the sorption properties of this type of material are mainly caused by the large hydroxyl group densities, groups that are available for linkage of heavy metal ions.

The core-shell systems consisting of Fe_3O_4 core/SiO_2 shell/secondary organic shell are used in various applications, the nanostructured silica imparting new properties to the magnetic ferrite core, the high specific surface area of silica being known and intensely exploited in a large number of applications. The organic shell used by Mahmoud et al. (2013) to remove heavy metal ions from water was triethylenetetraamine (TETA). In addition to the characteristic adsorption properties given by the silica shell, the amino groups of the organic moiety improved the capability of heavy metal ions removal. The synthesis was accomplished in three steps. First magnetic Fe_3O_4 nanoparticles were prepared by the coprecipitation method and were subsequently coated with a silica shell on treatment with sodium silicate solution and pH adjustment to 6. The silica shell was further functionalized by treatment with 3-chloropropyltrimethoxysilane to get nano-Fe_3O_4–SiO_2–Cl. The chlorine atoms were used for eventual functionalization with TETA when the nanosorbent nano-Fe_3O_4–SiO_2–TETA was obtained. The sorbent functionalized with amino groups has a high affinity for Cu(II), while the system without the organic shell is useful for the Pb(II) removal from water. A maximum Cu(II) sorption capacity of 480 μmol/g was determined in batch equilibrium studies carried out at pH 7 for the functionalized nanoadsorbent Fe_3O_4–SiO_2–TETA.

Due to their many uses, ferrite nanoparticles are a highly valuable commodity. In order to avoid environmental pollution, many researchers focused on the development of green synthetic methods, while others tried to convert environmentally pollutant sludges and waste materials into useful products. Kefeni et al. (2015) did just that by synthesizing Fe_3O_4 and $CoFe_2O_4$ nanoparticles starting with acid mine drainage, a waste fluid resulted from mining industry, and seeding it with pure ferrite nanoparticles in order to stimulate the growth of these particles from the mixture of components. This type of waste contains many types of minerals, all under various forms, salts, oxides, and other forms. The surface of the nanoparticles still contained traces of toxic impurities, which could be desorbed through carefully washing with water and organic solvents. This group demonstrated successfully the feasibility of the turning of environmentally polluting waste into precious chemical substances.

Savasari et al. (2015) developed and tested a core–shell system comprising of zero-valent iron nanoparticles and ascorbic acid shell and optimized the behavior of this system for industrial wastewater treatment via the response surface methodology. Through this technique they determined the optimum set of parameters in order to use this type of nanoadsorbent for industrial wastewater treatment. The following optimum set of parameters was found: a sorbent dosage of 2 g/L, an initial Cd(II) concentration of 15 mg/L, a contact time of 60 min and a pH value of 7, resulting in a removal efficiency of 79.68%. Optimization methods, although time consuming and laborious, are in many cases a necessity, and perhaps the only way of achieving the best possible throughput for a specific type of sorbent material.

Selective removal of Cd(II) from water was achieved by Luo et al. (2014) who used magnetic ion-imprinted fluorescent cadmium telluride quantum dots as a novel type of nanoadsorbent. The authors ingeniously applied a chemical etching strategy to obtain this composite nanomaterial. The synthesis was accomplished in a stepwise manner. Using a sol–gel method, magnetic Fe_3O_4 nanoparticles prepared by the coprecipitation method were coated with a shell of silica exposing functional thiol groups on its outer surface. To this end, the authors used teraethyl orthosilicate (TEOS) and 3-mercaptopropyl trimethoxysilane (MPTS) as reagents for functionalization of ferrite nanoparticles. In a separate synthetic step, 3-mercaptopropanoic acid (MPA)-capped cadmium telluride (CdTe) quantum dots were obtained using $CdCl_2$ as cadmium precursor, Na_2TeO_3 as tellurium source, $NaBH_4$ for fast reduction of Te^{4+} to Te^{2-}, and MPA as capping ligand. Next, functionalized magnetic Fe_3O_4/SiO_2-SH nanoparticles were combined with fluorescent MPA-capped CdTe quantum dots by means of thiol coordination leading to formation of the nanocomposite material denoted as Fe_3O_4@CdTe-MPA. Eventually, the Fe_3O_4@CdTe-MPA was chemically etched with EDTA according to the equilibrium reaction below. The MPA-CdTe complex was broken down by EDTA, MPA was detached and the Cd^{2+} ions on the surface of the Fe_3O_4@CdTe were removed by complexation with EDTA with concomitant loss of fluorescence. Thereby, specific Cd^{2+} recognition sites were created on the surface of the Fe_3O_4@CdTe nanocomposite resulting in the formation of the ion-imprinted nanomaterial Fe_3O_4@CdTe-IIP.

$$\begin{aligned}&Fe_3O_4@CdTe\text{-}MPA + HEDTA^{3-}\\&\rightleftharpoons Fe_3O_4@CdTe\text{-}IIP + MPA + Cd(HEDTA)^{-} + Te^{2-}\end{aligned}$$

When Cd^{2+} ions were added into the solution of the Fe_3O_4@CdTe-IIP, the above equilibrium was shifted to the left. The recognition sites created during the imprinting process were quickly reoccupied by the added Cd^{2+} ions, thereby restoring the fluorescence of the quantum dots and repairing the surface defects. The whole synthetic protocol is schematically illustrated in Fig. 16.3. The authors reported a selective adsorption capacity of 17.57 mg Cd(II)/g for the ion-imprinted quantum dots while the adsorption capacity of the nonimprinted Fe_3O_4@CdTe-MPA nanocomposite was only 8.03 mg/g. The method developed by this group for the selective removal of Cd(II) ions from water has also the advantage of an easily monitoring of the process due to regaining of fluorescence during Cd(II) adsorption.

Due to the known excellent adsorption properties, mesoporous silica–based materials have been extensively studied for their potential application in water decontamination. Chen et al. (2011) directly used mesoporous Ti-MCM-48 and tested its sorption properties for Cd(II) removal from water. The best adsorption results were obtained at pH 8.0, the authors reporting a maximum adsorption

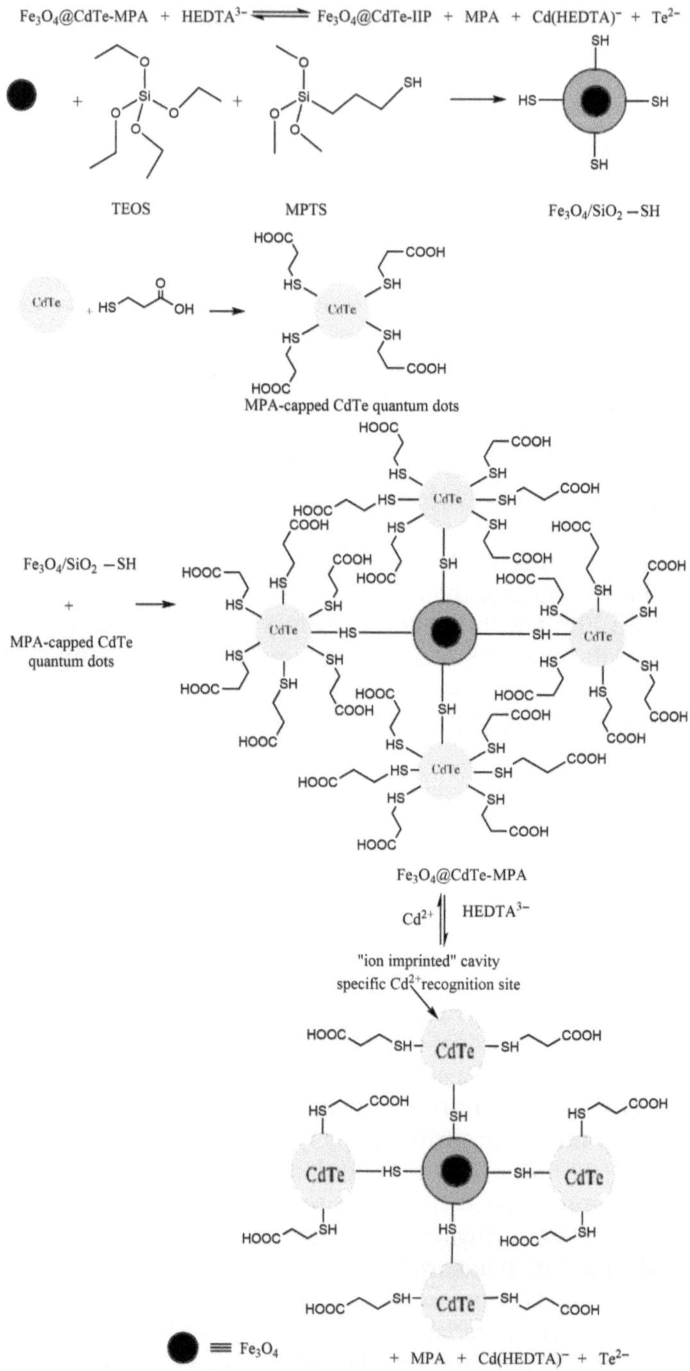

Figure 16.3. Schematic illustration of the synthesis of magnetic ion-imprinted fluorescent CdTe quantum dots for the selective removal of Cd^{2+} from water (Luo et al., 2014).

capacity of 83.57 mg/g. Wang et al. (2012) synthesized a mesoporous magnetic material and functionalized it the by the immobilization inside the pores of a pyrene-based receptor. The authors referred to it as a "multifunctional inorganic-organic hybrid nanomaterial" because it can be used for both detection and efficient removal of Hg^{2+} ions from water matrices. The improved fluorescence properties ensured high sensitivity and reproducibility of the detection process. A limit of detection of 1.72 ppb was reported. Moreover, another convenience feature of this material is its magnetic separability.

An environmental friendly method was used by Venkateswarlu et al. (2015) to prepare biogenic magnetite nanoparticles, which were afterward functionalized with MPA. This magnetic nanosorbent which again combines the sorption properties with the magnetic ones was used for selective sorption of Ni(II) from contaminated waters. In a batch adsorption experiment, the sorption equilibrium was reached in 60 min when operating at pH 6.0, an initial concentration of 80 mg/L, and a temperature of 30°C. The regeneration of this material was done exploiting its pH dependent sorption, by washing out the Ni(II)-loaded nanoadsorbent with HCl (pH 2) followed by rinsing it with double distilled water. The material preserved 75% of its initial sorption capacity after five cycles of adsorption–regeneration. The maximum adsorption capacity was found to be 42.01 mg/g at pH 6 and the optimum sorbent dosage was found to be 0.1 g/L, a very good set of characteristics.

ZnO nanoparticles synthesized by the gel combustion technique were chosen by Venkatesham et al. (2013) for removal of lead ions from water. The maximum sorbent capacity found by this group was 26.109 mg/g according to Langmuir isotherm and was highly pH dependent. The maximum sorption efficiency was reached at pH 4 and remained constant with further increasing pH. However, above the pH 10, $Pb(OH)_2$ might precipitate in the system. The specific surface area of the nanopowder was 80.425 m^2/g.

Magnetic Mn(II) containing Fe_3O_4 core–MnO_2 shell systems were studied by Chen et al. (2014) for the lead-removal properties it possesses. Although the specific surface area of the material this group obtained is not very high (113 m^2/g) the adsorption capacity for lead is high. Experiments carried out in the pH range 2–8 showed that the Pb removal rate increased with increasing pH but the adsorption capacity and kinetic studies were performed at pH 5.0 in order to avoid any hydroxide precipitation. A maximum sorption capacity of 261.1 mg/g was determined using the Langmuir model. Regeneration was achieved by treatment for 1 h with 0.1 M HCl and the nanoadsorbent is reusable its adsorption capacity remaining as high as 80% of the initial value after four cycles of regeneration.

Table 16.2 presents the most illustrative examples of cleaning up heavy metals from water using nanostructured metals and

Table 16.2 Performances of Nanostructured Metal and Metal Oxides Used for Decontamination of Heavy Metals from Drinking Water

Type of Adsorption Media	Adsorption (mg Adsorbed Species/Gram Adsorbent)	Other Advantages	References
Core–shell Fe_3O_4 nanoparticles functionalized with NH_2 groups	Cd(II) 2.683 mg/g	Real-life applicability, studies conducted with drilling fluid	Alimohammadi et al. (2013)
polyaniline–Fe^0 nanocomposite	As(V) 42.37 mg/g Cr(VI) 434.7 mg/g	Simpler reaction pathway, reduced costs of production	Bhaumik et al. (2014)
Zero-valent iron nanoparticles	—	Very rapid removal rate, 1–20 min; Cd(II), Cr(IV) and Pb(II) removal; studies conducted at neutral pH	Huang et al. (2013)
α-Fe_2O_3 nanofibers	Cr(VI) 90.9 mg/g	Synthesized via electrospinning method	Nalbandian et al. (2016)
Hydrated iron-oxide nanoparticles		Large-scale applications; marketed products	German et al. (2014)
Fe_3O_4 nanoparticles	Cr(VI) 2 μg/mg	Dynamic mode testing; real-water sample testing; Cr(VI) removal	Simeonidis et al. (2015)
α-Fe_2O_3 flower-like structures	Pb(II) 39 mg/g Cu(II) 48 mg/g	Low-cost, environmental-friendly method for synthesis; simultaneous removal of Pb(II), Cu(II), and Congo red	Zhu et al. (2013) and Cao et al. (2012)
Maghemite nanoparticles	Pb(II) 10.55 mg/g Zn(II) 4.49 mg/g Cd(II) 1.75 mg/g	Simultaneous removal of Pb(II), Zn(II), and Cd(II)	Ahmadi et al. (2014)
Fe_3O_4/ZnO microrods		Simultaneous removal of Co(II), Ni(II), Cu(II), Pb(II), Ag(I)	Xu et al. (2014)
Ni-doped maghemite	Cd(II) 65.46 mg/g Co(II) 43.30 mg/g Ni(II) 42.01 mg/g Cr(VI) 4.66 mg/g	Simultaneous removal of Cd, Co, Ni, and Cr	Lemine et al. (2014)
3D TiO_2 nanostructure	As(V) 59.70 mg/g Cr(VI) 21.92 mg/g	Simultaneous removal of As(V) and Cr(VI)	Guo et al. (2013)
Fe_3O_4/SiO_2 core–shell system	—	Simultaneous removal of Cu(II) and Pb(II)	Mahmoud et al. (2013)

Ferrite nanoparticles	—	Real samples (acid mine drainage) tested; simultaneous removal of Fe, Co, Ni, Mn, Zn, Al, Cr	Kefeni et al. (2015)
Zero-valent iron nanoparticles	—	Cd(II) removal from water samples; small sorbent dosage (2 g/L)	Savasari et al. (2015)
CdTe/Fe_3O_4 quantum dots	Cd(II) 17.57 mg/g	Highly selective adsorption of Cd(II)	Luo et al. (2014)
Ti-MCM-48	Cd(II) 83.57 mg/g	Highly selective adsorption of Cd(II)	Chen et al. (2011)
Magnetite nanoparticles	Ni(II) 42.01 mg/g	Environmental-friendly synthesis; rapid purification (60 min)	Venkateswarlu et al. (2015)
ZnO nanoparticles	Pb(II) 26.11 mg/g	pH reversible adsorption	Venkatesham et al. (2013)
Mn/MnO_2 core–shell systems	Pb(II) 261.1 mg/g	Pb removal from aqueous media, reversible adsorption, regeneration capabilities	Chen et al. (2014)

metal oxides along with the main characteristics of the nanoadsorbents and depollution methods.

6 Natural or Nature-Based Materials

The focus of the research community toward development of greener methods using less toxic or even biocompatible solvents, fewer reagents, precursors and catalysts, involving less steps in the synthetic pathway, all meet the same goal of pollution reduction and of usage of the natural resources in a more economical and environmental-friendly way. This goal materializes in the efforts to replace chemically synthesized materials with newer materials, either naturals, extracted from natural products using water or renewable solvents (such as ethanol, which can be renewed indefinitely through fermentation of fruits and vegetables), or with materials made from natural components. Others just take industrial waste materials and try to shape them into utilizable materials, like once the alchemists dreamt of changing iron into gold. Although an alchemist dream, today we are changing garbage into valuable chemicals through proper chemical modification and we have changed our perspective of the way we do things and why we do them, gradually adopting a more environmentally responsible behavior.

El-Sherif et al. (2013) developed a novel type of sorbent for Cr(III) that can successfully eliminate Cr(III) from tannery waste water. In this type of water Cr(III) can oxidize to Cr(VI) in a matter of hours, so just by simply storing these waste waters the toxicity increases with time through accumulation of toxic species on the account of nontoxic ones. The removal of Cr(III) from those waters before it gets a chance to oxidize is an important issue that was tackled by this group. Cysteine-modified polypyrrolepropylic acid nanofibers were prepared in two steps. First, electrochemical template-directed polymerization of pyrrolepropylic acid was performed using anodic alumina membrane as the working electrode (Tolani et al., 2010). Next, the carboxylic groups of the polymer were conjugated to the amino groups of cysteine through carbodiimide-mediated coupling. The functionalized nanowires exhibited a great capability for Cr(III) removal never reported before for any other material. After a thorough study, this team came with a set of optimized parameters for their material. The optimal nanofiber dosage and the exposure time were found to be 0.1 mg nanofibers per milliliter of tannery waste water with a pH of 5.5 and 45 min, respectively. Under these optimal conditions, a removal rate of 99.9% was obtained corresponding to a maximum adsorption capacity of 1.75 g of Cr(III) per gram of sorbent used, nearly double the amount of adsorbed species per gram of sorbent, which is never before encountered and a major accomplishment in the field of heavy metal ions removal.

Cellulose has a very high availability and it constitutes an ongoingly renewable source of materials. Maatar and Boufi (2015) used a natural source of cellulose, bleached eucalyptus pulp which was first oxidized and then converted to nanofibrils by means of high-pressure homogenization. Thereby, a highviscosity translucent gel was obtained. Nanofibrillated cellulose (NFC) gel was further subjected to freeze-drying in order to convert it into an aerogel. Using a grafting from strategy, grafted NFC aerogels were eventually fabricated from the pristine NFC aerogel by performing Fenton's reagent-initiated aqueous radical copolymerization of methacrylic acid (MAA) and maleic acid monomers (MA). The NFC-poly(MMA-*co*-MA) aerogel was tested for removal from water of the following heavy metal ions, Pb^{2+}, Cd^{2+}, Zn^{2+}, and Ni^{2+} and the following maximum adsorption capacyties 165, 135, 138, and 117 mg/g, respectively, were determined in batch experiments. Regeneration of the adsorbent was achieved with more than 98% efficiency by treating the aerogel with EDTA solution. This feature makes this material a very robust one, fit for longterm usage inside purification columns. The authors also used a column system wherein they deposited the nanomaterial and tested with promising results the capacity of heavy metal ions removal under dynamic conditions of continuous flow. This is a noteworthy aspect of the author's work that should be emphasized with respect to goal compliance (eventual industrial scale application).

Natural hydroxyapatite has a porous structure that makes it a suitable candidate for adsorption studies. However, since natural hydroxyapatite is difficult to purify, synthetic forms of the natural product seem to be an interesting subject in the field of nanomaterials. The compatibility of this type of material with the human bone tissue gives rise to new generations of prostethic devices and dental applications. But those applications should not end there, given the enormous amount of opportunities this nature wonder has to offer. He et al. (2016) tailored synthetic hydroxyapatite into nanowires for fluoride removal from water, but many other groups demonstrated that a large number of nanodevices can adsorb more than one species, thus being suited to larger applications, not just the one they were tested for.

As an example, nanocrystalline needle-shaped hydroxyapatite was used by Mobasherpour et al. (2012) in order to study the removal efficacy of this material on Pb(II), Cd(II), and Ni(II) from water matrices. The authors reported a very high value for the retention of lead ions, 1000 mg/g, but the values of 142.86 and 40 mg/g determined for Cd(II) and Ni(II) removal, respectively, are not to be neglected as well.

Mohammad et al. (2015) came up with an interesting strategy to combine hydroxyapatite with chitosan, which is extracted usually from shrimp and other crustacean shells with basic solutions, and usually de-acetylated up to a certain amount. This is just another example of how waste materials, like shrimp shells, which normally are discarded, can become precious commodities, chitosan being an important food supplement and not only, its importance in different kinds of sorbent being acknowledged by the scientific community through the many articles published using this material. The authors reported a one-pot synthesis of hydroxyapatite nanorods (nHAp)/chitosan (CS) composite material using the sol–gel method. The authors studied the effects of contact time, initial ion concentration, pH and sorbent dosage, in order to fully optimize the sorption of Pb(II) ions. for both pristine hydoxyapatite nanorods and the nHApCS composite. They observed that the sorption capacity of nHApCS overcame that of nHAp and suggested that the numerous amino groups present in the composite material were responsible for this result. The experimental data fitted well to Langmuir adsorption isotherms and a maximum sorption capacity of 196.1 mg/g was calculated for the nHApCS composite.

Gopalakannan and Viswanathan (2015) managed to develop a gelatin biocomposite embedded with nanohydroxyapatite and tested it for the Cr(VI) adsorption properties. The group tested the properties in batch mode and used real industrial water during their tests from a nearby industrial area. Their results were promising and add to those obtained by others in recommending nanocomposites-based on natural-like hydroxyapatite materials for adsorption of heavy metals from different types of polluted waters. Not many groups that tested sorbents actually used real samples taken from industrial areas, and even fewer groups tailored sorbents for a specific depollution behavior in a specific industrial area in order to overcome a threat. This group used real industrial water samples, which is noteworthy.

Zeolites are naturally occurring porous substrates which have a rather high specific surface area. Plotnikov et al. (2015) used a natural zeolite coming from "Holinskoe" mineral deposit (Russia, Republic of Buryatia) and modified it in order to make it more suitable for the removal of heavy metal ions from water samples. The zeolite was modified by deposition through the sol–gel method of nanofibers of aluminum oxyhydroxide and iron oxyhydroxide on the surface. The modifiers have been previously prepared using the electro-spark dispersion technique, the working solution being treated with a pulsed electric discharge for 5 min Encouraging results were obtained for extraction of Pb(II) and Cd(II) ions.

Polyaniline-grafted chitosan crosslinked with glutaraldehyde was proposed by Karthik and Meenakshi (2015) for removal of Pb(II) and Cd(II) ions from aqueous solutions. The sorption capacity was strongly pH-dependent and increased with increasing pH, the maximum removal capacity being achieved at pH 6.0. At this pH value, the surface of the polyaniline-grafted chitosan became negatively charged, thereby favoring the electrostatic fixation of the metal ions and subsequently metal coordination. At pH above 6.0, formation of lead and cadmium hydroxides occurred which impeded removal of the metal ions. The authors assumed that the purification process relies on an ion-exchange mechanism. The adsorption data obtained in batch experiments fitted well to the Freundlich isotherm model. Despite the fact that interference from coexisting metal ions (Ni^{2+}, Co^{2+}, Cu^{2+}, Zn^{2+}) reduced the removal percentage for both Cd(II) and Pb(II), sorption efficiencies as high as 98.21 and 94.92%, respectively, were determined using tap water samples. This was to be expected, being caused by the competitive adsorption of ions present in real-water samples.

Geopolymers is a quite controversial name and not always unitary in definition. The name of geopolymer comes from the fact that the starting substances are mainly of geological origin. Medpelli et al. (2015) synthesized such a material for the purpose of arsenic removal starting from kaolinite, a naturally occurring mineral. Onto this material iron oxide nanoparticles were synthesized aiming to give magnetic properties to the final material. The best results were obtained for a composite material containing about 20 wt % Fe whose adsorption capacity was 954 μg As/g dry media. Leaching tests on As(V) showed no significant leaching, proving that the material can be disposed of properly and that the used material does not pose a threat to the environment after use and can be safely stored in this fashion.

Yan et al. (2015) used biochar, a renewable resource material, magnetically modified and decorated with ZnS nanocrystals to remove lead ions from water. The material exhibited a high adsorption capacity for Pb(II) (367.65 mg/g).

Yang et al. (2012) took advantage of the rather unique properties of *Bacillus subtilis*, a spore forming bacterium to adsorb both As(III) and As(V) from the immediate environment. This group tested a novel approach in modifying the cellular membrane to contain iron by incubating those microorganisms with Fe(III). Only 30% of the cells remained viable, but the results are astonishing. At pH 3.0, after iron uptake, the cells selectively adsorbed only As(V) and not As(III), the adsorption of As(V) being three time as high as the one of the unmodified cells. Increasing pH to 10.0, the ferrated bacteria adsorbed both arsenic species and the adsorption capacity for As(V) was 11 fold higher than that of the genuine *B. subtilis*. This group presented yet another successful case in the application of natural solutions to industrial problems.

Same as Mohammad et al. (2015), Rozumova et al. (2015) also tackled the issue of taking waste materials and transforming them into commodities. Their material of choice was peanut husks, a waste material resulted from the food industry, from processed peanuts, and they managed to successfully turn it into a resource through minor modifications. The peanut husks have an inherent porous structure, which makes them ideal candidates for sorption studies. With the added qualities of the magnetically active iron oxide, the material becomes more valuable, due to the fact that it can be easily separated from the medium after an operation of successful extraction, regenerated and reused. The sorption capacity of this material is not great, only 28.3 mg/g for Pb^{2+} and 7.68 mg/g for Cd^{2+}, but this is not the main advantage of this material. The main advantage provided by this material is recycling of a waste material in a smarter, responsible way, instead of always trying to synthesize new materials that would in turn pollute the environment more than they are trying to clean it. This initiative is just one of the many which show that responsible chemistry is possible and should remain a priority of all the chemists worldwide.

Table 16.3 summarizes the nanoadsorbents based on natural materials.

7 Conclusions and Perspectives

The use of nanostructured materials over traditional ones has clear advantages, mainly due to the high surface area of the nanoscale materials, and, as the scale goes toward lower and lower values, the surface increases and the behavior modifies. In some cases, the nanoscale is not only responsible just for the increase in surface area, but some effects are to be explained by quantum effects that manifest themselves in this range, such as plasmonic coupling. Some authors found very high sorption rates for heavy metal ions removal [(El-Sherif et al., 2013) 1.75 g/g] while others got lower sorption rates, but through an environmentally friendly solution (Rozumova et al., 2015). Truth being said, both methods are to be encouraged by the scientific community. Sometimes a big sorption rate is required, especially in the case of large hazardous spillings where immediate action is required to minimize the environmental damage. In normal cases, such as getting clean water in the consumer's household, maybe a better strategy would be to use smarter materials, produced in an environmentally safe attitude.

Table 16.3 Performances of Natural-Based Nanoadsorbents

Type of Adsorption Media	Adsorption (mg Adsorbed Species/Gram Adsorbent)	Other Advantages	References
Cysteine-modified poly pyrrolepropylic acid nanofibers	Cr(III) 1.75 g/g	Huge purification potential; real sample application from tanning industry waste	El-Sherif et al. (2013)
Cellulose nanofibres	Pb(II) 165 mg/g Cd(II) 135 mg/g Zn(II) 138 mg/g Ni(II) 117 mg/g	Renewable materials, environmental friendly approach; reusable material	Maatar and Boufi (2015)
Hydroxyapatite	Pb(II) 1000.00 mg/g Cd(II) 142.86 mg/g Ni(II) 40.00 mg/g	Bicompatible, nontoxic, suitable for drinking water applications	Mobasherpour et al. (2012)
Hydroxyapatite/chitosan	Pb(II) 196.1 mg/g	Biocompatible, nontoxic, suitable for drinking water applications	Mohammad et al. (2015)
Chitosan/polyaniline	Pb(II) 27.57 mg/g Cd(II) 15.78 mg/g	High efficiency (more than 90% removal)	Karthik and Meenakshi (2015)
Magnetic-modified peanut husks	Pb(II) 28.30 mg/g Cd(II) 7.68 mg/g	Renewable source; environmental friendly	Rozumova et al. (2015)

Nogueira et al. (2015) focused their research toward the study of toxicological aspects resulting from wastewater treatment using nanomaterials. Tang et al. (2014) studied the potential interaction of nanomaterials with different components present in real samples, such as humic acid or fulvic acid. The study presented by Nogueira et al. (2015) clearly states that the effect of mass-scale nanoparticle production in the case of nano-TiO_2 or nano-Fe_2O_3 disturbs the activity of *Chironomus riparius*, a water midge with important activity in the metabolism of organic compounds in the sediment of the lakes. This study is not universally applicable and does not describe the fact that all nanomaterials are toxic or dangerous to the environment, but triggers a question mark on whether nanomaterials are sufficiently studied, are a good alternative, for both the environment and the human population, and presents a good argument for better solutions, greener chemistry, and a more responsible way of regarding the chemical synthesis and the nanomaterials in general. The interaction of nanomaterials with environmentally ubiquitous humic acid/fulvic acid may alter the stability, mobility, and transportation of heavy metal-loaded waste nanomaterials disposed in the environment resulting in possible enhancement of the bioavailability, bioaccumulation and toxicity of heavy metals (Tang et al., 2014). This raised an alarm toward responsible use of nanomaterials, pointing out that too many studies prepare sorbents and none of them are actually interested on what happens afterward with the used up sorbents or how those can be disposed properly after the regeneration cycles are complete and the sorption properties no longer fulfill the requirements.

Also not many studies are thorough enough, proposing very good solutions for distilled water samples, but lacking the transition toward real water samples. The studies that did this type of analysis clearly showed that interferences from sample matrix are real and have to be taken into account in real life situation and real water samples. Future studies should take this aspect into account, and focus their efforts more toward the goal of testing the sorbents they proposed in real samples situation, under dynamic flux conditions if they want to propose an industrial solution, or even small-scale applications when desirable. Although a high number of scientific articles exist in the field, few are thorough enough and maybe even fewer will be ever applied in real life situations. This only points out that there is still much to be done in this field and that the goal is set, pure water for the entire planet should not be a luxury but a service that everyone should enjoy.

References

Ahmadi, A., Heidarzadeh, S., Mokhtar, A.R., Darezereshki, E., Harouni, H.A., 2014. Optimization of heavy metal removal from aqueous solutions by maghemite (γ-Fe_2O_3) nanoparticles using response surface methodology. J. Geochem. Explor. 147, 151–158.

Alimohammadi, F., Gashti, M.P., Shamei, A., Kiumarsi, A., 2012. Deposition of silver nanoparticles on carbon nanotube by chemical reduction method: evaluation of surface, thermal and optical properties. Superlattice. Microst. 52, 50–62.

Alimohammadi, N., Shadizadeh, S.R., Kazeminezhad, I., 2013. Removal of cadmium from drilling fluid using nano-adsorbent. Fuel 111, 505–509.

Al-Rashdi, B.A.M., Johnson, D.J., Hilal, N., 2013. Removal of heavy metal ions by nanofiltration. Desalination 315, 2–17.

Awual, M.R., Hasan, M.M., 2014. Novel conjugate adsorbent for visual detection and removal of toxic lead(II) ions from water. Micropor. Mesopor. Mater. 196, 261–269.

Awual, Md.R., Yaita, T., El-Safty, S.A., Shiwaku, H., Suzuki, S., Okamoto, Y., 2013a. Copper(II) ions capturing from water using ligand modified a new type mesoporous adsorbent. Chem. Eng. J. 221, 322–330.

Awual, Md.R., Kobayashi, T., Shiwaku, H., Miyazaki, Y., Motokawa, R., Suzuki, S., Okamoto, Y., Yaita, T., 2013b. Evaluation of lanthanide sorption and their coordination mechanism by EXAFS measurement using novel hybrid adsorbent. Chem. Eng. J. 225, 558–566.

Awual, M.R., Yaita, T., Okamoto, Y., 2014. A novel ligand based dual conjugate adsorbent for cobalt(II) and copper(II) ions capturing from water. Sensors Actuat. B 203, 71–80.

Awual, M.R., Yaita, T., Suzuki, S., Shiwaku, H., 2015a. Ultimate selenium(IV) monitoring and removal from water using a new class of organic ligand based composite adsorbent. J. Hazard. Mater. 291, 111–119.
Awual, M.R., Yaita, T., Shiwaku, H., Suzuki, S., 2015b. A sensitive ligand embedded nano-conjugate adsorbent for effective cobalt(II) ions capturing from contaminated water. Chem. Eng. J. 276, 1–10.
Azamat, J., Sattary, B.S., Khataee, A., Joo, S.W., 2015. Removal of a hazardous heavy metal from aqueous solution using functionalized graphene and boron nitride nanosheets: insights from simulations. J. Mol. Graph. Model. 61, 13–20.
Bhaumik, M., Choi, H.J., McCrindle, R.I., Maity, A., 2014. Composite nanofibers prepared from metallic iron nanoparticles and polyaniline: high performance for water treatment applications. J. Colloid Interface Sci. 425, 75–82.
Cao, C.Y., Qu, J., Yan, W.S., Zhu, J.F., Wu, Z.Y., Song, W.G., 2012. Low cost synthesis of flowerlike α-Fe_2O_3 nanostructures for heavy metal ion removal: adsorption property and mechanism. Langmuir 28 (9), 4573–4579.
Chang, C.Y., Yu, H.Y., Chen, J.J., Li, F.B., Zhang, H.H., Liu, C.P., 2014. Accumulation of heavy metals in leaf vegetables from agricultural soils and associated potential health risks in the Pearl River Delta, South China. Environ. Monit. Asses. 186, 1547–1560.
Chen, J., Fang, K., Wu, L., Qian, Z., Chen, J., 2011. Removal of Cd(II) from aqueous by adsorption onto mesoporous Ti-MCM-48. Procedia Environ. Sci. 10, 2491–2497.
Chen, J., He, F., Zhang, H., Zhang, X., Zhang, G., Yuan, G., 2014. Novel core-shell structured Mn-Fe/MnO_2 magnetic nanoparticles for enhanced Pb(II) removal from aqueous solution. Ind. Eng. Chem. Res. 53 (48), 18481–18488.
Chen, S., Li, X., Sun, G., Zhang, Y., Su, J., Ye, J., 2015. Heavy metal induced resistance in bacterium LSJC7. Int. J. Mol. Sci. 16, 23390–23404.
Deng, J.H., Zhang, X.R., Zeng, G.M., Gong, J.L., Niu, Q.Y., Liang, J., 2013. Simultaneous removal of Cd(II) and ionic dyes from aqueous solution using magnetic graphene oxide nanocomposite as an adsorbent. Chem. Eng. J. 226, 189–200.
El-Safty, S.A., Awuala, Md.R., Shenashena, M.A., Shahat, A., 2013. Simultaneous optical detection and extraction of cobalt(II) from lithium ion batteries using nanocollector monoliths. Sensors Actuat. B 176, 1015–1025.
El-Sherif, I.Y., Tolani, S., Ofosu, K., Mohamed, O.A., Wanekaya, A.K., 2013. Polymeric nanofibers for the removal of Cr(III) from tannery waste water. J. Environ. Manage. 129, 410–413.
German, M., Seingheng, H., SenGupta, A.K., 2014. Mitigating arsenic crisis in the developing world: role of robust, reusable and selective hybrid anion exchanger (HAIX). Sci. Total Environ. 488-489, 547–553.
Ghaemi, N., Madaeni, S.S., Daraei, P., Rajabi, H., Zinadini, S., Alizadeh, A., Heydari, R., Beygzadeh, M., Ghouzivand, S., 2015. Polyethersulfone membrane enhanced with iron oxide nanoparticles for copper removal from water: Application of new functionalized Fe_3O_4 nanoparticles. Chem. Eng. J. 263, 101–112.
Gollavelli, G., Chang, C.C., Ling, Y.C., 2013. Facile synthesis of smart magnetic graphene for safe drinking water: heavy metal removal and disinfection control. ACS Sustain. Chem. Eng. 1 (5), 462–472.
Gopalakannan, V., Viswanathan, N., 2015. Development of nano-hydroxyapatite embedded gelatin biocomposite for effective chromium(VI) removal. Ind.Eng. Chem. Res. 54 (50), 12561–12569.
Gopalakrishnan, A., Krishnan, R., Thangavel, S., Venugopal, G., Kim, S.J., 2015. Removal of heavy metal ions from pharma-effluents using graphene-oxide

nanosorbents and study of their adsorption kinetics. J. Ind.Eng. Chem. 30, 14–19.

Gui, M., Ormsbee, L.E., Bhattacharyya, D., 2013. Reactive functionalized membranes for polychlorinated biphenyl degradation. Ind. Eng. Chem. Res. 52 (31), 10430–10440.

Gui, M., Papp, J.K., Colburn, A.S., Meeks, N.D., Weaver, B., Wilf, I., Bhattacharyya, D., 2015. Engineered iron/iron oxide functionalized membranes for selenium and other toxic metal removal from power plant scrubber water. J. Membr. Sci. 488, 79–91.

Guo, J., Cai, X., Li, Y., Zhai, R., Zhou, S., Na, P., 2013. The preparation and characterization of a three-dimensional titanium dioxide nanostructure with high surface hydroxyl group density and high performance in water treatment. Chem. Eng. J. 221, 342–352.

He, J., Matsuura, T., Chen, J.P., 2014. A novel Zr-based nanoparticle-embedded PSF blend hollow fiber membrane for treatment of arsenate contaminated water: material development, adsorption and filtration studies, and characterization. J. Membr. Sci. 452, 433–445.

He, J., Zhang, K., Wu, S., Cai, X., Chen, K., Li, Y., Sun, B., Jia, Y., Meng, F., Jin, Z., Kong, L., Liu, J., 2016. Performance of novel hydroxyapatite nanowires in treatment of fluoride contaminated water. J. Hazard. Mater. 303, 119–130.

Henriques, B., Goncalves, G., Emami, N., Pereira, E., Mercedes, V., Marques, P.A.A.P., 2016. Optimized graphene oxide foam with enhanced performance and high selectivity for mercury removal from water. J. Hazard. Mater. 301, 453–461.

Hu, H., Jin, Q., Kavan, P., 2014. A study of heavy metal pollution in China: current status, pollution-control policies and countermeasures. Sustainability 6, 5820–5838.

Huang, P., Ye, Z., Xie, W., Chen, Q., Li, J., Xu, Z., Yao, M., 2013. Rapid magnetic removal of aqueous heavy metals and their relevant mechanisms using nanoscale zero valent iron (nZVI) particles. Water Res. 47, 4050–4058.

Jaishankar, M., Tseten, T., Anbalagan, N., Mathew, B.B., Beeregowda, K.N., 2014. Toxicity, mechanism and health effects of some heavy metals. Interdiscip. Toxicol. 7 (2), 60–72.

Karthik, R., Meenakshi, S., 2015. Removal of Pb(II) and Cd(II) ions from aqueous solution using polyaniline grafted chitosan. Chem. Eng. J. 263, 168–177.

Kefeni, K.K., Msagati, T.M., Mamba, B.B., 2015. Synthesis and characterization of magnetic nanoparticles and study their removal capacity of metals from acid mine drainage. Chem. Eng. J. 276, 222–231.

Kim, Y.M., Chung, J.Y., An, H.S., Park, S.Y., Kim, B.G., Bae, J.W., Han, M., Cho, Y.J., Hong, Y.S., 2015. Biomonitoring of lead, cadmium, total mercury and methylmercury levels in maternal blood and in umbilical cord blood at birth in South Korea. Int. J. Environ. Res. Publ. Health 12, 13482–13493.

Lemine, O.M., Ghiloufi, I., Bououdina, M., Khezami, L., M'hamed, M.O., Hassan, A.T., 2014. Nanocrystalline Ni doped α-Fe_2O_3 for adsorption of metals from aqueous solution. J. Alloy. Compd. 588, 592–595.

Liu, X., Wang, M., Zhang, S., Pan, B., 2013. Application potential of carbon nanotubes in water treatment: a review. J. Environ. Sci. 25 (7), 1263–1280.

Luo, X., Guo, B., Wang, L., Deng, F., Qi, R., Luo, S., Au, C., 2014. Synthesis of magnetic ion-imprinted fluorescent CdTe quantum dots by chemical etching and their visualization application for selective removal of Cd(II) from water. Colloid. Surf. A 462, 186–193.

Ma, S., Zhan, S., Jia, Y., Zhou, Q., 2015. Highly efficient antibacterial and Pb(II) removal effects of Ag–$CoFe_2O_4$–GO nanocomposite. ACS Appl. Mater. Interfaces 7 (19), 10576–10586.

Maatar, W., Boufi, S., 2015. Poly(methacylic acid-*co*-maleic acid) grafted nanofibrillated cellulose as a reusable novel heavy metal ions adsorbent. Carbohydr. Polym. 126, 199–207.
Mahmoud, M.E., Abdelwahab, M.S., Fathallah, E.M., 2013. Design of novel nano-sorbents based on nano-magnetic iron oxide-bound-nano-silicon oxide-immobilized-triethylenetetraamine for implementation in water treatment of heavy metals. Chem. Eng. J. 223, 318–327.
Medpelli, D., Sandoval, R., Sherrill, L., Hristovski, K., Seo, D.K., 2015. Iron oxide-modified nanoporous geopolymers for arsenic removal from ground water. Resour. Efficient Technol. 1, 19–27.
Michalak, I., Chojnacka, K., Witek-Krowiak, A., 2013. State of the art for the biosorption process—a review. Appl. Biochem. Biotechnol. 170 (6), 1389–1416.
Mobasherpour, I., Salahi, E., Pazouki, M., 2012. Comparative of the removal of Pb^{2+}, Cd^{2+} and Ni^{2+} by nano crystallite hydroxyapatite from aqueous solutions: adsorption isotherm study. Arab. J. Chem. 5, 439–446.
Mohammad, A.M., Salah Eldin, T.A., Hassan, M.A., El-Anadouli, B.E., 2015. Efficient treatment of lead-containing wastewater by hydroxyapatite/chitosan nanostructures, Arabian Journal of Chemistry.
Nalbandian, M.J., Zhang, M., Sanchez, J., Choa, Y.H., Nam, J., Cwiertny, D.M., Myung, N.V., 2016. Synthesis and optimization of Fe_2O_3 nanofibers for chromate adsorbtion from contaminated water sources. Chemosphere 144, 975–981.
Nogueira, V., Lopes, I., Rocha-Santos, T., Goncalves, F., Pereira, R., 2015. Toxicity of solid residues resulting from wastewater treatment with nanomaterials. Aquat. Toxicol. 165, 172–178.
Pamphlett, R., Jew, S.K., 2013. Heavy metals in locus ceruleus and motor neurons in motor neuron disease. Acta Neuropathol. Commun. 81 (1), 1–15.
Plotnikov, E., Martemianova, I., Martemianov, D., Zhuravkov, S., Voronova, O., Korotkova, E., Vladimir, S., 2015. Water purification with natural sorbents: effect of surface modification with nano-structured particles. Procedia Chem. 15, 219–224.
Qu, X., Alvarez, P.J.J., Li, Q., 2013. Applications of nanotechnology in water and wastewater treatment. Water Res. 47, 3931–3946.
Rozumova, L., Zivotsky, O., Seidlerova, J., Motyka, O., Safarik, M., Safarikova, M., 2015. Magnetically modified peanut husks as an effective sorbent of heavy metals. J. Environ. Chem. Eng. 21, 1403–1409.
Savasari, M., Emadi, M., Bahmanyar, M.A., Biparva, P., 2015. Optimization of Cd(II) removal from aqueous solution by ascorbic acid-stabilized zero-valent iron nanoparticles using response surface methodology. J. Ind. Eng. Chem. 21, 1403–1409.
Shahat, A., Awual, M.R., Naushad, M., 2015. Functional ligand anchored nanomaterial based facial adsorbent for cobalt(II) detection and removal from water samples. Chem. Eng. J. 271, 155–163.
Simeonidis, K., Kaprara, E., Samaras, T., Angelakeris, M., Pliatsikas, N., Vourlias, G., Mitrakas, M., Andritsos, N., 2015. Optimizing magnetic nanoparticles for drinking water technology: the case of Cr(VI). Sci. Total Environ. 535, 61–68.
Sousa, V.S., Teixeira, M.R., 2015. Silver nanoparticles separation from the water using nanofiltration membranes: the role of mono–divalent salts and NOM. Separ. Purif. Technol. 149, 165–173.
Tang, W.W., Zeng, G.M., Gong, J.L., Liang, J., Xu, P., Zhang, C., Huang, B.B., 2014. Impact of humic/fulvic acid on the removal of heavy metals from aqueous solutions using nanomaterials: a review. Sci. Total Environ. 468-469, 1014–1027.

Thatai, S., Khurana, P., Boken, J., Prasad, S., Kumar, D., 2014. Nanoparticles and core–shell nanocomposite based new generation water remediation materials and analytical techniques: a review. Microchem. J. 116, 62–76.

Tofighy, M.A., Mohammadi, T., 2015. Copper ions removal from water using functionalized carbon nanotubes-mullite composite as adsorbent. Mater. Res. Bull. 68, 54–59.

Tolani, S., Mugweru, A., Craig, M., Wanekaya, A.K., 2010. Rapid and efficient removal of heavy metal ions from aqueous media using cysteine-modified polymer nanowires. J.Appl. Polym. Sci. 116 (1), 308–313.

Venkata Ramana, D.K., Yu, J.S., Seshaiah, K., 2013. Silver nanoparticles deposited multiwalled carbon nanotubes for removal of Cu(II) and Cd(II) from water: surface, kinetic, equilibrium, and thermal adsorption properties. Chem. Eng. J. 223, 806–815.

Venkatesham, V., Madhu, G.M., Satyanarayana, S.V., Preetham, H.S., 2013. Adsorption of lead on gel combustion derived nano ZnO. Procedia Eng. 51, 308–313.

Venkateswarlu, S., Kumar, S.H., Jyothi, N.V.V., 2015. Rapid removal of Ni(II) from aqueous solution using 3-mercaptopropionic acid functionalized bio magnetite nanoparticles. Water Res. Ind. 12, 1–7.

Wang, Y., Li, B., Zhang, L., Li, P., Wang, L., Zhang, J., 2012. Multifunctional magnetic mesoporous silica nanocomposites with improved sensing performance and effective removal ability toward Hg(II). Langmuir 28 (2), 1657–1662.

Xu, M., Li, Q., Fan, H., 2014. Monodisperse nanostructured Fe_3O_4/ZnO microrods using for waste water treatment. Adv. Powder Technol. 25, 1715–1720.

Yan, L., Kong, L., Qu, Z., Li, L., Shen, G., 2015. Magnetic biochar decorated with ZnS nanocrystals for Pb(II) removal. ACS Sustain. Chem. Eng. 3 (1), 125–132.

Yang, T., Chen, M.L., Liu, L.H., Wang, J.H., Dasgupta, P.K., 2012. Iron(III) modification of *Bacillus subtilis* membranes provides record sorption capacity for arsenic and endows unusual selectivity for As(V). Environ. Sci. Technol. 46 (4), 2251–2256.

Zhong, Y.J., You, S.J., Wang, X.H., Zhou, X., Gan, Y., Ren, N.Q., 2013. Synthesis of carbonaceous nanowire membrane for removing heavy metal ions and high water flux. Chem. Eng. J. 226, 217–226.

Zhu, D., Zhang, J., Song, J., Wang, H., Yu, Z., Shen, Y., Xie, A., 2013. Efficient one-pot synthesis of hierarchical flower-like α-Fe2O3 hollow spheres with excellent adsorption performance for water treatment. Appl. Surf. Sci. 284, 855–861.

17

APPLICATION OF NANOTECHNOLOGY, NANOFILTRATION, AND DRINKING AND WASTEWATER TREATMENT—A VISION FOR THE FUTURE

Sukanchan Palit
University of Petroleum and Energy Studies, Department of Chemical Engineering, Energy Acres, Dehradun, Uttarakhand, India

1 Introduction

The world of environmental engineering science is moving steadily and steadfastly through drastic challenges and insurmountable barriers. Global drinking water crisis is of immense significance (Drioli et al., 2002; Cheryan, 1998). Industrial wastewater treatment is moving through difficult terrains. Environmental sustainability is at a disastrous state in our present day human civilization. In such a crucial juncture of history and time, the importance of nanotechnology comes to the forefront. Application of nanotechnology, importance of nanofiltration (NF) and the wide world of drinking water treatment are the prime objectives of this instinctive treatise. The author carefully delineates the importance of nanotechnology in drinking water treatment and the immense innovative scientific endeavor behind the application. Provision of clean drinking water and successful environmental sustainability has an unsevered umbilical cord in our present day human civilization. History of mankind is repeating itself in every branch of human scientific endeavor. Environmental disaster and ecological imbalance is at its helm. So the answers to environmental sustainability are replete with scientific vision and scientific justification. Drinking water treatment is a primordial issue

Water Purification. http://dx.doi.org/10.1016/B978-0-12-804300-4.00017-4

in man's scientific research pursuit. In such a critical vein, nanotechnology will surely usher in a new era and a promising domain in the domain of drinking water treatment (Drioli et al., 2002; Cheryan, 1998).

Nanotechnology and its application in drinking water treatment is inevitably opening up new doors of scientific intuition and scientific innovation in years to come. Mankind's urge to excel is surely ushering in a new era of visionary direction. Application of NF to drinking water treatment is befitting to the domain of environmental sustainability and its immense emancipation. Environmental engineering paradigm is moving in a new direction of scientific fortitude. History of science and technology is ushering in a new era. Nanotechnology in today's world is steadfastly propelling to a new era of immense scientific vision and indomitable scientific truth. Application of nanotechnology in drinking water treatment, the visionary domain of NF and the futuristic vision of science and technology are the torchbearers of tomorrow's world of scientific research pursuit. In such a similar vein, advancement of science and technology, the future vision of NF and the immense visionary road ahead will lead to a new era of scientific understanding and scientific truth (Drioli et al., 2002; Ismail and David, 2001; Vrijenhoek et al., 2001; Cheryan, 1998).

1.1 Vision of This Present Treatise

Water process engineering and environmental engineering science are domains of engineering which are moving at a rapid pace. Drinking water treatment and industrial wastewater treatment are moving toward a newer visionary domain. History of human mankind, history of science and technology and the urge to excel in scientific endeavor has propelled civilization to rise toward greater heights. Zero-discharge norms and stringent environmental restrictions have urged the scientists and engineers to devise new and visionary technologies. In this treatise, the author delineates with cogent insight and deep comprehension the application and importance of nanotechnology in drinking water treatment and industrial wastewater treatment. The author also describes lucidly the relevant and far-reaching scientific endeavor in the field of NF of industrial wastewater and also in the application of nanotechnology in drinking water treatment. Science and engineering, as well as mankind's prowess will be emboldened if scientific truth and scientific validation is realized at its utmost. Nanotechnology is the next generation science and engineering. In such a critical vein, man's cognizance and mankind's prowess will be directed toward newer vision of nanotechnology. NF is another branch of nanotechnology, which has immense potential in treating drinking water and industrial wastewater. The author with immense insight brings forward to the scientific domain the visionary importance and the splendid details of the science and engineering of NF (Drioli et al., 2002; Ismail and David, 2001; Vrijenhoek et al., 2001; Cheryan, 1998).

1.2 Purpose and Aim of the Study

History of human civilization and history of science and technology are facing drastic challenges in today's world. Future of global drinking water crisis is at a difficult crossroads. Man's vision, a scientist's prowess and global concern for drinking water provision has plunged human civilization to a definitive challenge. In such a crucial juxtaposition, the environmental regulations and strict environmental restrictions have urged scientists and scientific generation to devise visionary techniques for provision of clean drinking water. The author with immense insight and deep comprehension delineates the future of global water crisis and the progress of technologies concerning provision of clean drinking water. A commendable effort has revealed the immense potential of NF and other techniques of nanotechnology in provision of clean drinking water and alleviating drinking water crisis. The reader can glean and unravel the visionary implication of the application of nanotechnology in drinking water treatment. Future of application of NF in drinking water provision and drinking water treatment is immensely bright and instinctive. Progress of science and technology, the urge to excel and the path toward scientific progress are surpassing scientific frontiers and overcoming scientific hurdles. Scientific endeavor and scientific vision is in the path of glory in today's human civilization. The study takes a visionary look into the application of NF and application of nanotechnology in the primordial and vexing issue of global drinking water crisis (Drioli et al., 2002; Mohammad et al., 2012; Schafer et al., 2003; Nghiem et al., 2005; Cheryan, 1998).

2 Global Drinking Water Crisis, Difficulties, and Success of Novel Separation Processes

Global drinking water crisis is at a devastating and deep peril. The success of scientific endeavor rests on the successful sustainable development and focusing on environmental sustainability. Science today stands in the backbone of nanotechnology and NF. History of science and engineering is witnessing a new dawn of human endeavor. Challenges are surmounted and frontiers surpassed. The crucial juncture is revealing a newer age and newer domain of environmental engineering science (Van der Bruggen et al., 2008; Childress and Elimelech, 2000; Lau et al., 2012; Vrouwenvelder et al., 2009; Cheryan, 1998).

The world of challenges and the path toward progress in the application of novel separation processes to drinking water treatment are wide and many. The difficulties, barriers, and hurdles are many yet the vision is inspiring. Science and technology is having a new beginning. Drinking water crisis and paradigm needs to be reenvisioned and reshaped. It is then the question of environmental sustainability realized. History of human mankind will then have a definitive and purposeful vision. The world of challenges will then be surpassed. Man's challenges, man's vision, and the drastic path toward emancipation are the torchbearers toward a definite futuristic target and future path. Global drinking water crisis is also in a similar vein in the path of greater scientific fortitude and scientific rejuvenation. Advancement of science and technology is moving toward a newer visionary domain. Environmental engineering science is gearing up with intense and vehement way toward the realization of successful sustainability.

Environmental engineering is in today's world in the path of immense scientific vision. Road to progress, immense scientific cognizance and the power to achieve will all lead to a greater emancipation in the path of application of environmental engineering science and membrane science. Membrane science is an effective chemical engineering and environmental engineering tool, which surpasses visionary frontiers of rationalization, optimization, and effectivity. NF, one of the membrane separation processes, stands upright with immense scientific vision, scientific understanding, and scientific justification (Luo and Wan, 2013; Alsvik and Hagg, 2013; Cheryan, 1998).

In the frame of Chemical Engineering world and Chemical Engineering applications, this branch of science today faces tremendous challenge from the domain of successful sustainability (Drioli et al., 2002). Sustainable solutions are needed to serve the increasing demand of raw materials, energy, and tailormade products. History of science and technology is moving toward a newer direction with the evolution of novel separation processes, such as membrane science (Opio, 2012; The RBC Harris Centre Drinking Water Applied Research Fund, 2009; Cheryan, 1998).

2.1 Access to Clean Drinking Water and Progress of Civilization

Challenges of advancement of science and technology are immense, as well as visionary. The vision of alleviating global drinking water crisis should reach out to the common mass. Progress of human civilization is on the path toward new dimension and newer direction. Developed, as well as developing countries are gearing toward the new challenge of groundwater heavy metal contamination. Groundwater remediation is the ultimate vision of tomorrow. Provision of clean drinking water is the vision for the future. Membrane science, NF, and other novel separation processes are the hallmarks of newer generation and newer scientific realm. NF and nanotechnology are surpassing visionary frontiers. The inevitable scientific truth, issues of environmental sustainability and the vexing question of application of nanotechnology will all lead a long way in rebuilding the history of environmental engineering science.

2.2 Global Initiative in Drinking Water Provision to Common Mass

Global initiative in clean drinking water provision to the common people is the primordial issue of tomorrow. The world is moving toward a newer dimension of future scientific pursuit in environmental engineering science. Environmental science, advancement of science and engineering and the immense path toward progress all will lead to a new future for global water initiative. The human civilization is steadily and drastically moving toward new dimensions of scientific research endeavor. Developed and developing nations are moving toward a new visionary realm of provision of clean drinking water. Global water initiative will be the hallmark of tomorrow's human progress. Challenges, scientific cognizance and the difficult path toward progress are the pallbearers of tomorrow.

2.3 Environmental Sustainability and the Progress of Scientific Endeavour

Environmental sustainability, provision of clean drinking water and the road toward progress are the vision of tomorrow. Successful environmental sustainability and global water initiative are in the process of new scientific vision. Provision of clean drinking water stands as a major prerogative of successful sustainable development in developed and developing countries. Environmental pollution control, industrial wastewater treatment, and veritably drinking water treatment in today's world are at a disastrous state with devastating end. History of science needs to be restructured at this crucial juncture. Environmental sustainability is in a disastrous state of immense distress. History of human civilization is at veritable crossroads. Man's vision, as well as a scientist's progress is in the process of reenvisioning with every step of progress.

3 Advancement of Science and Technology and the Vision of Membrane Science

Advancement of science and technology is befitting to the progress of environmental engineering science. Membrane science and separation and purification phenomena are in the path of visionary progress. A scientist's vision, mankind's yearning, and urge to excel and future of science and technology today stands in the midst of immense optimism. Vision of application of membrane science is in the definitive path. NF is one of the major membrane separation processes covering a wide domain of environmental engineering science. Today's technology of membrane separation process is applicable to industrial wastewater treatment and drinking water treatment. Advancement of science and engineering are unfolding the hidden scientific truth in industrial wastewater treatment and drinking water treatment. This vision is absolutely wide and far-reaching opening up new vistas in environmental sustainability.

3.1 Membrane Science—The Vision for the Future of Global Drinking Water Crisis

Membrane science and technology today stands in the midst of immense scientific vision and deep scientific understanding. The answers to environmental sustainability and the issues of environmental engineering science are revealed with indomitable scientific wisdom. Membrane science and membrane separation phenomenon in today's scientific world have an unsevered umbilical cord. Scientific research pursuit, deep comprehension, and immense forays are the hallmarks of a newer scientific generation.

The primary role of a membrane is to act as a selective barrier. It should allow passage of certain components and retain certain other components of a mixture (Cheryan, 1998). By implication, vision, and understanding, either the permeating stream or the retained phase should be enriched in one or more components. In its broadest and widest meaning, a membrane should be defined as "a region of discontinuity interposed between two phases," or as a "phase that acts as a barrier to prevent mass movement but allows restricted and/or regulated passage of one or more species through it." By these definitions, a membrane can be gaseous, liquid, or solid or combinations of these. Membranes can be further classified by (1) nature of membrane-natural versus synthetic; (2) structure of the membrane-porous versus nonporous, its morphological characteristics, or as liquid membranes; (3) application of the membrane-gaseous phase separations, gas–liquid, liquid–liquid; and so on; (4) mechanism of membrane action-adsorptive versus diffusive, ion-exchange, osmotic, or nonselective(inert) membranes (Drioli et al., 2002; Ray and Jain, 2011; Onjala et al., 2013; Cheryan, 1998).

The major membrane separation processes—reverse osmosis (RO), NF, UF (UF), microfiltration (MF), dialysis, electrodialysis (ED), and pervaporation (PV)—cover a wide range of particle/molecular sizes and applications. Osmosis is the transport of solvent through a semipermeable membrane from the dilute solution side to the concentration solution side of the membrane. With an ideal semipermeable membrane, only water should permeate through the membrane. The common laboratory technique of dialysis, on the other hand, is primarily a technique for purifying macromolecules, such as desalting of proteins, and the primary driving force is the difference in concentration of the permeable species between the solution in the dialysis bag and outside the bag. ED relies primarily on voltage or electromotive force and the application of ion-selective membranes (Cheryan, 1998).

3.2 Definition and Classification of Membrane Science

Filtration is defined as the separation of two or more components from a fluid stream based primarily on size differences (Drioli et al., 2002). In conventional terms, it usually refers to the separation of solid immiscible particles from liquid or gaseous streams (Drioli et al., 2002). Membrane filtration extends this application further to include the separation of dissolved solutes in liquid streams and for separation of gas mixtures (Table 17.1). Membrane separation processes domain in today's world reshaped and restructured. Scientific innovation, scientific advancements, and scientific doctrine are the pallbearers to a new scientific generation of vision and truth (Drioli et al., 2002; Cheryan, 1998).

What distinguishes the more common pressure-driven processes—MF, UF, and NF, and RO—is the application of hydraulic pressure to speed up the transport processes (Cheryan, 1998). However, the nature of the membrane itself controls which components permeate and which are retained. In its ideal definition, RO retains all components other than the solvent (e.g., water) itself, while UF retains only macromolecules or particles larger

Table 17.1 Visionary Scientific Endeavor in the Field of Membrane Science and NF

Research Workers	Visionary Scientific Research Pursuit
Drioli et al. (2002)	Hybrid membrane desalination systems
Ismail and David (2001)	A cogent insight and comprehensive review of carbon membranes for gas separation
Vrijenhoek et al. (2001)	Colloidal fouling of nanofiltration and reverse osmosis membranes
Mohammad et al. (2012)	A deep insight on UF in food industry
Schafer et al. (2003)	Application of reverse osmosis and NF in environmental engineering science (removal of natural hormone estrone from aqueous solution)
Nghiem et al. (2005)	Pharmaceutical retention by NF membranes (retention mechanisms of three pharmaceuticals, such as sulfamethoxazole, carbamazepine, and ibuprofen by NF membranes
Van der Bruggen et al. (2008)	Drawbacks of NF and methods to overcome them
Childress and Elimelech (2000)	NF membrane performance to membrane charge characteristics
Lau et al. (2012)	Recent progress in TFC membrane
Vrouwenvelder et al. (2009)	A critical flux to avoid biofouling of spiral wound NF and RO membranes
Luo and Wan (2013)	Effect of pH and salt on NF
Alsvik and Hagg (2013)	Pressure retarded osmosis and forward osmosis membranes, the vision behind materials and methods

than about 10–200 Angstrom units (about 0.001–0.002 μm). When first devised and developed in the 1960s, UF and RO and later joined by their sister pressure driven membrane processes, MF and NF—constituted the first continuous molecular separation processes that do not involve a phase change or interphase mass transfer (Drioli et al., 2002). Science and technology is moving at a rapid and drastic pace since then. Academic and scientific rigor, path toward progress and the vision to excel are propelling science toward newer future direction (Cheryan, 1998).

3.3 Historical Developments in the Path Toward Progress in Membrane Science

The phenomenon of osmosis, which is the transport of water or solvent through a semipermeable membrane (defined as a membrane that is permeable to solvent and impermeable to solutes), has been known about since 1748, when Abbe Nollet observed that water diffuses from a dilute solution to a more concentrated form when separated by a semipermeable membrane (Drioli et al., 2002; Cheryan, 1998). In 1855, Fick developed the first synthetic membrane, made apparently of nitrocellulose. Two years later, Traube also prepared artificial membranes, while Pfeffer reported in 1877 the successful manufacture of membranes made by precipitating copper ferrocyanide in the pores of porcelain. The first quantitative measurements of diffusion, phenomena and osmotic pressure were made using these early membranes (Cheryan, 1998).

The period of 1870–1920 saw the rapid development of theories of thermodynamics of solutions, most notably those of van't Hoff and his theory of dilute solutions and Gibbs, whose work led to the primary relationship between osmotic pressure and other thermodynamic properties. History of science and technology was moving in a definite and visionary direction. Advancements in science, vision to excel and the world of challenges and difficulties are the forerunners to a greater scientific vision and truth (Drioli et al., 2002).

Membrane filters became commercially available in 1927 from the Sartorius Company in Germany, manufactured using the Zsigmondy process. In 1931, Elford developed methods for sterilizing membrane filters using ultraviolet radiation (Cheryan, 1998). Up until 1945, membrane filters were used primarily for removal of microorganisms and particles from liquid and gaseous streams, for diffusion studies, and for sizing of macromolecules. German scientists also developed methods for culturing bacterial cells on membranes. History of membrane science and the wide road surpassing frontiers is moving toward a visionary direction and visionary realm. Scientific barriers and the vision of application of technology were leading toward newer dimension (Drioli et al., 2002; Cheryan, 1998).

In the early 1950s, Samuel Yuster, of the University of California, Los Angeles, had predicted that, based on the Gibbs adsorption isotherm, it should be possible to produce fresh water from brine (Drioli et al., 2002; Cheryan, 1998). Srinivasa Sourirajan, who also worked at the same university, reported and envisioned some success with this concept using commercially available homogeneous membranes (homogeneous from an ultrastructural point of view). Around the same time, some scientists independently obtained promising results using a homogeneous cellulose acetate membrane (Drioli et al., 2002).

At this juncture, the general conclusion by most researchers in the RO area were that, to obtain commercially feasible flux or dewatering rates, the most practical route would be to reduce the thickness of the membrane (Table 17.1). From 1958 to 1960, Sourirajan, now joined by Sidney Loeb, attempted to modify commercial cellulose acetate membranes by heating them under water, in the visionary hope that this would expand the pores and that the pores would remain open when the membrane cooled, thus increasing flux (Drioli et al., 2002). But exactly the opposite happened: heating contracted the pores. When the heat treatment was performed with commercially available UF membranes, it caused the pores to shrink, which increased the rejection of salts but also resulted in much higher flux than before. Science and technology were then moving in a visionary direction. These were the initial years of membrane science (Drioli et al., 2002). Advancement in science and technology, the immense vision and urge to excel paved a new era of membrane separation techniques and the future vision of NF. Drinking water treatment witnessed and scientifically validated the future vision of science and technology (Drioli et al., 2002; Cheryan, 1998).

3.4 Membrane Chemistry, Structure, and Function

Membrane chemistry, structure, and function are the inherent backbones of successful separation phenomenon of a membrane (Table 17.1). Scientific achievements, advancements in engineering, and the immense effective potential of membrane separation technique have urged human scientific endeavor to aim for visionary realms. In today's realm, environmental engineering science is moving rapidly and progressively with scientific progress (Drioli et al., 2002; Cheryan, 1998).

4 Man's Vision, a Scientist's Prowess, and the Future of Industrial Wastewater Treatment

Drinking water treatment and industrial wastewater treatment are the primordial issues of our present day human civilization. Human civilization is moving toward a new dawn and a newer vision. Man's vision, as well as a scientist's prowess is moving through difficult terrains. So, the human urge to excel needs to be reshaped and reenvisioned. Rigorous and stringent environmental engineering norms are the order of the day. Human scientific endeavor is at a disastrous state. History of science and technology is catastrophic as environmental disasters engulf human civilization. The immediate and imminent concern which is awe inspiring in today's human civilization is to restore ecological balance and realizing successful environmental sustainability. The future of industrial wastewater treatment is wide, bright, and groundbreaking. History of human civilization is slowly and steadily moving out of the deep and disastrous crisis. Application of nanotechnology is a challenge and a vehement torchbearer toward a newer vision (Drioli et al., 2002; Cheryan, 1998).

Intense scientific research pursuit, a scientist's vision and the future road to progress are the intense challenges for the future. Drinking water treatment scenario and the issue of provision of clean drinking water are the focal points of tomorrow's human progress. The vision of tomorrow's scientific progress is inspiring, as well as far-reaching. Scientific validation and scientific truth will go a long way in evolving new dimensions of scientific research pursuit. The vision still stands in the midst of immense optimism environmental engineering science in today's world is merged with drinking water treatment and industrial wastewater treatment. Scientific cognizance and scientific vision is moving toward a newer direction of drinking water treatment (Drioli et al., 2002; Cheryan, 1998).

4.1 Application of Nanotechnology and the Science of Nanofiltration

Application of nanotechnology and the future of NF is ushering in a new era of membrane science. Global drinking water challenges and the scientific urge to excel are the pallbearers to a new dawn of environmental engineering science. The question of nanoinspired vision has left a shining legacy to the wide world of drinking water treatment. History is repeating itself in every sphere of scientific urge and scientific vision of tomorrow. Nanotechnology is the visionary science of tomorrow. Barriers need to be reenvisioned and reshaped as indomitable frontiers realized. Science of NF will be witnessing a new beginning as futuristic vision of drinking water treatment realized (Drioli et al., 2002; Cheryan, 1998). The domain of nanotechnology and NF in the present day human civilization is moving toward a new path of greater emancipation and improved introspection in years to come (Drioli et al., 2002; Cheryan, 1998).

4.2 Challenges and Scientific Vision in Drinking Water Treatment

Immense challenges and monstrous scientific vision ushers in a new era of drinking water treatment. Human civilization is on the verge of new visionary future direction. Provision of clean drinking water is a primordial issue and a focal point in human progress.

Groundwater decontamination and groundwater remediation is similarly of immense importance. The world of challenges, the fetters surrounding science and man's scientific progress are going a long way in evolving new dimensions of future endeavor in drinking water treatment. Challenges, barriers, and scientific validation are the keywords of tomorrow in the quest for scientific pursuit. Drinking water treatment is in the verge of a new era (Bhalme and Nagarnaik, 2012; Report on Drinking Water Treatment, 2006; Casey and O'Reilly, 2009).

Scientific cognizance, scientific doctrine, and the immense difficulties ahead are the pallbearers, which will propel our human civilization to scale new heights. Science and engineering is moving toward a greater and visionary realm. Scientific validations along with environmental restrictions are the keywords toward a greater human destiny and improved human progress.

5 Nanofiltration, Global Drinking Water Crisis, and the Vision for Future

Challenge, barriers, and definitive vision are the pallbearers of the environmental engineering science today. NF and membrane science are the next generation science for drinking water treatment. Scientific endeavor in today's world is moving toward a visionary direction. NF is today linked by an umbilical cord with drinking water treatment. Drinking water treatment and drinking water disinfection is directly linked to the health of a human being. In such a critical juxtaposition, disinfection techniques, and industrial wastewater treatment are of utmost importance. Scientific endeavor in the field of drinking water treatment is visionary and far-reaching. Challenges, scientific vision, and scientific truth will lead a long way in the emancipation of application of nanotechnology in drinking water treatment and industrial wastewater treatment (Drioli et al., 2002; Cheryan, 1998).

Global drinking water treatment scenario and global drinking water crisis in the path toward greater progress and greater optimism. The vision for the future is splendid and ground-breaking. History of science and technology is moving toward a newer scientific generation and newer scientific vision. Man's vision, a scientist's progress, and human progress are the focal points and primordial issues toward engineering science today. Scientific cognizance and scientific invention in today's world are retructured and redefined with every step of history and time. Global drinking water crisis is in the path of newer vision and in the progress of indomitable challenges. Groundwater contamination, groundwater remediation, and the application of environmental engineering science are all moving in the path of newer vision. Scientific revelation, the future vision of environmental science and drastic scientific vision are paving way to a new decade and a new era of innovation (Drioli et al., 2002; Cheryan, 1998).

Scientific cognizance, scientific doctrine, and scientific foray into environmental engineering science are the hallmarks and focal points of a new generation of future thoughts and future vision. Innovation of science needs to be reshaped and reenvisioned with every step of human progress (Drioli et al., 2002; Cheryan, 1998).

5.1 Vision of Application of Membrane Science in Industrial Wastewater Treatment

The future of human civilization and the vision of science and technology are drastically and steadfastly moving toward a newer frontier. The world of challenges, urge to excel in scientific endeavor and the application of scientific techniques will lead a long way in realizing the vision of provision of clean drinking water. Industrial wastewater treatment and drinking water treatment today stands in the midst of immense scientific truth and scientific forbearance. History of science and technology is moving toward a definitive direction. Scientific challenges, scientific progress, and the immense urge to excel are powering scientific generations toward a newer realm.

6 Purposeful and Instinctive Vision in Tackling Global Drinking Water Crisis

Purposeful and instinctive vision will lead a long way in tackling global drinking water crisis. Health issues propel the ultimate vision of drinking water crisis. Human civilisation's progress and health issues of mankind are the ultimate focal points of the primordial issue of drinking water treatment and drinking water disinfection. A scientist's vision is vastly emboldened with every step of human history and time. Global drinking water crisis is moving toward a newer visionary realm. Challenges, difficulties, and barriers are befitting to the progress of human civilization. Mankind's prowess and man's utmost vision will lead a long way in alleviating global water shortage crisis.

6.1 Vision of Science and the Application of Nanotechnology

Science and technology of today's human civilization is moving drastically toward a newer age of immense challenge. Nanotechnology is propelling human civilization toward a new future direction. The needs of human civilization are emboldened with man's step into scientific endeavor in basic human needs. Provision of clean drinking water is of utmost importance to human progress. In such a critical juncture of global drinking water crisis, nanotechnology is propelling civilization toward newer dimension. The vision and road to future in the application of nanotechnology in drinking water treatment is far-reaching and unimaginable. History of science and engineering is surpassing wide and visionary frontiers. Fruits of science in today's wide civilization are reaching the teeming millions. History of science and technology stands in the midst of definitive visionary moment. Nanotechnology today is moving toward a visionary era and a visionary century. Nanoscience and its veritable progress are the harbingers of a new realm (Drioli et al., 2002; Ismail and David, 2001).

6.2 The World of Challenges and the Unsurpassed Frontiers of Science of Membranes

Challenges in drinking water treatment are surpassing wide and visionary frontiers. History of science and technology is moving toward a newer vision of environmental sustainability. Successful sustainable development has an umbilical cord in today's world with provision of clean drinking water. In such a vexing situation challenges needs to be restructured and redefined. The crucial juxtaposition of optimism and vision will go a long way in unveiling the scientific truth of drinking water treatment. Unsurpassed frontiers of science engulf environmental engineering. In a similar vein drinking water treatment and industrial wastewater treatment also are the torchbearers today toward a greater emancipation of environmental engineering science.

Membrane science today is in the path of immense scientific understanding and scientific vision. History of science and technology is slowly and drastically moving toward newer vision and newer scientific truth and justification. Unsurpassed frontiers are indomitable yet deeply comprehensive. Man's vision, scientist's prowess, and the world of scientific challenges are opening up new doors of innovation and scientific instinct in years to come. Some recent research endeavor in carbon membranes are delineated in this section.

Ismail and David (2001) delineated in a review latest developments of carbon membranes for gas separation. Inorganic membranes have been developed before 1945. The earlier application of inorganic membranes was primarily concentrated on military applications. This chapter reviewed the development of carbon membranes in the last 30 years and give a clear future understanding and future direction in research for carbon membranes. Nonnuclear applications of inorganic membranes started in the beginning of the 1980s with Membralox produced by Ceraver (now SCT), Carbosep produced by SFEC (now TECHSEP), and Ceraflo produced by Norton (now SCT). The potential of inorganic membranes was not widely recognized until high quality porous ceramic membranes were produced for industrial usage on a large scale. Membrane separation process in today's world is moving in a new and visionary direction with the innovation of carbon membranes. Carbon membranes revolutionized the membrane separation scenario. Although inorganic membranes are more expensive than organic polymeric membranes, they possess advantages of: temperature and wear resistance, well defined stable pore structure and chemical inertness. At present, interest in the development of porous inorganic membranes providing better selectivity, thermal stability, and chemical stability than polymer membranes has grown with immense proportions.

Vrijenhoek et al. (2001) delineated influence of membrane surface properties on initial rate of colloidal fouling of RO and NF membranes. This treatise investigates and identifies the most influential membrane properties governing colloidal fouling rate in RO/NF membranes. Colloidal fouling is one of the barriers to successful separation processes.

6.3 Scientific Endeavor in the Field of Nanofiltration of Drinking Watertreatment

Scientific research endeavor in the field of drinking water treatment is far-reaching and visionary. The success of environmental sustainability and ecological balance directly depends on visionary research pursuit. Today's human civilization is in the midst of immense optimism. Stringent environmental restrictions have propelled the human scientific mind to scale greater insurmountable heights. The author skillfully brings forward the recent scientific endeavor and scientific truth in the field of NF. NF is the next generation science and engineering veritably linked with clean drinking water provision and drinking water treatment. The immense challenges, indomitable barriers and the plausible solutions are the focal points of today's industrial wastewater treatment. Solutions to drinking water treatment are few and far-reaching. Scientific vision needs to be restructured in order to tackle drinking water purification phenomenon.

6.4 Intense Research Pursuit in the Field of Nanofiltration of Industrial Wastewater Treatment

NF of industrial wastewater treatment is slowly and steadily moving from one drastic challenge over another. History of human scientific progress is at a disastrous state. Man's vision, as well as a scientist's prowess needs to be reshaped and restructured. Science and technology is moving at a rapid pace toward a newer vision of scientific grit and scientific fortitude. Recent research pursuit is moving toward a definitive and purposeful direction. Future directions are opening up new dimensions of thought and new directions of scientific vision.

Among the various technical options offered by Chemical Engineering for a successful sustainable growth, membrane science, and technology today provide some of the most effective and desirable solutions for design, rationalization, and optimization of the visionary production cycle. The attractive prospects are related to the possibility of integrating various membrane units in different industrial cycles, with important overall benefits in terms of product quality, plant compactness, environmental impact, and energy concern and vision.

Intense research pursuit will go a long way in opening up new vistas in the domain of industrial wastewater treatment. Scientific cognizance, scientific grit, and determination are ushering in new dimensions of research and newer vision.

Drioli et al. (2002) was in the forefront of membrane science research. The phenomenol research area is vehemently dealt with in this research endeavor. The authors dealt incisively on the field of hybrid membrane desalination systems. In today's world, membrane science and technology provide with interesting and concrete solutions for design, rationalization, and optimization of innovative production chemical engineering cycles. Drioli's research work has surpassed visionary frontiers. Membrane science is ushering in a new emancipation of indigenous research and immense vision. The chapter holistically deals with the wide world of application of membrane science and membrane filtration. The author skillfully delineates the wide vision of membrane science integrated with into industrial procedures and hybridized with commercial vistas. Among the hybrid membrane desalination systems, the integration of thermal desalination, and membrane systems as hybrid desalination systems are of veritable and crucial importance.

Mohammad et al. (2012) in a definite insight delineated in a review paper UF in food processing industry. The authors reviewed application, membrane fouling, and fouling control. UF process has been widely applied in food processing industry for the last 20 years due to its advantages over conventional separation processes, such as gentle product treatment, high selectivity, and lower energy consumption. UF procedure today stands as an essential part in food technology and as a tool for separation and concentration. Science of UF is moving toward a newer direction and newer vision. However, membrane fouling comprises the benefits of UF as fouling significantly reduces the performance and thereby increases the cost of UF. In this review, the authors deeply investigate the fouling mechanisms of UF membranes, as well as the latest procedures to counter membrane fouling.

Schafer et al. (2003) delineated removal of natural hormone estrone from aqueous solutions with the help of RO and NF. The ability of a variety of NF and RO membranes to retain the natural hormone estrone is examined here with cogent insight as a function of solution conditions. While size exclusion dominates retention with tighter membranes both size exclusion and adsorptive effects appear to be instrumental in maintaining higher retention on NF membranes that otherwise exhibit relatively low ion retentions. These adsorptive effects may be driven by hydrogen bonding between estrone and the membrane. The results reported here indicate that while open NF membranes may be effective in retaining estrone under some conditions, the extent of retention may be susceptible to maintenance of adsorptive capacity at the membrane surface and depend on solution chemistry.

Nghiem et al. (2005) delineated in details pharmaceutical retention by NF mechanisms. This study investigates the retention mechanism of three pharmaceuticals—sulfamethoxazole, carbamazepine, and ibuprofen—by NF membranes. Results show that retention of pharmaceuticals by a tight NF membrane is dominated by steric exclusion, whereas both electrostatic repulsion and steric exclusion govern the retention of ionizable pharmaceuticals by a loose NF membrane. In the latter case, speciation of pharmaceuticals may lead to a dramatic change in retention as a function of pH, with much greater retention observed for ionized, negatively charged pharmaceuticals. Scientific vision in the NF of pharmaceuticals has reached visionary proportions. Pharmaceutical industries also create industrial pollution. The authors with deep and immense insight delineate the imminent importance of NF as a permanent solution to treat pharmaceutical wastewater. Science and technology is moving toward a newer scientific vision. Pharmaceutical industries are facing immense challenges.

In such a critical juncture, the authors unveil the hidden truth of industrial wastewater treatment in pharmaceutical industries.

Van der Bruggen et al. (2008) discussed in lucid details the primordial issue of drawbacks of NF and how to overcome them. As the author gleans over this issue, history and application of NF surpasses visionary frontiers. Difficulties, barriers, and vision are the ultimate torchbearers of the progress of life in earth. Environmental restrictions and immense environmental barriers are opening up new door of scientific innovation and scientific instinct in years to come. The authors delineated how still unresolved issues are vexing to the separation phenomenon of NF technique. This chapter identifies six difficult challenges for NF where solutions are scarce and minimum. The utmost and visionary challenges are:

1. Avoiding membrane fouling and vision to remediate it.
2. Improving the separation between solutes that can be achieved.
3. Further treatment of concentrates.
4. Chemical resistance and limited lifetime of membranes.
5. Insufficient rejection of pollutants in drinking water treatment.
6. The need for modeling and simulation tools.

The authors delineate with improved vision the path toward progress in separation phenomenon of NF technique. Human civilization's prowess is taking a decisive turn with the culmination of improved separation phenomenon of environmental engineering tools. Drawbacks should be surpassed and scientific validation should be realized.

Childress and Elimelech (2000) delineated in precise details the related phenomenon relating NF membrane performance to membrane charge (electrokinetic) characteristics. The performance (i.e., water flux and solute rejection) of a thin-film composite aromatic polyamide NF membrane and its relation to membrane surface charge (electrokinetic) characteristics were investigated. Membrane performance and streaming potential measurements were carried out as a function of pH for several solution chemistries, including an indefinite electrolyte, humic acid, and anionic and cationic surfactants. Performance results for the membrane were interpreted by relating the water flux and salt/ion rejection to the membrane charge characteristics.

Lau et al. (2012) reviewed the recent progress in thin film composite (TFC) membrane. The major breakthrough in TFC membrane via interfacial polymerization technique has resulted in tremendous achievements in producing a membrane with a right combination of flux and salt rejection and generated immense interest in industrial sector. Science and technology are emboldened with visionary impact to mankind and society. Application of membrane science in environmental engineering science has reached its zenith with every positive scientific endeavor. This review investigates the intensive and continuous development in research endeavor in TFC membrane. The authors deeply investigate the membrane productivity and selectivity, as well as its tolerance against chlorine, solvent fouling and so on. This informative chapter deeply delineates the recent research progress of the TFC membrane science and technology particularly in the field of water related separation processes. The vision of the review is absolutely definitive and purposeful. The authors address with deep insight to the readers the immense hidden application of TFC membrane. Reviewing and investigating the research progress is imperative since scientific endeavor opens up new doors of scientific innovation and scientific endurance. Scientific breakthrough in developing ultrathin film onto microporous substrate is equivalent to the historic announcement of Loeb–Sourirajan asymmetric membrane in 1960. Water and wastewater treatment scenario is emboldened due to the tremendous progress in production of RO and NF membranes by the innovative coating technique. Scientific breakthrough after the phenomenal achievement of Loeb–Sourirajan discovery ushered in a new dimension and newer direction in membrane science and technology. Challenge, vision, and purpose ushered in a new era of membrane technology.

Vrouwenvelder et al. (2009) discussed in deep comprehensive details on the issue of critical flux to avoid biofouling of spiral wound NF and RO membranes. The authors skillfully merged fact and fiction in the development of membrane technology. The relation between biofouling and membrane flux in spiral wound NF and RO membranes in drinking water stations with extensive pretreatment, such as UF has been studied in details. The authors put forward with immense vision and scientific understanding the utmost importance of critical flux in biofouling phenomenon.

Luo and Wan (2013) reviewed the effects of pH and salt on NF. The vision of the study reviews the reports of NF performance at different pH and salt conditions focusing on the mechanisms behind the varied separation phenomena induced by pH and salt.

Alsvik and Hagg (2013) dealt lucidly on pressure retarded osmosis and forward osmosis membranes with special emphasis on materials and methods. In the last 4 decades membrane development has occurred on the solid ground of demand of pressure driven processes. However, in the last decade the interest in osmotically driven processes, such as forward osmosis and pressure-retarded osmosis has increased. The vision for the future of membrane science lies in forward osmosis. Preparation of TFC membranes with a customized polysulfone support (PSf), electrospun support, TFC membranes on hydrophilic support, and hollow fiber membranes are discussed in details in this treatise.

The world of challenges in scientific revelation and scientific validation is immense and inspiring. Advancements in science and technology are the vision of tomorrow. Present direction and present dimension of research in membrane science and NF is absolutely far-reaching and have an umbilical cord with global water initiative and drinking water treatment. Challenges, vision, and purpose will go a long way in evolving newer emancipation in the future of global drinking water scenario. Developed and developing countries are gearing toward newer challenges and newer vision in alleviating global water needs and global water crisis. Environmental sustainability and global water initiative are the hallmarks of a greater holistic vision of provision of clean drinking water to the common mass. Advancements, scientific validation, and the emancipation of vision will go a long way in the path toward progress in holistic civilization's development.

7 Science, Technology, and the Vision of Domain of Drinking Water Treatment

Science, technology, and engineering are moving toward a new visionary domain with drastic pace. Human civilization's veritable needs should be fulfilled and environmental paradigm needs to be realized and envisioned. History of human civilization in today's world is witnessing new resurgence and reshaping. Provision of clean drinking water is of utmost importance. Science and technology is moving very rapidly. The question of environmental sustainability is of utmost importance in our present day human civilization. Drinking water treatment is moving in a new direction and new dimension of immense scientific understanding and scientific vision. Scientific cognizance and scientific doctrine are in the process of reshaping and restructuring at every step of human progress. Science and technology in today's world is progressing at a drastic pace toward visionary frontiers. Unsurpassed frontiers are the hurdles of scientific progress. Drinking water treatment and provision of clean drinking water today stands as a major hurdle to advancement of science. In such a crucial juncture of history and civilization, the world of challenges should be targeted toward basic human needs.

7.1 Global Vision and Initiative Over Drinking Water Treatment

Environmental engineering science, drinking water treatment, and global water crisis are in the brink of inevitable disaster. Grave concern and concerted effort will go a long and effective way in diminishing the disastrous crisis. Developed and developing nations are unveiling the scientific truth behind drinking water crisis.

Opio (2012) gives a deep comprehensive report to the scientific community the building of a scientific drinking water management system in Rural Africa. The CIGI (Centre for International Governance)–Africa Discussion Paper Series presents policy relevant, peer reviewed, field based research that addresses visionary issues in the areas of conflict resolution, energy, food security, health, migration, and climate change. The aim of the series propounds discussion and advanced knowledge on issues relevant to the policy makers and opinion leaders in Africa. The need to provide clean drinking water is widely recognized as a priority in rural Sub-Saharan Africa. There is lack of specific data on water quality to build effective drinking water management system. The discussion paper reviews and investigates a water quality study undertaken in Northern Uganda, to test the potability and potential contamination of water taken from wells, open water sources and households.

In 2009, upon receiving funds from the RBC Foundation's Blue Water fund, the Harris Centre in Canada created The RBC Harris Centre Drinking Water Applied Research Fund (The RBC Harris Centre Drinking Water Applied Research Fund, 2009). The report investigates historic issues in New Foundland and Labrador. The authors identified the visible knowledge gaps in the path toward progress. As a developed nation, Canada stands way ahead in the vision and urge to excel in drinking water treatment and provision.

Ray and Jain (2011) discussed in a comprehensive review drinking water treatment technology along with a comparative analysis. According to the treatise, water treatment technologies have evolved over the past few centuries to protect public health from pathogens and chemicals. The authors discuss in a deeply comprehensive paper the innovative technology of solar distillation. The study gleans over solar pasteurization, which depends upon solar energy for purifying small quantities of water for individual or family use. Various innovative techniques are discussed with cogent details to treat drinking water. Case studies of natural (riverbank and lakebank) filtration, membrane filtration, solar distillation, and solar pasteurization are presented with lucid details.

Onjala et al. (2013) reviewed risk perception, choice of drinking water, and water treatment in Kenyan towns. The developing country perspective is deeply reviewed. The Environment for Development provides insight into an international environmental economics policy. It focuses on international research collaboration, policy advice, and academic and practical training. It supports centers in Central America, China, Ethiopia, Kenya, South Africa, and Tanzania. This study delineates and uses household survey data from four Kenyan towns to examine the effect of households' characteristics and risk perceptions on their decision to treat/filter water, as well as their choice of main drinking water source. Drinking water crisis is gravest in African continent. A seemingly unrelated bivariate probit model is estimated. This chapter presents a study on decisions about drinking water sources and in-home water treatment behavior, collecting household data in Kenyan towns. The aim was to understand how people perceive the riskness of different water sources when they are choosing their drinking water and what their risk-averting behavior envisages.

A major devastating scenario in provision of drinking water is the effect of chlorine in drinking water. WHO (1996) report effectively envisioned guidelines for drinking water quality. The presence of chlorine in drinking water is a disaster and an unimaginable catastrophe to the future of human progress. The report delineates with absolute lucidity identity, physicochemical properties of chemicals, and organoleptic properties of chemicals, major use and environmental fate of chlorine. Presence of substantial amount of chlorine in water spells disaster in human health scenario and the holistic environmental engineering scenario.

Bhalme and Nagarnaik (2012) reviewed analysis of drinking water of different places. The study is based on the analysis of drinking water parameters in an educational institute situated in Hingna MIDC area, Nagpur, India. In this chapter different authors' paper are summarized for water analysis and their treatment process in different regions, which will be helpful to know the different treatment scenarios. Increase in societal urbanization, industrialization, agricultural activity, and various human activities have led to an increase in the pollution of surface water and groundwater.

The treatment of water and its vision is of utmost importance in the path toward human progress. Water should be free from the various contaminants, such as organic and inorganic pollutants, heavy metals, pesticides, and so on, as well as all its parameters, such as pH, electrical conductivity, calcium, magnesium, total hardness, carbonate, bicarbonate, chloride, total dissolved solids, alkalinity, sodium, potassium, nitrate, and DO should be within permissible limits. Science and technology is moving at a rapid pace and scientific frontiers surpassed. Human progress and forays of environmental engineering science are the hallmarks of the present decade.

The Report on Drinking Water Treatment (2006) on Cooperative Research Centre (2006–07) delineates the assistance of producing high quality water in Australia. Community confidence and safety issues are the pillars of the highly rated research endeavor. Innovation to meet global water challenges are opening up wide dimensions of future research. The report is an example of the future of global water quality.

Casey and O'Reilly (2009) lucidly delineates the present status of water quality in Ireland. The discussion centers around process engineering issues related to the use of chlorine and its compounds for the production of pathogen free drinking water.

Khayet (2010) deeply delineates with cogent comprehension the technique of desalination by membrane distillation. The futuristic vision of water science and technology is moving toward newer paradigm and surpassing visionary frontiers. Membrane distillation is a nonisothermal membrane separation process, which is leading a long way in alleviating global water shortage.

8 Application of Nanotechnology in Drinking Watertreatment: a Critical Overview

In today's human civilization, the target and focus of human scientific endeavor will be toward maintaining environmental and ecological balance (Drioli et al., 2002). A critical approach will reveal the immense potential and visionary implications of advanced oxidation processes, novel separation processes and groundwater decontamination and remediation. Nanotechnology is the next generation science and engineering. Technological and scientific validation will go a long way in focusing and targeting scientific vision. A critical review of the recent technological developments will surely bring in success and usher in an immense potential of the science of nanotechnology. Drinking water treatment scenario in today's world stands in the midst of scientific optimism and immense scientific vision. Vision, greatness, and scientific mission are the torchbearers of tomorrow (Drioli et al., 2002; Cheryan, 1998).

Man's vision, scientific cognizance, and the scientific struggle toward progress are the torchbearers to a newer visionary future. Environmental pollution control in the present day human civilization is at a disastrous yet at an optimistic state. The visionary realms in the field of application of nanotechnology need to be vehemently realized. Scientific determination and scientific justification are the order of the day. Science and technology are witnessing wide frontiers and visionary realms. Drinking water provision should be in a definitive visionary direction. The immense challenges are the pallbearers toward a definite vision (Drioli et al., 2002; Cheryan, 1998).

Scientific vision and scientific judgement in the domain of drinking water treatment is immense and indomitable. Cognizance, doctrine, and future progress of human civilization are scaling new heights. Man's, as well as a scientist's vision is emboldened with the passage of time and history. Application of nanotechnology in drinking water treatment is ushering in a new and progressive era. Science and technological applications will surely evolve new scientific visionary dimensions for the future (Drioli et al., 2002).

9 Arsenic Groundwater Crisis in South Asia, Developed and Developing Economies, and the Vision Ahead

Arsenic groundwater crisis in South Asia is in the midst of deep disaster. Incomprehensible devastation due to arsenic groundwater contamination has plunged our human civilization between devil and the deep sea (Drioli et al., 2002). Enigma, disaster, and catastrophe are the pallbearers to a human disaster of unimaginable proportions. Developed, as well as developing nations are facing such inimitable crisis. Scientific endeavor and effective participation of civil society will enhance the veritable solutions concerning this burning issue (Cheryan, 1998).

Global groundwater crisis today stands in the midst of chaos, difficulties, and scientific barriers. Scientific vision is at a devastating future. Man's vision is truncated at every step of history and time. The vision of tomorrow in drinking water treatment is far-reaching and surpassing visionary frontiers. Challenges, barriers, and vision in provision of clean drinking water today stands in a difficult and devastating position. The future needs to be reenvisioned (Cheryan, 1998).

9.1 Arsenic Groundwater Contamination and the Future of Human Civilization

Global groundwater contamination crisis is propelling toward an inevitable disaster. Catastrophe, challenges, and progress of science and engineering will all go a long way in evolving new dimensions of future scientific endeavor. Arsenic groundwater contamination scenario is in the midst of immense pessimism (Drioli et al., 2002). The future of human civilization today depends upon scientific fortitude and scientific forbearance. South Asia is in deep danger with the immense groundwater contamination scenario and the arsenic and heavy metal groundwater contamination. Science and technology needs to be restructured and re-envisioned with every step of human progress (Drioli et al., 2002).

9.2 Groundwater Remediation, Provision of Clean Drinking Water, and the March of Mankind Ahead

Groundwater remediation is the burning and vexing issue of present day human civilization. Provision of clean drinking water stands in the midst of immense scientific endeavor. March of mankind, progress of civilization, and the road for the future will all go a long way in evolving new future dimensions of research endeavor. Groundwater contamination and groundwater remediation will be the primordial issues of the future. Teeming millions in developing countries are on the verge of an unimaginable and indomitable human disaster. Progress of science, technological prowess, and fruits of engineering will eventually go a long way in opening new vistas of scientific pursuit and effective scientific judgement (Drioli et al., 2002).

10 Progress of Human Civilization, Progress of Science and Engineering, and Basic Human Needs

Human civilization is in great distress. Environmental and ecological imbalances are the vexing questions of our time. The world of serious and inevitable challenges in the provision of clean drinking water is the pallbearers of future scientific vision. Basic human needs are at a disastrous position. Man's, as well as a scientist's vision is at a devastating position. At this critical juncture of mankind's history and time, basic human needs, need to be addressed (Drioli et al., 2002).

Progress of human civilization is at a devastating stake. The world of purpose and vision in groundwater remediation needs to be reshaped. Provision of basic human needs, such as drinking water provision is at a state of distress. Man's vision, mankind's prowess, and progress of science and technology will surely lead human civilization toward a newer vision of scientific understanding (Drioli et al., 2002; Ismail and David, 2001; Vrijenhoek et al., 2001; Mohammad et al., 2012).

10.1 Doctrine of Science and the Inevitable Domain of Drinking Water Provision

Doctrine of science and technology needs to be reenvisioned and reemphasized. Challenges, difficulties, and barriers are the torchbearers to the newer scientific endeavor and newer scientific generation. Drinking water provision is a primordial issue and in such a similar vein, science, and engineering endeavor should be directed toward that vision.

Doctrine of science and technology is befitting to the progress of human civilization. Drinking water provision is an essential need to the human progress. History of science and technology, the urge to excel in the road toward progress and the vision ahead will all go a long way in the emancipation of human mankind from the clutches of disaster. Environmental pollution control scenario is at deep distress and inevitable disaster. The challenges need to be reshaped. Drinking water provision scenario is at deep crisis. Effective drinking water treatment and the alleviation of ecological balance will surely usher in a new era and a newer visionary domain.

10.2 Drinking Water Crisis and Forays of Science and Technology

Drinking water crisis is in a devastating situation. Forays into science and technology and scientific research pursuit today stand in the midst of immense optimism and vision. History of science and technology is repeating itself in a vehement and vicious way in the primordial issue of drinking water crisis. Crisis is surmounting day by day. The foray into newer dimensions of science and technology are befitting to the progress of human civilization. Drinking water treatment and industrial wastewater treatment in today's world stands in the midst of deep comprehension and immense introspection. The world of immense challenges are ushering in a new dimension in drinking water treatment.

11 Scientific Vision, Scientific Truth, and Introspection in the Domain of Drinking Water Treatment

Scientific vision and scientific truth are the torchbearers of visionary future of environmental pollution control and drinking water treatment. Zero-discharge norms and stringent environmental regulations are opening up vast doors of innovation in drinking water treatment. The challenge and vision stands indomitable. Scientific vision and scientific understanding are the torchbearers of tomorrow. Scientific struggle and successful environmental sustainability are the torchbearers of tomorrow. The introspection into science and technology is befitting to the future course of action. Man's vision and a scientist's foray are emboldened with the crucial passage of history and time (Drioli et al., 2002; Ismail and David, 2001; Vrijenhoek et al., 2001; Mohammad et al., 2012; Schafer et al., 2003).

Challenge and deep comprehension will evolve into new dimensions of drinking water treatment. Civilization is at a deep distress. The future of human mankind needs to be reshaped and reenvisioned. Scientific vision needs to be restructured. Introspection into science and engineering will surely lead to new areas of human development and scientific research pursuit. Successful environmental sustainability stands as a major backbone of human mankind. Energy and environmental sustainability are the orders of the day. Scientific truth and scientific empowerment are the visionary scenarios of research pursuit of tomorrow.

11.1 Scientific Validation and the Future of Drinking Water Treatment

The challenge of scientific validation stands today in the midst of immense optimism and deep scientific truth (Drioli et al., 2002). Man's vision, mankind's prowess, and scientific justification are the torchbearers of tomorrow's science and engineering. Science and technology in today's world is experiencing tremendous restructuring. In a similar vein, science behind drinking water treatment is surpassing visionary and indomitable frontiers. Scientific validation is the order of today's science. Frontiers are surpassed and barriers are crossed with every turn of the decade (Drioli et al., 2002).

12 The Future of Groundwater Remediation and the Issue of Drinking Water Treatment

Groundwater decontamination and groundwater remediation are in the brink of an imminent and inevitable disaster in today's world of environmental engineering science. Research focus, urge to excel and the vision forward are the utmost focal points of the future. In developing countries, groundwater contamination is a vexing issue. This has tremendous effect on the fate and result of the future of clean drinking water provision. World's water crisis is slowly moving positively. History of technology is witnessing a new beginning (Drioli et al., 2002; Cheryan, 1998).

12.1 Arsenic Groundwater Remediation and the Progress of Scientific Endeavor

Arsenic groundwater contamination with subsequent remediation is a vexing and primordial issue in developing South East Asian countries, such as India and Bangladesh. Groundwater contamination is linked to death defying and infectious diseases. History of human civilization today stands in the midst of greatness, challenges, and immense pessimism. Arsenic groundwater contamination is in a deep disaster and indomitable state. Yet, scientific frontiers are surpassed and history revisited. The world of challenges are befitting to the progress of civilization ahead (Drioli et al., 2002; Cheryan, 1998).

12.2 Groundwater Contamination and the Future of Mankind's Progress

Groundwater contamination is propelling our human civilization to a deep disaster and indomitable catastrophe (Drioli et al., 2002). Future of mankind's progress is at a disastrous stake. Scientific fetters are broken and vision realized. Man's prowess and mankind's cognizance are the torchbearers to a newer scientific world and newer scientific generation. Groundwater remediation is the target for the future. History of water resources engineering is witnessing a new dawn and a newer frontier. Mankind's progress is directly related to the advancement of science and technology. Clean drinking water provision marks as a primordial issue in successive decades (Drioli et al., 2002; Cheryan, 1998).

Groundwater contamination and groundwater remediation are in the verge of inimitable and immense disaster. Arsenic groundwater contamination in South Asia, as well as developing and developed economies is a burning, as well as a primordial issue. The question of global drinking water crisis can never be put in the foreground with the passage of history and time. Scientific endeavor should reach exotic and visionary heights for the emancipation of groundwater remediation. Future of mankind's progress will surely usher in a new era and new dimension if global water crisis and groundwater contamination is vehemently addressed (Drioli et al., 2002).

13 Effective Novel Separation Processes and Far-Reaching Environmental Engineering Techniques

Novel separation processes stands today in the environmental engineering scenario in the midst of immense scientific understanding and scientific optimism. Zero-discharge norms and stringent norms are ushering in a new visionary era. Environmental engineering norms are the futuristic vision of tomorrow. Challenges for the betterment of tomorrow, the vision to excel and the road ahead greatly enhances the impact of visionary technologies. History of environmental engineering science and water resources engineering is visibly ushering in a new era (Drioli et al., 2002).

13.1 Environmental Engineering Tools and Drinking Water Treatment

Environmental engineering tools, drinking water treatment, and successful environmental sustainability are the torchbearers of tomorrow's visionary world (Drioli et al., 2002). Future dimension and future perspectives are in the verge of newer vision and newer tomorrow. The challenges need to be restructured and reenvisaged (Drioli et al., 2002).

Environmental engineering tools, such as novel separation processes and advanced oxidation processes are the new generation technologies of scientific endeavor. History of science and technology and history of human civilization is moving toward a newer vision and newer scientific understanding. Provision of clean drinking water is the primordial issue of human society and human progress today. Science and engineering is emboldened at every turn of decade. The crucial juxtaposition of environmental engineering tools and drinking water treatment is befitting toward the progress of science.

13.2 Environmental Engineering Techniques and Industrial Wastewater Treatment

In today's human civilization, environmental engineering tools and industrial wastewater treatment have an unsevered and visionary umbilical cord. History of science and technology is in the brink of an appalling and unmitigated disaster. Disaster management needs to be reenvisioned. Challenges need to be redrawn (Drioli et al., 2002). Environmental engineering techniques is unraveling the scientific hurdles of drinking water treatment and industrial wastewater treatment. Industrial wastewater treatment today stands in the midst of scientific disaster and veritable catastrophe. The state of environment is shocking and awesome. In such a critical juxtaposition, vision powered by urge to excel and redefine will lead a long way in emancipation of water pollution control and environmental sustainability (Drioli et al., 2002).

14 Nanotechnology and the Vision for the Future

Nanotechnology and the application of NF to drinking water treatment are the futuristic and the ultimate vision of tomorrow's science. Drinking water treatment is in the brink of imminent disaster and unmitigated catastrophe with the evolution of ecological imbalance and environmental catastrophes (Drioli et al., 2002). Application of NF can create feasible wonders. Man's prowess needs to be reenvisaged at every step of scientific endeavor. The vision for the future in the domain of nanotechnology is absolute and definitive. Scientific cognizance, deep understanding, and comprehension are the scientific orders of the day (Drioli et al., 2002).

14.1 Future Directions, Future Dimensions, and the World of Intense Challenges

Future directions and future dimensions in today's world of intense challenges needs to be reshaped and reenvisaged. Mankind's history is in the brink of disaster with the passage of time in view of the monstrous ecological imbalance and veritable environmental calamities. Intense research endeavor needs to be directed toward successful sustainability. History of human mankind and history of science are in the brink of inimitable disaster in view of the global water crisis. Emancipation of science and road toward progress will go a long way ahead toward a newer scientific generation and newer scientific vision. Groundwater remediation is on the verge of scientific and research validation with the advent of new technologies. Intense and vehement challenges are shaping mankind's future. Future directions and future challenges will move toward a new direction in decades to come (Drioli et al., 2002).

14.2 Vision, Barriers, and Introspection in Application of Nanotechnology in Drinking Water Treatment

Vision, barriers, and hurdles are far and wide in the drinking water scenario of human civilization. Nanotechnology is widely diversifying itself in new and visionary ranges. Provision of clean drinking water today stands in the midst of immense scientific vision and scientific optimism. Introspection into the scientific endeavor involves intense vision. Scientific vision is winning over all odds in the visionary domain of drinking water treatment and the imminent vision of industrial wastewater treatment (Drioli et al., 2002).

14.3 Visionary Challenges and Effective Solutions

History of human civilization is moving toward an immense disaster. Science, technological advancements, and path toward progress are the torchbearers toward a greater vision of tomorrow. Groundwater contamination and groundwater remediation are the primordial issues of tomorrow's visionary era. Validation of science is another major issue. Entanglement of science with common mass will be the utmost vision of tomorrow's human progress. Environmental engineering science is in the new path of progress and vision. Global drinking water crisis is in deep disaster. So, the present and inevitable utmost importance is to go forward toward newer technologies and new vision (Drioli et al., 2002; Cheryan, 1998).

14.4 Man's Scientific Progress and Mankind's Prowess in Drinking Water Treatment

Man's scientific progress, mankind's vision, and prowess are the futuristic thoughts of tomorrow. History of civilization and history of science and technology are drastically moving toward an indomitable devastation due to global drinking water crisis. In such a situation, the world of challenges should be toward alleviation of drinking water issues and successful sustainable development. Environmental sustainability is in the path toward progress, as well as toward devastation in today's scenario. History of mankind is repeating itself with every step toward progress.

15 Future of Civilization and Future of Environmental Sustainability

The world of challenges are surmounting day by day. History of human civilization is moving through one crisis over another. Environmental sustainability thus stands in the midst of indomitable crisis and insurmountable deep comprehension. Successful sustainable development whether it is environmental or energy, question of futuristic vision is vexing. A scientist's vision needs to be reshaped with great challenges. History of mankind is in the verge of entry into a new era. Environmental sustainability is the order of the day (Drioli et al., 2002; Cheryan, 1998).

15.1 Future of Environmental Engineering Science and Future of Global Drinking Water Crisis

Industrial wastewater treatment, global drinking water crisis, and the human mankind's urge to excel are in a disastrous and declining situation. Environmental disasters befit successful sustainability and successful target toward alleviation of global drinking water crisis. The vision of tomorrow is immensely definitive and targeted toward a greater emancipation of global environmental sustainability. Drastic challenges, futuristic progress, and innovative changes are the torchbearers of tomorrow (Drioli et al., 2002; Ismail and David, 2001; Vrijenhoek et al., 2001; Cheryan, 1998).

15.2 Future Vision of Mankind's Progress and Provision of Clean Drinking Water

Future vision of drinking water treatment is absolutely inspiring and groundbreaking. History of science and engineering, technological advancements and the immense urge to excel has propelled the future of mankind toward new vision. The challenges toward environmental regulations are befitting toward the future of human civilization. Provision of clean drinking water and industrial wastewater treatment are in the verge of a new visionary dimension. Mankind's cognizance and in the similar vein scientific vision will go a long way in opening up new doors of innovation in years to come (Drioli et al., 2002; Cheryan, 1998).

16 Conclusions

History of human mankind and history of science and technology is steadily moving toward a new dimension of scientific vision and scientific determination. Challenges, fetters, and hurdles are the pallbearers of the next generation environmental engineering science. Sustainability and successful sustainable development are the order of the day. Application of nanotechnology in drinking water treatment will surely witness a new dawn of human civilization. The question of successful environmental sustainability arises at every step of human endurance. In such a crucial juncture, scientific validation of research is of utmost importance. Ecological balance along with provision of clean drinking water are of primordial importance. History of human mankind will surely witness a new era in decades to come. Environmental engineering science and water process engineering is moving toward a new dimension of future endeavor. The world of challenges are befitting to the human progress. In such a situation drinking water, provision is of utmost importance.

The impact of scientific endeavor to our present day civilization is immense and tremendous. Challenges to drinking water treatment are indomitable yet can be surpassed. Science and technology of nanotechnology needs to be reenvisioned with the progress of human needs and human progress. Application of NF in drinking water treatment will open up new vistas of research and the paradigm of scientific endeavor in today's world is far-reaching. Provision of drinking water is an immense parameter to mankind's growth. Engineering science will usher in a new era if scientific truth and scientific vision is realized. The visionary domain of nanotechnology is no more an illusion.

Acknowledgments

The author wishes to acknowledge the contribution of Chancellor, Vice-Chancellor, Staff, Students, and Management of University of Petroleum and Energy Studies, Dehradun, India without whom this writing project would not have been completed.

References

Alsvik, I.L., Hagg, M.B., 2013. Pressure retarded osmosis and forward osmosis membranes, materials and methods. Polymer 5, 303–327.

Bhalme, S.P., Nagarnaik, P.B., 2012. Analysis of drinking water of different places-a review. Int. J. Eng. Res. Appl. 2 (3), 3155–3158.

Casey, T.J., O'Reilly, P., 2009. Drinking water disinfection using chlorine theory and practice. Report, Aquavarra Research Limited, Ireland.

Cheryan, M., 1998. Ultrafiltration and Microfiltration Handbook. Technomic Publishing Co. Inc., Lancaster, Pennsylvania, USA.

Childress, A.E., Elimelech, M., 2000. Relating nanofiltration membrane performance to membrane charge (electrokinetic) characteristics. Environ. Sci. Technol. 34, 3710–3716.

Drioli, E., Curcio, E., Fontananova, E., 2002. Mass transfer operations: hybrid membrane processes. In: Pohorecki, R. Bridgwater, M., Molzahn, M., Gani, R., Gallegos, C. (Eds.), Chemical Engineering and Chemic Process Technology, vol. 2. Encyclopedia of Life Support Systems, United Nations Educational, Scientific and Cultural Organization, USA.

Ismail, A.F., David, L.I.B., 2001. A review on the latest development of carbon membranes for gas separation. J. Membr. Sci. 193, 1–18.

Khayet, M., 2010. Desalination by membrane distillation. In: Encyclopedia of Life Support Systems (EOLSS), water and wastewater treatment technologies.

Lau, W.J., Ismail, A.F., Misdan, N., Kassim, M.A., 2012. A recent progress in thin film composite membrane : a review. Desalination 287, 190–199.

Luo, J., Wan, Y., 2013. Effects of pH and salt on nanofiltration—a critical review. J. Membr. Sci. 438, 18–28.

Mohammad, A.W., Ng, C.Y., Lim, Y.P., Ng, G.H., 2012. Ultrafiltration in food processing industry: review on application, membrane fouling and fouling control. Food Bioproc. Technol. 5 (4), 1143–1156.

Nghiem, D.L., Schafer, A.I., Elimelech, M., 2005. Pharmaceutical retention mechanisms by nanofiltration membranes. Environ. Sci. Technol. 39 (19), 7698–7705.

Onjala, J., Ndiritu, S.W., Stage, J., 2013. Risk perception, choice of drinking water and water treatment: Evidence of Kenyan towns, Environment for Development (EFD), Discussion Paper, EFD.

Opio, C., 2012., Building effective drinking water management policies in rural Africa—lessons from Northern Uganda, CIGI Africa Initiative Discussion Paper Series, Published by the Africa Initiative and The Centre for International Governance Innovation.

Ray, C., Jain, R., 2011. Drinking Water Treatment Technology-Comparative analysis, Drinking Water Treatment, Strategies for Sustainability. Springer Science and Business Media B.V., Springer Publishing, Heidelberg, Germany.

Report on Drinking Water Treatment, 2006. The Cooperative Research Centre for Water Quality and Treatment, Australia.

Schafer, A., Nghiem, L.D., Waite, T.D., 2003. Removal of natural hormone estrone from aqueous solutions using nanofiltration and reverse osmosis. Environ. Sci. Technol. 37, 182–188.

The RBC Harris Centre Drinking Water Applied Research Fund, 2009. Drinking water quality research summary and suggested priorities report.

Van der Bruggen, B., Manttari, M., Nystrom, M., 2008. Drawbacks of applying nanofiltration and how to avoid them—a review. Sep. Purif. Technol. 63, 251–263.

Vrijenhoek, E.M., Hong, S., Elimelech, M., 2001. Influence of membrane surface properties on initial rate of colloidal fouling of reverse osmosis and nanofiltration membranes. J. Membr. Sci. 188, 115–128.

Vrouwenvelder, J.S., van Paassen, J.A.M., van Agtmaal, J.M.C., van Loosdrecht, M.C.M., Kruithof, J.C., 2009. A critical flux to avoid biofouling of spiral wound nanofiltration and reverse osmosis membranes: fact or fiction? J. Membr. Sci. 326, 36–44.

WHO, 1996. Chlorine in drinking water, Background document for development of WHO guidelines for drinking water quality. (WHO/SDE/WSH/03.04/45).

18

MICROBIAL BIOFILMS FROM THE AQUATIC ECOSYSTEMS AND WATER QUALITY

Liliana Neagu**,†, Doina Maria Cirstea*, Carmen Curutiu**, Magda Mihaela Mitache‡,§, Veronica Lazăr**, Mariana Carmen Chifiriuc**

**National Institute of Public Health, Institute of Biology, Bucharest, Romania; **University of Bucharest, Research Institute of the University of Bucharest, Bucharest, Romania; †Politehnica University of Bucharest, Department of Biomaterials and Medical Devices, Faculty of Medical Engineering, Bucharest, Romania; ‡Titu Maiorescu University, Faculty of Medicine, Bucharest, Romania; §Public Health Department Bucharest, Bucharest, Romania*

1 Introduction

In nature, bacteria generally exists in two major states: the free planktonic form, characteristic to a minority of bacteria, in which the unicellular organisms float or swim independently in a fluid environment, and respectively, the majoritary one, in which cells firmly attach to each other and also to a surface forming a solid structure. The last one could be regarded as a survival mechanism because inside this structure, bacteria are protected from toxic compounds, stressors, and so on (Johnson, 2007; Costerton et al., 1987; Flemming, 2009). In aquatic environments, biofilms develop on all exposed solid surfaces and interfaces and their structure depends on the species composition of the microbiota of the respective area, their microbial activity, and environmental conditions. The aquatic ecosystems biofilms are complex and dynamic communities with a considerable microbial biodiversity, composed of different species of bacteria, diatoms, and protozoa embedded in a protective polymer matrix. A major concern for water retailers is to maintain the quality of drinking water from the treatment plant to the tap of consumers. From the biological point of view, water must be characterized by a stable bacterial growth. However, distribution systems for drinking water are continuously

Water Purification. http://dx.doi.org/10.1016/B978-0-12-804300-4.00018-6

exposed to a variable flow of water, containing biodegradable organic substances and microorganisms, which originate either in the natural biofilms developed inside the water-treatment plants or from some incidents appeared in the distribution network. Some of these microorganisms (especially heterotrophic bacteria) can grow in this environment and can colonize the entire distribution system of drinking water, the highest density of microorganisms being found on the surface of the pipe wall, organized in dispersed microcolonies, mixed with corrosion products and inorganic precipitates (Block et al., 1993; September et al., 2007; Husham et al., 2012). The purpose of this chapter is to present the characteristics of microbial biofilms found in the aquatic ecosystems with a special focus on their both negative and beneficial roles for the water treatment and quality.

2 Definition of Biofilm

Biofilm represents a structured microbial community adherent to an inert or living surface embedded in a self-secreted polymeric matrix (Zarnea and Popescu, 2011). It has a heterogeneous structure, the cells embedded in this matrix exhibiting a modified phenotype, especially concerning the growth rate and gene transcription, and a mono- or, the most often, a multispecific composition, regulated by multiple signaling mechanisms dependent on the cellular density, known as quorum sensing and response (QS) systems (Costerton, 2009; Donlan, 2002; Lazăr and Chifiriuc, 2010a). In QS, the communication between the cellular components of a biofilm is achieved by chemical signal micromolecules with different structures represented by peptides in Gram-positive bacteria and L-homoserin-lactones in Gram-negative species (Davey and O'Toole, 2000; Davies et al., 1993, 1998; Chifiriuc et al., 2014). The QS regulation mechanism is involved in various physiological processes (biofilm formation, antibiotic production, bioluminescence, genetic competence, sporulation, swarming motility), as well as in the expression of virulence potential (Lazăr and Chifiriuc, 2010b). It has been shown that the QS-mutant strains have a decreased potential to develop mature biofilms, forming only thin and uniform structures (Delden and Iglewski, 1998). When the signaling molecules reach a threshold concentration which reflects a certain density of the microbial cells inside biofilm, the gene expression of 40–60% of the bacterial genome is changed having a consequence that the modification of biofilm cells phenotype in comparison with that of their planktonic counterparts (Israil and Chifiriuc, 2009).

Bacterial biofilms appeared about 3.25 billion years ago and represent a growth mode which protects cells and allows them to

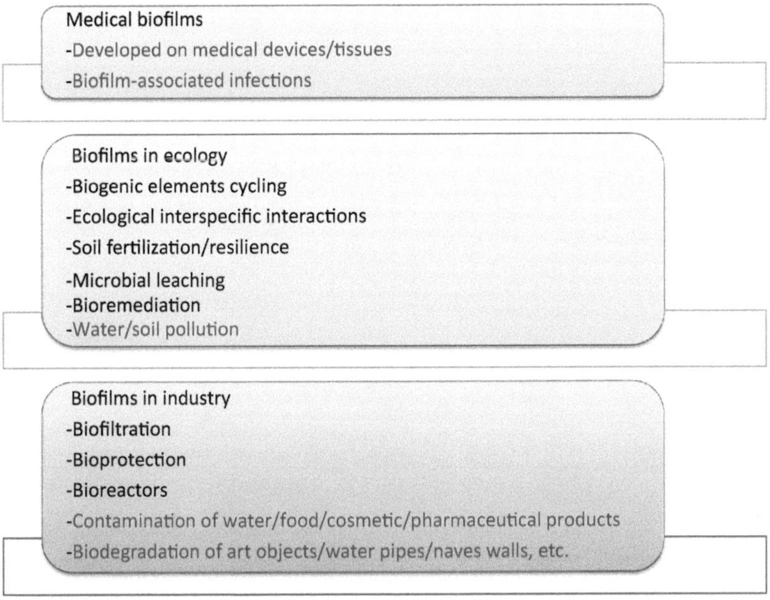

Figure 18.1. The beneficial and detrimental roles of biofilms for the medical, ecological, and industrial fields.

survive in hostile environments (Hall-Stoodley et al., 2004). So, one could state that biofilm development is a primitive form of bacterial life cycle and also, a key factor for their survival in various environments. Biofilms can be found everywhere in nature (Hall-Stoodley et al., 2004): in rainwater, thermal waters, marine waters and oceans, on various solid natural media (stone, wood, soil), and synthetic surfaces also as pipes, medical catheters, and so on, and have a major impact both on people's health and on the ecological and industrial processes (Chifiriuc et al., 2011) (Fig. 18.1).

3 Stages of Biofilm Development

Biofilm development represents a growth mode that confers protection of microbial cells, survival in hostile environments but also possibility of cells to detach from the biofilm, switch to the planktonic phenotype, disperse, and colonizes new niches. Microbial biofilm development mechanism (Fig. 18.2) undergoes several stages. The initial phase starts with a reversible attachment of free microorganisms to a surface. These first species that colonize a surface poorly adhere by van der Waals forces. This step is followed by the irreversible attachment of bacteria to the surface—if the primary colonists are not immediately removed from the surface; they can firmly anchor using cell-adhesion structures,

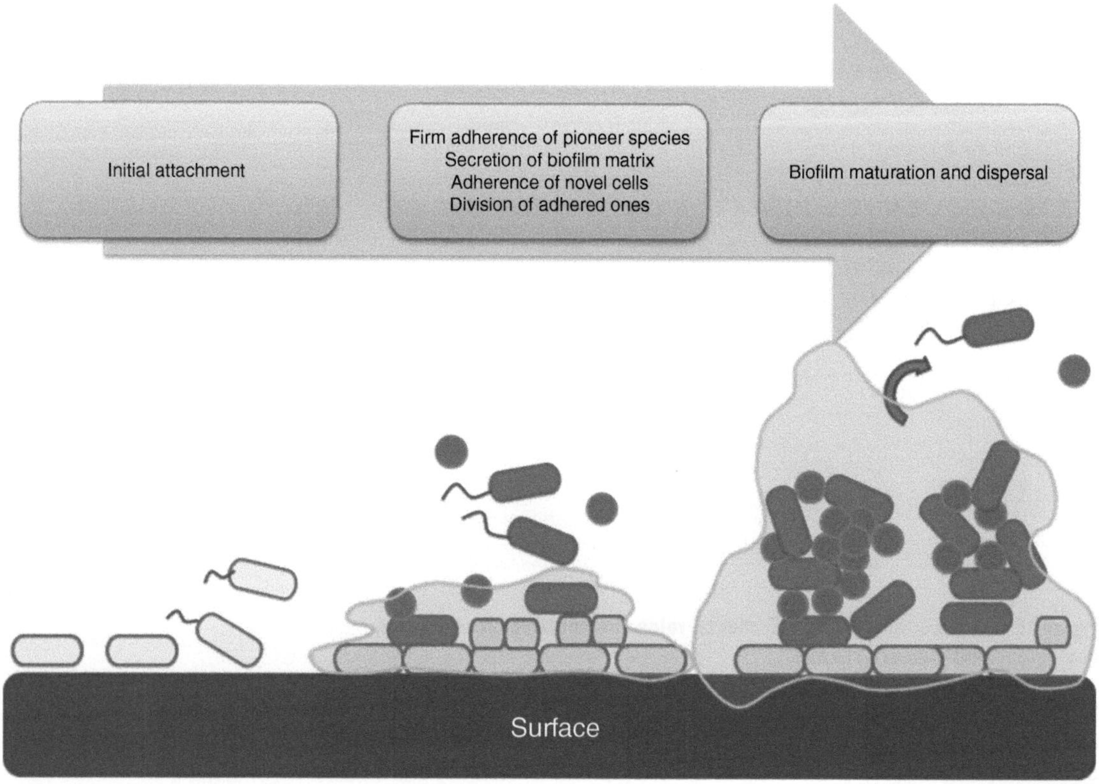

Figure 18.2. The main stages of biofilm development.

as extracellular polymeric substances or structural components as pilli. The first bacteria facilitates the adherence of other bacterial species by providing various adhesion sites and secretion of the polymeric matrix. Some species are not able to attach through their surfaces, but often are able to anchor to the matrix or directly to the first adherent bacteria. During the colonization these cells are able to communicate through QS molecules (Chifiriuc et al., 2014). Once colonization is initiated, biofilm grows through a combination of processes, including cell division and recruitment of new planktonic cells. This stage of biofilm development is known as maturation, in which the biofilm is stable and may suffer changes only in size and shape. In the final stage, the dispersal of cells from the biofilm occurs, allowing them to spread and colonize new surfaces (Finkel and Mitchell, 2011).

The biofilm is supported and protected by a matrix of polymeric compounds synthesized by its components. This matrix protects the cells and facilitates communication among them through biochemical signals. Almost every microbial species has the specific mechanisms through which they can adhere to surfaces, and to each other.

Microorganisms belonging to the biofilm are physiologically distinct from their free counterparts called planktonic cells. When a cell switches to biofilm growth, it undergoes phenotypic changes, the genes expression being differently regulated (An and Parsek, 2007). The microorganisms which form biofilms respond to many factors, which may include: recognition of specific or nonspecific adherence sites to a surface, nutritional index, or in some cases to the presence of harmful substances gradients (e.g., subinhibitory antibiotic concentrations, biocides, heavy metals) (Karatan and Watnick, 2009).

4 Microbial Biofilms in the Aquatic Environment

The number of adherent bacteria from the aquatic microbiota of flowing water is 500–500,000 times higher than that of planktonic bacteria, because the growth in biofilms ensures the survival and rapid adaptation of bacteria to changes of the environmental conditions, the proximity of the cells favoring the transfer of genetic material by conjugation, and selection of bacteria resistant/multiresistant to antibiotics and heavy metals (Lazăr et al., 2015).

Biofilms are formed not only in natural environments, but also in drinking water distribution systems (which may contain a very small number of bacteria); thus, major problems could arise in the context of public health, the distribution of water to consumer being realized through pipes in which water circulates slowly and could favor the biofilm development; from biofilms, planktonic cells are released periodically and could recontaminate water (*Pseudomonas aeruginosa*, *Legionella pneumophila*), and this cycle can be repeated over and over (Lazăr et al., 2015).

For example, water pipes from dental units are an ideal environment for growth and development of bacterial biofilms with bacteria originated from the water-distribution network (Walker et al., 2000). The instruments used by dentists are connected to the water-distribution system through a network of very-fine plastic tubules through which water and a stream of air flowing enact or cool the tools. In these very-narrow tubules, the column of water will flow into an axial stream, forming a thin film of immobile fluid on the internal surface of the tubules; this stagnant water will create favorable conditions for the development of a biofilm (Meiller et al., 1999; Pederson et al., 2002). It has been shown that the wastewater from specific equipments from the offices of dentists contains alarming levels of microorganisms belonging to some opportunistic species (*P. aeruginosa, L. pneumophila, Mycobacterium* spp., *Candida* spp.) (Szymanska et al., 2008) usually higher than those recorded in the distribution of drinking water (Douglas and van Noort, 1993; Fayle and Pollard, 1996; Martin, 1987). There are clinical studies that have shown contamination in this way of some immunosuppressed patients with *P. aeruginosa* (Schulze-Röbbecke et al., 1995). In one study, *L. pneumophila* was detected in 86.7% of the dental units at the beginning of the working day, before flushing, the microbial charge being reduced to 40% by flushing the waterlines. The highest contamination level was recorded in the periodontology and prosthodontics units (Ma'ayeh et al., 2008). In California, in 1994 it was reported death of a dentist as result of aerosol inhalation of *L. pneumophila* cells when handling dental instruments. Another fatal case of endocarditis was put into connections to the water during a simple operations of teeth cleaning (Lazăr et al., 2015).

In biofilms from drinking water there were identified six representative species (representing about 80% of all isolated and identified species), namely, *Methylobacterium* sp., *Mycobacterium mucogenicum*, encapsulated *Sphingomonas, Staphylococcus* sp., *Burkholderia cepacia, Acinetobacter calcoaceticus* (Simões et al., 2011; Lazăr et al., 2015).

A well-developed biofilm may compromise the chance of detecting the presence of fecal coliforms. High level of coliforms may indicate some deficiencies in the distribution system (e.g., corrosion of pipes). Bacteria involved in biological corrosion are ferroxidant and sulfate-reducing bacteria. Ferroxidant bacteria oxidize soluble iron to form insoluble precipitates in the pipeline. Bacteria that oxidize the iron/steel surfaces can lead to the formation of oxides of iron and manganese which accumulate in large amounts. In addition, sulfate-reducing bacteria (*Pseudomonas* spp., *Proteus* spp.—species that coupled this process with the oxidation of organic compounds) produce H_2S responsible for the specific smell of rotten eggs and for accelerate corrosion; nitrification bacteria corrode copper. Various species produce polymers that constitute matrix of biofilms and also could coupled with other pipe materials and modify redox potential of the surface and accelerate corrosion (Lazăr et al., 2015).

Growth in biofilms in the aquatic environments is thus leading to impaired water quality (organoleptic changes) and also promotes proliferation of trematodes, nematodes, and bacteriovorus protozoa which feed with bacteria from biofilms. Actinobacteria and algae produce geosmin and 2-methyl-liso-borneol conferring bad smell of water (such as a soil mold, http://www.who.int/water_sanitation_health/dwq/cyanobacteria_in_drinking-water.pdf). *Pseudomonas* sp. generates odor of sulfur and *Hyphomicrobium* sp. gives water a black color (Cullimore, 1978; Schulze and Mooney, 1993). Bacteria that oxidize iron and sulfur-oxidizing bacteria are lowering the pH of the water, and cause metallic taste and smell of sulfur, increased turbidity, and coloring water (through the development of dark-colored biofilms, called slime on the internal surface of water basins), produce obstruction of the piping, and reduce the flow of water and the efficiency of heat transfer (Schulze and Mooney, 1993). Biofilms have a negative role also in wastewater-treatment plants (WWTP), because they could also develop on biological filters and aeration devices. Also, WWTPs biofilms provide the perfect environment for the enrichment, recombination, and selection of antibiotic-resistant superbugs (Guardabassi and Dalsgaard, 2004). The membrane bioreactors increasingly used in wastewater treatment could facilitate the transfer of resistance genes among bacterial biofilm cells. It has been shown that the transfer frequency of the RP4 plasmid in these facilities was 2.76×10^{-5} per recipient, which was greater than the transfer frequency in wastewater and bacterial sludge reported previously, the recipient strains being very diverse, belonging *to Shewanella* spp., *Photobacterium* spp., *Pseudomonas* spp., *Proteus* spp., and *Vibrio* spp. genera (Yang et al., 2013).

In such a confined space, bacteria reach high densities to the high content of organic substances and share resistant genetic materials, and remaining antibiotics and other stressors, such as heavy metals may select multidrug resistant bacteria. The antibiotic resistant bacteria could eventually be discharged into streams and lakes, which further appear to play an important role in the emergence and dissemination of resistance in human population, highlighting the necessity for strategies of water-quality improvement (Lupo et al., 2012).

5 Microbial Biofilms in the Water Distribution Systems

With the development of human civilization, the drinking water distribution systems technology evolved. These distribution systems represent new aquatic environments that bacterial biofilms have colonized in time, from the time of Roman aqueducts to the most modern systems of the present. Biofilms formed on water pipes are formed by Gram-negative bacteria (species of *Pseudomonas*, *Sphingomonas*, *Flavimonas*, *Flavobacterium*,

Acinetobacter, Aeromonas, Legionella, coliforms), some atypical mycobacteria (*Aquabacterium*), viruses, amoebas, and parasites, such as *Cryptosporidium* and *Giardia* (Squinazi, 2004). Biofilms formed in drinking water systems consist most often in a combination of several species, including bacteria, fungi, algae, protozoa, but also debris particle and products resulted from corrosion.

The phenomenon of biofilm accumulation (biofouling) occurs without exception in all production, storage, and distribution of water systems (Fig. 18.3), in spite of a relatively unfavorable media, due to oligotrophy, low temperatures, presence of disinfectants, and hydraulic conditions (Moritz, 2011). Deposits had the origin in suspensioned matter carried by the water, in the microbial activity and physicochemical reactions from the interface and the mass of water (Echverría et al., 2009).

The presence of bacterial biofilms adhered on the surface of water-supply pipes in water-distribution networks could represent a source of nonconformity with standard regulations for drinking water quality and can be a suitable structure for the constitution of a reservoir of potentially pathogenic bacteria.

Bacterial dynamic in distribution systems is complex because it depends on different parameters, such as the biodegradable fraction of organic carbon, the presence of a residual disinfectant,

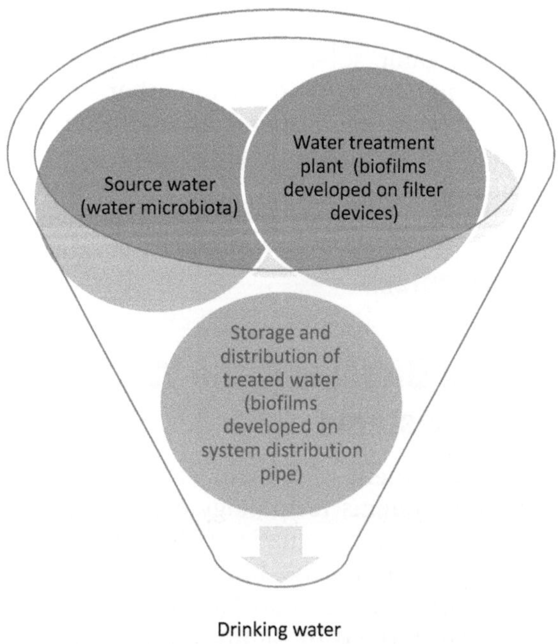

Figure 18.3. Biofilms from the treatment and distribution system.

materials, and condition of the pipe wall. Therefore, the development of biofilms in water distribution systems can be regarded as a dynamic and evolving process where compounds and microorganisms interact with each other. The proliferation and detachment of biomass from the surface materials that come in contact with potable water can lead to microbial contamination of the water distribution systems (Lechevallier et al., 1996). The microbiological quality of drinking water may be degraded significantly during transport in distribution networks, mainly due to bacteria dispersal from biofilms developed on the inner surface of the pipes.

For decreasing the microbiological contamination it is necessary to use chemical disinfectants, such as chlorine, which acts on microorganisms or of new technologies, which eliminate the risk of bacterial contamination by decreasing the concentration of organic substances from the distribution network. Moreover, in order to prevent any risk regarding bacterial growth in the distribution network, it should be kept permanently a residual disinfectant (chlorine, chlorine dioxide, monochloramine). However, some bacteria can adapt to this environment and even more, proliferate.

6 Other Strategies of Bacterial Survival in the Aquatic Environments

6.1 Starvation-Survival Lifestyle

In unfavorable environmental conditions (especially lack of energy required for growth and reproduction), both sessile and planktonic bacterial cells are undergoing morphological (e.g., reduced size from very small to ultramicrocells) and physiological changes characteristic for the starvation-survival state (Kjelleberg, 1993; Morita, 1997). The low size improves the surface/volume ratio favoring a more efficient sequestration of nutrients to confront nutrient limitation. Bacteria survive by degrading the reserve carbon polymers, such as polysaccharides, glycogen, and poly-β-hydroxybutyric acid in the endogenous metabolism. Bacteria respond to specific nutrient limitation by producing transport systems with increased affinities for the preferred nutrients but also for alternative nutrients, that could be relatively more abundant in the stressful environment (Leclerc, 2003). The stress response was well documented in *Escherichia coli*, in which it starts by the production of a guanosine 3′,5′-bispyrophosphate nucleotide acting as a signal for triggering a stringent stress response involving the activation of the alternative starvation sigma factor σs which activates a large number of stress proteins (Cst proteins) (Nyström 1993; Jones 1997;

Hengge-Aronis, 2000). Bacteria entering starvation-survival state become more resistant to various environmental stresses. However, the entrance in this state requires the synthesis of a great number of enzymes for the utilization of organic matter, which is most probable to happen in interactive community of bacteria, such as the biofilm's consortia) (Morita, 1997; Leclerc, 2003).

6.2 Viable but Noncultivable Cells

It has been found that a number of bacterial species (particularly Gram-negative bacteria) can exist in a different state, distinct from the starvation-survival one, called viable but nonculturable cells (VBNC). This form of metabolic latency is caused by exposure to stress conditions, for example, lack of nutrients (starvation state) or low temperatures, suggesting that this state is an adaptation strategy for long-term survival of bacteria in unfavorable environmental conditions (Oliver, 1993). In this state, the bacterial cells (smaller in size than usual), lose the ability of growing on solid media, fact that impede the detection of bacteria by conventional techniques (viability can be demonstrated by flow cytometry only) (Lehtinen, 2007). When stress conditions are removed, these bacteria have the ability to reverse to vegetative state. The ability of bacterial species to enter in VBNC state represents a risk for the human health because the total viable bacteria from a sample will be underestimated, if this sample is tested using colony forming units count. Moreover, if all the bacteria from the sample are in nonculturable state, the sample can be considered sterile, which is clearly a false negative result. The risk for public health is amplified by the fact that bacteria are nonvirulent in VBNC state, but after their reversion to cultivable state, the respective strains regain their virulence (Du et al., 2007; Li et al., 2014).

7 Microbiological Quality Assurance in Drinking Water Distribution Network

The main objective of water distribution networks is to deliver water to consumers in sufficient quantity and at any point in the network, at their request. For a long period of time, managing these networks was equal with simple management of water transport and quantitative aspects were privileged comparing with quality aspects, thus resulting the oversize of the distribution networks. Today, the notion of quantity is incompatible with the quality objective set by the regulations; water must meet the required quality conditions for human consumption.

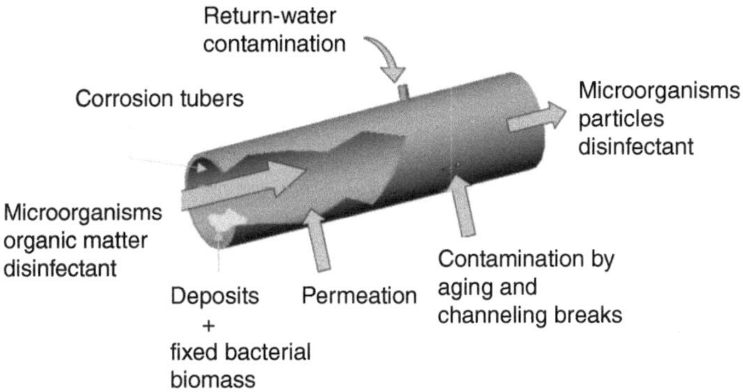

Figure 18.4. Diagram of a distribution network type reactor.

Indeed, a drinking water distribution system should not be considered only as an inert system, but as a complex reactor (Fig. 18.4), in which takes place physicochemical and biological interactions.

These reactions are related to the interaction of hydraulic conditions within the network, the type of materials used and the quality of water flowing. Reactions important for the quality of water entering the distribution network are: (1) the type and the amount of oxidant used (the generation of secondary products of disinfection); (2) content of biodegradable dissolved organic substances (that sustain bacterial growth); (3) smell (chlorine, sulfur, and so on); (4) the amount of mineral and organic particles (algae, bacteria); (5) corrosives conditions (http://www.who.int/water_sanitation_health/dwq/en/watreatpath.pdf; https://www.epa.ie/pubs/advice/drinkingwater/Disinfection2_web.pdf; http://www.who.int/water_sanitation_health/resourcesquality/waterqualmonitor.pdf; Schutte, 2006).

Water quality could be affected also during distribution. Drinking water resulted from the treatment station contains particulate matter that may come from: (1) the treatment plant itself, by releasing fine particles of carbon; (2) raw water used; (3) drinking water distribution system through the release of particles from metal corrosion or precipitation of dissolved elements (Fe, Ca, Mn, and so on); (4) the growth of microorganisms; (5) the tank, if water is in contact with the air.

The accumulation of these particles in the distribution network is directly related to sedimentation and transport profile (size, density, shape) (Gauthier, 2001). The particles present in drinking water distribution systems can accumulate in deposits as follows: (1) particle sedimentation from produced drinking water; (2) solid substances that can form particles in case of precipitation, flocculation, and organic growth; (3) particles from material resulting from corrosion; (4) particle sedimentation from external contamination.

The particles that remain in suspension can have significant effects on water quality (increased turbidity, color change), chemical effects (high concentrations of metals), and microbiological ones (an increase of bacteria number) (Gauthier, 2001). Water circulated through distribution networks carries large amounts of minerals (iron, silicates, aluminum, calcium, manganese, primarily as oxides) and in smaller quantities organic matter. Although consists of essentially dissolved elements, a small fraction of sediment can form deposits in the distribution system (Gauthier, 2001).

After treatment of drinking water, the water is not sterile. The aim of these treatments is to make it tasty to drink and more secure in terms of health. Therefore, several populations of organisms are still living in water distribution systems. These organisms can be classified into four groups (Block et al., 1993): (1) native species, represented mainly by microorganisms from soil, food, such as bacteria, fungi, and yeasts; (2) unexpected species to be found on tap water, such as protozoa or macroinvertebrates. Accidentally introduced in the network, they can survive and multiply at a rate of thousands per cubic meter; (3) pest species that can cause or can worsen corrosion, or determine flavoring compounds (e.g., actinomycetes). Another problem associated with these species, in case if their multiplication increased, is the impossibility to detect potentially pathogenic species; (4) dangerous species in terms of health. Some species pathogenic to humans can be detected in distribution networks. Their presence is the result of accidental contamination during the distribution of the water.

The bacterial biomass is found predominantly into the fuel tank and on alimentation pipes. Indeed, the pipes are largely colonized by microorganisms, being porous media that allow the attachment and growth of bacteria (Gauthier, 2001).

Deposits provide organic nutrients for microorganisms. Thus, they can be considered as reservoirs of biomass in drinking water distribution systems. The inorganic fraction mainly acts as a support for bacteria, being observed a greater number of bacteria in these systems. However, the nature of these mineral elements appears to influence survival and development of microorganisms.

In drinking water distribution systems, the biomass can be found in two states: (1) a fixed state, directly on the pipe surface or tubers, resulting from the process of corrosion. The term used for describe this state is the biofilm; (2) circulating state, bacteria being classified as plankton.

In general, the density of bacteria is higher in biofilm than in the aqueous phase. Thus, if the density of the microorganisms found in the flow before entering a pipe is 10^4–10^8 bacteria L^{-1} and 10^4 protozoa L^{-1}, inside a pipe with a diameter of less than 10 cm, the number of bacteria adhered and forming a biofilm is in average of 50–100 times higher (i.e., 10^3–10^7 bacteria cm^{-2} and 10^3 protozoa cm^{-2}) than that of the circulating bacteria 10^5–10^8 bacteria L^{-1} and 10^4–10^5 protozoa L^{-1}) (Block et al., 1993).

Studies realized in drinking water distribution networks, allowed the observation of changes in the microbiological quality of water transported along the distribution system.

The prolonged stagnation of water in the network is accompanied by increased bacterial density (except when are present chlorination systems in the network), reflecting a bacterial dynamic. This growth is the result of biological instability in drinking water distribution networks.

Drinking water distribution networks, like any other aqueous medium can be colonized by biofilms, in any conditions: absence of light, the presence of a circulating biocide agent, low concentration of nutrients (oligotrophic environments).

The level of microbiological contamination of the water network is based on the flow cell from the treatment plant. Once the bacteria start circulating in the pipes, they multiply on their inner surfaces, and then form a biofilm.

8 The Consequences of the Presence of Biofilms in Water Distribution Network

Natural water microbiota consists of native species, characteristic for the respective ecosystem, but also of nonindigenous, potentially pathogenic species, originated from soil and drilled by rainwater, or from different contamination sources represented, by wastewater from households, rich in sulfates originating from detergents, wastewater from livestock producers, chemical, mining, or metallurgy industry, accumulation of excess fertilizers, organic substances, and minerals from soil erosion. Groundwater is partially contaminated with heavy metals (from extractive/oil industry, oil–phenol-like products from refineries, drilling installations, agricultural activities) (Lazăr et al., 2015).

Various problems can be directly linked to the formation and development of biofilms on the walls of pipes for drinking water. These consequences can affect the distribution network itself, or may have negative effect on the consumption of contaminated water network.

These consequences are multiple and cover both bacterial populations and the physical characteristics of the network: (1) bacteria accumulated in biofilms are the first link in a trophic chain thus supporting the development of macroorganisms; (2) some types of bacteria can induce an increase turbidity, change of taste, and smell of water as result of their presence or their metabolic activity; (3) some bacteria can accelerate corrosion; (4) a reduced capacity of the distribution network due to an increase of resistance forces induced by the presence of the biofilm; (5) an increase in the number of nonconformities in relation with microbiological quality criteria for water for consumer. Nitrification bacteria development in anoxic zones, can also lead to nonconformities by exceeding maximum permissible concentrations for nitrates (Gauthier, 2001).

The presence of biomass in the constant flow within water distribution systems and its proliferation raise a problem in terms of public health. Indeed, the system is continuously inoculated with bacteria most of which being unknown, but potentially pathogenic organisms. Moreover, the conditions from the distribution network may allow maintaining or increasing level of coliforms and thus the standards for drinking water will be unfulfilled.

From the 300–500 million cases of acute diarrheal disease recorded annually in the world, over 60% are affecting children from rural areas of underdeveloped countries, where cholera, typhoid fever, parasitosis cause 20 million deaths, and it is estimated that in the entire world over 25,000 people die every day due to lack of drinking water (Ali et al., 2012; Cabulea, in press, http://www.who.int/water_sanitation_health/takingcharge.html#Foreword). Moreover, WHO estimates that the officially reported cases represent only a small proportion (5–10%) of the real number of cases occurring annually worldwide (Lanata et al., 2002). There are many worries lately regarding the widespread degradation of water's quality through uncontrolled discharge (sewage and other wastewater from industrial or agricultural activities and livestock).

Biological pollution of water, especially of surface water can be achieved directly from the pathogens (fecaloid pollution, or by waters from which the sick animals drink, bathing, or water filled with dead animals) or indirectly (through air, soil, food, in case of a flood, rain, earthquakes, cataclysms, the inadequacy/ineffectiveness decontamination, or through the use of unhygienic water sources).

Biological water pollution could favor: (1) development of a nutritional support for pathogens; (2) water demineralization process failure by blocking the ion exchange used as filter material in water-treatment installations; (3) a slower deferrization and demanganization processes (used as means of water treatment); (4) higher corrosion of the pipes from the water treatment, transport, and distribution plants; (5) changes of the organoleptic properties; (6) occurrence of potentially carcinogenic halogenated compounds in the process of water chlorination; (7) the presence of organic compounds in small quantities that may be toxic or carcinogenic; (8) high costs of water treatment.

The survival of microorganisms in aquatic ecosystems is dependent on physicochemical properties of the aquatic habitat. Survival and maintenance of pathogenic bacteria virulence is related to a complex of exogenous factors (treatment efficiency from the treatment plants, the volume and frequency of fecaloid-contaminated water discharge) and endogenous factors (water temperature, pH, oxygen concentration, the presence of organic compounds, solar irradiation, competition with indigenous microorganisms, the presence of bacteriophage, and so on).

A single drop of contaminated water contains more than 50 million of bacteria and other microorganisms, including pathogens that could be etiologic agents of some infectious diseases transmitted through water.

The risk of transmission through water of a pathogen depends on: (1) microbial species; (2) the degree of contamination of the water source; (3) the degree of the pathogens resistance in the aquatic environment; (4) the nature of the water source (e.g., fountain located near latrines); (5) the chemical composition of the water, pH, temperature, oxygenation level of the water; (6) the level of solar radiation; (7) the composition of saprotrophic microbiota; (8) the bacteriophages density; (9) the bacteriovore protozoa density; (10) the degree and frequency of refreshing of the water layer; (11) the frequency of water-source decontamination and nature of decontamination; (12) the sanitary conditions; (13) the epidemiological surveillance measures (Lazăr et al., 2015).

Limiting biological instability in water distribution systems, as well as the growth of microorganisms in biofilms represents point of interest both for users and distributors, underlying the necessity of water microbiological quality control during distribution.

Measures to eradicate biofilm or decrease its development on drinking water pipes include: (1) flushing the pipes with water currents and cleaning at regular intervals (current strong water jets of water at 95°C); (2) control and nutrient removal (preozone treatment, biological filtration, activated carbon, nanofiltration, reverse osmosis); (3) the use of disinfectants with residual activity; (4) materials and equipment contamination control; (5) control and reducing the problems related to hydraulic systems; (6) limitating the areas of very low flow speed of the water and "dead ends" with total stagnation of water; (7) control of supply lines and prevent water discharge lines; (8) pipeline corrosion control; (9) maintaining a constant positive pressure in the distribution system; (10) control of the water storage tanks (Lazăr et al., 2015).

9 Beneficial Roles of Microbial Biofilm in Water Treatment

Besides the negative impact on human health, it is important to note that biofilms can have beneficial applications, namely: their use in drinking water/wastewater treatment, or detoxification of toxic and hazardous waste, and so on. Bacterial biofilms can also be used by in vitro engineering for specific biotechnological applications. For over a century, biofilms have been successfully used in water treatment and wastewater treatment. The biological filters are used to reduce the concentration of biodegradable organic carbon that enters into the water distribution systems. Biofilms containing phototrophic cyanobacteria can be used to remove additional nutrients from secondary effluents of sewage-treatment stations. An increase in pH due to the biofilm photosynthetic activity causes the precipitation of dissolved phosphate (Roeselers et al., 2008). In this way removal of phosphorus helps to reduce the possibility of eutrophication. Immobilized biofilms can be used as bio barriers for capturing the contaminants from groundwater (De Beer and Stoodley, 2006).

Biofilms are biological barriers in the process of water treatment, being able to retain particles (including pathogens) and remove nutrients (Fonseca et al., 2001; Långmark et al., 2004; LeChevallier and Au, 2004; Tellen et al., 2010). Sessile bacteria from biofilms exhibit bioaccumulation and biodegradation properties higher than their planktonic counterparts (Araya et al., 2003).

In a distribution network it is estimated that about 95% of microorganisms are attached to the some surfaces, while only 5% of them are free on the water. Therefore, a small percentage can be collected and detected in routine analysis (Flemming et al., 2002).

Quantifying the effects of pollution on different components of the biofilm could offer a mean to directly evaluate the effects of pollutants on most-sensitive communities and the indirect consequences on the rest of biofilm components. Biofilms are relevant indicators, for environmental quality; therefore they can provide an early warning signal in case of contamination by pesticides or other toxic substances.

9.1 Use of Biofilm in Bioremediation of Wastewaters

If biofilms are a source of contamination in areas, such as food or medical fields, they can also be used positively in wastewater-treatment processes. Bioremediation is the application of biological processes to remove harmful pollutants from the environment, and represents in situ or ex situ use of microorganisms to clean a contaminated site (Kumar et al., 2011; Sharma, 2012; Gianfreda and Rao, 2004; Husham et al., 2012). There are successful examples of positive use of biofilms for the bioremediation of wastewater (Das et al., 2012). They confer several advantages including protection of embedded cells against hazardous effects of toxic pollutants. Reactors based on biofilms are commonly used for the treatment of high quantities of industrial and municipal wastewater (Borkar et al., 2013).

Over the past decades, these reactors have become attractive for bioremediation of xenobiotic compounds. Biofilms have the potential to remove heavy metals and other xenobiotic compounds from the wastewater and natural water containing low levels of pollutants. The study of biofilm communities and the genes transfer within biofilm facilitate the development of better techniques for bioremediation of polluted sites and wastewater (Das et al., 2012).

Biofilm can also be used for remediation of waste plastic or synthetic dyes removal from wastewater (Ong et al., 2011). These processes use fixed or mobile supports that may develop biofilms involved in treating pollution. For urban wastewater treatment, this type of process can replace the activated sludge process by reducing the size of the aeration tanks and removing sedimentation stage.

The biofilter is a filter in which on the surface of the filter material develops biological films (biofilms) consisting of bacteria which feed with certain pollutants from the water. In addition to the phenomenon of reducing biological pollutants, developed biofilms shrink the pores of filter of the material so that the quality of filtration of biofilters is improved, reaching in microfiltration and ultrafiltration domain (Gimbel et al., 2006). Thus, in one multilayer, biofilter can be reduced concentrations of a wide range of pollutants. The process of pollutants biofiltration from drinking water was highlighted by chance in some stations of potable water treatment when in certain periods reagents have not been added in chemical treatment step and washing filters was made at greater intervals of time, but a positive evolution of processed water parameters in time was still observed.

The development of biofilm on the solid surface in contact with the water has four stages: (1) conditioning film formation; (2) initial adherence of microorganisms at the surface; (3) biofilm development; (4) biofilm maturation—an equilibrium state between accumulation and detachment (Dunne, 2002).

Bacteria found in the transport and processing sections of drinking water are related to the substances dissolved in water, that along with the carbon are used as nutrients. These are: (1) bacteria involved in nitrogen cycle and in particular the nitrification and denitrification ones; (2) bacteria involved in cycle of iron and manganese; (3) bacteria from the sulfur cycle (Duarte, 2007).

Depending on the purpose, the biofilter can operate as a nitrification or denitrification filter. Filter materials constituting the biofilter (e.g., quartz sand) are capable also to retain other pollutants from water, such as iron and manganese in a higher oxidized form after the preoxidation step of water. In the biofilter, at the interface between reduced forms of iron and manganese (dissolved) and oxidized forms-precipitated, biofilms produce extracellular polymers which engulf iron and manganese oxides, keeping them with great efficiency.

In the biofiltration step, conversion of ammonium ions to nitrates takes place, by enabling conditions for nitrification, followed by another step in which nitrates are reduced to nitrogen gas through a process called denitrification (Chaudhary et al., 2003).

10 Conclusions

Using biofilters in order to significantly reduce pollutants (iron ions, hydrogen sulfide, and arsenic; manganese, ammonia, nitrites, nitrates, humic acids, pesticides, herbicides, and petroleum products; residues of pharmaceuticals or cosmetics) from water sources designed to provide potable water with lower energy consumption, represents a significant example for the beneficial exploitation of water biofilms.

References

Ali, M., Lopez, A.L., Young, A., Young, E.K., Binod, S., Brian, M., Clemens, J., 2012. The global burden of cholera. Bull. World Health Organ. 90, 209–218.

An, D., Parsek, M.R., 2007. The promise and peril of transcriptional profiling in biofilm communities. Curr. Opin. Microbiol. 10 (3), 292–296.

Araya, R., Tani, K., Takagi, T., Yamaguchi, N., Nasu, M., 2003. Bacterial activity and community composition in stream water and biofilm from an urban river determined by fluorescent in situ hybridization and DGGE analysis. FEMS Microbiol. Ecol. 43, 111–119.

Block, J.C., Haudidier, K., Paquin, J.L., Miazgaet, J., Lévi, Y., 1993. Biofilm accumulation in drinking water distribution systems. Biofouling 6, 333–343.

Borkar, R.P., Gulhane, M.L., Kotangale, A.J., 2013. Moving bed biofilm reactor—a new perspective in wastewater treatment. J. Environ. Sci. Toxicol. Food Technol. 6 (6), 15–21.

Cabulea, P.C., in press. Available from: https://ro.scribd.com/doc/100935059/Infectii-hidrice.

Chaudhary, D.S., Vigneswaran, S., Ngo, H.H., Shim, W.G., Moon, H., 2003. Biofilter in water and wastewater treatment. Korean J. Chem. Eng. 20 (6), 1054–1065.

Chifiriuc, M.C., Mihaescu, G., Lazar, V., 2011. Medical Microbiology and Virology. University of Bucharest House, Bucharest, 745 p..

Chifiriuc, M.C., Grumezescu, A.M., Lazar, V., 2014. Quorum sensing inhibitors from the sea: lessons from marine symbiotic relationships. Curr. Org. Chem. 18 (7), 823–839.

Costerton, J.W., 2009. Marine and industrial biofouling. In: Springer Series on Biofilms, vol. 4. Springer Verlag, Berlin Heidelberg, pp. 3–11.

Costerton, J.W., Cheng, K.J., Gessey, G.G., Ladd, T.I., Nickel, J.C., Dasgupta, M., Marrie, T.J., 1987. Bacterial biofilms in nature and disease. Annu. Rev. Microbiol. 41, 435–464.

Cullimore, D.R., 1978. The identification, cultivation and control of iron bacteria in ground water. In: Skinner, F.A., Shewan, J.M. (Eds.), Aquatic Microbiology. Academic Press, New York.

Das, N., Basak, L.V.G., Salam, J.A., Abigail, M.E.A., 2012. Application of biofilms on remediation of pollutants. J. Microbiol. Biotech. Res. 2 (5), 783–790.

Davey, M.E., O'Toole, G.A., 2000. Microbial biofilms: from ecology to molecular genetics. Microbiol. Mol. Biol. Rev. 64, 847–867.

Davies, D.G., Chakrabarty, A.M., Geesey, G.G., 1993. Exopolysaccharide production in biofilms: substratum activation of alginate gene expression by *Pseudomonas aeruginosa*. Appl. Environ. Microbiol. 59, 1181–1186.

Davies, D.G., Parsek, M.R., Pearson, J.P., Iglewski, B.H., Costerton, J.W., Greenberg, E.P., 1998. The involvement of cell-to-cell signals in the development of a bacterial biofilm. Science 280, 295–298.

De Beer, D., Stoodley, P., 2006. Microbial biofilms. In: Dworkin, M. (Ed.), The Prokaryotes. Springer, New York, pp. 904–937.

Delden, C.V., Iglewski, B.H., 1998. Cell-to-cell signaling and *Pseudomonas aeruginosa* infections. Emerg. Infect. Diseases 4 (4), 551–558.

Donlan, R.M., 2002. Biofilms: microbial life on surfaces. Emerg. Infect. Diseases 8 (9), 881–890.

Douglas, C.W., van Noort, R., 1993. Control of bacteria in dental water supplies. Br. Dent. J. 174 (5), 167–174.

Du, M., Chen, J., Zhang, X., Li, A., Li, Y., Wang, Y., 2007. Retention of virulence in a viable but nonculturable *Edwardsiella tarda* isolate. Appl. Environ. Microbiol. 73 (4), 1349–1354.

Duarte, M.C.M., 2007. Biofilm dynamics and drinking water stability: effects of hydrodynamics and surface materials. Dissertation submitted for the degree of Doctor in Sciences of Engineering.

Dunne, W.M., 2002. Bacterial adhesion: seen any good biofilms lately? Clin. Microbiol. Rev. 15 (2), 155–166.

Echverría, F., Castaño, J.G., Arroyave, C., Peñuela, G., Ramírez, A., Morató, J., 2009. Caracterization of deposits formed in a water distribution system. Ingeniare Revista Chilena de Ingeniería 17, 275–281.

Fayle, S.A., Pollard, M.A., 1996. Decontamination of dental unit water systems: a review of current recommendations. Br. Dent. J. 181 (10), 369–372.

Finkel, J.S., Mitchell, A.P., 2011. Genetic control of *Candida albicans* biofilm development. Nat. Rev. Microbiol. 9, 109–118.

Flemming, H.C., 2009. Why microorganisms live in biofilms and the problem of biofouling. In: Marine and Industrial Biofouling. In: Springer Series on Biofilms, vol. 4. Springer, Berlin Heidelberg. pp. 3–12.

Flemming, H.C., Percival, S.L., Walker, J.T., 2002. Contamination potential of biofilms in water distribution systems. Water Sci. Technol. 2 (1), 271–280.

Fonseca, A.C., Summers, R.S., Hernandez, M.T., 2001. Comparative measurements of microbial activity in drinking water biofilters. Water Res. 35 (16), 3817–3824.

Gauthier, F., 2001. Biofilm et qualité de l'eau potable. University of Picardie, Amiens.

Gianfreda, L., Rao, M.A., 2004. Potential of extra cellular enzymes in remediation of polluted soils. Enzyme Microb. Technol. 35, 339–354.

Gimbel, R., Graham, N., Collins, R., 2006. Recent Progress in Slow Sand and Alternative Biofiltration Processes. IWA Publishing, London.

Guardabassi, L., Dalsgaard, A., 2004. Occurrence, structure, and mobility of Tn1546-like elements in environmental isolates of vancomycin-resistant enterococci. Appl. Environ. Microbiol. 70, 984–990.

Hall-Stoodley, L., Costerton, J.W., Stoodley, P., 2004. Bacterial biofilms: from the natural environment to infectious diseases. Nat. Rev. Microbiol. 2, 95–108.

Hengge-Aronis, R., 2000. The general stress response in *Escherichia coli*. In: Storz, G., Hengge-Aronis, R. (Eds.), Bacterial Stress Responses. ASM Press, Washington, DC, pp. 161–178.

Husham, T.I., He, Q., Wisam, S., Al-Rekabi, Yang, Q., 2012. Improvements in biofilm processes for wastewater treatment. Pak. J. Nutr. 11 (8), 610–636.

Israil, A.M., Chifiriuc, M.C., 2009. Bacterial communication: new concepts in the antimicrobial therapy. Asclepius Publishing House, Bucharest, Romania, 110p.

Johnson, L., 2007. Microcolony and biofilm formation as a survival strategy for bacteria. J. Theor. Biol. 251, 24–34.

Jones, K., 1997. Strategies for survival. In: Kay, D., Fricker, C. (Eds.), Coliforms and *E. coli*. Problem or Solution?. The Royal Society of Chemistry, Cambridge.

Karatan, E., Watnick, P., 2009. Signals, regulatory networks, and materials that build and break bacterial biofilms. Microbiol. Mol. Biol. Rev. 73 (2), 310–347.

Kjelleberg, S., 1993. Starvation in Bacteria. Plenum Press, New York.

Lanata, C.F., Mendoza, W., Black, R.E., 2002. Improving Diarrhoea Estimates. World Health Organization, Geneva.

Långmark, J., Storey, M.V., Ashbolt, N.J., Stenström, T.A., 2004. Artificial groundwater treatment: biofilm activity and organic carbon removal performance. Water Res. 38, 740–748.

Lazăr, V., Chifiriuc, M.C., 2010a. Architecture and physiology of microbial biofilms. Roum. Arch. Microbiol. Immunol. 69 (3), 92–98.

Lazăr, V., Chifiriuc, M.C., 2010b. Medical significance and new therapeutical strategies for biofilm associated infections. Roum. Arch. Microbiol. Immunol. 69 (3), 121–127.

Lazăr,V., Chifiriuc, M.C., Curuțiu, C., Mitache, M.M., Marinescu, F., Croitoru, C., Mateescu, L., Ma˘ruțescu L., 2015. Methods and Standards for the Microbiological Control Laboratories. University from Bucharest, Romania. p. 196.

LeChevallier, M.W., Au, K.K., 2004. Water Treatment and Pathogen Control. IWA Publishing, London, Available from: http://www.who.int/water_sanitation_health/dwq/en/watreatpath.pdf.

Lechevallier, M.W., Welch, N.J., Smith, D.B., 1996. Full-scale studies of factors related to coliform regrowth in drinking water. Appl. Environ. Microbiol. 62, 2201–2211.

Leclerc, H., 2003. Relationships between common water bacteria and pathogens in drinking-water World Health Organization (WHO). In: Bartram, J., Cotruvo, J., Exner, M., Fricker, C., Glasmacher, A. (Eds.), Heterotrophic Plate Counts and Drinking-Water Safety. IWA Publishing, London.

Lehtinen, J., 2007. Improvements in the assessment of bacterial viability and killing. *SARJA—SER. A I OSA—TOM. 37* ASTRONOMICA–CHEMICA—PHYSICA—MATHEMATICA TURUN YLIOPISTO. TURUN YLIOPISTON JULKAISUJA ANNALES UNIVERSITATIS TURKUENSIS. Painosalama Oy – Turku, Finland, pp. 11–17.

Li, L., Mendis, N., Trigui, H., Oliver, J.D., Faucher, S.P., 2014. The importance of the viable but non-culturable state in human bacterial pathogens. Front. Microbiol. 5, 258.

Lupo, A., Coyne, S., Berendonk, T.U., 2012. Origin and evolution of antibiotic resistance: the common mechanisms of emergence and spread in water bodies. Front. Microbiol. 26 (3), 18.

Ma'ayeh, S.Y., Al-Hiyasat, A.S., Hindiyeh, M.Y., Khader, Y.S., 2008. *Legionella pneumophila* contamination of a dental unit water line system in a dental teaching centre. Int. J. Dent. Hyg. 6 (1), 48–55.

Martin, M.V., 1987. The significance of the bacterial contamination of dental unit water systems. Br. Dent. J. 163 (5), 152–154.

Meiller, T.F., Depaola, L.G., Kelley, J.I., Baqui, A.A., Turng, B.F., Falkler, W.A., 1999. Dental unit waterlines: biofilms, disinfection and recurrence. J. Am. Dent. Assoc. 130 (1), 65–72.

Morita, R.Y., 1997. Bacteria in Oligotrophic Environments. Starvation Lifestyle. Chapman and Hall, New York, 529 pp..

Moritz, M.M., 2011. Integration of hygienically relevant bacteria in drinking water biofilms on domestic plumbing materials. Teza˘ de Doctorat. Biofilm Centre. Duisburg-Essen University, Germania, 2; pp. 21–24; pp. 123–126; pp. 134–137.

Nyström, T., 1993. Global systems approach to the physiology of the starved cell. In: Kjelleberg, S. (Ed.), Starvation in Bacteria. Plenum Press, New York, pp. 129–146.

Oliver, J.D., 1993. Formation of viable but nonculturable cells. In: Kjelleberg, S. (Ed.), Starvation in Bacteria. Plenum Press, New York, pp. 239–272.

Ong, S.T., Keng, P.S., Lee, W.N., Ha, S.T., Hung, Y.T., 2011. Dye waste treatment. Water 3 (1), 157–176.

Pederson, E.D., Stone, M.E., Ragain, Jr., J.C., Simecek, J.W., 2002. Waterline biofilm and the dental treatment facility: a review. Gen. Dent. 50 (2), 190–195.

Roeselers, G., Loosdrecht, M.C.V., Muyzer, G., 2008. Phototrophic biofilms and their potential applications. J. Appl. Phycol. 20 (3), 227–235.

Schulze, E.D., Mooney, H.A., 1993. Biodiversity and Ecosystem Function. Springer-Verlag, pp. 88–90.

Schulze-Röbbecke, R., Feldmann, C., Fischeder, R., Janning, B., Exner, M., Wahl, G., 1995. Dental units: an environmental study of sources of potentially pathogenic mycobacteria. Tuber. Lung Dis. 76 (4), 318–323.

Schutte, F., 2006. The Water Research Commission, The Water Institute of Southern Africa.In: Water Utilisation Division, Department of Chemical Engineering, University of Pretoria. TT 265/06 March 2006.

September, S., Elsa, F., Venter, S., Brozel, V., 2007. Prevalence of bacterial pathogens in biofilms of drinking water distribution systems. Water Health 5, 219–227.

Sharma, S., 2012. Bioremediation: features, strategies and applications. J. Pharm. Life Sci. 2, 202.

Simões, L.C., Simões, M., Vieira, M.J., 2011. The effects of metabolite molecules produced by drinking water-isolated bacteria on their single and multispecies biofilms. Biofouling (7), 685–699.

Squinazi, F., 2004. Biofilm et matériaux des canalisations des réseaux de distribution d'eau. Laboratoired'hygiène de la ville de Paris.

Szymanska, J., Sitkowska, J., Dutkiewicz, J., 2008. Microbial contamination of dental unit waterlines. Ann. Agric. Environ. Med. 15, 173–179.

Tellen, V., Nkeng, G., Dentel, S., 2010. Improved filtration technology for pathogen reduction in rural water supplies. Water 2, 285–306.

Tool Kumar, A., Bisht, B.S., Joshi, V.D., Dhewa, T., 2011. Bioremediation of polluted environment: a management. Int. J. Environ. Sci. 1 (6).

Walker, J.T., Bradshaw, D.J., Bennett, A.M., Fulford, M.R., Martin, M.V., Marsh, P.D., 2000. Microbial biofilm formation and contamination of dental-unit water systems in general dental practice. Appl. Environ. Microbiol. 66 (8), 3363–3367.
Yang, D., Wang, J., Qiu, Z., Jin, M., Shen, Z., Chen, Z., Wang, X., Zhang, B., Li, J.W., 2013. Horizontal transfer of antibiotic resistance genes in a membrane bioreactor. J. Biotechnol. 167 (4), 441–447.
Zarnea, G., Popescu, O.V., 2011. Dictionnaire of general microbiology and molecular biology. Romanian Academy Publishing House, Bucharest, Romania, pp. 187–188.

19

REMOVAL OF LEAD IONS FROM WATER AND WASTEWATERS ELECTROCHEMICALLY

Isaiah A. Oke*,[a], Salihu Lukman,†, Abubakar Ismail†, Ezekiel O. Fehintola‡, Justinah S. Amoko‡**

**Department of Civil Engineering, Obafemi Awolowo University, Ile-Ife, Nigeria; **Department of Civil Engineering, University of Hafr Al-Batin, Hafr Al-Batin, Saudi Arabia; †Department of Water Resources and Environmental Engineering, Ahmadu Bello University, Zaria, Nigeria; ‡Department of Chemistry, Adeyemi College of Education, Ondo, Obafemi Awolowo University, Ile-Ife, Nigeria*

1 Introduction

Lead is among heavy metals in group IVB and period VI of the periodic table. Lead occurs in nature as galena, and as lead sulfide. It is distributed widely in the earth's crust. It is used in the manufacturing of water and gas pipes, lead sheet for roofing and as alloys for engineering uses. Lead is used in the manufacturing of electrical accumulators and paints. Lead in waters originated from wastes from storage and battery manufacture environment. Lead is cumulative poison, it damages the brain of young children and may cause death, it causes severe malfunctioning of the alimentary tracts, general weakness, and malaise impairs the functioning of the nervous system and sometimes paralysis of the exteriors muscles of the forearm.

Literature classifies inorganic chemicals found in water into four categories, namely:

1. *Type A inorganic chemicals*: these are chemicals found in wastewaters, which are known to be toxic to man. They are arsenic, chromium, mercury asbestos, copper, nitrate, nitrite, barium, fluoride, selenium, lead, and cadmium.
2. *Type B inorganic chemicals*: these are parameters that have been examined by the health authorities. They are rare in

[a]Senior author, currently on Sabbatical leave at Department of Civil Engineering, Elizade University, Ilara- Mokin, Ondo State, Nigeria.

Water Purification. http://dx.doi.org/10.1016/B978-0-12-804300-4.00019-8

drinking waters, and their toxicity is of limited concern. They are aluminum, nickel, sodium, silver, zinc, sulfate, molybdenum, and cyanide.

3. *Type C inorganic chemicals*: they are parameters that may have a high level of occurrence in wastewaters and waters, but are safe at these concentrations. They are expected, but their occurrence is rare and extremely limited. They are calcium, silicon (as silica), magnesium, carbon dioxide, manganese, bromine, chlorides, dissolved oxygen, bromide, iron, chlorine, lithium, phosphate, iodine, iodide, Ozone, and potassium.
4. *Type D inorganic chemicals*: these are inorganic parameters that are likely to be found in very low concentrations in wastewaters and waters with no toxicity at the low concentrations.

These contaminants (inorganic chemicals) originate from different sources in the community, namely: industrial and commercial sources, stormwater, infiltration from cracked pipes and leaky manhole covers.

Sources of lead in the environment: human beings, animals, and soil are the ultimate recipients of the lead particulate. It sometimes exists in the soil as lead (II) sulfate ($PbSO_4$). Concentrations of about 100–1000 ppm have been recorded depending on the nature of the activities carried out in a particular area (Galadima et al., 2010, 2011; Garba et al., 2010; Galadima and Garba, 2011, 2012). Other anthropogenic sources include mining and metallurgic industries, manufacture of batteries, sheet, ammunition, pipe, cable sheeting, solder, saint, and trash incineration. The principal route of exposure for people in the general population is food and lead in contaminated drinking water, working and hand to mouth activities of young children living in polluted environments and the lead dust brought home by industrial workers on their clothes and shoes. Paints containing lead are the most common high dose sources of lead exposure for school and preschool children. Most of them contain up to 50% of lead in the form of lead sulfide (PbS). Children can get seriously lead poisoned when renovations, modeling, and construction activities take place in a house or class that contains lead paints. Inhalation or swallowing of debris of the paints during regular playing causes the accumulation of the metal in the children body.

Effect of lead on man: lead exposure in young children has been linked to learning disabilities and children with blood lead concentrations greater than 10×10^{-2} mg/L are in danger of developmental disabilities. Increased blood lead level in children has correlated with decreases in intelligence quotient (IQ), nonverbal reasoning, short-term memory, attention, reading and arithmetic ability, fine motor skills, emotional regulation, and social engagement. The

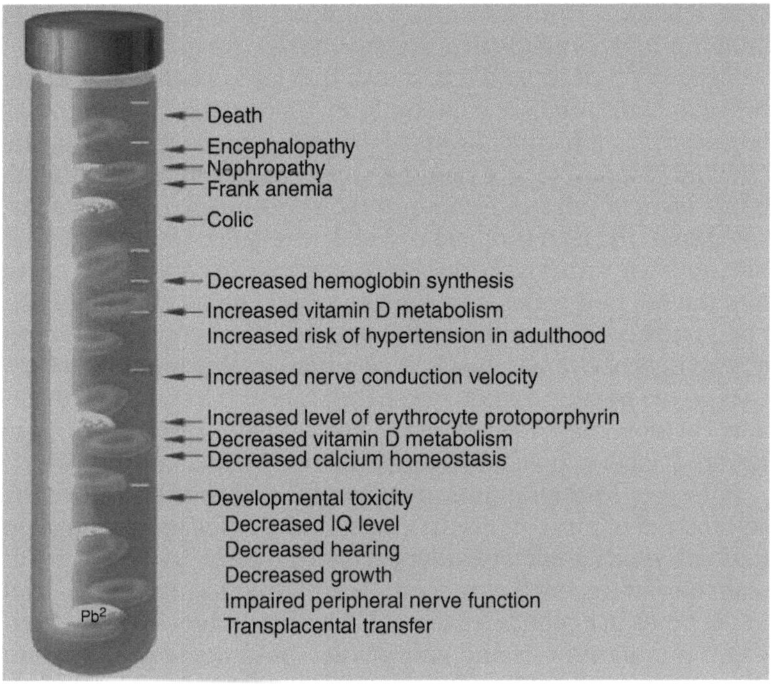

Figure 19.1. Paediatric effects of lead at various blood lead levels. WHO 2010 reproduced with ID 202358.

effect of lead on children's cognitive abilities takes place at very low levels. There is apparently no lower threshold to the dose–response relationship. Reduced academic performance has been associated with lead exposure even at blood lead levels lower than 5×10^{-2} mg/L. Blood lead levels below 10×10^{-2} mg/L have been reported to be associated with lower IQ and behavior problems, such as aggression, in proportion with blood lead levels (Wikipedia, 2015a). Fig. 19.1 shows pediatric effects of lead in the blood. The major health effects of lead in adults are manifest in three organ systems: the haematological system, the central nervous system, and the renal system failures. Lead also interferes with the last stage of heme synthesis, the incorporation of iron into protoporphyrin, catalyzed by hemesynthetase (ferrochelatase).

This step takes place in the inner matrix of the erythroid cell mitochondria in the bone marrow. It inhibits the transmitochondrial transfer of iron. This effect results in the accumulation of protoporphyrin in the mitochondria, its incorporation into the globin molecule in place of haem and the presence of elevated levels of free erythrocyte protoporphyrin in the peripheral blood. The principal organs affected are the central and peripheral nervous system and the cardiovascular, gastrointestinal, renal, endocrine, immune, and hematological systems (WHO, 2010)

Effects of lead on soil: It is known that lead accumulates in the soil with a high organic content. Lead deposited on the ground is transferred to the upper layers of the soil surface, where it may be retained for many years. In undisturbed soil, organic matter in the upper layer of soil surface retains the atmospheric lead). In cultivated soils, the lead is mixed with soil to a depth of 25 cm (within the root zone). Atmospheric lead in the soil will continue to move into the microorganisms and grazing food chains until equilibrium is reached. Literature highlighted that the uneven distribution of lead in ecosystems can displace other metals from the binding soils on the organic matter, which may hinder the chemical breakdown of inorganic soil fragments. Soluble leads in the soil are more readily available to be taken up by plants (Wikipedia, 2015a,b).

Effects of lead on plants: plants tend to absorb lead from the soil and retain most of this in their roots. Plant foliage may take up the lead, which moved to other parts of the plant. Some species of plant have the capacity to accumulate high concentrations of lead in the pores in a plant's leaves which let the carbon dioxide needed for photosynthesis and emit oxygen from the leaves. Lead on the surface of the leaf and reduces the amount of light reaching it. This result in stunting the growth or killing the plants by reducing the rate of photosynthesis, inhibiting respiration, encouraging an elongation of plant cells influencing root development; by causing premature aging. Lead can affect population genetics. The pattern and degree of lead accumulation on plants are largely influenced by the state of growth of the vegetation (Wikipedia, 2015a,b).

Effects of lead on microorganisms: literature has shown that lead at the concentrations occasionally found near roadsides (10,000–40,000 ppm dry weight), can wipe out populations of bacteria and fungi on leaf surfaces and in soil. This can have a significant impact, given that many of these microorganisms are an essential part of the decomposing food chain. The microorganism populations affected are likely to be replaced by others of the same or different species, although these may be less efficient at decomposing organic matter. It has been suggested that microorganisms can make the lead more soluble and hence more easily absorbed by plants (Wikipedia, 2015a).

Effects of lead on animals: lead affects the central nervous system of animals and inhibits their ability to synthesize red blood cells. Lead blood concentrations of above 40×10^{-2} mg/L can produce observable clinical symptoms in domestic animals. Calcium and phosphorus can reduce the intestinal absorption of lead. The literature reported that a regular diet of 2–8 mg of lead per kilogram of body weight per day, over an extended period, will cause death in most animals. Grazing animals are directly affected by the consumption of forage and feed contaminated by airborne lead and somewhat indirectly by the uptake of lead through plant roots. Invertebrates may also accumulate lead at levels toxic to their predators. Lead shot and lead weight can severely affect individual organisms and threaten ecosystems. After 3–10 days of waterfowl ingesting lead shot, the poison will reach the bloodstream and be carried to major organs, like the heart, liver, and kidneys. By the 17th–21st day, the bird falls into a coma and dies. Following the ingestion of lead shot, lead toxicosis has been observed in Magpie geese, Black swans, several species of duck (Black duck and Musk duck), and Hardhead species. Organic lead is much more readily taken up by birds and fish. Aquatic organisms take up inorganic lead through a transfer of lead from water and sediments; this is a relatively slow process. Organic lead is rapidly taken up by aquatic organisms from water and sediment. Aquatic animals are affected by lead at water concentrations lower than previously thought safe for wildlife. These concentrations occur often, but the impact of atmospheric lead on specific sites with high aquatic lead levels is not clear (Wikipedia, 2015a).

Other species: humans are not alone in suffering from lead's effects; plants and animals are also affected by lead toxicity to varying degrees depending on species. Animals experience many of the same effects of lead exposure as humans do, such as abdominal pain, peripheral neuropathy, and behavioral changes, such as increased aggression. Animals are used to test the effects of treatments (to provide information on the pathophysiology of lead and how it is absorbed and distributed in the body). Farm animals, such as cows and horses, as well as pet animals are also susceptible to the effects of lead toxicity. Sources of lead exposure in pets can be the same as those that present health threats to humans sharing the environment, such as paint and blinds, and there is sometimes lead in toys made for pets. Lead poisoning in a pet dog may indicate that children in the same household are at increased risk for elevated lead levels (Wikipedia, 2015a).

Wildlife: lead in water and wastewaters are one of the leading causes of toxicity in waterfowl, has been known to cause die-offs of wild bird populations. It has been reported that when hunters use lead shot, waterfowl, such as ducks can ingest the spent pellets later and be poisoned; predators that eat these birds are also at risk (Ferreyra et al., 2009). Lead shot-related waterfowl poisonings were first documented in the 1880s and by 1919, the spent lead pellets from waterfowl hunting were positively identified as the source of waterfowl deaths. Lead shot has been banned for hunting waterfowl in several countries. Other forms of lead that are threats to wildlife in the environment include lead paint, sediment from lead mines and smelters, and lead weights from fishing lines. Lead in some fishing gear has been banned in several countries. The critically endangered California condor has also been affected by lead poisoning. As scavengers, condors eat carcasses of the game that have been shot but not retrieved, and with them the fragments from lead bullets; this increases their lead levels. Among condors around the Grand Canyon, lead poisoning due to eating lead shot is the most frequently diagnosed cause of death. In an effort to protect these species, in areas designated as the California condor's range the use of projectiles containing lead has been banned (from hunting deer, feral pigs, elk, pronghorn antelope, coyotes, ground squirrels, and other nongame wildlife). Also, conservation programs exist which routinely capture condors, check their blood lead levels, and treat cases of poisoning (Wikipedia, 2015a). More on lead poisoning can be found in literature, such as Susan et al. (2015), Adie et al. (2010), and Eman et al. (2016).

Lead removal techniques: the techniques for lead ion removal ranges from physical and chemical treatments through membrane and irradiation to electrochemical, which can be described as follows: distillations are simply processes of evaporation followed by condensation whereby separation of volatile materials can be optimised by controlling both evaporation temperature and the condenser's temperature. Distillation separates miscible liquids for solvent reclamation. There are four most common distillation processes. They are batch and fractional distillations, steam stripping and thin film evaporation. Batch distillation can be used to purify waste containing a relatively high percentage of volatile organics. Batch distillation applies to wastes that have both volatile and nonvolatile components. Batch distillation can be applied to spent solvent wastes where the wastes are highly concentrated in the solvent and yielded significant amounts of recoverable materials upon separation. Fractional distillation applies to wastes containing greater than 7% organics. It is designed to achieve the highest degree of distillate purity of the separated components. It can be operated to produce multiple product streams for recovery of more than one organic constituent from the waste. It generates relatively small amounts of residue to be disposed of. In general, this physical treatment process (fractional distillation) is used where recovery of multiple constituents is desired and where the waste contains a minimum amount of suspended solids.

Chemical treatment processes are the addition of a chemical (liquid, solid, or gas) to wastewaters in order to remove some pollutants from wastewater or reduce the concentration of selected pollutants in the wastewater. The literature describes chemical methods as those processes used for the treatment of wastewater in which changes are brought about through chemical addition. The chemical treatment process is used to effect a chemical transformation of the waste to an innocuous or less toxic form. The amount of chemical that must be added to the wastewaters during chemical treatment vary with the following key factors: the quality of wastewater; the degree of treatment (removal) required; treatment conditions (pH adjustment, oxidation, reduction, and hydrolysis) and the quantity of the wastewater to be treated. Chemical treatment processes depend on the chemical properties of the impurities or utilize the chemical properties of added reagents. The addition of lime, alum, or ferric chloride to either effluent from a biological treatment process or untreated wastewater can achieve 50–95% removal of phosphorus. Some of the processes commonly employed as chemical treatment processes are adsorption, dehalogenation (dechlorination and defluorination), hydrolysis, neutralization, oxidation, reduction, chemical disinfections, and precipitation. The advantages of chemical treatment processes are well documented in the literature. They include the following:

1. Reliability of the process with proper operational control (it is capable of producing consistently high effluent quality).
2. The flexibility of the process (treatment efficiencies can be varied as needed).
3. The process is suitable for treatment or pretreatment of wastewaters containing toxic or otherwise refractory chemicals.

The disadvantages of chemical treatment processes also include the following:

1. The processes are of the relatively high cost if operated on a full-time basis.
2. The processes increase the amount of sludge generated thereby increasing sludge treatment and sludge disposal costs.
3. The processes generate high concentrations of chemical and or heavy metals or other toxic materials in the sludge may render the sludge unsatisfactory for use as a fertilizer and or soil conditioner and skilled labor is required for operation and maintenance of systems.
4. Certain mechanical equipment and chemicals required in the process may not readily available in developing countries.

Adsorption has been defined as the accumulation of dissolved particles from a solvent onto the surface of an absorbent. There are two types of adsorption, namely: chemical and physical adsorptions. Physical adsorption results from molecular condensation in the capillaries of the solid. Chemicals adsorption results in the formation of a monomolecular layer of adsorbate on the surface through forces of residual valence from the surface molecules. The adsorptive capacity of any material for a solute depends on both the material and the solute. The overall rate of absorption is controlled by the rate of diffusion of the solute molecules within the capillary pores of the adsorbent particles. The rate of adsorption varies reciprocally with the square of the particle diameter and contact time. It increases with increasing concentration of solute and temperature. Adsorption decreases with increasing molecular weight of the solute and with the pH of the solution. Adsorption is important in the removal of volatile organic compounds from the gas stream. It has been documented (Olarinoye et al., 2012) that there are three basic mathematical models of adsorption that relate the adsorption capacity to the equilibrium concentration of solute remaining in solution. The models are Brunauer, Emmett, and Teller (BET), Langmuir and Freundlich isotherms. The other models include Temklin and Redlich-Peterson isotherms.

Langmuir isotherm was developed from a theoretical consideration of adsorption based on the concept of equilibrium in a monomolecular surface layer. Langmuir isotherm can be expressed mathematically as shown in Eq. (19.1).

$$\frac{X_{adb}}{M_{adb}} = \frac{abC_{adb}}{1 + aC_{adb}} \tag{19.1}$$

Where a is the absorption constant; b is the absorption constant; C_{adb} is the concentration of the solute at equilibrium (mg/L); M_{adb} is the mass of adsorbent (mg) and X_{adb} is the mass of solute adsorbed (mg).

It has been reported that Langmuir isotherm has been modified for competitive adsorption of two substances as shown in Eq. (19.2).

$$\frac{X_a}{M_{adb}} = \frac{a_a b_a C_a}{1 + a_a C_a + a_b C_b} \tag{19.2}$$

Where a_a is the absorption constants for solute A; b_a is the absorption constants for solute A; C_a is the concentration of the solute A remaining at equilibrium (mg/L) and K_{adb} is the constant for adsorption.

The Freundlich equation is an empirical relationship, which often gives a more satisfactory model for experimental data. It can be expressed as Eq. (19.3).

$$\frac{X_{adb}}{M_{abd}} = K_{adb} C_{adb}^{1/N_{adb}} \tag{19.3}$$

Where K_{adb} is the constant for adsorption and N_{adb} is the imperial number for adsorption

Freundlich isotherm has been modified for multiple components adsorption isotherm. BET isotherm assumes that layers of molecules are adsorbed on top of previously adsorbed molecules. It can be expressed as:

$$\frac{C_{adb}}{C_{sa}-C_{adb}}\left(\frac{M_{adb}}{X_{adb}}\right)=\frac{1}{AX_m}+\left(\frac{A-1}{AX_m}\right)\frac{C_{adb}}{C_{sa}} \tag{19.4}$$

Where A is the concentration of reactant A (mg/L); C_{sa} is the saturation concentration of adsorbate (mg/L), and X_m is the number of moles of adsorbent adsorbed in forming a complete monolayer on the adsorbent surface

The study of adsorption dynamics describes the solute uptake rate, and evidently this rate controls the residence time of adsorbate uptake at the solid–solution interface. The kinetics of Cr (VI) adsorption on the activated carbons have been analyzed using pseudo-first-order, pseudo second-order, Elovich and intraparticle diffusion kinetic models. The conformity between experimental data and the model predicted values was expressed by the correlation coefficients (R^2, values close or equal to 1). A relatively high R^2 value indicates that the model successfully describes the kinetics of Cr (VI) adsorption. The pseudo-first-order equation is expressed as follows:

$$\frac{dq}{dt}=k_1(q_e-q_t) \tag{19.5}$$

Where k_1 is the rate constant of pseudo-first-order adsorption (l/h); q_e and q_t are the adsorption capacity at equilibrium and at time t respectively (mg/L·g),

After integration and applying boundary conditions $t=0$ to $t=t$ and $q_t=0$ to $q_t=q_t$, the integrated form of Eq. (19.5) becomes:

$$\log_e(q_e-q_t)=\log_e(q_e)-\frac{k_1}{2.303}t \tag{19.6}$$

The values of $\log_e(q_e-q_t)$ can be linearly correlated with t. The plot of $\log_e(q_e-q_t)$ against t should give a linear relationship from which k_1 and q_e can be determined from the slope and intercept of the respectively. The pseudo-second-order adsorption kinetic rate equation is expressed as:

$$\frac{dq}{dt}=k_2(q_e-q_t)^2 \tag{19.7}$$

Where k_2 is the rate constant of pseudo-second-order adsorp (g/mg·h) for the boundary conditions $t = 0$ to $t = t$ and $q_t = 0$ to $q_t = q_t$, the integrated form of Eq. (19.8) becomes:

$$\frac{1}{q_e - q_t} = \frac{1}{q_e} + k_2 t \tag{19.8}$$

which is the integrated rate law for a pseudo-second-order reaction. Eq. (19.8) can be rearranged to obtain Eq. (19.9), which has a linear form as follows:

$$\left(\frac{t}{q_t}\right) = \frac{1}{k_2 q_e^2} + \frac{1}{q_e} t \tag{19.9}$$

if the initial adsorption rate, h_{adb} (mg/g·h) is $h_{adb} = k_2 (q_e)^2$ then Eqs. (19.8) and (19.9) become:

$$\left(\frac{t}{q_t}\right) = \frac{1}{h_{adb}} + \frac{1}{q_e} t \tag{19.10}$$

the plot of (t/q_t) and t of Eq. (19.10) should give a linear relationship from which q_e and k_2 can be determined from the slope and intercept of the plot respectively.

The Elovich model equation is generally expressed as follows:

$$\frac{dq_t}{dt} = \alpha \exp(-\beta q_t) \tag{19.11}$$

where α is the initial adsorption rate (mg/g·h) for Elovich model; β is the desorption constant (g/mg) during any one experiment and exp is the exponential.

In order to simplify the Elovich equation, It has been assumed $\alpha\beta t >> t$ and by applying the boundary conditions $q_t = 0$ at $t = 0$ and $q_t = q_t$ at $t = t$ Eq. (19.11) becomes:

$$q_t = \frac{1}{\beta} \log_e(\alpha\beta) + \frac{1}{\beta} \log_e(t) \tag{19.12}$$

if Cr(VI) adsorption fits the Elovich model, a plot of qt versus $\log_e(t)$ should yield a linear relationship with a slope of $(1/\beta)$ and an intercept of $(1/\beta)\log_e(\alpha\beta)$. The intraparticle diffusion model is expressed as follows:

$$R_{adb} = k_{id} (t)^{ai} \tag{19.13}$$

a linearized form of the equation is followed by $\log R = \log k_{id} + a \log(t)$.

Over the years, the role of adsorption in wastewaters and water treatment has not often been underestimated (Oke et al., 2014).

Membrane: membrane treatment processes are those processes used in the treatment of wastewater in which change is brought about using the semipermeable substance. Membrane treatment processes are used on a large scale to produce water from seawater by reverse osmosis, to clean industrial wastewaters and to recover valuable constituents by electrodialysis, to fractionate macromolecular solutions in the wastewater by ultrafiltration. Although membrane treatment processes are very different in their mode of operations, in the structures used as separating barriers, and in the driving forces used for the transport of the different chemical components, they have several common separation features. These separation features of membranes make them attractive as separation techniques. Membrane treatment processes are faster, more efficient and more economical than conventional separation techniques. Another advantage of membranes over conventional techniques is that their separation techniques are usually performed at ambient temperature, thus allowing sensitive temperature pollutants to be removed without the constituents being damaged or chemically altered. The design of an energy-effective membrane always involves a trade-off between two major factors: selectivity and permeability.

Dialysis is one of the several membrane separation techniques. It consists, in essence, a barrier, which will preferentially pass certain components of wastewater. The driving force that causes such transfer of solvent to take place is the concentration gradient. Dialysis usually involves the flow of the waste stream containing solutes across one face of the membrane to come out from the second face. The membrane retains colloids and solutes of higher molecular weight. The transferring of the solvent through the membrane is by diffusion rather than by the hydrodynamic flow that would occur through a porous medium. Requirements for a membrane in dialysis process are good transfer rates, suitable mechanical strength, durability, resistance to chemical degradation, and low cost.

Ultrafiltration is a membrane treatment process with the ability to separate molecules in solution by size. It acts as a selective barrier. Ultrafiltration refers to treatment process in which the membranes have pore diameters in the range of 20–1000 Å($\times 10^{-9}$). The separation is achieved by concentrating the large molecules present in the waste stream on one side of the membrane, while the solvent and micromolecules are depleted as they pass through the membrane to the other side. Most ultrafiltration membranes are made from polymers, by the phase inversion process. Ultrafiltration membranes may also be formed dynamically (deposition of hydrous zirconium oxide on a porous medium under controlled conditions of flow and pressure). A few ultrafiltration membranes are prepared from ceramic materials. Ultrafiltration membranes are characterized based on pore size and porosity.

Ion exchange is a process of exchanging certain undesirable cations and anions of the wastewater for sodium or hydrogen ions in a resinous material. The resins, both natural and artificial commonly used are referred to as zeolites. In ion exchange, wastewater is passed through beds of cationic and anionic resins, which have been selected for the particular application and the deionized wastewater is then collected at the other end of the resins. The deionizing reactions in the resins are as follows:

$$\begin{matrix}\mathrm{Ca}\\ \mathrm{Mg}\end{matrix}\ \begin{matrix}(\mathrm{HCO_3})_2\\ \mathrm{SO_4}\\ \mathrm{Cl_2}\end{matrix} + \mathrm{Na_2Z} \longrightarrow \begin{matrix}\mathrm{Ca} & \mathrm{Z}\\ \mathrm{Mg} & \mathrm{Z}\end{matrix} + \begin{matrix}\mathrm{Na}\\ \\ \mathrm{Na_2}\end{matrix}\ \begin{matrix}(\mathrm{HCO_3})\\ \mathrm{Cl}\\ \mathrm{SO_4}\end{matrix} \qquad (19.14)$$

irradiation treatment processes are those processes used in the treatment of wastewater in which change is brought about through the application of radiation from solar energy or any other sources of rays. Electron beam radiation, gamma radiation, and ultraviolet radiation are the major application of radiation treatment processes in pollution control. Literature reviewed and stated some relevant applications of irradiation treatment processes.

Gamma radiation is a technology for purifying infectious waste through the use of ionizing radiation. The basic principle of the treatment process is that gamma rays from transition elements, which have unstable isotopes, such as cesium and cobalt, are passed through the solution. The electrical energy from this radiation destroys living organisms by altering the colloidal nature of the cell contents. The advantages of ionizing radiation as a treatment process are nominal electricity requirement, no stream requirement, no residual heat in treated waste, and the performance of the system is high. The principal disadvantages of gamma radiation include: high capital cost, it requires a highly trained operator and support personnel, it requires large space and the problem of ultimate disposal of the decayed radiation source.

Electron beam irradiation is radiation treatment process in which high-energy electron beam is pass through the waste to bring about change in the waste quality. The process of electron beam irradiation is best understood in aqueous solution in which sizable quantities of free radicals, such as hydrogen ion, hydroxyl ions, and the more stable oxidant hydrogen peroxide are produced. This highly reactive process decomposes the organic contaminants to produce carbon dioxide, water, and salts, which are no longer hazardous. Electron beam technology has potential application in the treatment of aqueous matrices, such as solutions and sludge (ultimate disposal of hazardous organic contaminants). Electron beam destruction has advantages over existing industrial waste treatment processes because it is not sensitive to the waste quality. This treatment process is particularly attractive in the treatment of halocarbons and waste streams containing up to 10% solids. Efficiency obtained by this treatment process depends only on the dose of radiation applied.

Electrical treatment processes are those processes used in the treatment of wastewater in which changes are brought about through the application of direct current or alternating current through electrodes to the cell, which contains the electrolyte. Electrodialysis, electrocoagulation, electron precipitation, and electrochemical treatment processes are among the major applications of electrical treatment processes. Electrodialysis is an electrical treatment process in which charged membranes are used to separate ions from aqueous solutions under the driving force of an electrical potential difference. The process utilizes an electrodialysis stack, built on the filter-press principle and containing several hundred of individual cells formed by a pair of anion and cation exchange membranes. The cells are arranged in an alternating pattern between an anode and a cathode to form individual cells. The basic principle of the process is that the solution is pumped through these cells and an electrical potential established between anode and cathode. The cations (positively charged ions) migrate toward the cathode, and the anions (negatively charged ions) migrate likewise to the anode. The cations pass easily through the negatively charged cation-exchange membrane but are retained by the positively charge anion-exchange membrane.

Electrocoagulation is the process of destabilizing suspended, emulsified, or dissolved contaminants in an aqueous medium by introducing an electrical current into the medium. In the electrocoagulation process, the electrical current is introduced into wastewater through electrodes constructed with metals that are selected to optimize the removal process. The most commonly used metallic materials for electrocoagulation are iron and aluminum. By Faraday's law, metal ions of these electrodes will be split off into the liquid medium, tend to form metal oxides that electromechanically attract to the contaminants that have been destabilized. Electrocoagulation treatment process utilizes direct current to cause sacrificial electrode ions to remove undesirable contaminants either by chemical reaction and precipitation or by causing colloidal materials to coalesce and then be removed by electrolytic flotation. The electrocoagulation system has proven to be able to cope with a variety of wastewaters. Limitations of electrocoagulation as a treatment process are that sacrificial electrodes are required, it treats waste that can conduct electricity, and the sludge volume has been reported to be higher.

Electroprecipitation treatment process is an electrical treatment process in which the basic device is the cell. The cell consists of an anode and a cathode immersed in an electrolyte. When direct current is applied, the dissolved metals in the electrolyte migrate to the cathode for electrode reaction. These metal ions are reduced and deposited at the cathode. As the process removes metals from solutions, which can be used for many other purposes, the process is called electrowinning. Electroprecipitation as a treatment process is not effective in removing all contaminants. Electroprecipitation is effective in removing the noble metals, such as gold and silver because these metals have high electrode potentials and are easily reduced and deposited.

An electrode is a conductor through which current enters or leaves electrolytes. There are two main types of electrodes, namely: anode (positive electrode, through which current enters) and cathode (negative electrode through which current leaves electrolyte). Current through electrode are characterized by the following properties: chemical properties of the metals are not altered; current is carried by electrons and increased in temperature increases resistance of the conductor. Literature reports that wetness of electrolyte, permeability, and conductivity of the electrode and the binding agent, porosity, surface area, and particles size affect the effectiveness of any electrode. The nature of electrode can be defined as an active electrode (takes part in the electrochemical reaction) or as a nonactive electrode (passive, does not take part in the electrochemical reaction). Carbon (graphite), titanium, titanium oxide, and platinum are the major nonactive electrodes, while aluminum, zinc, copper, lead, iron, and silver are active electrodes. Carbon is the most frequently used nonactive electrode during electrochemical treatment. It is known that carbon has been used from the beginning of fuel cell research as a support material for electrodes because of the following favorable properties, namely: large surface area, high electric conductivity, electrochemical and chemical stability, easy modification of surface properties, and large availability and low price. Carbon as an element is found in the form of pure carbon as diamond and impure form as graphite. Graphite exists as black, slippery hexagonal crystals. It is found naturally as plumbago and is manufactured artificially by heating coke to a very high temperature in the electric furnace (Acheson process). It is a good electrical conductor. It is attacked by potassium chlorate and nitric acid. It has been reported that when carbon (graphite) is to be used without an additional catalyst, it should be subjected to an activation treatment during which the chemical surface structure of the carbon is changed. The activation processes of carbon are achieved by one or several of the following procedures, namely: thermal treatment in different cases, treatment with oxidising mineral acid (nitric acid), modification of the carbon surface by ionizing radiation, and electrochemical modification. Optimum carbonizing and activating temperatures for palm kernel shell appeared to be in the range of 500–600°C and 800–900°C respectively. Literature, such as Rajkumar et al. (2001), Oke et al. (2012a,b,c,d), Li et al. (2014), and Zheng et al. (2015) reported the trend of electrochemical treatment of water and wastewater in environmental pollution control.

2 Previous Laboratory Setups of Electrochemical Treatment of Water and Wastewaters

Previous studies on electrochemical treatment of water and wastewaters revealed that there are various setups that are possible during electrochemical treatment of water and wastewaters. Chen (2004) stated that there are quite a few types of reactors, which found applications in metal recovery, from very basic reactors, such as tank cells, plate and frame cells, rotating cells, to complicated three-dimensional reactor systems like a fluidized bed, packed bed cell, or porous carbon packing cells. Fig. 19.2 presents basic electrochemical reactor used between 1973 and 2004. Tank cells, (Fig. 19.3) are one of the simplest and hence the most popular designs. The main advantages of tank cells are that it can be easily scaled up or down depending on the load of

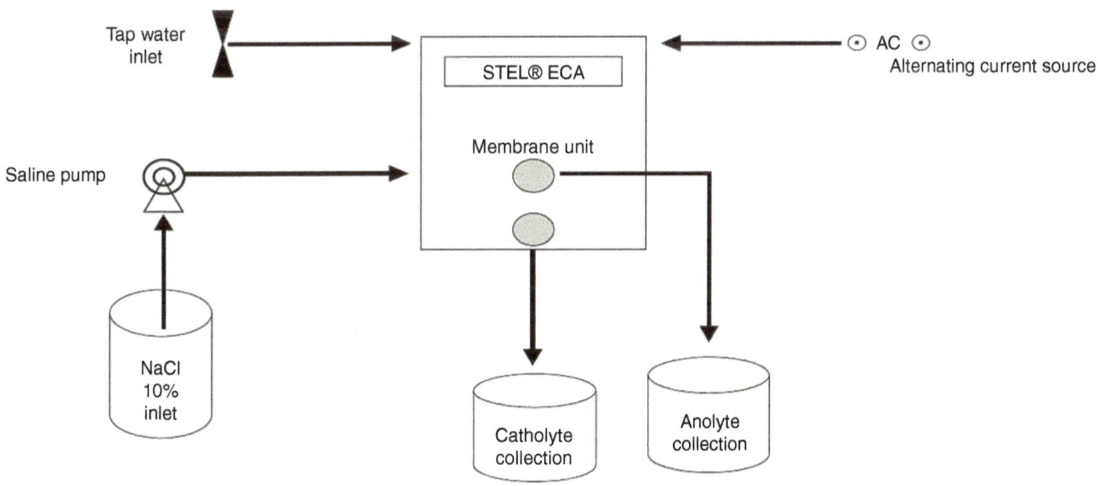

Figure 19.2. (A) Schematic diagram of electrochemical activation (ECA) for phenol degradation;

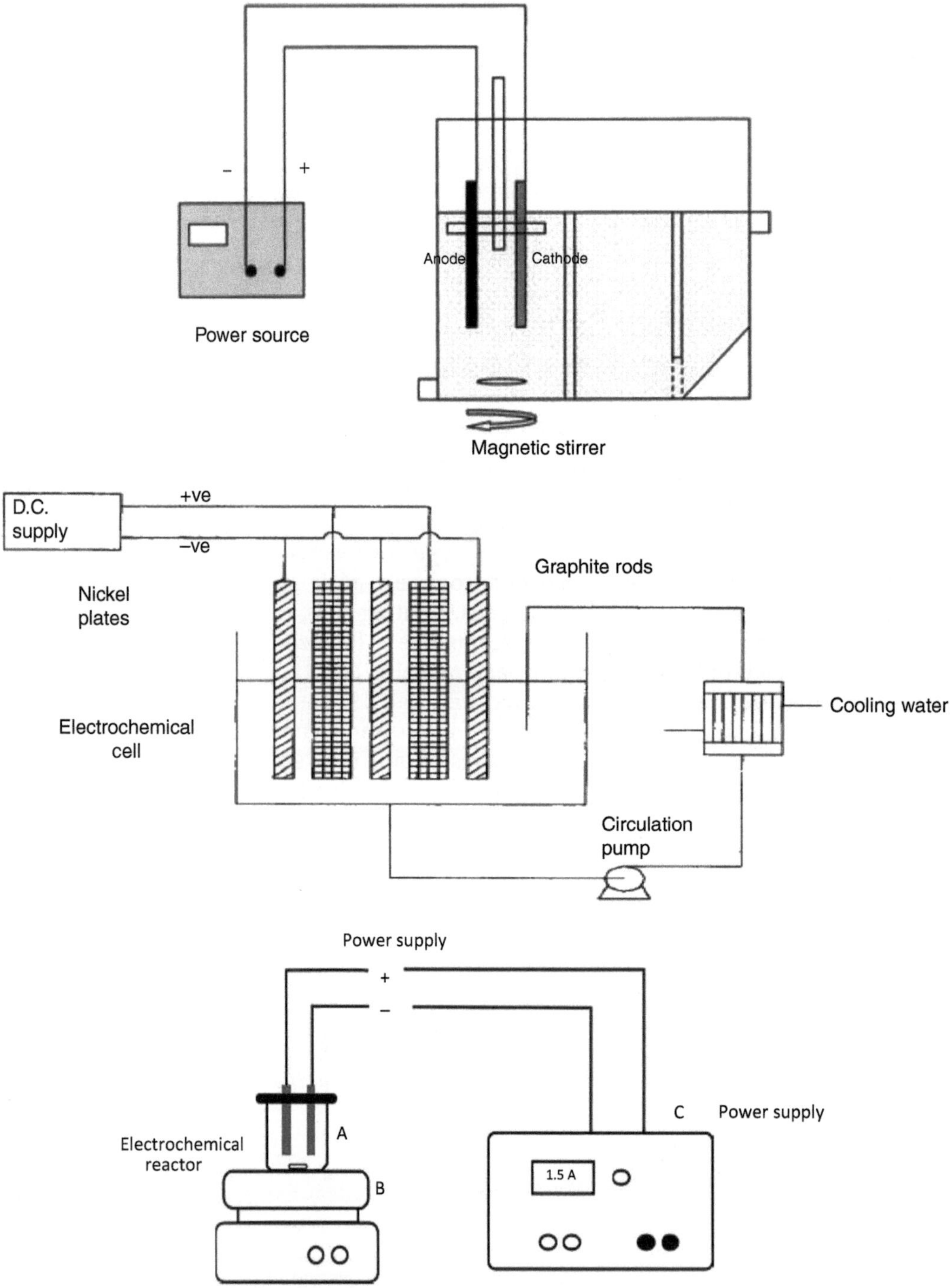

Figure 19.2. (*Continued*) (B) schematic diagram of electrochemical oxidation in continuous flow system; (C) schematic diagram of electrochemical oxidation process; (D) schematic diagram of basic electrochemical oxidation system;

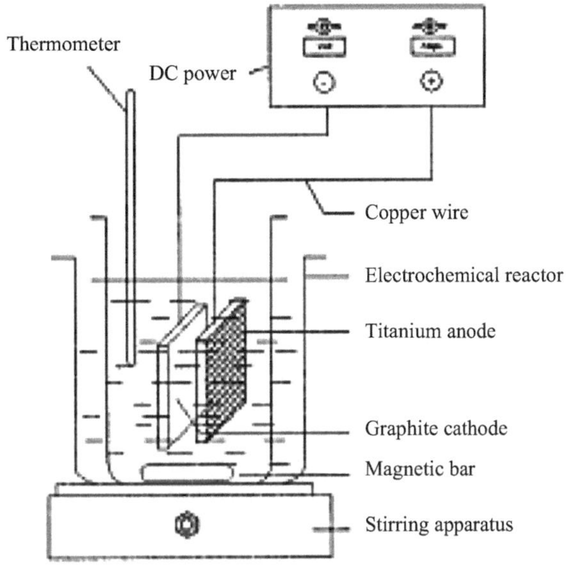

Figure 19.2. (*Continued*) (E) schematic diagram of electrochemical oxidation of catechol in chloride medium. Part A: Azni and Katayon (2002). Part B: Catalino et al. (2004). Part C: Allen et al. (1995). Part D: Inazaki et al. (2004). Part E: Rajkumar et al. (2001).

a process (Chen, 2004). The electrode can be arranged in monopolar (Fig. 19.4) or bipolar mode (Fig. 19.5). The main application of this type of reactor system is the recovery of metals from high concentration process streams, such as effluents from the electroplating baths, ethants, and eluates of an ion-exchange unit and the number of electrodes in a stack may vary from 10 to 100. The wastewater flow from the industry is usually induced by gravity.

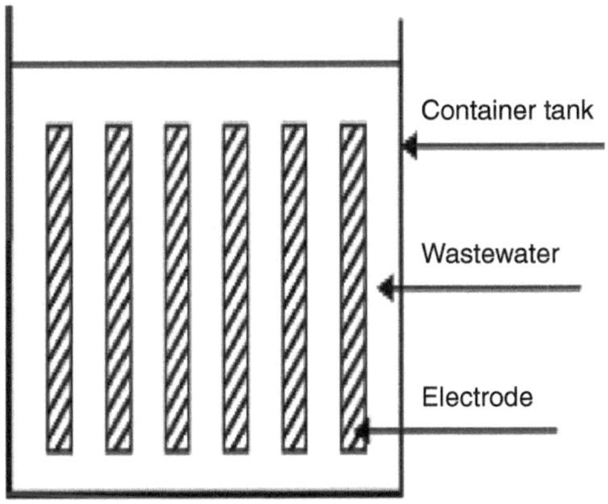

Figure 19.3. Schematic diagram of electrochemical oxidation tank cell. Chen (2004).

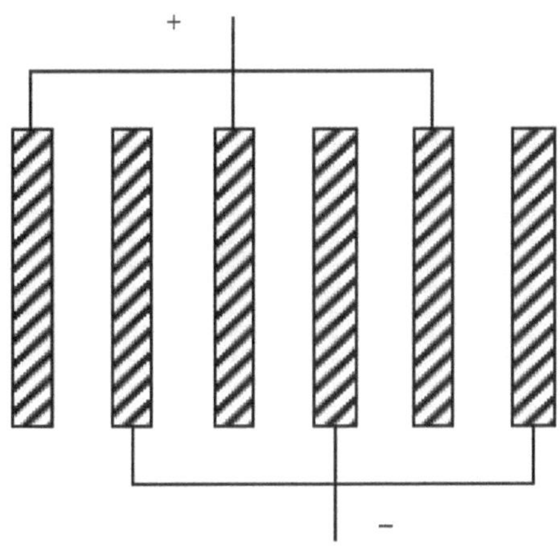

Figure 19.4. Schematic diagram of electrochemical oxidation mono polar system. Chen (2004).

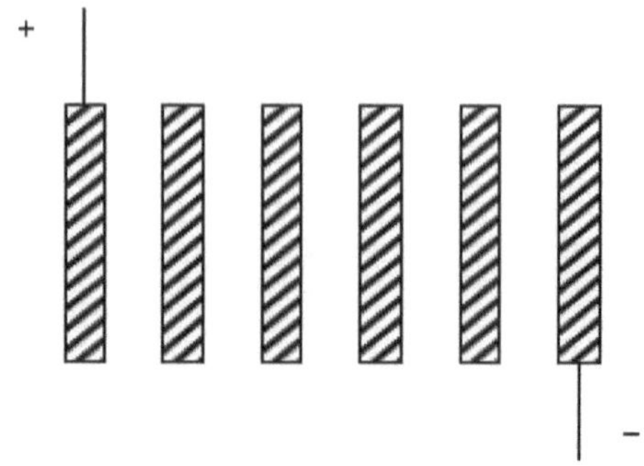

Figure 19.5. Schematic diagram of electrochemical oxidation bi polar system.
Chen (2004).

The plate and frame cell (called filter press, Fig. 19.6) is one of the most popular electrochemical reactor designs. Filter press conveniently houses units with an anode, a cathode, and a membrane in one module. This module system makes the design, operation, and maintenance of the reactor relatively simple. In order to enhance mass transfer from the bulk to the electrode surface and to remove the deposited metal powders from the cathode, the rotating cathode cell was designed and employed (Fig. 19.7). It was found that filter press can reduce copper content from 50 to 1.6 mg/L (ppm) by using the systems in a cascade version. The pump cell is another variant of rotating cathode cell (Fig. 19.8). A static anode and a rotating disk cathode, the narrow spacing between the electrodes in the pump cells allow the entrance of the effluent and effective wastewater treatment. The metals were electrically won

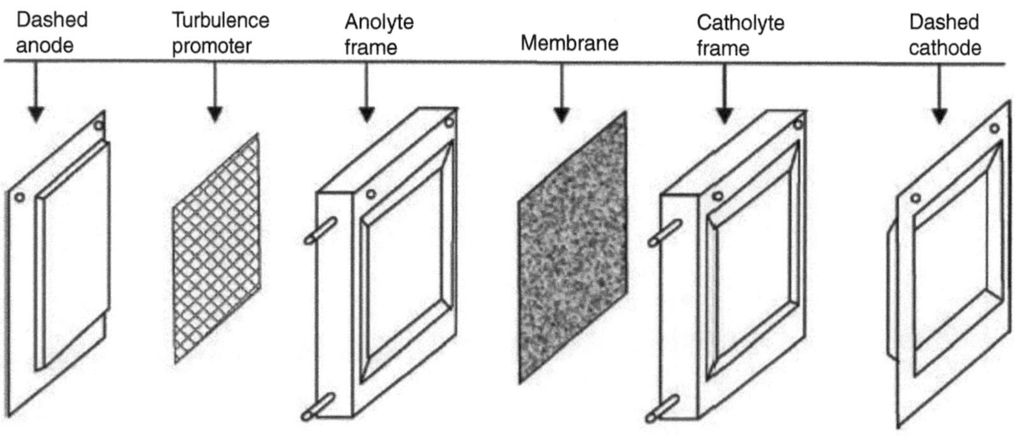

Figure 19.6. Schematic diagram of electrochemical oxidation filter press system.
Chen (2004).

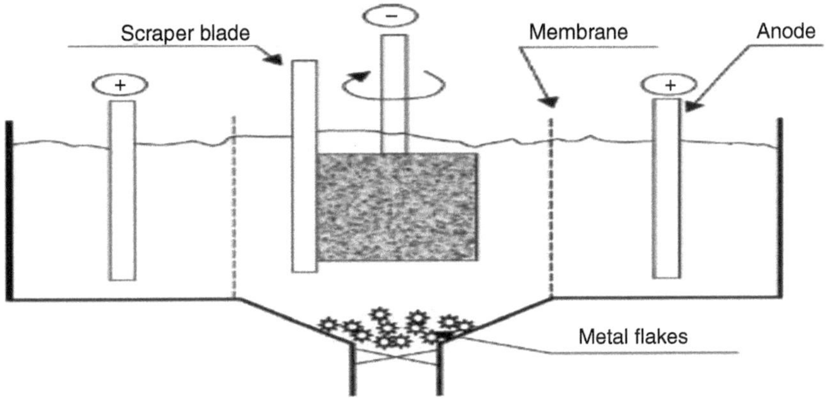

Figure 19.7. Schematic diagram of electrochemical oxidation with rotating cylinder electrode system. Chen (2004).

and scraped as powders. Another design employs rotating rod cathodes in between inner and outer anodes. Besides metal recovery, it is possible to have the anodic destruction of cyanides and the metal deposition happens at the surface of the cathode (increase the specific surface area in order to improve the space–time yield). In the design of fluidized bed (Fig. 19.9), the cathode was made of conductive particles in contact with a porous feeder electrode. The electrode can give a specific area of 200 m^2/m^3 (based on the fluidization of the particles by the water flow in the fluidized bed). The electrical contact in the fluidized bed is not always

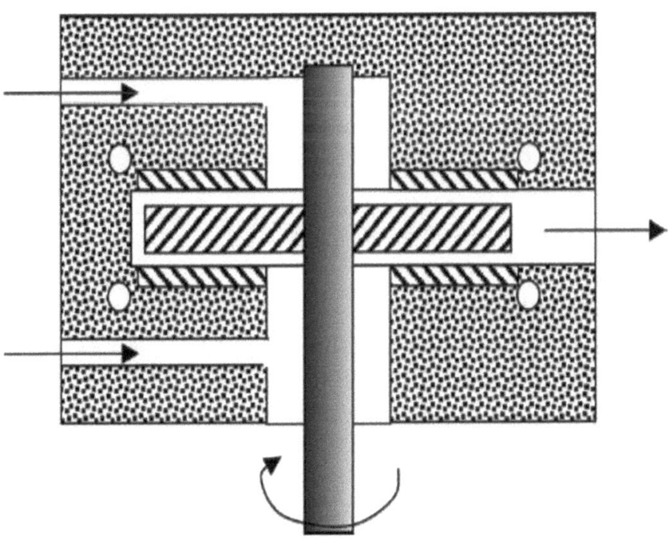

Figure 19.8. Schematic diagram of electrochemical oxidation with pump system. Chen (2004).

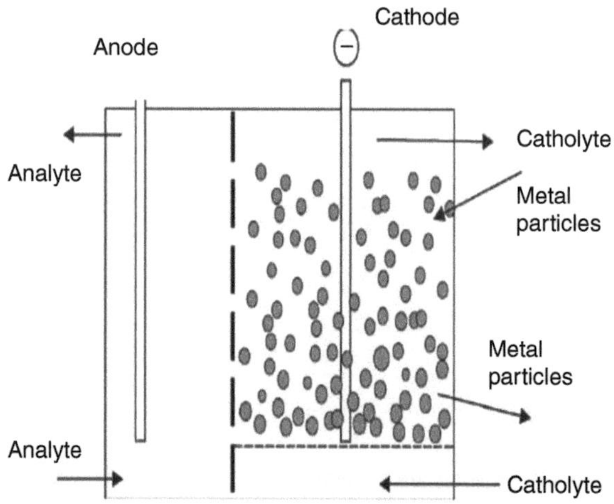

Figure 19.9. Schematic diagram of electrochemical oxidation with fluidized bed reactor system. Chen (2004).

maintained thus the current distribution is not always uniform, and the ohmic drop within the cell is high (in order to improve the contact between the electrode feeder and cathode particles, a large number of additional rod feeders are used). Inert particles can be employed in fluidized bed reactor to improve mass transfer rate in a ChemElec commercial design (Chen, 2004). With various reports on lead poisoning across the globe, there is the need to provide technique for lead removal from the environment and make the environment free from lead poisoning at all levels. The main of this study is to investigate effectiveness of electrochemical in removing lead ion from water and wastewaters with the aim of removing lead ions from water and wastewaters thus support the production and distribution of lead free water in the environment.

3 Materials and Method

3.1 Materials

The main materials and facilities used for the electrochemical treatment process in this research work were the electrolyzing unit, electrodes, synthetic wastewaters, and typical industrial wastewaters. An electrolyzing unit is the equipment that was used to convert alternating current to direct current with variable voltage output ranging from 0 to 60 V. Fig. 19.10A–B shows constructed electrolyzing equipment, some of the accessories used and laboratory setup. The equipment was constructed using local materials and calibrated using digital and analogue voltmeters. More

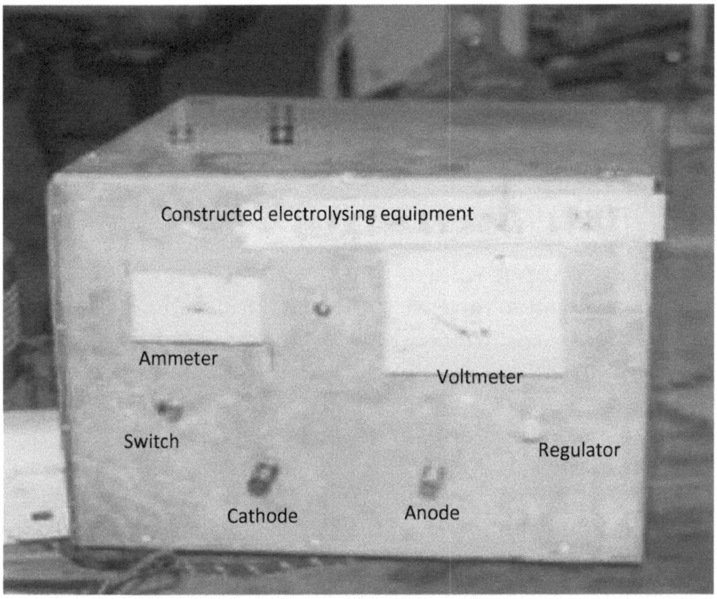

(A)

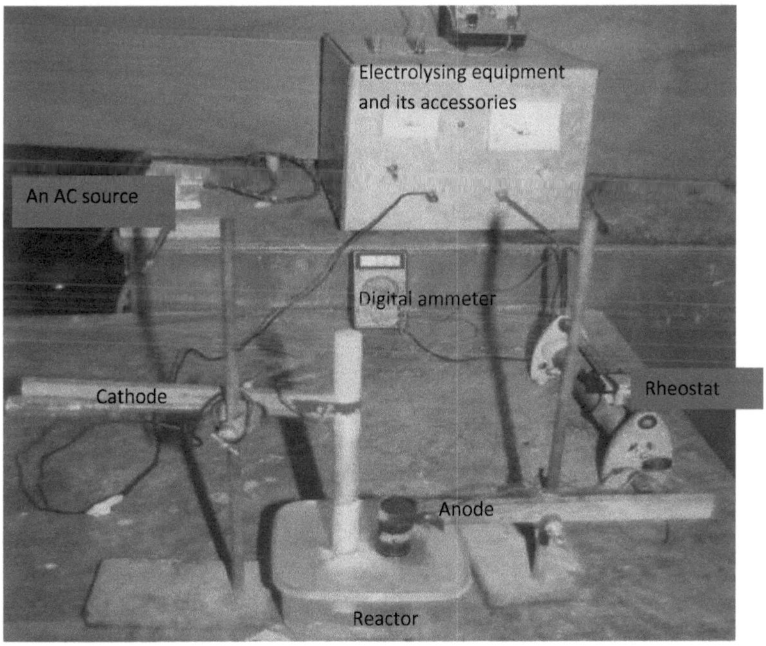

(B)

Figure 19.10. Constructed electrolyzing equipment. (A) Electrolyzing equipment and some of the accessories used and (B) laboratory setup for the studies.

on the electrolyzing equipment can be found in literature, such as Oke and Ogedengbe (2007). Apart from the electrolyzing unit, electrodes are another major part of the electrochemical treatment process because they are the devices that carry the current into and out of the wastewaters. Carbon (graphite) and aluminum electrodes were used in the research work. A Rheostat was used to regulate the current flow while an ammeter and a voltmeter were used to measure the current and voltage respectively. A Variac was used to regulate input voltage into the electrolyzing equipment. Synthetic wastewaters, such as sodium chloride solution and lead nitrate solution (wastewaters deliberately prepared by dissolving selected salts in distilled water) were used to determine the stability of the electrodes. Factors that affect the stability of the electrodes, factors that influence the efficacy of electrochemical treatment and efficacy of the treatment method in removing selected pollutants (lead ion, total solid, chromium, cadmium, etc.) were evaluated. Results of development of the electrodes, the stability of the electrodes and the efficacy of the electrodes are presented in Oke et al. (2007a,b,c, 2008, 2010, 2012e) and Oke (2009a,b). Effects of these factors (time, current, concentration of Pb^{2+}, etc.) on the performance of the treatment process were evaluated based on ability to remove Pb^{2+} from the aqueous solutions.Preparations of synthetic wastewaters are as follows:

1. *Preparation of synthetic waters*: synthetic wastewaters were prepared by dissolving known weights of selected salts in water (samples of known volume, resulting in synthetic wastewaters of known concentrations). Synthetic wastewaters (some containing single pollutants, others containing combinations of pollutants) were preserved as specified in APHA (2005). Synthetic wastewaters were employed in the electrochemical treatment studies at the first stage of the research work to avoid fluctuation of the wastewater characteristics and to obtain information on the accuracy of the treatment process. FMENV (2002) documented concentrations of some heavy metals and salts found in selected industries in Nigeria. According to FMENV (2002), maximum concentrations of chromium, chloride ion, sulfate ion, oil and grease, zinc, cadmium, nickel, and lead found in selected industrial wastewaters in Kano were 80.0, 13,000, 1,500, 2.2, 83, 2.6, 2.4, and 3.8 mg/L respectively. Ramgulam et al. (1999) reported maximum concentrations of cadmium, chromium, zinc, nickel, and lead found in textile wastewaters in Mauritius as 0.2, 0.7, 1.5, 0.32, and 2.2 mg/L respectively. In the preparation of synthetic wastewaters, these maximum concentrations of pollutants found in some industrial wastewaters documented in FMENV (2002) were targeted

as concentration of synthetic wastewaters (individually or as combinations) and their preparations are as described in APHA (2005) as follows:

a. *Chloride wastewater*: stock solution of chloride ion was prepared by dissolving 69.40 g of KCl in 100 mL of distilled water and was made up to 1000 mL mark with distilled water (APHA, 2005) to produce a known concentration of 33,000 mg/L of chloride ion (stock solution). The stock solution was preserved and discarded after 2 weeks if not used. The working solution of 13,000 mg/L of chloride ion was prepared daily from the stock solution.

b. *Sulfate wastewater*: stock solution of sulfate ion was prepared by dissolving 5.91 g of Na_2SO_4 in 100 mL of distilled water and was made up to 1000 mL mark with distilled water (APHA, 2005) to produce a known concentration of 4000 mg/L of sulfate ion. The stock solution was preserved and discarded after two weeks if not used. The working solution of 1500 mg/L of sulfate ion was prepared daily from the stock solution.

c. *Zinc wastewater*: stock solution of zinc ion was prepared by dissolving 15.00 g of zinc nitrate in 100 mL of distilled water and was made up to 1000 mL mark with distilled water (APHA, 2005) to produce a known concentration of 1000 mg/L of zinc ion. The stock solution was preserved and discarded after two weeks if not used. The working solution of 100 mg/L of zinc ion was prepared daily from the stock solution.

d. *Cadmium wastewater*: stock solution of cadmium ion was prepared by dissolving 2.74 g of $Cd(NO_3)_2{\cdot}4H_2O$ in 100 mL of distilled water and was made up to 1000 mL mark with distilled water (APHA, 2005) to produce a known concentration of 4.0 mg/L of cadmium ion. The stock solution was preserved and discarded after two weeks if not used. The working solution of 2.6 mg/L of cadmium ion was prepared daily from the stock solution.

e. *Chromium wastewater*: stock solution of chromium ion was prepared by dissolving 28.28 g of $K_2Cr_2O_4$ in 100 mL of distilled water and was made up to 1000 mL mark with distilled water (APHA, 2005) to produce a known concentration of 100 mg/L of chromium ion. It was preserved and discarded after two weeks if not used. The working solution of 80.00 mg/L of chromium ion was prepared daily from the stock solution.

f. *Lead wastewater*: stock solution of lead ion was prepared by dissolving 1.598 g of $Pb(NO_3)_2$ in 100 mL of distilled water and was made up to 1000 mL mark with distilled water (APHA, 2005) to produce a known concentration of 100 mg/L of lead ion. It was preserved and discarded after two weeks if not used. The working solution of 3.80 mg/L of lead ion was prepared daily from the stock solution.

g. *Nickel wastewater*: stock solution of nickel ion was prepared by dissolving 1.735 g of nickel chloride in 200 mL of distilled water and was made up to 1000 mL mark with distilled water (APHA, 2005) to produce a known concentration of 10 mg/L of nickel ion. It was preserved and discarded after two weeks if not used. The working solution of 2.80 mg/L of nickel ion was prepared daily from the stock solution.

h. *Oil and grease wastewater*: stock solution of oil and grease was prepared by dissolving 1.273 g of nonpolar oil, such as engine oil of type SAE 40 in 200 mL of acetone and was made up to 1000 mL mark with distilled water to produce a known concentration of 1273 mg/L of grease and oil. It was preserved and discarded after two weeks if not used. The working solution of 28.0 mg/L of oil and grease was prepared daily from the stock solution.

i. *Composite wastewater*: stock solution of the composite wastewater was prepared by dissolving a mixture of two or more of 1.598 g of $Pb(NO_3)_2$, 2.828 g of $K_2Cr_2O_4$, 2.744 g of $Cd(NO_3)_2{\cdot}4H_2O$, 1.500 g of nitrate, 1.753g of nickel chloride, 22.179 g of Na_2SO_4, 1.273 g of nonpolar oil, such as engine oil of type SAE 40 and 69.40 g of KCl in 100 mL of distilled water and was made up to 1000 mL mark with distilled water to produce a known concentration of the pollutant as the case may be. It was preserved and discarded after 2 weeks if not used. The working solution of composite water was prepared daily from the stock solution. Each of these prepared synthetic wastewaters was subjected to electrochemical treatment to monitor the efficacy of the treatment of the process in removing selected pollutants.

2. *Preparation and electrochemical treatment of synthesized selected industrial wastewaters*: selected industrial wastewaters, such as textile and tannery were prepared using some analytical reagents. Compositions of the wastewaters were based on stated concentration in literature, such as Orhon et al. (2000). These synthesized wastewaters were subjected to electrochemical treatment using optimal operational factors.
3. *Collection and electrochemical treatment of typical industrial wastewaters*: industrial wastewaters were collected from selected industrial sources according to Standard Methods for Water and Wastewaters Examination (APHA, 2005) within the

country, specifically Kano (Kano state), Kaduna (Kaduna state), Lagos (Lagos state), Ibadan (Oyo state), Ilesha (Osun state), and Oyo (Oyo state), Nigeria. Each of the industrial wastewaters collected was subjected to electrochemical treatment under specified treatment conditions (electrode depth, contact area voltage and current, volume of the liquid, and treatment period) within a maximum period of 24 h. Selected contaminants (chloride ion, sulfate ion, oil and grease, and selected heavy metals) removal was monitored as specified in APHA (2005).

4. *Analytical procedures*: in the analysis of selected pollutants concentrations (chloride, sulfate, cadmium, zinc, nickel, and others), Standard Methods for the Examination of Water and Wastewaters (APHA, 2005; Marks et al., 2016) was followed. However, in the analysis of heavy metal concentration Atomic Adsorption Spectrometer (AAS) facility available in Central Science Laboratory was used after its calibration using standard solutions and computation of the pollutant removed were conducted as follows:
   a. *Chloride concentration determination*: in the determination of chloride ions removed, procedures for the determination of chloride concentration in wastewater as specified in the Standard Methods for the Examination of Water and Wastewaters (APHA, 2005) was followed. Chloride ion concentrations removed was calculated using Eq. (19.15)

$$\text{Chloride removed (mg/L)} = Cl_i - Cl_f \qquad (19.15)$$

   where Cl_i and Cl_f are the initial and final concentrations of chloride respectively.
   b. *Dissolved solids concentration determination*: similarly, determination of dissolved solids removed from wastewaters (raw and treated wastewaters) was as specified in the Standard Methods for the Examination of Water and Wastewaters (APHA, 2005) was followed. Solid concentrations removed was computed using Eq. (19.16):

$$\text{Solids removed (mg/L)} = SlO_i - SlO_f \qquad (19.16)$$

   where SlO_i and SlO_f are the initial and final concentrations of solid respectively.
   c. *Grease and oil concentration determination*: grease and oil determination was carried out as outlined in APHA. Grease and oil concentrations removed were calculated using Eq. (19.17).

$$\text{Grease and oil removed (mg/L)} = Gre_i - Gre_f \qquad (19.17)$$

where Gre_i and Gre_f are the initial and final concentrations of grease and oil respectively.

d. *Sulfate concentration determination*: in the determination of sulfate ion removed from wastewaters, procedures for the determination of sulfate ion concentration in wastewater as specified in the Standard Methods for the Examination of Water and Wastewaters (APHA, 2005) was followed. Calculation of sulfate ion concentrations removed was conducted as follows:

$$\text{Sulphate ion removed (mg/L)} = SO_i - SO_f \tag{19.18}$$

where SO_i and SO_f are the initial and final concentrations of sulfate respectively.

e. *Cadmium concentration determination*: in the determination of heavy metals in water and wastewater there are various methods and approaches, but APHA (2005) recommended atomic absorption spectrometric method because of its reliability and consistency. Therefore in the determination of cadmium concentration in both raw wastewater samples and treated wastewater samples, atomic absorption spectrometric method was used. The equipment was calibrated with a standard cadmium solution prepared from the stock solution (1.500 g of cadmium nitrate was dissolved in 200 mL of distilled water and was diluted to 1000 mL with distilled water). Atomic absorption spectrometric facilities were utilized and cadmium concentrations removed was calculated using expression:

$$\text{Cadmium removed (mg/L)} = Cd_i - Cd_f \tag{19.19}$$

where Cd_i and Cd_f are the initial and final concentrations of cadmium respectively.

f. *Chromium concentration determination*: determination of chromium concentration in both raw wastewater samples and treated wastewater samples was carried out using atomic absorption spectrometric method (APHA, 2005). The equipment was calibrated with a standard chromium solution prepared from a stock solution (2.828 g of anhydrous potassium dichromate ($K_2Cr_2O_7$) was dissolved in 200 mL of water, 15.0 mL of concentrated tetraoxonitrate (V), (HNO_3) acid was added and was diluted to 1000 mL with distilled water). Atomic absorption spectrometric facility was utilized. Chromium removed due to the treatment was calculated using expression

$$\text{Chromium removed (mg/L)} = Cr_i - Cr_f \tag{19.20}$$

where Cr_i and Cr_f are the initial and final concentrations of Chromium respectively.

g. *Nickel concentration determination*: in the determination of nickel concentration in both raw wastewater samples and treated wastewater samples, atomic absorption spectrometric method was used as specified in APHA (2005). The equipment was calibrated with a standard nickel solution prepared from a stock solution (1.753 g of nickel chloride was dissolved in 200 mL of distilled water and was diluted to 1000 mL with distilled water). Atomic absorption spectrometric facility was utilized and nickel removed by the treatment received was calculated using expression:

$$\text{Nickel removed (mg/L)} = Ni_i - Ni_f \quad (19.21)$$

where Ni_i and Ni_f are the initial and final concentrations of Nickel respectively.

h. *Zinc concentration determination*: the determination of zinc concentration in both raw wastewater samples and treated wastewater samples, atomic absorption spectrometric method was followed as instructed in APHA (2005). The equipment was calibrated with a standard zinc solution prepared from a stock solution (dissolve 1.500 g of zinc nitrate was dissolved in 200 mL of distilled water and was diluted to 1000 mL with distilled water). Atomic absorption spectrometric facility was utilized. Zinc removed electrochemically was calculated using expression:

$$\text{Zinc removed (mg/L)} = Zn_i - Zn_f \quad (19.22)$$

where Zn_i and Zn_f are the initial and final concentrations of Zinc respectively.

i. *Lead concentration determination*: determination of lead concentration in both raw wastewater samples and treated wastewater samples, atomic absorption spectrometric method was systematically followed as indicated in APHA (2005). The equipment was calibrated with a standard lead solutions of lead from prepared stock solution [1.598 g of anhydrous lead nitrate, $Pb(NO_3)_2$] was dissolved in 200 mL of water, 15.0 mL of concentrated tetraoxonitrate (V), (HNO_3) acid was added and was diluted to 1000 mL with distilled water (APHA, 2005). Atomic absorption spectrometric facility was utilized. Lead removed electrochemically was calculated using expression

$$\text{Lead removed (mg/L)} = Pb_i - Pb_f \quad (19.23)$$

where Pb_i and Pb_f are the initial and final concentrations of lead respectively.

5. *Computation of power and energy consumed*: computations of power and energy used during electrochemical treatment of these wastewaters were determined using Eqs. (19.24) and (19.25) respectively.

$$\text{Power} = IV \tag{19.24}$$

$$\text{Energy} = IVT \tag{19.25}$$

where I is the current through the wastewaters (A); V is the voltage across the electrolyte (V); and T is the treatment time (second).

6. *Electro chemical treatment of synthetic wastewaters*: prepared synthetic wastewater samples were subjected to electrochemical treatment with specified electrode configuration (electrode space, depth of electrode, contact area and type, Oke et al., 2012a,b,c,d) specified voltage from electrolysis equipment and fixed detention period for the removal of selected contaminants, which include chloride ions, sulfate ions, dissolved solids, cadmium, lead, zinc, nickel, grease and oil, and chromium concentrations.

4 Result and Discussion

4.1 Relationship Between Efficacy Electrochemical Treatment and Selected Factors

Results of the effect of selected factors on efficacy electrochemical treatment method in removing lead ion concentration from water and wastewaters are as presented in Figs. 19.11–19.15. Fig. 19.11 presents the relationship between the selected factors and efficacy (response) of the treatment process. Fig. 19.12 shows effects of current through the electrodes on the efficiency of the process in removing lead ion. From the figure, it can be seen that as current increases at a fixed time, the amount of lead ion removed increases. Similarly, when the time decreases at a fixed current through the electrodes, the lead ion removed decreases. When both the current through the electrodes and the time increase it can be seen that lead ion removed increases exponentially (Fig. 19.12). This result indicates that the higher the current, the higher the efficiency and confirms the results from factorial experiments and optimization (Oke et al., 2007c). Fig. 19.13 presents influences of the volume of wastewater on the efficiency of the process. Like Fig. 19.12, it can be seen from the figure that as volume increases the amount of lead

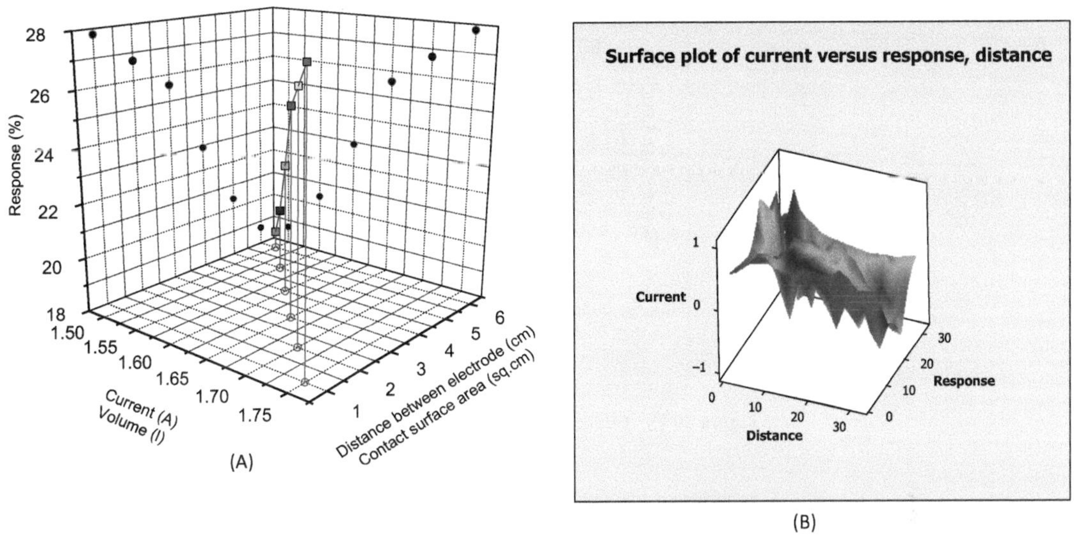

Figure 19.11. Relationship between factors and response. (A) In 3D with trajectory line and (B) surface plot of current against response

ion removed decreases (a negative factor). When the volume of the wastewater decreases, it can be seen that lead ion removed increases dramatically (Fig. 19.13). This result shows that the higher the volume, the lower the efficiency and confirms the results from factorial experiments and optimization. It can be explained further that efficiency of the process is a function of discharge and retention time

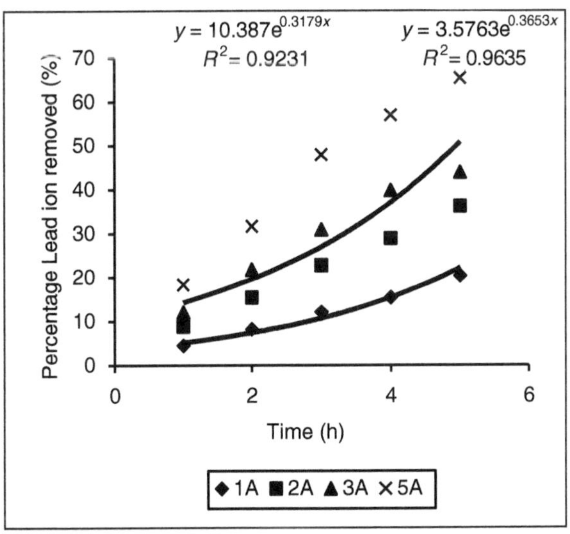

Figure 19.12. Effect of current on lead ion removal.

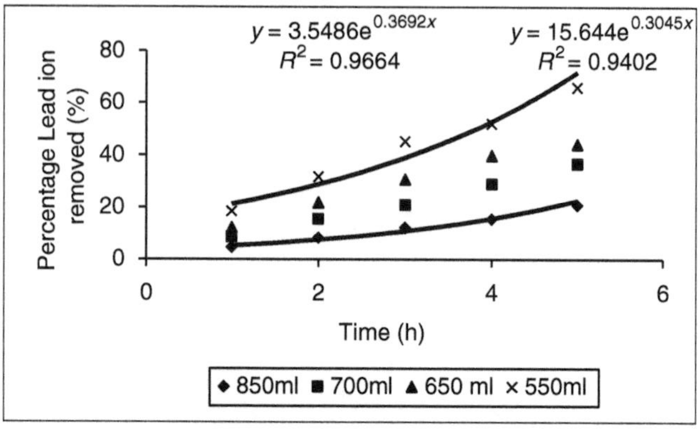

Figure 19.13. Effect of volume of wastewater used on lead ion removal.

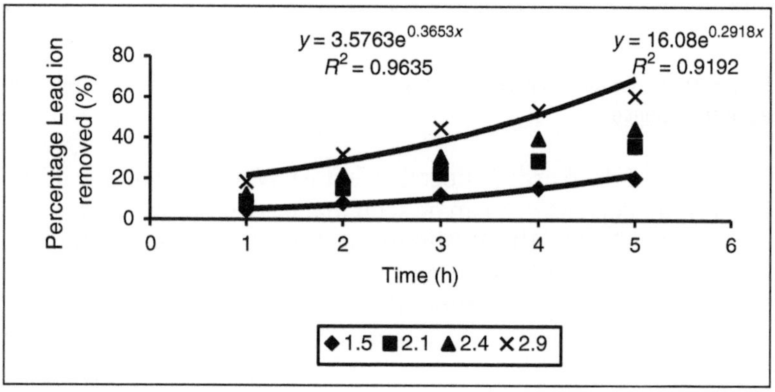

Figure 19.14. Effect of electrode's surface contact area on lead ion removal.

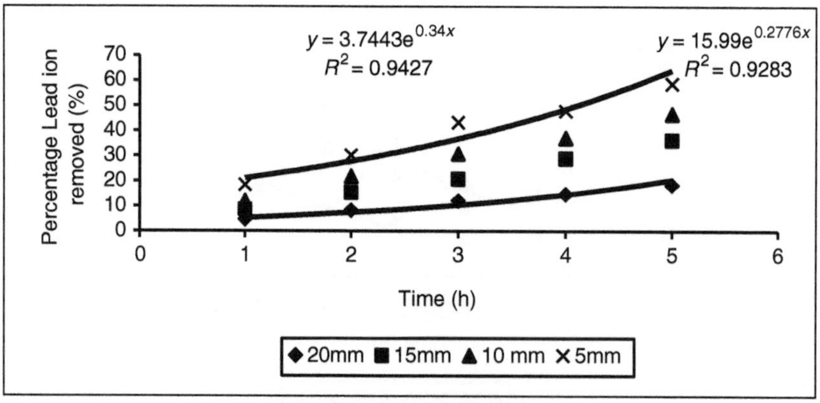

Figure 19.15. Effect of separation distances between the electrodes on lead ion removal.

[as volume can be expressed as presented in Eq. (19.26)]. This result indicates that efficiency of the process decreases with increasing flow rate and increases with increasing retention time. In addition to all these, it means that Faraday's first law [removal of dissolved substance is proportional to the quantity of electricity passed) can be written mathematically as Eq. (19.27)]. Eqs. (19.28) and (19.29) show that the higher the volume, the higher the time and current required for effective treatment (higher efficiency).

$$V = Qt \tag{19.26}$$

$$M = kIt = Qt\rho \tag{19.27}$$

From Eq. (19.27) the volume of wastewater treated at a particular time can be expressed as follows:

$$M = kIt = V\rho \tag{19.28}$$

$$V = \frac{k}{\rho} It \tag{19.29}$$

Fig. 19.14 shows the influence of contact surface area of the electrodes on the efficiency of the process, which indicates that the higher the contact surface area, the higher the efficiency. This observation also agrees with the results from factorial experiments and optimization. This observation indicates that efficiency of the process is a function of depth and width of the electrodes (higher depth or/and width of the electrodes) will increase the efficiency of the process. Fig. 19.15 represents the influence of separation distance between the electrodes on the efficiency of the process, which indicates that the higher the separation distances between the electrodes, the lower the efficiency. This result confirms the results of factorial experiments and optimization (Oke et al., 2007a). This result can be explained further that the higher the distance between the electrodes the lower the efficiency, as more energy will be required to overcome the resistance of the wastewater (lower conductivity of the wastewater). Tables 19.1a–c present analysis of variance of effect of the selected factors on efficacies of electrochemical treatment of water and wastewaters (removal of lead ion concentration electrochemically). From the tables, it was revealed that time ($F_{20,4}$ = 19.589; p = 1.08 × 10^{-16}); Space between the electrodes ($F_{10,2}$ = 15.821; p = 0.0008) current ($F_{24,4}$ = 277.963; p = 1.01 × 10^{-19}); coions ($F_{10,5}$ = 10.214; p = 0.0011) and concentration of Pb^{2+} ($F_{20,5}$ = 8.069; p = 0.000265) were significant factors in Pb^{2+} removal electrochemically at 95% confidence level. Pb^{2} Effects of pH of the solution ($F_{24,6}$ = 2.029; p = 0.1008)were found to be not significant at all level.

Table 19.1a Effect of Concentration of Lead Ion and Time on Efficacy of the Treatment Method

Source of Variation	Sum of Squared	Degree of Freedom	Mean Squared	F-Value	P-Value
Concentration of lead	2198.695	5	439.739	8.069099	0.000265
Time	4270.155	4	1067.539	19.58906	1.08E-06
Error	1089.933	20	54.49667		
Total	7558.783	29			

Table 19.1b Effect of Coion Concentration and Space Between the Electrodes on Efficacy of the Treatment Method

Source of Variation	Sum of Squared	Degree of Freedom	Mean Squared	F-Value	P-Value
Coion	1425.856	5	285.1712	10.21403	0.001108
Space	883.4311	2	441.7156	15.82101	0.000799
Error	279.1956	10	27.91956		
Total	2588.483	17			

Table 19.1c Effect of pH of the Wastewaters and Current Through the Electrodes on Efficacy of the Treatment Method

Source of Variation	Sum of Squared	Degree of Freedom	Mean Squared	F-Value	P-Value
pH	73.324	6	12.22067	2.029441	0.100845
Current	6695.218	4	1673.805	277.9625	1.01E-19
Error	144.5206	24	6.021692		
Total	6913.063	34			

4.2 Heavy Metal Ions Removal From Synthetic Wastewaters

Synthetic wastewaters containing known concentrations of lead ion (from 10.59 to 444.3 mg/L) were subjected to electrochemical treatment. From the results, removal of lead ions increases with increasing initial concentration, time and current flow through the electrolyte (Fig. 19.16). This result indicates that lead ion removal is a function of current, time, and initial concentration. It was observed that the efficacy of the treatment process was 100% (from 106.7–451.9 mg/L to least than 0.01 mg/L) in all the individual systems and reduced to between 52% and 62.5% for combination system. Lead ion reductions by the treatment process can be attributed to three basic reactions during the treatment process, namely: electrolysis process, electrochemical/electroprecipitation process, and adsorption. In electrolysis process lead ions were reduced (accepted electrons) at the cathode to form lead atom and finally deposited as sludge, slurry, or precipitates [Eq. (19.30)]. During electrochemical/electroprecipitation process hydroxyl ions (a product during electrolysis) from anode combined with lead ion to form a metallic compound [lead hydroxide, Eq. (19.27)] which were deposited as sludge. In the case of adsorption process spent carbon particles from anode adsorbed the pollutant.

$$Pb^{2+} + 2\bar{e} \rightarrow Pb \tag{19.30}$$

$$Pb^{2+} + 2OH^{-} \rightarrow Pb[OH]_2 \tag{19.31}$$

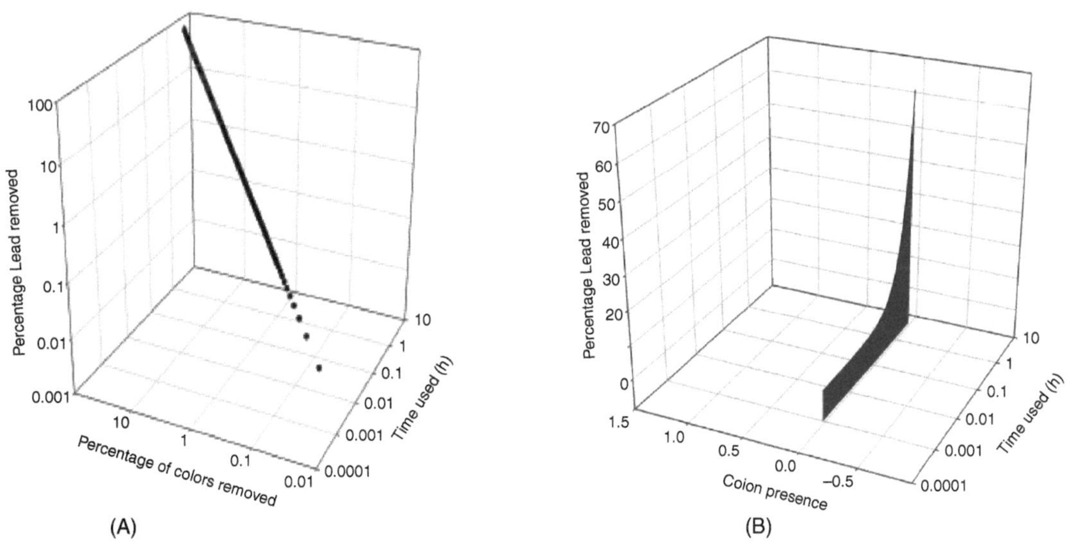

Figure 19.16. Relationship between lead ion removal, treatment time, and presence of coion. (A) In multicomponent solution and (B) in mono-component solution.

Also, 100% removal of lead ions from the individual system and lower removal from the multicomponent system might have been attributed to position lead ion in the electrochemical table, periodic table and electronic configuration effect of coions and ionic strength, which are major factors in selective removal of pollutants (cations) at the cathode. The results of total and partial removal of lead ions from mono- and multicomponent systems are similar to Panayotov et al. (2001) description of lead removal from solutions. Fig. 19.16A–B presents the relationship between lead ion removal from solutions in respect to coions and treatment time.

4.3 Treatment of Synthesized Industrial Wastewaters

Literature, such as Orhon et al. (2000); UNIDO (2004) listed the compositions of some industrial wastewaters (tannery, textile, and pharmaceutical industries). Based on their report synthesized tanning and texturing wastewaters were prepared and subjected to three possible treatment options (electrochemical treatment, chemical only, electrochemical and chemical). Efficacies (based on ability to reduce selected pollutants) of electrochemical treatment, electrochemical and chemical, and chemical (using calcium hypochlorite)

Table 19.2 Treatments of Synthesized Textile Wastewater

Treatment Processes	Time (h)	Cur (A)	SS	% SS Removed	Cr	% Cr Removed	Col.	% Col. Removed
Electrochemical	0	1.85	2370	0.00	0.22	0.00	520	0.00
	1	2.3	1680	29.11	0.1	54.55	470	9.62
	2	2.45	1160	51.05	0.06	72.73	440	15.38
	3	2.25	880	62.87	0.04	81.82	420	19.23
	4	2.03	120	94.94	0	100.00	380	26.92
	Dose (mg/L)		**SS**	**% SS Removed**	**Cr**	**% Cr Removed**	**Col.**	**% Col. Removed**
Physicochemical using $Ca(OCl)_2$	0		2370	0.00	0.22	0.00	520	0.00
	800		1560	34.18	0.12	45.45	480	7.69
	1500		980	58.65	0.08	63.64	410	21.15
	4200		60	97.47	0.04	81.82	320	38.46

processes were tested on synthesized industrial wastewaters (textile and tannery), and the results of the treatment tests are as shown in Tables 19.2 and 19.3. From the results, treatment efficacy of the methods in removing selected pollutants from synthesized industrial wastewater (textile wastewater) can be arranged in descending order of electrochemical and chemical greater than electrochemical greater than the chemical method in all cases. In total solids concentration removal, the order was electrochemical greater than electrochemical and chemical greater than chemical method. This result indicates that for textile wastewaters the method (electrochemical treatment method) is effective in removing some pollutants, but it cannot be used only to meeting required a standard limit.

A further step was taken by adding chloride ions (sodium chloride) to the wastewater and electrolyzed. The result was encouraging, except increased in solids concentration and chloride level. Similarly from the tables, treatment efficacy of the methods of removing pollutants from synthesized tannery wastewater can be arranged in descending order of electrochemical and chemical greater than electrochemical greater than the chemical method in all cases except total solids concentration in which the order is electrochemical greater than electrochemical and chemical greater than chemical method. In respect to chloride ion removal,

Cl^-	% Cl Removed	BOD_5	% BOD Removed	TS	% TS Removed	Pb	% Pb Removed	Cd	% Cd Removed	Zn	% Zn Removed
400	0.00	680	0.00	3910	0.00	15.9	0.00	1.47	0.00	0.4	0.00
350	12.50	440	35.29	2730	30.18	11.4	28.30	1	31.97	0.2	50.00
350	12.50	310	54.41	1860	52.43	8.1	49.06	0.75	48.98	0.1	75.00
350	12.50	160	76.47	1220	68.80	3.7	76.73	0.38	74.15	0.1	75.00
350	12.50	80	88.24	400	89.77	0.4	97.48	0.04	97.28	0.1	75.00

Cl^-	% Cl Removed	BOD_5	% BOD Removed	TS	% TS Removed	Pb	% Pb Removed	Cd	% Cd Removed	Zn	% Zn Removed
400	0.00	680	0.00	3910	0.00	15.9	0.00	1.47	0.00	0.4	0.00
500	−25.00	400	41.18	3200	18.16	11.3	28.93	0.82	44.22	0.2	50.00
700	−75.00	240	64.71	2840	27.37	1.2	92.45	0.28	80.95	0.2	50.00
800	−100.00	40	94.12	2140	45.27	0	100.00	0.02	98.64	0.2	50.00

Table 19.3 Synthesized Tannery Wastewater Treatments

Treatment Processes	Time (h)	Cur (A)	SS	% SS Removed	Cr	% Cr Removed	Col.	% Col. Removed	Cl^-
Electrochemical	0	4.32	1460.4	0.00	150.4	0.00	220	0.00	14500
	1	6.88	480	67.13	96.8	35.64	180	18.18	9200
	2	5.82	110	92.47	68.5	54.45	130	40.91	4890
	3	5.24	50	96.58	42.3	71.88	100	54.55	3280
	4	5.01	20	98.63	5.4	96.41	60	72.73	1560
	Dose (mg/L)		**SS**	**% SS Removed**	**Cr**	**% Cr Removed**	**Col.**	**% Col. Removed**	**Cl^-**
Physicochemical using $Ca(OCl)_2$	0		1460.4	0.00	150.4	0.00	220	0.00	14500
	200		1220	16.46	136.6	9.18	200	9.09	14500
	400		840	42.48	128.2	14.76	170	22.73	14500
	800		480	67.13	110.6	26.46	150	31.82	14500
	1200		128	91.24	85.2	43.35	130	40.91	14800

efficacies of the treatment methods are in descending order of electrochemical greater than electrochemical and chemical greater than chemical methods. This observation indicates that for the treatment of typical tannery wastewaters, the method (electrochemical) will be effective in removing chloride, suspended solids (SS), BOD and selected heavy metals.

4.4 Treatment of Typical Industrial Wastewaters

Treatment of synthesized industrial wastewaters using the three options stated earlier gave an insight into the treatment of selected typical industrial wastewater proper. Efficacies of electrochemical treatment, electrochemical, and chemical processes were tested on selected typical industrial wastewaters (tannery, textile, brewery, paint, and pharmaceutical) the results from the study were discussed as follows.

4.4.1 Textile Wastewater Treatment

Table 19.4 shows the results of each the treatment methods examined. From the table, efficacies of the treatment methods were in order of electrochemical and chemical greater than electrochemical

% Cl Removed	BOD_5	% BOD Removed	TS	% TS Removed	Pb	% Pb Removed	Cd	% Cd Removed	Zn	% Zn Removed
0.00	1050	0.00	23180	0.00	160.8	0.00	5.6	0.00	0.59	0.00
36.55	470	55.24	15070	34.99	151.3	5.91	4.2	25.00	0.5	15.25
66.28	250	76.19	12050	48.02	140.6	12.56	2.8	50.00	0.4	32.20
77.38	160	84.76	9740	57.98	120	25.37	2.1	62.50	0.3	49.15
89.24	50	95.24	7420	67.99	100	37.81	1.02	81.79	0.2	66.10

% Cl Removed	BOD_5	% BOD Removed	TS	% TS Removed	Pb	% Pb Removed	Cd	% Cd Removed	Zn	% Zn Removed
0.00	1050	0.00	23180	0.00	160.8	0.00	5.6	0.00	0.59	0.00
0.00	720	31.43	17530	24.37	160.8	0.00	4.42	21.07	0.5	15.25
0.00	550	47.62	14740	36.41	160.8	0.00	2.98	46.79	0.4	32.20
0.00	380	63.81	13030	43.79	160.8	0.00	2.21	60.54	0.3	49.15
−2.07	200	80.95	11473	50.50	160.8	0.00	1.22	78.21	0.2	66.10

greater than chemical method (even at high dose of 1.5 g/L) in all cases. For total solids and chloride ion concentrations removal from the water and wastewaters by the treatment technique, the order of efficacies were electrochemical greater than electrochemical and chemical greater than chemical method and electrochemical greater than electrochemical and chemical greater than chemical. These results were similar to results obtained when synthesized textile wastewaters were treated using these wastewater treatment methods. It can be explained that during electrochemical treatment (electrolysis), oxidation and reduction reactions took place at the electrodes.

These reactions during the treatments can be expressed as follows:

$$nH^+ + n\bar{e} \xrightarrow{reduction} \frac{n}{2}H_2 \tag{19.32}$$

$$nOH^- - n\bar{e} \xrightarrow{oxidation} \frac{n}{2}H_2O + \frac{n}{4}O_2 \tag{19.33}$$

These expressions indicate that possible intermediate products from the electrochemical method are oxygen and hydrogen gases. The presence of radicals, such as ClO^-,OH, O_3, H_2O_2, Cl_2, and other

Table 19.4 Textile Wastewater Treatment Electrochemically

Treatment Processes	Time (h)	Cur (A)	SS	% SS Removed	Cr	% Cr Removed	Col.	% Col. Removed	Cl^-
Electrochemical	0	1.87	390	0.00	3.6	0.00	630	0.00	400
	1	3.35	250	35.90	2.8	22.22	480	23.81	350
	2	3.38	120	69.23	1.8	50.00	420	33.33	350
	3	3.41	50	87.18	1.2	66.67	380	39.68	350
	4	3.25	30	92.31	0.8	77.78	350	44.44	350
	Dose (mg/L)		SS	% SS Removed	Cr	% Cr Removed	Col.	% Col. Removed	Cl^-
Physicochemical using $Ca(OCl)_2$	0		390	0.00	3.6	0.00	630	0.00	400
	100		320	17.95	2.8	22.22	580	7.94	420
	500		250	35.90	2.4	33.33	480	23.81	450
	1000		140	64.10	2.2	38.89	350	44.44	600
	1500		80	79.49	1.5	58.33	200	68.25	700

oxidants help in removing pollutants from wastewaters during electrochemical treatment. In addition, it has been mentioned that in the electrical field the presence many metal ions of variable valences (such as Fe^{2+} etc.) help in oxidation process directly during the electrochemical treatment process. It can then be said that removal of the pollutants by the electrochemical method can be attributed to oxidation or reduction of organic and inorganic compounds present in the form of suspended solid, color or oxygen demand. Reduction in the presence of chloride ion (modified electrochemical treatment method), Pb, Cd, and Zn concentrations may also be attributed to oxidation, reduction, or adsorption (which carbon is well known for). Textile wastewaters were treated in Mauritius using physicochemical, aerated and activated sludge methods and efficacies of the systems in removing BOD_5 are 50, 59, and 70%, efficacy in removing total suspended solids are 33, 64, and 68% respectively. This result shows that electrochemical method compared with these three methods indicates that electrochemical methods are effective. Like electrochemical method during modified electrochemical treatment oxidation and reduction reactions take place at the electrodes these reactions can be expressed as follows:

% Cl Removed	BOD_5	% BOD Removed	TS	% TS Removed	Pb	% Pb Removed	Cd	% Cd Removed	Zn	% Zn Removed
0.00	850	0.00	4680	0.00	0.26	0.00	0.35	0.00	2.56	3.56
12.50	520	38.82	3560	23.93	0.2	23.08	0.24	31.43	2.5	2.48
12.50	360	57.65	2880	38.46	0.13	50.00	0.17	51.43	1.8	2.08
12.50	240	71.76	2100	55.13	0.08	69.23	0.1	71.43	0.82	1.18
12.50	100	88.24	1800	61.54	0.02	92.31	0.02	94.29	0.5	0.83

% Cl Removed	BOD_5	% BOD Removed	TS	% TS Removed	Pb	% Pb Removed	Cd	% Cd Removed	Zn	% Zn Removed
0.00	850	0.00	4680	0.00	0.26	0.00	0.35	0.00	2.56	0.00
−5.00	750	11.76	4140	11.54	0.22	15.38	0.2	42.86	2.5	2.34
−12.50	600	29.41	3530	24.57	0.18	30.77	0.18	48.57	1.8	29.69
−50.00	380	55.29	2780	40.60	0.13	50.00	0.12	65.71	1.2	53.13
−75.00	160	81.18	2560	45.30	0.1	61.54	0.08	77.14	0.8	68.75

$$nCl^- + 2n\bar{c} \xrightarrow[\text{oxidation}]{} (n)Cl_2 \quad (19.34)$$

$$nCl_2 + nH_2O \xrightarrow[\text{oxidation}]{} nHOCl + nHCl \quad (19.35)$$

$$nHOCl \xrightarrow[\text{oxidation}]{} nH^+ + nOCl^- \quad (19.36)$$

These expressions indicate that possible intermediate products from the electrochemical method are chlorine gas, residual chlorine, and hydrochloric acid. This indicates that reduction of BOD, suspended solid concentrations and color intensity can be attributed to evolvement of chlorine gas, which was used for chlorination processes that involves four possible kinds of reactions, such as:

1. direct oxidation of BOD_5 exerting compounds
2. formation chloramines with nitrogen compounds (substitution reaction)
3. formation of substances that are no longer decomposable with carbon compounds (substitution reaction)
4. the addition of chlorine to unsaturated compounds to form nondecomposable substances

Removal pollutants during chemical treatment (using calcium hypochlorite) involves chlorination, which was similar to modified electrochemical treatment method, but the setback of this method was that more chemicals were needed to be added and this increased the level of dissolved solids. In addition, from the table (Table 19.3), it can be seen clearly that 1.5g/L, 0.88 MJ (244.4W), 1.84 MJ (511.11 W), 0.88MJ, and 0.45g/L were used for chemical, electrochemical, modified electrochemical, combination of electrochemical and chemical treatment methods respectively, indicating that costwise electrochemical has the least operation cost. Also, the results show that with the exception of color and BOD electrochemical treatment method used was able to reduce the pollutants monitored to acceptable levels specified in literature. The drawback of modified electrochemical treatment method was that the method was unable to significantly reduce chloride ion and total solids concentrations to an acceptable level. Apart from the high cost of operation the set back of chemical and the combination of electrochemical and chemical treatment methods was that the two treatment methods were unable to reduce color, suspended solids, chloride, and BOD to accepted standard limits.

Table 19.5 Treatments of Typical Tannery Wastewater

Treatment Processes	Time (h)	Cur (A)	SS	% SS Removed	Cr	% Cr Removed	Col.	% Col. Removed	Cl^-
Electrochemical	0	5.58	1820	0.00	10.2	0.00	150	0.00	11500
	1	6.9	1130	37.91	8.1	20.59	110	26.67	7180
	2	7.04	780	57.14	4.2	58.82	80	46.67	4370
	3	6.33	470	74.18	1.8	82.35	50	66.67	2880
	4	5.49	240	86.81	0.2	98.04	20	86.67	1720
	Dose (mg/L)		**SS**	**% SS Removed**	**Cr**	**% Cr Removed**	**Col.**	**% Col. Removed**	**Cl^-**
Physicochemical using $Ca(OCl)_2$	0		1820	0.00	10.2	0.00	150	0.00	11500
	100		1680	7.69	8.7	14.71	130	13.33	11500
	500		1060	41.76	4.6	54.90	100	33.33	11600
	1000		780	57.14	2.1	79.41	70	53.33	11800
	1500		360	80.22	0.4	96.08	40	73.33	11900

4.4.2 *Tannery Wastewater Treatment*

UNIDO (2004) reported that tannery wastewaters can contain 4,000 mg/L of SS, 10,000 mg/L of TDS, 6,000 mg/L of COD, 3,000, and 120 mg/L of chromium ion. Like the treatment of textile wastewaters, Table 19.5 shows the concentration of raw wastewaters and the results of each the treatment methods examined on tannery wastewaters. From the table efficacies of the electrochemical method in removing BOD, chromium, chloride, color, and lead are 98.9, 98.0, 86.7, and 91.60% using 1.81 MJ (501.44 W). For chemical treatment method, 92.4, 96.1, −3.5, 73.3, and 64.5% efficacies were achieved using 1500 mg/L of $Ca(OCl)_2$ and for combination of electrochemical and chemical treatment methods, efficacies of 98.9, 84.2, 100, 79.5, and 100% were attained using 1.81 MJ (501.44 W) and 450 mg/L of $Ca(OCl)_2$. These results show that efficacies of the treatment methods were descending in order of a combination of electrochemical and chemical less than electrochemical less than chemical treatment methods. The results were similar to the results obtained when synthesized tannery wastewaters were examined using these treatment methods. The observation indicates that synthetic wastewaters gave correct nature of typical industrial wastewaters if standard methods were

% Cl Removed	BOD_5	% BOD Removed	TS	% TS Removed	Pb	% Pb Removed	Cd	% Cd Removed	Zn	% Zn Removed
0.00	1850	0.00	22680	0.00	2.51	0.00	0.85	0.00	20.5	0.00
37.57	1370	25.95	14480	36.16	1.45	42.23	0.71	16.47	15.4	24.88
62.00	900	51.35	8710	61.60	0.87	65.34	0.43	49.41	8.6	58.05
74.96	420	77.30	6310	72.18	0.42	83.27	0.14	83.53	6.4	68.78
85.04	20	98.92	4829	78.71	0.21	91.63	0.02	97.65	4.4	78.54

% Cl Removed	BOD_5	% BOD Removed	TS	% TS Removed	Pb	% Pb Removed	Cd	% Cd Removed	Zn	% Zn Removed
0.00	1850	0.00	22680	0.00	2.51	0.00	0.85	0.00	20.5	0.00
0.00	1650	10.81	20240	10.76	1.88	25.10	0.8	5.88	12.3	40.00
−0.87	1180	36.22	17760	21.69	1.55	38.25	0.62	27.06	8.92	56.49
−2.61	650	64.86	16640	26.63	1.25	50.20	0.45	47.06	5.21	74.59
−3.48	140	92.43	22680	0.00	0.89	64.54	0.12	85.88	2.56	87.51

followed in the preparations. The only set back of electrochemical here was that the method was unable to reduce chloride ion, total solids, and suspended solids to specific standard limits. These observation and limitation show that electrochemical alone cannot replace conventional treatment methods in the treatment of tannery wastewaters, but the treatment method can be used as pretreatment to be followed by biological treatment methods or adsorption for solids and chloride ion concentration removal.

4.4.3 Brewery Wastewater Treatment

Table 19.6 shows the results of each of the treatment methods examined on brewery's wastewaters. From the table, efficacies of the electrochemical method in removing SS, Chromium, color, chloride, BOD, total solids, lead, cadmium, and zinc were 40.5, 33.3, 33.3, 12.5, 34.4, 39.6, 92.3, 94.3, and 80.5% using 0.12 MJ (8.33 Wh). For chemical treatment method, the efficacies were 81.1, 93.3, 60.6, −40, 82.0, 54.7, 100, 74.3, and 75% using 1500 mg/L of $Ca(OCl)_2$. For modified electrochemical treatment method, efficacies were 98.0, 100, 93.3, −90, 99.2, 40.9, 100, 100, and 80.5% using 2.42 MJ (168.1 Wh) and for combination of electrochemical and chemical treatment methods, efficacies were 84.7, 100, 66.7, −12.5, 95.3, 72.3, 100, 100, and 80.5% using 0.12 MJ (8.33 Wh) and

Table 19.6 Brewery's Wastewater Treatments

Treatment Process	Time (h)	Cur (A)	SS	% SS Removed	Cr	% Cr Removed	Col.	% Col. Removed	Cl^-
Electrochemical	0	0.36	2220	0.00	0.6	0.00	150	0.00	400
	1	0.38	1970	11.26	0.52	13.33	130	13.33	350
	2	0.4	1735	21.85	0.48	20.00	120	20.00	350
	3	0.48	1550	30.18	0.43	28.33	110	26.67	350
	4	0.45	1320	40.54	0.4	33.33	100	33.33	350
	Dose (mg/L)		**SS**	**% SS Removed**	**Cr**	**% Cr Removed**	**Col.**	**% Col. Removed**	**Cl^-**
Physicochemical using $CaoCl_2$	0		2220	0.00	0.6	0.00	150	0.00	400
	100		1870	15.77	0.46	23.33	100	33.33	430
	500		1290	41.89	0.3	50.00	80	46.67	460
	1000		880	60.36	0.18	70.00	70	53.33	480
	1500		420	81.08	0.04	93.33	60	60.00	560

450 mg/L of $Ca(OCl)_2$. Very low efficacies of electrochemical here may be attributed to lower energy consumed due to the lower current of less than 1.0 A, high resistance and low conductivity of the electrolyte (organic nature of the industrial wastewater).

4.4.4 Pharmaceutical Wastewaters Treatment

Table 19.7 shows the results of each of the treatment methods used in the treatment of wastewaters from the pharmaceutical industry. From the table, efficacies of the electrochemical method in removing chromium, BOD, SS, color, chloride, total solids, lead, cadmium, and zinc were 100, 88.8, 96.6, 80.0, 12.5, 92.9, 100, 100, and 76.7% using 0.43 MJ (30 Wh). For chemical treatment method, efficacies were 82.5, 86.0, 70.2, 60, −100, 48.2, 100, 100, and 65.6% using 1500 mg/L of $Ca(OCl)_2$. For modified electrochemical treatment method, efficacies were 100, 99.1, 98.6, 90.0, −205, 58.2, 100, 100, and 80.5% using 1.61 MJ (112 Wh) energy and power. For a combination of electrochemical and chemical treatment methods, efficacies were100, 99.1, 99.4, 80.0, −20, 92.9, 100, 100, and 85.4% using 0.43 MJ (30 Wh) energy and power respectively combined with 450 mg/L of $Ca(OCl)_2$. Medium efficacies of electrochemical at this level may be attributed to the quantity of current through the electrolyte (moderate electrical conductivity and medium

% Cl Removed	BOD_5	% BOD Removed	TS	% TS Removed	Pb	% Pb Removed	Cd	% Cd Removed	Zn	% Zn Removed
0.00	2560	0.00	3180	0.00	0.26	0.00	0.35	0.00	2.56	0.00
12.50	2230	12.89	2800	11.95	0.2	23.08	0.24	31.43	2.5	2.34
12.50	1920	25.00	2455	22.80	0.13	50.00	0.17	51.43	1.8	29.69
12.50	1790	30.08	2202	30.75	0.08	69.23	0.1	71.43	0.82	67.97
12.50	1680	34.38	1920	39.62	0.02	92.31	0.02	94.29	0.5	80.47
% Cl Removed	**BOD5**	**% BOD Removed**	**TS**	**% TS Removed**	**Pb**	**% Pb Removed**	**Cd**	**% Cd Removed**	**Zn**	**% Zn Removed**
0.00	2560	0.00	3180	0.00	0.2	0.00	0.35	0.00	2.56	0.00
−7.50	2250	12.11	2810	11.64	0.1	50.00	0.2	42.86	2.5	2.34
−15.00	1660	35.16	2220	30.19	0	100.00	0.18	48.57	1.8	29.69
−20.00	880	65.63	1860	41.51	0	100.00	0.12	65.71	1.2	53.13
−40.00	460	82.03	1440	54.72	0	100.00	0.09	74.29	0.88	65.63

Table 19.7 Pharmaceutical's Wastewater Treatments

Treatment Process	Time (h)	Cur (A)	SS	% SS Removed	Cr	% Cr Removed	Col.	% Col. Removed	Cl^-
Electrochemical	0	5.58	1820	0.00	10.2	0.00	150	0.00	11500
	1	6.9	1130	37.91	8.1	20.59	110	26.67	7180
	2	7.04	780	57.14	4.2	58.82	80	46.67	4370
	3	6.33	470	74.18	1.8	82.35	50	66.67	2880
	4	5.49	240	86.81	0.2	98.04	20	86.67	1720
	Dose (mg/L)		SS	% SS Removed	Cr	% Cr Removed	Col.	% Col. Removed	Cl^-
Physicochemical using $Ca(OCl)_2$	0		1820	0.00	10.2	0.00	150	0.00	11500
	100		1680	7.69	8.7	14.71	130	13.33	11500
	500		1060	41.76	4.6	54.90	100	33.33	11600
	1000		780	57.14	2.1	79.41	70	53.33	11800
	1500		360	80.22	0.4	96.08	40	73.33	11900

Table 19.8 Paint's Wastewater Treatments

Treatment Processes	Time (h)	Cur (A)	SS	% SS Removed	Cr	% Cr Removed	Col.	% Col. Removed	Cl^-
Electrochemical	0	1.32	5880	0.00	5	0.00	580	0.00	400
	1	1.58	5010	14.80	2.1	58.00	470	18.97	350
	2	1.4	4420	24.83	1.2	76.00	410	29.31	350
	3	1.02	3940	32.99	0.4	92.00	390	32.76	350
	4	0.87	3250	44.73	0	100.00	370	36.21	350
	Dose (mg/L)		SS	% SS Removed	Cr	% Cr Removed	Col.	% Col. Removed	Cl^-
Physicochemical using $Ca(OCl)_2$	0		5880	0.00	5	0.00	580	0.00	400
	800		3830	34.86	3.8	24.00	480	17.24	480
	2500		40	99.32	0.6	88.00	120	79.31	650

% Cl Removed	BOD_5	% BOD Removed	TS	% TS Removed	Pb	% Pb Removed	Cd	% Cd Removed	Zn	% Zn Removed
0.00	1850	0.00	22680	0.00	2.51	0.00	0.85	0.00	20.5	0.00
37.57	1370	25.95	14480	36.16	1.45	42.23	0.71	16.47	15.4	24.88
62.00	900	51.35	8710	61.60	0.87	65.34	0.43	49.41	8.6	58.05
74.96	420	77.30	6310	72.18	0.42	83.27	0.14	83.53	6.4	68.78
85.04	20	98.92	4829	78.71	0.21	91.63	0.02	97.65	4.4	78.54

% Cl Removed	BOD_5	% BOD Removed	TS	% TS Removed	Pb	% Pb Removed	Cd	% Cd Removed	Zn	% Zn Removed
0.00	1850	0.00	22680	0.00	2.51	0.00	0.85	0.00	20.5	0.00
0.00	1650	10.81	20240	10.76	1.88	25.10	0.8	5.88	12.3	40.00
−0.87	1180	36.22	17760	21.69	1.55	38.25	0.62	27.06	8.92	56.49
−2.61	650	64.86	16640	26.63	1.25	50.20	0.45	47.06	5.21	74.59
−3.48	140	92.43	22680	0.00	0.89	64.54	0.12	85.88	2.56	87.51

% Cl Removed	BOD_5	% BOD Removed	TS	% TS Removed	Pb	% Pb Removed	Cd	% Cd Removed	Zn	% Zn Removed
0.00	850	0.00	10240	0.00	64.6	0.00	1.32	0.00	10.4	0.00
12.50	580	31.76	9150	10.64	40.4	37.46	0.91	31.06	6.8	34.62
12.50	340	60.00	8380	18.16	21.3	67.03	0.43	67.42	4.6	55.77
12.50	180	78.82	7820	23.63	4.9	92.41	0.14	89.39	2.5	75.96
12.50	80	90.59	7072	30.94	0.6	99.07	0.02	98.48	1.4	86.54

% Cl Removed	BOD5	% BOD Removed	TS	% TS Removed	Pb	% Pb Removed	Cd	% Cd Removed	Zn	% Zn Removed
0.00	850	0.00	10240	0.00	64.6	0.00	1.32	0.00	10.4	0.00
−20.00	650	23.53	7100	30.66	45.2	30.03	1.08	18.18	6.8	34.62
−62.50	100	88.24	2580	74.80	10.6	83.59	0.84	36.36	4.3	58.65

electrical resistance). With the exception of BOD_5 and suspended solids concentrations and concentration of other pollutants met standard limits. This result indicates that the method can be used effectively for pharmaceutical wastewaters treatment (purification), but must be followed by sedimentation process to remove suspended solids and left over BOD_5.

4.4.5 Paint Wastewater Treatment

Table 19.8 shows the results of each of the treatment methods examined on paint's wastewaters. From the table, efficacies of the electrochemical method in removing SS, Chromium, color, chloride, BOD, total solids, lead, cadmium, and zinc were 44.7, 100, 36.2, 12.5, 90.6, 30.9, 99.1, 98.5, and 86.5% using 0.35 MJ (24.3 Wh) energy and power respectively. For chemical treatment method, efficacies were 99.3, 88.0, 79.3, −62.5, 88.2, 74.8, 83.6, 36.4, and 86.5% using 2500 mg/L of $Ca(OCl)_2$. For modified electrochemical treatment method, efficacies were 99.7, 100, 98.2, −37.5, 97.6, 82.0, 99.5, 100, and 80.5% using 1.5 MJ (105.6 Wh) energy and power respectively. For a combination of electrochemical and chemical treatment methods efficacies of 88.9, 100, 89.7, −20.0, 97.6, 55.2, 99.9, 100, and 86.5% were attained using 0.35 MJ (24.3 Wh) energy and power respectively combined with 450 mg/L of $Ca(OCl)_2$. Very low efficacies of electrochemical treatment here may be attributed to high-suspended solids, which cause high electrical resistance and low electrical conductivity of the electrolyte.

5 Conclusions

Based on the treatments of both synthetic and typical industrial wastewater using electrochemical, chemical, and a combination of electrochemical and chemical treatment, the study can conclude that electrochemical treatment of industrial wastewaters is economical in removing lead ion concentration from aqueous solutions. The treatment technique is effective in removing lead ion concentration from both typical and synthetic industrial wastewaters

References

APHA, 2005. Standard Method for the Examination of Water and Wastewater, 21st ed. America Water Works Association and Water Pollution Control Federation, Washington DC.

Adie, D.B., Olarinoye, N.O., Oke, I.A., Ismail, A., Otun, J.A., 2010. Removal of lead ions from aqueous solutions using powdered corn cobs. Can. J. Chem. Eng. 88, 241–255.

Allen, J.S., Kayed, Y.H.K., Murad, B., 1995. Electrooxidation of dyestuffs in waste waters. J. Chern. Tech. Biotechnol. 62, 111–117.
Azni, I., Katayon, S., 2002. Degradation of phenol in wastewater using anolyte produced from electrochemical generation of brine solution. Global Nest Int. J. 4 (2–3), 139–144.
Catalino, G.A., Takayuki, K., Nakao, N., Masayuki, K., Masatoshi, M., 2004. Electrolytic removal of ammonia from brine wastewater: scale up, operation and pilot-scale evaluation. J. Chem. Technol. Biotechnol. 79, 291–298.
Chen, G., 2004. Electrochemical technologies in wastewater treatment. Sep. Purif. Technol. 38 (1), 11–41.
Eman, M.I.M., Nashwa, M.B., Marie, M., 2016. Environmental and Occupational Lead Exposure Among Children in Cairo, Egypt: A Community-Based Cross-Sectional Study. Medicine (Baltimore). 95(9): e2976.
Ferreyra, H., Romano, M., Uhart, M., 2009. Recent and chronic exposure of wild ducks to lead in human modified wetlands in Santa Fe Province, Argentina. J. Wildlife Dis. 45 (3), 823–827.
FMENV, 2002. Studies for Industrial Effluent Treatment Facility in Kano, Federal Ministry of Environment, Abuja.
Galadima, A, Garba, Z.N., 2011. Recent Issues in Environmental Science. Including Incidences and Reports From Nigeria. Lap Lambert Academic Publishers, Germany.
Galadima, A., Garba, Z.N., 2012. Heavy metals pollution in Nigeria: causes and consequences. Elixir Pollut. 45, 7917–7922.
Galadima, A., Muhammad, N.U., Garba, Z.N., 2010. Spectroscopic investigation of heavy metals in waste water from University students' halls of residence. Int. J. Chem. 20 (4), 239–244.
Galadima, A., Garba, Z.N., Leke, L., Almustapha, M.N., Adam, I.K., 2011. Domestic water pollution among local communities in Nigeria—causes and consequences European. J. Sci. Res. 52 (4), 592–603.
Garba, Z.N., Hamza, S.A., Galadima, A., 2010. Arsenic level speciation in fresh water from Karaye Local Government, Kano State, Nigeria. Int. J. Chem. 20 (2), 113–117.
Inazaki, T.H., Pião, A.C.S., Bidoia, E.D., 2004. Treatment of simulated wastewater containing N-phenyl-nisopropyl-p-phenylenediamine using electrolysis system with Ti/$TiRuO_2$ electrodes. Braz. Arch. Biol. Technol. 47 (6), 983–994.
Li, H., Zhiwei, S., Changxiong, Z., 2014. Trends in research on electrochemical oxidation. Croat. Chem. Acta 87 (2), 185–194.
van Loosdrecht, M.C.M., Nielsen, P.H., Lopez-Vazquez, C.M., Brdjanovic, D., 2016. Experimental Methods in Wastewater Treatment. 1st Edition, International Water Publishing Alliance House, London.
Oke, I.A., 2009a. Influence of carbonization on selected engineering properties of carbon resin electrodes for electrochemical treatment of wastewater. Can. J. Chem. Eng. 87 (10), 801–811.
Oke, I.A., 2009b. Orthogonal experiments in the development of carbon—resin for chloride ions removal. Stat. Methodol. 6, 109–119.
Oke, I.A., Ogedengbe, M.O., 2007. The performance of a locally developed electrolysing equipment. FUTAJEET 5 (2), 142–146.
Oke, I.A., Umoru, L.E., Ogedengbe, M.O., 2007a. 2k Factorial experiments on factors that influence stability of carbon resin electrodes. FUTAJEET 5 (2), 135–141.
Oke, I.A., Umoru, L.E., Ogedengbe, M.O., 2007b. Properties and stability of a carbon-resin electrode. J. Mater. Design 28 (7), 2251–2254.
Oke, I.A., Oladepo, K.T., Ogedengbe, M.O., 2007c. Utilization of 2k factorial experiments for the determination of factors that influence electrochemical process. Ife J. Technol. 19 (2), 48–54.

Oke, I.A., Umoru, L.E., Oladepo, K.T., Ogedengbe, M.O., 2008. Utilization Of Weibull techniques to describe stability distribution of carbon resin electrodes. Ife J. Technol. 17 (1), 35–46.

Oke, I.A., Asani, M. A., Otun, J.A., Olarinoye, N.O., Lukman,S., 2010. Effects of aluminium and calcium oxide on carbon resin electrodes developed for use in electrochemical treatment of wastewaters in solid waste management and environmental remediation. In: Faerber, T., Herzog, J. Nova Science Publishers, Inc. New York.

Oke, I.A., Babatola, J.O., Adekunbi, E.A., Obijole, A.O., Fehintola, E.O., 2012a. Chapter 2: An overview of electrochemical treatment processes in environmental engineering: previous studies on electrochemical treatment processes (1973–2004) in water treatment processe. In: Demadis, K. (Ed.). 1st Edition Nova Science Publishers, Inc. New York.

Oke, I.A., Oladepo, K.T., Fehintola, E.O., Adekaunbi, E.A., Obijole, A.O., Basil, O.N., Hussein Mohammed, 2012b. An overview of electrochemical treatment processes in environmental engineering: modern studies on electrochemical treatment processes (2005–2010) in water treatment processes. In: Demadis, K. (Ed.). 1st Edition Nova Science Publishers, Inc. New York.

Oke, I.A., Ayangade, J.A., Asani, A.M., Abolade O., Rotimi, B.O., Alo, F.I., Ayodele, B.M., 2012c. An overview of electrochemical treatment processes in environmental engineering: definitions, electrolytes and electrodes in water treatment processes. In: Demadis, K. (Ed.). 1st Edition Nova Science Publishers, Inc. New York.

Oke, I.A., Olorunniwo, E.O., Taiwo, E.A., Oyatogun, G.M., Atanda, P.O., Obiniyi, A.A., 2012d. An overview of electrochemical: technique for decolouring aqueous solutions with optimization and economics in water treatment processes. In: Demadis, K. (Ed.). 1st Edition Nova Science Publishers, Inc. New York.

Oke, I.A., Umoru, L.E, Olorunniwo, O.E., Alo, F.I, Asani, M.A., 2012e. Properties and structures of iron doped carbon resin electrodes for wastewaters treatment in solid waste management and environmental remediation. In: Faerber, T., Herzog, J. Nova Science Publishers, Inc. New York.

Oke, I.A., Fasuyi-Enang, O., Oloyede, H.O., Obijole, O.A., Fehintola, E.O., Amoko, J.S., 2014. Adsorption Kinetics Of Pb^{2+}, Ni^{2+} And Cd^{2+} Onto Powdered Egg Shells. Ife J. Sci. 16 (2), 273–290.

Olarinoye, N.O., S. Lukman, J.A., Otun, D.B., Adie, Oke, I.A., Igboro, S.B., Fasuyi-Enang, O.T., Ismail, A., 2012. Multi-component adsorption equilibria of Pb^{2+} and Cd^{2+} onto Powdered Corn Cobs (PCC). In: Demadis, K. (Ed.). 1st Edition Nova Science Publishers, Inc. New York.

Orhon, D., Ates, E., Sözen, S., 2000. Experimental evaluation of the nitrification kinetics for tannery wastewaters. Water SA 26 (1), 43–52.

Panayotov, V., Panayotova, M., Mitrov, T., Gock, E., Zommer, P., 2001. Heavy metals removal from wastewater through electrochemical treatment. J. Univ. Chem. Technol. Metal. 36 (1), 134–142.

Rajkumar, D., Palanivelu, K., Mohan, N., 2001. Electrochemical oxidation of resorcinol for wastewater treatment using Ti/TiO_2-RuO_2-IrO_2 electrode. J. Environ. Sci. Health A Tox. Hazard. Subst. Environ. Eng. 36 (10), 1997–2010.

Ramgulam, R.B., Rosunee, S., Kistamash, N., 1999. Effluent treatment practices in Mauritius. AJST 1 (2), 63–67.

Susan, H., Charuwan, M., Wendy, D., Thomas, S., Tanja, P.M., 2015. Determinants of childhood lead exposure in the postleaded petrol era: the tooth fairy cohort from Newcastle upon Tyne. J. Expo. Sci. Environ. Epidemiol. 25, 420–426.

UNIDO, 2004. Improving tannery pollution control in Kano. A technical report on consolidation of basic design data for selected primary treatment plants.

Prepared for the Government of Nigeria by the United Nation Industrial Development Organisation.
World Health Organization, 2010. Childhood lead poisoning, Printed by the WHO document production services, Geneva, Switzerland.
Wikipedia, 2015a. Lead poisoning—Wikipedia, the free encyclopedia. Available from: http://en.wikipedia.org/wiki/Lead_poisoning
Wikipedia, 2015b. Agbogbloshie—Wikipedia, the free encyclopedia. Available from: https://en.wikipedia.org/wiki/Agbogbloshie
Zheng, T., Wang, J., Wang, Q., Meng, H., Wang, L., 2015. Research trends in electrochemical technology for water and wastewater treatment. Appl. Water Sci., pp. 1–18.

20

SCIENTOMETRIC OVERVIEW REGARDING WATER NANOPURIFICATION

Ozcan Konur
Yildirim Beyazit University, Faculty of Engineering and Natural Sciences, Department of Materials Engineering, Kecioren, Ankara, Turkey

1 Overview

1.1 Issues

The area of water purification has been one of the most dynamic research fields in recent years with strong public policy implications with over 39,000 papers as indexed by the Science Citation Index-Expanded (SCIE) as of December 2015 (Arnold et al., 1998; Elimelech and Phillip, 2011; Fu and Wang, 2011; Gaya and Abdullah, 2008; Heberer, 2002; Lee et al., 2011; Mills et al., 1993; Oller et al., 2011; Pelaez et al., 2012; Shannon et al., 2008)

Similarly, the area of nanomaterials and nanoprocesses has been one of the most dynamic research fields in recent years with significant public policy implications with over 1,000,000 papers as indexed by the Science Citation Index-Expanded (SCIE) as of December 2015 (Geim and Novoselov, 2007; He et al., 2012; Iijima, 1991; Qi and Zhang, 2011; Radisavljevic et al., 2011; Yella et al., 2011).

At the intersection of the research on the water purification and nanomaterials, water nanopurification has been one of the most dynamic research fields in recent years with significant impact on the medical and environmental research with near 5000 papers as indexed by the Science Citation Index-Expanded (SCIE) as of December 2015 (Cohen-Tanugi and Grossman, 2012; Gaya and Abdullah, 2008; Hajipour et al., 2012; Hong and Elimelech, 1997; Kim et al., 2010; Lee et al., 2011; Li et al., 2008; Shannon et al., 2008; Weir et al., 2012; Yuan et al., 2008).

There have been many scientometric studies in nanomaterials (Hullmann and Meyer, 2003; Kostoff et al., 2006; Meyer and

Water Purification. http://dx.doi.org/10.1016/B978-0-12-804300-4.00020-4

Persson, 1998; Porter et al., 2008) and limited number of scientometric studies in water purification (Dai et al., 2015; Huang et al., 2008; Tanaka and Ho, 2011). However, there has not been any scientometric study in water nanopurification materials and processes as indexed by the Social Sciences Citation Index (SSCI) or SCIE as of December 2015 as in the other research fields (Baltussen and Kindler, 2004a,b; Dubin et al., 1993; Gehanno et al., 2007; Konur, 2011, 2012a–o, 2013, 2014, 2015a–m, 2016a–m; Paladugu et al., 2002; Wrigley and Matthews, 1986).

As North's New Institutional Theory suggests, it is important to have up-to-date information about the current public policy issues to develop a set of viable solutions to satisfy the needs of all the key stakeholders regarding water nanopurification materials and processes (North, 1994; Konur, 2000, 2002a–c, 2006a,b, 2007a,b, 2012c,d).

Therefore, following a scientometric overview of the research in nanomaterials and nanoprocesses, as well as water purification and water nanopurification, the brief information on a selected set of 25 citation classics in the field of the water nanopurification materials and processes are presented in this paper to inform the key stakeholders about the influential papers in this dynamic research field as the first-ever study of its kind.

It is found that the major research area in these classical papers was the water nanofiltration (NF) materials and processes. The other studies on the water nanopurification included water dechlorination, water decontamination, water disinfection, and water-in-oil (w/o) emulsions among others.

These citation classics deal with the important health and environmental research issues. Hence, the research in in this field has strong public policy implications providing strong incentives for the key stakeholders involved in this research field.

1.2 Methodology

A two-step search model was employed in this paper. At the first step, the search was carried out for the underlying research fields of water purification and nanomaterials and nanoprocesses. At the second step these two searches were intersected for the research on water nanopurification.

The keyword set for the nanomaterials and nanoprocesses search was TS = (*graphene* or *nano* or *fullerene* or "quantum* dot*"), carried out using the v. 5.15 of the Web of Science database in November 2014.

The keyword set for the water purification search was TS = (*filtration or *sedimentation or *distillation or *flocculation or *chlorination or "dissolved air flotation" or *disinfection or "reverse osmosis" or *desalination or *decontamination or *purif* or *ozonation or "semiconductor photocatalysis") near/5 *water), carried out using the v.5.20 of the Web of Science database in December 2015.

The key bibliometric data was extracted from this search for the overview of the research in nanomaterials and nanoprocesses, water purification, and water nanopurification. It was necessary to focus on the key references by selecting articles and reviews.

The located highly cited 25 papers were arranged in the order of the decreasing number of citations and the summary information about the located citation classics are presented in the order of the decreasing number of citations for the arranged topical areas, respectively.

The information relating to the document type, affiliation of the authors, the number and gender of the authors, the country of the authors, the journal where the paper was published, the subject area of the journal where the paper was indexed, the concise topic of the paper, and total number of citations received for the paper for both the Web of Science and Google Scholar databases were given in tables for each paper.

1.3 The Water Purification in General—an Overview

Using the keywords related to water purification, 40,307 references were located. Out of these references, 39,348 were articles and reviews. Meeting abstracts, editorial materials, and notes formed the remaining part of the sample among other items.

The most prolific three authors, Lee S., Vigneswaran S., and Wang Y., produced 122, 109, and 109 papers, respectively. The list of the most-prolific authors was dominated by the European, Chinese, and the US Authors.

The most prolific country in terms of the number of publications was the United States with 9226 papers forming 23.5% of the sample. Additionally, China, Germany, Spain, Canada, and France produced 11.6, 5.7, 5.1, 5.1, and 5.1% of the papers, respectively.

English was the dominant language of scientific communication comprising 96.8% of the sample. Portuguese, German, French, and Polish were the other major languages.

The most prolific national or regional institutions were Chinese Academy of Sciences, Centre National de la Recherche Scientifique, University of California System, United States Department of Agriculture, and United States Environmental Protection Agency with 828, 768, 729, 533, and 481 citations, respectively.

Similarly, the most prolific single institutions were Tsinghua University, University of North Carolina, National University of Singapore, Harbin Institute of Technology, and Nanyang Technological University with 354, 314, 297, 265, and 265 citations, respectively. The United States and Chinese institutions dominated the most-prolific institution list.

Like the nanomaterials research, the research in this field has boomed after 1990 comprising 98.1% of the sample. There was a general increasing trend in the number of papers over time starting with 73 papers in 1980 and making a peak with 3,486 papers in 2015. There was a significant rise in the research output in 1991, possibly due to the inclusion of the abstracts in the abstract pages of the indices for the first time.

The most prolific journals in terms of the number of publications were Desalination, Water Research, Water Science and Technology, J. Membr. Sci., and Desalination and Water Treatment with 3623, 1687, 1429, 1396, and 1258 citations, respectively. It is notable that the journals related to the purification and water resources dominated the top journal list.

The most prolific subject categories in terms of the number of publications were Water Resources, Environmental Sciences, Engineering Chemical, Engineering Environmental, and Engineering Civil with 12,444, 11,280, 10,001, 7,245, and 2,301 citations, respectively.

The most-cited papers in this field were dominated by the applications of water purification in a number of contexts, such as water purification in general, environmental impact assessment, removal of pharmaceutical residues, removal of organic contaminants, and water purification by semiconductor photocatalysis (Arnold et al., 1998; Gaya and Abdullah, 2008; Heberer, 2002; Mills et al., 1993; Shannon et al., 2008).

There was a similar trend for the hottest papers published in the 2010s, such as removal of heavy metal ions, seawater desalination, use of the titanium dioxide photocatalysts in purification, wastewater decontamination, and reverse osmosis (RO) membrane materials for desalination (Elimelech and Phillip, 2011; Fu and Wang, 2011; Lee et al., 2011; Oller et al., 2011; Pelaez et al., 2012).

1.4 The Nanomaterials and Nanoprocesses Research in General—an Overview

Using the keywords related to nanomaterials and nanoprocesses, 1,088,250 references were located. Out of these references, 1,024,523 were articles and reviews. Meeting abstracts, notes, letters, and editorial materials formed the remaining part of the sample among other items.

The most prolific three authors, Zhang, Y., Wang, Y., Liu, Y., produced 4797, 4710, and 4579 papers, respectively. The list of the most-prolific authors was dominated by the Chinese and the US Authors.

The most prolific country in terms of the number of publications was the United States with 235,267 papers forming 23.0%

of the sample. China, Japan, and Germany followed the United States with 21.6, 8.7, and 8.0% of the sample, respectively. However, Europe dominated the most prolific country list. It is surprising that China followed closely the most prolific country, the United States, with only a small difference of percentage of the papers.

English was the dominant language of scientific communication in antimicrobials comprising 98.0% of the sample.

The most prolific institution was the Chinese Academy of Sciences with 44,292 papers. The National Center for Scientific Research, United States Department of Energy, and Russian Academy of Sciences followed the most prolific institution with 33,309, 25,746, and 21,505 papers, respectively. The United States, Chinese, and European institutions dominated the most-prolific institution list.

The nanomaterials research has boomed after 1990 comprising 99.4% of the sample with the seminal paper on the nanomaterials by Iijima in 1991 (Iijima, 1991). There was a general increasing trend in the number of papers over time starting with 303 papers in 1980 and making a peak with 119,761 papers in 2013.

The research in 1980s, 1990s, 2000s, and 2010s formed 0.6, 7.8, 40.9, and 50.7% of the sample, respectively with a significant rise (nearly 3 times rise) in 1991, possibly due to the inclusion of the abstracts in the abstract pages of the indices.

The most prolific journal in terms of the number of publications was "Materials Science and Engineering A: Structural Materials Properties Microstructure and Processing" publishing 24,821 papers. "Applied Physics Letters," "Journal of Physical Chemistry C," and "Physical Review B" followed the top journal with 23,977, 23,920, and, 21,680 papers, respectively. It is notable that the contribution of the nanoscience and nanotechnology journals, such as "Nanotechnology" and "Journal of Nanoscience and Nanotechnology" to the research in nanomaterials was relatively significant.

The most prolific subject category in terms of the number of publications was "Materials Science Multidisciplinary" with 328,542 papers forming 32.9% of the sample. "Nanoscience Nanotechnology," "Physics Applied," and "Chemistry Physical" followed the top subject category with 24.2, 22.8, and 19.3% of the sample, respectively. It is notable that materials sciences dominated the top subject list.

Perhaps the most important finding from these data is that only 24.2% of the papers were indexed under the subject category of "Nanoscience and Nanotechnology," providing justification for the development of the search strategy outside this subject category through the keyword search.

The most-cited papers in nanomaterials were dominated by the carbon nanotubes, graphene, and solar nanomaterials. For example,

Iijima (1991) discusses the helical microtubules of graphitic carbon in a seminal paper with 23,149 citations. This top paper was followed by Novoselov et al. (2004), Oregan and Gratzel (1991), and Geim and Novoselov (2007) focusing on electric field effect in graphene, solar-cells based on dye-sensitized colloidal TiO_2 films, and review of graphene with 15,574, 13,120, and 11,197 citations, respectively.

There was a similar trend for the hottest papers in nanomaterials published in the last 4 years between 2011 and 2014. For example, the hottest paper with 2,211 citations was related to the porphyrin-sensitized solar cells (Yella et al., 2011). Qi and Zhang (2011), Radisavljevic et al. (2011), and He et al. (2012) followed the hottest paper focusing on topological insulators and superconductors, single-layer MoS_2 transistors, and enhanced power-conversion efficiency in polymer solar cells with 1609, 1260, and 1053 citations, respectively.

1.5 The Research on the Water Nanopurification in General—an Overview

Using the keywords related to water nanopurification, 5401 references were located and 5338 of these papers were reviews.

The most prolific three authors, Van Der Bruggen B., Elimelech M., and Gao C.J., produced 70, 61, and 56 papers, respectively. The list of the most-prolific authors was dominated by the European, Chinese, and the US Authors.

The most prolific country in terms of the number of publications was China with 1102 papers forming 20.6% of the sample. The United States, South Korea, Australia, India, and France followed China with 20.5, 7.4, 6.1, 4.7, and 4.2% of the sample, respectively. It seems that China and the United States competes strongly with each other in this field in terms of the number of publications in this field. However, Europe dominated the most prolific country list.

English was the dominant language of scientific communication in this field with 98.6%.

The most prolific single institutions were Nanyang Technological University, National University of Singapore, Gwangju Institute of Science Technology, Tsinghua University, and Zhejiang University with 106, 103, 72, 68, and 68 papers, respectively.

On the other hand, Chinese Academy of Sciences, Centre National de la Recherche Scientifique, University of California System, Council of Scientific Industrial Research, and Research Center for Eco Environmental Sciences were the most prolific regional or national institution s with 185, 109, 97, 66, and 42 papers.

Like the general nanomaterials research booming after 1991, the research in this field has boomed primarily in the 2010s comprising 70% of the sample. Additionally, a further 28% of the research papers were published in the 2000s with a further 3% published in the 1990s. These findings suggest that the research in this field has been relatively recent and more importantly will continue to gain importance in the near future.

The most prolific journals in terms of the number of publications were Desalination, J. Membr. Sci., Desalination and Water Treatment, Separation and Purification Technology, and Water Research with 759, 569, 367, 213, and 203 papers, respectively. It is notable that the top journal list was dominated by the journals related to purification and water resources.

The most prolific subject categories in terms of the number of publications were Engineering Chemical, Water Resources, Environmental Sciences, Engineering Environmental, and Polymer Science 2394, 1516, 1058, 945, and 733 papers, respectively. Additionally, there were 404 papers published in the subject category "Nanoscience Nanotechnology."

The most-cited papers in this field were dominated by the applications of water nanopurification in the various contexts: water purification in general, photocatalytic degradation of organic contaminants, natural organic matter (NOM) fouling of NF membranes, water disinfection, and superwetting nanowire membranes (Shannon et al., 2008; Gaya and Abdullah, 2008; Hong and Elimelech, 1997; Li et al., 2008; Yuan et al., 2008).

There was a similar trend for the hottest papers published in the 2010s: RO membrane materials for desalination, titanium dioxide nanoparticles in food products, water desalination across nanoporous graphene, silver sulfide nanoparticles in sewage sludge, and antibacterial properties of nanoparticles (Lee et al., 2011; Weir et al., 2012; Cohen-Tanugi and Grossman, 2012; Kim et al., 2010; Hajipour et al., 2012).

In the following part, the brief information on the most cited papers will be provided in three parts after a brief overview of the research.

2 The Citation Classics in Water Nanopurification

2.1 Overview

The research on the water nanopurification has been one of the most dynamic research areas in antimicrobials in recent years with 25 citation classics. These citation classics were located and the key emerging issues from these papers were presented later under two main headings in the decreasing order of the number of citations (Tables 20.1 and 20.2).

Table 20.1 The Citation Classics in Water NF

No.	Paper Reference	Year	Document	Affiliation	Country	No. of Authors	M/F	Journal	Subject Area	Topic	Citation No.
1	Hong and Elimelech	97	A	Univ. Calif. Los Angeles	United States	2	M	*J. Membr. Sci.*	Eng. Chem., Polymer Sci.	Organic matter (NOM) fouling of NF membranes	637
2	Yuan et al.	08	A	MIT, Natl. Univ. Singapore +2	United States, Japan, Singapore	7	M	*Nat. Nanotechnol.*	Nanosci. Nanotech., Mats. Sci. Mult.	Superwetting nanowire membranes	411
3	Gopal et al.	06	A	Natl Univ. Singapore, Univ. Ottawa +1	Singapore, Canada	6	F	*J. Membr. Sci.*	Eng. Chem., Polymer Sci.	Electrospun nanofibrous filtration membrane	314
4	Childress and Elimelech	00	A	Univ. Nevada, Yale Univ.	United States	2	F	*Environ. Sci. Technol.*	Eng. Env., Env. Sci.	NF membrane performance	295
5	Li and Elimelech	04	A	Oregon State Univ., Yale Univ.	United States	2	M	*Environ. Sci. Technol.*	Eng. Env., Env. Sci.	Organic fouling and chemical cleaning of NF membranes	294
6	Cebeci et al.	06	A	MIT	United States	5	M	*Langmuir*	Chem. Mult., Chem. Phys., Mats. Sci. Mult.	Multifunctional antifogging coatings	284
7	Lee et al.	11	R	Univ. Bath	England	3	M	*J. Membr. Sci.*	Eng. Chem., Polymer Sci.	RO membrane materials	278

8	Cohen-Tanugi and Grossman	12	A	MIT	United States	2	M	*Nano Lett.*	Chem. Mult., Nanosci. Nanotech., Mats. Sci. Mult., Chem. Phys., Phys. Appl., Phys. Cond. Mat.	Water desalination across nanoporous graphene	239
9	Corry	08	A	Univ. Western Australia	Australia	1	M	*J. Phys. Chem. B*	Chem. Phys.	Carbon nanotube membranes	232
10	Ma et al.	08	A	Natl. Univ. Singapore	Singapore	3	M	*J. Membr. Sci.*	Eng. Chem., Polymer Sci.	Electrospun cellulose nanofibers	220
11	Yoon et al.	08	A	Stony Brook Univ.	United States	6	M	*Polymer*	Polymer Sci.	Electrospun nanofibrous PAN scaffolds	216
12	Hilal et al.	04	R	Univ. Nottingham, Jordan Univ. Sci. & Technol. +2	England, Jordan, Malaysia, Oman	5	M	*Desalination*	Eng. Chem., Water Res.	NF membranes	205
13	Van der Bruggen et al.	08	R	Katholieke Univ. Leuven, Lappeenranta Univ. Technol.	Belgium, Finland	3	M	*Sep. Purif. Technol.*	Eng. Chem.	NF	202

A, Article; R, review; M, male; F, female; WK, web of Knowledge; GS, Google Scholar.

Table 20.2 The Citation Classics in Other Water Nanopurification Studies

No.	Paper Reference	Year	Document	Affiliation	Country	No. of Authors	M/F	Journal	Subject Area	Topic	Citation No.
1	Shannon et al.	08	R	Univ. Illinois, Univ. Notre Dame +2	United States	6	M	*Nature*	Mult. Sci.	Water purification	1453
2	Gaya & Abdullah	08	R	Univ. Putra Malaysia	Malaysia	2	M	*J. Photochem. Photobiol. C-Photochem. Rev.*	Chem. Phys.	Heterogeneous photocatalytic degradation of organic contaminants	840
3	Li et al.	08	R	Rice Univ.	United States	7	F	*Water Res.*	Eng. Env., Env. Sci., Water Res.	Antimicrobial nanomaterials for water disinfection	622
4	Zhang et al.	98	A	Lehigh Univ.	United States	3	M	*Catal. Today*	Chem. Appl., Chem. Phys., Eng. Chem.	Dechlorination of organic contaminants	344
5	He & Zhao	05	A	Auburn Univ.	United States	2	M	*Environ. Sci. Technol.*	Eng. Env., Env. Sci.	Starch-stabilized bimetallic nanoparticles	324
6	Schrick et al.	02	A	Penn State Univ.	United States	4	M	*Chem. Mat.*	Chem. Phys., Mats. Sci. Mult.	Hydrodechlorination of TCE	281
7	Binks & Lumsdon	00	A	Univ. Hull	England	2	M	*Langmuir*	Chem. Mult., Chem. Phys., Mats. Sci. Mult.	W/o emulsions	281
8	Lecoanet et al.	04	A	Rice Univ., Univ. Aix Marseille 3	France, United States	3	F	*Environ. Sci. Technol.*	Eng. Env., Env. Sci.	Mobility of nanomaterials in porous media	271

9	Schrick et al.	04	A	Penn State Univ.	United States	4	F	*Chem. Mat.*	Chem. Phys., Mats. Sci. Mult.	Zerovalent metal nanoparticles	267
10	He et al.	07	A	Auburn Univ.	United States	4	M	*Ind. Eng. Chem. Res.*	Eng. Chem.	Enhanced transport and dechlorination of TCE	260
11	Weir et al.	12	A	Arizona State Univ., ETH +1	Swit-zerland, Norway, United States.	5	M	*Environ. Sci. Technol.*	Eng. Env., Env. Sci.	Titanium dioxide nanoparticles	248
12	Tratnyek & Johnson	06	A	Oregon Hlth. & Sci. Univ.	United States	2	M	*Nano Today*	Chem. Mult., Nanosci. Nanotech., Mats. Sci. Mult.	Nanotechnologies for environmental cleanup	241

A, Article; R, review; M, male; F, female.

Elimelech M. was the most prolific authors with four papers. Blough J.L., He F., Li Q.L., Ma Z.W., Mallouk T.E., Ramakrishna S., Schrick B., and Zhao D.Y. were the other prolific researchers with two papers each. In total, 80 authors contributed to these classical papers.

The United States was the most prolific country with 17 papers. England, Singapore, and Malaysia were the other prolific countries with three, three, and two papers, respectively.

Massachusetts Institute of Technology, National University of Singapore, Yale University, Auburn University, and Penn State University were the most prolific single institutions with four, three, three, two, and two papers, respectively. In total, 43 national, regional, or single institutions contributed to these classical papers.

2008 was the most prolific publication year with six papers. 2004, 2006, 2000, 2005, and 2012 were the other prolific publication years with four, four, two, two, and two papers, respectively.

The most prolific journal was Environmental Science Technology with five papers. J. Membr. Sci., Chemistry of Materials, and Langmuir were the other prolific journals with four, two, and two papers, respectively. In total, these papers were published in 17 journals.

The most prolific subject categories were Chemistry Physical and Engineering Chemical with eight papers, each. Materials Science Multidisciplinary, Engineering Environmental, Environmental Sciences, and Polymer Science were the other prolific subject categories with seven, six, six, and five papers, respectively. Additionally, the subject category of Nanoscience Nanotechnology had further three papers.

The citation impact of these classical papers were significant with 374 citations per paper and H-index of 25.

There was a significant gender deficit in this field as there were only five papers with a female first author out of 25 papers.

The papers were presented under two major headings: water NF and other water nanopurification studies.

Organic matter fouling of NF membranes, superwetting nanowire membranes, polyvinylidene fluoride nanofibers electrospun NF membrane for liquid separation, polyamide NF membrane performance, fouling and cleaning of NF membranes, superhydrophilic behavior of nanoporous thin films, development of electrospun regenerated cellulose nanofiber membranes, NF membranes, and NF processes were the key research areas studied under the heading of water NF. Additionally, there were three papers in this area with a focus on desalination: RO membrane materials for desalination, water desalination with nanoporous graphene, and water desalination with carbon nanotube membranes.

The use of nanoscale bimetallic particles in the dechlorination of organic contaminants, use of the bimetallic nanoparticles for the dechlorination of hydrocarbons, hydrodechlorination of organic contaminants, and zerovalent iron nanoparticles for the dehalogenation of chlorinated hydrocarbons in groundwater and soils were the research areas in the second major category with a focus on the dechlorination studies.

Photocatalytic degradation of organic contaminants and remediation of contaminated groundwater using nanoparticles containing zerovalent iron were the research areas in the second category with a focus on the decontamination studies.

Water disinfection by antimicrobial nanomaterials, stabilization of w/o emulsions by silica nanoparticles, water purification in general, mobility of nanomaterials in porous media, human exposure to the titanium dioxide nanoparticles in food and personal care products were the other key research areas under the second heading.

2.2 The Most-Cited Papers in Water Nanofiltration

2.2.1 General Studies

Hong and Elimelech (1997) discuss NOM fouling of NF membranes in a paper originating from the United States with 637 citations. They find that membrane fouling increases with increasing electrolyte concentration, decreasing solution pH, and addition of divalent cations. Additionally, NOM fouling rate increases substantially with increasing initial permeation rate. They show that NOM fouling of NF membranes involves coupling between physical and chemical interactions. Thus, the addition of a strong chelating agent (EDTA) to feed water reduces NOM fouling significantly by removing free and NOM-complexed calcium ions, as well as improves the cleaning efficiency dramatically.

Yuan et al. (2008) discuss superwetting nanowire membranes in a paper originating from Singapore, Japan and the United States with 411 citations. They present a self-assembly method for constructing thermally stable, freestanding nanowire membranes that exhibit controlled wetting behavior ranging from superhydrophilic to superhydrophobic. They find that the nanowires that form the membrane structure can be resuspended in solutions and subsequently reform the original paper-like morphology over many cycles.

Gopal et al. (2006) discuss polyvinylidene fluoride nanofibers electrospun NF membrane for liquid separation in a paper originating from Canada and Singapore with 314 citations. They find that these electrospun membranes were successful in rejecting more than 90% of the microparticles from solution. These membranes could be used for pretreatment of water prior to RO or as prefilters to minimize fouling and contamination prior to NF.

Childress and Elimelech (2000) discuss the polyamide NF membrane performance in a paper originating from the United States with 295 citations. They find that water flux and salt passage were maximal at the membrane pore isoelectric point (pH 5). Ion rejection is directly related to the membrane pore charge. At low pH, negative rejection of protons was observed. Humic acid and sodium dodecyl sulfate had very little and significant effects on the short-term performance of the membrane, respectively.

Li and Elimelech (2004) discuss the fouling and cleaning of NF membranes in a paper originating from the United States with 294 citations. They find a significant correlation between the measured adhesion forces and the fouling and cleaning behavior of the membrane under various solution chemistries. Divalent calcium ions greatly enhance NOM fouling by complexation and subsequent formation of intermolecular bridges among organic foulant molecules. They further find that efficient chemical cleaning was achieved only when the calcium ion bridging was eliminated as a result of the interaction between the chemical cleaning agent and the fouling layer.

Cebeci et al. (2006) discuss superhydrophilic behavior of nanoporous thin films in a paper originating from the United States with 284 citations. They find that stable superhydrophilic wetting characteristics were obtained only after a critical number of bilayers were deposited onto a surface. The assembly conditions, such as solution pH and nanoparticle concentration, as well as the choice of nanoparticle size strongly influenced film properties. They assert that the superhydrophilic behavior of these films is driven by the rapid infiltration of water into a 3D nanoporous network created under specific assembly conditions.

Ma et al. (2005) discuss the development of electrospun regenerated cellulose (RC) nanofiber membranes in a paper originating from Singapore with 220 citations. The RC nanofiber membrane was further surface functionalized with a general affinity dye ligand (CB) for separation of many biomolecules. They find that The CB derived RC nanofiber membrane has a CB content of 130 μmol/g, and capture capacity of 13 mg/g for bovine serum albumin (BSA) and 4 mg/g for bilirubin. The membrane showed reusability after regeneration with elution buffer.

Yoon et al. (2006) discuss NF membranes in a paper originating from the United States with 216 citations. They show a high flux NF membrane based on an electrospun nanofibrous polyacrylonitrile

(PAN) scaffold coupled with a thin top layer of chitosan. This membrane exhibited a flux rate that is an order magnitude higher than commercial NF membranes, while maintaining the same rejection efficiency for oily wastewater filtration.

Hilal et al. (2004) discuss the NF membranes in water treatment in a review paper originating from England, Jordan, Malaysia, and Oman with 205 citations. They note that these membranes remove turbidity, microorganisms and hardness, as well as a fraction of the dissolved salts resulting in a much more energy-efficient process. They focus on the applications of NF in treating water and in the pretreatment process for desalination, the mechanism and minimization of NF membrane fouling problems.

Van der Bruggen et al. (2008) discuss NF processes in a review paper originating from Belgium and Finland with 202 citations. They identify six challenges for NF, such as avoiding membrane fouling, and possibilities to remediate, improving the separation between solutes that can be achieved, further treatment of concentrates, chemical resistance and limited lifetime of membranes, insufficient rejection of pollutants in water treatment, and the need for modeling and simulation tools. They note that these challenges are to some extent interrelated and offer possible pathways to solutions.

2.2.2 Desalination Studies

Lee et al. (2011) discuss RO membrane materials for desalination in a review paper originating from England with 278 citations. They consider both the first generation of asymmetric polymeric membranes and the nanostructured membrane materials including zeolite membranes, thin film nanocomposite membranes, carbon nanotube membranes, and biomimetic membranes. They argue that these nanomembranes would result for enhanced RO desalination performance.

Cohen-Tanugi and Grossman (2012) discuss water desalination with nanoporous graphene in a paper originating from the United States with 239 citations. They find that the graphene membrane's ability to prevent the salt passage depended critically on pore diameter with adequately sized pores allowing for water flow while blocking ions whilst commonly occurring hydroxyl groups can roughly double the water flux thanks to their hydrophilic character. However, the increase in water flux comes at the expense of less consistent salt rejection performance.

Corry (2008) discuss water desalination with carbon nanotube membranes in a paper originating from Australia with 232 citations. They show that ions face a large energy barrier and will not pass through the narrower tubes studied but can pass through the wider nanotubes. Water, however, faces no such impediment due to the formation of stable hydrogen bonds and crosses all of the tubes studied at very large rates. They affirm that these membranes can achieve a high degree of desalination at flow rates far in excess of existing membranes.

2.3 The Most-Cited Papers in Other Water Nanopurification Studies

2.3.1 Dechlorination Studies

Zhang et al. (1998) discuss the use of nanoscale bimetallic particles in the dechlorination of organic contaminants in a paper originating from the United States with 344 citations. They achieve rapid and complete dechlorination of several chlorinated organic solvents and chlorinated aromatic compounds by using the nanoscale bimetallic particles (Pd/Fe, Pd/Zn, Pt/Fe, Ni/Fe). They reason that one metal (Fe, Zn) serves primarily as electron donor while the other as catalyst (Pd, Pt).

He and Zhao (2005) discuss the use of the bimetallic nanoparticles for the dechlorination of hydrocarbons [trichloroethylene (TCE) and polychlorinated biphenyl (PCB)] in water in a paper originating from the United States with 324 citations. They use starched palladized iron (Fe–Pd) nanoparticles. The starched nanoparticles displayed much less agglomeration but greater dechlorination power than those prepared without a stabilizer. These particles showed good reactivity when used for dechlorination of TCE or PCBs in water as these particles destroyed 98% of TCE and transformed over 80% of PCBs.

Schrick et al. (2002) discuss the hydrodechlorination of organic contaminants using bimetallic Ni–Fe nanoparticles in a paper originating from the United States with 281 citations. These particles reduced TCE from an aqueous solution as the reaction occurred by nickel-catalyzed hydrodechlorination. They reason that as the iron actively corroded, the cathodically protected nickel surface chemisorbed hydrogen ions where the C—H bond formation was rate-determining and toxic-side products did not accumulate.

Schrick et al. (2004) discuss zerovalent iron nanoparticles for the dehalogenation of chlorinated hydrocarbons in groundwater and soils in a paper originating from the United States with 267 citations. They find that the Fe/C and Fe/PAA [poly(acrylic acid)] particles form colloidal suspensions that settle very slowly in water. Their anionic surface charge facilitates transport through soil- and sand-packed columns. They further find that the nanoparticle diffusion is the dominant filtration mechanism.

He et al. (2007) discuss dechlorination of organic contaminants using stabilized Fe–Pd nanoparticles in a paper originating from the United States with 260 citations. They use sodium carboxymethyl cellulose (CMC) as a stabilizer. Compared to nonstabilized Fe–Pd particles, they find that the CMC-stabilized nanoparticles displayed markedly improved stability against aggregation, chemical reactivity, and soil transport. CMC molecules were adsorbed to iron nanoparticles primarily through the carboxylate groups through monodentate complexation whilst –OH groups in CMC were also involved in interactions with iron particles.

2.3.2 Decontamination Studies

Gaya and Abdullah (2008) discuss the photocatalytic degradation of organic contaminants in wastewater in a review paper originating from Malaysia with 840 citations. They note the methods to improve the efficiency of photodecomposition over titania by combining the photoprocess with either physical or chemical operations. The resulting combined processes revealed a flexible line of action for wastewater treatment technologies where the choice of treatment method usually depends upon the composition of the wastewater. They focus on the use of the TiO2 photocatalyst for remediation and decontamination of wastewater.

Tratnyek and Johnson (2006) discuss the remediation of contaminated groundwater using nanoparticles containing zerovalent iron (nZVI) in a paper originating from the United States with 241 citations. They note the difficulties to engineer applications for optimal performance or to assess the risk to human or ecological health.

2.3.3 Water Disinfection Studies

Li et al. (2008) discuss the water disinfection by antimicrobial nanomaterials in a review paper originating from the United States with 622 citations. They cite a number of nanomaterials with strong antimicrobial properties, such as TiO_2, ZnO and fullerol, peptides, chitosan, carboxyfullerene, carbon nanotubes, ZnO and silver nanoparticles, aqueous fullerene nanoparticles, and chitosan. They discuss the antimicrobial mechanisms of these nanoparticles and highlight research needs to utilize these novel nanomaterials for water treatment applications.

2.3.4 Sedimentation Studies

Binks and Lumsdon (2000) discuss the stabilization of w/o emulsions by silica nanoparticles in a paper originating from England with 281 citations. They find that emulsions of are completely stable to coalescence as a result of an adsorbed layer of particles at the oil–water interface. Their stability to sedimentation increases with particle concentration due to network formation of the particles in the continuous oil phase. They further find that the system hydrophile–lipophile balance is determined by the particle wettability.

2.3.5 Other Studies

Shannon et al. (2008) discuss the water purification in a review paper originating from the United States with 1453 citations. They note the importance of eater scarcity and urge for research to identify robust new methods of purifying water at lower cost and with less energy, while at the same time minimizing the use of chemicals and impact on the environment. They focus on the disinfection, decontamination of water, and desalination of sea and brackish water.

Lecoanet et al. (2004) discuss mobility of nanomaterials in porous media, such as groundwater aquifers and water treatment plant filters in a paper originating from France and the United States with 271 citations. They show that these nanomaterials exhibit widely differing transport behaviors. Fullerene-based nanomaterials that had been functionalized to facilitate dispersal in water displayed the highest mobilities whilst colloidal aggregates of C_{60} were among the least mobile of the nanomaterials evaluated.

Weir et al. (2012) discuss the human exposure to the titanium dioxide nanoparticles in food and personal care products in a paper originating from Switzerland, Norway, and the United States with 248 citations. They note that while many white-colored food and personal care products contained titanium, it was not a prerequisite. Although several of these product classes contained low amounts of titanium, their widespread use and disposal down the drain and eventually to wastewater treatment plants deserves attention. They identify children as having the highest exposures because TiO_2 content of sweets is higher than other food products and they find a typical exposure for a US adult on the order of 1 mg Ti per kilogram body weight per day.

3 Conclusions

The area of water purification has been one of the most dynamic research fields in recent years with strong public policy implications with over 39,000 papers. Similarly, the areas of nanomaterials and nanoprocesses have been one of the most dynamic research fields in recent years with significant public policy implications with over 1,000,000 papers.

At the intersection of the research on the water purification and nanomaterials, the field of water nanopurification materials and processes has been one of the most dynamic research fields in recent years with significant impact on the medical and environmental research with near 5000 papers.

Following a scientometric overview of the research in nanomaterials and nanoprocesses, as well as water purification, the brief information on a selected set of 25 citation classics in the field of the water nanopurification materials and processes were presented in this paper to inform the key stakeholders about the influential papers in this dynamic research field as the first-ever study of its kind.

It was found that the major research area in these classical papers was the water NF materials and processes. The other studies on the water nanopurification included water dechlorination, water decontamination, water disinfection, and w/o emulsions.

As these citation classics deal with the important health and environmental research issues, the research in this field has strong public policy implications providing strong incentives for the key stakeholders involved in this research field.

Further research is recommended for the detailed studies including scientometric studies and citation classic studies for each of these topical areas, as well as the underlying research fields.

References

Arnold, J.G., Srinivasan, R., Muttiah, R.S., Williams, J.R., 1998. Large area hydrologic modeling and assessment—Part 1: model development. J. Am. Water Resour. Assoc. 34 (1), 73–89.

Baltussen, A., Kindler, C.H., 2004a. Citation classics in anesthetic journals. Anesth. Analg. 98, 443–451.

Baltussen, A., Kindler, C.H., 2004b. Citation classics in critical care medicine. Intensive Care Med. 30, 902–910.

Binks, B.P., Lumsdon, S.O., 2000. Catastrophic phase inversion of water-in-oil emulsions stabilized by hydrophobic silica. Langmuir 16 (6), 2539–2547.

Cebeci, F.C., Wu, Z.Z., Zhai, L., Cohen, R.E., Rubner, M.F., 2006. Nanoporosity-driven superhydrophilicity: a means to create multifunctional antifogging coatings. Langmuir 22 (6), 2856–2862.

Childress, A.E., Elimelech, M., 2000. Relating nanofiltration membrane performance to membrane charge (electrokinetic) characteristics. Environ. Sci. Technol. 34 (17), 3710–3716.

Cohen-Tanugi, D., Grossman, J.C., 2012. Water desalination across nanoporous graphene. Nano Lett. 12 (7), 3602–3608.

Corry, B., 2008. Designing carbon nanotube membranes for efficient water desalination. J. Phys. Chem. B 112 (5), 1427–1434.

Dai, Y.R., Song, Y.H., Gao, H.J., Wang, S.Y., Yuan, Y., 2015. Bibliometric analysis of research progress in membrane water treatment technology from 1985 to 2013. Scientometrics 105 (1), 577–591.

Dubin, D., Hafner, A.W., Arndt, K.A., 1993. Citation-classics in clinical dermatological journals: Citation analysis, biomedical journals, and landmark articles, 1945–1990. Archiv. Dermatol. 129, 1121–1129.
Elimelech, M., Phillip, W.A., 2011. The future of seawater desalination: energy, technology, and the environment. Science 333 (6043), 712–717.
Fu, F.L., Wang, Q., 2011. Removal of heavy metal ions from wastewaters: a review. J. Environ. Manag. 92 (3), 407–418.
Gaya, U.I., Abdullah, A.H., 2008. Heterogeneous photocatalytic degradation of organic contaminants over titanium dioxide: a review of fundamentals, progress and problems. J. Photochem. Photobiol. C 9 (1), 1–12.
Gehanno, J.F., Takahashi, K., Darmoni, S., Weber, J., 2007. Citation classics in occupational medicine journals. Scand. J. Work, Environ. Health 33, 245–251.
Geim, A.K., Novoselov, K.S., 2007. The rise of graphene. Nat. Mater. 6, 183–191.
Gopal, R., Kaur, S., Ma, Z.W., Chan, C., Ramakrishna, S., Matsuura, T., 2006. Electrospun nanofibrous filtration membrane. J. Membr. Sci. 281 (1-2), 581–586.
Hajipour, M.J., Fromm, K.M., Ashkarran, A.A., de Aberasturi, D.J., de Larramendi, I.R., Rojo, T., et al., 2012. Antibacterial properties of nanoparticles. Trends Biotechnol. 30 (10), 499–511.
He, F., Zhao, D.Y., 2005. Preparation and characterization of a new class of starch-stabilized bimetallic nanoparticles for degradation of chlorinated hydrocarbons in water. Environ. Sci. Technol. 39 (9), 3314–3320.
He, F., Zhao, D.Y., Liu, J.C., Roberts, C.B., 2007. Stabilization of Fe-Pd nanoparticles with sodium carboxymethyl cellulose for enhanced transport and dechlorination of trichloroethylene in soil and groundwater. Ind. Eng. Chem. Res. 46 (1), 29–34.
He, Z.C., Zhong, C.M., Su, S.J., Xu, M., Wu, H.B., Cao, Y., 2012. Enhanced power-conversion efficiency in polymer solar cells using an inverted device structure. Nat. Photonics 6, 591–595.
Heberer, T., 2002. Occurrence, fate, and removal of pharmaceutical residues in the aquatic environment: a review of recent research data. Toxicol. Lett. 131 (1-2), 5–17.
Hilal, N., Al-Zoubi, H., Darwish, N.A., Mohammad, A.W., Abu Arabi, M., 2004. A comprehensive review of nanofiltration membranes: treatment, pretreatment, modelling, and atomic force microscopy. Desalination 170 (3), 281–308.
Hong, S.K., Elimelech, M., 1997. Chemical and physical aspects of natural organic matter (NOM) fouling of nanofiltration membranes. J. Membr. Sci. 132 (2), 159–181.
Huang, Y., Yang, Q., Ao, X.L., 2008. Bibliometric analysis of pentachlorophenol remediation methods during the period of 1994 to 2005. Scientometrics 77 (1), 177–186.
Hullmann, A., Meyer, M., 2003. Publications and patents in nanotechnology—an overview of previous studies and the state of the art. Scientometrics 58, 507–527.
Iijima, S., 1991. Helical microtubules of graphitic carbon. Nature 354, 56–58.
Kim, B., Park, C.S., Murayama, M., Hochella, M.F., 2010. Discovery and characterization of silver sulfide nanoparticles in final sewage sludge products. Environ. Sci. Technol. 44 (19), 7509–7514.
Konur, O., 2000. Creating enforceable civil rights for disabled students in higher education: an institutional theory perspective. Disabil. Soc. 15, 1041–1063.
Konur, O., 2002a. Assessment of disabled students in higher education: current public policy issues. Assess. Eval. High. Educ. 27, 131–152.
Konur, O., 2002b. Access to employment by disabled people in the UK: is the Disability Discrimination Act working? Int. J. Discr. Law 5, 247–279.

Konur, O., 2002c. Access to nursing education by disabled students: rights and duties of nursing programs. Nurse Educ. Today 22, 364–374.
Konur, O., 2006a. Participation of children with dyslexia in compulsory education: current public policy issues. Dyslexia 12, 51–67.
Konur, O., 2006b. Teaching disabled students in Higher Education. Teach. High. Educ. 11, 351–363.
Konur, O., 2007a. A judicial outcome analysis of the Disability Discrimination Act: a windfall for the employers? Disabil. Soc. 22, 187–204.
Konur, O., 2007b. Computer-assisted teaching and assessment of disabled students in Higher Education: the interface between academic standards and disability rights. J. Comput. Assist. Learn 23, 207–219.
Konur, O., 2011. The scientometric evaluation of the research on the algae and bio-energy. Appl. Energy 88, 3532–3540.
Konur, O., 2012a. 100 citation classics in energy and fuels. Energy Educ. Sci. Technol. Part A 30 (si 1), 319–332.
Konur, O., 2012b. What have we learned from the citation classics in energy and fuels: a mixed study. Energy Educ. Sci. Technol. Part A 30 (si 1), 255–268.
Konur, O., 2012c. The gradual improvement of disability rights for the disabled tenants in the UK: the promising road is still ahead. Social Polit. Econ. Cult. Res. 4, 71–112.
Konur, O., 2012d. The policies and practices for the academic assessment of blind students in higher education and professions. Energy Educ. Sci. Technol. Part B 4 (si 1), 240–244.
Konur, O., 2012e. Prof. Dr. Ayhan Demirbas' scientometric biography. Energy Educ. Sci. Technol. Part A 28, 727–738.
Konur, O., 2012f. The evaluation of the research on the biofuels: a scientometric approach. Energy Educ. Sci. Technol. Part A 28, 903–916.
Konur, O., 2012g. The evaluation of the research on the biodiesel: a scientometric approach. Energy Educ. Sci. Technol. Part A 28, 1003–1014.
Konur, O., 2012h. The evaluation of the research on the bioethanol: a scientometric approach. Energy Educ. Sci. Technol. Part A 28, 1051–1064.
Konur, O., 2012i. The evaluation of the research on the microbial fuel cells: a scientometric approach. Energy Educ. Sci. Technol. Part A 29, 309–322.
Konur, O., 2012. The evaluation of the research on the biohydrogen: a scientometric approach. Energy Educ. Sci. Technol. Part A 29, 323–338.
Konur, O., 2012. The evaluation of the biogas research: a scientometric approach. Energy Educ. Sci. Technol. Part A 29, 1277–1292.
Konur, O., 2012. The scientometric evaluation of the research on the production of bio-energy from biomass. Biomass Bioenerg. 47, 504–515.
Konur, O., 2012. The evaluation of the global energy and fuels research: a scientometric approach. Energy Educ. Sci. Technol. Part A 30, 613–628.
Konur, O., 2012. The evaluation of the biorefinery research: a scientometric approach. Energy Educ. Sci. Technol. Part A 30 (si 1), 347–358.
Konur, O., 2012. The evaluation of the bio-oil research: a scientometric approach. Energy Educ. Sci. Technol. Part A 30 (si 1), 379–392.
Konur, O., 2013. What have we learned from the research on the International Financial Reporting Standards (IFRS)? A mixed study. Energy Educ. Sci. Technol. Part D 5, 29–40.
Konur, O., 2014. The periodical research on the experiences of the college students with ADHD: a mixed study. Int. J. Res. Teacher Educ. 5, 12–33.
Konur, O., 2015a. Algal biosorption of heavy metals from wastes. In: Kim, S.K., Lee, C.G. (Eds.), Marine Bioenergy: Trends and Developments. CRC Press, Boca Raton, Florida, pp. 597–626.

Konur, O., 2015b. Algal economics and optimization. In: Kim, S.K., Lee, C.G. (Eds.), Marine Bioenergy: Trends and Developments. CRC Press, Boca Raton, Florida, pp. 691–716.

Konur, O., 2015c. Algal high-value consumer products. In: Kim, S.K., Lee, C.G. (Eds.), Marine Bioenergy: Trends and Developments. CRC Press, Boca Raton, Florida, pp. 653–682.

Konur, O., 2015d. Algal photobioreactors. In: Kim, S.K., Lee, C.G. (Eds.), Marine Bioenergy: Trends and Developments. CRC Press, Boca Raton, Florida, pp. 81–108.

Konur, O., 2015e. Algal photosynthesis, biosorption, biotechnology, and biofuels. In: Kim, .S.K. (Ed.), Springer Handbook of Marine Biotechnology. Springer, Berlin, Heidelberg, Germany, pp. 1131–1161.

Konur, O., 2015f. Current state of research on algal biodiesel. In: Kim, S.K., Lee, C.G. (Eds.), Marine Bioenergy: Trends and Developments. CRC Press, Boca Raton, Florida, pp. 487–512.

Konur, O., 2015g. Current state of research on algal bioelectricity and algal microbial fuel cells. In: Kim, S.K., Lee, C.G. (Eds.), Marine Bioenergy: Trends and Developments. CRC Press, Boca Raton, Florida, pp. 527–556.

Konur, O., 2015h. Current state of research on algal bioethanol. In: Kim, S.K., Lee, C.G. (Eds.), Marine Bioenergy: Trends and Developments. CRC Press, Boca Raton, Florida, pp. 217–244.

Konur, O., 2015i. Current state of research on algal biohydrogen. In: Kim, S.K., Lee, C.G. (Eds.), Marine Bioenergy: Trends and Developments. CRC Press, Boca Raton, Florida, pp. 393–422.

Konur, O., 2015. Current state of research on algal biomethane. In: Kim, S.K., Lee, C.G. (Eds.), Marine Bioenergy: Trends and Developments. CRC Press, Boca Raton, Florida, pp. 273–302.

Konur, O., 2015. Current state of research on algal biomethanol. In: Kim, S.K., Lee, C.G. (Eds.), Marine Bioenergy: Trends and Developments. CRC Press, Boca Raton, Florida, pp. 327–370.

Konur, O., 2015. The scientometric study of the global energy research. In: Prasad, R., Sivakumar, S., Sharma, U.C. (Eds.), Energy Science and Technology. V.1. Opportunities and Challenges. Studium Press LLC, Houston, TX, pp. 475–489.

Konur, O., 2015. The review of citation classics on the global energy research. In: Prasad, R., Sivakumar, S., Sharma, U.C. (Eds.), Energy Science and Technology. V.1. Opportunities and Challenges. Studium Press LLC, Houston, TX, pp. 490–526.

Konur, O., 2016a. Scientometric overview regarding antimicrobial nanobiomaterials. In: Grumezescu, A.M. (Ed.), Applications of Nanobiomaterials. Volume VI: Nanobiomaterials in Antimicrobial Therapy. Elsevier, Amsterdam.

Konur, O., 2016b. Scientometric overview regarding surface chemistry of nanobiomaterials. In: Grumezescu, A.M. (Ed.), Applications of Nanobiomaterials. Volume 3: Surface Chemistry of Nanobiomaterials. Elsevier, Amsterdam.

Konur, O., 2016c. Scientometric overview regarding nanobiomaterials in dentistry. In: Grumezescu, A.M. (Ed.), Applications of Nanobiomaterials. Volume 11: Nanobiomaterials in Dentistry. Elsevier, Amsterdam.

Konur, O., 2016d. The scientometric overview in nanobiodrugs. In: Grumezescu, A. (Ed.), Nanoarchitectonics for Smart Delivery and Drug Targeting. Elsevier, Amsterdam.

Konur, O., 2016e. The scientometric overview in cancer targeting. In: Grumezescu, A. (Ed.), Nanoarchitectonics for Smart Delivery and Drug Targeting. Elsevier, Amsterdam.

Konur, O., 2016f. Scientometric overview regarding water nanopurification. In: Grumezescu, A. (Ed.), Nanotechnology in Agri-Food Industry: Water Purification. Elsevier Academic Press, San Diego, USA.
Konur, O., 2016g. The hot papers in the surface engineering of nanobiomaterials. In: Grumezescu, A. (Ed.), Therapeutic Nanostructures. Volume 1: General Approach. Elsevier, Amsterdam.
Konur, O., 2016h. The recent citation classics in antimicrobial nanobiomaterials. In: Grumezescu, A. (Ed.), Therapeutic Nanostructures. Volume 3: Antimicrobials. Elsevier, Amsterdam.
Konur, O., 2016i. Glycoscience: the current state of the research. In: Kim, S.K. (Ed.), Marine Glycobiology: Principles and Applications, Part 1. Introduction to Marine Glycobiology. CRC Press, Boca Raton, Florida.
Konur, O., 2016. Algal Omics: The current state of the research. In: Kim, S.K. (Ed.), Marine Omics: Principles and Applications. Part 1. Introduction to Marine Omics. CRC Press, Boca Raton, Florida.
Konur, O., 2016. Scientometric overview regarding nanoemulsions. In: Grumezescu, A. (Ed.), Nanotechnology in Agri-Food Industry: Emulsions. Elsevier Academic Press, San Diego, USA.
Konur, O., 2016. The state of the research in algal genomics. In: Kim, S.K. (Ed.), Encyclopedia of Marine Biotechnology. Wiley-Blackwell, New Jersey, USA.
Konur, O., 2016. The state of the research in algal drugs. In: Kim, S.K. (Ed.), Encyclopedia of Marine Biotechnology. Wiley-Blackwell, New Jersey, USA.
Kostoff, R.N., Stump, J.A., Johnson, D., Murday, J.S., Lau, C.G.Y., Tolles, W.M., 2006. The structure and infrastructure of the global nanotechnology literature. J. Nanopart. Res. 8, 301–321.
Lecoanet, H.F., Bottero, J.Y., Wiesner, M.R., 2004. Laboratory assessment of the mobility of nanomaterials in porous media. Environ. Sci. Technol. 38 (19), 5164–5169.
Lee, K.P., Arnot, T.C., Mattia, D., 2011. A review of reverse osmosis membrane materials for desalination-development to date and future potential. J. Membr. Sci. 370 (1-2), 1–22.
Li, Q.L., Elimelech, M., 2004. Organic fouling and chemical cleaning of nanofiltration membranes: measurements and mechanisms. Environ. Sci. Technol. 38 (17), 4683–4693.
Li, Q.L., Mahendra, S., Lyon, D.Y., Brunet, I., Liga, M.V., Li, D., et al., 2008. Antimicrobial nanomaterials for water disinfection and microbial control: potential applications and implications. Water Res. 42 (18), 4591–4602.
Ma, Z.W., Kotaki, M., Ramakrishna, S., 2005. Electrospun cellulose nanofiber as affinity membrane. J. Membr. Sci. 265 (1-2), 115–123.
Meyer, M., Persson, O., 1998. Nanotechnology—interdisciplinarity, patterns of collaboration and differences in application. Scientometrics 42, 195–205.
Mills, A., Davies, R.H., Worsley, D., 1993. Water-purification by semiconductor photocatalysis. Chem. Soc. Rev. 22 (6), 417–425.
North, D., 1994. Economic-performance through time. Am. Econ. Rev. 84, 359–368.
Novoselov, K.S., Geim, A.K., Morozov, S.V., Jiang, D., Zhang, Y., Dubonos, S.V., et al., 2004. Electric field effect in atomically thin carbon films. Science 306, 666–669.
Oller, I., Malato, S., Sanchez-Perez, J.A., 2011. Combination of advanced oxidation processes and biological treatments for wastewater decontamination—a review. Sci. Total Environ. 209 (20), 4141–4166.
Oregan, B., Gratzel, M., 1991. A low-cost, high-efficiency solar-cell based on dye-sensitized colloidal TiO_2 films. Nature 353, 737–740.

Paladugu, R.M., Chein, S., Gardezi, S., Wise, L., 2002. One hundred citation classics in general surgical journals. World J. Surg. 26, 1099–1105.
Pelaez, M., Nolan, N.T., Pillai, S.C., Seery, M.K., Falaras, P., Kontos, A.G., et al., 2012. A review on the visible light active titanium dioxide photocatalysts for environmental applications. Appl. Catal. B 125, 331–349.
Porter, A.L., Youtie, J., Shapira, P., Schoeneck, D.J., 2008. Refining search terms for nanotechnology. J. Nanopart. Res. 10, 715–728.
Qi, X.L., Zhang, S.C., 2011. Topological insulators and superconductors. Rev. Modern Phys. 83, 1057–1110.
Radisavljevic, B., Radenovic, A., Brivio, J., Giacometti, V., Kis, A., 2011. Single-layer MoS_2 transistors. Nat. Nanotechnol. 6, 147–150.
Schrick, B., Hydutsky, B.W., Blough, J.L., Mallouk, T.E., 2004. Delivery vehicles for zerovalent metal nanoparticles in soil and groundwater. Chem. Mater. 16 (11), 2187–2193.
Schrick, B., Blough, J.L., Jones, A.D., Mallouk, T.E., 2002. Hydrodechlorination of trichloroethylene to hydrocarbons using bimetallic nickel-iron nanoparticles. Chem. Mater. 14 (12), 5140–5147.
Shannon, M.A., Bohn, P.W., Elimelech, M., Georgiadis, J.G., Marinas, B.J., Mayes, A.M., 2008. Science and technology for water purification in the coming decades. Nature 452 (7185), 301–310.
Tanaka, H., Ho, Y.-S., 2011. Global trends and performances of desalination research. Desalin. Water Treat. 25 (1–3), 1–12.
Tratnyek, P.G., Johnson, R.L., 2006. Nanotechnologies for environmental cleanup. Nano Today 1 (2), 44–48.
Van der Bruggen, B., Manttari, M., Nystrom, M., 2008. Drawbacks of applying nanofiltration and how to avoid them: a review. Sep. Purif. Technol. 63 (2), 251–263.
Weir, A., Westerhoff, P., Fabricius, L., Hristovski, K., von Goetz, N., 2012. Titanium dioxide nanoparticles in food and personal care products. Environ. Sci. Technol. 46 (4), 2242–2250.
Wrigley, N., Matthews, S., 1986. Citation-classics and citation levels in geography. Area 18, 185–194.
Yella, A., Lee, H.W., Tsao, H.N., Yi, C.Y., Chandiran, A.K., Nazeeruddin, M.K., et al., 2011. Porphyrin-sensitized solar cells with Cobalt (II/III)-based redox electrolyte exceed 12 percent efficiency. Science 334, 629–634.
Yoon, K., Kim, K., Wang, X.F., Fang, D.F., Hsiao, B.S., Chu, B., 2006. High flux ultrafiltration membranes based on electrospun nanofibrous PAN scaffolds and chitosan coating. Polymer 47 (7), 2434–2441.
Yuan, J.K., Liu, X.G., Akbulut, O., Hu, J.Q., Suib, S.L., Kong, J., et al., 2008. Superwetting nanowire membranes for selective absorption. Nat. Nanotechnol. 3 (6), 332–336.
Zhang, W.X., Wang, C.B., Lien, H.L., 1998. Treatment of chlorinated organic contaminants with nanoscale bimetallic particles. Catal. Today 40 (4), 387–395.

SUBJECT INDEX

FSC
www.fsc.org
MIX
Papier | Fördert
gute Waldnutzung
FSC® C083411

Kolibri 360 GmbH
Ferdinand-Jühlke-Straße 7
99095 Erfurt, Deutschland
produktsicherheit@kolibri360.de